Lecture Notes in Electrical Engineering

Volume 255

For further volumes:
http://www.springer.com/series/7818

Zengqi Sun · Zhidong Deng
Editors

Proceedings of 2013 Chinese Intelligent Automation Conference

Intelligent Automation & Intelligent Technology and Systems

Editors
Zengqi Sun
Zhidong Deng
Department of Computer Science
Tsinghua University
Beijing
People's Republic of China

ISSN 1876-1100 ISSN 1876-1119 (electronic)
ISBN 978-3-642-38459-2 ISBN 978-3-642-38460-8 (eBook)
DOI 10.1007/978-3-642-38460-8
Springer Heidelberg New York Dordrecht London

Library of Congress Control Number: 2013939563

Printed on acid-free paper

Springer is part of Springer Science+Business Media (www.springer.com)

Preface

The 2013 Chinese Intelligent Automation Conference (CIAC2013) was Sponsored by Intelligent Automation Committee, Chinese Association of Automation and organized by Yangzhou University in Yangzhou, Jiangsu Province, China; 23–25, August 2013. The objective of CIAC2013 was to provide a platform for researchers, engineers, academicians as well as industrial professionals from all over the world to present their research results and development activities in Intelligent Control, Intelligent Information Processing and Intelligent Technology and Systems. This conference provided opportunities for the delegates to exchange new ideas and application experiences face-to-face, to establish research or business relations and to find partners for future collaboration.

We have received more than 800 papers. The topics include adaptive control, fuzzy control, neural network-based control, knowledge-based control, hybrid intelligent control, learning control, evolutionary mechanism-based control, multi-sensor integration, failure diagnosis, and reconfigurable control, etc. Engineers and researchers from academia, industry, and government can gain an inside view of new solutions combining ideas from multiple disciplines in the field of Intelligent Automation. All submitted papers have been subject to a strict peer-review process; 285 of them were selected for presentation at the conference and included in the CIAC2013 proceedings. We believe the proceedings can provide the readers a broad overview of the latest advances in the fields of Intelligent Automation.

CIAC2013 has been supported by many professors (see the name list of the committees); we would like to take this opportunity to express our sincere gratitude and highest respect to them. At the same time, we also express our sincere thanks for the support of every delegate.

Zengqi Sun
Chair of CIAC2013

Committees

Honorary Chairs

Yanda Li, Tsinghua University, China
Bo Zhang, Tsinghua University, China

General Chair

Zengqi Sun, Tsinghua University, China

General Co-Chairs

Hongxin Wu, Beijing Institute of Control Engineering
Qidi Wu, Tongji University
Jue Wang, Institute of Automation, Chinese Academy of Science
Wenli Xu, Tsinghua University
Xingyu Wang, East China University of Science and Technology
Yixin Yin, University of Science and Technology, Beijing
Zushu Li, Chongqing University
Limin Jia, Beijing Jiaotong University
Zhidong Deng, Tsinghua University
Junping Du, Beijing University of Posts and Telecommunications

Technical Program Committee Co-Chairs

Zhidong Deng, Tsinghua University
Yixin Yin, University of Science and Technology, Beijing
Limin Jia, Beijing Jiaotong University
Junping Du, Beijing University of Posts and Telecommunications
Tianping Zhang, Yangzhou University

Organizing Committee Co-Chairs

Bin Li, Yangzhou University
Tianping Zhang, Yangzhou University
Yun Li, Yangzhou University
Yuequan Yang, Yangzhou University
Hua Ye, Southeast University

Contents

Part I
Intelligent Automation

Chapter 1
Application of Production Rule in Intelligent Collaboration System of Information Appliances

Huazhong Liu, Jihong Ding and Anyuan Deng

Abstract The intelligent collaboration among information appliances based on production rule is researched to enable them to achieve mutual recognition, mutual communication and mutual cooperation. An intelligent collaboration platform of information appliances is built and the intelligent collaborative process is discussed. And then all the facts in information appliances are showed with abstract representation, all the rules are showed with production rule, and they are stored in the knowledge bases. Finally, an inference algorithm is designed combining production rule and expert system. The experimental results and application show that the intelligent collaborative system of information appliances based on production rule have the advantage of convenient control, clear rule description, intuitive knowledge representation and good inference effectiveness.

Keywords Information appliances · Production rule · Intelligent inference · Autonomous learning

1.1 Introduction

Since the concept of Smart Home is introduced, it gains rapid development, which focuses on the intelligent and humanized research. Information appliance is a digital equipment belonged to smart home product, combined the computer technology, communication technology and electronic technology on the basis of traditional appliance [1]. Owing to the types of equipment have become more diverse and the function of equipment have become more powerful, the more convenient interconnection and more intelligent collaboration have become the key of research [2].

H. Liu (✉) · J. Ding · A. Deng
School of Information Science and Technology, Jiujiang University, Jiujiang, China
e-mail: sharpshark_ding@163.com

Z. Sun and Z. Deng (eds.), *Proceedings of 2013 Chinese Intelligent Automation Conference*, Lecture Notes in Electrical Engineering 255,
DOI: 10.1007/978-3-642-38460-8_1, © Springer-Verlag Berlin Heidelberg 2013

The research on intelligent control among information appliances can be divided into two stages [3]. The first stage is mechanical control, we can control appliances by manipulating the terminals in virtue of communication network, in which, each appliance itself is an intelligent product, but these appliances can not communicate and collaborate with each other. The second stage is intelligent control, each appliance not only can be operated according to the master's will, but also all the appliances can communicate and collaborate with each other because all the information appliances form a network, and they can react intelligently according to the master's habits and outside information obtained by terminals, moreover, they have the capacity of independent learning. Currently, the control way has changed from mechanical control to intelligent control.

1.2 Intelligent Collaboration System of Information Appliances

1.2.1 Construction of Platform

From a structural point of view, the intelligent collaboration system of information appliances (for short ICSIA) is composed of user terminal, intelligent gateway, information appliances and communication network [4], which is shown in Fig. 1.1. In this model, people can send operating command to intelligent gateway via the communication network by manipulating the user terminal such as remote control, the remote telephone, PDA and the computer connected to the network, etc. Intelligent gateway will manipulate the corresponding information appliance according to the command. At the same time, intelligent gateway will detect the signal of each sensor in real time and react automatically according to the rules set by the master.

1.2.2 Collaborative Process

The intelligent gateway, is the core equipment in ICSIA, manage uniformly all information appliances of the smart home, in which, some modules are running

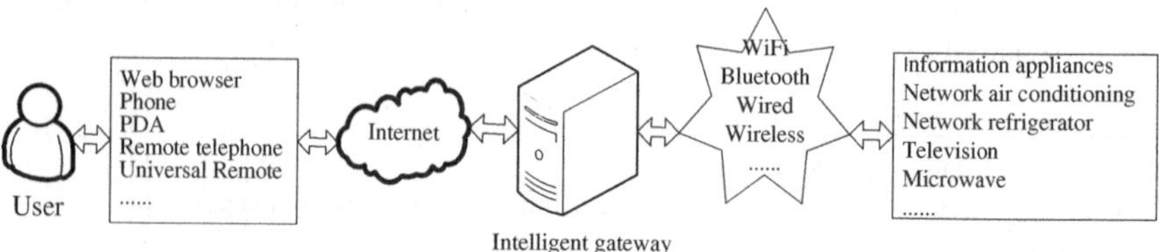

Fig. 1.1 Structure of intelligent collaboration platform

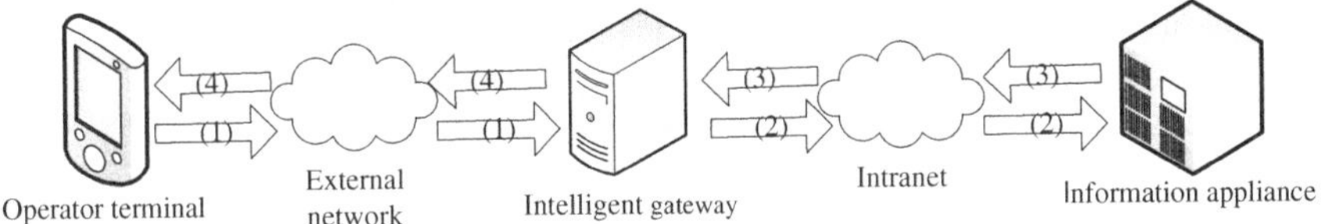

Fig. 1.2 Collaborative process of information appliances

such as embedded database, Information appliances interface definition language (IAIDL) compiler [5], device adapter, intelligent collaboration module and network communication module, etc. In embedded database, the whole registration information, status information and statistics data of all information appliances are stored. In intelligent collaboration module, some rules can be input by user and these rules can be improved step by step through autonomous learning, and the intelligent collaborative program is running to achieve unified management to a variety of information appliances.

If the user intend to manipulate the appliance, the user can send a service request to the intelligent gateway by manipulating the terminal, the intelligent gateway determine whether the service request can be processed or not according to the security policy. Only the legitimate user possess corresponding permission, will the intelligent gateway forward the service request to the corresponding information appliance, and then the information appliance respond to the request and send the result message to the intelligent gateway, finally, the intelligent gateway forward the results of the service response to the user terminal [6], the process is shown in Fig. 1.2.

1.3 Production Rule

Production system, is proposed by the American mathematician E. Post in 1943, and is introduced into the field of artificial intelligence by A. Newell and E. A. Simon in 1972 [7]. Now, production system has been widely used in artificial intelligence and has made great progress.

The general form of Production Rule is as follows:

IF <X> THEN <Y> λ.

In the formula, X represents precondition, Y represents conclusion, and λ represents credibility. When the precondition <X> is true, the credibility of the conclusion <Y> is λ, and the precondition <X> can be the conjunction of the facts or assertions [8], which can be shown in Fig. 1.3.

Production rule can express not only the certain knowledge but the uncertain knowledge, not only the heuristic knowledge but the procedural knowledge.

If production rule represents the certain knowledge, it is generally expressed as: IF <A> THEN <B>, simplified as $A \rightarrow B$.

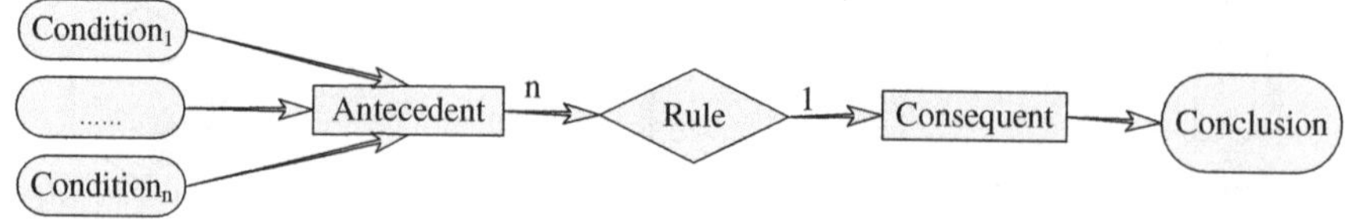

Fig. 1.3 Production rule model

It means that if A is true, then B will be executed. In the study of ICSIA, production rule express certain knowledge and has the following characteristics:

The same condition can draw different conclusions. For instance, if the dinner is prepared (A), then the light of the dining-room is set to dining-light mode (B) and the background music of dining-room is played (C). The knowledge representation of the case using production rule is: IF A THEN B, IF A THEN C, simplified as $A \rightarrow B$, $A \rightarrow C$.

The same conclusion can be drawed by different conditions. For instance, if air-conditioner is open (A), then the windows will be shut (C); or if it is rain (B), the windows will also be shut (C). The knowledge representation of the case using production rule is: IF A OR B THEN C, simplified as $A \vee B \rightarrow C$.

The conclusion can be drawed only when a plurality of conditions are met simultaneously. For instance, if the door is opened (A), and the light is less than set value (B), then the lamps will be turned on (C). The knowledge representation of the case using production rule is: IF A AND B THEN C, simplified as $A \wedge B \rightarrow C$.

The conclusion of one rule could be the condition of another rule. For instance, if the temperature detected by the detector is higher than set value (A), then the air-conditioning will be opened (B); and if the air-conditioning is open (B), the windows will be closed (C). The knowledge representation of the case using production rule is: IF A THEN B, IF B THEN C, simplified as $A \rightarrow B$, $B \rightarrow C$.

1.4 Intelligent Collaboration System of Information Appliances Based on Production Rule

1.4.1 Detailed Structure of System

The core of the ICSIA is the knowledge base and the inference engine. Firstly, all the knowledge of information appliances are represented and stored in the knowledge base on the basis of automata theory. Secondly, the rule bases are established according to the owner's habits. Finally, the inference algorithm is implemented combined production rule and expert system. Its structure is shown in Fig. 1.4.

In the process of knowledge acquisition, the expert knowledge will be input into the knowledge base of the expert system in specific representation form, and the

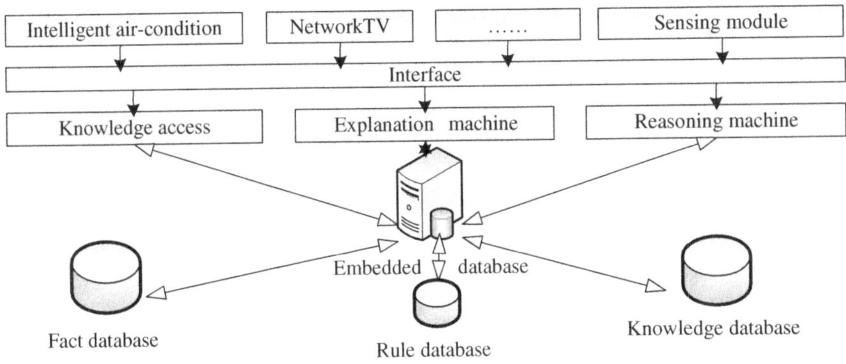

Fig. 1.4 Detailed structure diagram

knowledge engineer need to translate the expert knowledge into specific knowledge representation form if experts do not well in computer knowledge.

1.4.2 Abstract Representation of Fact

In ICSIA, all the facts need to be represented in a unified way, an abstract representation is used in this paper, which is expressed as X_{ij}.

X represents the type of device, it is the abbreviation of the corresponding device, i represents serial number of device, the legal value is 1–9, a–z and A–Z. j represents the status of device, the legal value is 0–9, a–z and A–Z.

Each information appliance must belong to a certain type, and there may be many devices for the same type, and there may be a lot of status for the same device belonged to the same type. For instance, lamp belong to a type of device, it can be represented by L, and there are a number of lamps in home, it can be distinguished from 1–9, a–z and A–Z, so there are total 61 kinds of representation can be expressed, and 0 is used to represent all the devices belonged to the same type, and assumed that each lamp has two status of ON and OFF, it can be expressed by 1 and 0.

If the lamp NO. 1 is on, it can be expressed as L11, if the lamp NO. 2 is off, it can be expressed as L20; if all the lamps are on, it can be expressed as L01. For each abstract representation such as L11, it can represent the status of the device, and it can also represent the fact in the condition base and the conclusion base, moreover, it can also represent an operation in the conclusion base which can call the corresponding API function to trigger the hardware device to response the appropriate operation.

1.4.3 Construction of Knowledge Base

In order to represent the collaborative relationship of information appliances and store all the knowledge and production rules, five databases are designed in the paper.

1. Basic fact table
   In order to store the whole factual information, a fact table is designed and its structure is as follows:
   Fact (Fact_Id, Fact_Value, Fact_Description).
   Fact_Id represents the record number of the table, and it is the primary key of the table. Fact_Value represents various facts, adopting the above abstract representation like X_{ij}. Fact_Description represents the description of the fact.
2. Temporary fact table
   In order to store the initial facts and the middle facts generated in the inference process, a temporary fact table is designed. In the implementation process, a linked list is used because the size of the temporary fact is unpredictable.
3. Model fact table
   In ICSIA, detector generally just detect the device state, it does not necessarily match with some specific fact in the fact table. So, in order to judge whether a particular fact is occurred, a model fact table is also designed to store the fact to be judged and the corresponding model algorithm. The structure is as follows:
   Fact_Model (Model_ID, Fact_Id, Model, Effect, Valid).
   Model_ID represents the record number of the table, it is the primary key of the table. Fact_Id represents the identifier of the current fact, it is the foreign key reference to Fact table. Model represents the function or algorithm which used to judge whether the fact was occurred. Effect represents whether the fact is valid or not. Valid represents whether the fact was established or not.
4. Rule condition table
   In order to store all the rules, the condition and conclusion in each rule expression will be decomposed, and all the rule conditions are stored in the rule condition table, and all the rule conclusion are stored in the rule conclusion table. The separate store way can eliminate data redundancy and avoid insertion and deletion anomalies to enhance the stability and flexibility of the database.
   Rule_Condition (Condition_Id, Conclusion_Id, Fact_Id, Used, Prompt).
   Condition_Id represents the record number of the table, it is the primary key of the table. Fact_Id represents the identifier of the fact, it is the foreign key reference to Fact table. Conclusion_Id represents the identifier of the conclusion in the Rule_Conclusion table, it is the foreign key reference to Rule_Conclusion table. Used represents whether the condition have been visited or not, it is to prevent the current tuple from operating repeatedly to improve the efficiency of inference, the initial value is 0. Prompt represents descriptive instructions.
   Rule condition table is used to store the conditions of all rules. The pre-condition in some rules may be the conjunction of several conditions after

preprocessing, that is, the conclusion of the rule will be occur only when all the conditions of the rule are satisfied simultaneously. The table structure is as follows:

5. Rule conclusion table
   Rule conclusion table is used to store the conclusion of corresponding rule, the conclusion of the rule will be executed only when all the conditions of the rule are satisfied simultaneously. The table structure is as follows:
   Rule_Conclusion (Conclusion_Id, Fact_Id, CondNum, CondLack).
   Conclusion_Id represents the record number of the table, it is the primary key of the table. Fact_Id represents the identifier of the fact which is triggered by the conclusion, it is the foreign key reference to Fact table. CondNum represents the total number of conditions for current rule. CondLack represents the missing number of conditions for current rule.
   In which, the initial value of CondNum and CondLack is equal, the value of CondLack will minus one when one of the conditions satisfies the rule.

1.4.4 Autonomous Learning

There probably exist some rules is contradictory and unreasonable in the rule base, so some detection methods need to used to eliminate these rules such as contradictory rule and cyclic rule [9].

Contradictory rule is that the same conditions can infer the opposite conclusion directly or indirectly. The detection method is to construct the inference chain for every same condition of all rules respectively, and if any conflict rule is found in the process of construction, it indicates that the contradictory rule exist.

Cyclic rule is that the conditions and conclusions of a set of rules form a cyclic chain in the inference process. For instance, A → B, B → C, C → A is a cyclic chain, and they will go into an infinite loop in the inference process. The detection method is to establish a two-dimensional table, the column represents the condition and the row represents the conclusion, and then mark out all the rules in the table. Find one rule from the table in which the conclusion is the condition of the other rule (A → B, B → C), and then find all the other rules according to the above method starting from the rule and form a rule chain (A → B → C → E → F), if there exist the same part along the rule chain (A → B → C→ E → B, the same part is B), it is obvious that there exist cyclic rule.

1.4.5 Intelligent Inference Process

The inference engine is the “thinking unit” and the core part of the intelligent collaboration system, its function is to infer the issues raised by users to the

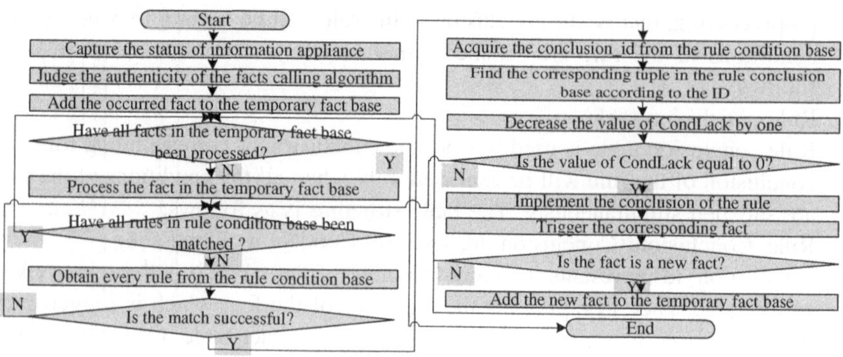

Fig. 1.5 Inference flowchart

corresponding conclusion by means of the knowledge of the knowledge base according to the inference strategy.

There usually exist three inference methods, event-driven forward inference, goal-driven backward inference and the hybrid inference [10]. Forward inference is to infer the conclusion starting from the conditions. In the inference process, judge whether the conditions in the rule meet it through search rule base one by one, and the rule will be executed only when the whole facts of the rule occur synchronously. Backward inference is to trace back to the facts supported the condition starting from the conclusion, that is, assume that the conclusion is correct, then verify whether the conditions meet the rule, if all conditions are met it, it is confirmed that the conclusions is correct. Hybrid inference is the combination of these two inference methods.

Backward inference is often more effective than forward inference for these problem whose conclusion is knowable, but the conclusion of the system is unknowable, so the event-driven forward inference is selected in the inference engine of the system.

The inference flowchart is shown in Fig. 1.5. Firstly, the inference machine read the current status of information appliances through the intelligent collaboration module in gateway, and judge whether the fact is occurred through the corresponding model algorithm in Fact_Model table, if the fact is true, it will be added to the temporary fact table, and these facts constitute the initial facts. Secondly, the inference machine judge whether every fact in the temporary fact table would meet which rule in the rule condition table, and record its Conclusion_Id of the tuple in the Rule_Condition table only if the processing fact is consistent with the Fact_Id of the current tuple. Finally, find the corresponding tuple in the Rule_Conclusion table according to above Conclusion_Id, and decrease the value of CondLack by one, if the value of CondLack is unequal to 0, and continue to process the next fact in the temporary fact table, if the CondLack value is equal to 0, it is obvious that all the facts lead to the conclusion have been occurred, and the conclusion of the rule will be implemented and the corresponding fact will be triggered. If the triggered fact is a new fact, it will be appended to the temporary fact table. In the inference process, the first-match strategy is used for conflict resolution, that is, the rule will be executed once the rule is matched in the problem-solving process.

1.5 Implementation of Inference Algorithm

The hardware platform of the experiment is built adopting the S3C2410X microprocessor and expanding other information acquisition device. The software system of the experiment is developed taking Linux as the operating system and taking Sqlite as the database, meanwhile, the inference algorithm is written in C language.

Algorithm Intelligent inference algorithm of information appliances based on production rule.
Input The status of information appliances.
Output Activated fact in the inference process.
Process

Step 1 Create and initialize all the data tables in advance, mainly for the fact table, the fact model table, the rule condition table and the rule conclusion table.

Step 2 Traverse the rule conclusion table, set m_iCondLack and m_iCondNum is equal, traverse rule condition table and clear Used flag, traverse fact model table and clear Valid flag and Effect flag.

Step 3 Acquire the status of information appliances and call model algorithm to judge whether the facts is occurred, set Effect flag and Valid flag in fact model table according to the judgement result, if it is uncertain, set Valid flag to 0, if it is false, set Valid flag to 1 and set Effect flag to 0, if it is true, set Valid flag and Effect flag to 1, and add the corresponding fact code to the temporary fact table.

Step 4 For every fact f_i in the temporary fact table, complete the following step 5 to step 8.

Step 5 For each tuple t_i in the rule condition table, if the Used flag is 0 and the Fact_ID match f_i, extract the Conclusion_ID n_i and complete the following steps.

Step 6 Find the corresponding tuple t_j in the rule conclusion table according to the Conclusion_ID n_i.

Step 7 Decrease the CondLack of the tuple t_j by one, if the value of CondLack is unequal to 0, then turn to step 5, if it is equal 0, then execute the current rule and implement the corresponding conclusion fact f_j.

Step 8 Call corresponding API function to respond to the fact f_j; if it is a new fact, append it to the temporary fact table, set m_iCondLack and m_iCondNum is equal in the rule conclusion table and turn to step 4.

1.6 Conclusion

In this paper, in order to enable all the information appliances to run collaboratively, knowledge representation and knowledge storage is introduced and inference algorithm is designed combined production rules and the expert system on the basis of IAIDL. And the experimental results and application of the intelligent collaboration system of information appliances based on production rule show that it has the advantage of convenient control, intuitive knowledge representation and good inference effectiveness.

References

1. Man S, Yang H, Peng Y, Wang X (2010) Design of embedded wireless smart home gateway based on ARM9. J Comput Appl 7:2541–2544 (in Chinese)
2. Intelligent building to wisdom city (2012) http://www.qianjia.com/forum/
3. Li L, Luo J (2007) Smart home and its development trend. Comput Modernization 11:18–21 (in Chinese)
4. Liu H (2009) The Research and implementation of the collaborative model of information appliances. Hunan Normal University, Changsha, pp 22–30 (in Chinese)
5. Yang L (2007) The Research and design of the interface definition language of information appliance and its compiler. Hunan Normal University, Changsha, pp 26–30 (in Chinese)
6. Peng H (2008) The research and design of the versatile controller of information appliance in the intelligent house. Hunan Normal University, Changsha, pp 26–30 (in Chinese)
7. Li L, Gao T (2006) The theory and realization of production rules expert systems. Microcomput Appl 27:631–633 (in Chinese)
8. Shang T, Tao J (2010) Production rule-based assessment system of built environment. Eng J Wuhan Univ 6:809–811 (in Chinese)
9. Suo H (2011) Research and analysis of production rules in expert system. J Weinan Teach Univ 26:63–65 (in Chinese)
10. Min H, Gan X (2012) A real-time reasoning mechanism of intelligent robot. Comput Eng 22:141–144 (in Chinese)

Chapter 2
Research on Intelligent Fault Diagnosis of Engine Based on MOBP Neural Network

Heda Zhang, Jiantong Song, Jialin Han, Fang Fang and Wanqiang Ren

Abstract Facing the demands of precise and intelligent diagnosis for engine faults, this paper applies the artificial intelligent theory, studies and realizes the application of the MOBP neural network on the intelligent diagnosis for engine faults. As is proved by the experiments, the training time based on the MOBP neural network is shorter than traditional methods, and the reasoning efficiency is significantly improved. It is confirmed that the study of the intelligent fault diagnosis for engine based on the MOBP neural network is practical.

Keywords Engine diagnosis · MOBP algorithm · Neural network

2.1 Introduction

The engine is the core component of the vehicle, and every part influences and associates each other. Because of the flexible and terrible operation condition, there are more chances of fault in the engine, about 40 % faults among the vehicle. With the widely used of the electrical engine, the amount electric elements and sensors, self-diagnosis function was essential in the electrical engines. But this was restricted in the common fault diagnosis, not efficient in the flexible fault diagnosis. So the study for the intelligent diagnosis methods for the electrical engine is more essential [1].

At present, the artificial intelligence based on neural network technology is widely used in many fields [2]. The neural network based momentum back

H. Zhang (✉) · J. Han
Institute of Industrial Engineering, Beijing Institute of Technology, Beijing, China
e-mail: zhang_heda@126.com

H. Zhang · J. Song · F. Fang · W. Ren
Institute of Automotive Engineering, Beijing Polytechnic, Beijing, China

Z. Sun and Z. Deng (eds.), *Proceedings of 2013 Chinese Intelligent Automation Conference*, Lecture Notes in Electrical Engineering 255,
DOI: 10.1007/978-3-642-38460-8_2,

propagation (MOBP) algorithm can make the way of diagnosis of engine fault intelligent [3]; in addition, the application of the fast adaptive algorithm of the learning rate shorten the training time than the traditional way. The reasoning ability of the neural network may improve obviously, and effectively heighten the ability of intelligent diagnosis for electrical engine.

2.2 The Structure and Principle of BP Neural Network

Back propagation neural network is a multilayer feed forward network with error back propagation algorithm training, consisting of two process of the information forward propagation and error back propagation, which is one of the most widely used neural network models [4].

BP neural network topology includes the input layer, hidden layer, and output layer. Each neuron of the input layer is responsible for receiving the input information from the outside, and pass them to the each neural element of intermediate layer; The intermediate layer which is responsible for the information conversion is an internal information processing layer; in addition, the intermediate layer can be designed as a single hidden layer or multi-hidden-layers structure according to the requirement of information variety; When the information transported from the last hidden layer to each neural element of output layer was further disposed, one of the forward propagation learning processes is completed, and the final result is produced from output layer to outside [5].

If the actual output does not match the expected output, then the error back propagation stage begin, and every layer weighted value is corrected by the way of error gradient descent from the input layer to the hidden layer, layer by layer. Cycle of information forward propagation process and error back propagation process, which is not only the process of each layer's weighted value constantly adjusting, but also the training process of the neural network learning [6].

2.3 The Intelligent Fault Diagnosis System of Engine

In designing the engine intelligent fault diagnosis neural network, make fault symptoms as the input node of the network, the cause of the fault as the output node [7]. Firstly train the network with fault samples to determine the structure of the network, the transfer function of the middle layer, the number of neurons and neuron weights and threshold value [8]. To make up for the slow convergence rate and local minimum value of the standard BP algorithm, such as drawback using MOBP algorithm, that is based on the momentum BP neural network model. Greater learning rate can be used in momentum back propagation algorithm, without causing the divergence of the learning process.

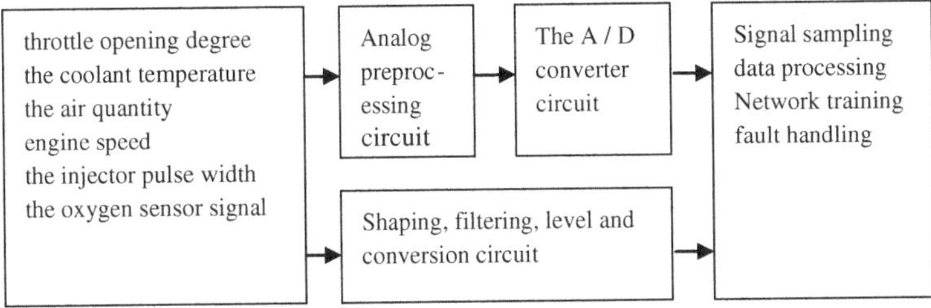

Fig. 2.1 Diagram of the test system

2.3.1 The Framework of Test System

Combined with the actual test conditions, the network input layer includes six input variables, Fig. 2.1 is a block diagram of the test system.

2.3.2 The MOBP Algorithm of Neural Network

MOBP algorithm reduces the sensitivity of the network to the error surface, effectively suppresses the network falling into local minimum. As follows is the expression of the algorithm:

$$\Delta x(k+1) = \eta \Delta x(k) + \alpha(1-\eta)\frac{\partial E(k)}{\partial x(k)} \tag{2.1}$$

$$x(k+1) = x(k) + \Delta x(k+1) \tag{2.2}$$

where, $x(k)$ is connection weights vector or threshold vector at the k times;
η is the momentum factor $(0<\eta<1)$;
E_k is the total error function at the k times iteration.

To make the algorithm stable as much as possible, and ensure that the learning step is as large as possible, this test supplemented fast adaptive learning rate algorithm, Making a corresponding adjustment of the learning rate according to the local error surface. Adaptive learning rate adjustment formula is as follows:

$$\alpha(k+1) = \begin{cases} 1.05\alpha(k)E(k+1)<E(k) \\ 0.7\alpha(k)E(k+1) > 1.04E(k) \\ \alpha(k) \end{cases} \tag{2.3}$$

In this algorithm, each time the last correction result influences this correction value. If the last correction value was excessive, this one will reduce, and this algorithm plays a role in reducing vibrations; On the other hand, if the last correction value was too low, this one will increase, and this algorithm plays a role in accelerating correction.

2.3.3 Training the Network and Analyzing the Results

The test object of this paper was Passat 3.0-liter V-6 cylinder FSI engine, and we collected a great deal of real-time data involving Idle condition, start state, accelerate condition, and every common load condition, so that the neural network obtained good generalization ability.

The testing process had been simplified to four common faults including 1 cylinder injector fault, Oxygen sensor fault, the air quantity meter fault and coolant temperature sensor fault. As Table 2.1 shown is fault model identification of training network, Table 2.2 shows part of the samples data in the network training. In practical application, in order to make the engine fault diagnosis system more accurate, that was best to have a choice of more training samples.

Among all the samples of the network training, the different operating parameters have different units and magnitude, and every data differ greatly in different dimensions, so all the data should be normalized before network training by the conversion processing the input and output data of the network is limited within the interval (0, 1), so that each input component is given equal importance from the beginning. If the processing is done with original data, it will result in the magnified data and dropped astringency, or even no astringency. Final output of the network must also be anti-normalization processing, so it will be able to show the actual significance.

The paper adopted linear normalization method as follows:

$$\overline{a_{ji}} = \frac{a_{ji} - a_{j\min}}{a_{j\max} - a_{j\min}}, \quad j \leq m,\ i \leq n \tag{2.4}$$

Among the linear normalization method, a_{ji}, $a_{j\min}$ are respectively the normalized value, the actual value, the maximum value and the minimum value for each group sample data; m is the samples number; n is the nods number. This method not only ensures the data in the interval of (0, 1), but also amplify the differences between the various data in the same set of samples. The training process of the neural network is the process of obtaining the right value and the

Table 2.1 The fault pattern recognition of network training

Number	Pattern type	Vector output
1	No fault	000
2	The injector fault of no. 1 cylinder	001
3	Oxygen sensor fault	010
4	Air mass meter fault	011
5	Coolant temperature sensor fault	100

Table 2.2 Part of the samples data in the network training

Sample no.	Engine speed / rpm	Throttle opening / mV	Air mass / V	Coolant temperature / V	Oxygen sensor /mV	Fuel injection pulse /ms	Desired output
1	906	506.5	1.716	0.375	39.1	3.3	000
2	899	503.9	1.758	0.378	39.5	3.5	100
3	875	504.8	1.742	0.356	39.6	3.2	011
4	904	506.8	1.733	0.324	39.7	3.2	010
5	878	527.5	1.798	0.395	39.5	3.5	001

Table 2.3 Dates of testing samples and the simulation results

Sample no.	Engine speed /rpm	Throttle degree /mV	Air mass / V	Coolant temperature / V	Oxygen sensor /mV	Injection pulse /ms	Desired output
1	1150	605.2	1.753	0.356	39.5	3.2	000
2	860	586.5	1.745	0.365	39.4	3.3	100
3	968	512.5	1.762	0.332	39.6	3.1	011
4	1520	652.2	1.712	0.325	39.5	3.1	010
5	1250	625.5	1.758	0.398	39.3	3.8	001

threshold values, edit the network training program through MATLAB (Table 2.3).

After repeated training and comparison, the network structure finalized one hidden layer structure as 6-13-5 involving six input nods, three hide nods, and three output nods. The transfer function of the hide layer adopts the S type tansig function, the transfer function of output layer adopts the S type logsig function, the network training function adopt trainlm function, the final error is 0.01. After the transient training, the target of the network reached, and the best result was conducted as the model network.

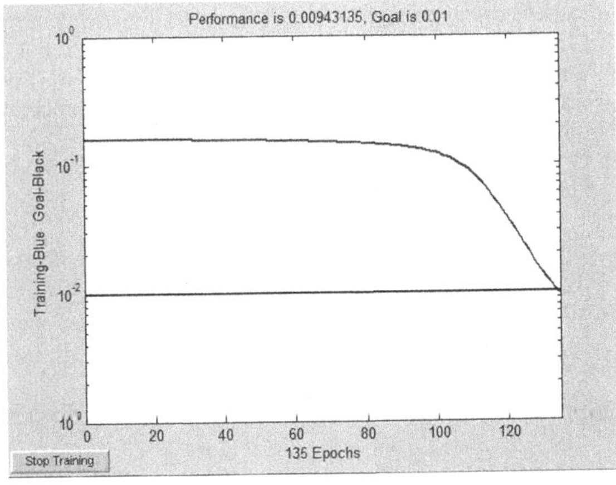

Fig. 2.2 The training error curve of the network

The training time was set 500 times in the neural network for engine intelligent fault diagnosis system, but the convergence was completed after 135 times, and the training accuracy was obtained, as the Fig. 2.2 shows is the training error curve of network.

2.4 Conclusion

Through the practical tests for the real engine faults, the neural network based MOBP algorithm realized the precise mapping and the accuracy is up to 0.01. The total operating time for engine faults diagnosis is merely about 9 ms. and the real diagnosis could be realized. It's proved by the experiment that the intelligent fault diagnosis system of engine based on MOBP neural network algorithm is workable and effective.

References

1. Benkaci M, Hoblos G (2012) Feature selection combined with neural network for diesel engine diagnosis. In: Proceedings of the ICINCO 2012 9th international conference on informatics in control, automation and robotics, p 317–24
2. Calise AJ, Hovakimyan N (2001) Adaptive output feedback control of nonlinear system using neural networks. Automatica 37:126–179
3. Wang Z, Yugeng S, Zhang Q, Qin J, Sun X, Shen H (2006) Research on fuzzy neural network algorithms for nonlinear network traffic predicting. Opto-Electron Lett 2(5) (in Chinese)
4. Shi L, Wang XC (2011) Application of probabilistic neural network in engine fault diagnosis. Process Autom Instrum 32(3):33–35 March (in Chinese)
5. Wang Y, Yanfeng X, He H (2010) An intelligent approach for engine fault diagnosis based on wavelet pre-processing neural network model. In: Proceedings of the 2010 international conference on information and automation (ICIA 2010) 576–581 (in Chinese)
6. Zhang Q, Chen N, Huang J, Meng Z (2012) Application of expert system fuzzy BP neural network in fault diagnosis of piston engine. In: Proceedings of the 2012 international conference on computer science and electronics engineering (ICCSEE 2012) 604–607 (in Chinese)
7. Chen T, Sun J, Hao Y (2006) Neural network and Dempster-Shafter theory based fault diagnosis for aeroengine gas path . Acta Aeronautica Et Astronautica Sinica 1014–1017 (in Chinese)
8. Li G, Yang Q (2007) FNN-based intelligent fault diagnosis system for vehicle engine. J Syst Simul 1034–1037 (in Chinese)

Chapter 3
Feedback Linearization Optimal Control for Bilinear Systems with Time-Delay in Control Input

Dexin Gao

Abstract This paper considers the optimal control problem for bilinear system with time-delay in control input based on state feedback. Firstly, we change a bilinear system with time-delay in control action model to a time-delay pseudo linear system model by local linearization. Through the Artstein transformation, a linear controllable system without delay is obtained. Then based on the theory of linear quadratic optimal control, a optimal controller is designed by solving the Riccati equation. At last, the simulation results show the effectiveness of the method.

Keywords Bilinear systems · Optimal control · Feedback linearization · Time-delay

3.1 Introduction

Bilinear system is a special nonlinear system, during the processes of the engineering, social economy and ecology, there are so many objects can be described by bilinear systems. Bilinear system is close to linear system in the aspects of form, so some theory of linear systems can be used for bilinear systems. Meanwhile, because of bilinear systems can be approximated as many nonlinear systems, it is more accurate than the traditional linear approximation. Therefore, the study of bilinear systems is becoming particularly important. At present, some research results about the bilinear systems have been obtained. For example, Aganovic proposed a method of global successive approximation about bilinear system [1]; DISOPE approximate algorithm based on bilinear model is presented

D. Gao (✉)
College of Automation and Electronic Engineer, Qingdao University of Science & Technology, No 53 ZhengZhou Road, Shangdong 266042 QingDao, China
e-mail: qdgaodexin@126.com

Z. Sun and Z. Deng (eds.), *Proceedings of 2013 Chinese Intelligent Automation Conference*, Lecture Notes in Electrical Engineering 255,
DOI: 10.1007/978-3-642-38460-8_3,

by Li [2]; Tang has studied the optimal control of the discrete bilinear system [3–5]. The optimal iterative algorithm based on quadratic performance index about bilinear system is given in the Ref. [6–8], etc. In many bilinear process controls, time delay phenomenon exists widely in the transmission process of control signal, which can not be neglected in some accurate control systems.

This paper proposed an optimal control design method for time-delay bilinear systems based on state-feedback linearization and model transformation. Firstly, the model of the control time-delay bilinear system is given in this paper and changed to the nonlinear system mode; Secondly, a complex nonlinear system model is changed to a easy pseudo linear system model by the differential homeomorphism; then through the Artstein transformation, the time-delay pseudo linear system is converted to an easy pseudo linear system; At last, a optimal controller is designed by solving the Riccati equation.

3.2 Problem Statement

3.2.1 System Description

Consider bilinear systems with control time-delay described by the following difference equations

$$\begin{aligned} \dot{x} &= \tilde{A}x + \{Nx\}u(t-\tau) + \tilde{B}u(t-\tau) \\ y &= h(x) \\ x(t_0) &= x_0 \\ \{Nx\} &= \sum_{j=1}^{n} N_j x_j(t) \end{aligned} \tag{3.1}$$

where $x(t)$ is the state vector; $u(t)$ is the control vector; $y(t)$ is the output vector; $\tilde{A}$, $\tilde{B}$, N_j are scalar matrixes of appropriate dimensions; x_j is the j-th component of state vector; $\{Nx\}u$ is the bilinear term; $h(x)$ is the scalar function of x. $\tau > 0$ is the known control delay, let $u(t) = 0, t \in [-\tau, 0)$

Through transformation, change the control time-delay bilinear system to the general expression of control time-delay nonlinear system

$$\begin{aligned} \dot{x} &= f(x) + g(x)u(t-\tau) \\ y &= h(x) \end{aligned} \tag{3.2}$$

Where $f(x) = \tilde{A}x$, $g(x) = \{Nx\} + \tilde{B}$.

Through exact linearization and delay-free transformation, we can change the time-delay nonlinear system (3.2) to an easy pseudo linear system (3.3)

$$\dot{\eta} = A\eta + Bu(t) \tag{3.3}$$

where $z = [z_1\ z_2\ \ldots z_r]^T$ is the new state vector,

$$A = \begin{bmatrix} 0 & 1 & 0 & \cdots & 0 & 0 \\ 0 & 0 & 1 & \cdots & 0 & 0 \\ \vdots & \vdots & \vdots & \vdots & \vdots & \vdots \\ 0 & 0 & 0 & \cdots & 0 & 1 \\ c_1 & c_2 & c_3 & \cdots & c_{r-1} & c_r \end{bmatrix} \quad B = e^{-A\tau}\begin{bmatrix} 0 \\ 0 \\ \vdots \\ 0 \\ m \end{bmatrix}$$

Then we can get the optimal control law based on the theory of linear quadratic optimal control.

3.2.2 System Transformation

Consider the control time-delay nonlinear systems described by the general model, assuming that the relative degree of the output y with respect to the input $u(t)$ is r, we get

$$z_i = L_f^{i-1}h(x) \quad i = 1, 2, \ldots r \tag{3.4}$$

$$L_f^r h(x) = \sum_{i=1}^{r} c_i L_f^{i-1} h(x) \tag{3.5}$$

$$\begin{aligned} &L_g L_f^{i-1} h(x) = 0 \quad i = 1, 2, \ldots r-1 \\ &L_g L_f^{r-1} h(x) = m \end{aligned} \tag{3.6}$$

Let:

$$z = \phi(x) = \left[h(x)\ \ L_f h(x) \ldots L_f^{r-1} h(x)\right]^T, r \le n \tag{3.7}$$

where $\phi(x)$ is the partial differential homeomorphism. We obtain

$$\begin{aligned} \dot{z}_1 &= \frac{\partial h(x)}{\partial x}\dot{x} = \frac{\partial h(x)}{\partial x}(f(x) + g(x)u(t-\tau)) \\ &= L_f h(x) + L_g L_f^0 h(x) u(t-\tau) \\ &= L_f h(x) = z_2 \end{aligned} \tag{3.8}$$

$$\dot{z}_2 = z_3 \tag{3.9}$$

Then get

$$\dot{z}_{r-1} = L_f^{r-1} h(x) + L_g L_f^{r-2} h(x) u(t-\tau) = z_r \tag{3.10}$$

$$\begin{aligned}\dot{z}_r &= L_f^r h(x) + L_g L_f^{r-1} h(x) u(t-\tau) \\ &= \sum_{i=1}^{r} c_i z_i + m u(t-\tau)\end{aligned} \tag{3.11}$$

Equations (3.8)–(3.11) can be written as

$$\begin{aligned}\dot{z}(t) &= Az(t) + B_1 u(t-\tau) \\ z(0) &= z_0 \\ y(t) &= h(x) = z_1\end{aligned} \tag{3.12}$$

where $z = [z_1\ z_2 \ldots z_r]^T$ is the new state vector.

$$A = \begin{bmatrix} 0 & 1 & 0 & \cdots & 0 & 0 \\ 0 & 0 & 1 & \cdots & 0 & 0 \\ \vdots & \vdots & \vdots & \vdots & \vdots & \vdots \\ 0 & 0 & 0 & \cdots & 0 & 1 \\ c_1 & c_2 & c_3 & \cdots & c_{r-1} & c_r \end{bmatrix} \quad B_1 = \begin{bmatrix} 0 \\ 0 \\ \vdots \\ 0 \\ m \end{bmatrix} \tag{3.13}$$

Written Eq. (3.12) as the form of integral equation

$$\begin{aligned}z(t) &= e^{At} z_0 + \int_0^t e^{A(t-h)} [B_1 u(h-\tau)] dh \\ &= e^{At} z_0 + \int_0^t e^{A(t-h)} [e^{-A\tau} B_1 u(h)] dh - \int_{t-\tau}^t e^{A(t-h)} e^{-A\tau} B_1 u(h) dh\end{aligned} \tag{3.14}$$

let

$$\begin{aligned}\eta(t) &= z(t) + \int_{t-\tau}^t e^{A(t-h)} B u(h) dh \\ B &= e^{-A\tau} B_1\end{aligned} \tag{3.15}$$

The system (3.12) can be converted to an easy pseudo linear system (3.16)

$$\begin{aligned}\dot{\eta} &= A\eta(t) + Bu(t) \\ \eta(0) &= z_0 \\ z(t) &= \eta(t) - \int_{t-\tau}^t e^{A(t-h)} B u(h) dh\end{aligned} \tag{3.16}$$

where $\eta(t)$ is the new state vector.

3.3 Optimal Controller Design

3.3.1 Quadratic Performance Index Analysis

The optimal control problem of the control time-delay bilinear system (3.1) could be transformed into to the l linear system described by (3.16).

For the finite horizon optimal control problem, select the quadratic performance index as:

$$J = \frac{1}{2}\eta^T(t_f)Q_f\eta(t_f) + \frac{1}{2}\int_{t_0}^{t_f} [\eta^T(t)Q\eta(t) + u^T(t)Ru(t)]dt \tag{3.17}$$

where Q is a $n \times n$ positive semi-definite matrix, R is a $m \times m$ positive definite symmetric matrix. t_f is the end time known.

For the infinite horizon optimal control problem, select the quadratic performance index as:

$$J = \frac{1}{2}\int_0^{\infty} [\eta^T(t)Q\eta(t) + u^T(t)Ru(t)]dt \tag{3.18}$$

The optimal control problem is to find a control law $u^*(t)$, which minimizes the quadratic performance index J described by (3.17), (3.18).

3.3.2 Optimal Control Design of Finite Time

Theorem 3.1 *Consider the optimal control problem of system* (3.16) *with the quadratic performance index* (3.17). *The optimal control law is existent and unique and its form as follows*:

$$u^*(t) = -R^{-1}B^TP(t)\eta(t) \tag{3.19}$$

$P(t)$ is the unique positive definite solution of the Riccati matrix difference Eq. (3.20)

$$-\dot{P}(t) = A^TP(t) + P(t)A - P(t)SP(t) + Q \tag{3.20}$$

where $S = BR^{-1}B^T$.

Proof Let

$$\lambda(t) = P(t)\eta(t) \tag{3.21}$$

According to (3.21), get(3.22)

$$\dot{\lambda}(t) = [\dot{P}(t) + P(t)A - P(t)SP(t)]\eta(t) \tag{3.22}$$

$$-\dot{\lambda}(t) = Q\eta(t) + A^T\lambda(t) \tag{3.23}$$

the transversal condition

$$\lambda(t_f) = Q_f\eta(t_f) \tag{3.24}$$

Adding Eqs. (3.22) and (3.23), *Riccati* Eq. (3.20) is obtained, and the optimal control law is obtained as follow

$$u^*(t) = -R^{-1}B^TP(t)(z(t) + \int_{t-\tau}^{t} e^{A(t-h)}Bu(h)dh) \tag{3.25}$$

Combined above derived process, the optimal control law of control time-delay bilinear system in infinite domain has been obtained. The proof is complete.

3.3.3 Optimal Control Design of Infinite Time

Theorem 3.2 *Consider the optimal control problem of system* (3.16) *with the quadratic performance index* (3.18)

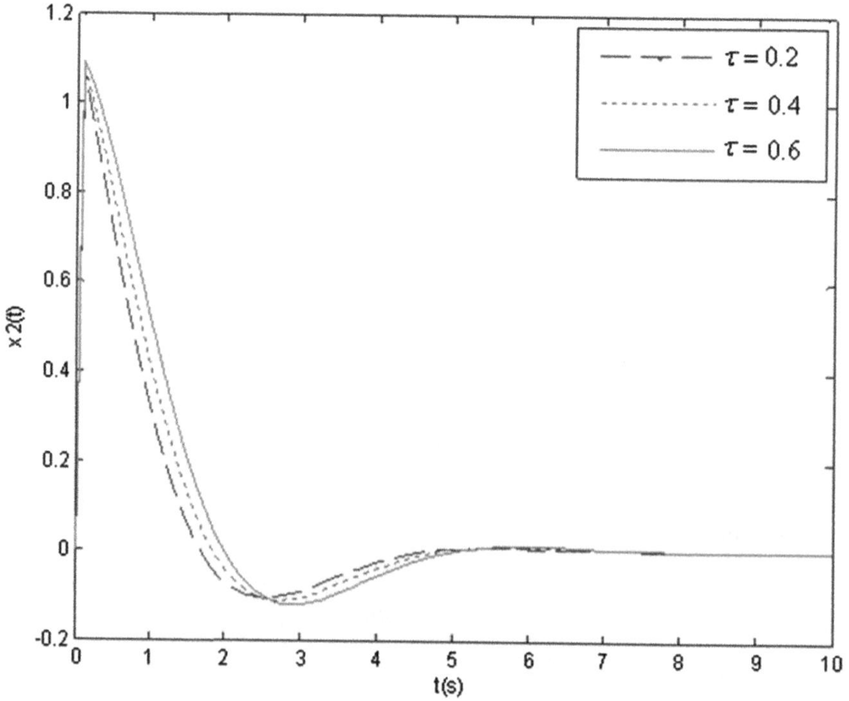

Fig. 3.1 State vector $x_2(t)$

$$u^{*}(t) = -R^{-1}B^{T}P(z(t) + \int_{t-\tau}^{t} e^{A(t-h)}Bu(h)dh) \tag{3.26}$$

P is the unique positive definite solution of the Riccati matrix Eq. (3.27)

$$A^{T}P + PA - PSP + Q = 0 \tag{3.27}$$

The derived process is the same as the finite time optimal control design.

3.4 Simulation Example

In order to illustrate the effectiveness and feasibility of this method, consider a control time-delay bilinear system as follows

$$\begin{aligned} \dot{x} &= \tilde{A}x + \{Nx\}u(t-\tau) + \tilde{B}u(t-\tau) \\ y &= x_2 \end{aligned} \tag{3.28}$$

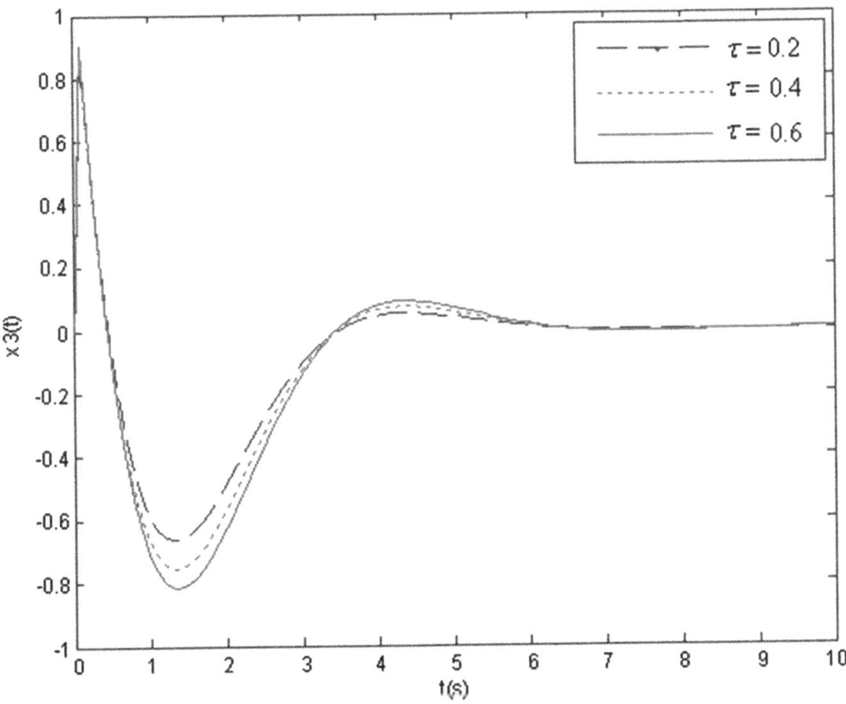

Fig. 3.2 State vector $x_3(t)$

$$\tilde{A} = \begin{bmatrix} 1 & 1 & 1 \\ 0 & 0 & 1 \\ 0 & -1 & 0 \end{bmatrix} \quad N = \begin{bmatrix} 1 & 0 & 0 \\ 0 & 0 & 0 \\ 0 & 0 & 0 \end{bmatrix} \quad \tilde{B} = \begin{bmatrix} 1 \\ 0 \\ 1 \end{bmatrix}.$$

According to calculation $r = 2$, through local linearization, we can get a time-delay pseudo linear system

$$\dot{z} = Az + B_1 u(t - \tau) \tag{3.29}$$

where $A = \begin{bmatrix} 0 & 1 \\ -1 & 0 \end{bmatrix}$, $B = \begin{bmatrix} 0 \\ 1 \end{bmatrix}$. Using model transformation, change the system with time-delay to a linear controllable system without delay

$$\dot{\eta} = A\eta(t) + Bu(t) \tag{3.30}$$

where $A = \begin{bmatrix} 0 & 1 \\ -1 & 0 \end{bmatrix}$, $B = e^{-A\tau} B_1$. Select $R = 1.0$, $Q = diag(1, 1)$, Consider the different effects of time-delay, we respectively select $\tau = 0.2, 0.4, 0.6$. The simulation curves of the state component $x_2(t)$, $x_3(t)$ and the control variable $u^*(t)$ are presented in Figs. 3.1, 3.2, 3.3.

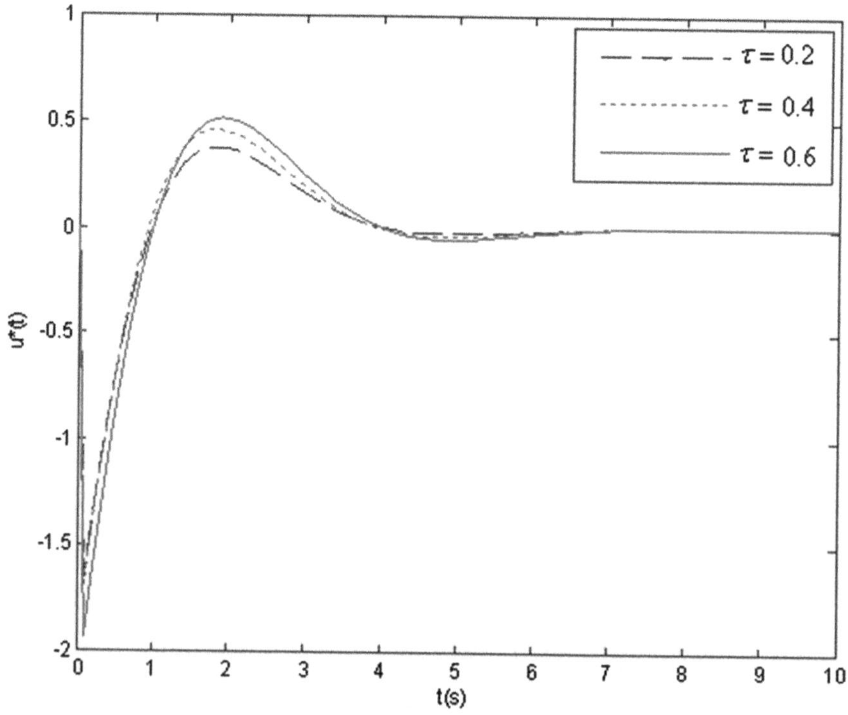

Fig. 3.3 The control vector $u(t)$

It is shown that the optimal control design method for bilinear system with time-delay in control input based on state feedback is effectiveness in different effects of time-delay. The proposed approach, which uses model transformation to devise the optimal control law of control time-delay bilinear systems, is valid.

3.5 Conclusion

This paper concentrates on the solution of the optimal control problem for time-delay bilinear systems based on state-feedback linearization and model transformation. Firstly, we change a bilinear system with time-delay in control action model to a time-delay pseudo linear system model by local linearization. Through the Artstein transformation, a linear controllable system without delay is obtained. Then based on the theory of linear quadratic optimal control, a optimal controller is designed by solving the Riccati equation. At last, the simulation results show the effectiveness and the feasibility of the method.

Acknowledgments This work was supported in part by the National Natural Science Foundation of China (60804005), and by the Natural Science Foundation of Shandong Province Grant ZR2011FQ006), by the Natural Science Foundation of Qingdao City (12-1-4-3-(17)-jch).

References

1. Aganovic Z, Gajic Z (1994) The successive approximation procedure for finite-time optimal control of bilinear systems. IEEE Trans Autom Control 39(9):1932–1935
2. Li J-M, Sun Y-P, Liu Y, Wang Y-L (2007) Convergent iterative algorithm of optimal control for discrete-time bilinear-quadratic problem. Numer Math J Chin Univ 29(3):216–225 (in Chinese)
3. Tang G-Y, Ma H, Zhang B-L (2005) Successive-approximation approach of optimal control for bilinear discrete-time systems. IEEE Proc Control Theor Appl 152(6):639–644 (in Chinese)
4. Tang G-Y, Zhao Y-D, Hu N-P (2007) Design of optimal tracking controllers with observer for bilinear systems. Control Theor Appl 24(3):503–507 (in Chinese)
5. Tang G-Y, Lei J, Sun L (2009) Observer-based optimal disturbance-rejection for linear systems with time-delay in control action. Control Theor Appl 26(2):209–214 (in Chinese)
6. Xu D-X, Liu D, Qian F-C (2005) Two-Level algorithm of solving discrete bilinear system optimal control. J Xi'an Univ Technol 21(01):73–76 (in Chinese)
7. Xu D-X, Qian F-C (2006) Optimal control of discrete bilinear system based on the hopfield neural network. Inf Control 35(1):90–92 (in Chinese)
8. Ma H, Tang G-Y (2007) Suboptimal control for bilinear discrete systems with time delay. Periodical Ocean Univ China 37(5):852–856 (in Chinese)

Chapter 4
L_∞ Dynamic Surface Control for a Class of Nonlinear Pure-Feedback Systems with Finite-Time Extended State Observer

Guofa Sun, Xuemei Ren and Dongwu Li

Abstract This paper presents a novel dynamic surface control (DSC) design for a class of nonlinear pure-feedback system without using adaptive approach. By introducing a set of alternative state variables and the corresponding transform, state-feedback control of the pure-feedback can be viewed as output-feedback control of a canonical system. To estimate unknown states of the newly derived canonical system and the lumped overall uncertainty, a finite-time extended state observer (ESO) is adopted. The closed-loop stability and L_∞ convergence of the tracking error to a small compact set around zero are proved. Comparative simulations are included to verify the reliability and effectiveness of the proposed approach.

Keywords Dynamic surface control · Pure-feedback system · Finite-time extended state observer · L_∞ stability

4.1 Introduction

The output feedback problem in backstepping or DSC is preminarily studied in the latest work by introducing a state observer [1, 2, 3]. However, in the mentioned results, only linear state observers are employed and ultimately uniformly bound

G. Sun (✉) · X. Ren (✉) · D. Li
School of Automation, Beijing Institute of Technology,
5 South Zhongguancun Street, Haidian, Beijing, People's Republic of China
e-mail: sungf@bit.edu.cn

X. Ren
e-mail: xmren@bit.edu.cn

D. Li
Institute 23 of China Aerospace Science and Industry Corporation, 52 Yongding Road, Haidian, Beijing, People's Republic of China
e-mail: lidongwu2002@163.com

Z. Sun and Z. Deng (eds.), *Proceedings of 2013 Chinese Intelligent Automation Conference*, Lecture Notes in Electrical Engineering 255,
DOI: 10.1007/978-3-642-38460-8_4,

(UUB) of all closed-loop signals are proved without considering transient tracking performance.

In this paper, considering the practical implementation issue of DSC designed for pure-feedback nonlinear system, an L_∞ output tracking control scheme is proposed based on finite-time nonlinear extend state observer (ESO). In recent years, ESO is successfully applied to lots of practical systems since it is first produced by Han [4] in 1995, for example, in speed regulation problem of PMSM [5], uncertain multivariable systems with time-delay [6] and the problem of secure data transmission [7]. Motivated by the idea of finite-time observer, a finite-time ESO will be employed before the L_∞ DSC is designed.

The contributions of this paper are briefly summarized as follows.

1. A simplified DSC method without function approximation is employed as the adopted control technique. It is combined with ESO to address the output-feedback control problem, which is more applicable comparing with existing state-feedback control schemes in practical systems. Furthermore, the L_∞ Stability is proved to make sure that tracking signals maintain good transient performance.
2. Since the considered pure-feedback nonlinear models include a large class of SISO nonlinear systems, the proposed method can be used for strict-feedback systems and other pure-feedback systems with few modifications.
3. A finite-time ESO is employed to address the output-feedback problem that the desired feedback controllers are asserted to exist but immeasurable. As a result, the proposed approach can be easily implemented in practical plants, without the sophisticated controller expression and long time online computing.

4.2 Coordinate Transformation and Problem Formulation

Consider the following pure-feedback continuous-time nonlinear system

$$\begin{cases} \dot{x}_i = f_i(\bar{x}_{i+1}) + x_{i+1} \\ \dot{x}_n = f_n(\bar{x}_n) + u \\ y = x_1 \end{cases} \tag{4.1}$$

where $\bar{x}_i = (x_1, \ldots, x_i)^T \in R^i$, $i = 1, \ldots, n$ denote the system states, $u \in R, y \in R$ are the control input and output of the system, and $f_i(\bullet), i = 1, \ldots, n$ are smooth nonlinear functions.

Applying the mean value theorem [8], system (4.1) can be transformed into

$$\begin{cases} \dot{x}_i = f_i(\bar{x}_i) + g_i(x_{i+1})x_{i+1}, & g_i(x_{i+1}) = \bar{g}_i(\bar{x}_i, x_{i+1}^{\gamma_i}) + 1, \ 1 \le i \le n+1 \\ \dot{x}_n = f_n(\bar{x}_n) + g_n(x_n, u)u, & g_n(x_n, u) = \bar{g}_n(\bar{x}_n, u^{\gamma_n}) + 1 \end{cases} \tag{4.2}$$

By defining alternative state variables as $z_1 = x_1$, $z_i = \dot{z}_{i-1}$, $y = z_1 = x_1$, $i = 2, \ldots, n$, by induction [9], the derivative of z_i, $i = 1, \ldots, n-1$ can be written in the following form of $\dot{z}_i = \alpha_i(\bar{x}_i) + \beta_i(\bar{x}_{i+1})x_{i+1}$ with

$$\alpha_i(\bar{x}_i) = \sum_{j=1}^{i-1}\left(\frac{\partial f_{i-1}}{\partial x_j} + \frac{\partial g_{i-1}}{\partial x_j}\right)(f_j + g_j x_{j+1}) + \left(\frac{\partial g_{i-1}}{\partial x_i}x_i + g_{i-1}\right)f_i,\ \beta_i(\bar{x}_{i+1})$$
$$= \left(\frac{\partial g_{i-1}}{\partial x_i}x_i + g_{i-1}\right)g_i.$$

Then, system (4.1) can be represented to be

$$\begin{cases} \dot{z}_i = z_{i+1}, \quad i = 1, \ldots, n-1, \quad y = z_1 \\ \dot{z}_n = \alpha_n(\bar{x}_n) + \beta_n(\bar{x}_n, u^{\gamma_n})u, \quad 0<\gamma_n<1 \end{cases} \tag{4.3}$$

It is shown that with the newly defined states and coordinate transform, pure-feedback system (4.1) is now reformulated as a Brunovsky system.

In the following, we will consider the output-feedback control design for (4.3) to achieve tracking control of (4.1).

4.3 Output DSC Controller Design

4.3.1 Finite-Time Extended State Observer Design

Since only the input and output signals are available, a state observer must be constructed to estimate the unmeasured variables z_i, $i = 1, \ldots, n$. To end this, the structure of the finite-time extended state observer (FTESO) is in the following form.

$$\begin{cases} \dot{\xi}_1 = \xi_2 - \kappa_1[\eta_1]^{\lambda}, \\ \dot{\xi}_i = \xi_{i+1} - \kappa_2[\eta_1]^{i\lambda-(i-1)}, \\ \dot{\xi}_n = \xi_{n+1} - \kappa_n[\eta_1]^{n\lambda-(n-1)} + \beta_0 u \\ \dot{\xi}_{n+1} = -\kappa_{n+1}[\eta_1]^{(n+1)\lambda-n} \end{cases}$$
$$\begin{cases} \eta_1 = \xi_1 - z_1, \\ \dot{\eta}_1 = \eta_2 - \kappa_1[\eta_1]^{\lambda}, \\ \dot{\eta}_i = \eta_{i+1} - \kappa_i[\eta_1]^{i\lambda-(i-1)}, \ i = 2, \cdots, n-1 \\ \dot{\eta}_{n+1} = -\kappa_{n+1}[\eta_1]^{(n+1)\lambda-n} \end{cases} \tag{4.4}$$

where $[x]^{\lambda} = |x|^{\lambda}\mathrm{sgn}(x)$ for all $x \in R$ and for $\lambda > 0$. The errors $\eta_i = \xi_i - z_i$, $i - 1, \ldots, n$ satisfy

$$\dot{\eta} = \psi(\lambda, \eta) \tag{4.5}$$

Lemma 1 *Set the gains* $(\kappa_1, \ldots, \kappa_{n+1})$ *to be Hurwitz. There exists* $\varepsilon \in [1 - \frac{1}{n-1}, 1)$ *such that, for all* $\lambda \in (1 - \varepsilon, 1)$*, system (4.3) is globally finite-time stability. For all* $\lambda \in (1 - \varepsilon, 1)$*, the Lyapunov candidate function is chosen as:*

$$V_o(\lambda, \eta) = \zeta^T P \zeta \tag{4.6}$$

Where ζ *is defined as* $\zeta = ([\eta_1]^q \; [\eta_2]^{1/(nq)} \; \cdots \; [\eta_n]^{1/([n\lambda-(n-1)]q)} \; [\eta_{n+1}]^{1/([(n+1)\lambda-n]q)})^T$ *with* $q = \Pi_{i=1}^n((i-1)\lambda - (i-2))$ *and P is the solution of Lyapunov equation* $A_o^T P + PA_o = -I$*.*

The detailed proof is similar to the proof of theorem 10 in Ref. [7].

According to the above lemma, by properly choosing the gain parameters $\kappa_i, \; i = 1, \ldots, n+1$, the observer error can be made converge to the origin asymptotically and the inequality $\dot{V}_o \leq -\|\zeta_i\|^2$ holds.

4.3.2 Dynamic Surface Controller Design

In the following development, based on the above finite-time ESO, a L_∞ DSC will be derived for (4.1).

Step 1 The first error surface is defined as $e_1 = \xi_1 - x_r$ with x_r as the desired trajectory. Taking (4.4) into consideration, one obtains the derivative of e_1 as

$$\dot{e}_1 = \xi_2 - \kappa_1[\eta_1]^\lambda + \eta_2 - \dot{x}_r \tag{4.7}$$

A virtual control α_1 is chosen as

$$v_2 = -k_1 e - e_1 + \kappa_1[\eta_1]^\lambda - \eta_2 + \dot{x}_d \tag{4.8}$$

Here, the quadratic form Lyapunov candidate $V_1 = \frac{1}{2}e_1^2$ is considered. Its derivative is obtained as

$$\dot{V}_1 = -k_1 e_1^2 + e_1(\xi_2 - v_2) \tag{4.9}$$

Let v_2 pass through a first-order filter with time constant τ_2 to obtain ϑ_2, i.e., $\tau_2\dot{\vartheta}_2 + \vartheta_2 = v_2, \; \vartheta_2(0) = v_2(0)$. Then, $e_2 = \xi_2 - \vartheta_2$ becomes the second error surface and the subsequent steps from 2 to n-1 can be performed in the same fashion recursively.

Step i $(2 \leq i \leq n-1)$: The ith error surface is defined to be $e_i = \xi_i - \vartheta_i$. Its derivative is obtained as

$$\dot{e}_i = -\kappa_i[\eta_1]^{i\lambda-(i-1)} + \xi_{i+1} - \dot{\vartheta}_i \tag{4.10}$$

Similarly to step 1, the Lyapunov candidate is chosen as $V_i = \frac{1}{2}e_i^2$ and the virtual control signal is selected to be

$$v_{i+1} = -k_i e_i + \kappa_i[\eta_1]^{i\lambda-(i-1)} + \dot{\vartheta}_i \tag{4.11}$$

Then, it can be readily checked that

$$\dot{V}_i \leq -k_i e_i^2 + e_i(\xi_{i+1} - v_{i+1}) \tag{4.12}$$

Introduce a set of new state variables ϑ_{i+1} and let v_{i+1} pass through a first order filter with time constant τ_{i+1} to obtain ϑ_{i+1}, i.e., $\tau_{i+1}\dot{\vartheta}_{i+1} + \vartheta_{i+1} = v_{i+1}$, $\vartheta_{i+1}(0) = v_{i+1}(0)$.

Step n The last error surface is defined to be $e_n = \xi_n - \vartheta_n$, whose derivative is $\dot{e}_n = -\kappa_n[\eta_1]^{n\lambda-(n-1)} + \xi_{n+1} - \dot{\vartheta}_n$. The Lyapunov candidate function of the nth subsystem is given as $V_n = \frac{1}{2}e_n^2$. Finally, the actual control signal is chosen as

$$u = k_n e_n + \kappa_n[\eta_1]^{n\lambda-(n-1)} - \xi_{n+1} + \dot{\vartheta}_n \tag{4.13}$$

which results in $\dot{V}_n \leq -k_n e_n$.

4.4 L$_\infty$ Stability Analysis

Theorem *Considering the closed-loop system consisting of the plant* (4.1), *with finite-time ESO* (4.4), *the virtual control laws* v_i $i = 2, \ldots, n$ *given by* (4.8), (4.11) *and actual control law* u *given by* (4.13). *For any given constants* $\mu > 0$, $R_0 > 0$, *if* $V(0) \leq \mu$, $x_r + \dot{x}_r + \ddot{x}_r \leq R_0$, *there exist design parameters* k_i, τ_{j+1}, $i = 1, \ldots, n$, $j = 1, \ldots, n-1$ *and* κ_i *such that all signals of the overall closed-loop system are uniformly bounded. Moreover, by properly choosing the design parameters, the* L$_\infty$ *norm of tracking errors can converge to a residual set that can be made arbitrarily small.*

Proof In order to facilitate the validation procedure, for the ith subsystem, the following notations representing the filter error at each step of the controller design procedure are needed.

$$\begin{aligned} E_2 &= \vartheta_2 - v_2 = \vartheta_2 + k_2 e_2 - \kappa_2[\eta_1]^{2\lambda-1} - \dot{\vartheta}_2 \\ E_i &= \vartheta_i - \alpha_{i-1} = \vartheta_i + k_i e_i - \kappa_i[\eta_1]^{i\lambda-(i-1)} - \dot{\vartheta}_i, \quad i = 3, \ldots, n \end{aligned} \tag{4.14}$$

It is not difficult to obtain that

$$\dot{\vartheta}_i = -\frac{E_2}{\tau_2}, \quad \xi_i - \alpha_i = e_i + E_i, \quad i = 2, \ldots, n \tag{4.15}$$

substituting which into (4.14) yields

$$\left|\dot{E}_i + \frac{E_i}{\tau_i}\right| \leq \chi_i(e_1, \ldots, e_{i+1},\ E_2, \ldots, E_{i+1},\ x_r,\ \dot{x}_r,\ \ddot{x}_r,\ \eta_1), \quad i = 2, 3, \cdots, n \tag{4.16}$$

where χ_i, $i = 2, \ldots, n$ are continuous functions. Hence, it can be readily checked that $E_i\dot{E}_i \leq -\frac{1}{\tau_i}E_i^2 + \chi_i|E_i|$, $\quad i = 2, \ldots, n$. Consider the Lyapunov function candidate to be defined by

$$V = V_o + \sum_{i=1}^{n} V_i + \frac{1}{2}\sum_{i=2}^{n} E_i \tag{4.17}$$

Its derivative statisfies

$$\dot{V} = -\sum_{i=1}^{n} k_i e_i + \sum_{i=1}^{n-1} e_i(\xi_{i+1} - v_{i+1}) + V_o + \sum_{i=2}^{n} E_i\dot{E}_i \tag{4.18}$$

Substituting (4.14) into (4.18) yields

$$\dot{V} = -\sum_{i=1}^{n} k_i e_i + \sum_{i=1}^{n-1} e_i(e_i + E_i) + V_o + \sum_{i=2}^{n} E_i\dot{E}_i \tag{4.19}$$

Considering (4.16) obtains the following in equations

$$\dot{V} \leq -\sum_{i=1}^{n} k_i e_i + \sum_{i=1}^{n-1} e_i(e_i + E_i) + V_o + \sum_{i=2}^{n}\left(-\frac{1}{\tau_i}E_i^2 + \chi_i|E_i|\right) \tag{4.20}$$

Here, in order to proceed the proof, an Young' in equation as $e_i(e_{i+1} + E_{i+1}) \leq \frac{1}{2}e_i^2 + e_{i+1}^2 + E_{i+1}^2$ is needed. Then, the inequality (4.20) can be rewritten as

$$\begin{aligned}\dot{V} &\leq -\sum_{i=1}^{n} k_i e_i^2 + \sum_{i=1}^{n-1}\left(\frac{1}{2}e_i^2 + e_{i+1}^2 + E_{i+1}^2\right) + V_o + \sum_{i=2}^{n}\left(-\frac{1}{\tau_i}E_i^2 + \chi_i|E_i|\right) \\ &\leq -\left(c_1 - \frac{1}{2}\right)e_1^2 - \sum_{i=2}^{n}\left(c_i - \frac{3}{2}\right)e_i^2 + V_o - \sum_{i=2}^{n}\chi_i|E_i|\end{aligned} \tag{4.21}$$

Since for any $R_0 > 0$ and $\mu > 0$, the following sets are compact on R^3 and R^{3i-1}, respectively. $\Omega := \{(x_r, \dot{x}_r, \ddot{x}_r) : x_r + \dot{x}_r + \ddot{x}_r \leq R_0\}$, $\Omega_i := \Big\{\Sigma_{j=1}^{i} e_j^2 + \Sigma_{j=1}^{i} v_j^T P_j v_j + \Sigma_{j=2}^{i} E_j^2 \leq 2\mu\}$, $\quad i = 1, \ldots, n$. Then $\Omega \times \Omega_i$ is also compact in R^{2i+2}. Therefore, $|\chi_{i+1}|$ has a maximum B_{i+1}, on $\Omega \times \Omega_i$. Thus, for any constant $\delta > 0$, $\chi_i|E_i| \leq \frac{\chi_i^2 E_i^2}{2\delta} + \frac{\delta}{2} \leq \frac{B_i^2}{2\delta}E_i^2 + \frac{\delta}{2}$, holds, which together with (4.16), implies that

$$\dot{V} \le -\left(k_1-\frac{1}{2}\right)e_1^2-\sum_{i=2}^{n}\left(k_i-\frac{3}{2}\right)e_i^2+V_o-\sum_{i=2}^{n}\left(\frac{1}{\tau_i}-1E_i^2\right)+\sum_{i=2}^{n}\left(\frac{B_i^2}{2\delta}E_i^2+\frac{\delta}{2}\right) \le -\left(k_1-\frac{1}{2}\right)e_1^2$$
$$-\sum_{i=2}^{n}\left(k_i-\frac{3}{2}e_i^2\right)-\|v_i\|^2-\sum_{i=2}^{n}\left(\frac{1}{\tau_i}-1-\frac{B_i^2}{2\delta}E_i^2\right)+\frac{(n-1)\delta}{2} \tag{4.22}$$

Letting $\rho=\min_{2\le i\le n}\left\{k_1-\frac{1}{2},\,k_i-\frac{3}{2},\,\frac{1}{\tau_i}-1-\frac{B_i^2}{2\delta},\,\frac{1}{\lambda_{\max}(P)}\right\}$, we have the following inequality.

$$\dot{V}\le 2\rho V+\varsigma,\quad \varsigma=\frac{(n-1)\delta}{2} \tag{4.23}$$

If the design parameters are chosen such that $\rho>\varsigma/(2\mu)$, then $\dot{V}<0$ on $V=\mu$. It implies that $V\le\mu$ is an invariant set, i.e., $V(0)\le\mu$, then $V(t)\le\mu$ for all $t\ge0$. Moreover, if the initial conditions of (4.4) are set to be zero except $\xi_1(0)=z_1(0)$, $x_r(0)=z_1(0)$, we have $E_i(0)=0,\quad i=2,\ldots,n,\quad e_j(0)=0$, $j=1,\ldots,n$. Hence, in view of V, $V(0)$ can be computed as $V(0)=\zeta^T(0)P\zeta(0)$. Since the initial conditions of the (4.4) are set to be zero except for $\xi_1(0)=z_1(0)$, from the definition of ξ, we have $\xi(0)=[z_1(0)\ ,\quad 0,\ \cdots\quad 0]^T$. Therefore, $\eta(0)=[0,-z_2(0),\cdots-z_{n+1}(0)]^T$ which together with the expression of ζ implies that

$$V(0)=\zeta^T(0)P\zeta(0)\le\lambda_{\max}(P)\sum_{i=2}^{n+1}\left([z_i(0)]^{\frac{1}{(i\lambda-(i-1))q}}\right)^2\le\frac{1}{\rho}\sum_{i=2}^{n+1}\left([z_i(0)]^{\frac{1}{(i\lambda-(i-1))q}}\right)^2 \tag{4.24}$$

Substituting (1.12) into $V(t)$ arrive at

$$V(t)\le\frac{1}{\rho}\sum_{i=2}^{n+1}\left([z_i(0)]^{\frac{1}{(i\lambda-(i-1))q}}\right)^2=\frac{\varsigma_0}{2\rho},\quad \varsigma_0=2\sum_{i=2}^{n+1}\left([z_i(0)]^{\frac{1}{(i\lambda-(i-1))q}}\right)^2 \tag{4.25}$$

As a result, it follows that $\|e_i\|_\infty\le\sqrt{2\|V(t)\|_\infty}\le\sqrt{\frac{\varsigma_0}{\rho}},\quad i=1,\ldots,n$. Then it can be concluded that the L_∞ norms of the tracking errors can be made arbitrarily small with a sufficiently large ρ.

4.5 Simulation Results

In this section, a simulation example is provided to show the effectiveness of the developed methods. Consider the one-link robot system with the inclusion of motor dynamics whose model is given by $D\ddot{q}+B\dot{q}+N\sin q=\tau,\quad M\dot{\tau}+H\tau=u-K_m\dot{q}$, which can be expressed in the pure-feedback form by noting that

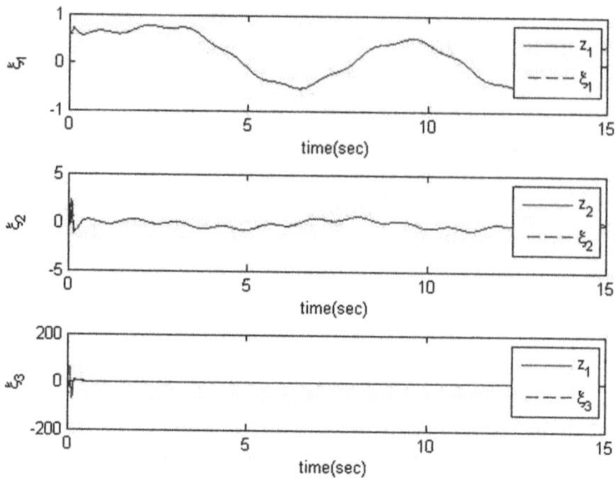

Fig. 4.1 Performance of ESO for the considered system

$$x_1 = q,\ x_2 = \dot{q},\ x_3 = \tau,\ f_1 = 0,\ \ g_1 = 1,$$
$$f_2 = (-N\sin x_1 - Bx_2)/D,\ g_2 = 1/D,\ f_3 = (-K_m x_2 - Hx_3)/M,\ \ g_3 = 1/M$$

The parameter values with appropriate units are given to be $D = 1,\ M = 0.05,\ B = 1,\ K_m = 10,\ H = 0.5,\ N = 10$. The sampling time is 0.01 s. The reference trajectory is $x_r(t) = r_1(t)/(s+1)^2,\quad r_1(t) = \sin 5t + \sin t$. The control objective is to make x_1 follow the reference trajectory x_r. According to Theorem 1, the controller and observer parameters are chosen as $k_1 = 7,\ k_2 = 23,\ k_3 = 31,\ \beta_1 = 100,\ \beta_2 = 500,\ \beta_3 = 2{,}000,\ \beta_4 = 12{,}000,\ \tau_2 = \tau_3 = 0.01$.

The initial conditions of system states are $x_1(0) = x_2(0) = x_3(0) = 0$ and the initial condition of reference signal is $x_r = 0.6$.

The simulation results are depicted in Figs 4.1, 4.2 and 4.3. From Fig. 4.1, we observe that good estimating performance is achieved and that the estimating

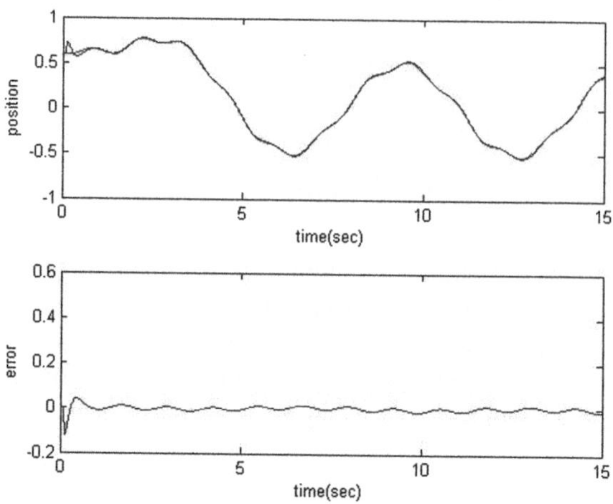

Fig. 4.2 Tracking performance of L_∞ DSC nonlinear system

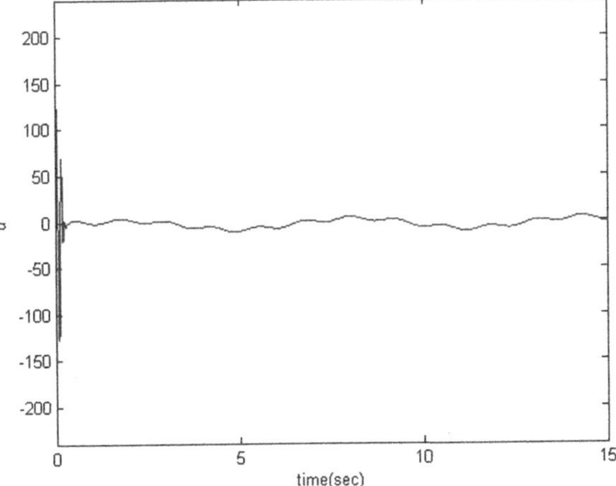

Fig. 4.3 Closed loop control signal with the proposed L_∞ DSC scheme

errors converge to a small neighborhood of zero in less than one period of oscillation. At the same time, tracking error is kept bounded, as seen in Fig. 4.2. It is noted that, due to the initialization technique, the L_∞ performance of control signal with small amplitude of impulse is maintained, Fig. 4.3.

4.6 Conclusion

In this paper, an output-feedback DSC has been developed for a class of pure feedback nonlinear system. By state transformation, the state-feedback control problem of a pure-feedback system was transformed into an output control problem of a canonical system. A finite-time ESO is applied to estimate unknown states of the transformed system and the lumped uncertainty. The proposed scheme is able to eliminate the algebraic loop problem in traditional backstepping and guarantee the L_∞ performance of the tracking errors by introducing the initialization technique. Simulations were performed to illustrate the effectiveness of the proposed scheme. Future work will be focused on the experiments on practical systems to further demonstrate the reliability of the proposed approach.

Acknowledgments This work is supported by National Natural Science Foundation of China (No. 61273150 and No.60974046), and the Research Fund for the Doctoral Program of Higher Education of China (No. 20121101110029).

References

1. Zhou J, Wen C, Zhang Y (2006) Adaptive output control of nonlinear systems with uncertain dead-zone nonlinearity. IEEE Trans Autom Control 51(3):504–511

2. Tong SC, Li YM (2012) Adaptive fuzzy output feedback tracking backstepping control of strict-feedback nonlinear systems with unknown dead zones. IEEE Trans Fuzzy Syst 20(1):168–180
3. Wang CL, Lin Y (2012) Decentralised adaptive dynamic surface control for a class of interconnected non-linear systems. IET Control Theory Appl 6(9):1172–1181
4. Han JQ (2009) From PID to active disturbance rejection control. IEEE Trans Industr Electron 56(3):900–906
5. Liu HX, Li SH (2012) Speed control for PMSM servo system using predictive functional control and extended state observer. IEEE Trans Industr Electron 59(2):1171–1183
6. Xia YQ, Shi P, Liu GP, Rees D, Han JQ (2007) Active disturbance rejection control for uncertain multivariable systems with time-delay. IET Control Theory Appl 1(1):75–83
7. Perruquetti W, Floquet T, Moulay E (2008) Finite-time observers: application to secure communication. IEEE Trans Autom Control 53(1):356–360
8. Ge SS, Fan H, Tong Heng L (2004) Adaptive neural control of nonlinear time-delay systems with unknown virtual control coefficients. IEEE Trans Syst Man Cybern Part B Cybern 34:499–516
9. Park JH, Kim SH, Moon CJ (2009) Adaptive neural control for strict-feedback nonlinear systems without backstepping. IEEE Trans Neural Networks 20(7):204–1209

Chapter 5
Cross Function Projective Synchronization of Liu System and the New Lorenz System with Known and Unknown Parameters

Yuqing Sun, Wuneng Zhou and Yuan Du

Abstract In this paper, we focuses on a new way of the function projective synchronization (FPS) of the hyperchaotic Liu system and the hyperchaotic New Lorenz system. We call this new method the cross function projective synchronization (CFPS). Within the two systems, we achieved the CFPS at the first place through a proper control scheme. Furthermore, by designing the parameter update law, the time of reaching projective synchronization could be adjustable. Eventually, several numerical simulations are presented to verify the feasibility and effectiveness of the method.

Keywords Cross function projective synchronization · Adaptive control · Hyperchaotic system · Lyapunov stability theorem

5.1 Introduction

In 1990, Pecora and Carroll successfully synchronized two identical chaotic systems with different initial conditions [1]. Since then, the research of synchronization has attracted much attention from scientists and engineers. So chaos synchronization has become a very hot topic in the past three decades due to its potential applications in vast fields, especially in secure communication, neural networks and biological networks. With the further studies of the research, various methods have been proposed to achieve chaos synchronization, such as Ott-Grebogi-Yorke (OGY) method [2], Phase Control (PC) method [1], nonlinear control [3], adaptive method [4], impulse control [5] and coupling control [6], etc. Based on these control methods, many synchronization schemes have been studied

Y. Sun (✉) · W. Zhou · Y. Du
College of Information Science and Technology, Donghua University,
28 Building, No 300 WenHui Road, ShangHai City 5006 SongJiang, China
e-mail: departedcliff@sina.cn

Z. Sun and Z. Deng (eds.), *Proceedings of 2013 Chinese Intelligent Automation Conference*, Lecture Notes in Electrical Engineering 255,
DOI: 10.1007/978-3-642-38460-8_5,

such as complete synchronization (CS) [1], phase synchronization (PS) [7], lag synchronization (LS) [8], generalized synchronization [8] and so on.

Since first introduced by Mainieri and Rehacek [9], projective synchronization has gained wide attention. Later, by making the master and the slave system synchronize to a scaling factor α, a new synchronization called generalized projective synchronization (GPS) was observed by Yan and Li [10]. On the basis of GPS, modified projective synchronization (MPS) has been proposed by Li [11], where the responses of the synchronized dynamical states synchronize up to a constant scaling matrix. Moving a step further, Park proposed an AMPS scheme [12], called adaptive modified projective synchronization, where the synchronization of two systems was achieved with unknown parameters. Then Tang et al. [13] extended MPS and forwarded to a new projective synchronization, called function projective synchronization (FPS), where master and slave systems could be synchronized up to a desired scaling function rather than a constant. Combining Park's and Tang's results, Du introduced parameter adaptive scheme into FPS [14], which exclude the problems of uncertainty of parameters and realizes Adaptive Function Projective Synchronization (AFPS). This could be used to increases the security during communication.

Recently, Wang introduced a fourth state vector into Liu chaotic system and transformed it into hyperchaotic Liu system [15]. Also, by introducing a state feedback to the traditional Lorenz system [16], Jia proposed the new Lorenz hyperchaotic system. Based on this, we propose a new method to achieve the cross function projective synchronization of the hyperchaotic Liu system and the hyperchaotic New Lorenz system.

First, on the basis of Lyapunov stability theorem, an adaptive controller is designed. By using the controller, the cross function projective synchronization of the hyperchaotic Liu system and the hyperchaotic New Lorenz system can be easily achieved. Meantime, an impact factor is added in the controller. By changing the value of impact factor, the time of reaching projective synchronization could be adjustable, expanding the scope of chaos synchronization and communication security effect. By this method, one can achieve hyperchaotic synchronization and identify the unknown parameters simultaneously.

The organization of this paper is as follows: In Sect. 5.2, the definition of the cross function projective synchronization is presented and the characters of the two hyperchaotic systems are studied. In the Sect. 5.3, by means of active control, we realize the CFPS of Liu hyperchaotic system and the new Lorenz hyperchaotic system. In Sect. 5.4, we proposed CFPS of the two systems with unknown parameters, which realizes AFPS of the two systems. In both of Sects. 5.3 and 5.4, numerical simulations are given to manifest the effectiveness of the methods. The conclusion is finally drawn in Sect. 5.5.

5.2 Conception Definition and System Description

In this section, we give definition of CFPS and the descriptions of the two hyperchaotic systems.

5.2.1 Definition of CFPS

The master and slave hyperchaotic system of the cross function projective synchronization can be illustrated as follows:

$$\begin{cases} \dot{x}_1 = f_1(x_1, x_2) \\ \dot{x}_2 = f_2(x_1, x_2) \end{cases} \tag{5.1}$$

$$\begin{cases} \dot{y}_1 = g\,(y_1, y_2) + u\,(x_1, x_2, y_1, y_2) \\ \dot{y}_2 = g\,(y_1, y_2) + u\,(x_1, x_2, y_1, y_2) \end{cases} \tag{5.2}$$

where $x_1, x_2, y_1, y_2 \in R^n$ are state vectors, $f, g : R^n \to R^n$ are nonlinear vector functions, and $u(x, y)$ is the vector controller.

In order to achieve the cross function projective synchronization of the master and slave system, we define synchronization errors as follows:

$$\begin{cases} e_1 = y_2 - h_1(t, x_1, x_2)x_1 \\ e_2 = y_1 - h_2(t, x_1, x_2)x_2 \end{cases} \tag{5.3}$$

where $e \in R^n$ is the error vector, and $h(t, x) = (h_1, h_2, \ldots, h_n)$, $(h_i \neq 0)$ are called scaling function vector which represents the proportion of projection. Then the synchronization error dynamic system can be obtained as follows:

$$\begin{cases} \dot{e}_1 = \dot{y}_2 - \dot{h}_1\,(t, x_1, x_2)\,x_1 - h_1\,(t, x_1, x_2)\,\dot{x}_1 \\ \dot{e}_2 = \dot{y}_1 - \dot{h}_2\,(t, x_1, x_2)\,x_2 - h_2\,(t, x_1, x_2)\,\dot{x}_2 \end{cases} \tag{5.4}$$

Definition 5.1 For drive system (5.1) and response system (5.2), it is acknowledged that the system (5.1) and the system (5.2) are function projective synchronization (FPS), if we can find a scaling function h(t,x) making

$$\begin{cases} \lim\limits_{t\to\infty} \|e_1\| = \lim\limits_{t\to\infty} \|y_2 - h_1(t, x_1, x_2)x_1\| = 0 \\ \lim\limits_{t\to\infty} \|e_2\| = \lim\limits_{t\to\infty} \|y_1 - h_2(t, x_1, x_2)x_2\| = 0 \end{cases} \tag{5.5}$$

come into existence.

The vector controller implies that the error dynamic system (5.4) between the drive system and the response system is globally asymptotically stable.

5.2.2 System Description

In this paper, we use Liu hyperchaotic system and new Lorenz system to realize CFPS and AFPS. The Liu hyperchaotic system is described as:

$$\begin{cases} \dot{x}_1 = a(x_2 - x_1) \\ \dot{x}_2 = bx_1 - kx_1x_3 + x_4 \\ \dot{x}_3 = -cx_3 + gx_1^2 \\ \dot{x}_4 = -dx_1 \end{cases} \tag{5.6}$$

where $(x_1, x_2, x_3, x_4) \in R^4$ and $a, b, c, d, k, g \in R^1$ are parameters of system (5.6). When $a = 10,\ b = 40,\ c = 2.5,\ g = 4,\ d = 10.6$, there are two positive Lyapunov exponents: $\lambda_1 = 1.14,\ \lambda_2 = 0.13$. The hyperchaotic attractor is shown in Fig. 5.1.

The new Lorenz hyperchaotic system is obtained as:

$$\begin{cases} \dot{y}_1 = -a_1(y_1 - y_2) + y_4 \\ \dot{y}_2 = -y_1y_3 + r_1y_1 - y_2 \\ \dot{y}_3 = y_1y_2 - b_1y_3 \\ \dot{y}_4 = -y_1y_3 + d_1y_4 \end{cases} \tag{5.7}$$

where $(y_1, y_2, y_3, y_4) \in R^4$, and $a_1, b_1, r_1, d_1 \in R^1$ are parameters of system (5.7). When $a_1 = 10,\ b_1 = \frac{8}{3},\ r_1 = 28,\ d_1 = 1.3,$ there are two positive Lyapunov exponents: $\lambda_1 = 0.3985,\ \lambda_2 = 0.2418$. The hyperchaotic attractors are shown in Fig. 5.2.

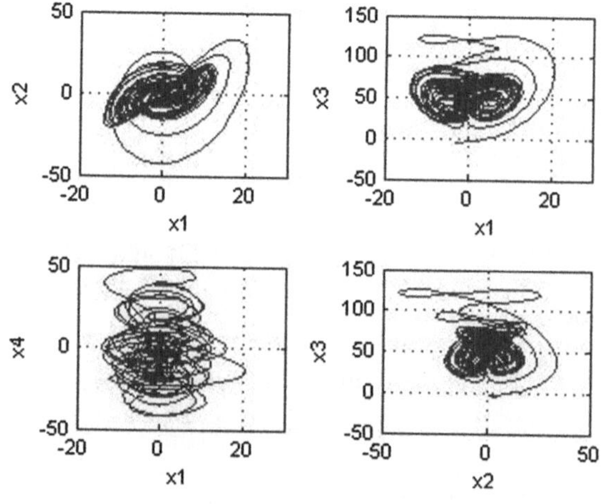

Fig. 5.1 Phase portraits of Liu hyperchaotic system in **a** the x1 − x2 space; **b** the x1 − x3 space; **c** the x1 − x4 space; **d** the x2 − x3 space

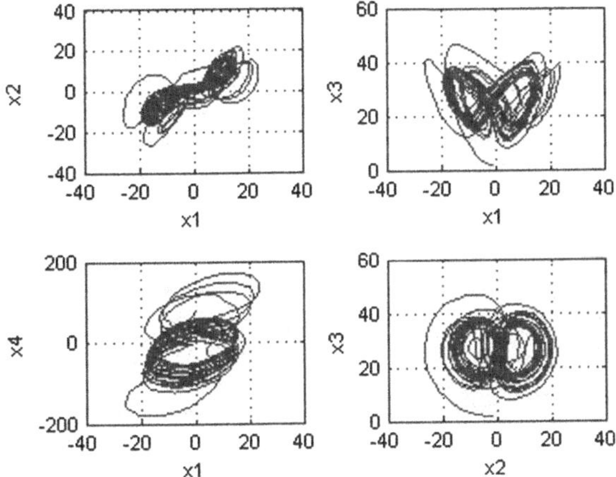

Fig. 5.2 Phase portraits of the new Lorenz hyperchaotic system in **a** the x1 − x2 space; **b** the x1 − x3 space; **c** the x1 − x4 space; **d** the x2 − x3 space

5.3 Active Synchronization of Liu Hyperchaotic System and the New Lorenz System

In this section, we design a controller using active control method to obtain CFPS of the hyperchaotic Liu system and the New Lorenz system. The feasibility and effectiveness of the control scheme will be verified by Lyapunov stability theory. We choose Liu system as the master system and the new Lorenz system as the slave system. So the master system is (5.6) and the slave system can be given as follows:

$$\begin{cases} \dot{y}_1 = -a_1(y_1 - y_2) + y_4 + u_1 \\ \dot{y}_2 = -y_1y_3 + r_1y_1 - y_2 + u_2 \\ \dot{y}_3 = y_1y_2 - b_1y_3 + u_3 \\ \dot{y}_4 = -y_1y_3 + d_1y_4 + u_4 \end{cases} \tag{5.8}$$

In the slave system, u_1, u_2, u_3, u_4 are the linear or nonlinear control functions. According to (5.4), subtracting the slave system (5.8) by the product of the projection law and the master system (5.6), we obtain the error dynamical system as follows:

$$\begin{cases} \dot{e}_1 = \dot{y}_2 - \dot{x}_1 = -y_1y_3 + r_1y_1 - y_2 + u_2 - \dot{h}_1x_1 - h_1a\,(x_2 - x_1) \\ \dot{e}_2 = \dot{y}_3 - \dot{x}_2 = y_1y_2 - b_1y_3 + u_3 - \dot{h}_2x_2 - h_2\,(bx_1 - kx_1x_3 + x_4) \\ \dot{e}_3 = \dot{y}_4 - \dot{x}_3 = -y_1y_3 + d_1y_3 + u_4 - \dot{h}_3x_3 - h_3(-cx_3 + gx_1^2) \\ \dot{e}_4 = \dot{y}_1 - \dot{x}_4 = -a_1\,(y_1 - y_2) + y_4 + u_1 - \dot{h}_4x_4 + h_4(-dx_1) \end{cases} \tag{5.9}$$

Due to the error dynamical system, we choose following control law:

$$\begin{cases} u_1 = a_1(y_1 - y_2) - y_4 + \dot{h}_4 x_4 + h_4 d x_1 - k_4 e_4 \\ u_2 = y_1 y_3 - r_1 y_1 + y_2 + \dot{h}_1 x_1 + h_1 a(x_2 - x_1) - k_1 e_1 \\ u_3 = b_1 y_3 - y_1 y_2 + \dot{h}_2 x_2 + h_2(b x_1 - k x_1 x_3 + x_4) - k_2 e_2 \\ u_4 = y_1 y_3 - d_1 y_4 + \dot{h}_3 x_3 - h_3(c x_3 - g x_1^2) - k_3 e_3 \end{cases} \tag{5.10}$$

where $k_1,\ k_2,\ k_3,\ k_4$ are constant and $k_1,\ k_2,\ k_3,\ k_4 \geq 0$

Theorem 5.1 *For any initial condition, if we choose the controller satisfies control law* (5.10) *and* $k_1,\ k_2,\ k_3,\ k_4$, *then the master system* (5.6) *and the slave system* (5.8) *will asymptotically achieve CFPS.*

Proof Choose the Lyapunov function as follows:

$$V = \frac{1}{2}\left(e_1^T e_1 + e_2^T e_2 + e_3^T e_3 + e_4^T e_4\right) \tag{5.11}$$

The time derivative of V along the trajectories of (5.9) is

$$\begin{aligned} \dot{V} &= \frac{1}{2}(\dot{e}_1^T e_1 + e_1^T \dot{e}_1 + \dot{e}_2^T e_2 + e_2^T \dot{e}_2 + \dot{e}_3^T e_3 + e_3^T \dot{e}_3 + \dot{e}_4^T e_4 + e_4^T \dot{e}_4) \\ &= e_1 \dot{e}_1 + e_2 \dot{e}_2 + e_3 \dot{e}_3 + e_4 \dot{e}_4 \\ &= e_1(-y_1 y_3 + r_1 y_1 - y_2 + u_2 - \dot{h}_1 x_1 - h_1 a(x_2 - x_1)) \\ &\quad + e_2(y_1 y_2 - b_1 y_3 + u_3 - \dot{h}_2 x_2 - h_2(b x_1 - k x_1 x_3 + x_4)) \\ &\quad + e_3(-y_1 y_3 + d_1 y_3 + u_4 - \dot{h}_3 x_3 - h_3(-c x_3 + g x_1^2)) \\ &\quad + e_4(-a_1(y_1 - y_2) + y_4 + u_1 - \dot{h}_4 x_4 + h_4(-d x_1)) \end{aligned} \tag{5.12}$$

Substitute for u_i in (5.12) with the controller (5.11), then after simplification, we could notice the equation turns into:

$$\dot{V} = -k_1 e_1^2 - k_2 e_2^2 - k_3 e_3^2 - k_4 e_4^2 \tag{5.13}$$

Since $k_1,\ k_2,\ k_3,\ k_4 \geq 0$,

$$\dot{V} = -k_1 e_1^2 - k_2 e_2^2 - k_3 e_3^2 - k_4 e_4^2 \leq 0 \tag{5.14}$$

On the basis of the Lyapunov stability theorem, the error vector $e(t)$ asymptotically tends to zero, which means that cross function projective synchronization between systems (5.6) and (5.8) is achieved.

According to the scheme mentioned above, numerical simulations are given to show the effectiveness of the controller (5.10). The parameters of the master system (5.6) and slave system (5.8) are $a = 10,\ b = 40,\ k = 1,\ c = 2.5,\ g = 4,\ d = 10.6$ and $a_1 = 10,\ b_1 = \frac{8}{3},\ r_1 = 28,\ d_1 = 1.3$. Choosing the function projection scaling vector $h = (\sin t, t, 3, x_4)$, then $\dot{h} = (\cos t, 1, 0, \dot{x}_4)$. Set the initial value $(x_1, x_2, x_3, x_4) = (-3, 5, -5, 3)$ and $(y_1, y_2, y_3, y_4) = (1, -2, 2, -3)$. Several simulation figures are obtained by using Fourth order Runge–Kutta

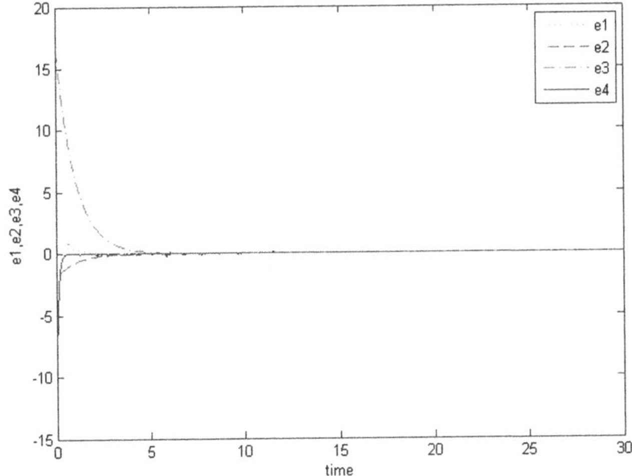

Fig. 5.3 The synchronization errors of four states versus time

method. Errors of synchronization of the two hyperchaotic system are shown in Fig. 5.3. From Fig. 5.3, we can see that they tend to zero after a sufficiently long time, the CFPS of the two systems has achieved.

5.4 Adaptive Function Projective Sychronization of Hyperchaotic Liu System and the New Lorenz System by Parameter Adaptive Control

Although cross function projective synchronization via active control is feasible and effective, we cannot ignore that there always exist parameter disturbance and mismatch in real world. So in order to solve the problem of parameter uncertainty, in this section, we will introduce a parameter adaptive control scheme. We still use system (5.6) and system (5.8) as the master and slave system. The error dynamical system is identical to (5.9). Since we do not know the precise value of the parameters of the master and slave system, we choose the control law and the parameter adapt law as follows:

$$\begin{cases} u_1 = \widehat{a}_1(y_1 - y_2) - y_4 + \dot{h}_4 x_4 + h_4 \widehat{d} x_1 - k_4 e_4 \\ u_2 = y_1 y_3 - \widehat{r}_1 y_1 + y_2 + \dot{h}_1 x_1 + h_1 \widehat{a}(x_2 - x_1) - k_1 e_1 \\ u_3 = \widehat{b}_1 y_3 - y_1 y_2 + \dot{h}_2 x_2 + h_2(\widehat{b} x_1 - \widehat{k} x_1 x_3 + x_4) - k_2 e_2 \\ u_4 = y_1 y_3 - \widehat{d}_1 y_4 + \dot{h}_3 x_3 - h_3(\widehat{c} x_3 - \widehat{g} x_1^2) - k_3 e_3 \end{cases} \tag{5.15}$$

where k_1, k_2, k_3, k_4 are constant and $k_1, k_2, k_3, k_4 \geq 0$. $\hat{a}, \hat{b}, \hat{c}, \hat{d}, \hat{f}, \hat{g}, \hat{a}_1, \hat{b}_1, \hat{d}_1, \hat{r}_1$ are the estimation of systems' parameters $a, b, c, d, f, g, a_1, b_1, d_1, r_1$.

At the same time, in order to make the estimation parameters gradually approach to the real value, we design following parameter update law:

$$\begin{aligned}
&\dot{\tilde{a}} = e_1h_1(x_2 - x_1) - k_5\tilde{a}, && \dot{\tilde{b}} = e_2h_2x_1 - k_6\tilde{b} \\
&\dot{\tilde{c}} = -h_3e_3x_3 - k_7\tilde{c}, && \dot{\tilde{d}} = -e_4h_4x_1 - k_8\tilde{d} \\
&\dot{\tilde{g}} = h_3e_3x_1^2 - k_9\tilde{g}, && \dot{\tilde{k}} = -e_2h_2x_1x_3 - k_{10}\tilde{k} \\
&\dot{\tilde{a}}_1 = e_1(y_1 - y_2) - k_{11}\tilde{a}_1, && \dot{\tilde{b}}_1 = e_3y_3 - k_{12}\tilde{b}_1 \\
&\dot{\tilde{d}}_1 = -e_4y_4 - k_{13}\tilde{d}_1, && \dot{\tilde{r}}_1 = -e_2y_1 - k_{14}\tilde{r}_1
\end{aligned} \tag{5.16}$$

where $k_1, k_2, k_3, k_4, k_5, k_6, k_7, k_8, k_9, k_{10}, k_{11}, k_{12}, k_{13}, k_{14}$ are constant and $k_1, k_2, k_3, k_4, k_5, k_6, k_7, k_8, k_9, k_9, k_{10}, k_{11}, k_{12}, k_{13}, k_{14} \geq 0$.

Theorem 5.2 *For any initial condition, the AMFP between the master system* (5.6) *and slave system* (5.8) *will occur by the control law* (5.15) *and parameter update law* (5.16).

Proof Applying control law (5.15) to the error dynamical system (5.9) yields the resulting system:

$$\begin{cases}
\dot{e}_1 = -y_1y_3 + \hat{r}_1y_1 - y_2 + u_2 - \dot{h}_1x_1 - h_1\hat{a}(x_2 - x_1) \\
\dot{e}_2 = y_1y_2 - \hat{b}_1y_3 + u_3 - \dot{h}_2x_2 - h_2(\hat{b}x_1 - \hat{k}x_1x_3 + x_4) \\
\dot{e}_3 = -y_1y_3 + \hat{d}_1y_3 + u_4 - \dot{h}_3x_3 - h_3(-\hat{c}x_3 + \hat{g}x_1^2) \\
\dot{e}_4 = -\hat{a}_1(y_1 - y_2) + y_4 + u_1 - \dot{h}_4x_4 + h_4(-\hat{d}x_1)
\end{cases} \tag{5.17}$$

where

$$\begin{aligned}
&\tilde{a} = a - \hat{a},\ \tilde{b} = b - \hat{b},\ \tilde{c} = c - \hat{c}, \\
&\tilde{d} = d - \hat{d},\ \tilde{g} = g - \hat{g},\ \tilde{k} = k - \hat{k} \\
&\tilde{a}_1 = a_1 - \hat{a}_1,\ \tilde{b}_1 = b_1 - \hat{b}_1 \\
&\tilde{d}_1 = d_1 - \hat{d}_1,\ \tilde{r}_1 = r_1 - \hat{r}_1
\end{aligned} \tag{5.18}$$

Choose the following Lyapunov function:

$$\begin{aligned}
V =& \frac{1}{2}(e_1^Te_1 + e_2^Te_2 + e_3^Te_3 + e_4^Te_4) \\
&+ \frac{1}{2}(\tilde{a}^T\tilde{a} + \tilde{b}^T\tilde{b} + \tilde{c}^T\tilde{c} + \tilde{d}^T\tilde{d} + \tilde{g}^T\tilde{g} + \tilde{k}^T\tilde{k}) \\
&+ \frac{1}{2}(\tilde{a}_1^T\tilde{a}_1 + \tilde{b}_1^T\tilde{b}_1 + \tilde{r}_1^T\tilde{r}_1 + \tilde{d}_1^T\tilde{d}_1)
\end{aligned} \tag{5.19}$$

Compute the time derivative of (5.19) and apply control (5.15) and parameter update law (5.16), after simplification of the equation, we could finally get:

$$\begin{aligned}\dot{V} = & -k_1e_1^2 - k_2e_2^2 - k_3e_3^2 - k_4e_4^2 \\ & -k_5a^2 - k_6b^2 - k_7c^2 - k_8d^2 - k_9g^2 - k_{10}k^2 \\ & -k_{11}a_1^2 - k_{12}b_1^2 - k_{13}d_1^2 - k_{14}r_1^2\end{aligned} \tag{5.20}$$

since $k_1, k_2, k_3, k_4, k_5, k_6, k_7, k_8, k_9, k_{10}, k_{11}, k_{12}, k_{13}, k_{14} \geq 0$

$$\begin{aligned}\dot{V} = & -k_1e_1^2 - k_2e_2^2 - k_3e_3^2 - k_4e_4^2 \\ & -k_5a^2 - k_6b^2 - k_7c^2 - k_8d^2 - k_9g^2 - k_{10}k^2 \\ & -k_{11}a_1^2 - k_{12}b_1^2 - k_{13}d_1^2 - k_{14}r_1^2 \leq 0\end{aligned} \tag{5.21}$$

Then according to Lyapunov stability theory, the AFPS of hyperchaotic system (5.6) and (5.8) is obtained under the chosen controllers (5.15) and the designed parameter update law (5.16).

Remark 5.1 When $h = (1, 1, 1, 1)$, AFPS is simplified to complete synchronization. When $h = (\alpha, \alpha, \alpha, \alpha)$, FPS converts to general projective synchronization. When $h = (\alpha_1, \alpha_2, \alpha_3, \alpha_4)$, where α_i is a scaling factor, modified projective synchronization will be realized.

Numerical simulations results are presented to demonstrate the effectiveness of the proposed synchronization methods. The unknown estimation parameters of the master system (5.6) are $a = 10,\ b = 40,\ c = 2.5,\ d = 10.6,\ g = 4,\ k = 1$ and the unknown estimation parameters of the slave system (5.8) are $a_1 = 10,\ b_1 = \frac{8}{3}$, $r_1 = 28,\ d_1 = 1.3$. Suppose the function projection scaling vector $h =$

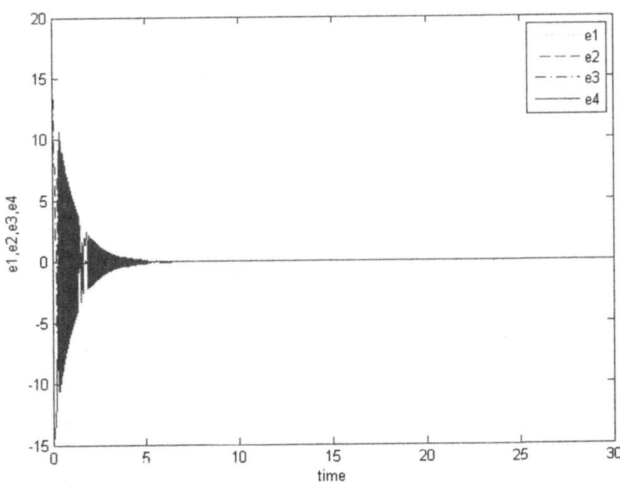

Fig. 5.4 The synchronization errors of four states versus time

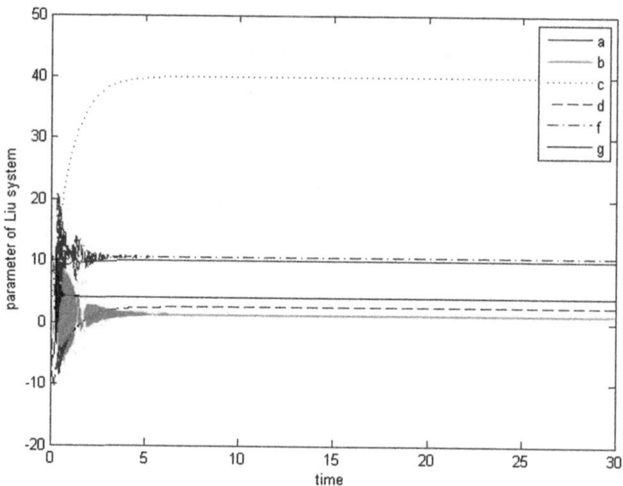

Fig. 5.5 Changing parameters a, b, c, d, f, g of Liu hyperchaotic system with time

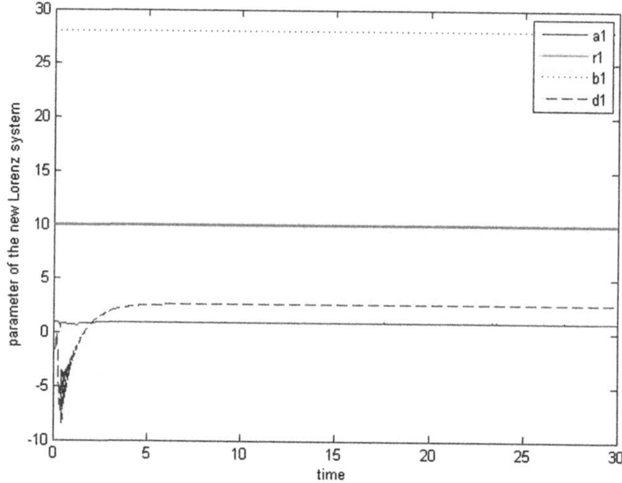

Fig. 5.6 Changing parameter a_1, b_1, d_1, r_1 of the New Lorenz hyperchaotic system with time

$(\sin t, t, 3, x_4)$, then $\dot{h} = (\cos t, 1, 0, \dot{x}_4)$. Make $(k_1, k_2, k_3, k_4, k_5, k_6, k_7, k_8, k_9, k_{10}, k_{11}, k_{12}, k_{13}, k_{14}) = (1, 1000, 0.1, 1,\ \ 1, 1, 1, 1, 1, 1, 5000, 3000, 1, 1)$. Let the initial values $(x_1(0), x_2(0), x_3(0), x_4(0)) = (-3, 5, -5, 3)$ and $(y_1(0), y_2(0), y_3(0), y_4(0)) = (1, -2, 2, -3)$. Furthermore, the initial values of estimated parameters are $(\hat{a}(0), \hat{b}(0), \hat{c}(0), \hat{d}(0), \hat{g}(0), \hat{k}(0), \hat{a}_1(0), \hat{b}_1(0), \hat{d}_1(0)\ \hat{r}_1(0)) = (8, 1, 9, 1, 1, 1, 30, 1, 1, 1)$. Synchronization of systems (5.8) and (5.9) under adaptive control law (5.16) are shown in Fig. 5.4. And the parameters changing curves are shown in Figs. 5.5 and 5.6. In Fig. 5.4, we can see the synchronization errors between systems (5.6) and (5.8) tend to zero after a sufficiently long time. Figures 5.5 and 5.6 indicate that the identified parameters $(a, b, c, d, g, k, a_1, b_1, d_1, r_1)$ approach the desired values $(10,\ 40, 1, 2.5, 10.6, 4, 10, \frac{8}{3}, 1.3, 28)$.

5.5 Conclusion

In this letter, we studied the cross function projective synchronization problem of Liu hyperchaotic system and the New Lorenz hyperchaotic system with known and unknown parameters. Proven by Lyapunov Stability Theorem, a novel parameters identification and control method have been confirmed suitable for the synchronization of the two systems. Numerical simulations are used to verify the feasibility and effectiveness of the proposed synchronization scheme.

Acknowledgments This work was supported by the National Natural Science Foundation of China (61075060), the Innovation Program of Shanghai Municipal Education Commission (12zz064) and the Open Project of State Key Labora-tory of Industrial Control Technology (ICT1231).

References

1. Pecora LM, Carroll TL (1990) Synchronization in chaotic systems. Phys Rev Lett 64:821–824
2. Ott E, Grebogi C, Yorke A (1990) Controlling chaos. Phys Rev Lett 64(11):1196–1199
3. Zhang Q, Lu J (2008) Chaos synchronization of a new chaotic system via nonlinear control. Chaos Solitons Fractals 37(1):175–179
4. Park JH (2005) Adaptive synchronization of a unified chaotic system with an uncertain parameter. Int J Nonlinear Sci Numer Simul 6(2):201–206
5. Stojanovski T, Kocarev L, Parlitz U (1996) Driving and synchronizing by chaotic impulses. Phys Rev Lett 43(9):782–785
6. Li D, Lu J-A, Wu X (2005) Linearly coupled synchronization of the unified chaotic systems and the Lorenz systems. Chaos Solitons Fractals 23:79–85
7. Rosenblum MG, Pikovsky AS, Kurths J (1996) Phase synchronization of chaotic oscillators. Phys Rev Lett 76:1804
8. Rosenblum MG, Pikovsky AS (1997) From phase to lag synchronization in coupled chaotic scillators. Phys Rev Lett 78:4193
9. Mainieri G, Rehacek J (1999) Projective synchronization in three-dimensional chaotic systems. Phys Rev Lett 82(15):3042–3045
10. Yan J, Li C (2005) Generalized projective synchronization of a unified chaotic system. Chaos Solitons Fractals 26:1119–1124
11. Li G (2007) Modified projective synchronization of chaotic system. Chaos Solution Fractals 32:1786–1790
12. Park JH (2008) Adaptive control for modified projective synchronization of a four-dimensional chaotic system with uncertain parameters. J Comput Appl Math 213:288–293
13. Tang XH, Lu JA, Zhang WW (2007) The FPS of chaotic system using backstepping design. China Dyn Control 0705:216

14. Du H, Zeng Q (2008) Function projective synchronization of different chaotic systems with uncertain parameters. Phys Rev Lett A 372:5402–5410
15. Wang F, Liu C (2006) Hyperchaos evolved from the Liu chaotic system. China Phys 15(5):963–968
16. Jia Q (2007) Projective synchronization of a new hyperchaotic Lorenz system. Phys Rev Lett A 370:40–45

Chapter 6
The Application of Multiwavelets to Chaotic Time Series Analysis

Zhihong Zhao, Shaopu Yang and Yu Lei

Abstract Multiwavelets offer symmetry, orthogonality, and short support, which is not possible with scalar wavelet. In this paper, we analyzed chaotic time series using multiwavelets. Two benchmark chaotic time series and the real-world data of Sunspots time series were analyzed. Three type of accuracy criteria: Mean square error, Root mean square error and Mean absolute error were used to measure the performance of the reconstruction. And the results were compared to the scalar wavelet. Experimental results indicate that multiwavelets were superior to scalar wavelet for chaotic time series analysis.

Keywords Multiwavelets · Chaotic time series · Lorenz time series

6.1 Introduction

Chaotic phenomenon is an irregular motion produced by determinate nonlinear dynamical system. It is widely existed in various fields such as physics, mathematics, and biology, etc. Chaotic time series provides useful information for analysis and interpretation of the physical system that produce it. Different methods have been used to analyze the chaotic time series. Among them wavelet

Z. Zhao (✉) · Y. Lei
School of Computing and Informatics, Shijiazhuang Tiedao University, Shijiazhuang 050043, China
e-mail: hb_zhaozhihong@126.com

Y. Lei
e-mail: leiyu@eyou.com

S. Yang
Institute of Traffic Environment and Safety Engineering, Shijiazhuang 050043, China
e-mail: yangsp@stdu.edu.cn

Z. Sun and Z. Deng (eds.), *Proceedings of 2013 Chinese Intelligent Automation Conference*, Lecture Notes in Electrical Engineering 255,
DOI: 10.1007/978-3-642-38460-8_6,

transform (WT) is widely used. WT works as a mathematical microscope on a specific part of time series to extract local structures and singularities. This makes the WT ideal for handling non-stationary time series [1].

Multiwavelets are new development of the wavelet theory. Multiwavelets have several advantages in comparison to scalar wavelet. Such features as short support, orthogonality, symmetry, and vanishing moments. A scalar wavelet cannot possess all these properties at the same time. While a multiwavelet system can simultaneously provide perfect reconstruction while preserving length (orthogonality), and a high order of approximation (vanishing moments) [2]. Multiwavelets are now widely used in various of science and engineering. The most application of multiwavelets are signal denoising [3], image processing [4], and machinery fault diagnosis [5]. Multiwavelets have proved to be superior performance for image processing applications, compared with scalar wavelets.

Although multiwavelets have a lot of applications, few applicated to chaotic time series analysis. The goal of this paper is to analyze chaotic time series using multiwavelets. The experimental chaotic time series that we analyzed are three well-known time series: the sunspot time series, the Mackey–Glass time series and Lorenz time series.

6.2 Multiwavelets Theory

The concept of multiresolution analysis can be extended from scalar case (when only one scaling function is assumed) to the general case of r scaling functions. Similar to the scalar case, r scaling functions also satisfy two-scale refinement equations:

$$\phi(t) = \sqrt{2}\sum_{k} G_k \phi(2t-k) \tag{6.1}$$

$$\psi(t) = \sqrt{2}\sum_{k} H_k \phi(2t-k) \tag{6.2}$$

Where $G_{k\in Z}$ are $r \times r$ scaling coefficients and $H_{k\in Z}$ are $r \times r$ wavelets coefficients.

In practice multiscaling and wavelet functions often have multiplicity $r = 2$. An important multiwavelet system was constructed by Geronimo, Hardin and Massopust, named as GHM multiwavelets [6].

The GHM scaling functions satisfy Eq. (6.1) with four coefficients as (6.3). The multiscaling functions are shown in Fig. 6.1.

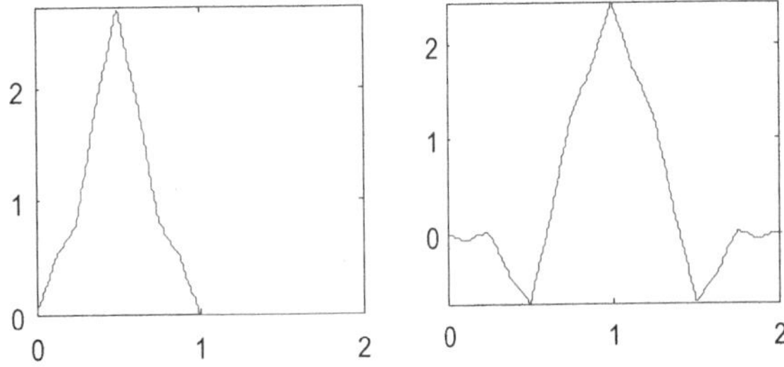

Fig. 6.1 Geronimo-Hardin-Massopust pair of scaling functions

$$H_0 = \begin{bmatrix} -\frac{1}{20} & -\frac{3}{10\sqrt{2}} \\ \frac{1}{10\sqrt{2}} & \frac{3}{10} \end{bmatrix}, \quad H_1 = \begin{bmatrix} \frac{9}{20} & -\frac{1}{\sqrt{2}} \\ -\frac{9}{10\sqrt{2}} & 0 \end{bmatrix}$$
$$H_2 = \begin{bmatrix} \frac{9}{20} & -\frac{3}{10\sqrt{2}} \\ \frac{9}{10\sqrt{2}} & -\frac{3}{10} \end{bmatrix}, \quad H_3 = \begin{bmatrix} -\frac{1}{20} & 0 \\ -\frac{1}{10\sqrt{2}} & 0 \end{bmatrix} \tag{6.3}$$

And the multiwavelets satisfy Eq. (6.2) with four coefficients as (6.4). The multiwavelet functions are shown in Fig. 6.2.

$$G_0 = \begin{bmatrix} \frac{3}{5\sqrt{2}} & \frac{4}{5} \\ -\frac{1}{20} & -\frac{3}{10\sqrt{2}} \end{bmatrix}, \quad G_1 = \begin{bmatrix} \frac{3}{5\sqrt{2}} & 0 \\ \frac{9}{20} & \frac{1}{\sqrt{2}} \end{bmatrix}$$
$$G_2 = \begin{bmatrix} 0 & 0 \\ \frac{9}{20} & -\frac{3}{10\sqrt{2}} \end{bmatrix}, \quad G_3 = \begin{bmatrix} 0 & 0 \\ -\frac{1}{20} & 0 \end{bmatrix} \tag{6.4}$$

GHM scaling functions have four remarkable properties showing that multiwavelets can combine more useful features than scalar wavelets [7]:

1. Both scaling functions have short supports [0,1] and [0,2]. In the scalar case one would expect support [0,3], for a scaling function satisfying a two-scale equation with 4 coefficients.
2. The system has second order of approximation.
3. Translates of the scaling functions and wavelets are orthogonal.
4. Both scaling functions and the wavelets are symmetric.

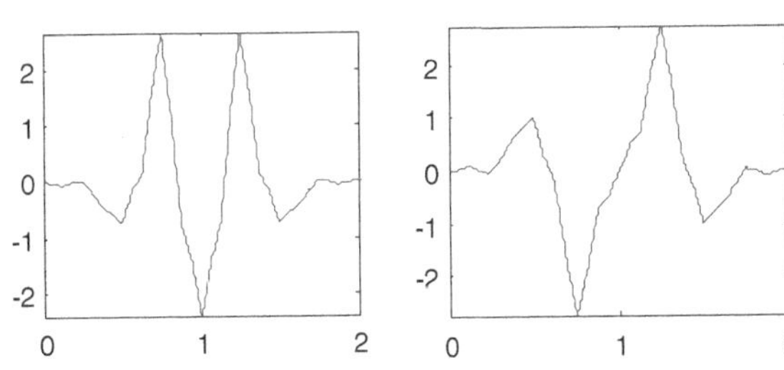

Fig. 6.2 Geronimo-Hardin-Massopust multiwavelets

In the scalar case, symmetry, orthogonality and approximation order higher than 1 cannot be combined. The GHM multiscaling and multiwavelet functions are also quite smooth-almost differentiable.

6.3 Applications and Results

Three chaotic time series experiments were carried out to assess the reconstruction performance of the multiwavelets method. We compared the performance of the reconstruction of the time series within three type of accuracy criteria: Mean square error (MSE), Root mean square error (RMSE) and Mean absolute error (MAE). GHM multiwavelets were also compared to scalar wavelet using Daubechies wavelet, LA(8) wavelet (i.e. a Least-Asymetric filter of length 8) and Coiflets 2 wavelet (Coif2).

6.3.1 The Sunspot Time Series

The data analysis here represents the annual average number of Sunspots during the period from 1700 to 1956. It contains 256 data points. Figure 6.3 illustrates the number of Sunspots for the given period of time.

The multiwavelets decomposition of the Sunspot data are shown in Fig. 6.4. The level of the decomposition is 3.

Table 6.1 shows the reconstruction accuracy results using multiwavelet and scalar wavelet method on the sunspot time series. As can be seen, GHM multiwavlet gives a lower MSE, RMSE and MAE than the scalar wavelet. This indicates that GHM multiwavelet can better represent the Sunspot time series than the scalar wavelet.

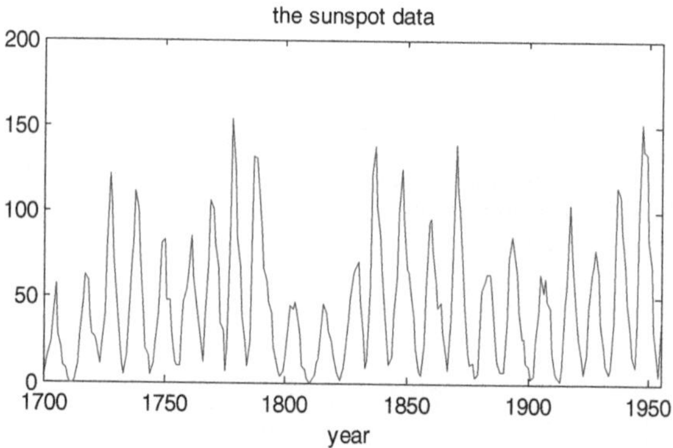

Fig. 6.3 The sunspot time series

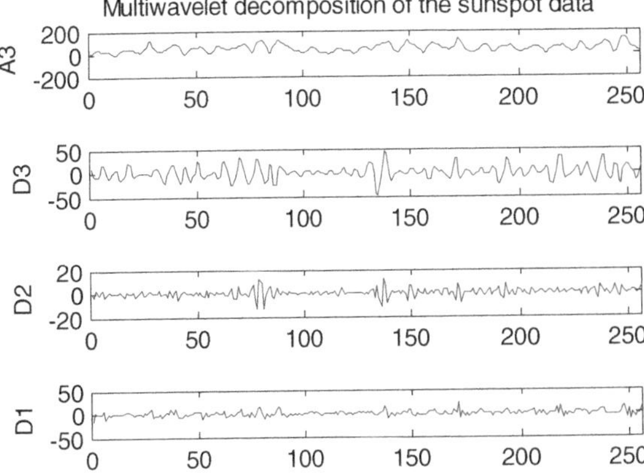

Fig. 6.4 The multiwavelet decomposition of the sunspot time series

Table 6.1 Reconstruction accuracy of the sunspot time series

	MSE	RMSE	MAE
GHM	1.8605e-028	1.3640e-014	8.7536e-015
D4	1.3115e-020	1.1452e-010	9.7181e-011
Coif2	1.0660e-018	1.0325e-009	8.7566e-010
LA8	1.1354e-024	1.0655e-012	9.4439e-013

6.3.2 The MacKey–Glass Time Series

The multiwavelets have also analysed on the chaotic series generated by the Mackey–Glass delay-differential equation [8]:

$$\frac{dx(t)}{dt} = -0.1x(t) + \frac{0.2x(t-\tau)}{1+x(t-\tau)^{10}} \tag{6.5}$$

The parameters of the series τ is 17. The time series were generated by numerical integration using a fourth order Runge–Kutta method. The Mackey–Glass time series is shown in Fig. 6.5.

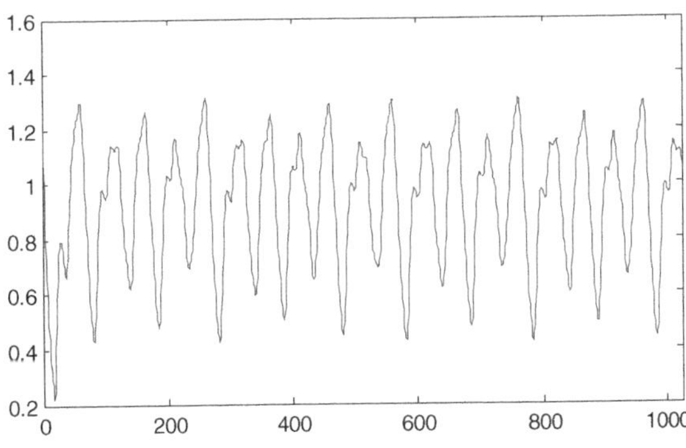

Fig. 6.5 The MacKey–Glass time series

Table 6.2 Reconstruction accuracy of the Mackey–Glass time series

	MSE	RMSE	MAE
GHM	5.8230e-032	2.4131e-016	1.8309e-016
D4	2.5443e-025	5.0441e-013	4.2758e-013
Coif2	9.2525e-024	3.0418e-012	2.5430e-012
LA8	3.8729e-028	1.9680e-014	1.9386e-014

Table 6.2 shows the reconstruction accuracy results using multiwavelet and scalar wavelets on the MacKey–Glass time series. Similar to the Sunspot time series, GHM multiwavlet gives a lower MSE, RMSE and MAE than the scalar wavelet on the Mackey–Glass time series. This indicates that GHM multiwavelet can better represent the Mackey–Glass time series.

6.3.3 Lorenz Time Series

We also considered the chaotic time series of the Lorenz differential equation [9]

$$\begin{aligned} \dot{x} &= \sigma(y - x) \\ \dot{y} &= rx - y - xz \\ \dot{z} &= xy - by \end{aligned} \tag{6.6}$$

In this paper parameters are set to be $\sigma = 10$, $b = \frac{8}{3}$, and $r = 28$ for a rich dynamical behavior. The fourth-order Runge–Kutta method is used to solve the equations. The initial state was {x, y, z} = {0.01, 0.01, 0.0}. The x component was taken as the Lorenz time series which is shown in Fig. 6.6.

Table 6.3 shows the reconstruction accuracy results using multiwavelets and scalar wavelets on the Lorenz time series. It can be seen from the table that multiwavelets perform better in all the accuracy criteria.

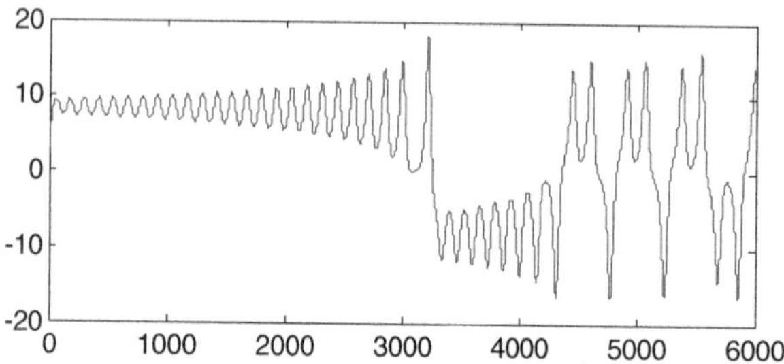

Fig. 6.6 The Lorenz time series

Table 6.3 Reconstruction accuracy of the Lorenz time series

	MSE	RMSE	MAE
GHM	4.4853e-030	2.1178e-015	1.5469e-015
D4	1.5106e-024	1.2290e-012	8.7898e-013
Coif2	3.7857e-023	6.1528e-012	4.3373e-012
LA8	2.9846e-026	1.7276e-013	1.6308e-013

6.4 Conclusion

We have examined the use of multiwavelets in chaotic time series reconstruction. Multiwavelets offer the advantages of combining symmetry, orthogonality, and short support, properties not mutually achievable with scalar wavelet systems. We applied multiwavelets to three chaotic time series reconstruction and compared with the scalar wavelet technique. All the chaotic time series reconstruction experiments show that the multiwavelets techniques are superior to the scalar wavelet. The results indicate that multiwavelets are better than the scalar wavelet in chaotic time series analysis. These results suggest that multiwavelets are worth of further investigation as a technique for chaotic time series analysis.

Acknowledgments This work was supported by National Natural Science Foundation of China (Grant No. 11172182, 11227201, and 11202141)

References

1. Murguía J, Campos-Cantón E (2006) Wavelet analysis of chaotic time series. Revista mexicana de física 52(2):155–162
2. Strela V et al (1999) The application of multiwavelet filter banks to image processing. IEEE Trans Image Proces 8(4):548–563
3. Wang X, Zi Y, He Z (2011) Multiwavelet denoising with improved neighboring coefficients for application on rolling bearing fault diagnosis. Mech Sys Signal Proces 25(1):285–304
4. Yang YF, Su ZX (2012) A novel medical image enhancement algorithm based on multiwavelet transform. Adv Sci Eng Med 4(6):545–549
5. Chen J et al (2012) Compound faults detection of rotating machinery using improved adaptive redundant lifting multiwavelet. Mech Sys Signal Proces
6. Geronimo J, Hardin D, Massopust PR (1994) Fractal functions and wavelet expansions based on several scaling functions. J Approximation Theor 78:373–401
7. Strela V, Walden AT (1998) Orthogonal and biorthogonal multiwavelets for signal denoising and image compression. Aerosp Def Sens Controls: Int Soc Opt Photon 3391:96–107
8. Mackey MC, Glass L (1977) Oscillation and chaos in physiological control systems. Science 197(4300):287–289
9. Lorenz EN (1963) Deterministic nonperiodic flow. J Atmos Sci 20(2):130–141

Chapter 7
Localization and Navigation of Intelligent Wheelchair in Dynamic Environments

Jingchuan Wang, Juncheng Wang and Weidong Chen

Abstract An intelligent wheelchair JiaoLong with multi-mode is developed for the handicapped and the elderly. JiaoLong is designed of two manipulate modes according to the user's disability and the environments for use. Based on the dynamic localizability matrix, an improved particle filter localization algorithm is proposed in this paper. The results of experiments show the practicability of the system design and the effectiveness provided by the improved localization method.

Keywords Intelligent · Wheelchair · Control modes · Localization · Algorithm

7.1 Introduction

Intelligent wheelchairs have been developed in a number of research labs over the past years. Conventional wheelchairs require the presence of healthcare stuff or family members who are burdened with actually operating the wheelchairs. In order to release the burden of staff in facilities such as hospitals and nursing rooms, fundamental functions such as: Avoid Obstacle, Follow Wall, and Pass Doorway have been proposed [1]. According to the past researches, safety and comfort have also been important aspects for intelligent wheelchair. On top of that, enhancement of independent mobility by a smart wheelchair can also help people rebuild confidence of social skills [2].

J. Wang (✉) · J. Wang · W. Chen
Department of Automation, Shanghai Jiao Tong University, and Key Laboratory of System Control and Information Processing, Ministry of Education of China, Shanghai 200240, China
e-mail: jchwang@sjtu.edu.cn

J. Wang · J. Wang · W. Chen
State Key Laboratory of Robotics and System (HIT), Harbin 150001, China

Z. Sun and Z. Deng (eds.), *Proceedings of 2013 Chinese Intelligent Automation Conference*, Lecture Notes in Electrical Engineering 255,
DOI: 10.1007/978-3-642-38460-8_7, © Springer-Verlag Berlin Heidelberg 2013

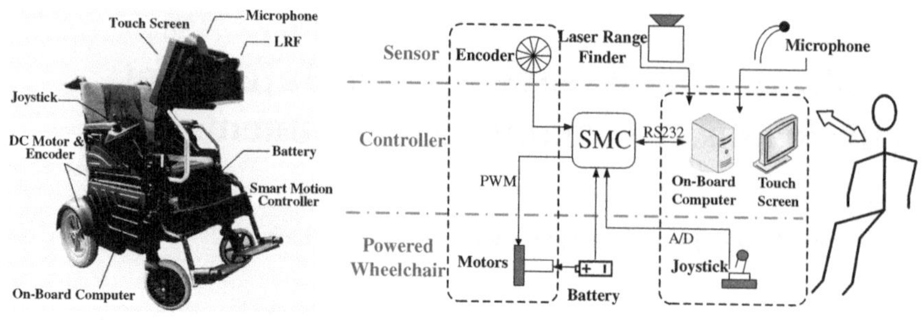

Fig. 7.1 Prototype and hardware structure of JiaoLong

However, a more important feature of smart wheelchair is functionality, to make the systems adaptable to the particular needs of each user according to the type and degree of handicap involved. In this aspect, the control system of a smart wheelchair should change the degree of autonomy according to the user's control ability. This user adapted idea is very useful especially when people are having rehabilitation training [3–5]. In this paper we introduce an intelligent wheelchair: JiaoLong with multi-modes, the aim of which is to provide an aid to mobility for different disabled and elderly people.

As shown in Fig. 7.1, based on the commercial powered wheelchair, JiaoLong is equipped with encoder and Laser Range Finder (LRF) [6], has been designed to meet a wide range of users through multi-mode control system design. This paper introduces the control system of JiaoLong, the realization of two modes aiming to meet the demands of users of different level of handicap, an improved particle filter localization algorithm we proposed to ensure safety travel, as well as the results of the evaluation experiments carried out in welfare institute environments.

7.2 Control Modes

According to the type and degree of different users' handicap, two control modes of JiaoLong are provided: Dynamic Shared Control Mode and Autonomous Navigation Mode. Control methods based on multi-mode make JiaoLong adaptable to the needs of different users in different environments [7].

7.2.1 Dynamic Shared Control Mode

Humans, especially old or disabled ones, are less precise in maneuver, do not preserve curvature well and sometimes have difficulty in perceiving their surroundings. It is significant to find out a method that can combine human control ability and machine control ability effectively [7]. Machine assistance should be

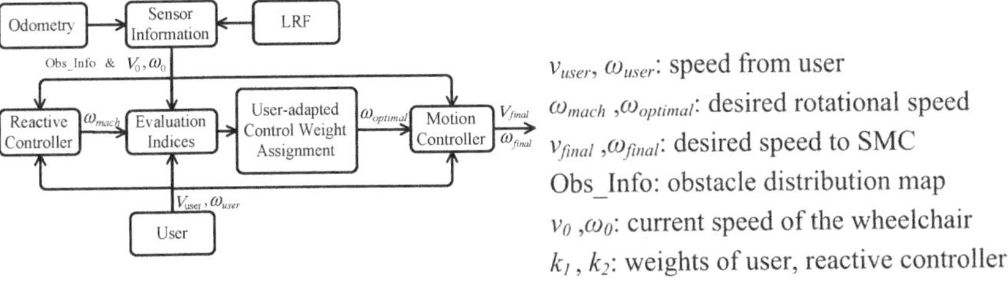

Fig. 7.2 Architecture of the dynamic shared control system

adaptable to the difference of the users' control abilities. A Dynamic Shared Control (DSC) algorithm rooted in this idea is proposed in this paper [8].

Figure 7.2 depicts the functional diagram of this algorithm. There are two key parts in this architecture: the reactive controller and the user-adapted control weight assignment. The reactive controller provides basic machine help (ω_{mach}) using MVFH&VFF methods. The user-adapted control weight assignment assigns control weights to user and the reactive controller. The minimum vector field histogram (MVFH) method and vector force field (VFF) method which the reactive controller adopts are developed by the University of Michigan [9, 10].

Three indices are proposed to evaluate the controller's performance: *safety*, *comfort* and *obedience*. Performance in maneuver is measured by evaluation indices and the weight optimizer calculates the user's control weight according to this performance. User-adapted control weight assignment is the center of dynamic shared-control mode. Through the weights dynamic assignment of the user and the reactive controller, the maximum indices is obtained.

The *safety* index needs to be able to reflect the possibility of a wheelchair colliding with obstacles. Three indices are defined as:

$$\begin{aligned} safety &= 1 - \exp(-\alpha \cdot dis) \\ comfort &= \exp(-\beta|\omega - \omega_0|) \\ obedience &= \exp(-\gamma|\xi - \xi^*|) \end{aligned} \tag{7.1}$$

Where, α is a constant *dis* represents the distance between the wheelchair and the nearest obstacle in its path, β is a constant, ξ^* is the orientation calculated from the user's input v_{user} and ω_{user}; ξ is the orientation determined by v and ω; γ is a constant.

7.2.2 Autonomous Navigation Mode

In some structured environment such as in hospitals and welfare institutes, Jiao-Long can carry users to navigate automatically in autonomous navigation mode. It adapts to the users who have difficulty and even no ability in powered wheelchair control, and lighten the burdens of nurses and the families of patients. During navigation JiaoLong can navigate to the destination automatically and avoid the dynamic obstacles [11]. In localization, the fundamental particle filtering method [12] is used. In path planning and trajectory following, the classical A* algorithm [13] and the Lane Curvature Method (LCM) [14] are used.

We proposed an improved particle filter localization algorithm [15] which is based on the dynamic localizability matrix. Framework of the improved particle filter localization based on localizability is shown in Fig. 7.3.

On one hand, this algorithm estimates the belief of laser range finder observations using the localizability matrix of observation model. On the other hand, it estimates the belief of the odometer data using the covariance matrix of prediction model. Then based on these two indicators, the predicted robot position is modified according to the observed information.

The dynamic localizability indicator was introduced in paper [15]. The structure of map, noisy of sensor and the effect of the unknown obstacles are considered comprehensively. As show as Eq. (7.2)

$$\boldsymbol{D}(\boldsymbol{P}) = \sum_{i}^{n} \left(\frac{1 - s_i}{\sigma_i^2} \begin{bmatrix} \frac{\Delta r_{iE}^2}{\Delta x^2} & \frac{\Delta r_{iE}^2}{\Delta x \Delta y} & \frac{\Delta r_{iE}^2}{\Delta x \Delta \theta} \\ \frac{\Delta r_{iE}^2}{\Delta x \Delta y} & \frac{\Delta r_{iE}^2}{\Delta y^2} & \frac{\Delta r_{iE}^2}{\Delta y \Delta \theta} \\ \frac{\Delta r_{iE}^2}{\Delta x \Delta \theta} & \frac{\Delta r_{iE}^2}{\Delta y \Delta \theta} & \frac{\Delta r_{iE}^2}{\Delta \theta^2} \end{bmatrix} \right) \tag{7.2}$$

s_i is the effecter of unknown obstacles; σ_i^2: is the variance of LRF measurement; $\frac{\Delta r_{iE}}{\Delta x}, \frac{\Delta r_{iE}}{\Delta y}, \frac{\Delta r_{iE}}{\Delta z}$ is the measurement change of LRF after robot's movement in Δx, Δy and Δz.

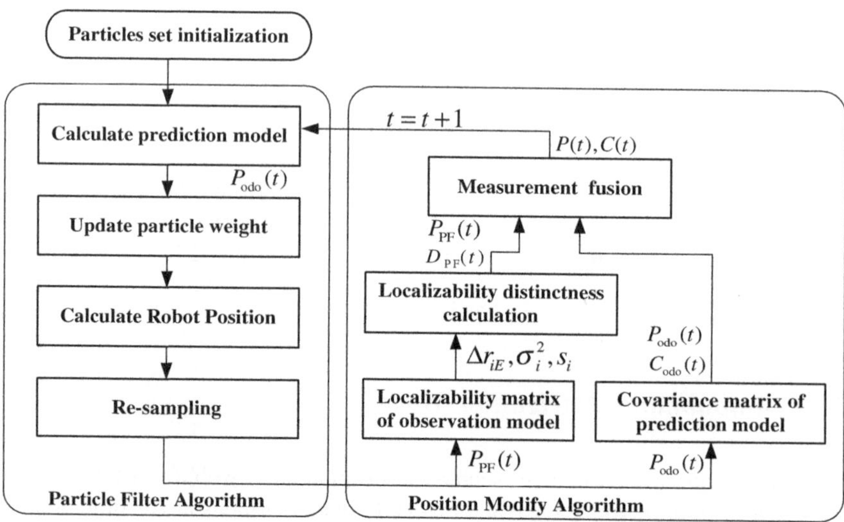

Fig. 7.3 Framework of improved particle filter localization

The localizability matrix $\boldsymbol{D}_{PF}(\boldsymbol{P})$ of observation model for particle filter algorithm can be defined as:

$$\boldsymbol{D}_{\mathrm{PF}}(\boldsymbol{P}) = k_{\mathrm{P}}\boldsymbol{D}(\boldsymbol{P}), \quad k_{\mathrm{P}} = 1 - \frac{1}{\log n_{\mathrm{P}}} \tag{7.3}$$

$\boldsymbol{P}$ is robot position (assume to the result of particle filter), n_{p} is the number of particle set.

At time t, based on prediction model robot position can be calculated as:

$$\boldsymbol{P}_{\mathrm{odo}}(t) = \boldsymbol{P}(t-1) + \boldsymbol{f}_{\mathrm{odo}}(\Delta u(t)) \tag{7.4}$$

$\boldsymbol{P}(t-1)$ is robot position at time $t-1$; $\Delta u(t)$ is the input of odometer at time $t-1$; f_{odo} is the kinematic model; $\boldsymbol{C}_{\mathrm{odo}}(t)$ is the covariance matrix of prediction model.

Based on prediction model, the new particle set $\{\boldsymbol{P}_{\mathrm{pf}}(t), w(t)\}$ will be update. $\boldsymbol{P}_{\mathrm{pf}}(t)$ is the position of particles; $w(t)$ is the weight of particles.

The robot position can be calculated through new particle set:

$$\boldsymbol{P}_{\mathrm{PF}}(t) = \sum_i w^i(z(t), M)\boldsymbol{p}^i_{\mathrm{PF}}(t) \tag{7.5}$$

$z(t)$: observation of LRF; $\boldsymbol{M}$: map information.

The predictive position is modified based on observation, it can be calculated:

$$\Delta\boldsymbol{P}_{\mathrm{obs}}(t) = \boldsymbol{P}_{\mathrm{PF}}(t) - \boldsymbol{P}_{\mathrm{odo}}(t) \tag{7.6}$$

At time t, the predictive position is modified based on the localizability matrix $\boldsymbol{D}_{PF}(\boldsymbol{P})$ of observation model and the covariance matrix of prediction model [16]:

$$\boldsymbol{P}(t) = \boldsymbol{P}_{\mathrm{odo}}(t) + \frac{\boldsymbol{D}_{\mathrm{PF}}(t)}{\boldsymbol{D}_{\mathrm{PF}}(t) + k_1\boldsymbol{C}^{-1}_{\mathrm{odo}}(t)}\Delta\boldsymbol{P}_{\mathrm{obs}}(t) \tag{7.7}$$

k_1: proportion between the localizability of observation model and the covariance matrix of prediction model.

After the re-sampling period, the particle set is modified in the differencing direction of robot position and predictive position:

$$\{\boldsymbol{p}'(t)\} = \{\boldsymbol{p}(t)\} + P(t) - P_{\mathrm{PF}}(t) \tag{7.8}$$

According to Eq. (7.7), in crowed environment, the localizability is lower. Robot localization relies mainly on the predictive information of odometer. In loose environment, the particle set is modified according to the predictive position of observation model. The accumulative error is eliminated because of the higher localizability.

7.3 Experiments

7.3.1 Door Passage Experiment

In laboratory experiments, volunteers were asked to pretend to have some kinds of disabilities like visual defects to pass through a door. α, β and γ in the three indices are set to 0.0005, 0.006 and 0.003 in this experiment.

The left part of Fig. 7.4 includes four variation curves. From top to bottom, they are user's control weight, user's desired rotational speed, machine's desired rotational speed and the combination of user's and machine's rotational speed. The right part illustrates the trajectory and the weight assignment of real-time in this test.

At the beginning there was no obstacle around the wheelchair and no collision was possible to happen. So, the *obedience* became the dominating index, and the user got a relatively high control weight. At the time point 25, if the wheelchair was still controlled completely under the user's will, there would be a collision. To preserve *safety* the algorithm started to decrease the user's control weight. As can be seen from the curve of the user's control weight, it was changed according to his control ability to prevent collision. This seamless cooperation ensured the movement of the wheelchair safe and smooth.

7.3.2 Autonomous Navigation Experiment

In order to verify the accuracy of localization, another experiment was defined: the wheelchair was operated in autonomous navigation mode and navigated to a same destination (a site in the hall) from corridor. This experiment was tested in different crowded environments (there is different quantity of pedestrians in the hall) with common particle filter algorithm and improved particle filter algorithm. The localization results were recorded in Fig. 7.5.

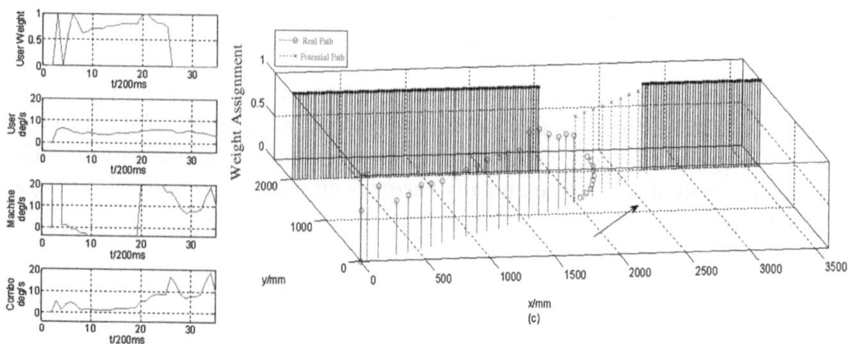

Fig. 7.4 User's control weight and trajectory in door passage

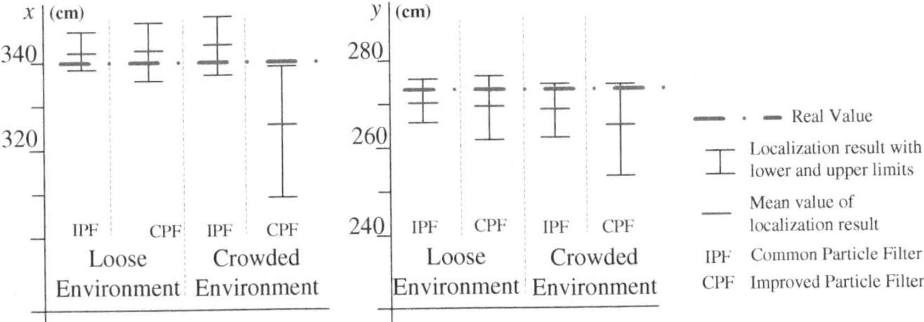

Fig. 7.5 Comparison of localization results

As shown as Fig. 7.5, in loose environment (3 pedestrians in the hall), the errors of localization results of common and improved particle filter algorithm were few. The variance of improved particle filter algorithm was lower. However, in crowded environment (8 pedestrians in the hall), the error of localization results of common particle filter algorithm was larger than that of improved particle filter algorithm. So, our improved particle filter algorithm showed effectiveness and accuracy especially in practical crowded environments compared with the common particle filter algorithm.

7.4 Test and Evaluations

JiaoLong was evaluated and test in two situations: the Shanghai 3rd Elderly Home.

In Shanghai 3rd Elderly Home, we invited five elderly people who are in normal states of mind but have some difficulty in mobility with ages ranging from 75 to 84. Besides, two of them have disabilities. One is suffering from weakness in limbs and the other is suffering from left-handed stroke.

All the subjects were asked to drive the wheelchair through three different tasks (HT: Hall Tour, DP: Door Passage, OA: Obstacle Avoidance) under two operating modes (EM: Electric Mode, DSC: Dynamic Shared-Control Mode).

There are two types of evaluation data: (1) Subjective data which records the subjects' feeling (TD: Task Difficulty, OD: Operating Difficulty and S: Satisfaction). (2) Objective data which records the task completion time, collision times, trajectory smoothness, trajectory fluency, trajectory length, user's control weight, and the three evaluation indices (*safety*, *comfort* and *obedience*).

Table 7.1 shows that the average speeds in different tasks have significant differences and the fluency and smoothness indices get a relative higher value under DSC mode, meaning that the trajectory quality was improved in DSC mode. These improvements are mainly caused by the restriction of the evaluation indices and the increase of the user's control performance as the tasks difficulty decreased.

Table 7.1 Mean and standard deviation of trajectory quality

MEAN/STD	HT		DP		OA	
	EM	DSC	EM	DSC	EM	DSC
Speed (m/s)	0.319/0.016	0.318/0.015	0.236/0.022	0.237/0.018	0.212/0.027	0.201/0.019
Fluency	0.648/0.081	**0.685/0.058**	0.578/0.023	**0.634/0.061**	0.636/0.087	**0.712/0.068**
Smoothness	0.698/0.057	**0.812/0.076**	0.668/0.055	**0.719/0.042**	0.550/0.24	**0.736/0.079**

Table 7.2 JiaoLong test questionnaires

	STD	TD	OD	S
HT	EM	0.5	1.5	1.5
	DSC	0.5	−3	3.5
DT	EM	2	3.5	−1
	DSC	2	−1.5	2
OA	EM	1.5	4	−1
	DSC	1.5	−2	2.5

Table 7.2 shows the questionnaires: the door passage task is the hardest one and the hall tour is the easiest. By using the DSC algorithm, user's satisfaction was improved and task like door passage becomes easier.

7.5 Conclusion

This paper shows the prototype, hardware architecture, improved algorithm in localization, experiments and tests results of the intelligent wheelchair called JiaoLong. Fundamental ideas and methods used in the dynamic shared control mode and the autonomous navigation mode are also proposed. The experiments in different modes show that the proposed algorithm and ideas can not only adapt the degree of autonomy to the user's control ability but also show the accuracy and effectiveness especially in crowded dynamic environments.

Acknowledgments This work is partly supported by ITER Research and Development Program under grant 2012GB102001 and 2012GB102002, the National High Technology Research and Development Program of China under grant 2012AA041403, the Natural Science Foundation of China under grant 60934006 and 61175088, the State Key Laboratory of Robotics and System (HIT), the Research Fund for the Doctoral Program of Higher Education under grant 20100073110018, and Research Councils UK under UK-China Science Bridge Grant No. EP/G042594/1.

References

1. Prassler E, Scholz J, Fiorini P (2001) A robotics wheelchair for crowded public environment. IEEE Robot Autom Mag 8(1):38–45
2. Bourhis G, Pino P (1996) Mobile robotics and mobility assistance for people with motor impairments: rational justification for the VAHM project. IEEE Trans Rehab Eng 4:7–11
3. Gignac MA, Cotta C, Badley EM (2000) Adaptation to chronic illness and disability and its relationship to perceptions of independence and dependence. J Gerontol B Psychol Sci Soc Sci 55(6):362–372
4. Lankenau A, Rofer T (2001) A versatile and safe mobility assistant. IEEE Robot Autom Mag 8(1):29–37
5. Bourhis G, Horn O, Habert O, Pruski A (2001) An autonomous vehicle for people with motor disabilities. IEEE Robot Autom Mag 88(1):20–28
6. Wang JC, Chen WD (2009) Design and Implementation of a multi-sensor based autonomous wheelchair (in Chinese). J Nanjing Univ Sci Technol (Nat Sci) 33:104–409
7. Gomi T, Griffith A (1998) Developing intelligent wheelchairs for the handicapped wheelchair. Assistive Technol Artif Intell 151–178
8. Li Q, Chen WD, Wang JC (2011) Dynamic shared control for human-wheelchair cooperation. In: Proceedings of 2011 IEEE international conference on robotics and automation, Shanghai, China, 9–13 May 2011
9. Levine SP, Bell DA, Jaros LA et al (1999) The NavChair assistive wheelchair navigation system. IEEE Trans Rehabil Eng 7(4):443–451
10. Borenstein J, Koren Y (1991) The vector field histogram-fast obstacle avoidance for mobile robots. IEEE Trans Robot Autom 7(3):278–288
11. Wang Y, Chen WD, Wang JC (2011) Hybrid map-based navigation for intelligent wheelchair. In: Proceedings of 2011 IEEE international conference on robotics and automation, Shanghai, China, 9–13 May 2011
12. Dellaert F, Burgard W, Fox D et al (1999) Monte Carlo localization for mobile robots. In: Proceedings of the IEEE international conference on robotics and automation, vol 2. Detroit, MI, May 1999, pp 1322–1328
13. Hart PE, Nilsson NJ, Raphael B (1968) A formal basis for the heuristic determination of minimum cost paths. IEEE Trans Sys Sci Cybern Ssc-4(2):100–107
14. Ko NY, Simmons RG (1998) The Lane-curvature method for local obstacle avoidance. In: Proceedings of 1998 IEEE international conference on intelligent robotics and systems, Victoria, BC, Canada
15. Wang W, Chen WD, Wang Y (2012) Localizability estimation for mobile robot with use of probabilistic grid map (in Chinese). Robot 34(4):485–491,512
16. Roecker JA, McGillem CD (1988) Comparison of two-sensor tracking methods based on state vector fusion and measurement fusion. IEEE Trans Aeros Electron Sys 24:447–449

Chapter 8
Adaptive Particle Filter with Estimation Windows

Peng Li and Jian Tang

Abstract Particle filter is well suited to estimate the state of non-linear non-Gaussian dynamic systems, which comes at the cost of higher computational complexity. But in many real time applications, it must deal with constraints imposed by limited computational resources. To deal with this question, we distribute the samples among the different observations arriving during a filter update, the novel algorithm represents densities over the state space by mixtures of sample sets. Another contribution of this paper is to increasing the efficiency of particle filters by adapting the size of sample sets during the estimation process. According to the relative entropy theory and particle number controller idea, we choose the number of samples, decrease computation overhead. A simulation of the classic HARD bearing only tracking problem is presented, the results show that the novel algorithm performs better than generic particle filter.

Keywords Particle filter · Estimation window · Relative entropy · Particle number controller

8.1 Introduction

Due to their sample-based representation, particle filters have been applied with great success to a variety of state estimation problems including object tracking, speech recognition, and mobile robotics. Unfortunately, the sample-based representation of particle filters comes at increased computational requirements, especially compared to the Kalman filter and its extensions, resulting in common situations where the rate of incoming sensor data is higher than the update rate of

P. Li (✉) · J. Tang
College of Information and Technology, Key Laboratory of Intelligent Computing and Information Processing, Ministry of Education, Xiangtan University, Yuhu, Xiangtan City, China
e-mail: pengli.hit@gmail.com.cn

Z. Sun and Z. Deng (eds.), *Proceedings of 2013 Chinese Intelligent Automation Conference*, Lecture Notes in Electrical Engineering 255,
DOI: 10.1007/978-3-642-38460-8_8,

the particle filter [1, 2]. We introduced adaptive particle filters to deal with such situations. Instead of discarding sensor readings, this technique distributes the samples among the different observations arriving during a filter update. In view of the problems which computational burden for the particle filter dependent on the number of particles and on the resampling calculation, A novel approach to determine the number of particles is also developed, using a relay and an integrator in a feedback system. Note that the controller is implemented in resampling step.

The organization of the paper is as follows. In Sect. 8.2, we introduce classical particle filter with estimation windows. The main results are given in Sect. 8.3, where the adaptive particle filter is completed, which works according to the results deduced from relative entropy and the controller of particle number. The adaptive particle filter with estimation window comes and the performance of the proposed algorithm is illustrated on a simulation about the hard nonlinear bearing only tracking problem in Sect. 8.4. This paper is finalized with conclusion in Sect. 8.5 and followed by acknowledgments.

8.2 Classical Particle Filter with Estimation Windows

We will now present a novel approach to dealing with limited computational resources. The key idea is to consider all sensor measurements by distributing the samples among the observations within an update window. Additionally, by weighting the different sample sets within a window, our approach focuses the computational resources (samples) on the most valuable observation [3]. Figure 8.1 illustrates the approach. As can be seen, instead of one sample set at time t, we maintain k smaller sample sets, one for each observation. At the end of each estimation window, algorithm determines the weights of the mixture belief such that the approximation error relative to the optimal filter process is minimal.

Let's introduce some notation, assume that observations arrive at time intervals Δ, which we call observation interval. Let n be the number of samples required by the particle filter. Assume that the resulting update cycle of the particle filter takes

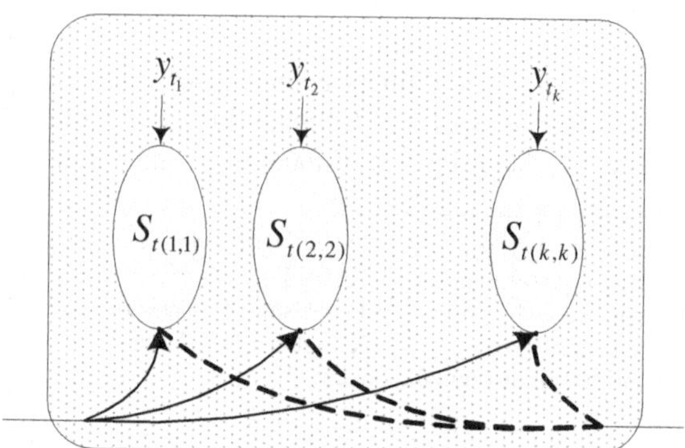

Fig. 8.1 Particle filter with estimation windows

$k\Delta$ and is called the estimation interval or estimation window. Accordingly, k observations arrive during one estimation interval. We call this number the window size of the filter, which is the number of observations obtained during a filter update. The ith observation and state within window t are denoted y_{t_i} and x_{t_i} respectively.

Now we show how the new algorithm works. In each estimation time windows t, the new algorithm only do once particle filter operation and calculate the corresponding weight for each sub-particle set $S_{t(i)}$ at time $t(i)$. At the other times $t(j)$ $(j = 1, 2, \ldots k : j \neq i)$, the state is updated according to the dynamic equations. At the end of the estimation windows t, the estimation of state is calculated by unscented particle filter, using all the particle from k particle sets, algorithm gets into next estimation window $t + 1$.

8.3 Adaptive Particle Filter

In this section, we propose a novel approach to change the particle number. To begin, Compute the probability density functions of two sets of particles, which get from the same probability distribution, we get the relative entropy of the two sets. Then by the relative and the particle number controller, we change the particle number in estimation.

8.3.1 Relative Entropy

Definition The relative entropy of probability density function $p(x)$ and $q(x)$ is

$$D(p\|q) = \sum_{x \in \aleph} p(x) \log \frac{p(x)}{q(x)} \tag{8.1}$$

Lemma 1 *Jensen's Inequation* [4]

If $p(x)$ *and* $q(x)$ *are probability density functions, then*

$$D(p\|q) \geq 0 \tag{8.2}$$

with equation iff $p(x) = q(x)$ *for arbitrary value of* x.

Lemma 2 Summation Inequation of Logarithm [4]

For non negative number $a_1, a_2 \ldots a_n$ *and* $b_1, b_2 \ldots b_n$, *we have*

$$\sum_{i=1}^{n} a_i \log \frac{a_i}{b_l} \geq \left(\sum_{i=1}^{n} a_i \right) \log \left(\sum_{i-1}^{n} a_i \Big/ \sum_{i-1}^{n} b_i \right) \tag{8.3}$$

with equation iff $\frac{a_i}{b_i}$ *is constant.*

Lemma 3 Convexity of Relative Entropy [4, 5]
If (p_1, q_1) *and* (p_2, q_2) *are two groups of probability density functions, and* $D(p\|q)$ *is convexity function of* (p, q), *then for all value of* λ *between* 0 *and* 1, *we have*

$$D((\lambda p_1 + (1-\lambda)p_2)\|(\lambda q_1 + (1-\lambda)q_2)) \le \lambda D(p_1\|q_1) + (1-\lambda)D(p_2\|q_2) \quad (8.4)$$

Lemma 4 Monotone Decreasing Characteristics of Empirical Distribute [4, 6]
Let $X_1, X_2, \cdots, X_n$ *be independent identical distribution functions, the density function is* $p(x)$, $\hat{p}(x)$ *is the empirical probability density function of* $X_1, X_2, \cdots, X_n$, *then*

$$E(D(\hat{p}_n\|p)) \le E(D(\hat{p}_{n-1}\|p)) \quad (8.5)$$

$$E(D(p\|\hat{p}_n)) \le E(D(p\|\hat{p}_{n-1})) \quad (8.6)$$

Theorem 1 *Let* m_1, m_2, n_1, n_2 *be positive integer, and they satisfies* $m_1 \ge m_2$ $n_1 \ge n_2$. $\hat{p}_{m_1}(x), \hat{p}_{m_2}(x), \hat{p}_{n_1}(x), \hat{p}_{n_2}(x)$ *are the empirical probability density function of four samples, which respective has the particle number of* m_1, m_2, n_1, *then*

$$E(D(\hat{p}_{m_1}\|\hat{p}_{n_1})) \le E(D(\hat{p}_{m_2}\|\hat{p}_{n_2})) \quad (8.7)$$

Proof According to the lemma 1, we could easy get

$$E(D(\hat{p}_{m_1}\|\hat{p}_{n_1})) \le E(D(\hat{p}_{m_2}\|\hat{p}_{n_1})) \quad (8.8)$$

by the mathematical induction. In the same way, we could prove that

$$E(D(\hat{p}_{m_2}\|\hat{p}_{n_1})) \le E(D(\hat{p}_{m_2}\|\hat{p}_{n_2})) \quad (8.9)$$

Combining these two equations, we obtain

$$E(D(\hat{p}_{m_1}\|\hat{p}_{n_1})) \le E(D(\hat{p}_{m_2}\|\hat{p}_{n_2})) \quad (8.10)$$

Thus the theorem is proved.

8.3.2 Particle Number Controller

The computational burden for the particle filter is dependent on the number of particles and on the resampling calculation. However, the resampling can be efficiently implemented using a classical algorithm. For sampling N ordered independent identically distributed variables. A novel approach is to apply a simple control structure according to Fig. 8.1 [7]. The number of particles needed is determined by the controller using the residual $\varepsilon_t = \left\|\mu_t^{(1)} - \mu_t^{(2)}\right\|$, where $\mu_t^{(1)}$ and $\mu_t^{(2)}$ are some statistical property from the particle filters, using different number of particles. Possible choices are for instance some relevant statistics, such

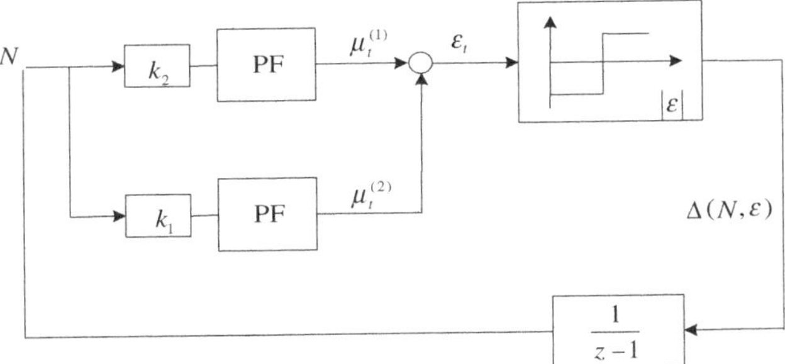

Fig. 8.2 Controller of particle number

as the mean estimate from the particle filter or utilization of the probability density or the cumulative density function. The control structure used is a nonlinear block consisting of a relay and an integrator using

$$\begin{cases} N_{t+1} = (1-\beta)N, & if|\varepsilon_t| < \Delta \\ N_{t+1} = (1+\alpha)N, & if|\varepsilon_t| > \Delta \end{cases} \tag{8.11}$$

For maneuvering targets in a tracking application the controller can reduce or increase the number of particles during the tracking envelope, and the controller is implemented in the resampling step (Fig. 8.2).

8.3.3 Adapt Particle Number

Combining the theorem 1 and particle number controller, we proposed the algorithm of adaptive choosing particle number as Fig. 8.3.

8.4 Novel Particle Filter and its Simulation

The novel particle filter is deduced by combined the classical particle filter with windows in Sect. 8.2 and the adaptive particle filter in Sect. 8.3. At the end, the performance of the novel algorithm is testing in target tracking on reentry.

8.4.1 Adaptive Particle Filter with Estimation Windows

The Adaptive Particle Filter with Estimation Windows comes as we combined the developments in Sects. 8.2 and 8.3. First, we represented posteriors as mixtures of sample sets, where each mixture component integrates one observation arriving

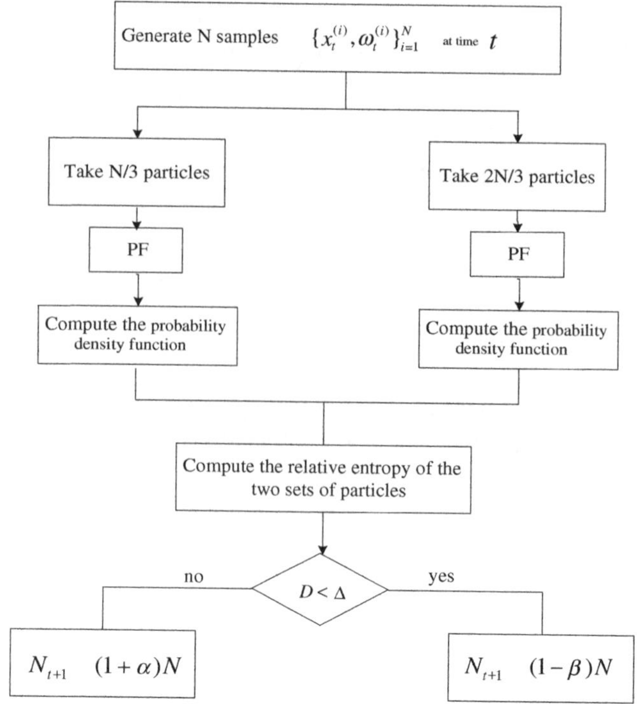

Fig. 8.3 Flowchart of the novel algorithm

during a filter update. Then, we achieved adaptive particle filter by computed the particle number in the resampling process, the number of particles needed is determined by the controller using the residual, that is some statistical property from the particle filters, here is the relative entropy

8.4.2 Simulation

The following state space model is used [7, 8]

$$X(k) = \begin{bmatrix} 1 & 1 & 0 & 0 \\ 0 & 1 & 0 & 0 \\ 0 & 0 & 1 & 1 \\ 0 & 0 & 0 & 1 \end{bmatrix} X(k-1) + \begin{bmatrix} 0.5 & 0 \\ 1 & 0 \\ 0 & 0.5 \\ 0 & 1 \end{bmatrix} V(k-1)$$

$$Y(k) = \arctan(X_3(k)/X_1(k)) + N(k)$$

Where the state vector is defined as the 2D position and velocity vector of the target, relative to a fixed external reference frame. $X_1(k)$ is x- position at time k, $X_2(k)$ is x- velocity at time k, $X_3(k)$ is y-position at time k, $X_4(k)$ is y-velocity at

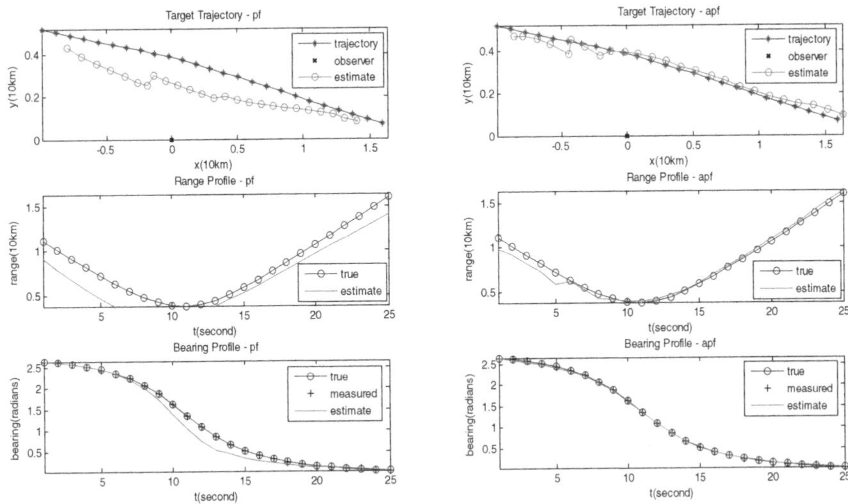

Fig. 8.4 Simulations of bearing tracking

time k, $Y(k)$ is the bearing angle (in radians) from the fixed observer towards the target. The state dynamics and the observations are driven by white Gaussian noise source.

The preliminary study has demonstrated the operation of the adaptive particle filter on a realistic guidance problem. The particle filter and adaptive particle filter with estimation windows were implemented with 25,000 state-space samples, illustrative results are presented for a single run. Figure 8.4 shows the position, the range and the bearing for this example. Simulations have shown significant performance improvement over particle filter for severely nonlinear problems.

8.5 Conclusion

The adaptive particle filter with estimation windows has been applied to hard nonlinear bearing only tracking problems. Simulations have shown significant performance improvement over particle filter for severely nonlinear problems. The great strength of adaptive particle filter is representing densities over the state space by mixtures of sample sets, dealing with constraints imposed by limited computational resources. A further advantage is adapting the size of sample sets during the estimation process, so the efficiency of particle filters is increasing.

Acknowledgments This research is supported by the Education Department of Hunan Province (CAREER grant number 11C1217), the Hunan Provincial Science & Technology Department (CAREER grant number 2012GK3141), the construct program of the key discipline in Hunan province, and the Key Laboratory of Intelligent Computing & Information Processing (Xiangtan University), Ministry of Education.

References

1. Doucet A, de Freitas N, Gordon N (eds) (2001) Sequential Monte Carlo methods in practice. Springer, New York
2. Doucet A, Godsill S, Andrieu C (2000) On sequential Monte Carlo sampling methods for Bayesian filtering. Stat Comput 10(3):752–763
3. Gustafsson F (2000) Adaptive filtering and change detection. Wiley, NY
4. Cover TM, Thomas JA (1991) Elements of information theory. Wiley Series in TelecommunicationsWiley, New York
5. Fox D (2003) Adapting the sample size in particle filters through KLD sampling. Int J Robot Res (IJRR) 22(12):1002–1029
6. Fox D (2001) KLD-sampling: adaptive particle filters and mobile robot localization. Adv Neural Inf Proces Sys (NIPS) 2001:2491–2503
7. Karlsson R, Gustafsson F (2001) Monte Carlo data association for multiple target tracking. Invited paper to IEEE workshop on target tracking, Eindhoven, NL, pp 97–102, 232–242
8. Gordon N, Salmond D, Ewing C (1995) Bayesian state estimation for tracking and guidance using the bootstrap filter. J Guidance Control Dyn 18(6):296–314

Chapter 9
Binary Encoding Differential Evolution with Application to Combinatorial Optimization Problem

Changshou Deng, Bingyan Zhao, Yanlin Yang and Hai Zhang

Abstract Differential Evolution algorithm is a new competitive heuristic optimization algorithm in the continuous field. The operators in the original Differential Evolution are simple; however, these operators make it impossible to use the Differential Evolution in the binary space directly. Based on the analysis of problems led by the mutation operator of the original Differential Evolution in the binary space, a new mutation operator was proposed to enable this optimization technique can be used in binary space. The new mutation operator, which is called semi-probability mutation operator, is a combination of the original mutation operator and a new probability-based defined operator. Initial experimental results of two different combinatorial optimization problems show its effectiveness and validity.

Keywords Combinatorial optimization problem · Differential evolution · Binary encoding · Semi-probability mutation operator

9.1 Introduction

Differential Evolution (DE) is a competitive optimization technique for numerical optimization problems with real parameter [1]. It is a simple yet powerful algorithm for global optimization over continuous spaces, which use the greedy selection criterion to determine which of the rivals to remain in the next generation. Since its invention, the DE algorithm has become quite popular in the machine intelligence and cybernetics. It has been successfully been applied to diverse fields of science and engineering, such as mechanical engineering design

C. Deng (✉) · B. Zhao · Y. Yang · H. Zhang
School of Information Science and Technology, Jiujiang University,
Jiujiang, Jiangxi Province, China
e-mail: csdeng@jju.edu.cn

Z. Sun and Z. Deng (eds.), *Proceedings of 2013 Chinese Intelligent Automation Conference*, Lecture Notes in Electrical Engineering 255,
DOI: 10.1007/978-3-642-38460-8_9, © Springer-Verlag Berlin Heidelberg 2013

[2], signal processing [3] and pattern recognition [4]. It has been proved to perform better than the Genetic Algorithm (GA) or the Particle Swarm Optimization (PSO) by numerical benchmarks experiments [5]. Despite the simplicity and successful application in many engineering fields, its application on the solution of optimization problems with binary decision variables is still unusual. One of the possible reasons for this lack is that DE cannot keep the closure when the original DE operators are used in discrete domain directly, for the operators designed in the original DE are designed only for continuous domain. A few works exploit its usage for discrete optimization problems, particularly the combinatorial optimization problem. The Differential Evolution with binary encoding in [6] may be the first version of binary Differential Evolution. In our recent work [7], a novel binary Differential Evolution without scale factor F was proposed. In this work, the mutation result of the original Differential Evolution was analyzed in depth and then a new semi-probability mutation operator was derived for the binary variables.

The remainder of this paper is structured as follows. Section 9.2 gives a brief introduction of original DE. A new binary encoding DE (BEDE) is presented in Sect. 9.3. Two different combinatorial optimization problems are used to evaluate this binary encoding DE in Sect. 9.4. Section 9.5 concludes this paper.

9.2 Differential Evolution

DE is a population-based stochastic optimizer that starts to explore the search space by sampling at multiple, randomly chosen initial points. It is a kind of float point encoding evolutionary optimization algorithm. AT present, there have been several variants of DE [1]. One of the most promising schemes, DE/RAND/1/BIN scheme of Storn & Price, is presented in great detail. In order to clarify the notation used in this paper, the minimization of the objective function $f(x)$ is referred.

9.2.1 Generation of Initial Population

The DE Algorithm starts with the initial target population $X = (x_{ij})_{m\times n}$ with the size m and the dimension n, which is generated by the following way.

$$x_{ij}(0) = x_j^l + rand(0, 1)(x_j^u - x_j^l) \tag{9.1}$$

where $i = 1, 2, \ldots, m$, $j = 1, 2, \ldots, n$, x_j^u denotes the upper constraints, and x_j^l denotes the lower constraints.

9.2.2 Mutation Operator

For each target vector $x_i(i = 1, 2, \ldots, m)$, a mutant vector is produced by

$$h_i(t+1) = x_{r1} + F(x_{r2} - x_{r3}) \tag{9.2}$$

where $i, r_1, r_2 \in \{1, 2, \ldots, m\}$ are randomly chosen and must be different from each other. And F is the scaling factor which has an effect on the difference between the individual x_{r1} and x_{r2}.

9.2.3 Crossover Operator

DE employs the crossover operator to add the diversity of the population. The approach is given below.

$$u_i(t+1) = \begin{cases} h_i(t+1), & rand \leq CR\, or\, j = rand(i) \\ x_i(t), & otherwise \end{cases} \tag{9.3}$$

where $i = 1, 2, \ldots, m, j = 1, 2, \ldots, n, CR \in [0, 1]$ is crossover constant and $rand(i) \in (1, 2, \ldots, n)$ is the randomly selected index. In other words, the trial individual is made up with some components of the mutant individual, or at least one of the parameters randomly selected, and some of other parameters of the target individual.

9.2.4 Selection Operator

To decide whether the trial individual $u_i(t+1)$ should be a member of the next generation, it is compared to the corresponding $h_i(t+1)$. The selection operator is based on the survival of the fitness among the trial individual and the corresponding one such that:

$$x_i(t+1) = \begin{cases} u_i(t+1), & if\, f(u_i(t+1)) < f(x_i(t)) \\ x_i(t), & otherwise \end{cases} \tag{9.4}$$

DE can adapt itself during the search process and find the optimum efficiently and effectively. The mutation operator can not be used in the binary space directly, while the crossover operator and selection operator can be used in binary space directly.

9.3 Binary Encoding Differential Evolution

A conclusion can be easily drawn from the mutation operator, denoted by the formula (9.2), that it can only keep the closure in the field of real numbers. However, the original DE cannot keep the closure when it is applied in discrete domain. Thus it cannot be used in discrete optimization problems directly. A Binary-coding DE with new mutation rules was proposed to expand DE into the binary space.

9.3.1 Semi-Probability Mutation Operator

A new binary mutation operator was derived from the table of original DE mutation results. Assumed that the binary encoding is used ant the value of F is 1. All the possible results of the original DE mutation operator in binary space directly are shown in Table 9.1.

It is easy to see that there are eight kinds of different combination of the mutually different x_{r1}, x_{r2}, and x_{r3}. And three fourths of the different kinds of combination of the three individuals can achieve binary number, shown in bold form in Table 9.1. Hence this six kinds of combination are eligible even the individual is in binary encoding. Furthermore, the following rules can be derived.

1. If x_{r1} equals to zero and x_{r2} equals to x_{r3}, or x_{r1} equals to one and x_{r2} equals to zero and x_{r3} equals to one, then the result of mutation equals to zero.
2. If x_{r1} equals to one and x_{r2} equals to x_{r3}, or x_{r1} equals to one and the three variables are identical, then the result of mutation equals to one.

The results of the rest two kinds of combination are not zero or one. Some modifications must be done to the original DE mutation operator. By further analyzing the remainder forms of the three individual's combination, probability-based rules can be proposed. A probability denoted by pr can be defined as (9.6).

$$pr = \frac{1}{1 + e^{-h_i^{(t+1)}}} \tag{9.5}$$

Table 9.1 Results of original DE operator on binary space

x_{r1}	x_{r2}	x_{r3}	F	Results
0	0	0	1	0
0	0	1	1	−1
0	1	0	1	1
0	1	1	1	0
1	0	0	1	1
1	0	1	1	0
1	1	0	1	2
1	1	1	1	1

The semi-probability Mutation operator in this paper is defined as follows.

$$h_i(t+1) = \begin{cases} x_{r1} + F(x_{r2} - x_{r3}), & if(h_i(t+1) = 0\, or\, h_i(t+1) = 1) \\ 0, & pr < rand \\ 1, & otherwise \end{cases} \tag{9.6}$$

9.3.2 Flowchart of the BEDE

The flowchart of the BEDE is given as Fig. 9.1.

9.4 Numerical Examples

The 0-1 knapsack problem and one-max problem are the typical combinatorial optimization problems. This section will use three knapsack problems with different size and One-Max problem to initially evaluate the BEDE.

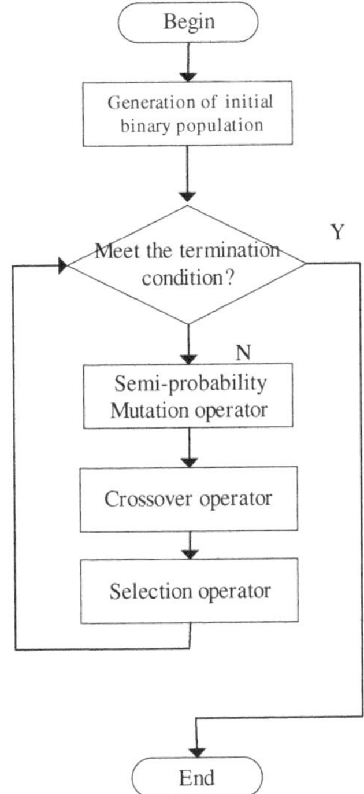

Fig. 9.1 The flowchart of binary encoding differential evolution

9.4.1 Knapsack Problems

The typical 0-1 knapsack problem (KP) is that there are n given items has to be packed in a knapsack of weight capacity V. Each item has a profit p_i and a weight w_i. The problem is to select a subset of the items whose total profit is a maximized, while whose total weight does not exceed the capacity V. Then the 0-1 knapsack problem can be formulated as (9.7) [8].

$$\begin{aligned} &Maximize \quad f = \sum_{i=1}^{n} p_i x_i w_i \\ &s.t. \quad \sum_{i=1}^{n} w_i x_i \le V \\ &x_i \in \{0, 1\} \quad (i = 1, 2, \ldots n) \end{aligned} \tag{9.7}$$

During the evolution of searching the best solution of the knapsack problem, infeasible solutions will appear which means that the conditions in the formula (9.7) cannot be met. When the total items outrange the capacity of the knapsack, the items with lower profit density will be discarded by priority. By this way, the remainder of the items is closer to the optimum than that of random way. A fixed-up operator is defined to solve this problem. Two steps are adopted in the fixed-up operation. Firstly, all the items are sorted in profit density ascending order. Then one solution is infeasible, the items are discarded in the order of profit density until it is feasible.

9.4.2 Numerical Experiment

The parameters of the BEDE and the NBDE [7] is presented in Table 9.2. For each of the three knapsack problems, 50 trials have been conducted and the best results (Best), average results (Avg), worst results (Worst) and standard deviations (Dev) of the BEDE and NBDE are shown in Table 9.3.

The results in Table 9.3 show us that BEDE can find the optimums of the three Knapsack Problems with comparable small size of population. The BEDE is better than the NBDE in Avg, Worst, and Dev terms.

Table 9.2 Parameters of binary encoding DE for the three KPs

Examples	Population size	CR	Maxgen
Kp1	20	0.5	50
Kp2	50	0.5	200
Kp3	50	0.5	1,000

Table 9.3 The statistical results of the three knapsack problems

Problems	Algorithm	Best	Avg	Worst	Dev
KP1	BEDE	1,042	1,041.8	1,037	1.8516
	NBDE	1,042	1,039	1,030	3.2482
KP2	BEDE	3,119	3,116.6	3,111	1.2898
	NBDE	3,119	3,114.4	3,102	4.7559
KP3	BEDE	26,559	26,555.3	26,535	5.1912
	NBDE	26,559	26,550	26,529	9.378

Table 9.4 The statistical results of the one-max problem

Algorithm	Best	Avg	Worst	Dev
BEDE	120	119.96	119	0.1979
NBDE	120	119.84	119	0.3703

9.4.2.1 One-Max Problem

The aim of a One-Max problem is simply to maximize the ones in a binary string. The fitness of a string is the number of ones it has. The string length 120 is used for this study, with optimum 120.

The parameters set in this study for the One-Max problem is as fellows. Population size is set to 50, $CR = 0.15$ and the maximum generation is 500.

For each of the One-Max problem with length 120, 50 trials have been conducted and the best results (Best), average results (Avg), worst results (Worst) and standard deviations (Dev) of BEDE and NBDE [7] are shown in Table 9.4.

The results in Talbe 9.4 show us that the proposed BEDE can find the optimum and the performances of Avg, Worst and Dev are slightly better than that of the NBDE.

9.5 Conclusion

DE is a recently developed algorithm that is very efficient for global optimization over continuous spaces. A binary encoding DE using semi-probability based mutation operator was proposed in this paper to extend the field of DE from the continuous domain to the binary field. The new mutation operator can be applied in binary space directly. Initial experiments on the three different sizes of Knapsack Problems and One-Max problem with size length 120 show that the proposed BEDE is an effective and efficient way to solve combinatorial optimization problems compared with the other one version of binary DE.

Acknowledgments This work is partially support by State Key Laboratory of Software Engineering under grant No. SKLSE2012-09-39 and the Science and Technology Foundation of Jiangxi Province, China under grant No.GJJ11616 as well.

References

1. Storn R, Price K (1997) Differential evolution—a simple and efficient heuristic for global optimization over continuous spaces. J Global Optim 11:241–354
2. Rogalsky T, Derksen RW, Kocabiyik S (1999) Differential evolution in aerodynamic optimization. In: Proceedings of 46th annual conference of Canadian Aeronautics and Space Institute, pp 29–36
3. Das S, Konar A (2006) Design of tow dimensional IIR filters with modern search heuristics: a comparative study. Int J Comput Intell Appl 6(3):176–185
4. Das S, Abraham A, Konar A (2008) Adaptive clustering using improved differential evolution algorithm. IEEE Trans Syst Man Cybern Part A 38(1):218–237
5. Versterstrom J, Thomsen R (2004) A comparative study of differential evolution, particle swarm optimization, and evolutionary algorithm on numerical benchmark problems. In: Proceedings of computation, CEC2004, vol 2. pp 1980–1987
6. Gong T, Andrew LT (2007) Differential evolution for binary encoding. Soft Comput Ind Appl ASC 39:251–262
7. Deng C, Zhao B, Yang Y et al (2010) Novel binary differential evolution without scale factor F. In: Proceedings of third international workshop on advanced computational intelligence, vol 8. pp 250–253
8. Liu Y, Liu C (2009) A schema-guiding evolutionary algorithm for 0-1 Knapsack problem. In: Proceedings of 2009 international association of computer science and information technology—spring conference, pp 160–164

Chapter 10
The Research on the Fuzzing

Lanlan Qi, Jiangtao Wen, Hui Huang, Xiong Wang and Zhiyong Wu

Abstract From Miller firstly introduced the fuzzing in 1990 and found failures in over 25 % of UNIX programs, to recent TaintScope system presented by Peking University and the discovery of 27 0day vulnerabilities in several popular software including Adobe Acrobat, the practical experiences and results have illuminated that fuzzing are effective for vulnerability mining. In this paper, fuzzing is studied and surveyed. First, new features of fuzzing are analyzed. And then, give the definition of Cd and Md. Cd and Md describe the ability to implement vulnerability mining on the vulnerability of the test case. Based on the Cd and Md, this paper also gives a new fuzzing architecture Feedback Fuzz.

Keywords Fuzzing · Mining · Accuracy · Deformity

10.1 Introduction

Fuzzing is an effective vulnerabilities mining, Miller [1, 2] firstly introduced the fuzzing in 1990 and crashed more than 25 % of UNIX programs. Aitel designed and implemented their own Fuzzing tools SPIKE [3, 4], then successfully found numbers of unknown vulnerabilities. Godefroid used SAGE [5, 6] discovered more than 20 0day vulnerabilities in large-scale Windows applications. In 2010, TaintScope system [7] presented by Peking University found 27 0day vulnerabilities in several popular software including Adobe Acrobat, Google Picasa,

L. Qi (✉) · J. Wen · H. Huang · X. Wang · Z. Wu
Department of Computer Science and Technology, Tsinghua University, Beijing, China
e-mail: qilanlan09@gmail.com

L. Qi · J. Wen · H. Huang · X. Wang · Z. Wu
Electronic Engineering Institute of PLA, Hefei, China

Z. Sun and Z. Deng (eds.), *Proceedings of 2013 Chinese Intelligent Automation Conference*, Lecture Notes in Electrical Engineering 255,
DOI: 10.1007/978-3-642-38460-8_10,

Microsoft Paint, ImageMagick and other applications. Fuzzing is an automated software testing technique based on fault injection. It uses a large number of semivalid data [8] as the input of application, then monitors and records any exception. The so-called semivalid data is that the necessary identifies of document and most of the data is valid, while the rest of the data is invalid. When the target application handles invalid data, it may cause the application to crash or trigger a security hole.

10.2 New Feature

After recent 20 years development, fuzzing has some new features.

10.2.1 Feedback Driven Fuzzing

Leverage the programs' internal information during execution to direct test data generation. As shown in Fig. 10.1, unsafe function information, runtime information, code coverage and other feedback information direct smartly the correct and malformed test data generation. It can greatly improve the efficiency of vulnerability mining, such as TaintScope [7], BuzzFuzz [9], FTSG [10] and other tools.

10.2.2 Integration of White-Box Fuzzing A

Symbolic-execution-based white-box fuzzing (such as EXE [11], KLEE [12]) can substitute all program inputs with symbolic values, gather input constraints on a program trace and generate new inputs that can drive program executions along different traces. These tools are able to provide good code coverage and have proven to highly improve the effectiveness of traditional fuzzing tools. They

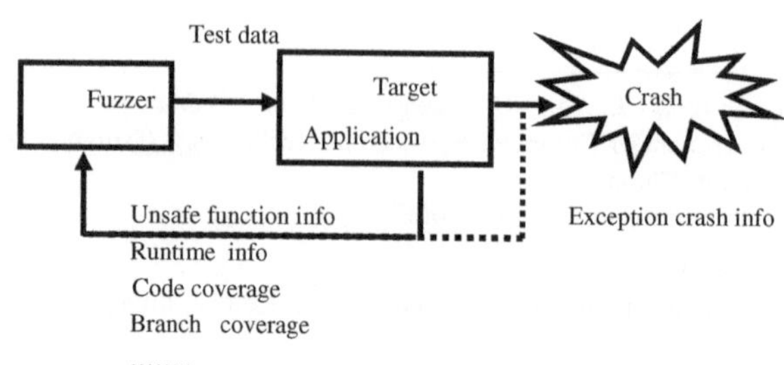

Fig. 10.1 Based-on feedback fuzzing

successfully detected serious bugs in GNU coreutils [12], large shipped Windows applications [13], and Linux file systems [11].

10.2.3 Model Inference Assisted Fuzzing

Many protocol reverser-engineering tools (Such as FTSG [10], Prospex [14]) can be used to guide fuzzing tests. These tools can extract format specifications for input data by analyzing network traffic, monitoring the execution of a program while it processes input data, or analyzing binary executables directly. The extracted protocol specifications can be further translated into fuzzing specifications. This technique can reduce the complexity of the test data structure and improve the code coverage [15].

10.3 Cd and Md

The following terms represent key concepts for fuzzing. Code coverage[15]—a technique for determining which pieces of code have run during a specific time period, often by instrumenting the code to be tested with special instructions that run at the beginning and end of a function. When the code runs, the special instructions record which functions were executed. Recording which code a test case covers can indicate how much additional testing is needed to cover all the code.

Cd enables the test cases cover the vulnerable position in the target application, and Md enables the test cases trigger the vulnerability in the target application. This paper gives the definitions of Cd and Md that describes the characteristics of generated test cases. Cr is the collection of constraints which have to be satisfied when test cases want to cover the vulnerable statement, and Ce is the collection of constraints which have to be satisfied when the test cases want to trigger the vulnerability after the vulnerable statement has been covered, and Cs is the collection constraints which have to be satisfied when test case can trigger the vulnerability.

There are some relationships between them:

$$\mathrm{Cr} \cap \mathrm{Ce} = \Phi \tag{10.1}$$

$$\mathrm{Cs} \subseteq \mathrm{Cr} \cup \mathrm{Ce} \tag{10.2}$$

When the test case can successfully trigger the vulnerability, if and only if $\mathrm{Cs} = \mathrm{Cr} \cup \mathrm{Ce}$; when the test case can not trigger the vulnerability, if and only if $\mathrm{Cs} \subset \mathrm{Cr} \cup \mathrm{Ce}$.

Fig. 10.2 Code with buffer overflow

```
int function(char *input_str)
    {
      char src[6];
      if (input_str[0] == 'W' &&
      input_str[1] == 'F')
        {
          strcpy(src, input_str + 2);
        }
      return 1;
    }
```

|Cr| means the number of the member of Cr, and |Cs| means the number of the member of Cs, and |Ce| means the number of the member of Ce.

Definition 10.1 Cd, the degree of correctness, it means the percentage of the Cr which is satisfied by a test case. And

$$\mathrm{Cd} = |\mathrm{Cs} \cap \mathrm{Cr}\ |/|\mathrm{Cr}| \tag{10.3}$$

Definition 10.2 Md, the degree of malformation, it means the percentage of the Ce which is satisfied by a test case. And

$$\mathrm{Md} = |\mathrm{Cs} \cap \mathrm{Ce}\ |/|\mathrm{Ce}| \tag{10.4}$$

Cd and Md describe the ability to implement vulnerability mining on the vulnerability of the test case. Cd is the percentage of achieved constraints of covering some vulnerable statement, while Md is the percentage of achieved constraints of triggering the exception of the vulnerability.

As in Fig. 10.2, the first constraint is c1 which requires that the first character of input_str is 'W', and c2 requires that the second character is 'F', and c3 requires that len(input_str) > 9. In which, $\mathrm{Cr} = \{\mathrm{c}1, \mathrm{c}2\}, \mathrm{Ce} = \{\mathrm{c}3\}$.

For example, there are four test cases as follows:

"WW", "FWFFFFFFFF", "WFW", "WFWWWWWWWW"
Obviously, "WW" satisfied with c1, but not c2 and c3, so Cd = ½, Md = 0.

And other test cases Cd and Md values are:

"FWFFFFFFFF": Cd = 0, Md = 1;
"WFW": Cd = 1, Md = 0;
"WFWWWWWWWW": Cd = 1, Md = 1.

By the definition and analysis, the Cd and Md of the test case to the vulnerability are independent, and $\mathrm{Cd} \in [0, 1]$, $\mathrm{Md} \in [0, 1]$. If and only if both their values are 1, the test case can successfully trigger the vulnerability.

10.4 Architecture of Feedback Fuzz

It is important that fuzzing is how to produce malformed data. So a good test data generation technique is very important. Traditionally, there are two main ways to generate test cases: generation-based and mutation-based [15]. However, the common drawback of traditional fuzzing techniques is that most malformed inputs are prematurely dropped. Based on improvement the Cd and Md, we have proposed Feedback Fuzz.

The early implementation of fuzzing [1, 2] adopts the simple model shown in Fig. 10.3. It includes engine and monitor. The function of engine is to generate needed data and send them to the target application. The function of monitor is to monitor the running status of target program and to capture exceptions and errors by simple scripts.

As the development of fuzzing, the above architecture can't express the complex mechanism of current fuzzing. So the paper gives a new architecture as shown in Fig. 10.4. The new architecture has three properties:

1. Engine is divided into four modules. It strengthens the reusability of codes and flexibility, then the users can quickly custom a Fuzzer according to need of target protocols.
2. The Agent is refined Improvements. With the agent, the fuzzing task can be divided into several parts and run simultaneously in different agents, so the efficiency is improved greatly. On the other hand, the agent can automatically save the dynamic execution information of the target application. The dynamic execution information can improves Cd and Md, also guides engine generating test data.
3. Format script and sample data are added. Format scripts give the relationships between input elements, the input elements' types, the lengths and so on. For example, SPIKE [4] use C—like scripting language and Peach [16] use XML

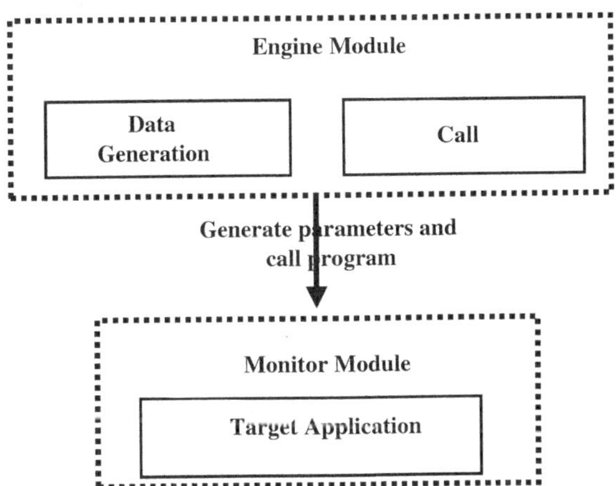

Fig. 10.3 Early fuzzing architecture

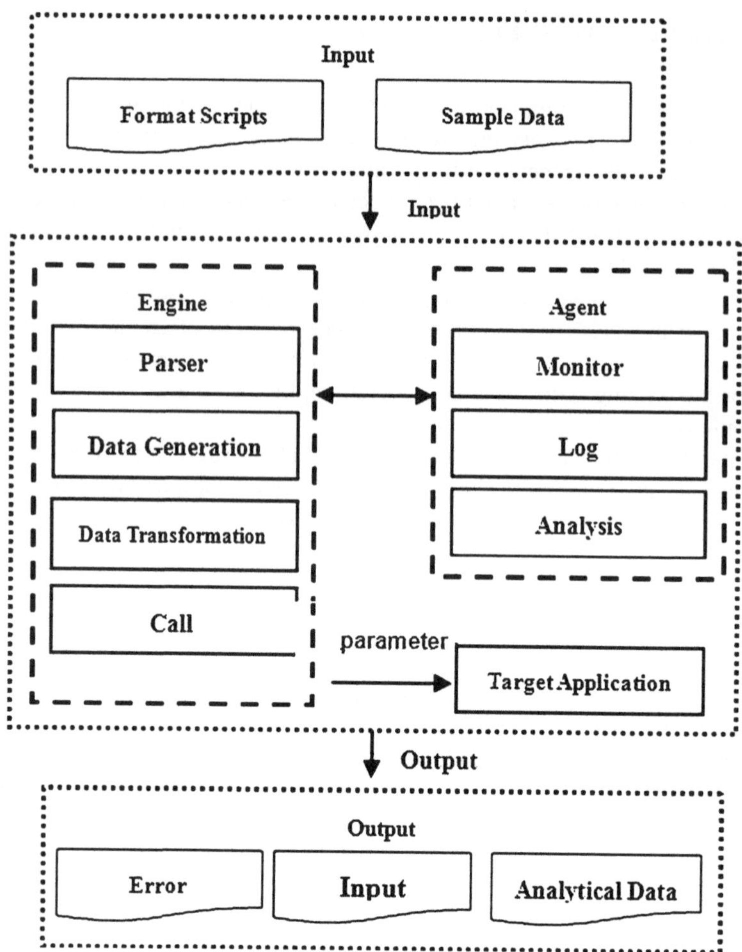

Fig. 10.4 Feedback fuzz architecture

language format scripts. The sample data is the base which new test cases are generated. Based on well-formed sample data, the generated test cases can cover more codes of target application. Introducing the two modules can avoid large amount of test cases generation with low correctness and improve the code coverage.

Acknowledgments This material is based upon work partially supported by the National Natural Science Foundation of China No 61070208 & 61170050.

References

1. Miller BP, Fredrikson L, So B (1990) An empirical study of the reliability of UNIX utilities. Comm ACM 33(12):32
2. Miller BP, Koski D, Lee CP, Maganty V, Murthy R, Natarajan A, Steidl J (1995) fuzzing revisited: a reexamination of the reliability of UNIX utilities and services. Technical report, University of Wisconsin Madison
3. Aitel D (2002) The advantages of block-based protocol analysis for security testing. Immunity Inc.
4. SPIKE [CP/OL] (2009) Web page: http://www.immuni-tysec.com/resources-freesoftware.shtml. Visited on Jun 2009
5. Godefroid P, Levin M, Molnar D (2008) Active property checking. In: Proceedings of EMSOFT'08, 19–24 Oct 2008
6. Godefroid P, de Halleux p, Nori AV, Rajamani SK, Schulte W, Tillmann N (2008) Automating software testing using program analysis. In: Proceedings of IEEE SOFTWARE, Sep/Oct 2008, pp 30–37
7. Wang TL, Wei T, Gu GF, Zou W (2010) TaintScope: a checksum-aware directed fuzzing tool for automatic software vulnerability detection. In: Proceedings of symposium on security and privacy, Oakland, CA, 16–19 May 2010
8. Oehlert P (2005) Violating assumptions with fuzzing. IEEE Comput Soc 3(2):58–62
9. Ganesh V, Leek T, Rinard M (2009) Taint-based directed whitebox fuzzing. In: Proceedings of 31st international conference on software engineering (ICSE 2009), Vancouver, Canada, 16–27 May 2009
10. Wu Z, William Atwood J, Zhu X (2009) A new fuzzing technique for software vulnerability mining. In: Proceedings of IEEE CONSEG'09, Chennai, India, 17–19 Dec 2009, pp 59-66
11. Cadar C, Ganesh V, Pawlowski PM, Dill DL, Engler DR (2006) Exe: automatically generating inputs of death. In: Proceedings of the 13th ACM conference on computer and communications security (CCS'06), pp 322–335
12. Cadar C, Dunbar D, Engler D (2008) Klee: unassisted and automatic generation of high-coverage tests for complex systems programs. In: USENIX symposium on operating systems design and implementation (OSDI'08), San Diego, CA, USA
13. Godefroid P, Kiezun A, Levin MY (2008) Grammar-based whitebox fuzzing. In: Proceedings of the ACM SIGPLAN conference on programming language design and implementation (PLDI'08), ACM
14. Comparetti PM, Wondracek G, Kruegel C, Kirda E (2009) Prospex: protocol specification extraction. In: Proceedings of IEEE symposium on security and privacy, IEEE Computer Society Press, USA
15. Oehlert P (2005) Violating assumptions with fuzzing. IEEE Secur Priv 3(2):58–62
16. Peach [CP/OL] (2009) Web page: http://www.PeachFuzzer.com; http://Peachfuzz.sourceforge.net; http://Pe-achfuzz@googlegroups.com. Visited on Jun 2009

Chapter 11
Attitude Control of Solar Sail Spacecraft Using Fractional-Order PID Controller

Yong Wang, Min Zhu, Yiheng Wei, Cheng Peng and Yang Zhang

Abstract In consideration of the extreme flexibility characteristics of film sail and large-scale booms of solar sail, a nonlinear rigid-flexible coupling dynamic model is presented. Based on the coupling dynamic model, a fractional order PID controller is designed to stabilize the system. The simulation results show the effectiveness of the proposed method, as the asymptotic tracking of the sail target attitude angle and robust vibration suppression of flexible structures are all achieved, and results is much better compared with the traditional PID controller.

Keywords Solar sail spacecraft · Rigid-flexible coupling · Fractional order PID control

11.1 Introduction

Solar sail spacecraft is a novel spacecraft propelled by the solar radiation pressure (SRP) [1]. Compared with the conventional spacecraft, the solar sail has the advantage of providing continuous propulsion and is needless to carry large amounts of fuel [2]. Consequently, the solar sail propulsion has broad prospect for interplanetary flight and deep space exploration, meanwhile, it has received more and more attention in recent years.

Y. Wang · M. Zhu (✉) · Y. Wei · C. Peng · Y. Zhang
Department of Automation, University of Science and Technology of China, Hefei, China
e-mail: zhumin@mail.ustc.edu.cn

Y. Wang
e-mail: yongwang@ustc.edu.cn

Z. Sun and Z. Deng (eds.), *Proceedings of 2013 Chinese Intelligent Automation Conference*, Lecture Notes in Electrical Engineering 255,
DOI: 10.1007/978-3-642-38460-8_11,

At present, there are two kinds of widely used solar sail models. One is the four tip-mounted control vanes model which is proposed by NASA [3], the other is the gimbaled boom control model which is proposed by DLR [4, 5]. Based on those two models, Bong Wie proposed the solar radiation pressure model and the rigid body attitude dynamics model, respectively [6–9]. However, the film sail and large-scale booms of solar sail have the extreme flexibility characteristics, thus, the attitude adjustment of solar sail will induce flexible structural vibration. Thus, the rigid body dynamic model of solar sail can not accurately describe the characteristics of its flexible structures. For this reason, the authors have presented a nonlinear rigid-flexible coupling dynamic model in [10, 11].

The fractional order calculus is the extension of integer order calculus and is a more accurate instrument than the integer system in describing the real world object [12]. Meanwhile, compared with traditional controller, using fractional order controller can enhance the performance of the control systems [13]. Among a great variety of fractional order controllers, the $PI^{\lambda}D^{\mu}$ controller proposed by Podlubny is widely used for control system analysis, where λ and μ are the fractional orders of the integral and derivative parts of the controller, respectively.

Based on the rigid-flexible coupling attitude dynamic model, a novel fractional order PID controller is proposed. Numerical simulation is presented to illustrate the effectiveness of the proposed method and the results shows that it can suppress the vibration of flexible structures induced by the attitude adjustment and track the sail target attitude angle well.

11.2 Problem Descriptions

Based on the DLR's gimbaled boom control model, Wie [6, 7] established the pitch axis attitude dynamic model of solar sail which is shown as Fig. 11.1. And the rigid dynamic equations were obtained as follows

$$\begin{aligned}(a_1 + a_2\cos\delta)\ddot{\alpha} + (a_2\cos\delta)\ddot{\delta} - a_2\left(\dot{\alpha} + \dot{\delta}\right)^2\sin\delta \\ = -a_3\cos\alpha\sin\alpha + a_8 f\cos\delta - T_g\end{aligned} \tag{11.1}$$

$$\begin{aligned}(a_4 + a_2\cos\delta)\ddot{\alpha} + a_4\ddot{\delta} + a_2\dot{\alpha}^2\sin\delta = & - a_5\cos\alpha\sin\alpha\cos\delta \\ & - a_6\cos^2\alpha\sin\delta \\ & - a_7\cos\alpha\sin\delta + a_9 f + T_g\end{aligned} \tag{11.2}$$

where α is the pitch angle of the solar sail, δ is the angle between the gimbaled boom and the roll axis, T_g is the gimbaled thruster torque, f is the reaction thruster force. $a_i (i = 1, \ldots 9)$ is the constant coefficient.

As is mentioned above, in order to suppress the vibration of flexible structures induced by the attitude adjustment, the authors present a nonlinear rigid-flexible

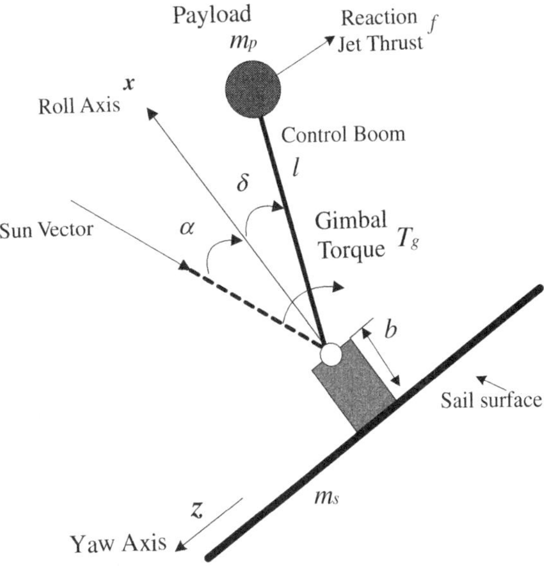

Fig. 11.1 The pitch axis dynamic model of solar sail

coupling dynamic model as Eqs. (11.3–11.5) by using the hybrid coordinate method.

$$(a_1 + a_2)\ddot{\alpha} + a_2\ddot{\delta} + \mathbf{p}^{\mathrm{T}}\ddot{\mathbf{q}} = -a_3 \cos\alpha \sin\alpha + a_8 f - T_g \tag{11.3}$$

$$\begin{aligned}(a_4 + a_2)\ddot{\alpha} + a_4\ddot{\delta} &+ \left(a_6 \cos^2\alpha + a_7 \cos\alpha\right)\delta + \mathbf{p}^{\mathrm{T}}\ddot{\mathbf{q}} \\ &= -a_5 \cos\alpha \sin\alpha + a_9 f + T_g\end{aligned} \tag{11.4}$$

$$\ddot{\mathbf{q}} + \mathbf{C}\dot{\mathbf{q}} + \mathbf{K}\mathbf{q} + \mathbf{p}\ddot{\alpha} = 0 \tag{11.5}$$

where $\mathbf{p} = [p_1, p_2, \ldots p_n]^T$ is the rigid-flexible coupling coefficient, $\mathbf{q} = [q_1, q_2, \ldots q_n]^T$ is the flexible mode vector, $\mathbf{C}$ and $\mathbf{K}$ are the relative matrix coefficient.

By setting the state vector as $z = [\alpha, \delta, q^T]^T \in \mathbb{R}^n$, the dynamic model shown in Eqs. (11.3–11.5) can be expressed as

$$M\ddot{z} + N\dot{z} + (E + E_z)z = Fu \tag{11.6}$$

where M, N, E and F are constant matrix, $M > 0$, E_z is the function of z, and $\|E_z\|_2 \le \varepsilon_1$.

By setting $x = \begin{bmatrix} z \\ \dot{z} \end{bmatrix} \in \mathbb{R}^{2n}$, the dynamic model of solar sail can be rewritten as

$$\dot{x} = (A + \Delta A)x + Bu \tag{11.7}$$

11.3 Attitude Control Using Fractional Order PID Controller

11.3.1 Fractional Calculus

There are three widely used definitions of fractional calculus which are the Riemann–Liouville's definition, the Caputo's definition and the GrÏnwald-Letnikov's definition [12]. These three definitions are equivalent if the considered function is relaxed at the origin. And in this paper the third definition is used which is defined as Eq. (11.8)

$$D^{\lambda}f(t) = \lim_{h\to 0} h^{-\lambda} \sum_{j=0}^{[t/h]} (-1)^j \binom{\lambda}{j} f(t-jh) \tag{11.8}$$

where λ is the fractional order and

$$\binom{\lambda}{j} = \frac{\Gamma(\lambda+1)}{\Gamma(j+1)\Gamma(\lambda-j+1)} \tag{11.9}$$

Assume that Laplace transformation of $f(t)$ is $F(s) = L\{f(t)\}$. Then the Laplace transformation for the fractional order integration and differentiation can be obtained as

$$L\{D^{\lambda}f(t)\} = s^{\lambda}F(s) \tag{11.10}$$

11.3.2 Fractional Order PID Controller

As is shown in Fig. 11.2, a unit negative feedback fractional order control system is considered. Where *r* is the reference input, *e* is the tracking error, *u* is the control input and *y* is the system output, *P*(*s*) is the control plant (either fractional order or integer order). *C*(*s*) is the fractional order PID controller where in this paper is the $PI^{\lambda}D^{\mu}$ controller, which is described as Eq. (11.11)

$$C(s) = K_p + \frac{K_i}{s^{\lambda}} + K_d s^{\mu} \tag{11.11}$$

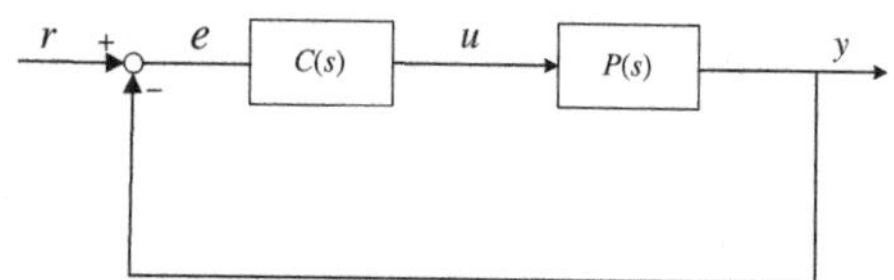

Fig. 11.2 Fractional order control system structure

where $\lambda,\ \mu \in (0,\ 2)$. Since this kind of controller has five parameters $(K_p,\ K_i,\ K_d,\ \lambda,\ \mu)$ to tune, it is more flexible than the traditional one. With the help of FOTF toolbox, we can easily implement the $PI^{\lambda}D^{\mu}$ controller by designing the fractional order operator block in simulink.

The purpose of our $PI^{\lambda}D^{\mu}$ controller design is to asymptoticly track the sail target attitude angle and suppress the robust vibration of flexible structures.

To achieve a good tracking performance, the ITAE criterion of the tracking error is used to optimize the controller which is defined as follows

$$J_{\mathrm{ITAE}} = \int_0^{\infty} t|e(t)|dt \tag{11.12}$$

where the tracking error $e = \alpha_d - \alpha$, here $\alpha_d = 35^\circ$ is the reference pitch attitude angle. By setting $X = (K_p,\ K_i,\ K_d,\ \lambda,\ \mu)$, which satisfies some bound restrictions denoted by $X_{\min}$ and $X_{\max}$, then the design problem of the fractional order PID controller can be converted to the following optimization problem

$$\begin{aligned} &\min \int_0^{\infty} t|\alpha_d - \alpha|dt \\ &s.t.\quad X_{\min} < X < X_{\max} \end{aligned} \tag{11.13}$$

11.4 Simulations

The simulation parameters which we use of the solar sail come from the 40×40 m ST7 solar sail configuration proposed by JPL [7] and is shown in Table 11.1. Only the first two order flexible modes are considered, the related parameters are shown in Table 11.2.

Based on the ITAE criterion, a traditional PID controller and an fractional order PID controller can be obtained as Eqs. (11.14) and (11.15), respectively,

$$C_{\mathrm{PID}} = 8.1841 + \frac{0.7146}{s} + 53.4295s \tag{11.14}$$

Table 11.1 Basic parameters of solar sail

Parameter	Value	Parameter	Value
m_s	40 kg	m_p	120 kg
b	0.5 m	l	2 m
J_s	3000 kg m^2	J_p	20 kg m^2
P	4.563×10^{-6} N/m^2	A	1400 m^2

Table 11.2 Parameters of first two order flexible modes of solar sail

	p_i	C_i	K_i
$i = 1$	54.4	4.51×10^{-4}	0.17
$i = 2$	12.4	6.31×10^{-6}	0.257

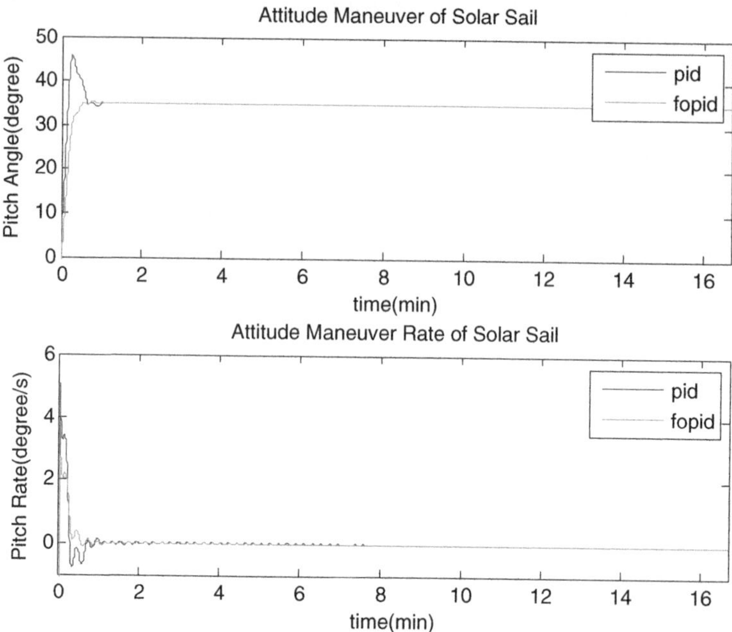

Fig. 11.3 The tracking carve of pitch angle and its rate

$$C_{PI^{\lambda}D^{\mu}} = 82.6572 + \frac{76.6311}{s^{0.1231}} + 54.2406s^{1.1439} \tag{11.15}$$

The curves of the system performance are shown in Figs. 11.3, 11.4 and 11.5. We can see from Figs. 11.3 and 11.4 that both the traditional and fractional PID controllers satisfy the design requirements as they both track the sail target attitude angle and suppress the robust vibration of flexible structures well.

Meanwhile, according to Figs. 11.3, 11.4, 11.5 and 11.6, we can easily find that the performance of the fractional PID controllers is better than the traditional one.

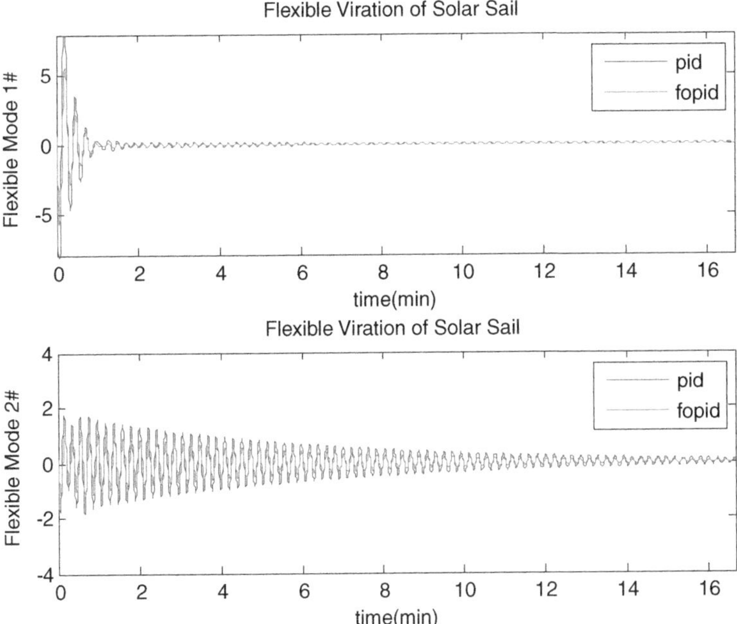

Fig. 11.4 The curve of first two order flexible modes

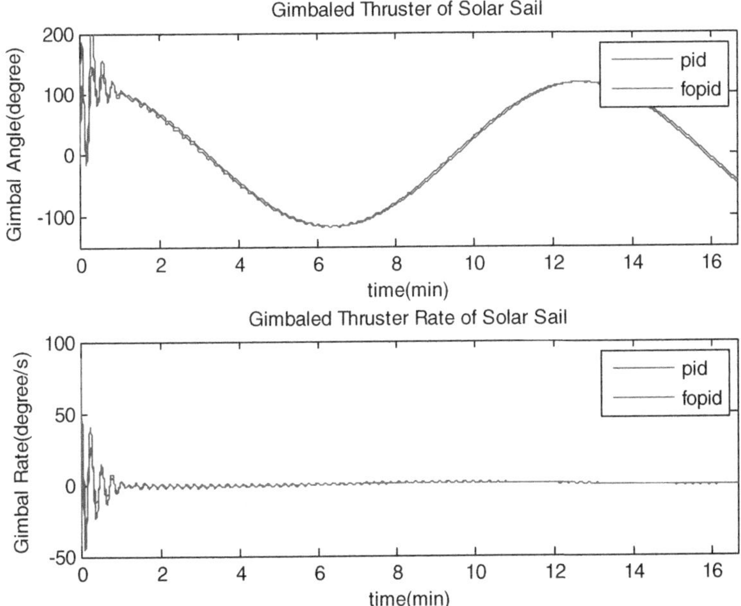

Fig. 11.5 The curve of gimbaled angle and its rate

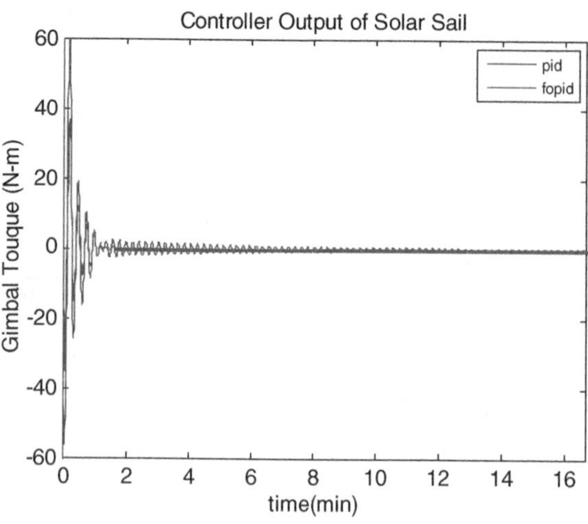

Fig. 11.6 Controller outputs

11.5 Conclusion

In this paper, based on the nonlinear coupling dynamic model of solar sail attitude control, a fractional order PID controller is designed to stabilize the system. The simulation results show the effectiveness of the proposed method, as the asymptotic tracking of the sail target attitude angle and robust vibration suppression of flexible structures are all achieved. Meanwhile, compared with the traditional PID controller, the performance of the proposed method is much better.

Acknowledgments This work was supported by the National Natural Science Foundation of China (no. 60974103&61004017), and the National 863 Project (no. 2011AA7034056).

References

1. McInnes CR (1999) Solar sailing: technology, dynamics, and mission applications. Springer, London
2. Frisbee RH (2003) Advanced propulsion for the 21st century. In: Proceedings of AIAA/ICAS international air and space symposium and exposition: the next 100 Y. Dayton, Ohio, 14–17 Jul 2003
3. Wie B (2002) Dynamic modeling and attitude control of solar sail spacecraft. NASA solar sail technology working group (SSTWG) final report
4. Leipold M, Eiden M et al (2003) Solar sail technology development and demonstration. Acta Astronaut 52(2–6):317–326
5. Leipold M, Garner CE, Freeland R, Herrmann A, Pagal GNM (1998) ODISSEE-a proposal for demonstration of a solar sail in earth orbit. Acta Astronaut 45(4):557–566
6. Wie B (2004) Solar sail attitude control and dynamics, part I. J Guidance Control Dyn 27 (4):526–535
7. Wie B (2004) Solar sail attitude control and dynamics, part II. J Guidance Control Dyn 27 (4):536–544

8. Wie B, Murphy D (2004) Robust attitude control systems design for solar sail, part 1: Propellantless primary ACS. In: Proceedings of AIAA guidance, navigation, and control conference and exhibit, providence, Rhode Island, 2004, pp 1–28
9. Wie B, Murphy D (2004) Robust attitude control systems design for solar sail, part 2: MicroPPT-based secondary ACS. In: Proceedings of AIAA guidance, navigation, and control conference and exhibit, providence, Rhode Island, 2004, pp 1–16
10. Zhang Y (2010) Attitude control and trajectory optimization of solar sail spacecraft. University of Science and Technology of China, Hefei (in Chinese 张洋. 太阳帆航天器姿态控制与轨迹优化[D]. 合肥, 中国科学技术大学. 2010.)
11. Zhu M, Zhang Y, Wei Y, Wang Y (2013) Attitude dynamic modeling and feedback LPV control of solar sail. Inf Control 42(2):196–201 (in Chinese 朱敏, 张洋, 卫一恒, 王永. 太阳帆航天器姿态动力学建模与反馈LPV控制[J]. 信息与控制. 已收录.)
12. Podlubny I (1999) Fractional differential equations. Academic press, New York
13. Chen YQ, Monje CM, Vinagre BM, Xue D, Feliu V (2010) Fractional-order systems and controls—fundamentals and applications. Springer, London

Chapter 12
Ubiquitous Visual Navigation for the Robotic Wheelchair

Liang Yuan

Abstract A new Robotic Wheelchair System (RWS) is presented. The new RWS departs completely from the conventional approaches of autonomous robots which rely on onboard sensors to map locally their environment for navigating progressively and autonomously. Instead, our RWS makes use of an innovative tracking and servoing system to achieve global and reliable navigation of robotic wheelchairs. The UVSS enables visual feedback control of robotic wheelchair in a global environment. Furthermore it coordinates with an onboard autonomous control system, whose purpose is to avoid the obstacles in the neighborhood of the wheelchair and to enable navigation when the global control from RWS is not available due to delays and packets dropouts in the communication channel. In addition to servoing and navigation, the RWS also provides continuous communication between the wheelchair riders and their supporters. To test the proposed system, experiments for servoing the robotic wheelchair in following a designed path under UVSS is performed. Experimental results validates the advantage of the new system.

Keywords Robotic wheelchair · Ubiquitous visual navigation · Tracking · Remote control

12.1 Introduction

Providing feasible mobility to elderly people or people with disabilities has been a research effort for decades. Today electric wheelchairs are recognized as a primary tool for this purpose. In general, electric wheelchairs can be divided into two

L. Yuan (✉)
College of Mechanical Engineering, Xinjiang University, 1230 Yanan Road, Urumqi 830046 Xinjiang, China
e-mail: ylhap@163.com

Z. Sun and Z. Deng (eds.), *Proceedings of 2013 Chinese Intelligent Automation Conference*, Lecture Notes in Electrical Engineering 255,
DOI: 10.1007/978-3-642-38460-8_12,

categories: (1) manually controlled; and (2) automatically controlled. Manually controlled wheelchairs are operated by their users with a joystick whereas automatically controlled wheelchairs can navigate themselves automatically. Since manually controlled wheelchairs are not suitable for those who are unable to operate them, research works have primarily focused on automatic wheelchairs used in the field [1].

To make automatic operation possible it is necessary to use sensors such as infrared, sonic, or visual sensors. Of all the sensors used, visual sensors are most popular. A common practice is to install visual sensors on the wheelchairs to mimic human's ability of sensing the surroundings. For example, in [2], the authors placed a camera on a wheelchair to find local image correspondences for baseline image matching such that the wheelchair can be localized. More recently, MIT has developed a smart wheelchair equipped with sensors that can perceive the wheelchair's surroundings [3]. In [4], Fine and Tsotsos used a wheelchair-mounted visual sensors are used for high-level control such as path planning. A camera locally fixed on the wheelchair captures the images of the surroundings, which are processed to extract the geometric features, and the path of the wheelchair is then planned accordingly [5]. As the speed of the visual sensor becomes faster, an exciting new technology called [6] has emerged which has attracted a great deal of attention in the robotics community due to a number of advantages including direct measurement, non-contact sensing, and operating in complicated environments [7]. A large number of works have applied visual feedback control to robot manipulators [8], while relatively few to mobile robots [9].

Unfortunately locally installed visual sensors even with visual feedback control have severe limitations. In this paper, we propose a completely new approach which utilizes the Ubiquitous Visual Sensor Systems (UVSS). We propose to utilize these systems together with Internet and wireless technologies to monitor, navigate, and control the robotic wheelchairs. The paper is organized as follows. In Sect. 12.2 we functionally describe the Ubiquitous Visual Sensor Systems. In Sect. 12.3, we present the key components of the Ubiquitous Visual Sensor Systems including real-time tracking and real-time networked control and communication. In Sect. 12.4 we present experimental results when the robotic wheelchair runs under the Ubiquitous Visual Sensor Systems. In Sect. 12.5 we discuss our proposed system and summarize our conclusions.

12.2 Ubiquitous Visual Sensor Systems

The UVSS is designed for the global monitoring, navigation, and control of the robotic wheelchairs. The RWS was inspired by the rapid deployment of visual sensors in many facets of our technologically-driven society. To enable the desired and optimum operation of RWS, we propose to develop the following innovative technologies:

1. Real-time tracking of the wheelchairs: For efficient and robust real-time tracking we propose to integrate the mean-shift algorithm with particle filtering. This is an innovative and promising approach and has not been tried in the past for video image tracking. With this approach the real-time location, velocity and acceleration of the wheelchairs will be computed and monitored in real-time. The availability of the real-time position, velocity and acceleration of the wheelchairs provides the feedback information required for their real-time control and monitoring [10].
2. Create distinct markers for feasible identification of robotic wheelchairs: Tracking of a wheelchair follows its identification inside the visual image. Robust identification is one of the most difficult problems in computer vision [11]. For the solution of this problem we propose to attach to each wheelchair special markers with distinct features. The choice of the special markets and features will facilitate reliable and fast target identification inside dynamically varying and complex environments.
3. Real-time networked control and communication: The proposed system uses a networked control mechanism to control multiple robotic wheelchairs instead of using dedicated links to each wheelchair. This networked control architecture makes full use of all communication channels while maintaining the required update data rates. The proposed network consists of either wired or wireless links which are transparent to the riders or supporters. We propose the integration of the local sensors and decision-making with the global navigation and control in order to minimize the effects of delays or packet dropouts in the wireless communication channels. Through the proposed network the UVSS will provide audio and video support for the riders and their supporters [12].

12.3 The Key of Components of Ubiquitous Visual Sensor Systems

12.3.1 Real-Time Tracking of the Robotic Wheelchair

The navigation and control of the robotic wheelchairs is based on their effective and robust image detection and tracking. The mean-shift tracking algorithm is to compute iteratively the mean-shift vector on the locally estimated density until convergence to a local maximum [13]. Since the convergence needs only a few iterations, this is a fast and effective tracking method and has been used successfully in real-time surveillance video tracking. The mean-shift algorithm, however, is not effective for tracking abrupt motions. In recent years particle filtering has emerged as the dominant method for the filtering and data association approach. The main advantage of particle filtering is its capability to handle effectively non-linear non-Particle filtering is more effective and reliable as compared to the mean-shift approach. Particle filtering, however, is difficult to use for real-time applications and especially for tracking multiple wheelchairs because of the large number of particles required for effective multi-target tracking. To achieve effective, reliable, and robust real-time tracking of multiple wheelchairs, we propose a novel approach which integrates particle filtering with environmental constraints into the mean-shift algorithm. The proposed algorithm will be implemented with the following steps:

1. Compute the mean-shift vector;
2. Compute the next location according to the local density estimation;
3. Iterate 1 and 2 for a limited number of steps to reach the local maximum; If not reached go to 4;
4. Sample particles according to the dynamic state function;
5. Calculate weights of particles according to the observation function;
6. Resample particles and calculate the posterior mean according to the weights;
7. Go back to 1.

In the first three steps, the mean-shift algorithm will track successfully multiple targets with smooth motions. When abrupt motions occur, particle filtering will be engaged in the last three steps to locate the lost object for the effective implementation of the mean-shift algorithm [14, 15].

12.3.2 Real-Time Networked Control and Communication

For the control of the wheelchairs, a real-time networked feedback system has been implemented as shown in Fig. 12.1. The video cameras send the images of all wheelchairs to the network server. The network server performs multiple target identification and tracking of the wheelchairs and sends driving control commands via a wireless network. For a real-time networked control system, we will consider two technical issues: (1) the bit rate constraints between network nodes, and (2) the effects of delay and packet dropouts. To overcome the bit rate constraints we propose the following format for controlling the wheelchair movements: the left motor velocity (16 bits) and right motor velocity (16 bits) which corresponds to 32 bits (4 bytes) of data. Sending one control command per millisecond requires a bandwidth of 32 Kbits/sec/wheelchair which corresponds to 32 Mbits/sec for handling 1,000 wheelchairs. The remaining bandwidth will be used to handle audio or video communications between the rider and the supporters.

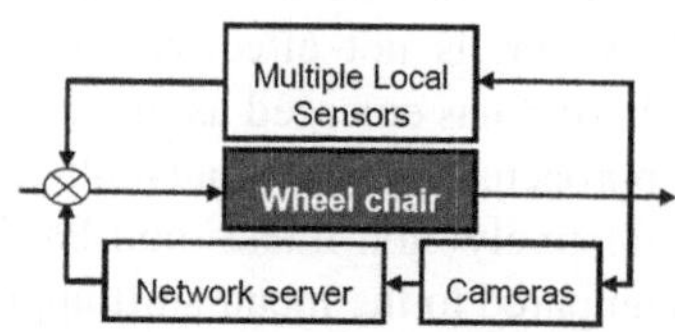

Fig. 12.1 Network feedback servoing of the robotic wheelchair

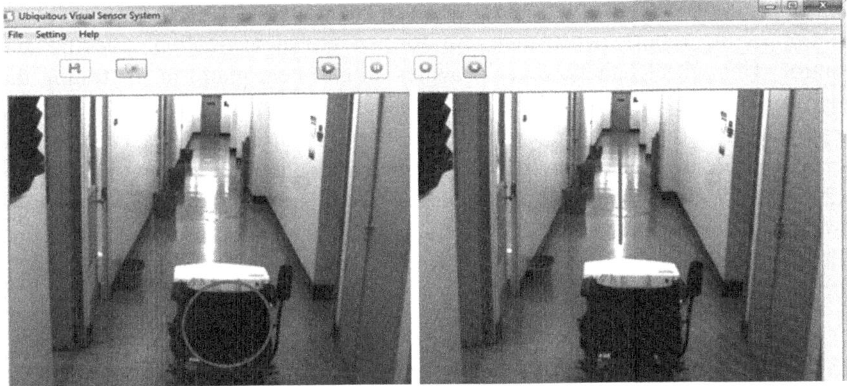

Fig. 12.2 Software interface on the server

To overcome the effects of delays and packet dropouts in the wireless communication channel, we propose to use a real-time dual feedback control system consisting of both local and global control as shown in Fig. 12.3. For the local control multiple sensors (IR, Camera, and possibly Lidar) are installed in the wheelchair. The purpose of these sensors is to detect the obstacles in the neighborhood of the wheelchair and to enable its autonomous navigation when the global control from the network server is not available due to delays and packet dropouts. The wheelchair will completely stop when the loss of the global control is too long to ensure a safe and reliable operation of the wheelchair. The proposed use of a dual local and global feedback control system minimizes the effects on the control of the wheelchairs caused by the delays and packet dropouts in wireless communication channel.

12.4 Experiments

We have tested the performance of our proposed UVSS in our laboratory. In this section we present the robotic wheelchair to follow a designed path under the UVSS.

12.4.1 Experimental Preparation

We implemented the integration of the proposed UVSS with a robotic wheelchair in our experiment. The experiment requires three major components: (1) wheelchair, (2) cameras, and (3) wireless network. This wheelchair was developed as an autonomous mobile robot equipped with a number of local sensors including sonic, infrared, and visual sensors. These sensors have been designed only for local control of the robotic wheelchair not for the global navigation, tracking and control. This wheelchair however provides an excellent platform for testing the approach. To realize the UVSS, cameras and wireless network need to be deployed. In this experiment, we use the high-definition cameras (The Source Imaging, Charlotte, NC) to track the wheelchair in the real-time. Its intensity is 1,280*960. We use the a wireless router (Cisco Linksys WRT54GL Wireless-G Broadband Router) to setup the wireless communication between the server and clients.

12.4.2 Experimental Results

The software interface of the server is shown in Fig. 12.2. There are two windows to monitor the wheelchair. The left window shows the real-time tracking. A red ellipse represents the tracking results. The right windows shows the status of the robotic wheelchair following the designed path in the hallway. The red line represents the designed path.

Path following algorithm under UVSS: The motion control of the robotic wheelchair is to control two variables including its linear velocity (v) and angular velocity (ω). In practice, the actual control for the robotic wheelchair is to directly drive the left and right motors. Equation (12.1) shows the relationship between the control variables and velocities of the left and right wheels

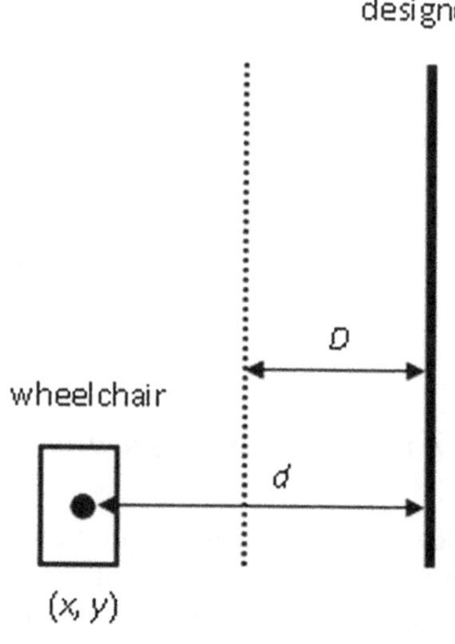

Fig. 12.3 Schematic diagram of following the designed path

$$\begin{aligned} v_l &= v - \frac{D}{2}\omega \\ v_r &= v + \frac{D}{2}\omega \end{aligned} \tag{12.1}$$

where v_r and v_m represent the velocities of the left and right motor, respectively; D represents the width of the wheelchair.

To follow the designed path, first of all, the wheelchair position need to be tracked. The distance (*d*) from the wheelchair to the designed path can be calculated. We assume a safe region which has a width of D, where the center line of the region is the designed path (Fig. 12.3). If the distance (*d*) of the wheelchair from the center line is larger than D, the angular velocity is equal to the maximum velocity. However, if $d < D$, the angular velocity is linearly decreased. This motion control strategy of the wheelchair can be modeled as follows:

$$\begin{aligned} &v = v_{\max} \\ &\omega = \begin{cases} r_{\max} & d > D \\ \frac{d}{D} r_{\max} & 0 \le d \le D \end{cases} \end{aligned} \tag{12.2}$$

where v_{max} and r_{max} represent the maximum linear and angular velocities, respectively.

12.5 Conclusion

In this paper, we have described a completely new idea for navigation of robotic wheelchairs in structured environments. The proposed Robotic Wheelchair System (RWS) makes use of the Ubiquitous Visual Sensor Systems (UVSS) to enable global navigation and monitoring of wheelchairs. The UVSS works together with the Internet and Wireless communication technologies such that the algorithm in the context of global navigation and monitoring becomes possible. The proposed technologies are utilized by a visual feedback control algorithm using solutions of mean-shift and particle filtering for robust non-linear object tracking. Integrated with the local sensors installed on the wheelchairs, the proposed mechanism is more robust than conventional ones which rely only on local sensing for progressive navigation. Consequently, the new RWS results in more reliable and safer operation of wheelchairs in structured environments such as hospitals, nursing homes, college campuses, museums and many others, where an UVSS can be installed. Experimental results have validated the proposed new approach.

Acknowledgments This work is supported by the grant to Liang Yuan from the National Natural Science Foundation of China (61262059/F020502).

References

1. Cahn D, Stephen P (1975) ROBNAV: a range-based robot navigation and obstacle avoidance algorithm. IEEE Trans Syst Man Cybern SMC-5(19):544–551
2. Goedeme T, Tuytelaars T, Van Gool L (2004) Fast wide baseline matching for visual navigation. In: proceedings IEEE computer society conference on computer vision and pattern recognition
3. http://rvsn.csail.mit.edu/wheelchair/
4. Fine G, Tsotsos J (2009) Examining the feasibility of face gesture detection using a wheelchair mounted camera. In: proceeding first ACM SIGMM international workshop on media studies and implementations that help improving access to disabled users, pp 19–27
5. Dunn E, van den Berg J, Frahm J-M (2009) Developing visual sensing strategies through next best view planning. In: proceeding international conference on intelligent robots and systems, pp 4001–4008
6. Hutchinson SA, Hager GD, Corke PI (1996) Adaptive visual servo control of robots. IEEE Trans Robot Autom 12(5):651–670
7. Weiss L, Sanderson A, Neuman C (1987) Dynamic sensor-based control of robots with visual feedback. IEEE J Robot Autom 3(5):404–417
8. de Luca A, Oriolo G, Paone L, Giordano P (2002) experiments in visual feedback control of a wheeled mobile robot. In: proceeding international conference on robotics and automation, pp 2073–2078
9. Xu L, Zheng YF (2001) Moving personal robots in real-time using primitive motions. Auton Robots 10:175–183
10. Balster E, Zheng YF, Ewing R (2005) Feature-based wavelet shrinkage algorithm for image denoising. IEEE Trans Image Process 14:2024–2039
11. Balster E, Zheng YF, Ewing R (2006) Combined spatial and temporal domain wavelet shrinkage algorithm for video denoising. IEEE Trans Circuits Syst Video Technol 16:220–230
12. Comaniciu D, Ramesh V, Meer P (2005) Kernal-based object tracking. IEEE Trans Pattern Anal Mach Intell 25:564–577
13. Comaniciu D, Meer P (2002) Mean-shift: a robust approach toward feature space analysis. IEEE Trans Pattern Anal Mach Intell 24:603–619
14. Ristic B, Arulampalam S, Gordon N (2004) Beyond the Kalman filter: particle filters for tracking applications. Artech House, Boston
15. Yuan L, Zheng YF, Zhu J, Wang L, Brown A (2012) Object tracking with particle filtering in fluorescence microscopy images: application to the motion of neurofilaments in axons. IEEE Trans Med Imaging 31:117–130

Chapter 13
Torque Distribution Control for Independent Wheel Drive Electric Vehicles with Varying Vertical Load

Guodong Yin, Jinxiang Wang and Dawei Pi

Abstract The performance of independent wheel drive electric vehicles (EVs) to prevent the generation of vehicle slipping is effective by distributing the four wheels' torque independently. The direct yaw moment generated by the tire force difference between the two sides of the vehicle is taken as the control input. Considering that the control performance is affected by varying vertical load, and the designed torque distribution controller presents the effective method to give practical benefits in motion control. Last computer simulation using Matlab/Simulink-Carsim is carried out to investigate the method of torque distribution based on the ratio of the vertical load of each wheel to the total vehicle load for EVs to improve the performance of handling and security.

Keywords Electric vehicle · Torque distribution control · Varying vertical load

13.1 Introduction

Electric vehicles equipped with in-wheel motors have appeared as future personal electric vehicles based on several advantages in the viewpoint of energy efficiency and motion control. Over the past few years, a great deal of research on motion

G. Yin (✉)
School of Mechanical Engineering, Southeast University, Nanjing, China
e-mail: ygd@seu.edu.cn

J. Wang
State Key Laboratory of Mechanical Transmission, Chongqing University, Chongqing, China

D. Pi (✉)
School of Mechanical Engineering, Nanjing University of Science and Technology, Nanjing, China
e-mail: 526547291@qq.com

Z. Sun and Z. Deng (eds.), *Proceedings of 2013 Chinese Intelligent Automation Conference*, Lecture Notes in Electrical Engineering 255,
DOI: 10.1007/978-3-642-38460-8_13,

controls, including traction control or yaw stability control, has been done utilizing independently driven in-wheel motors [1–3]. The purpose of these motion controls is to prevent unintended vehicle behavior through active vehicle control and assist drivers in maintaining the controllability and stability of vehicles. Four wheel independently actuated (FWIA) electric vehicles utilize four in-wheel (or hub) motors to drive the four wheels independently, thus the direct yaw-moment, which is usually employed to ensure the vehicle stability, can be easily generated by the motor torque difference between the left and right sides of the vehicle. The main goal of most motion control systems is to control the sideslip angle and yaw rate of the vehicles. In Cong et al. [4], direct yaw moment control based on sideslip angle estimation was proposed for improving the stability of in wheel- motor-driven electric vehicles (IWM-EVs). Fuzzy-rule based control and sliding-mode control algorithms for vehicle stability enhancement were proposed and evaluated through experiments [5, 6].

It is known that the vehicle parameters are subject to many types of uncertainties such as external disturbances, un-modeled dynamics, road roughness, wind gusts, and changes of payload. Thus, a robust stability problem for vehicle control has been raised, that is the vehicle controller has to cope with these uncertainties such that the maneuvering stability and the desired performance can be achieved. Modern robust control theories have been proved to be useful techniques in dealing with the aforementioned concerns. The H_2/H_∞ synthesis is a typical robust control method, and the performance of the robust controller can be further enhanced by the μ-analysis [7, 8]. This paper studies the stability control problem of FWIA vehicles and a μ-synthesis based robust controller is designed to yield the vehicle virtual direct yaw-moment for maintaining the vehicle motion stability in the presence of uncertainties. The proposed control method can improve the vehicle performance such as the robustness and lateral motion stability with respect to a given class of uncertainties including payload variations. Simulations using a high-fidelity, full-vehicle model in CarSim®, show that the FWIA vehicle with the proposed controller can achieve the desired performance and motion control with vertical load variations under different conditions.

13.2 Vehicle System Dynamics

Assuming the vehicle is running in a constant speed and a schematic diagram of a vehicle model is shown in Fig. 13.1, where we can find that the motions of the vehicle in the horizontal plane can be characterized by the following two equations:

$$\text{Lateral motion} \quad F_{xf}\sin\delta_f + F_{yf}\cos\delta_f + F_{yr} = F_y \tag{13.1}$$

$$\text{Yaw motion} \quad L_f F_{xf}\sin\delta_f + L_f F_{yf}\cos\delta_f - L_r F_{yr} + M = I_z\dot{\gamma} \tag{13.2}$$

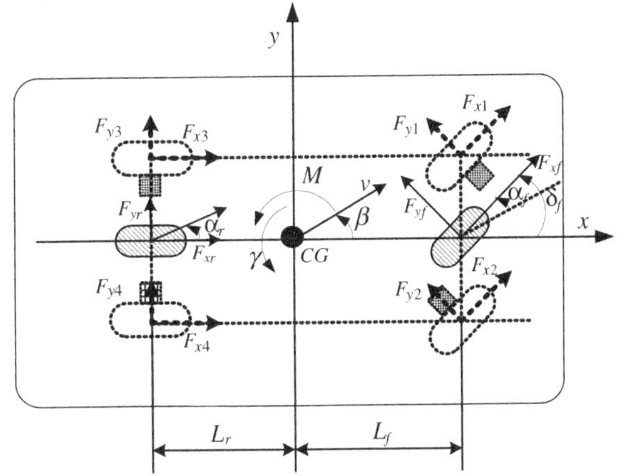

Fig. 13.1 Schematic diagram of a vehicle model

where δ_f is the front wheel steering angle, γ is the yaw rate. I_z is the yaw moment of inertia about its mass center z-axis, and m is the vehicle mass. F_x/F_y is the total tire longitudinal/lateral force represented by the summation of the tire longitudinal/lateral forces generated at all the four tires. F_{xf} is the total longitudinal force of the front tires, F_{yf}/F_{yr} is the total lateral forces of the front/rear tires, respectively. L_f/L_r is the distance between the center of mass and the front/rear axle. M is the direct yaw moment and will be taken as the control input to the vehicle. Note that the direct yaw moment can be generated by the motor torque difference between the two sides of the vehicle.

If all the tires are working in the linear tire force region, the dynamic Eqs. (13.1) and (13.2) can be expressed by a transfer function from the front wheel steering angle and direct yaw moment to yaw rate, which can be written as

$$G_{f-\gamma}(s) = \frac{b_1 s + b_2}{s^2 + a_1 s + a_2}, \quad G_{M-\gamma}(s) = \frac{b_3 s + b_4}{s^2 + a_1 s + a_2} \tag{13.3}$$

with $b_1 = \frac{2L_f K_f}{I_z}$; $b_2 = \frac{4K_f K_r (L_f + L_r)}{mvI_z}$; $b_3 = \frac{1}{I_z}$, $b_4 = \frac{2(K_f + K_r)}{mvI_z}$; $a_1 = \frac{2(L_f^2 K_f + L_r^2 K_r)}{I_z v} + \frac{2(K_f + K_r)}{mv}$; $a_2 = \frac{4K_f K_r (L_f + L_r)^2}{mv^2 I_z} + \frac{2(L_r K_r - L_f K_f)}{I_z}$ where K_f/K_r is the cornering stiffness for the front/rear tire, v is the vehicle longitudinal velocity. β is the vehicle sideslip angle.

13.3 Robust Control Design

The μ-synthesis control technique is used here to synthesize a controller for the AWD system. The designed controller has to be insensitive to un-modeled high-frequency dynamics, neglected nonlinearities, and variations in the vehicle system for the simple vehicle model. In other words, a vehicle robust stability controller which takes the additional yaw-moment generated by the tire force difference between the two sides of the vehicle is designed in this section.

13.3.1 Method to Distribute to the Wheels

The front wheel driving force F_f and the rear wheel driving force F_r are derived using relationships between the total driving force F_{car}. The load movement z occurs when driving operations are performed while running on flat roads.

The part on which wheels touch the surface of a road is pulled back with the driving force F_{car}. Thus, if an acceleration sensor is installed at the center of the car body, the acceleration α_{car} of the inertial force is detectable.

$$M \cdot \alpha_{car} = (T - T_d - J_w \dot{\omega}_w)/r - F_w \tag{13.4}$$

The driving force F_{car} is used to get the amount z of load movement generated when the driving pedal is stepped on as given by

$$z = \left| \frac{F_{car} \cdot H_{car}}{L_{car}} \right| \tag{13.5}$$

where G is the center of gravity, H_{car} is the height of the center of gravity, z is the amount of load movement, and L_{car} is the wheel base.

Assuming that the load is shift to the front wheel side by the amount z on driving, the driving forces and F_f and F_r, which can be generated for the front and rear wheel sides, are given by Eqs. (13.6) and (13.7), respectively, i.e.,

$$F_f = \mu(W_f - z) = \mu\left(W_f - \left|\frac{F_{car} \cdot H_{car}}{L_{car}}\right|\right) \tag{13.6}$$

$$F_r = \mu(W_r + z) = \mu\left(W_r + \left|\frac{F_{car} \cdot H_{car}}{L_{car}}\right|\right) \tag{13.7}$$

13.3.2 Nominal Models Design for μ Control

The objective of stability control is to improve the vehicle steadiness and transient response properties, enhancing vehicle handling performance and maintaining stability in those cornering maneuvers, i.e., the yaw rate r or sideslip angle β of the vehicle should be close to the desired vehicle responses [4].

To fit the system model into the μ control framework, the model is modified by changing Eq. (13.3) into an equivalent form considering the mass variation. $G_{f-\gamma}(m_0)$ and $G_{M-\gamma}(m_0)$ are the nominal front wheel angle and direct yaw moment transfer function blocks, respectively, which are obtained based on Eq. (13.3) by taking all parameters at their nominal values ($m = m_0 = 830$ kg). For simplicity, we treat all parameters of L_f and L_r as constants except the mass m since it dominates the uncertainty, and the velocity v is considered as a time-varying parameter. Thus, the blocks $\triangle_f$ and $\triangle_M$ in the controller design are the plant structured uncertainty perturbations to the nominal transfer functions $G_{f-\gamma}(m_0)$ and $G_{M-\gamma}(m_0)$, respectively. These uncertainty perturbations are added according to the multiplicative perturbation model of the μ-analysis. The weighting functions W_{sf} and W_{sM} represent the uncertainty frequency response of the corresponding transfer function blocks. Here, m_0 is used to denote nominal mass. Additional measurement noise, external disturbance, and uncertainty blocks can be added to the model depending on the situation and practical needs.

Based on the above regulations and vehicle control performance, we can calculate the weighting functions

$$W_{\rm sf} = \frac{0.007s^2 + 5.5s + 15}{s^2 + 18s + 82}; \; W_{\rm sM} = \frac{0.008s^2 + 9s + 24}{s^2 + 25s + 103}.$$

13.3.3 Structured Singular Value and μ-Synthesis

The structured singular value (μ) is an appropriate tool for analyzing the robustness and synthesizing of a system subjected to structured linear fraction transformers (LFT). Consider the feedback control system with the generalized plant $P(s)$, the controller $K(s)$ and the uncertainty block $\Delta(s)$. Here, δ_f is the exogenous input vector, e is the error output vector, y is the measured output vector and u is the control input vector to the generalized plant.

It is worth to note that P consists of a nominal vehicle model and weighting functions. Further, all the uncertainty elements have been pulled out of the vehicle system and placed in the Δ block, which is defined as

$$\Delta = diag\left[\Delta_f(s),\ \Delta_M(s)\right], \tag{13.8}$$

where the Δ block is stable and norm bounded, $\|\Delta\|_\infty \leqq 1$. The perturbations are now represented by $w = [d1 \bullet d2]^{\rm T}$ and $z = [z_f \bullet z_M]^{\rm T}$, respectively. The performance outputs $e = [e_\gamma \bullet e_\beta]^{\rm T}$ representing error signals are usually kept small in the control system. Consider an uncertainty with known structure, bounded value and belonging to the set $B\Delta$:

$$B\Delta = \{\Delta \in \Delta | \bar{\sigma}(\Delta) \leq 1\} \tag{13.9}$$

where $\bar{\sigma}(\Delta)$ is the maximum singular value of Δ. The controller K can be combined with P via a lower linear fractional transformation (LFT) to yield the transfer function matrix M, so the structural singular value (μ) is defined as:

$$\mu_\Delta^{-1}(M) = \min\{\bar{\sigma}(\Delta) : \Delta \in \Delta,\ \det(I - M\Delta) = 0\} \tag{13.10}$$

Furthermore, let

$$\|M\|_\mu = \sup_w \mu_\Delta(M(jw)).$$

Thus, $\mu_\Delta(M)$ is a measure of the smallest structured Δ that causes instability of the feedback loop. Given a desired uncertainty level, the purpose of the design is to look for a control law, which can reduce the μ level of the closed-loop system and ensure the stability of the system for all possible uncertainty descriptions.

Ideally, based on the μ-theory, the robust stability and performance hold for a given $M-\Delta$ configuration if and only if:

$$\inf_K \|M\|_\mu \leq 1.$$

In other words, the performance and stability assessment of the closed-loop system M is a μ test over frequency range and for the given uncertainty structure Δ. Using the performance robustness condition and the well-known upper bound for μ, the robust synthesis problem to be solved is reduced to the following:

$$\min_K \inf_D \sup_w \bar{\sigma}\left(DM(jw)D^{-1}\right) \tag{13.11}$$

D and K are iteratively solved (D-K iteration algorithm). Here D is any positive definite symmetric matrix with appropriate dimension and $\bar{\sigma}(.)$ denotes the maximum singular value of a matrix.

13.4 Simulation Results

Simulations of the robust controller based on a high-fidelity, full-vehicle model constructed in CarSim® are conducted. The vehicle parameters in the simulations are taken from an actual prototype FWIA electric vehicle developed in the authors' group at Southeast University.

A J-turn maneuvering is made in this simulation, and the front wheel steering angle is shown in Fig. 13.2. Note that this front ground wheel steering angle is in the reasonable range for the given vehicle speed. The controller assumes that the vehicle speed is 20 m/s. In order to test the robustness of the proposed controller, the vehicle speed is not kept at the constant value in the simulation. Instead the speed decreased from around 21.5–19 m/s.

Figure 13.3 displays the direct yaw moment required by the controller, from which we can see that the direct yaw moment is generated as soon as the vehicle started making the J-turn. This is because that the controller tries to control the vehicle such that the motions of the controlled vehicle can be the same with the ones described by reference model. The vehicle yaw rates and slip angles are shown in Figs. 13.4 and 13.5, respectively. We can see from these two figures that the performance of the controlled vehicle is very close to the reference model. In

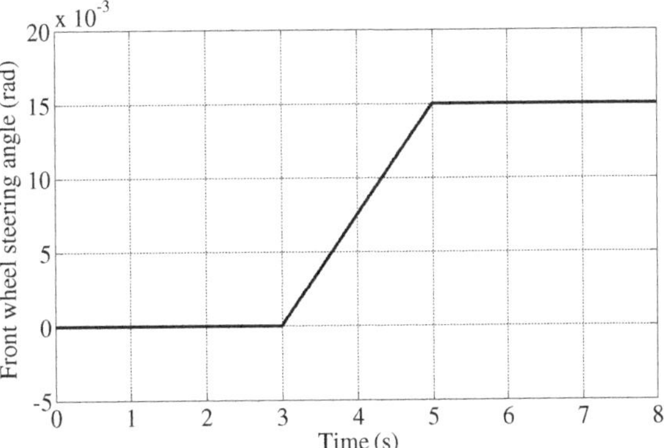

Fig. 13.2 Front wheel steering J-turn simulation

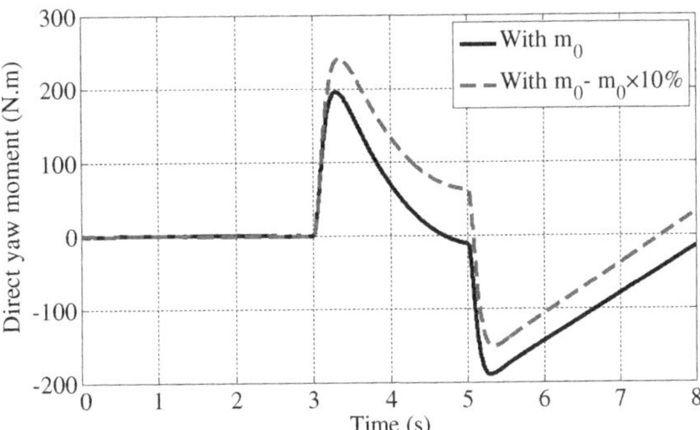

Fig. 13.3 Direct yaw moment with m variation

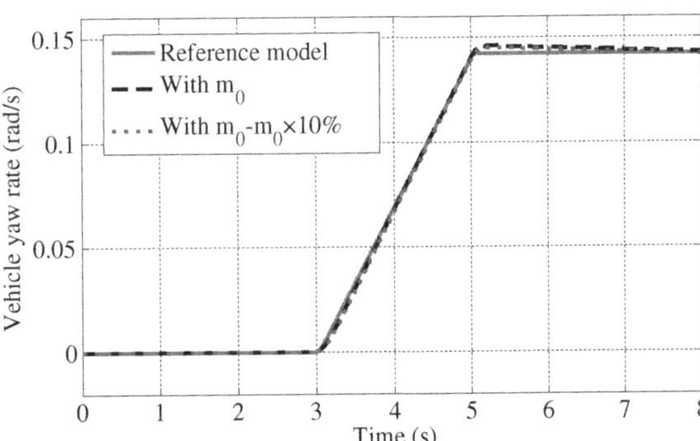

Fig. 13.4 Vehicle yaw rate with m variation

order to test the robustness of the proposed controller with respect to the mass (payload) change, the vehicle mass in the CarSim®, full-vehicle model is adjusted to be 90 percent of the nominal value. We can see again that the proposed controller controlled the vehicle well.

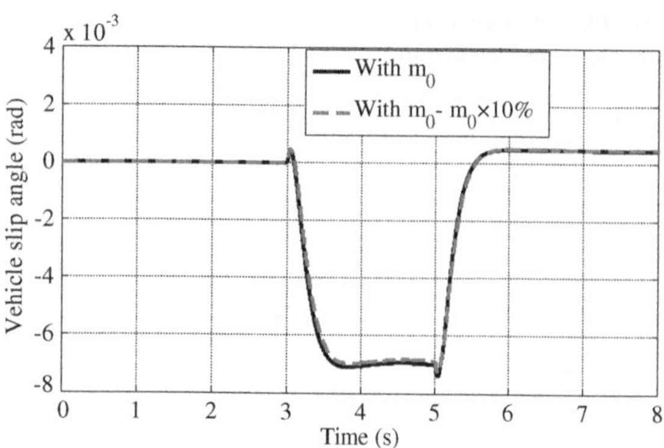

Fig. 13.5 Vehicle slid slip angle with m variation

13.5 Conclusion

A robust controller based vehicle stability control method is presented for AWDEV. Simulations using a high-fidelity, CarSim®, full-vehicle model evidence that the vehicle stability and desired maneuverability can be achieved with the help of the proposed robust control system.

Acknowledgments This research was supported by National Science Foundation of China (51105074, 51205158, 51205204) and Foundation of State Key Laboratory of Mechanical Transmission (SKLMT-KFKT-201206).

References

1. Hori Y (2004) Future vehicle driven by electricity and control-research on fourwheel-motored 'UOT Electric March II'. IEEE Trans Ind Electron 51(5):954–962
2. Magallan GA, Angelo CHD, Garcia GO (2011) Maximization of the traction forces in a 2WD electric vehicle. IEEE Trans Veh Technol 60(2):369–380
3. Sakai S, Sado H, Hori Y (1999) Motion control in an electric vehicle with four independently driven in-wheel motors. IEEE/ASME Trans Mechatron 4(1):9–16
4. Cong G, Mostefai L, Denai M, Hori Y (2009) Direct yaw-moment control of an in-wheel-motored electric vehicle based on body slip angle fuzzy observer. IEEE Trans Ind Electron 56(5):1411–1419
5. Kim K, Hwang S, Kim H (2008) Vehicle stability enhancement of four wheel-drive hybrid electric vehicle using rear motor control. IEEE Trans Veh Technol 57(2):727–735
6. Kim J, Park C, Hwang S, Hori Y, Kim H (2010) Control algorithm for an independent motor-drive vehicle. IEEE Trans Veh Technol 59(7):3213–3222
7. Packard A, Doyle J (1993) Complex structured singular value. Automatica 29:71–109
8. Castellanos R, Messina AR, Sarmiento H (2005) Robust stability analysis of large power systems using the structured singular value theory. Int J Electr Power Energy Syst 27(5–6):389–397

Chapter 14
Fault Diagnosis of Subway Auxiliary Inverter Based on PCA and WNN

Junwei Gao, Ziwen Leng, Yong Qin, Dechen Yao and Xiaofeng Li

Abstract Taken the nonlinearity of fault signals in subway auxiliary inverter and the diagnostic precision into consideration, the paper proposes the fault diagnosis method on the basis of principal component analysis (PCA) and wavelet neural network (WNN). Firstly, extract the initial feature vectors of fault signals by the decomposition and reconstruction of wavelet package, then use PCA to reduce the dimension of initial feature vectors, so as to eliminate redundant data information. Finally, the processed feature vectors will be taken as the input samples of wavelet neural network for the fault diagnosis. Experiment results have tested and verified the feasibility and effectiveness of the method. The proposed diagnostic method has higher precision and stronger convergence than the network directly using initial feature vectors.

Keywords Subway auxiliary inverter · PCA · Wavelet package · Wavelet neural network · Fault diagnosis

14.1 Introduction

As an important electrical component in subway train, auxiliary inverter takes the task of providing power to the electrical equipments apart from train traction. Therefore, the power quality and working stability of auxiliary inverter will directly affect the running efficiency and the safety of subway train. Fault signals in auxiliary

J. Gao (✉) · Z. Leng
College of Automation Engineering, Qingdao University,
No. 308 Ningxia Road, Qingdao 266071, China
e-mail: qdgao163@163.com

J. Gao · Y. Qin · D. Yao · X. Li
State Key Laboratory of Rail Traffic Control and Safety, Beijing Jiaotong University,
Beijing 100044, China

Z. Sun and Z. Deng (eds.), *Proceedings of 2013 Chinese Intelligent Automation Conference*, Lecture Notes in Electrical Engineering 255,
DOI: 10.1007/978-3-642-38460-8_14,

inverter usually have the characteristics of nonlinearity and non-stationarity. As a consequence, the accurate fault diagnosis of auxiliary inverter is of great significance to guarantee the safe operation of subway train.

Wavelet analysis is applicable to the processing of non-stationary signal, while it can achieve the local analysis of time domain and frequency domain [1, 2], and conduct refined and multi-level analysis of signal by scale and shift operation. Wavelet package [3] is an orthogonal decomposition method on the basis of the multi-solution analysis of wavelet transform, and it can conduct more elaborate decomposition and reconstruction to the signal. The paper takes the frequency band energy extracted by wavelet package to compose the feature vector, which has the shortcomings of high dimension and serious self-correlation among data. Principal component analysis (PCA) [4] is a data mining technique of multivariate statistics, and it can reduce the dimension of data which contain a large number of relevant information. After the dimension reduction processing, the new data save the main information of original data. Therefore, PCA is widely applied in signal processing and artificial neural network. Artificial neural network has the unique capabilities of self-learning, self-organization, associative memory and parallel processing [5]. Wavelet neural network is a feed-forward network which takes the wavelet basis function as the activation function [6]. It is widely used in fault diagnosis and has achieved better results in actual application.

This paper consists of five sections. Initial fault feature extraction by wavelet package and the dimension reduction of extracted feature vectors by PCA are described in next section. Basic theory of wavelet neural network is given in Sect. 14.3. Simulation results and conclusion are presented in Sects. 14.4 and 14.5.

14.2 Fault Feature Extraction

Fault diagnosis belongs to the field of pattern recognition, and the feature extraction is an important part of diagnosis.

14.2.1 Initial Feature Vector Extracted by Wavelet Package

Wavelet package can further decompose the high frequency parts, select corresponding frequency band to match the signal spectrum adaptively to improve the time–frequency resolution. Therefore, wavelet package is the deeper analysis of multi-resolution and will have stronger description of fault signal.

For the fault diagnosis of auxiliary inverter, the paper extracts the frequency band energy as the initial feature vector and the extraction steps are as follows.

1. Select appropriate scale to decompose fault signal. The paper decomposes the fault signal at the level 3, and extracts the coefficients of eight frequency bands from low-frequency to high-frequency of the 3rd level respectively.
2. Reconstruct the decomposed coefficients of wavelet package, and extract the signal of each frequency band. S_{ij} represents reconstructed signal at the ith layer of the jth node.
3. Extract the energy feature based on the reconstructed signal S_{3k}.

$$E_{3k} = \int |S_{3k}(t)|^2 dt = \sum_{i=0}^{N-1} |d_{ki}|^2, \quad k = 0, 1, \ldots, 7 \tag{14.1}$$

where i is sample point, d_{ki} is the discrete point amplitude of reconstructed signal S_{3k}, N is the total number of sample point.

4. Calculate the whole energy E and compose the normalized feature vector T.

$$E = \sum_{j=0}^{7} E_{3j} \tag{14.2}$$

$$T = [E_{30}/E, \ E_{31}/E, \ldots, \ E_{37}/E] \tag{14.3}$$

14.2.2 Feature Extraction Based on PCA

PCA can map the data from the original high-dimension space to a low-dimension vector space [7] and achieve the description of sample data by less feature values, so as to reduce the dimension, eliminate the overlap and redundancy of sample. Following steps achieve the dimension reduction and the final feature extraction.

1. Normalization of initial feature sample. Suppose $X = (x_{ij})_{n \times p}$ is the initial feature sample, n is the number of feature vector, p is the number of feature parameters. To avoid the network saturation caused by the excessive input sample, normalize the initial feature sample and obtain the processed sample X^*.

$$x_{ij}^* = (x_{ij} - \bar{x}_j)/S_j \tag{14.4}$$

where x_{ij}^* is the normalized data, $\bar{x}_j$ is the average value of sample, S_j is the standard deviation.

2. Calculate correlative coefficient matrix. Obtain the coefficient matrix $R = (r_{ij})_{p \times p}$ based on the normalized sample X^*.

$$r_{ij} = \frac{1}{n}\sum_{k=1}^{n} x_{ki}^{*} \times x_{kj}^{*} \tag{14.5}$$

where x_{ki}^{*} and x_{kj}^{*} are the ith row and the jth column of X^{*} respectively.

3. Calculate the feature values and feature vectors of correlative coefficient matrix. $\lambda_1 \geq \lambda_2 \geq \cdots \geq \lambda_p > 0$ are the feature values, with the corresponding vectors $C = [c_1, c_2, \cdots, c_p]$. The new feature sample is $Y_{n\times p} = CX^{*}$.
4. Calculate the contribution rate and the cumulative contribution rate of the main components. Feature value reflects the deviation degree of corresponding component in sample. When the cumulative contribution rate of the former m components reaches 95 %, selects corresponding components to constitute the new feature sample.

$$\eta_i = \lambda_i \Big/ \sum_{k=1}^{p} \lambda_k \quad (i = 1, 2, \cdots, p) \tag{14.6}$$

$$\varphi(\eta) = \sum_{j=1}^{m} \eta_j \Big/ \sum_{i=1}^{p} \eta_i \tag{14.7}$$

where η_i is the contribution rate of the ith component, $\varphi(\eta)$ is the cumulative contribution rate of the former m components, $m<p$.

14.3 Basic Theory of Wavelet Neural Network

Wavelet neural network is the organic combination of wavelet theory and artificial neural network [8], while it selects nonlinear wavelet basis function to replace the activation function *Sigmoid* of hidden layer and establishes the connection of wavelet transform and network parameters by affine transform. Wavelet neural network takes the advantage of multi-scale time–frequency analyzing capability of wavelet transform and the generalized ability of neural network, so it has stronger nonlinear approximation, fault-tolerance and pattern classification capabilities.

Because the three-layer forward neural network can approximate a nonlinear mapping at arbitrary precision, wavelet neural network selects the three-layer network with single hidden layer.

Suppose the input layer has N neurons, with hidden layer neurons H and output layer neurons M. Input sample is $X = (x_1, x_2, \cdots, x_N)^T$, actual network output is $\bar{Y} = (\bar{y}_1, \bar{y}_2, \cdots, \bar{y}_M)^T$, the desired output is $Y = (y_1, y_2, \cdots, y_M)^T$.

Activation function of hidden layer selects *Morlet* basis function:

$$\varphi(x) = \cos(0.75x)\exp(-x^2/2) \tag{14.8}$$

Activation function of output layer selects *Sigmoid* function:

$$g(x) = [1 + \exp(-x)]^{-1} \tag{14.9}$$

Output of wavelet neural network is:

$$\bar{y} = g\left[\sum_{j}^{H} w_{jk}\varphi_{a,b}\left(\sum_{i}^{N}\left(w_{ij}x_i - b_j\right)/a_j\right)\right] \tag{14.10}$$

where a_j, b_j are the scale factor and shift factor of the jth hidden layer neuron, w_{ij} is the connection weight between input layer and hidden layer, w_{jk} is the connection weight between hidden layer and output layer.

Define the network error E.

$$E = \sum_{k=1}^{M} (\bar{y}_k - y_k)^2 \tag{14.11}$$

where $\bar{y}_k$ is the actual output, y_k is the desired output.

Wavelet neural network uses the gradient descent algorithm, and takes the sum of error square of actual output and desired output as the learning target to adjust the network connection weight, scale factor and shift scale. However, gradient descent algorithm has the shortcomings of slow convergence when learning samples are larger. The paper introduces the momentum factor to improve the network learning efficiency.

The network connection weight, scale and shift factor are corrected as follows.

$$\begin{aligned} w_{ij}^{t+1} &= w_{ij}^{t} - \eta\frac{\partial E}{\partial w_{ij}^{t}} + \alpha\Delta w_{ij}^{t}; \quad w_{jk}^{t+1} = w_{jk}^{t} - \eta\frac{\partial E}{\partial w_{jk}^{t}} + \alpha\Delta w_{jk}^{t} \\ a_j^{t+1} &= a_j^{t} - \eta\frac{\partial E}{\partial a_j^{t}} + \alpha\Delta a_j^{t}; \quad b_j^{t+1} = b_j^{t} - \eta\frac{\partial E}{\partial b_j^{t}} + \alpha\Delta b_j^{t} \end{aligned} \tag{14.12}$$

where t is number of iterations, η is learning rate, α is momentum factor.

14.4 Simulation Experiments

Common types of faults in auxiliary inverter usually include voltage fluctuation, pulse transient and frequency variation. Collect 30 groups of data of each type with sample frequency 4,096 Hz, randomly select 18 groups of each fault with a total of 54 groups as training samples, the remaining 36 groups are taken as testing samples. The diagnostic labels of voltage fluctuation, pulse transient and frequency variation are (1 0 0), (0 1 0) and (0 0 1) respectively.

Table 14.1 Parts of initial feature samples

Fault type	E_0	E_1	E_2	E_3	E_4	E_5	E_6	E_7
Voltage fluctuation	0.8363	0.0426	0.0312	0.0415	0.0015	0.0049	0.0277	0.0145
	0.8446	0.0399	0.0312	0.0373	0.0013	0.0046	0.0277	0.0134
Pulse transient	0.5082	0.2783	0.0130	0.0816	0.0022	0.0290	0.0728	0.0149
	0.4849	0.2889	0.0194	0.0853	0.0050	0.0389	0.0662	0.0113
Frequency variation	0.3078	0.2190	0.1438	0.1718	0.0091	0.0208	0.0855	0.0422
	0.2977	0.2085	0.1462	0.1876	0.0104	0.0222	0.0843	0.0432

14.4.1 Fault Feature Extraction

Select wavelet basis 'db4' to decompose fault signals and get the eight-dimension feature vector, parts of initial feature samples are shown in Table 14.1.

Initial feature samples extracted by wavelet package transform are analyzed by PCA, as shown in Table 14.2.

Concluded from Table 14.2, the cumulative contribution rate of former 6 components has reached 96.08 %, which exceeds the prescribed 95 % and nearly contains whole information. Therefore, the paper selects the former 6 components to compose the final fault feature vector to achieve dimension reduction.

14.4.2 Simulation Results

Neurons of input layer and output layer of PCA-WNN are determined 6 and 3 based on the fault feature samples processed by PCA. Based on amounts of tests, neurons of hidden layer are determined 13. Learning rate is 0.01, momentum factor is 0.923 and target error is 0.01. Initial feature samples are taken as the input of wavelet neural network (WP-WNN) with network structure 8-13-3, while other parameters are the same as PCA-WNN. Training error curves of PCA-WNN and WP-WNN are shown in Figs. 14.1 and 14.2. Table 14.3 shows parts of output PCA-WNN, Table 14.4 is the diagnostic comparison of PCA-WNN and WP-WNN.

Concluded from Figs. 14.1 and 14.2, when the target error of wavelet neural network is 0.01, the network error of PCA-WNN converges to 0.0098 at 84 steps, while WP-WNN needs 218 steps of training to reach 0.0099. As is seen in Tables 14.3 and 14.4, actual outputs of PCA-WNN are closely approximating the desired outputs with the fault diagnostic accuracy 97.22 %, while the diagnostic accuracy of WP-WNN is 86.11 %.

Table 14.2 Analysis of initial feature samples by PCA

Main component	Feature value	Contribution rate (%)	Cumulative contribution rate (%)
1	3.8993	45.91	45.91
2	1.7660	20.79	66.70
3	0.9617	11.32	78.02
4	0.7462	8.79	86.81
5	0.4728	5.57	92.38
6	0.3144	3.70	96.08
7	0.2259	2.66	98.74
8	0.1074	1.26	100.00

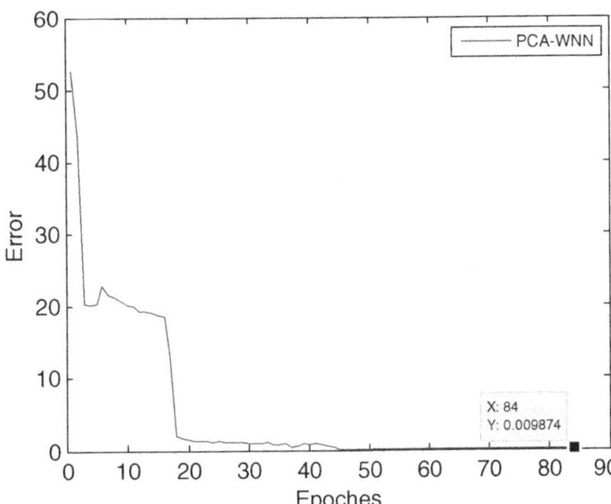

Fig. 14.1 Training error curve of PCA-WNN

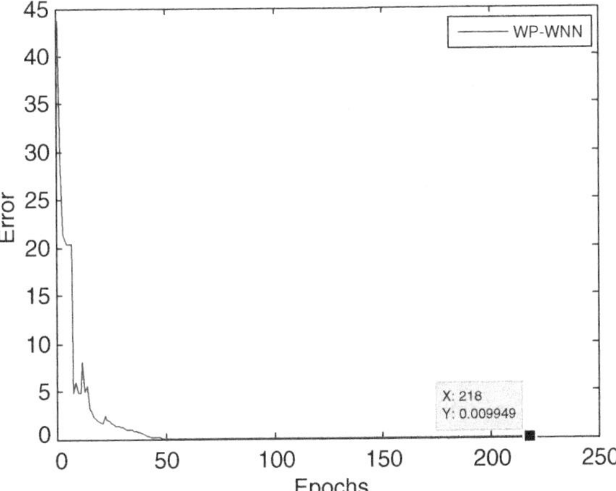

Fig. 14.2 Training error curve of WP-WNN

Table 14.3 Parts of network output of PCA-WNN

Fault type	Actual output			Desired output
Voltage fluctuation	0.9834	0.0116	0.0098	1 0 0
	0.9755	0.0184	0.0236	1 0 0
	0.9803	0.0165	0.0102	1 0 0
Pulse transient	0.0027	0.9963	0.0092	0 1 0
	0.0086	0.9909	0.0067	0 1 0
	0.0093	0.9891	0.0069	0 1 0
Frequency variation	0.0105	0.0128	0.9884	0 0 1
	0.0078	0.0097	0.9934	0 0 1
	0.0112	0.0088	0.9913	0 0 1

Table 14.4 Diagnostic comparison of PCA-WNN and WP-WNN

Model	Voltage fluctuation	Pulse transient	Frequency variation	Diagnostic accuracy
PCA-WNN	12/12	11/12	12/12	97.22 % (35/36)
WP-WNN	11/12	10/12	10/12	86.11 % (31/36)

14.5 Conclusion

The paper proposes the fault diagnosis method of auxiliary inverter based on the PCA and wavelet neural network. Initial feature vector extracted by wavelet package transform indicates the energy variation of fault signal, while PCA reduces the dimension of feature vector and simplifies the network structure. Simulation results show that, the proposed diagnostic method has fast convergence and classification precision, which is applicable to the diagnosis of auxiliary inverter.

Acknowledgments This work is partially supported by the National Key Technology R&D Program (2011BAG01B05), the Foundation of Shandong Province (BS2011DX008, ZR2011FQ012, ZR2011FM008), 863 Program (2011AA110501) and the State Key Laboratory of Rail Traffic Control and Safety Foundation (RCS2011K005, RCS2012K006) Beijing Jiaotong University.

References

1. Gao JW, Yu JP, Leng ZW et al (2013) The application of PSO-LSSVM in fault diagnosis of subway auxiliary inverter. ICIC Express Letters, Part B: Applications, 4(3):777–784
2. Mallat S (1989) A theory for multiresolution signal decomposition: the wavelet representation. IEEE Trans Pattern Anal Mach Intell 11(7):674–693
3. Gui ZH, Han FQ (2005) Neural network based on wavelet packet-characteristic entropy for fault diagnosis of draft tube. Proc CSEE 25(4):99–102 (In Chinese)
4. Wu Q, Cai HN, Huang LF (2011) Feature-level fusion fault diagnosis based on PCA. Comput Sci 38(1):268–270 (In Chinese)

5. Gao JW, Cai GC, Li QC et al (2011) Short-term urban road traffic flow forecasting based on ANFIS. ICIC Express Lett 5(10):3671–3675
6. Ma WJ (2008) Power transformers fault diagnosis based on the wavelet neural network. Control Theory Appl 27(12):17–19 (In Chinese)
7. Lu CH, Wang Y, Wang LF (2008) Fault diagnosis in analog circuit based on PCA and SVM. J Electron Meas Instrum 22(3):64–68 (In Chinese)
8. Yang C, Wang W (2011) Application of wavelet neural network optimized by genetic algorithm in traffic volume prediction. Comput Eng 37(14):149–151 (In Chinese)

Chapter 15
Ultrasonic Array Based Obstacle Detection in Automatic Parking

Huihai Cui, Jinze Song and Daxue Liu

Abstract An ultrasonic array for automatic parking is designed and accomplished in this paper. The details of software and hardware of the system were presented in the article, and the system was validated in a refitted vehicle by automatic parking. The detection result of the ultrasonic array was contrast with that of LADAR. The system showed the advantages in the robust, economy, environments irrespective, and easy to widely application.

Keywords Ultrasonic array · Automatic parking · Obstacle detection · LADAR

15.1 Introduction

Automatic parking is one of the important applications of unmanned vehicle autonomy technology. Automatic parking is usually activated when the vehicle is driven with low speed (about 5–15 km/h). The unoccupied parking place is detected firstly and the localization information is transmitted to the driving control system, the vehicle is then driven by the autonomous control system to the parking place. During the reverse, the sensors are going to detect the surrounding obstacles to ensure the safety.

Since the parking environment information is mainly represented by the distance and orientation of the obstacles, the performance of the range detection sensors is of essentially importance for the overall performance of the automatic parking system. So far, popular range detection sensors include: vision, laser, millimeter-wave radar, ultrasonic or infrared sensors. Ultrasonic sensors, which have been widely applied in the research of mobile [1, 2], have many merits compared to other

H. Cui (✉) · J. Song · D. Liu
College of Mechatronics Engineering and Automation, National University of Defense Technology, Deya Road, Changsha 410073 Hunan, China
e-mail: huihai_cs@163.com

Z. Sun and Z. Deng (eds.), *Proceedings of 2013 Chinese Intelligent Automation Conference*, Lecture Notes in Electrical Engineering 255,
DOI: 10.1007/978-3-642-38460-8_15,

popular range sensors, such as: cheap, accurate range readings, unaffected by light intensity, etc.

A single sonar return signal is usually represented by an arc centered at the position of the ultrasonic sensor, which is limited in the range of particular beam angle. Therefore, ultrasonic sensor suffers from bad angular resolution [3], and it is difficult to achieved high localization accuracy by single sensor. There is also blind area problem arise from the limited beam angle. To solve the problems, a multi ultrasonic sensor array is developed, and the accurate obstacle localization is acquired through the optimally process of the grouped sensor data, using the redundancy relationship.

15.2 Ultrasonic Sensor Array Design

In order to decrease the blind area, improve the localization accuracy of the obstacle detection system, a fixed ultrasonic sensor array is designed in this section. Firstly, the physical characteristic of the ultrasonic sensor is analyzed.

15.2.1 Beam Angle Analysis of Polaroid 600 Ultrasonic Sensor

Polaroid 600, a kind of transmitting-receiving integrated sensor unit, is selected to construct the ultrasonic sensor array. The valid range is from 0.4 to 10.6 m, the relative accuracy is ±1 %D, the beam angle is about 30°, which is shown in Fig. 15.1a [4].

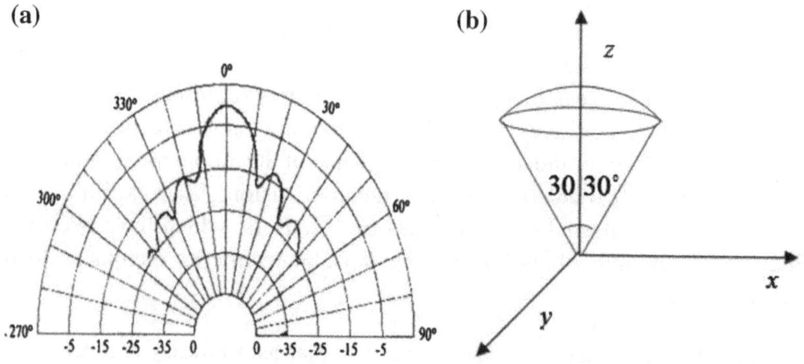

Fig. 15.1 Polaroid 600 sensor characteristic, (**a**) beam angle model (**b**) range detection limit

In the figure, the valid detection range of Polaroid 600 sensor is depicted by the area enclosed by the wave curve, which is made up of one major lobe and multiple minor lobes. The sonar pressure reaches the highest level at 0° angle offset the main sonar wave axis, and decreases with the offset angle increases. The beam angle is defined as the angle between the major axis and the point where the major beam function reaches 0.707 of the major maximum value, that is, the half power point. As we can see in the figure, the detection range is various with the non-uniform sonar pressure in the range of beam angle. On the observation that the furthest detection range of Polaroid 600 sensor, is 10.6 m, and the detection range required is limited in 5 m in the parking system, so we assume that the sonar pressure is uniformly distributed on equal distance spherical surface in the beam angle range as shown in Fig. 15.1b. The following ultrasonic array design is based on the spherical analysis model.

15.2.2 Safety Warning Area Segmentation

Ultrasonic sensors for obstacle detection are usually setup at the rear of the car. Given that the height of the ultrasonic sensor is $h_1 = 0.45$ m from ground, the minimum ground clearance of a sedan is $h_2 = 0.15$ m (the minimum ground clearance of a SUV is 0.2–0.25 m), Fig. 15.2 illustrates the range detection model of rear-setup ultrasonic sensor.

To keep the car away from collision with obstacles while reversing, the height of obstacle should be lower than the ground clearance. Therefore, it is determined as dangerous status when the height of the obstacle is beyond of h_2, otherwise safe when the height is lower than h_2. According to the triangular formula, the distance of the sonar wave from the ultrasonic sensor to ground surface, is as follows:

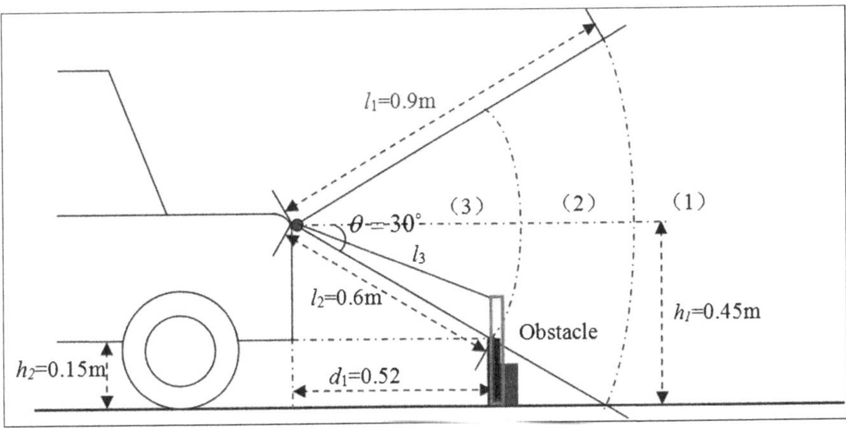

Fig. 15.2 Rear-setup ultrasonic sensor diagram

$$l_1 = \frac{h_1}{\sin \theta} = 0.9\,\text{m} \tag{15.1}$$

Therefore, the valid detection range of the ultrasonic sensor is a sector with 0.9 m radius, and 60° central angle. Suppose that the reversing velocity is 5 km/h, the time of sonar wave transmit through 0.9 m, together with the time of data processing, is about 0.007 s, during which the movement distance of the car is 0.0097 m, less than 1 cm. Therefore, even an obstacle, 0.9 m far away, is detected, which is outside of the sector range, it is determined as safe status.

According to the principle of ultrasonic range detection, only the nearest obstacle is detected. Therefore, for an obstacle lying in the sector range, only the nearest point on the object's surface is detected, the object's height is difficult to measure. In the automatic parking system, the known beam angle and the sector range edge are utilized together to estimate the height of the obstacle.

Suppose there is an obstacle A, with a height of 0.15 m (the same as the ground clearance of the car, shown in the Fig. 15.2 as a black rectangle), which is the limit to pass through safely. When the car is reversed toward the obstacle gradually, the distance detected by the ultrasonic sensor decreases gradually. When the car move to the position shown in Fig. 15.2, where the sector range edge of the ultrasonic sensor interacts with the upper edge of the obstacle, the distance detected by the sensor is as follows:

$$l_2 = l_1 - \frac{h_2}{\sin \theta} = 0.6\,\text{m} \tag{15.2}$$

The actual distance between the obstacle and the car is as follows:

$$d_2 = l_2 \times \cos \theta = 0.52\,\text{m} \tag{15.3}$$

Given the reversing velocity is 5 km/h, the sonar wave transmitting time and the processing time are estimated as 0.0047 s, the distance of the car movement is 0.0064 m, which means d_2 is a safe distance.

If the height of the obstacle is less than 0.15 m (shown as the blue rectangle in Fig. 15.2), the upper edge of the obstacle lies below the sector range edge, then the returns received by the ultrasonic sensor come from ground, l_2 is larger than 0.6 m, which means the obstacle can be pass through safely. If the height of the obstacle is larger than 0.15 m (shown as the red rectangle in Fig. 15.2), then the detected distance l_3 is less than 0.6 m, apparently the obstacle can't be pass through. Therefore we set $l_2 = 0.6$ as the limit to determine the height of the obstacle, and the ultrasonic sensor's detection range can be segmented to three parts: (1) $l_2 > 0.9$ m as safe distance, non-obstacle detected, the car keep reversing; (2) 0.6 m $\leq l_2 \leq 0.9$ m as middle area, obstacle has been detected, but the height can't be measured, keep reversing; (3) $l_2 > 0.6$ m as warning area, detected obstacle can't be pass through, stop reversing.

One should be noted that the analysis values above are relevant with the height h_1 of the ultrasonic sensor, larger h_1 would bring out larger warning distance and longer warning time.

15.2.3 Sensor Array Design and Range Analysis

The sensor array is designed to meet the requirement of information availability, efficiency, and time applicability [5]. The amount of the sensors is chosen to ensure the information acquisition, and minimized. Given the amount of the sensors, the parameter of time applicability is maximized as possible as we can, that is minimize the data latency.

Given the rear width of the car is 1.8 m. Figure 15.3 illustrates the ultrasonic sensor array planar model, in which 4 Polaroid 600 sensors are equally spaced on the rear of the car (0.6 m), with the setup height of 0.45 m. Because of the beam angle limit, there exist blind area where obstacle can't be detected between two sensors [6], illustrated as gray area shown in Fig. 15.3.

The detectable area by ultrasonic sensors is denoted by digital, representing the number of the sensor which can detect the area. For example, (1), (2), (3), (4) denote that the area is covered by sensor (1), sensor (2), sensor (3), sensor (4), and (1, 2), (2, 3), (3, 4) denote that the area is covered simultaneously by two sensors, whereas (1, 2, 3), (2, 3, 4) denote that the area is covered by three sensors.

According to the analysis on safety warning area segmentation, the system is going to warn when detected range is less than 0.6 m, and the car should be stopped. The above warning area is denoted by grid. As shown in the Fig. 15.3, the furthest distance of the blind area of the ultrasonic array is 0.52 m, which is less than the radius of the warning area, 0.6 m, that is the blind area is covered by the warning area. Therefore, the possibility of fail to detect arisen from the blind area

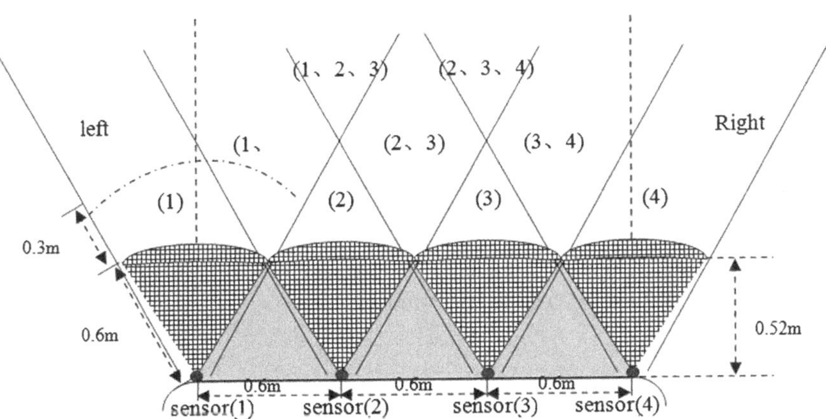

Fig. 15.3 Ultrasonic sensor array diagram

is eliminated through the design of the ultrasonic sensor array. Furthermore, there is physical blind area of the sensor arisen from the time of vibration stop of the ultrasonic wave impulse, which limits the minimum detection distance of the ultrasonic sensor. The minimum detection distance of Polaroid 600 is 0.4 m, less than the radius of the warning sector area. Therefore, the impact of the physical blind area is also eliminated from the obstacle detection process.

15.3 Obstacle Localization Based on Sensor Array

In order to make it easy to analysis the obstacle detection performance, the rear area of the car is segmented to three parts: left-rear, middle-rear, right-rear, as shown in Fig. 15.3. While reversing movement, the distance between the ultrasonic sensors and the obstacle vary dynamically from far to near position. When the sensor can't detect obstacles or the detected distance is larger than the maximum limit, the output of the sensors is set zero.

Given that only one sensor can receive return, the orientation of the obstacle can be coarsely determined according to the position of the sensor which receive returns. For example, if only sensor (1) get returns, obstacle is located in the left-rear area, similarly, if only sensor (4) gets return, obstacle is determined in the right-rear area, and the obstacle may be located in the middle part (3) area if sensor (3) gets return. Obstacles located in other area, usually be detected by two or more sensors simultaneously. It is of critical importance for obstacle localization to fuse multiple detected distance information arise from one obstacle.

15.3.1 Fusion of Pair Wise Range Information

If two ultrasonic sensors get returns simultaneously, there should be an obstacle located in the intersection of the two spherical sector areas which is correspondent with the beam angle range [7]. Therefore, we can accurately estimate the location of the object through the point of intersection of two spherical sectors, which can be simplified as a point of intersection of two arcs when the sector areas are projected onto the horizontal plane, shown as the Fig. 15.4a. Given the obstacle T, which global coordinate is (x_T, y_T), the coordinates of the two ultrasonic sensors are (x_1, y_1) and (x_2, y_2) respectively, the detected distance by sensor (1) is l_1, and the detected distance by sensor (2) is l_2, the distance between the two sensors is d, d can be figure out by the following formula:

$$d = \sqrt{(x_1 - x_2)^2 + (y_1 - y_2)^2} \tag{15.4}$$

In the Fig. 15.4a, given the lengths of the three edges of the triangle, the angle α can be computed according to the law of cosines:

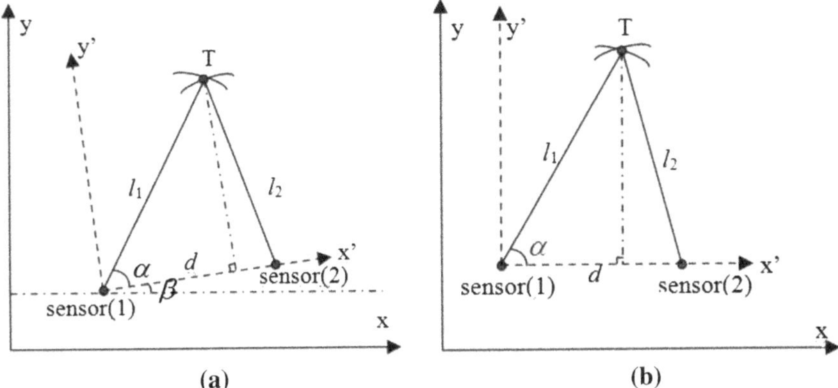

Fig. 15.4 Returns of a pair of sensor

$$\cos \alpha = \frac{d^2 + l_1^2 - l_2^2}{2 \times l_1 \times d} \tag{15.5}$$

$$\alpha = \arccos\left(\frac{d^2 + l_1^2 - l_2^2}{2 \times l_1 \times d}\right) \tag{15.6}$$

A new coordinate frame can be defined, with the origin is set as the location of the sensor (1), the X axis is set as the line pass through the two sensors. Then the new coordinates of the obstacle T is:

$$x'_T = \cos \alpha \times l_1 \tag{15.7}$$

$$y'_T = \sin \alpha \times l_1 \tag{15.8}$$

In the Fig. 15.4a, the new coordinate frame is not parallel with the global coordinate frame. And there is an angle β, which can be got by:

$$\beta = \arctan\left(\frac{y_2 - y_1}{x_2 - x_1}\right) \tag{15.9}$$

After transformed by rotation and translation, the coordinate of T in the global coordinate frame is:

$$x_T = x_1 + \left(x'_T \cos \beta - y'_T \sin \beta\right) \tag{15.10}$$

$$y_T = y_1 + \left(x'_T \sin \beta + y'_T \cos \beta\right) \tag{15.11}$$

In the Fig. 15.4b, $\beta = 0^\circ$, which means that the new coordinate frame is parallel with the global coordinate frame, so the global coordinate of T can be simplified as:

$$x_T = x_1 + x'_T \tag{15.12}$$

$$y_T = y_1 + y'_T \tag{15.13}$$

15.3.2 Fusion of Triple Range Information

From Fig. 15.3 we can see, in the areas of (1, 2, 3) and (2, 3, 4), there are three ultrasonic sensors detect obstacles. When process the triple returns, we can fuse the distance data got from pair wise sensors, following the method of dual distance data fusion, discussed above.

In ideal condition, the computation results from three fusions should be the same value. That is three arcs corresponding to the three distances should intersect at one point. However, because of the impact of various noise and measurement error of the sensors, there are usually three intersect points among the three arcs, as illustrated in Fig. 15.5.

To deal with the problem, one simple method is to compute the average of the three locations, T12, T13, T23, and regard the average as the estimated location of the obstacle. Another way is the circumcircle method [6]. The basic idea is to locate the center of the circumcircle, using three points of intersection, and estimate the location of the obstacle through the coordinates of the center of the circle (x_p, y_p), which can be solved from the following equation.

$$\begin{cases} (x_{T12} - x_p)^2 + (y_{T12} - y_p)^2 = r^2 \\ (x_{T13} - x_p)^2 + (y_{T13} - y_p)^2 = r^2 \\ (x_{T23} - x_p)^2 + (y_{T23} - y_p)^2 = r^2 \end{cases} \tag{15.14}$$

where (x_{T12}, y_{T12}), (x_{T13}, y_{T13}), (x_{T23}, y_{T23}) is the coordinate of T12, T13, T23 respectively, the detailed solution of the equation can be referenced in literature [1].

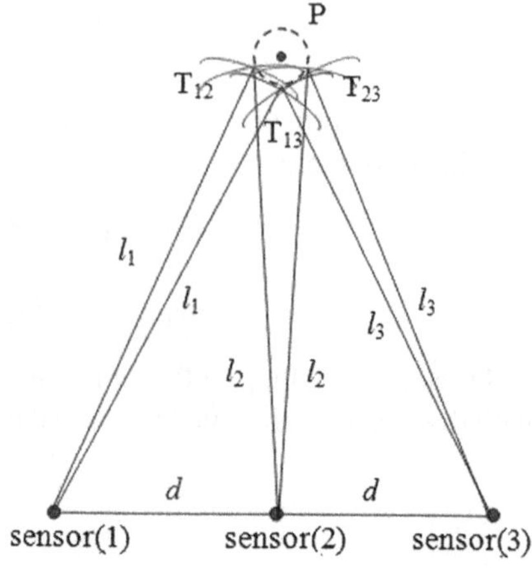

Fig. 15.5 Returns of three sensors

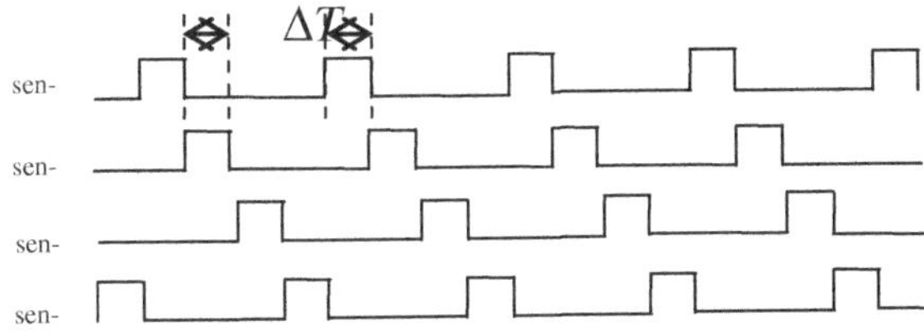

Fig. 15.6 Working sequence of the ultrasonic sensor array

15.4 The Working Mode of the Sensor Array

If the multiple sensors in the ultrasonic array are triggered simultaneously with the same cycle time, it is easy to form crosstalk, and result in wrong determination of the orientation of the obstacle. Furthermore, at the simultaneous trigger working mode, the blind area in the time domain is hardly be improved, the update of environment status is not good in real-time performance.

To cope with the above problems, a circularly scan-trigger working mode is adopted to the ultrasonic sensor array. In detail, after one sensor is triggered, another sensor is not triggered until delay a while, the sensors are triggered one by one, and the working sequence is illustrated in Fig. 15.6.

In the Fig. 15.6, the delay time for every sensor is ΔT, which has effect on the overall performance of the ultrasonic sensors. If ΔT is too small, it is easy to result in crosstalk and wrong localization, whereas ΔT is too large, it might be a large relative offset from the obstacle in one cycle of the sensor array, result in large localization error. In the parking system, ΔT is set as the working time of the sensor, T_w, the working duration time for every sensor is $3T_w$. When a sensor ends its detection process, the next sensor is triggered immediately. Because there is always sensor working, the blind area in the time domain is eliminated. The sensor working time T_w is going to increase with the detection range. Given T_w is set as the time to detect an obstacle 5 m away, the relative offset is only several centimeters, which can be ignored.

Besides, since there is intrinsic error in the detected distance, this kind of error can't be totally eliminated through fusion procedure, and multiple sonar returns might not be consistent, all of these problems are going to decrease the overall performance of the sensor array. Therefore, during the localization procedure of the ultrasonic sensor array, acquiring localization information of obstacle with more believability, need continuous fusion data.

15.5 Experiment Result

The ultrasonic sensor array based obstacle detection system is implemented on a refitted car. In the experiment, when an operator push the automatic parking switch button, starting to scout the parking place, the ultrasonic sensor array is activated to work, during which the car is controlled by a driver. In the Fig. 15.7, red points

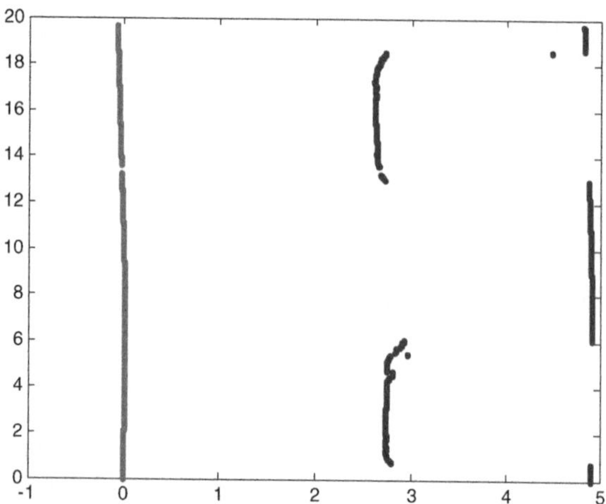

Fig. 15.7 Obstacles detection during park

represent the actual trajectory of the car, and blue points represent the obstacles detected by ultrasonic sensors. The origin is the position where the parking system is started, the Y axis is the heading direction of the car, and the X axis is the right-hand direction, the unit for both the x and y coordinates is m. As we can see from the distribution of the blue points, the ultrasonic sensors system successfully detected two cars parked aside the road, together with the obstacles.

In Fig. 15.8, When the parking place is found, the vehicle is going to direct with speaker, then the parking trajectory is planned, represented by red points, according to the obstacle information detected by the ultrasonic sensors. The purple points represent the trajectory of the car while searching the parking place, the red circle and the black circle represent the parking place detected. Because the planned trajectory is referenced with the right-rear wheel, red points offset half width of the car from the center position. The origin is the position where the

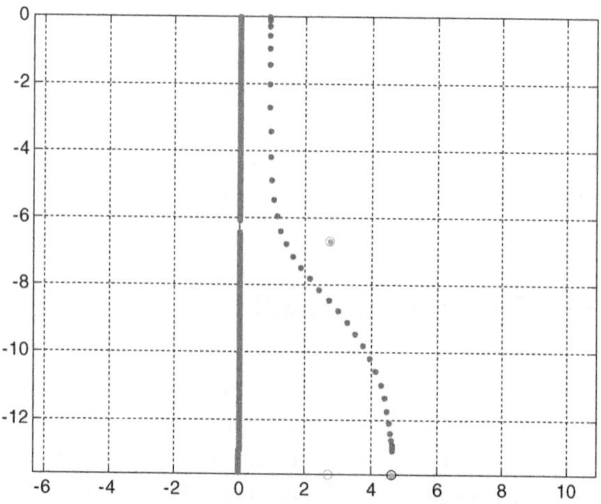

Fig. 15.8 The planned parking trajectory

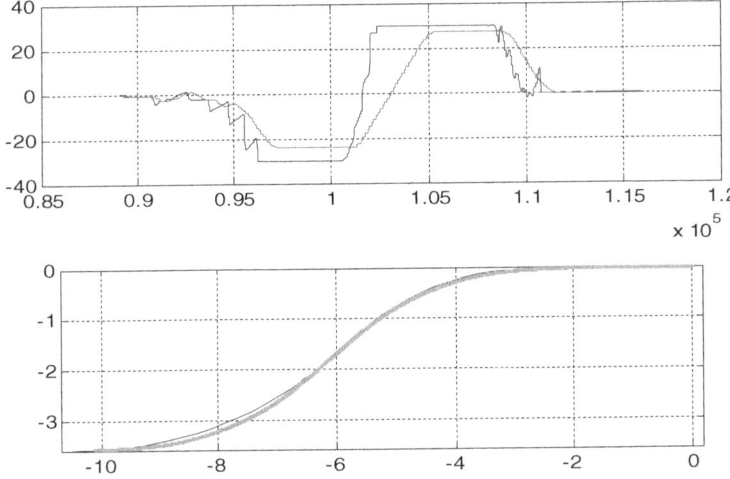

Fig. 15.9 The actual reverse trajectory and the control input

parking place is found, the Y axis is the heading direction of the car, and the X axis is the right-hand direction, the unit for both the x and y coordinates is m.

Figure 15.9 illustrates the actual automatic parking result. The upper diagram shows the expected steering angle (blue) and the actual steering angle (red), the longitudinal coordinate denotes the steering angle, unit as degree, the horizontal coordinate denotes time, unit as millisecond. The lower diagram shows the planned parking trajectory (blue) and the actual reversing trajectory (red). The origin is the start position, and the car was heading right. The unit for horizontal and longitudinal coordinates is meter.

In Fig. 15.10, the detection result by the ultrasonic sensors is compared with that of single line scan LADAR. The origin is the parking start position, the Y axis is the heading direction of the car, and the X axis is the right-hand direction, the

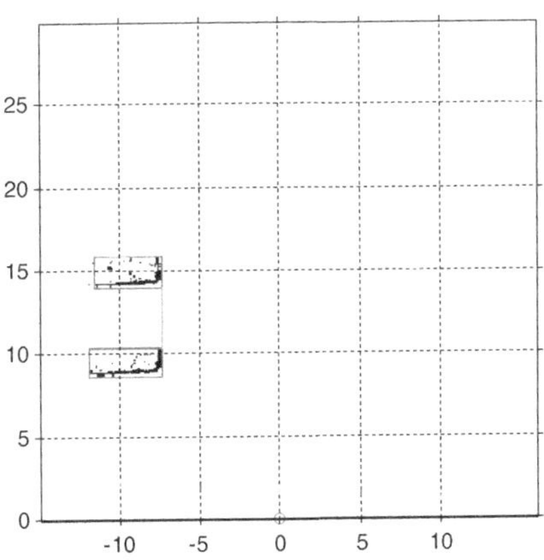

Fig. 15.10 The parking place detected by LADAR

unit for both the x and y coordinates is meter. During the experimental car move forward, the parked vehicles are continually detected. In the experimental environment, the parked vehicles are longitudinally orientated. As we can see in the Fig. 15.10, although the points detected by LADAR is much more than the ultrasonic sensors, more details may bring out more troubles for the vehicle detection, such as the points inside the vehicles which is generated when the laser scan through the window.

15.6 Conclusion

The ultrasonic sensor array based obstacle detection system achieves high accuracy, which can satisfy the requirement of automatic parking. Compared with LADAR, ultrasonic sensors have many merits such as robustness, cheap cost, environment adaptability, and convenient for large scale applications.

References

1. Wu L (2006) Research on ultrasonic obstacle detection technology applied in ALV. Master thesis (in Chinese)
2. Borenstein J, Shoval S (2001) Using coded signals to benefit from ultrasonic sensor crosstalk in mobile robot obstacle. In: IEEE international conference on robotics and automation 2001, pp 2879–2884
3. Sun H, Chen J, Wu L (2003) Multi-sensor information fusion and its application to the robot. J Transducer Technol 22(9):1–4 (in Chinese)
4. Liu X, Zhou Z (2003) Technology of robot's safe obstacle avoidance based on multi-ultrasonic sensor. Meas Control Technol 23(2):71–73 (in Chinese)
5. Yun W, Xi Y (1996) Information system of intelligent robot. ROBOT 18(5):257–263 (in Chinese)
6. Borenstein J, Koren Y (1989) Real time obstacle avoidance for fast mobile robots. IEEE Trans Syst Man Cybern 19(5):1179–1187
7. Luo X, Zhan X, Jiao J, Gao B, Geng X (2007) Multi-targets orientation system based on ultrasonic sensor. Sens World 09:25–29 (in Chinese)

Chapter 16
Approximate Dynamic Programming Based Controller Design Using an Improved Learning Algorithm with Application to Tracking Control of Aircraft

Xiong Luo, Yuchao Zhou and Zengqi Sun

Abstract The strategy using approximate/adaptive dynamic programming (ADP) has been widely used to design a learning controller for complex systems of higher dimension in recent years. This paper aims at handling an important problem in the design of ADP learning controllers, which is the improvement of learning algorithm for its convergence performance. We analyze ADP controller implementation framework according to the requirement of tracking control task, with emphasis on providing an improved weight-updating gradient descent approach in optimizing connection weights in network structures. A comparison of the proposed method and classic ADP design for tracking and controlling pitch angle of aircraft is presented. It verifies the feasibility in the design of the proposed ADP based controller.

Keywords Approximate dynamic programming · Controller · Learning algorithm · Aircraft control

---

X. Luo (✉) · Y. Zhou
School of Computer and Communication Engineering,
University of Science and Technology Beijing (USTB), 30 Xueyuan Road,
Haidian District, Beijing 100083, China
e-mail: xluo@ustb.edu.cn

X. Luo · Y. Zhou
Beijing Key Laboratory of Knowledge Engineering for Materials Science,
30 Xueyuan Road, Haidian District, Beijing 100083, China

Z. Sun
Department of Computer Science and Technology, Tsinghua University,
Haidian District, Beijing 100084, China

Z. Sun and Z. Deng (eds.), *Proceedings of 2013 Chinese Intelligent Automation Conference*, Lecture Notes in Electrical Engineering 255,
DOI: 10.1007/978-3-642-38460-8_16,

16.1 Introduction

It is well known, on the issue of designing a stable controller for nonlinear dynamic system, dynamic programming (DP) method does not completely overcome the challenge of "curse of dimensionality". Then approximate/adaptive dynamic programming (ADP) was presented by means of a combination of the adaptive critics as well as the actor critic techniques, which has shown great promise to handle complex control problem for high-dimensional systems by introducing approximating function structures to estimate and update the control policy [1, 2]. Among a series of adaptive critic design approaches for ADP, the neural network-based implementation is popular and has been a focus in recent years for its potential scalability to address complex control problems [3].

ADP has recently demonstrated many successful applications especially in the field of system control. The aircraft control has always been an active area, which is the application we discuss in this paper. There are many examples of successful application [4–8]. In these applications, an important issue in fulfilling the control task is how to design a learning controller with satisfied time performance. It plays an key role in some applications of real-time control. Generally, under the neural network based implementation framework for ADP, the standard gradient descent algorithm is adopted to update the connection weights of neural network. Although it is one of the simplest ways to implement learning control in ADP, the convergence performance is not guaranteed when dealing with some complex control problems. Thus, it is necessary to improve the standard gradient-based learning algorithm. For example, in [9], a special recursive Levenberg–Marquardt (LM) method, i.e., a combination of vanilla gradient descent and Gauss–Newton iteration, was investigated and incorporated into the weight-updating scheme of ADP design to improve the learning robustness and convergence. However, the implementation of this algorithm is slightly complicated.

In this paper, we will discuss some easy improvements on existing gradient descent method, inspired by the optimization learning strategies for back propagation (BP) neural network. After introducing the momentum term and adopting an accelerated learning strategy, an improved weight-updating learning algorithm is proposed and embodied in the ADP design. Moreover, the proposed ADP-based learning controller is developed to address a special control problem, which is the pitch tracking control of aircraft.

16.2 Statement of Approximate Dynamic Programming

For a nonlinear system given by

$$X(t+1) = F(X(t), U(t)), \tag{16.1}$$

where $X \in \mathbb{R}^n$ and $U \in \mathbb{R}^m$ represent the state vector and the control action, and F denotes the system function. Suppose that it is to find a control sequence $U(t)$, so that the system (16.1) can follow a trajectory starting from an initial state while minimizing the cost function J through the principle of optimality, given by

$$J^*(X(t)) = \min_{U(t)}\{\psi(X(t), U(t)) + \alpha \cdot J^*(X(t+1))\}, \tag{16.2}$$

where $J^*(X(t))$ is the optimal cost from time t, ψ represents the utility function, and $\alpha \in (0, 1)$ denotes the discount factor. ADP is designed to obtain the approximate solutions of (16.2) by means of approximation structures of neural network during the iterative updating process of control policy.

16.3 Approximate Dynamic Programming Based Controller Design

16.3.1 System Framework

A design framework of the proposed ADP learning controller is shown in Fig. 16.1. In this figure, $J(t)$ represents the cost-to-go function, $X(t)$ represents the system states, and $U(t)$ represents the control action. $U_c(t)$ represents the desired ultimate cost objective, and it is set to 0 in this design paradigm. $X_d(t)$ represents the desired system reference states. The discount factor is $\alpha \in (0, 1)$. The key components here are the same with the ones in the classic online model-free ADP [10]. There are two neural networks, namely, the critic network and the action network. They are both a nonlinear multilayer feedforward network with only one hidden layer. The action network is trained to estimate the control action $U(t)$. The critic network is trained to approximate the cost-to-go objective function $J(t)$.

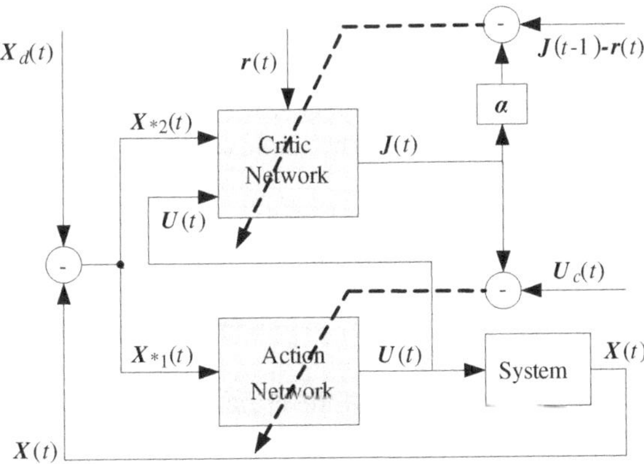

Fig. 16.1 The design framework of ADP learning controller used in tracking control task

Moreover, $r(t)$ is the reinforcement signal. It is generated from the external environment. Generally, $r(t)$ is defined as a simple delayed feedback of binary signal form, in which "0" or "-1" is provided when the objective is met or not met, respectively [10]. Because there is more explicit state information in the flight control task of aircraft [6], the reinforcement signal can be set using instantaneous state and desired state with a more informative quadratic regulator form

$$r(t) = -\sum_{k}\left(\left(x^{(k)}(t) - x_d^{(k)}(t)\right)/x_{\max}^{(k)}(t)\right)^2, \tag{16.3}$$

where $x^{(k)}(t)$ represents the k-th state of the state vector $X(t)$, $x_d^{(k)}(t)$ represents the k-th desired reference state, and $x_{\max}^{(k)}(t)$ represents the nominal maximum state.

Compared to the classic ADP learning controller, there is a different treatment for input system states in Fig. 16.1, considering the requirement of tracking control task. In [11], the inputs to the training network, $X_{*\{1,\,2\}}(t)$, are a normalization of value $(X(t){-}X_d(t))$. The input to the action network, $X_{*1}(t)$, is designed to achieve the tracking of desired state, and the input to the critic network, $X_{*2}(t)$, is used to improve the performance of approximation for objective function $J(t)$. Note that they may take different implementation forms according to control requirement.

16.3.2 Improved Learning Algorithm

In the critic network, the prediction error and the objective function to be minimized are defined as

$$e_c(t) = \alpha J(t) - [J(t-1) - r(t)]. \tag{16.4}$$

$$E_c(t) = e_c^2(t)/2. \tag{16.5}$$

In general, the weights learning rule for the critic network is a traditional gradient descent algorithm [10]. Although it is one of the simple and popular learning methods, there are some drawbacks during its learning process, e.g. its slow convergence rate. To address those problems, inspired by the adjustment strategy of momentum term with variable coefficients [12], and the accelerated learning method for weights [13], we propose a learning algorithm to improve the learning performance. The improved weight update rule for the critic network is

$$\begin{cases} \mathbf{w}_c(t+1) = \mathbf{w}_c(t) + \Delta\mathbf{w}_c(t) \\ \Delta\mathbf{w}_c(t) = \tau_c(t)\left[-\frac{\partial E_c(t)}{\partial \mathbf{w}_c(t)} - \frac{\partial E_c(t-1)}{\partial \mathbf{w}_c(t-1)} + \beta_c(t)\Delta\mathbf{w}_c(t-1)\right] \end{cases}, \tag{16.6}$$

where, at time t, $\mathbf{w}_c(t)$ represents the weight vector in the critic network, $\tau_c(t)$ represents the learning rate of the critic network, and $\beta_c(t)$ represents the momentum coefficient of the critic network. And we have the following definition

$$\begin{cases} \tau_c(t) = \tau_c(t-1)(1+\cos\phi_c(t)/2) \\ \beta_c(t) = \left\|\frac{\partial E_c(t)}{\partial \mathbf{w}_c(t)}\right\| \Big/ \|\Delta \mathbf{w}_c(t-1)\| \end{cases}, \tag{16.7}$$

where $\phi_c(t)$ represents the angle between the current gradient $(-\partial E_c(t)/\partial \mathbf{w}_c(t))$ and the previous update $(\Delta \mathbf{w}_c(t-1))$, at time t. The gradient $(\partial E_c(t)/\partial \mathbf{w}_c(t))$ can be calculated based on the critic network structure, which was discussed in [10].

In (16.6), by introducing the current gradient $(-\partial E_c(t)/\partial \mathbf{w}_c(t))$, the previous gradient $(\partial E_c(t)/\partial \mathbf{w}_c(t))$, and the momentum term, it can serve the purpose of improving training speed [12].

In the action network, the prediction error and the objective function are

$$e_a(t) = J(t) - U_c(t). \tag{16.8}$$

$$E_a(t) = e_a^2(t)/2. \tag{16.9}$$

Similarly, the weight update rule for the action network is designed by

$$\begin{cases} \mathbf{w}_a(t+1) = \mathbf{w}_a(t) + \Delta \mathbf{w}_a(t) \\ \Delta \mathbf{w}_a(t) = \tau_a(t)\left[-\frac{\partial E_a(t)}{\partial \mathbf{w}_a(t)} - \frac{\partial E_a(t-1)}{\partial \mathbf{w}_a(t-1)} + \beta_a(t)\Delta \mathbf{w}_a(t-1)\right] \end{cases}, \tag{16.10}$$

where, at time t, $\mathbf{w}_a(t)$ represents the weight vector in the action network, $\tau_a(t)$ represents the learning rate of the action network, and $\beta_a(t)$ represents the momentum coefficient of the action network. And we have the following definition

$$\begin{cases} \tau_a(t) = \tau_a(t-1)(1+\cos\phi_a(t)/2) \\ \beta_a(t) = \left\|\frac{\partial E_a(t)}{\partial \mathbf{w}_a(t)}\right\| \Big/ \|\Delta \mathbf{w}_a(t-1)\| \end{cases}, \tag{16.11}$$

where $\phi_a(t)$ represents the angle between the current gradient $(-\partial E_a(t)/\partial \mathbf{w}_a(t))$ and the previous update $(\Delta \mathbf{w}_a(t-1))$, at time t. The gradient $(\partial E_a(t)/\partial \mathbf{w}_a(t))$ is calculated based on the action network structure, which was discussed in [10].

In Fig. 16.1, there are two inputs, i.e., $X_{*1}(t)$ of the action network, and $X_{*2}(t)$ of the critic network. As mentioned in Sect. 3.1, these inputs may be different. In the pitch tracking control application of aircraft, they can be defined as

$$\begin{cases} X_{*1}(t) = (X(t) - X_d(t))/\|X(t) - X_d(t)\|_1 \\ X_{*2}(t) = (X_{*1}(t))^2 \end{cases}. \tag{16.12}$$

In so doing, it will result in significant performance improvements over results in traditional applications [11].

16.4 Simulation Result

Here, under our proposed controller design scheme, we implement the pitch control for a simplified aircraft. Under a steady cruise state with a constant velocity and altitude, the system used in the pitch control is the same as the one in [14, 15].

$$\begin{cases} d\alpha/dt = -0.313\alpha + 56.7q + 0.232\eta, \\ dq/dt = -0.0139\alpha - 0.426q + 0.0203\eta, \\ d\theta/dt = 56.7q, \end{cases} \tag{16.13}$$

where α, θ, and η is the attack, pitch, and elevator deflection angle, respectively. q is the pitch rate. The goal of this control example is to control the elevator angle η to track a desired pitch angle θ. Although the above dynamic equations are simplified linear model, it is not easy to design a satisfying learning controller. Because the adjustment of the elevator angle η will result in the slow change of α until it settles. So it is nontrivial.

When the pitch angle $\theta(t)$ is outside the range of $[-0.5, 0.5]$ rad, we set the reinforcement signal $r(t)$ to -1; otherwise, we set $r(t)$ according to (16.3). In our simulation the desired reference pitch angle $\theta_d = 0.1$ rad, and the nominal maximum pitch angle $\theta_{\max} = 0.5$ rad. Here, the controller action $\eta(t)$ is limited in the range of $[-1.4, 1.4]$ rad, for all t. In the initial state of system, the angle of attack $\alpha(0) = 0$ rad, the pitch rate $q(0) = 0$ rad/s, and the pitch angle $\theta(0) = 0$ rad.

The number of the hidden layer nodes in the critic network and the action network are set to 6. The discount factor is set to $\alpha = 0.95$. The internal cycle of the critic network and the action network are set to 80 and 100, respectively. The internal training error threshold for the critic network and the action network are set to 0.05 and 0.005, respectively. The initial learning rates of two networks are 0.3 and they are decreased by 0.05 every 5 time steps until they reach 0.001. In the simulation, 100 runs have been performed to obtain the results. Each run consists of 1,000 trials. If the last trial of the run has lasted 6,000 time steps, it is considered successful. The comparison results between our proposed learning controller and the classic online ADP controller proposed in [10], are illustrated in Table 16.1. All the runs are successful for this pitch control task. In terms of the average number of trials needed to learn to implement the pitch control for aircraft, the result of the proposed controller with an improved learning algorithm is less than the one in [10]. It verifies the feasibility and reliability in the design of the proposed ADP controller under the implementation of aircraft pitch control.

Table 16.1 Performance comparison of two learning controllers

Controller	Success rate (%)	Average number of trials
The classic online ADP controller	100	6.5
The proposed ADP controller	100	4.5

16.5 Conclusion

In this paper, we presented a learning method in the ADP controller design. Starting from the existing gradient-based weight-updating method used in the classic ADP structure, we adopted an accelerated learning approach to improve the learning performance in optimizing the connection weights of the critic network and the action network. This paper has successfully applied the improved ADP to track the pitch angle of aircraft. Results presented here showed the proposed ADP learning controller was able to achieve the control goal with faster convergence rate than the classic online ADP controller. The proposed improved learning method can be also extended to the general control system design.

Acknowledgments This work was jointly supported by the National Natural Science Foundation of China (Grants Nos. 61174103, 61074066, 61174069, 61004021), the Fundamental Research Funds for the Central Universities (Grant No. FRF-TP-11-002B), and 2012 Ladder Plan Project of Beijing Key Laboratory of Knowledge Engineering for Materials Science (Grant No. Z121101002812005).

References

1. Barto AG, Sutton RS, Anderson CW (1983) Neuron-like adaptive elements that can solve difficult learning control problems. IEEE Trans Syst Man Cybern 13(5):834–847
2. Si J, Barto A, Powell W, Wunsch D (2004) Handbook of learning and approximate dynamic programming. Wiley, New York
3. Werbos PJ (1989) Neural networks for control and system identification. Proc IEEE Conf Decis Control 1:260–265
4. van Kampen E, Chu QP, Mulder JA (2006) Continuous adaptive critic flight control aided with approximated plant dynamics. Proc AIAA Guidance Navig Control Conf 5:2989–3016
5. Ferrari S, Stengel RF (2004) Online adaptive critic flight control. J Guid Control Dyn 27(5):777–786
6. Enns R, Si J (2003) Helicopter trimming and tracking control using direct neural dynamic programming. IEEE Trans Neural Netw 14(4):929–939
7. Wang FY, Zhang HG, Liu DR (2009) Adaptive dynamic programming: an introduction. IEEE Comput Intell Mag 4(2):39–47
8. Powell WB (2011) Approximate dynamic programming: solving the curses of dimensionality, 2nd edn. Wiley-Blackwell, New York
9. Fu J, He H, Zhou X (2011) Adaptive learning and control for MIMO system based on adaptive dynamic programming. IEEE Trans Neural Netw 22(7):1133–1148
10. Si J, Wang YT (2001) On-line learning control by association and reinforcement. IEEE Trans Neural Netw 12(2):264–276
11. Enns R, Si J (2002) Apache helicopter stabilization using neural dynamic programming. J Guid Control Dyn 25(1):19–25
12. Chan LW, Fallside F (1987) An adaptive training algorithm for back propagation networks. Comput Speech Lang 2(3–4):205–218
13. Zhang R, Han J (1998) An accelerated learning algorithm for neural networks by gradient power method. Syst Eng Theory Pract 18(6):97–101 (In Chinese)

14. Messner B, Tilbury D (1996) Digital control tutorial, University of Michigan, Ann Arbor. Available: http://www.engin.umich.edu/group/ctm/digital/digital.html
15. Waegeman T, Wyffels F, Schrauwen B (2012) Feedback control by online learning an inverse model. IEEE Trans Neural Netw Learn Syst 23(10):1637–1648

Chapter 17
New Necessary and Sufficient Conditions for Schur D-Stability of Matrices

Jinfang Han

Abstract In this paper, the Schur D-stability and vertex stability of matrices are investigated by means of the matrix eigenvalue theory and spectral radius approach. Four new necessary and sufficient conditions are obtained which guarantee the Schur D-stability of matrices. That the conditions limit of tridiagonal matrix and non-negative matrix in the previous literatures are abandoned. These results are wider applicable and less conservative than those in recent criteria. Three equivalence relations between the Schur D-stability, Schur stability and vertex stability are established.

Keywords Matrices · Schur D-stability · Eigenvalue theory · Spectral radius · Schur stability · Vertex stability · Equivalence relations

17.1 Introduction

It is well-known that the stability of a system is directly related to the stability of its state matrix, especially the stability of linear system or linear dynamic interval systems can be determined completely by the stability of its state matrix. Therefore, the researches on various stability of matrix and interval matrix has been regarded as an important subject in control theory for a long time, which also attract much attention of the numerous cybernetics experts and mathematicians et al. all over the world. The research papers on this field appear repeatedly in

J. Han (✉)
Institute of Engineering Mathematics Hebei University of Science ans Technology, No 26 Yuxiang Street, Shijiazhuang, China
e-mail: JFHanemail@126.com; JFHan@hebust.edu.cn

J. Han
4–1–1 Kitakaname, Hiratsuka, Kanagawa 259–1292, Japan

Z. Sun and Z. Deng (eds.), *Proceedings of 2013 Chinese Intelligent Automation Conference*, Lecture Notes in Electrical Engineering 255,
DOI: 10.1007/978-3-642-38460-8_17,

various international and domestic publications and never lose attractiveness. But it is worth mentioning that the D-stability of the matrices, as Wang and Wang in note [1] say: "is still an open problem that is not solved thoroughly".

The D-stability (Hurwitz and Schur) of the general matrix and the linear systems (including the singular linear systems) has been discussed respectively in [1–4], [2–5], etc. Some corresponding meaningful results are presented in the papers as well. However, the research on relationship between Schur stability, Schur D-stability and vertex stability of matrices, and the research on necessary and sufficient conditions for Schur D-stability of matrix. These are rarely touched. In [6], Fleming et al. only for nonnegative matrix presents the necessary and sufficient condition about Schur stability and Schur D-stability. That is, if $A \geq 0$, then A is Schur D-stable if and only if A is Schur stable. And for the special case of tridiagonal matrix, they gives the necessary and sufficient condition about vertex stability and Schur D-stability. That is, if A be a tridiagonal matrix. Then A is Schur D-stable if and only if A is vertex stable. Where A with operator norm.

Under the enlightenment of paper [1, 6–8], the author tries to consider the relationship between Schur stability, Schur D-stability and vertex stability of general matrices, to get rid of the conditions about none matrix and tridiagonal matrix. In terms of matrix eigenvalues theory and properties of spectral radius, we shall propose some new necessary and sufficient conditions (criteria) which guarantee the matrix is Schur D-stable. This results is shown to be less conservative than those in recent literatures [6–9]. In addition, the author expects the paper can be early published for the first time, so that to be discussed with the people of the same trade and experts.

This paper is organized as follows. In Sect. 17.2, Ideas of Schur D-stability, Schur stability and vertex stability and lemmas are given. In Sect. 17.3, four necessary and sufficient conditions for the Schur D-stability of matrices, and three equivalence relations between the Schur D-stability, Schur stability and vertex stability are derived. Finally, the conclusion is provided in Sect. 17.4.

17.2 Preliminaries

In this section we will first give some basic nomenclatures and lemmas which will be used in the following sections.

Given a square complex matrix $A = (a_{ij}) \in C^{n\times n}$, $|A| = (|a_{ij}|)_{n\times n}$ denotes the modulus matrix of A, where $|a_{ij}|$ denotes the modulus (absolute value) of a_{ij}. For a real matrix $A = (a_{ij}) \in R^{n\times n}$ is nonnegative (positive) if its entries $a_{ij} \geq 0$ $(a_{ij} > 0)$ and $A \geq B \geq 0 (A > B > 0)$ means that the entries of A and B satisfy $a_{ij} \geq b_{ij} \geq 0$ $(a_{ij} > b_{ij} > 0), i,j \in N = \{1, 2, \cdots, n\}$, or $A - B \geq 0 (A - B > 0)$. $\rho(A)$ the spectral radius of matrix A and I the identity matrix. $\lambda(A)$ denotes any eigenvalue of A. The set of all eigenvalues of A, denoted by $\sigma(A)$, is called the spectrum of A. the operator norm or spectral norm of A is denoted by $||A||$. In the next sections, we assume A with operator norm.

Lemma 17.1 [6] *Let* $A=(a_{ij})\in C^{n\times n}$, $B=(b_{ij})\in C^{n\times n}$, *If* $0\le A\le B$, *then* $\rho(A)\le\rho(B)$.

Lemma 17.2 [6, 7] *For any matrices* $A=(a_{ij})\in R^{n\times n}(C^{n\times n})$, $B=(b_{ij})\in R^{n\times n}(C^{n\times n})$ *and* $C=(c_{ij})\in R^{n\times n}(C^{n\times n})$, *if* $|A|\le C$, *then*

1. $\rho(A)\le\rho(|A|)\le\rho(C)$;
2. $\rho(AB)\le\rho(|A||B|)\le\rho(C|B|)$;
3. $\rho(A+B)\le\rho(|A+B|)\le\rho(|A|+|B|)\le\rho(C+|B|)$.

Definition 2.1 [6] A real or complex square matrix A is said to be Schur stable if $\rho(A)<1$. A real or complex square matrix A is said to be Schur D-stable if $\rho(AD)<1$ for every real diagonal D with $|D|\le I$. A real or complex square matrix A is said to be vertex stable if $\rho(AD)<1$ for all real diagonal D with $|D|=I$. Where I denote the identity matrix of size appropriate for the context.

17.3 Main Results

In this section, we prove some general results about matrices A is schur D-stable, and presents several necessary and sufficient conditions for A is schur D-stable. We start our discussions by introducing the following generalizations.

Theorem 17.1 *Let A be a generic* $n\times n$ *matrix with operator norm, then A is Schur D-stable if and only if A is Schur stable. i.e. the Schur D-stability and Schur stability of the A is equivalent.*

Proof The necessity is obvious. Indeed, if a square matrix A is Schur D-stable i.e., $\rho(AD)<1$ for all real diagonal D with $|D|\le I$, particularly, take $D=I$ we have $\rho(A)=\rho(AI)=\rho(AD)<1$, i.e., A is Schur stable.

On the contrary, if A is Schur stable, i.e., $\rho(A)<1$, then for all real diagonal D with $|D|\le I$, by the lemma 17.1 and lemma 17.2, and note that there is only one difference in sign between eigenvalues of A and $-A$, i.e. if $Ax=\lambda x, \lambda\in\sigma(A)$, then $-Ax=-\lambda x$. So we can get

$$\rho(AD)\le\rho(|AD|)\le\rho(|A||D|)\le\rho(|A|I)=\rho(|A|)=\rho(\pm A)<1$$

That is, the A is Schur D-stable. This completes the proof.

Theorem 17.2 *Let A be a generic* $n\times n$ *matrix with operator norm, then A is Schur D-stable if and only if A is vertex stable. i.e. that Schur D-stability and vertex stability of the A is equivalent.*

Proof The necessity is obvious. Indeed, if a square matrix A is Schur D-stable, i.e., $\rho(AD) < 1$ for all real diagonal D with $|D| \leq I$, particularly, for $|D| = I$ also holds.

To prove sufficiency, if a square matrix A is vertex stable, i.e., $\rho(AD) < 1$ for all real diagonal D with $|D| = I$, then for all real diagonal $\bar{D}$ with $|\bar{D}| \leq I$, by virtue of lemma 17.1 and lemma 17.2, we have

$$\rho(A\bar{D}) \leq \rho(|A\bar{D}|) \leq \rho(|A||\bar{D}|) \leq \rho(|A|I) = \rho[|A||D|] = \rho(\pm AD), \qquad (17.3.1)$$

we denote $B = AD$, Because of that there is only one difference in sign between eigenvalues of B and $-B$, so we have $\rho(B) = \rho(-B)$, i.e., $\rho(+AD) = \rho(-AD)$, thus, if A is vertex stable, then for all real diagonal D with $|D| = I$,we have

$$\rho(AD) = \rho(+AD) = \rho(-AD) < 1, \qquad (17.3.2)$$

Combining (17.3.1, 17.3.2), we can get $\rho(A\bar{D}) < 1$, that is, the A is Schur D-stable. The theorem is proved.

Theorem 17.3 *Let A be a generic $n \times n$ matrix with operator norm, then A is Schur stable if and only if A is vertex stable. i.e. the Schur stability and vertex stability of the A is equivalent.*

Proof The necessity is obvious. Indeed, suppose square matrix A is Schur stable, i.e., $\rho(A) < 1$, then for all real diagonal D with $|D| = I$, by the Lemma 17.2, and note that there is only one difference in sign between eigenvalues of A and $-A$, we can get $\rho(AD) \leq \rho(|AD|) \leq \rho(|A||D|) = \rho(|A|I) = \rho(|A|) = \rho(\pm A) < 1$, that is, A is vertex stable.

To prove sufficiency. Suppose square matrix A is vertex stable, i.e., $\rho(AD) < 1$ for all real diagonal D with $|D| = I$, then by the Lemma 17.2 and properties of spectral radius, we have

$$\rho(A) = \rho(AI) = \rho(A|D|) = \rho[A(\pm D)] = \rho[(\pm A)D] = \rho(\pm AD), \qquad (17.3.3)$$

We denote $B = AD$, Because of that there is only one difference in sign between eigenvalues of B and $-B$, so we have $\rho(B) = \rho(-B)$, i.e., $\rho(+AD) = \rho(-AD)$, thus, if A is vertex stable, then we have

$$\rho(AD) = \rho(+AD) = \rho(-AD) < 1, \qquad (17.3.4)$$

Combining (17.3.3, 17.3.4), we can get $\rho(A) < 1$, that is, the A is Schur stable. The proof is completed.

Theorem 17.4 *Let A be a generic $n \times n$ matrix with operator norm. If $AD \geq 0$ for all real diagonal D with $|D| \leq I$, then A is Schur D-stable if and only if $I - AD$ is nonsingular and $(I - AD)^{-1} \geq 0$.*

Proof The necessity is obvious. We prove this only for sufficiency. If A satisfy $AD \geq 0$ for all real diagonal D with $|D| \leq I$, and $I - AD$ is nonsingular and

$(I - AD)^{-1} \geq 0$. Then for any eigenvalue λ of AD, it's eigenvector $x \neq 0$, we have $ADx = \lambda x$, and note that consistency, we can obtain

$$\begin{aligned} AD|x| = |AD||x| \geq |\lambda||x| &\Rightarrow (I - AD)|x| \leq (1 - |\lambda|)|x| \\ &\Rightarrow |x| \leq (I - AD)^{-1}(1 - |\lambda|)|x| \end{aligned}$$

Because of $x \neq 0$ and $(I - AD)^{-1} \geq 0$, so $|\lambda| < 1$. By the arbitrariness of λ, we know that $\rho(AD) < 1$, i.e., A is Schur D-stable. This completes the proof.

17.4 Conclusion

In this paper, we discuss the Schur D-stability problem of matrices. By using the matrix eigenvalue theory and spectral radius approach, four new necessary and sufficient conditions (criteria) for guaranteeing the Schur D-stability of matrices are derived and three equivalence relations between the Schur D-stability, Schur stability and vertex stability are provided.

Acknowledgments This work was supported in part by NNSFC under Grant Nos. 70271006, 60674107.

References

1. Wang H, Wang C (1998) The relations for some stable matrices. Chinese J Eng Math 15(1):109–112. (in Chinese)
2. Yu L (2001) Robustness of D-stability for linear systems. Acta Automatica Sinica 27(6):860–862. (in Chinese)
3. Henrion D et al (2001) D-stability of polynomial matrices. Int J Control 74(8):845–856
4. Hu G, Xie X-S (2003) Robust control for uncertain discrete-time singular systems with D-stability. Acta Automatica Sinica 29(1):142–147. (in Chinese)
5. Cain BE (1984) Inside the D-stable matrices. Linear Alg Appl 56:237–243
6. Fleming R et al (1998) On schur D-stable matrices. Linear Alg Appl 279(1-3):39–50
7. Su T-J, Shyr W-J (1994) Robust D-stability for linear uncertain discrete time-delay systems. IEEE Trans Automat Contr 39(2):425–428
8. Berman A (1984) Characterization of a cyclic D-stable matrices. Linear Alg Appl 58:17–31
9. Han J, Qiu J, Liu Z (2006) Schur D-stability analysis for interval matrices. Adv Syst Sci Appl 1:57 –61

Chapter 18
Aerial Cooperative Combination Formation Method of Manned/Unmanned Combats Agents

Lujun Wan and Peiyang Yao

Abstract Aerial cooperative combination formation (ACCF) become a new problem when the manned/unmanned combat agents task coalition performing the air–air task. The problem of ACC formation focuses on the method to allocate the weapon units, guidance units and targets in a time-slack in order to maximize the effectiveness of the task coalition. According to the operational context, a constrained optimization model was proposed for the problem and a novel nested genetic algorithm (NGA) was designed to solve the model. In NGA, the outer-loop of GA searched for the best weapon-guidance combination and the inner-loop of GA searched for the best weapon-target allocation. Aiming to discrete characteristics of the problem, the coding rules, crossover operators and mutation operators was specially designed. Experimental results show that the proposed algorithm can solve the three-dimensional ACCF problem effectively.

Keywords Manned/unmanned combat agents · Cooperative combination · Nested genetic algorithm · Constraint optimization

18.1 Introduction

In future distributed networked operations, the unmanned combat agent (UCA) with integrative function of perceive, decision and action will be utilized more and more widely. The UCA, for example land robots, unmanned sea vessel and unmanned aerial vehicles, with the advantage of high manipulation, low cost and stealth [1, 2], can execute tasks in extremely execrable survival environment. In this way, the operator can avoid of dangerous. In the recent past, considerable

L. Wan (✉) · P. Yao
School of Information and Navigation, Air Force Engineering University,
No 1, FengHao Road, LianHu, Xi'an, China
e-mail: pandawlj@126.com

Z. Sun and Z. Deng (eds.), *Proceedings of 2013 Chinese Intelligent Automation Conference*, Lecture Notes in Electrical Engineering 255,
DOI: 10.1007/978-3-642-38460-8_18,

efforts focus on the problem of platform assign to target task and the scheduling demanded in multi-vehicle systems [3, 4]. But the UCA has to affront the shortage of intelligent and autonomous control when it executes dependent mission. Sometimes the mission or operation is too complex to afforded by UCA independently, through cooperating of unmanned combat agent and manned combat agent (MCA) mixed grouping, the MCA/UCA formation would improve mission executing effectiveness in dynamic environment [5].

With high performance of air sensor, cooperative guidance become an important tactical way to enhance battle survivability. In future, when MCA and UCA cooperative execute mission or task, UCA was supposed to make aerial guidance for the weapon emitted by MCA, in such way, it can guarantee MCA's safety. In order to realize above assumption, it demand to form an time-slack combat combination with one target, one MCA and one UCA, viz. aerial cooperative combination formation (ACCF) problem. The high countering aerial environment require the combat agents' quick and consistency decision making. In such a dynamic and stochastic setting, agents are difficult to seek consensus based distributed communication owing to low cognitive and deficient negotiation mechanism. Accordingly, for the sake of satisfying the time effect, it is necessary to set an aerial coordinate center to charge formation tactical decision making. Select one MCA to be the coordinate center, which establish centralized planning in a decision making time-slack considering the whole coalition's benefit. ACCF problem focus on the method to assign the appropriate target to the appropriate MCA while fix on the cooperation relationship between MCA and UCA that providing guidance for the former. Substantially, the ACCF problem belongs to task allocation category, we can figure out it by modeling and optimizing.

18.2 The Constraint Optimized Model of ACCF Problem

For convenient, it need to give several hypothetical conditions before modeling.

Assumption 1: After weapon emit, there is a smooth connected process to deal with the MCA and UCA cooperative guidance, assume that the success probability of smooth connect is 100 %;

Assumption 2: The combat situation can be shared where and who necessary, then assume the constructed command and control communication relation structure between multiple agents can realize points to points and blackboard communication freely and reliably;

Assumption 3: Time axes of combat course can be divided into several decision making time-slack window, MCA and UCA satisfies launch and guidance demand in single decision making time-slack window respectively;

Assumption 4: The weapon range and guidance device can both cover targets, and assume the situation information obtain by MCA and UCA is consistent.

18.2.1 Objective Function

Note $T = \{t_1, t_2, \cdots, t_{N_T}\}$ is a set of counter targets that performing intercept task, $M = \{m_1, m_2, \cdots, m_{N_M}\}$ is a set of MCA and $A = \{a_1, a_2, \cdots, a_{N_A}\}$ is a set of UCA, the former charging for launching weapon, and the latter guiding for the MCA. ACC_i denote as $\langle t_i, m_j, a_k \rangle$ is an aerial cooperative combination, where $1 \le i \le N_T$, $1 \le j \le N_M$, $1 \le k \le N_A$, in cooperative combination ACC_i, the MCA m_j charge for attacking target t_i and the UCA a_k provide guidance control for a weapon emitted by m_j. Air counter situation of multiple combat agents can be depicted by Fig. 18.1.

In Fig. 18.1, LOS is target line, D_{ji} represents the space distance between MCA m_j and target t_i, x_j and v_j is respectively the axes or velocity vector of m_j, ε_{ji} represents the abaxial emission angle of target t_i relative to m_j. Then thr_{ij} which is threaten level of target t_i versus MCA m_j can be defined as follow:

$$thr_{ij} = \left(\omega_1 P_{ij}^D P_{ij}^\varepsilon + \omega_2 P_{ij}^v + \omega_3 P_{ij}^E\right) \cdot \xi_{ij} \tag{18.1}$$

Here, ω_1, ω_2 and ω_3 are weight coefficient, $\omega_1 + \omega_2 + \omega_3 = 1$; the distance threaten variable $P_{ij}^D \in [0, 1]$, the angle threaten variable $P_{ij}^\varepsilon \in [0, 1]$, the velocity threaten variable $P_{ij}^v \in [0, 1]$, the combat capability threaten variable $P_{ij}^E \in [0, 1]$, and ξ_{ij} is intercept intention probability of target t_i versus MCA m_j.

The MCA influenced by UCA come down to two points: kill probability of one weapon assault one target and emitted weapon number. For a match combination $ACC_i = \langle t_i, m_j, a_k \rangle$, suppose the independent probability of Nth weapon assault target t_i is $p_{ijk}(N)$, for instance, if the weapon number is 3, the probability ps_{ijk} of target t_i successfully destroy by combination of m_j and a_k as follows:

$$ps_{ijk} = \left[1 - \left(1 - p_{ijk}(1)\right) \times p_{ijk}(2)\right] \times p_{ijk}(3) \tag{18.2}$$

Matching the ACC to make the task coalition combat efficiency or the loss of oppose maximize, so the objective function is built as follows:

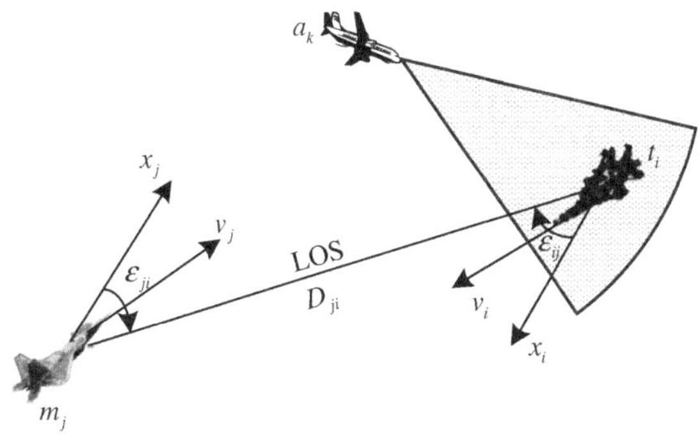

Fig. 18.1 Air counter situation of combat agents

$$\max J = \sum_{i,j}\left\{thr_{ij} \times \left[1 - \prod_{k=1}^{N_A}\prod_{j=1}^{N_M}(1 - ps_{ijk})^{x_{ijk}}\right]\right\} \quad (18.3)$$

where $x_{ijk} \in \{0, 1\}$ is decision making variable, $x_{ijk} = 1$ represents the t_i, m_j, a_k make up of temporary ACC_i, otherwise the combination isn't existence.

18.2.2 Constraint Conditions

1. Each target must assign an MCA, $\forall t_i \in T$, if MCA $m_j \in ACC_i$, then must assign a guidance UCA to t_i and m_j to cooperate, the Eq. (18.4) should tenable:

$$\sum_{k=1}^{N_A}\sum_{j=1}^{N_M} x_{ijk} = 1 \quad \& \quad \sum_{k=1}^{N_A} x_{ijk} = 1 \quad (18.4)$$

2. The number of targets assigned to MCA m_j no more then the number of weapons N_W; Each UCA can provide guidance for several weapons, which may emitted by one MCA or different MCA. So the weapon number with UCA's guidance is limited to the maximal guidance capacity N_F; Assume the guidance total capacity in model can meet each weapon's requirement of cooperative guidance, see in Eq. (18.5):

$$\sum_{k=1}^{N_T} x_{ijk} \le N_W \quad \& \quad \sum_{i=1}^{N_T}\sum_{j=1}^{N_M} x_{ijk} \le N_F \quad \& \quad N_M \cdot N_W \le N_A \cdot N_F \quad (18.5)$$

18.3 Designing of Optimized Algorithm

18.3.1 Model Solution Analysis

ACCF mathematical model is three-dimension decision making nonlinear integral programming, which is belong to generally nondeterministic polynomial time NP-hard. The search spaces will exponential increased along with problem scale added, generally it can not find the best solution no other than to employ enumerated method. In modern aerial combat, compute and obtain ACCF problem solution in seconds is not practical because of dynamic and complicated battle environment, but we can get satisfied solution or inferior solution in control problem scale or limited time. To solve the ACCF constraint optimized problem, firstly, it should deal with constraint conditions that transform constraint model into non-constraint problem; secondly, it need to choose appropriate algorithm which can locate the best field of fitness function. GA is especially suitable for discontinuous optimization problems with huge search, and it a suitable candidate for our problem [6, 7]. In order to solve the problem more efficiently, one intuitive way is to separate the problem into two sub problems: targets assignment and guidance platform matching. Therefore, we employ an Nested Genetic Algorithms (NGA) to solve above problem by internal and external GA respectively.

The NGA consists of two loops, namely: (1) the outer-loop GA (OLG); and (2) the inner-loop GA (ILG). The OLG seek to find best match of m_j and a_k, then the ILG is responsible for finding optimized allocation of m_j and t_i. Concrete steps of method as follows:

Step 1: Initialization of all the parameter uploaded by MCA and UCA, create threaten level matrix of all the targets and kill probability matrix of MCA and UCA match combination;
Step 2: Utilized the mode of extended natural number coding, initialize of outer-loop population Y and inner-loop population X separately which satisfied the constraint condition;
Step 3: Carry out arithmetic crossover and nonuniform mutation operators in population Y and X, then make use of list identify policy to decode chromosome;
Step 4: Calculate the fitness value of population Y and X, reserve elitist individual based on the rank of its fitness value, and carry out select operators in remain populations to create new generation;
Step 5: Judge whether the algorithm satisfied maximal iterative times or not, if reached, stop optimization, output currently optimized ACCF scheme; else continue algorithm running.

18.3.2 Nested Genetic Algorithm Design

18.3.2.1 Chromosome Representation

A chromosome representation scheme is determined by the structure of the problem. Good representation can greatly improve the performance of GA. For the m_j and a_k matching problem, which is basically a combinatorial problem, neither binary nor floating number representations are efficient, since there is too much redundancy in the search space. Specifically, a chromosome coding mode satisfied constraint condition is employed in OLG, chromosome code C_j expressed as decimal number, $N_M \times N_W$ is code length, $i = 1, 2, \ldots, N_M$, $j = 1, 2, \ldots, N_W$. The initialization of the population is *pop_size*, single locational code C_j is randomly generated from 1 to N_A. The ILG code is similar as OLG, and the code length is also $N_M \times N_W$, single locational code C'_j is randomly generated from 1 to N_T.

18.3.2.2 Decoding Policy and Fitness Function

If some of the randomly generated chromosomes are not feasible, it need to decode the infeasible chromosomes in order to obtain the code's preimage. We design the list label method to decode infeasible chromosomes into preimage, hence, we can calculate the fitness value. The decoding steps are as follows:

Step 1: Build a code label linker list→*List*, in *List* from head to end is corresponding to label of 1 to N_A, all the labels are initialization to be 0;

Step 2: Polling the *pop_size* chromosomes randomly, read each code of chromosome, once the number appear, the corresponding label of *List* value $List\_num(k) + 1$; if $List\_num(k) \geq N_F$, find the minimum and shortest Hamming range label point in *List* to instead of the former label value, if $List\_num(k) < N_F$, read the next point code.

Step 3: Repeating above steps, until decoding all the chromosomes.

During each generation, chromosomes are evaluated using a measure of fitness. The following three major steps are included in the evaluation phase: (1) convert chromosomes to the MCA-task assignment matrix and the MCA-UCA matching matrix; (2) obtain the MCA-task assignment matrix and the MCA-UCA matching matrix by running the ILG and OLG respectively; (3) calculate the objective function values for these tentative solutions. OLG and ILG utilize the same decoding policy, then fitness value can be computed by Eq. (18.6):

$$f_{fitness}(x) = \sum_{i,j} \left\{ thr_{ij} \times \left[1 - \prod_{k=1}^{N_A} \prod_{j=1}^{N_M} \left(1 - ps_{ijk}\right)^{x_{ijk}} \right] \right\} \tag{18.6}$$

18.3.2.3 Crossover and Mutation Operators

There are two basic considerations when designing crossover operators, namely: (1) make fewer changes when crossing over so as to inherit the parents' features as much as possible; and (2) make more changes when crossing over so as to explore a new pattern of allocation and thereby enhance the search's ability to find a global optimum. The mutation operators alter one parent by changing one or more variables in some way, or by some random amount, to form an offspring. According to the code structure characteristic of chromosomes, the crossover operators we use for OLG and ILG are multipoint arithmetic crossover, and the mutation operators we employ in our algorithm are nonuniform random mutation.

1. Multipoint Arithmetic Crossover

Suppose$0 \leq a \leq 1$, $C^i(1)$ and $C^i(2)$ are the randomly selected *i*th genes of two parents, $C^{i+1}(1)$ and $C^{i+1}(2)$ are the *i*th genes of two children, suppose κ is a random point of ith chromosomes' crossover quarter, then the ith genes of two children are given by:

$$C_{\kappa}^{i+1}(1) = \begin{cases} \lceil aC_{\kappa}^{i}(1) + bC_{\kappa}^{i}(2) \rceil, & C_{\kappa}^{i}(1) > C_{\kappa}^{i}(2) \\ \lfloor aC_{\kappa}^{i}(1) + bC_{\kappa}^{i}(2) \rfloor, & \text{otherwise} \end{cases} \tag{18.7}$$

$$C_{\kappa}^{i+1}(2) = \begin{cases} \lceil bC_{\kappa}^{i}(1) + aC_{\kappa}^{i}(2) \rceil, & C_{\kappa}^{i}(1) > C_{\kappa}^{i}(2) \\ \lfloor bC_{\kappa}^{i}(1) + aC_{\kappa}^{i}(2) \rfloor, & \text{otherwise} \end{cases} \tag{18.8}$$

where $b = 1 - a$, $\lceil x \rceil$ is the smallest integer greater than or equal to x, and $\lfloor x \rfloor$ is the largest integer less than or equal to x.

Follow above method, reassembling the different chromosomes' crossover points randomly by arithmetic crossover operator.

2. Nonuniform Random Mutation

To enhance the local search capability of NGA algorithm, randomly selects variable $B_{i\kappa}(l)$ of the ith genes' individual $C_i(l)$ and sets it equal to a nonuniform random integer number (u_1, u_2), then the new mutation point code are obtained:

$$B'_{i\kappa}(l) = \begin{cases} \lceil B_{i\kappa}(l) + (u_2 - B_{i\kappa}(l))f(G) \rceil, \ r_1 < 0.5 \\ \lfloor B_{i\kappa}(l) - (u_1 + B_{i\kappa}(l))f(G) \rfloor, \ r_1 \geq 0.5 \end{cases} \tag{18.9}$$

where, $f(G) = \left(r_2 \cdot \left(1 - {}^{G}\!/_{G_{\max}}\right)\right)^s$.

Where r_1 and r_2 are uniformly distributed random numbers in $[0, 1]$, G is the current generation, $G_{\max}$ is the maximum number of generations, and s is a shape parameter.

18.3.2.4 Select Policy

There are several schemes for determining and assigning the selection procedure, e.g. roulette wheel selection, scaling selection, ranking selection, and crowding selection for probabilistic selection; and tournament selection and elitist models for deterministic. We adopted the normalized geometric ranking approach associated with elitist policy as the selection procedure for the NGA. The chromosomes are ordered according to their fitness value. The new population is produced in such way that part of it from the old population with the highest fitness value, and the remaining is from other old population crossover and mutation operators. The probability of selecting a chromosome is based on its rank:

$$P(r) = \frac{q}{1 - (1 - q)^{pop\_size}} \cdot (1 - q)^{r-1} \tag{18.10}$$

where q is the probability of selecting the best individual, r is the rank of the individual, pop_size is the population size.

18.3.2.5 Termination Criterion

When the GA reaches a prespecified number of generation, generally, the ILG needs less number of generation than the OLG. Output the current best chromosome individual which is regarded as optimized solution.

18.4 Illustrative Simulation

In this section, the algorithm is implemented using MATLAB, while all the computational results are obtained on an Intel Pentium IV 3.0 GHz computer. We consider an illustrative example of designing an Aerial cooperative combination formation process, in which aerial combat there are 4 MCA execute fire attacking and 5 UCA charge for guidance, viz. $N_M = 4$, $N_A = 5$. Set the target number $N_T = 6$, the weapon number of each MCA $N_W = 2$, the maximal guidance capacity of UCA $N_F = 2$, penalty factor is 2. The parameters of OLG and ILG are same, set as: *pop_size* = 30, $q = 0.1$, $a = 0.25$.

Utilize the presented code, crossover, mutation and decode policy, keep the population size the same for each iteration, consider 3 different cases: (a) 50 % of the new population is from the best 50 % of the previous population, 20 % of the new population is from crossover, 30 % of the new population is from mutations; (b) 30 % of the new population is from elitism, 30 % is from crossover, 40 % is from mutation; (c) 10 % of the new population is from elitism, 40 % is from crossover, 50 % is from mutation. Run NGA algorithm 10 times, set maximum generations as 50, we can obtain the calculated result evolution curve in Fig. 18.2.

By generating the new population synthetically, we can maintain both elitism and diversity in the population. However, the origin of new population in different proportion, the algorithm would represent vary performance. From Fig. 18.2, we notice that if the elitism proportion is relatively high, the NGA can possess good convergence except for finding global solution hardly (e.g. case a). If the reduce the elitism proportion, algorithm's convergence speed would slow down, whereas it can guarantee to search the global optimized solution (e.g. case c). A good algorithm parameters' setting has to achieve a balance between convergence and global searching. For general efficiency, the parameters of algorithm set as case b is up to the standard mostly. The NGA can find the best solution in about 40 generations, which testify it can figure out small or middle scale ACCF problem (Fig. 18.2, Table 18.1).

Assume that size of OLG population is n_1, size of ILG population is n_2, iterative times is m, when the NGA algorithm solve $(N_M \times N_W) \times N_T \times (N_A \times N_F)$ size ACCF problem model, the algorithm compute complexity compose of 5 segments. We notice that the worst compute complexity of proposed algorithm is quantity

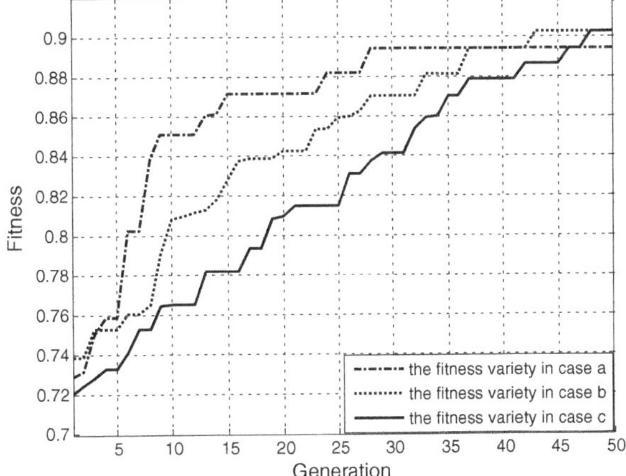

Fig. 18.2 The fitness of NGA algorithms in different case

Table 18.1 The optimized ACCF result calculated by NGA

MA	1		2		3		4	
Weapon	1	2	1	2	1	2	1	2
Target	3	6	2	5	6	1	4	1
UCA	1	1	2	5	3	3	2	4
Fitness		0.9024						

level of $O\left(n_1 \cdot N_T \cdot (N_M \times N_W)^2\right)$ or $O\left(n_2 \cdot (N_A \times N_F) \cdot (N_M \times N_W)^2\right)$, viz. the algorithm complexity is three cubed rank of problem size. So the algorithm run time increasing would not along with increasing of problem scale.

18.5 Conclusion

In this paper, we proposed aerial cooperative combination formation concept against a background of MCA/UCA cooperative air combat. In future, when MCA and UCA cooperates on executing mission or task, the UCA is supposed to make aerial guidance for the weapon emitted by MCA. Different from previous researched task/target assignment problem, the built ACCF model is dynamic three-dimension decision making nonlinear integral programming. Based on the mathematical model's characteristic, a novel nested genetic algorithm is designed. In order to solve the problem effectively, the coding rules, crossover operators, mutation operators and selecting policy are devised, which can decrease the temporal complicated degree and find the best solution in limited generations.

In actually weapon attacking process, the match of MCA and UCA must consider the problem of spationtemporal constraint, viz. the position of UCA and

MCA, the maximum guidance effect distance, antenna array direction, azimuth angle and elevation angle are fit for pointing and launching. Future work needs to focus on introducing spationtemporal constraint into ACCF modeling. Furthermore, extending the design approach to include flexibility and robustness criteria in the design process is also our critical job.

Acknowledgments This work was financially supported by the National Natural Science Foundation of China (70771157), Innovation Program of Air Force Engineering University Doctor Commission (KDY2011-002).

References

1. Shi ZF, Yang B, Liu HY (2010) Modeling and simulation of UCAV swarm cooperative task assignment. In: Proceedings of the third international conference on information and computing
2. Mohsen RG, Hossein H, Lucas C (2009) Discrete invasive weed optimization algorithm: application to cooperative multiple task assignment of UAVs. Joint 48th IEEE conference on decision and control, Shanghai
3. Laura H, Kelly C, Steven R (2010) Application of proper orthogonal decomposition and artificial neural networks to multiple UAV task. AIAA 2010–8439
4. Edison E, Shima T (2008) Genetic algorithm for cooperative UAV task assignment and path optimization. AIAA 2008–6317
5. Valenti M, Schouwenaars T, Kuwata Y (2004) Implementation of a manned vehicle-UAV mission system. AIAA 2004–5142
6. Yu FL, Tu F, Pattipati KR (2006) A novel congruent organizational design methodology using group technology and a nested genetic algorithm. IEEE Trans Syst Man, Cybern-Part A: Syst Hum 36(1):5–18
7. Mu L, Xiu B X, Huang JC (2009) The optimization design of C2 organization communication network based on nested genetic algorithm. In: Proceedings of 8th international conference on machine learning and cybernetics, Baoding

Chapter 19
Finite-Time Adaptive Dynamic Surface Control of Dual Motor Driving Systems

Dongwu Li, Guofa Sun, Xuemei Ren and Xiaohua Lv

Abstract This paper is focused on finite-time adaptive control of dual motor driving servo system. By introducing a set of alternative state variables and the corresponding transform, a state space description of dual motor driving servo system is yields. Based on the obtained state space model of dual motor driving servo system, a dynamic surface control (DSC) scheme is proposed, which is able to eliminate the "explosion of complexity" problem inherent in traditional backstepping design and guarantee the finite convergence performance of the tracking errors. Simulation results are given to illustrate the effectiveness of the proposed approach.

Keywords Dual motor driving servo system · Finite-time convergence · Dynamic surface control · Neural network (NN)

19.1 Introduction

With the rapid development of industry, dual motor driving servo system is required in many applications. Like most of the mechanical systems, there exists non-smooth nonlinear parts such as gears to transmit motor torque to the load.

D. Li (✉) · G. Sun · X. Ren
School of Automation, Beijing Institute of Technology,
5 South Zhongguancun Street, Haidian, Beijing, China
e-mail: lidongwu2002@163.com

G. Sun
e-mail: sungf@bit.edu.cn

X. Ren
e-mail: xmren@bit.edu.cn

D. Li · X. Lv
Institute 23 of China Aerospace Science and Industry Corporation,
52 Yongding Road, Haidian, Beijing, China
e-mail: lv_xiaohua@yahoo.cn

Z. Sun and Z. Deng (eds.), *Proceedings of 2013 Chinese Intelligent Automation Conference*, Lecture Notes in Electrical Engineering 255,
DOI: 10.1007/978-3-642-38460-8_19,

Such nonlinear parts induce dead zone, backlash or hysteresis, which negatively affect control performance. Although there are lots of controller design techniques with widely applications, few papers have been published on dual motor drive servo systems based on these techniques.

Itoh [1] dealt with a control technique of eliminating the transient vibration of a twin-drive geared mechanical system with backlash. Based on dual motor driving servo system with pre-loaded torque bias, backstepping was studied by Zhao and Hu in [2] and a function link neural network (FLNN) expended by trigonometric polynomials was introduced in [3]. In [4], Zhao and Zhou proposed a model for dual motor driving servo system with unknown backlash and designed an adaptive controller with state feedback. Gawronski et al. [5] explained the shaping principles of the circuit and show how the circuits can be modified to improve the antenna dynamics under external disturbances. It is not difficult to see that control approaches applied to dual motor driving servo systems are very limited. Even the methods have been tried to use are bothered by the controller complexity.

Recently, much research work have focused on adaptive DSC method. It is proposed by Yip in [6] as a new algorithm for adaptive backstepping control of nonlinear uncertain systems. Following this idea, it is applied to strict nonlinear systems with dead zone in [7], time delay in [8] or in other forms in [9, 10]. In this paper, we have applied the adaptive DSC to dual motor driving servo systems.

19.2 Problem Formulation

Consider the following dual motor driving servo system with backlash described by dead zone

$$\begin{cases} J_i\ddot{\theta}_i + b_i\dot{\theta}_i = u(t) - \tau_i(t) - (-1)^i\omega \\ J_m\ddot{\theta}_m + b_m\dot{\theta}_m(t) = \Sigma_{i=1}^2\tau_i(t) \end{cases} \tag{19.1}$$

where θ_i, $(i = 1, 2)$ and θ_m represent the angle of driven motors and the load, $\dot{\theta}_i(i = 1, 2)$ and $\dot{\theta}_m$ are the velocity of the motors and the load, J_i and b_i are the inertia and friction coefficient of the motors, J_m and b_m are the inertia and friction coefficient of the load, u is the system input torque, ω is torque bias, $\tau_i(i = 1, 2)$ is the transmission torque of the motors and the load.

Under the effect of backlash nonlinearity of the gears, τ_i can be expressed as:

$$\tau_i(t) = kf(z_i(t)) + cf(\dot{z}_i(t)) \tag{19.2}$$

where k and c are the torsional coefficient and the damping coefficient, $f(z_i(t))$ is the deadzone function, which can be expressed as:

$$f(z_i(t)) = \begin{cases} z_i(t) + \alpha, & z < -\alpha \\ 0, & -\alpha < z < \alpha \\ z_i(t) - \alpha, & z > \alpha \end{cases} \tag{19.3}$$

where $z_i(t) = \theta_i(t) - \theta_m(t)$. Because the dead zone is not differentiable, it is impossible to derive the control signal. Motivated by the work of [4], we instead the dead zone model by a smooth, continuous and differentiable function as follow:

$$f(z_i(t)) = z_i - \alpha\left(\frac{2}{1 - e^{-rz_i}} - 1\right). \tag{19.4}$$

Then the transmission force can be expressed as:

$$\tau_i(t) = k\left(z_i - \alpha\left(\frac{2}{1 - e^{-rz_i}} - 1\right)\right) + c\dot{z}_i\left(1 - 2r\alpha\frac{e^{-rz_i}}{(1 + e^{-rz_i})^2}\right). \tag{19.5}$$

Before transforming the plant model into state space representation, we need the following notations:

$$x_1 = \theta_m, x_2 = \dot{\theta}_m, x_{3i} = z_i - \alpha\left(\frac{2}{1 + e^{-rz_i}} - 1\right), x_{4i} = \dot{z}_i\left(1 - 2r\alpha\frac{e^{-rz_i}}{(1 + e^{-rz_i})^2}\right) \tag{19.6}$$

Then we have $\tau_i(t) = kx_{3i} + cx_{4i}$. Choose $x_1,\ x_2,\ x_{3i},\ x_{4i}$ as the state variables, the state space expression of the plant is

$$\begin{cases} \dot{x}_1 = x_2, y = x_1, \dot{x}_2 = -a_1 x_2 + a_3 \sum_{i=1}^{2} \tau_i, \\ \dot{x}_{31} = x_{41}, \dot{x}_{41} = -a_{01}x_{41} - (a_{01} - a_1)x_2\rho + a_{21}(u_1(t) - \tau_1 + \omega)\rho - a_3\rho\sum_{i=1}^{2}\tau_i + 2r^2\alpha\dot{z}_1\bar{\omega}_1 \\ \dot{x}_{32} = x_{42}, \dot{x}_{42} = -a_{02}x_{42} - (a_{02} - a_1)x_2\rho + a_{22}(u_2(t) - \tau_2 - \omega)\rho - a_3\rho\sum_{i=1}^{2}\tau_i + 2r^2\alpha\dot{z}_2\bar{\omega}_2 \end{cases} \tag{19.7}$$

where $a_{0i} = \frac{b_i}{J_i}, a_1 = \frac{b_m}{J_m}, a_{2i} = \frac{1}{J_i}, a_3 = \frac{1}{J_m}, \rho_i = 1 - 8r\alpha\frac{e^{-rz_i}}{(1+e^{-rz_i})^2}, \bar{\omega}_i = \frac{e^{-r\alpha}(1-e^{-rz_i})}{(1+e^{-rz_i})^3}$.

To facilitate the controller design procedure, the dual motor drive servo system is redescribed as

$$\begin{cases} \dot{x}_1 = x_2 \quad \dot{x}_2 = f_2(x) + a_3 x_{4i}, \quad y = x_1 \\ \dot{x}_{31} = x_{41}, \quad \dot{x}_{41} = f_{41}(x_1) + a_{21}\rho_1 u_1(t) \\ \dot{x}_{32} = x_{42}, \quad \dot{x}_{42} = f_{42}(x_2) + a_{22}\rho_2 u_2(t) \end{cases} \tag{19.8}$$

where $f_2 = -a_1 x_2 + a_3 \sum_{i=1}^{2} \tau_i - a_3 x_{3i},\ f_{4i} = -a_{0i}x_{4i} - (a_{0i} - a_1)x_2\rho + a_{2i}\rho(-\tau_i + \omega) - a_3\rho\sum_{j=1}^{2}\tau_j + 8r^2\alpha\dot{z}_i\bar{\omega}_i$

The high order neural networks (HONN) will be used in this paper takes the form $f(X) = W^T S(X) + \varepsilon^*, X \in \Omega$ with $|\varepsilon^*| \leq \varepsilon_m$. Denote the components of $S(X)$ by $s_k(X),\ k = 1\ ,\ldots,L$ with

$$s(X_j) = \frac{a}{b + e(-X_j/c)} + d. \tag{19.9}$$

19.3 Controller Design

In this section, the adaptive neural network DSC technique proposed in [6] will be applied to the dual motor drive servo system described by (19.8). According to the relative order of the considered model, the recursive design procedure contains 3 steps for each driving motor.

Case 1: In this case, an adaptive robust finite-time DSC control scheme is designed for motor 1 in detail.

Step 1: At this step, we consider the first equation in (19.8), i.e., $\dot{x}_1 = x_2$. Let $e_1 = x_1 - x_r$ which is called the error surface with $x_r$as the desired trajectory. Then the time derivative of e_1 along (19.8) is $\dot{e}_1 = x_2 - \dot{x}_r$. Choose a virtual control α_2 as follows

$$\alpha_2 = -k_1 e_1 - c_1 |e_1|^r sgn(e_1) + \dot{x}_r. \tag{19.10}$$

Introduce a new state variable z_2 and let α_2 pass through a first-order filter with time constant τ_2 to obtain z_2 as

$$\tau_2 \dot{z}_2 + z_2 = \alpha_2, z_2(0) = \alpha_2(0). \tag{19.11}$$

Step 2: Consider the second equation of plant (19.8) as $\dot{x}_2 = f_2(x) + a_3 x_{3i}$. Given a compact set $\Omega_x \in R^4$, let W_1^* and ε_1^* be such that $f_2(x) = W_1^{*T} S_1(x) + \varepsilon_1^*$ for $x = [x_1 x_2 x_{31} x_{41} x_{32} x_{42}]^T \in \Omega_x$ with $|\varepsilon_1^*| \leq \varepsilon_m$. Let $e_2 = x_2 - z_2$ be the second error surface. Then the time derivative of e_2 is derived to be $\dot{e}_2 = a_3 x_{41} + f_2 - \dot{z}_2$. Choose a virtual control α_{31} to be

$$\alpha_{31} = \frac{1}{a_3}\left(-\hat{W}_1^T S_1(x) - k_2 e_2 - c_2 |e_2|^r sgn(e_2) + \dot{z}_2\right) \tag{19.12}$$

where the estimation of W_1^* is updated as

$$\dot{\hat{W}}_1 = \Gamma_1 S_1(x) e_2 - \sigma_1 \Gamma_1 \hat{W}_1. \tag{19.13}$$

with any constant matrix $\Gamma_1 = \Gamma_1^T > 0$. Introduce a new state variable z_3 and let α_{31} pass through a first order filter as

$$\tau_3 \dot{z}_3 + z_3 = \alpha_3, z_3(0) = \alpha_3(0). \tag{19.14}$$

Step 3: The final control law for driving motor 1 will be derived in this step. The forth equation of controlled plant (19.8) is $\dot{x}_{41} = f_{41}(X_1) + a_{21}\rho_1 u_1(t)$. Define the third error surface e_{31} of motor 1 to be $e_3 = x_{41} - z_3$. Considering (19.8), its derivative is given as $\dot{e}_3 = f_{41} + a_{21}\rho_1 u_1(t) - \dot{z}_3$. Given a compact set $\Omega_{X_1} \in R^4$, let W_2^* and ε_2^* be such that $f_{41}(X_1) = W_2^{*T} S_2(X_1) + \varepsilon_2^*$. For any $X_1 = (x_1 x_2 x_{31} x_{41})^T \in \Omega_{X_1}$ with $\varepsilon_2^* \le \varepsilon_m$. Here, it is noted that, taking the synchronous speed error, we need the following definition $e_s = \dot{\theta}_1 - \dot{\theta}_2 = \dot{z}_1 - \dot{z}_2 = x_{41} - x_{42} = e_3 - e_4$. Where e_4 will be defined in later. Finally, let the control signal to be

$$u_1 = \frac{1}{a_{21}\rho_1}\left(-\hat{W}_2^T S_2(X_1) - k_{31}e_{31} - k_4 e_s - c_{31}|e_{31}|^r sgn(e_3) + \dot{z}_{31}\right) \tag{19.15}$$

where the estimation of W_2^* is updated as

$$\dot{\hat{W}}_2 = \Gamma_2 S_2(x) e_3 - \sigma_2 \Gamma_2 \hat{W}_2 \tag{19.16}$$

Case 2: In this case, an adaptive robust finite time DSC control scheme is designed for motor 2.

The first two steps are the same to case 1, i.e.

$$\begin{aligned} \alpha_2 &= -k_1 e_1 - c_1|e_1|^r sgn(e_1) + \dot{y}_r \\ \alpha_{32} &= \frac{1}{a_3}\left(-\hat{W}_1^T S_1(X) - k_2 e_2 - c_2|e_2|^r sgn(e_2) + \dot{z}_2\right) \end{aligned} \tag{19.17}$$

Step 3: Define the third error surface e_{32} of motor 1 to be $e_4 = x_{42} - z_3$.

Then the third coordinate of error space is $\dot{e}_4 = x_{42} + a_{22}\rho_2 u_2(t) - \dot{z}_3$. Given another compact set $\Omega_{X_2} \in R^4$, let W_3^* and ε_3^* be such that $f_{42}(X_2) = W_3^{*T} S_3(X_1) + \varepsilon_3^*$. For any $X_2 = [x_1 x_2 x_{32} x_{42}]^T \in \Omega_{X_2}$ with $\varepsilon_3^* \le \varepsilon_m$. Finally, let the control signal u_2 be as follows

$$u_2 = \frac{1}{a_{22}\rho_2}\left(-\hat{W}_3^T S_3(x_1) - k_{32}e_4 + k_4 e_s - c_4|e_4|^r sgn(e_4) + \dot{z}_{32}\right) \tag{19.18}$$

where $\hat{W}_3$ is updated to be

$$\dot{\hat{W}}_3 = \Gamma_3 S_3(X_2) e_4 - \sigma_3 \Gamma_3 \hat{W}_3 \tag{19.19}$$

19.4 Stability Analysis

In this section, two theorems are provided to the finite time stability of all signals and finite-time stability of system (19.8).

Lemma *Suppose that there exists continuous, positive definite function $V(t)$ satisfying differential inequality as*

$$\dot{V}(t) + \alpha V(t) + \beta V^{\xi}(t) \le 0,\ t \ge t_0,\ V(t_0) \ge 0 \tag{19.20}$$

where $\alpha, \beta > 0,\ 0 < \zeta < 1$ *are constants. Then, for any given* t_0, $V(t)$ *satisfies the for linequality:* $V^{1-\varsigma} \leq (\alpha V^{1-\varsigma}(t_0) + \beta)\exp^{-\alpha(1-\varsigma)(t-t_0)} - \beta, t_0 \leq t \leq t_s$ *and* $V(t) \equiv 0,\ \forall\ t \geq t_s$ *with* t_s *given by*

$$t_s = t_0 + \frac{1}{\alpha(1-\varsigma)} \ln \frac{\alpha V^{(1-\varsigma)}(t_0) + \beta}{\beta}. \tag{19.21}$$

Theorem 1 *Consider the closed loop dual motor driving servo system described by* (19.8) *with the final controller signals* (29), *the modified virtual stabilizing functions* (30) *and parameter adaptive laws* (19.13), (19.16) *and* (19.19). *If the parameters* $\delta_i,\ i = 1,2,3$ *are chosen as* (19.22), *the closed-loop error system can converge to its equilibrium in finite time. Moreover, since the equilibrium is 0, the system output signal of* (19.8) *will track the reference signal precisely in finite time.*

$$\delta_1 \geq \varepsilon_m,\quad \delta_2 \geq a_3\varepsilon_m + \eta_1,\quad \delta_3 \geq \eta_1,\quad \delta_4 \geq \eta_1,\quad |\varepsilon_i| \leq \varepsilon_m,\quad i = 1,\ 2 \tag{19.22}$$

$$\begin{aligned} \alpha_2 &= -k_1 e_1 - c_1|e_1|^r sgn(e_1) - \delta_1 sgn(e_1) + \dot{y}_r \\ \alpha_{3i} &= \frac{1}{a_3}\left(-\hat{W}_1^T S_1(x) - k_2 e_2 - e_1 e_2 - c_2|e_2|^r sgn(e_2) - \delta_2 sgn(e_2) + \dot{z}_2\right),\quad i = 1,2 \end{aligned} \tag{19.23}$$

$$\begin{aligned} u_1 &= \frac{1}{a_{21}\rho_1}\left(-\hat{W}_2^T S_2(X_1) - k_3 e_3 - e_2 e_3 - c_3|e_3|^r sgn(e_3) - \delta_3 sgn(e_3) + \dot{z}_{31}\right) \\ u_2 &= \frac{1}{a_{21}\rho_2}\left(-\hat{W}_3^T S_3(X_2) - k_4 e_4 - e_2 e_4 - c_4|e_4|^r sgn(e_4) - \delta_4 sgn(e_4) + \dot{z}_{32}\right) \end{aligned} \tag{19.24}$$

Proof From (19.9), we can conclude that the sigmoid function $s_i(X)$ is bounded by $0 \leq s_i(X) \leq L_0, i = 1,\ldots,L_1$, with $L_0 = \max\left\{|\frac{a}{b}|, |\frac{a}{b+1} + d|\right\}$, therefore $s_i(X)$ is bounded by $\|s(X)\| \leq L_0\sqrt{L_1}$, where $\|.\|$ denotes the Euclidean norm of a vector, $S(X) = [s_1(X),\ldots,s_L(X)]^T$. Denote $\|W_F\| \leq \bar{W}$. According the property of Forensics norm, we can obtain the following in equation $\|\tilde{W}^T S(X)\|_F \leq \|\tilde{W}\|_F \|S(X)\|$.. Moreover, we have the following in equation as $\|\tilde{W}^T S(X)\|_F \leq \eta_1$ where $\eta_1 = \bar{W} L_0 \sqrt{L_1}$. According to the controller design procedure, the error dynamic system is

$$\begin{aligned} &\dot{e}_1 = x_2 - \dot{y}_r,\ \dot{e}_2 = a_3 x_{41} + f_2 - \dot{z}_2(t),\ \dot{e}_3 = f_{41} + a_{21}\rho_1 u_1(t) - \dot{z}_{31}(t), \\ &\dot{e}_4 = x_{42} + a_{22}\rho_2 u_2(t) - \dot{z}_{32}(t) \end{aligned}$$

Substituting the modified virtual stabilizing functions (27) and the final controller signals (19.24) into the above error system yields

$$
\begin{aligned}
\dot{e}_1 &= -k_1 e_1 - c_1|e_1|^r sgn(e_1) - \delta_1 sgn(e_1) + \epsilon_1, \\
\dot{e}_2 &= -\tilde{W}_1^T S_1(X) - k_2 e_2 - c_2|e_2|^r sgn(e_2) - \delta_2 sgn(e_2) + a_3\epsilon_2 \\
\dot{e}_3 &= -\tilde{W}_2^T S_2(X_1) - k_3 e_3 - c_3|e_3|^r sgn(e_3) - \delta_3 sgn(e_3) \\
\dot{e}_4 &= -\tilde{W}_3^T S_3(X_2) - k_4 e_4 - c_4|e_4|^r sgn(e_4) - \delta_4 sgn(e_4)
\end{aligned}
\tag{19.25}
$$

Consider the Lyapunov function candidate to be $V_1 = \frac{1}{2}\Sigma_{i=1}^4 e_i^2$. The derivative of V_1 is $\dot{V}_1 = \Sigma_{i=1}^4 e_i \dot{e}_i$. Substituting Eq. (19.25) into $\dot{V}_1$ yields

$$
\begin{aligned}
\dot{V}_1 &= -\Sigma_{i=1}^4 \left(k_i e_i^2 + c_i \parallel e_i \parallel^{r+1} + \delta_i \mid e_i \mid\right) + a_3\epsilon_2 e_2 - \Sigma_{j=1}^3 \tilde{W}_j^T S_j(X_{j-1}) e_j + \epsilon_1 e_1 \\
&\le -\Sigma_{i=1}^4 \left(k_i e_i^2 + \parallel e_i \parallel^{r+1} + \delta_i \parallel e_i \parallel\right) + \Sigma_{j=1}^3 \eta_1 \mid e_j \mid + \mid \epsilon_1 \parallel e_1 \mid + a_3 \mid \epsilon_2 \parallel e_2 \mid
\end{aligned}
\tag{19.26}
$$

Using Young's in equation yields $\dot{V}_1 \le -\Sigma_{i=1}^4 \left(k_i e_i^2 + c_i \parallel e_i \parallel^{r+1}\right) \le -2kV_1 - 2cV_1^{\frac{r+1}{2}}$ where $k = \min\{k_i\}, i = 1,2,3,4$ and $c = \min\{c_i\}, i = 1,2,3,4$. Then we obtain the inequation $\dot{V}_1 + \bar{k}V_1 + \bar{c}V_1^{\bar{r}} \le 0$, where $\bar{k} = 2k, \bar{c} = 2c$ and $\bar{r} = \frac{r+1}{2}$. According to Lemma 1, it can be concluded that the error dynamic system can converge to the equilibrium point within a finite time t_1 given by $t_1 = \frac{1}{\bar{k}(1-\bar{r})} \ln \frac{\bar{k}V_1^{1-\bar{r}}(t_0) + \bar{c}}{\bar{c}}$.

19.5 Simulation Studies

In order to show the superior tracking performance of the proposed scheme, three different control schemes, including the single motor driving PID control, the dual motor driving PID control, and the dual motor driving FARC control are performed in simulations. The plant parameter values are chosen as $J_1 = J_2 = 0.026, J_m = 0.9, b_1 = b_2 = 0.4, b_m = 0.7, k = 376, c = 30, w = 25, \alpha = 0.2$.

Then, three simulation examples are performed for fair comparison of three different controllers, single motor driving PID Control, dual motor driving PID Control and dual motor driving Finite-time Dynamic Surface Control (FDSC). The simulation results are depicted in Figs. 19.1, 19.2, 19.3. For fair comparison, the initial states of the system are set the same and the parameters of NN are chosen to be $(x(0), \dot{x}(0)) = (0,0), \Gamma = 0.05, a = 2, b = 10, c = 1$ and $\mathrm{d} = -10$.

The expression of PID controller is given as $u = k_P e + k_I \int_0^t e(t)dt + k_D \dot{e}$, where the design parameters $\mathrm{k}_P = 10$, $\mathrm{k}_I = 12$ and $K_D = 3$ are determined by using a heuristic tuning approach for a given reference signal, e.g. $x_d = 0.85\sin(2\pi t/5)$. For the FDSC controller, the parameters are chosen to be

$$
k_1 = 3, k_2 = 2.3, k_3 = 7, k_4 = 5, c_1 = 3.5, c_2 = 2.7, c_3 = 6, c_4 = 9, \delta_1 = 0.2, \\
\delta_2 = 0.3, \delta_3 = 0.2, \delta_4 = 0.16
$$

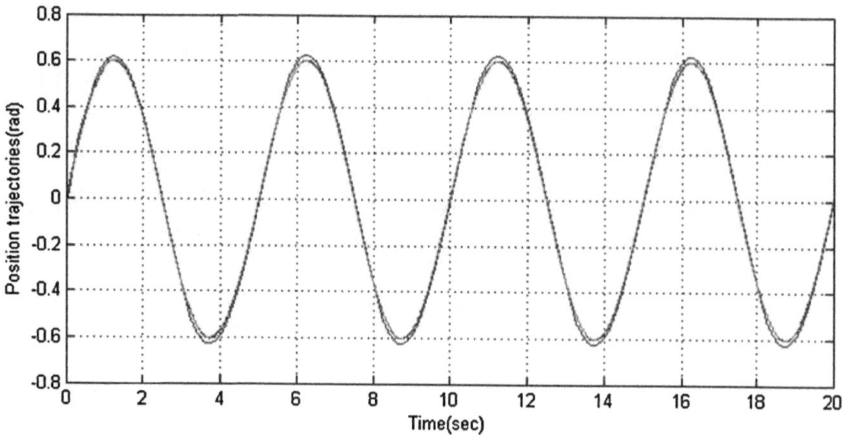

Fig. 19.1 Tracking performance of system output with three different methods

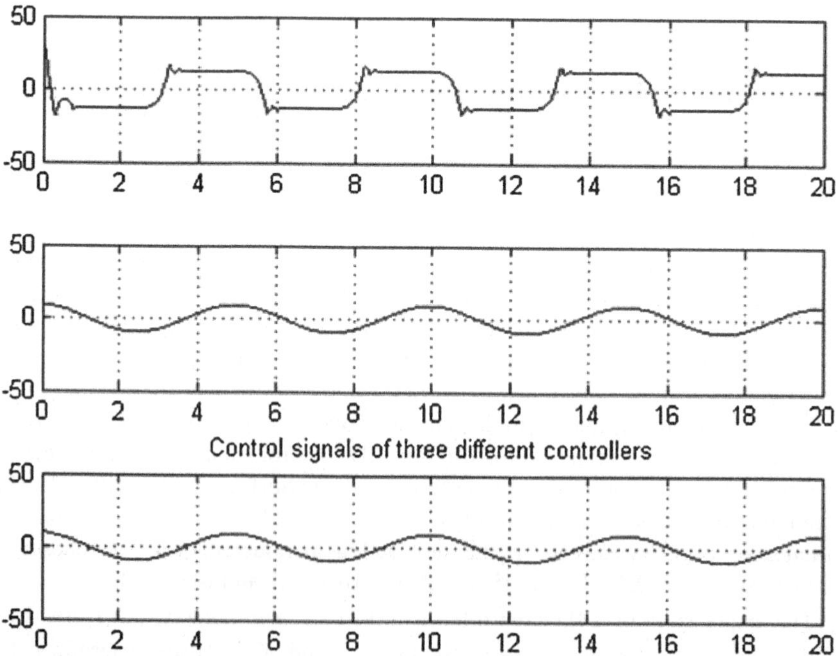

Fig. 19.2 Control signal of three control schemes

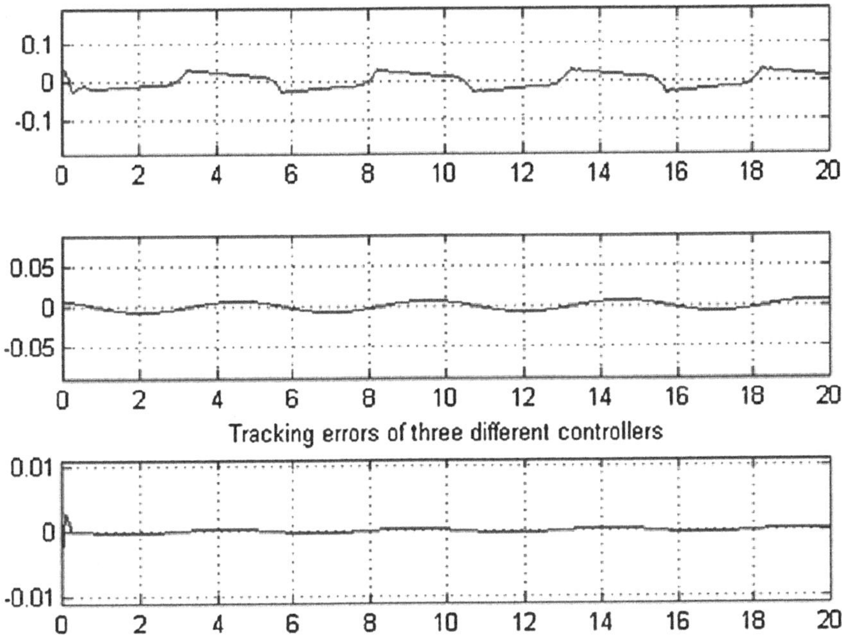

Fig. 19.3 Tracking error of the closed loop system with three control schemes

From the simulation results, it is clear that even through the backlash phenomenon is severe, the proposed adaptive NN controller guarantees both stability and good tracking performance of the closed loop dual motor driving serve system.

19.6 Conclusion

In this paper, finite-time adaptive dynamic surface control scheme is proposed for dual motor driving servo system. Smooth continuous function is utilized to approximate unknown dead zone caused by backlash nonlinearity. As a result, original plant is transformed into uncertain state space model and high order neural network is applied to approximate unknown functions. The tuning laws of HONN parameters are derived by DSC technique. The stability analysis is shown by Lyapunov method and all signals of the closed loop dual motor driving servo system are proved first to be uniformly ultimate bounded and then finite-time convergence. Simulation results illustrate our proposed control scheme.

Acknowledgments This work is supported by National Natural Science Foundation of China (No. 61273150 and No.60974046), and the Research Fund for the Doctoral Program of Higher Education of China (No. 20121101110029).

References

1. Itoh M (2008) Torsional vibration suppression of a twin-drive geared system using model based control. The 10th IEEE international workshop on advanced motion control, pp 176–181
2. Zhao G, Hu W (2005) Backlash nonlinearity control of servo system driven by two motors. Electric Drive 35(2):24–27 (In Chinese)
3. Fan W, Zhao G, Chen Q, Hu W (2006) Design of neural network controller of dual-motors driving servo system. Electr Mach Contr 10(3):260–268 (In Chinese)
4. Zhao H, Zhou X (2011) Backstepping adaptive control of dual-motor driving servo system. Control Theor Appl 28(5):745–751 (In Chinese)
5. Gawronski W, Beech-Brandf JJ, Ahlstrom Jr HG, Maneri E (2000) Torque-bias profile for improved tracking of the deep space network antennas. IEEE Antennas Propag Mag 42(6):35–45
6. Patrick Yip P, Karl Hedrick J (1998) Adaptive dynamic surface control: a simplified algorithm for adaptive backstepping control of nonlinear systems. Int J Contr 71(5):959–979
7. Aung O, Lin Y, Zhao Q (2010) Fuzzy based adaptive dynamic surface control for a class of nonlinear systems with unknown dead zone. In: Proceedings of the 2010 IEEE international conference on mechatronics and automation, pp 687–692
8. Wang D, Huang J (2005) Neural network-based adaptive dynamic surface control for a class of uncertain nonlinear systems in strict-feedback form. IEEE Trans Neural Netw 16(1):195–202
9. Yoo S, Park J, Choi Y (2007) Adaptive dynamic surface control for stabilization of parametric strict-feedback nonlinear systems with unknown time delays. IEEE Trans Autom Contr 52(12):2360–2365
10. Song B, Hedrick JK (2004) Observer-based dynamic surface control for a class of nonlinear systems: an LMI approach. IEEE Trans Autom Contr 49(11):1995–2001

Chapter 20
Hybrid Reinforcement Learning and Uneven Generalization of Learning Space Method for Robot Obstacle Avoidance

Jianghao Li, Weihong Bi and Mingda Li

Abstract This paper introduces a hybrid reinforcement learning algorithm for robot obstacle avoidance. This algorithm is based on SARSA (λ), and mix with the supervised learning. This hybrid learning algorithm can reduce the learning time obviously which is demonstrated by the simulations. In reinforcement learning, generalization of learning space is important for learning efficiency. An uneven generalization model is designed for improving the learning efficiency. The simulations show that the uneven model can not only reduce the learning time, but also the moving steps.

Keywords Reinforcement learning · Supervised learning · Space generalization · Obstacle avoidance

20.1 Introduction

As an important machine learning method, the reinforcement learning is widely used in many fields, such as autonomous control [1], behavior generation [2], and multi-agents collaboration [3], etc.

Since the foundation date of the reinforcement learning, several algorithm were designed, such as dynamic program, Monte Carlo etc. In recent years, some researchers still research the algorithm improving and parameters optimization

---

J. Li (✉) · W. Bi
College of Information Science and Engineering, Yanshan University,
The Key Laboratory for Special Fiber and Fiber Sensor of Hebei Province,
No. 438 West Hebei Avenue, Qinhuangdao City, China
e-mail: ljh@ysu.edu.cn

M. Li
Health School of Qinhuangdao, No. 400 Yanshan Avenue,
Qinhuangdao City, China

Z. Sun and Z. Deng (eds.), *Proceedings of 2013 Chinese Intelligent Automation Conference*, Lecture Notes in Electrical Engineering 255,
DOI: 10.1007/978-3-642-38460-8_20,

[4, 5]. In the area of generalization, the even tile coding method was widely used [6, 7]. However, its learning efficiency is low.

The contents of this paper are organized as follow: Sect. 20.2 introduces the robot model and learning task. Section 20.3 presents the hybrid reinforcement learning, and Sect. 20.4 introduces the generalization of the learning space. The simulations and analysis are shown in Sect. 20.5, followed by the conclusion.

20.2 Robot Model and Learning Task

As shown in Fig. 20.1, the mobile robot posture is presented as

$$\xi = (x, y, \theta)^{\mathrm{T}} \tag{20.1}$$

where x, y represents the position, and θ represents the wheels' orientation. The learning space size is 150×150, which consists of the mobile robot, the target and some obstacles. The target position is located at (140, 140). The prototype was presented in Ref. [8].

The mobile robot must move or steer step by step. And each step has 4 possible actions: move forward, move backward, turn left and turn right. The distance or steering angle of each step is equal. The translational motion can only change the robot's position (x, y), while the steering can only change the robot's orientation (θ).

The learning task of the mobile robot is to find an optimal path to arrive at the target using less time and less moving steps. Meanwhile, the positioning error which is defined as the distance between the target and the robot final stop position is as little as possible. During the path finding process, the mobile robot must avoid the obstacles.

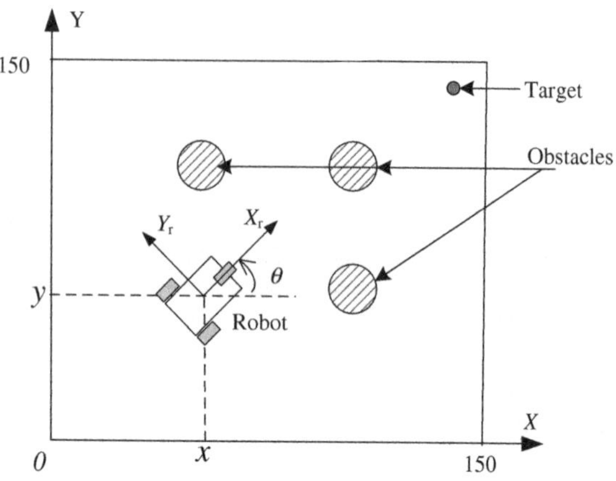

Fig. 20.1 The mobile robot in the learning space

20.3 Hybrid Reinforcement Learning Algorithm

20.3.1 Reinforcement Learning Algorithm

In the robot obstacle avoiding task, one reinforcement learning algorithm named as the linear, gradient-descent Sarsa (λ) with ε-greedy policy is employed. This algorithm is a kind of function approximation method which was introduced detailed in a Sutton's work.

The approximate function is the most important fact in this algorithm, which is shown below

$$V_t(s) = \vec{\theta}_t^T \vec{\phi}_s = \sum_{i=1}^{m} \theta_t(i)\phi_s(i) \tag{20.2}$$

where V_t is a linear function of a vector $\vec{\theta}_t$, and $\vec{\phi}_s = (\phi_s(1), \phi_s(1), \ldots, \phi_s(m))^T$, is a column vector of features, corresponding to every state s.

The method is mainly based on Sutton's SARSA (λ) algorithm, it employs tile coding function approximation and replacing eligibility traces principles.

20.3.2 Supervised Learning Algorithm

The reinforcement learning algorithm mixed a supervised learning method. During the robot moving process, the supervised learning always detects the new state of the robot named S'. So the system can know whether the robot bumps with the obstacles. If the collision happens, the supervised learning may begin to work. The system will assign an action a^{-1} to the robot. The action a^{-1} represents the inverse action of the former action which makes the collision. It is a wrong action which should be corrected.

20.3.3 Reward Function Design

As described in above chapter, for avoiding the robot bumping the obstacles, the supervised learning method is mixed into the reinforcement learning. Besides this, when the robot bumps the obstacles, it will be punished with the reward function.

The reward function for the reinforcement learning is designed as below:

$$R_{eward} = \begin{cases} -100: & \text{Bumpingobstacle} \\ -10: & \text{Normal steps} \\ 0: & \text{Achieving target} \end{cases} \tag{20.3}$$

The reward is always a negative value on all the time until the robot arrives at the target. In this learning task, the ε-greedy policy is used and the factor ε equals zero. So the negative reward can make the states visited by robot are valued worse than those unvisited states. This policy drives the robot to explore new states continually, until finding a proper path to the target.

The negative rewards between the bumping obstacle and normal steps are different. For normal step, the negative value is only −10. For bumping obstacle, it is −100 which is 10 multiple than the normal step situation. The reason of this design that the great larger punishment for bumping is to avoid the robot bumps the obstacle in next learning process.

20.4 Generalization of Learning Space

20.4.1 Even Generalization Method

As shown in formula (20.2), the features could be constructed from the states in many ways. The tile coding is well suited for efficient on-line learning, which is called the learning state generalization.

The most common tile coding method for state generalization is named as even tile coding, as is shown in Fig. 20.2. It uses many grids to cover the whole learning space. The size of each grid is same and the distribution of them is evenly.

The virtual of the even tile coding schedule is easy to design and to realize. Its principle is simply to understand. However, its drawback is evenly generalization resolution. It is unnecessary and will decrease the learning efficiency.

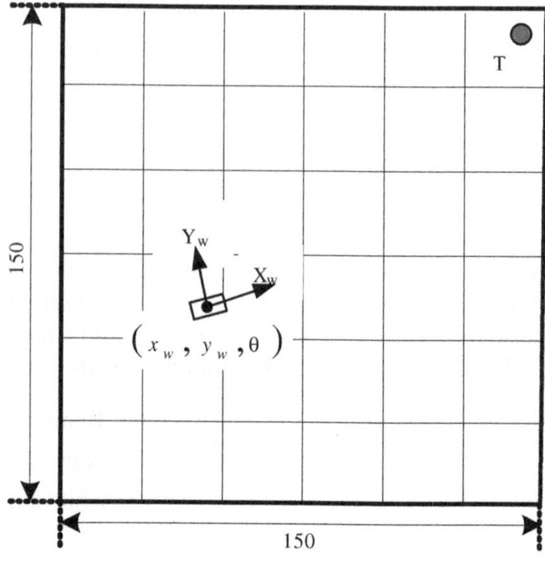

Fig. 20.2 Evenly tile coding

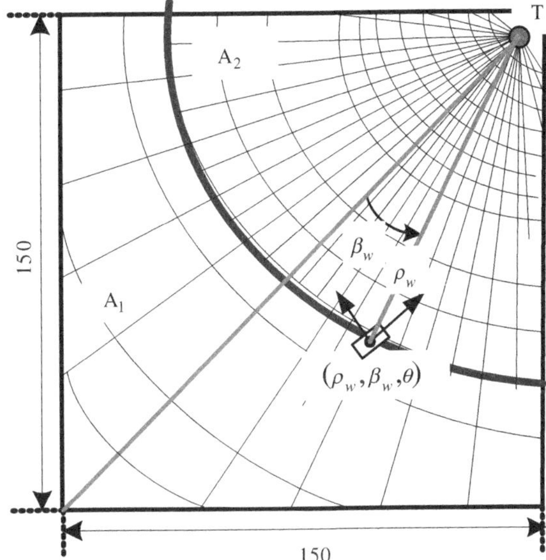

Fig. 20.3 Cobweb tile coding schedule

20.4.2 Uneven Generalization Method

A better generalization method is shown in Fig. 20.3. It is called cobweb tile coding schedule, because it looks like a cobweb [8].

In this schedule, the whole learning space is divided into two areas. The tile's shape is sector. The sector's density in each area is different.

The area which contains the target position has higher sector density. Furthermore, the density in one area is also gradual change. So this makes the gradual change of the generalization resolution from the start point to the target point in the whole learning space.

The cobweb tile coding schedule is reasonable and will improve the learning efficiency.

20.5 Simulations of Robot Obstacles Avoiding

20.5.1 Comparison Between Single and Hybrid Reinforcement Using Even Tile Coding

Only with reinforcement learning, the robot can avoid the obstacle and achieve the target, as shown in Fig. 20.4. The learning process convergences when learning about 1,300 times by the Fig. 20.4a. In Fig. 20.4b, the inner black circle represents the obstacle, while the outer red circle represents the danger area which the robot can not bump.

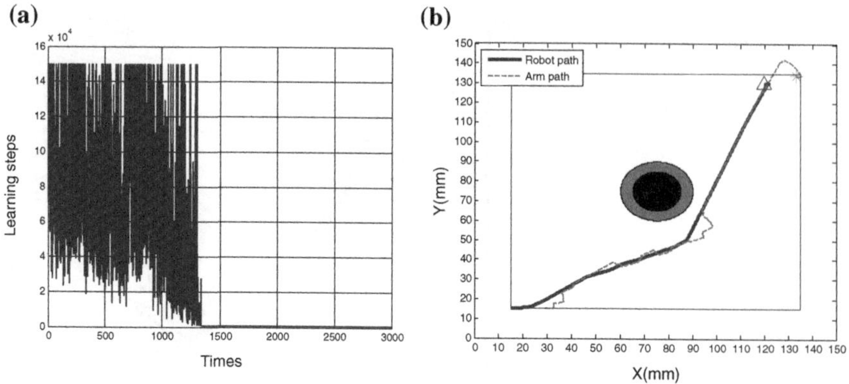

Fig. 20.4 The single reinforcement learning curves using even tile coding. **a** Learning convergence. **b** CurveRobot movement curve

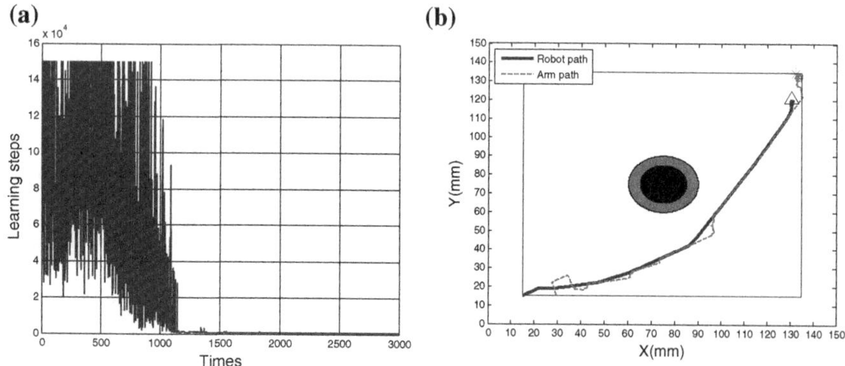

Fig. 20.5 The hybrid learning curves using even tile coding. **a** Learning convergence curve. **b** Robot movement curve

Figure 20.5 shows the learning situation of the hybrid learning algorithm. From (a), it is obviously that the supervised learning improves the convergence speed. When learning about 1,100 times, the learning curve is converged.

The concrete data of the two simulations above are listed in Table 20.1 for comparison detailed. The hybrid method is better than single method from the two aspect of learning time and precision. But it also needs more steps (893 steps) than the single method (862 steps).

Table 20.1 Learning results comparison between single reinforcement and hybrid method

Learning method	Learning time	Movement steps	Precision
Single reinforcement	1,875 s	862	0.29
Hybrid reinforcement	1,568 s	893	0.21

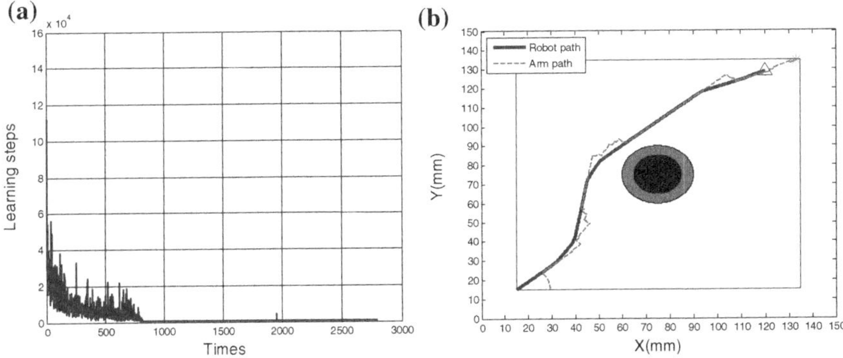

Fig. 20.6 The hybrid learning curves using cobweb tile coding for one obstacle. **a** Learning convergence curve. **b** Robot movement curve

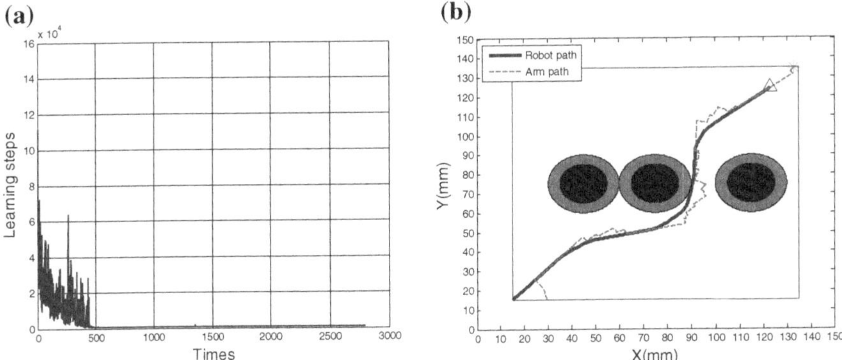

Fig. 20.7 The hybrid learning curves using cobweb tile coding for three obstacles. **a** Learning convergence curve. **b** Robot movement curve

20.5.2 Obstacle Avoiding Using Cobweb Tile Coding

The hybrid learning algorithm using cobweb tile coding is also employed to calculate the obstacle avoiding path. The cases of one obstacle and three obstacles are shown in Figs. 20.6 and 20.7 respectively. The learning can converge fast.

The learning data results are listed in Table 20.2. It is obviously that the cobweb method achieves less learning time, less movement steps, and higher precision than the even tile method.

Table 20.2 Learning results of hybrid reinforcement using cobweb tile coding

Obstacles quantity	Learning time	Movement steps	Precision
One	188 s	817	0.19
Three	183 s	854	0.09

20.6 Conclusion

As a major artificial learning method, the reinforcement learning can apply to the obstacle avoiding of robot. When it merges with supervised learning, the learning speed can be improved.

Furthermore, the uneven generalization of learning space is better than even generalization. An uneven model, which is called cobweb tile coding, is designed. The simulations demonstrate it can improve the learning efficiency for the obstacle avoiding task.

Acknowledgments This work is supported by Hebei Province Natural Science Foundation for Youth (No. F2011203065), Major Basic Research Program of Applied Basic Research Project of Hebei Province Science and Technology R&D Project (No. 11963545D), and Qinhuangdao City Science and Technology R&D Project (No. 2012021A31).

References

1. Maravall, D., de, L. J., & Martín, H.J.A. (2009). Hybridizing evolutionary computation and reinforcement learning for the design of almost universal controllers for autonomous robots. Neurocomputing, 72(4-6), 887-894
2. Duan Y, Liu Q, Xu X (2007) Application of reinforcement learning in robot soccer. Eng Appl Artif Intell 20(7):936–950
3. Mabu S, Hatakeyamay H, Thu MT (2011) Genetic network programming with reinforcement learning and its application to making mobile robot behavior. IEEJ Transactions on Electronics, Information and Systems 126(8):1009–1015
4. Grefenstette JJ, Moriarty DE, Schultz AC (2011) Evolutionary algorithms for reinforcement learning. Journal Of Artificial Intelligence Research 67(11):241–276
5. Csaba S (2010) Algorithms for reinforcement learning. Synthesis Lectures on Artificial Intelligence and Machine Learning 4(1):1–103
6. Stone P, Sutton RS, Kuhlmann G (2005) Reinforcement learning for RoboCup-soccer keepaway. Adaptive Behavior 13(3):165–188
7. Tokarchuk L, Bigham J, Cuthbert L (2006) Fuzzy and tile coding function approximation in agent coevolution. In: Deved V (ed) Proceedings of the 24th IASTED International Conference on Artificial Intelligence and Applications. ACTA Press, Anaheim, pp 353–358
8. Li JH, Li ZB, Chen JP (2011) Microassembly path planning using reinforcement learning for improving positioning accuracy of a 1cm3 omni-directional mobile microrobot. Applied Intelligence 34(2):211–225

Chapter 21
The Effects of Public Expenditure on Pro-poor Growth in Rural China: A General Equilibrium Simulation Approach

Yibing Wang and Xiaoyun Ma

Abstract This paper analyzes the effects of public expenditure on poverty in different regions of rural China. A general equilibrium simulation is used to examine the effects of government actions on the rural poor populations' welfare. The data used in this paper are available from the China's National Bureau of Statistics. The results of this study indicate that the economic growth in China is pro-poor no matter what simulation scenario is adopted. The public expenditures from the governments made significant contribution to the reduction of poverty severity in rural China. In other words, the poor in rural China benefited from fruits of the economic growth induced by the public investments. However not all the regions have profited uniformly from the economic growth.

Keywords Public expenditure · Economic growth General · Equilibrium simulation

21.1 Introduction

Poverty used to be narrowly defined as a state wherein a person is below a minimum threshold of supply in food, income or consumption [1]. This narrow view is obsolete today for the specialists in social sciences as well as for organizations and development economists. More stress is placed on the way how the poor view their own situation and the way poverty is conceived within different

Y. Wang (✉)
School of Business, Shandong University at Weihai, Weihai, China
e-mail: wang_yibing@yahoo.com

X. Ma
School of Economics and Management, Ningxia University, Yinchuan, China
e-mail: fatyma@foxmail.com

Z. Sun and Z. Deng (eds.), *Proceedings of 2013 Chinese Intelligent Automation Conference*, Lecture Notes in Electrical Engineering 255,
DOI: 10.1007/978-3-642-38460-8_21,

cultures. It is widely accepted that the poor is those who do not meet the perceived minimum conditions within societies throughout the world. Ravallion and Datt [2] noticed that such a definition takes into account the public interventions in primary education, basic medical cares and road maintenance etc. This alternative definition of poverty allows a researcher to make rigorous analysis and recommend proper intervention policies.

Kakwani [3] analyzed the impact of poverty among the target groups. It was found that rational public expenditures and optimal distributions appear to be efficient in reducing the inequality and poverty of populations. The debate today lies in the nature and the relevance of the government intervention. How can public expenditures reduce poverty through economic growth? Barro [4] pointed out there is a positive correlation between the public expenditures and the economic growth in a region. However, is this growth beneficial to the poor? Is a pro-poor growth established in rural China?

This study intends to examine the effects of government actions in poverty reduction via economic growth. Some economists made relevant researches previously. Kraay [5] found that a public expenditure leads to a qualified growth of pro-poor when it is more beneficial to the poor than to the wealthy or when its contribution to the poor is percentage higher than its contribution to the national income. From then on the pro-poor growth is defined as better distribution of the unequal public expenditures and termination of expenses only beneficial to wealthy groups.

21.2 The Poverty and Inequalities in China

There are 122.38 millions of inhabitants in rural China lives in poverty [6]. Table 21.1 shows the evolution of poverty from 2002 to 2011 in rural China. The poverty reduction in rural China displays a saddle pattern since 2002. The rural poverty population increased from 28.2 million in 2002 to 29 million in 2003. Thanks to the increase of public expenditure in rural China since 2004 the rural poverty population declined rapidly from 26.1 million in 2004 to 14.79 million in 2007. The poverty standard was revised upwards in China in 2008. The rural poverty population continued to decline until 2010. In 2011 the poverty population has jumped to 122.38 million from 26.88 million, an increase of more than 300 %. The incidence of poverty in rural China displays a similar pattern. Estimated at 3 % on average in 2002, the population of poor people reaches its peak in 2011 [6].

The data about income inequalities evolution in rural China are calculated from the data provided by China's National Bureau of Statistics (NBS). The income inequalities in rural China increased significantly. The Gini coefficient has increased 4.17 % from 0.29864 in 2002 to 0.3111 in 2011. The Gini coefficient shows an upward trend in the short run and could easily surpass the warning line of

Table 21.1 Distribution of winners

Region	SIM1 (κ_α)	SIM2 (κ_α)
The north	0.18	0.22
The north–east	−0.26	0.08
The north–west	−0.07	−0.01
The south	1.25	1.48
The south–west	0.01	0.94
The east	1.84	1.91
The middle	0.02	1.13
The middle-west	0.94	1.03
The whole	1.01	1.21

0.4, which could bring serious negative effects on social development and economic growth.

The income inequality in rural China displays a widening trend. The income ratio of high income over low income group has soared along with the increase of Gini coefficient. The income ratios of almost all other income groups over low income group display a widening trend. The income ratios of middle income, mid-high income and high income group over low income group have increased from 1.95, 2.63 and 4.88 in 2002 to 5.45, 7.53 and 14.35 in 2011 respectively. The only exception is the income ratio of mid-low income group over low income group, which decreased slightly from 1.47 in 2002 to 1.40 in 2011. In sum, income inequalities between the poor and the rich are reinforced while the income of mid-low income group is falling down rapidly close to that of low income group.

Income inequalities in China go up despite the continuous and regular rise in governmental expenditures. The governmental expenditures' share in GDP increases rapidly. The Chinese current expenditures amounted to 18.33 % of the gross domestic product (GDP) in 2002 against 23.1 % in 2011 (NBS, 2012). It appears that governmental expenditures do reduce absolute poverty but not relative poverty.

21.3 The Literature Review

A pro-poor growth promotion—which improves the poor people's ability to participate in the economic activity and to get an advantage from it—will be essential to get the poor out of poverty and reach the millennium objectives for the development [7]. Why has growth succeeded in reducing poverty in some countries and not in others? How can the poor participate in the economic growth process and take advantage of it? Why the growth favorable to the poor is important and how can the backers promote it? The answers to those questions constitute the essential of the literature review.

The theories that treat the public expenditures as the source of the economic growth do not win unanimous support among the economists. Many researchers [8] clearly claim that public expenditures explain the economic growth in a positive way. However authors [9] concluded in their works that public expenditures do not make much contribution to the improvement of the growth rate in the developing countries.

Dévarajan [10] was one of the first to use the concept of unproductive and productive public expenditures. This classification permits to characterize the expenditures likely to affect in a positive and significant way the growth of the gross domestic product in the short-and-medium-term. Public expenditures which are described as unproductive in the short-term are considered productive in the long-term. Gunter [11] explained a more global analysis about the public interventions and their ability to reduce the numerical index of poverty. Gunter [11] explained that sole use of public expenditures could not guarantee a growth profitable to the poor. They think that in order to fight efficiently against poverty, it is better to combine the actions and economic policies in several domains. The infrastructure projects relating to education, accommodation and health hardly ever reach their term. Gupta and Mitra [12] confirmed that public interventions aimed at poverty reduction remain marginal and inefficient.

Concerning China, many economists have been interested in the phenomenon of urban and rural poverty. Ravallion [13] found economic growth spurred by public expenditures reduced poverty in China.

As it can be noticed, the public expenditures do not necessarily lead to a pro-poor growth. That's the reason why this paper tries to analyze the impact of government actions on poverty in rural China.

21.4 The Methodological Approach

The simulated general equilibrium approach is inspired by the works of Pereira [14]. Ravallion [13] defined the pro-poor growth rate (PPGR) as being the area located below the growth effect curve up to poverty rate. If $g_t(p)$ is considered as the real consumption growth rate for the percentile p of the distribution and the poverty rate at the initial point in time (t), then the PPGR is:

$$PPGR = \int g_t(p)dp/H_t \tag{21.1}$$

The pro-poor growth rate is therefore the average growth of the population consumption featuring below the poverty line. If P_α is a measure of poverty, μ the average income, η_α the elasticity growth of income and ε_α the elasticity growth of public expenditures, then the proportional variation of the households' poverty is:

$$dP_\alpha/P_\alpha = \eta_\alpha(d\mu/\mu) + \varepsilon_\alpha(dG/G) \tag{21.2}$$

From that, the elasticity of global poverty δ_α is deduced by dividing the above equation by the average income growth rate.

$$\delta_\alpha = (dP_\alpha/P_\alpha)/(d\mu/\mu) = \eta_\alpha + \varepsilon_\alpha(dG/G)/(d\mu/\mu) = \eta_\alpha + \varepsilon_\alpha\lambda \qquad (21.3)$$

The parameter δ_α measures poverty change percentage further to a variation of the growth rate by one percent. The parameter λ measures the Gini change percentage further to a variation of the growth rate by one percent. Thus when λ is positive (negative) the growth process is followed by an increase (reduction) of inequalities. In other words the growth will be qualified as pro-poor when λ is negative. Kakwani [3] deduced the pro-poor index from it, which is defined as:

$$\kappa_\alpha = \delta_\alpha/\eta_\alpha = 1 + \lambda(\varepsilon_\alpha/\eta_\alpha) \qquad (21.4)$$

This index corresponds to the ratio of the elasticity of global poverty added to the elasticity of poverty growth. It indicates the existence of a pro-poor growth when $\kappa_\alpha > 1$. In other words, the results of growth as regards redistribution are proportionally more profitable to the poor than to the non-poor. The growth is not strictly pro-poor even if a reduction of poverty effect is witnessed when $1 > \kappa_\alpha > 0$. Finally economic growth leads to an increase in the state of poverty when $\kappa_\alpha < 0$. The economic growth has no impact on poverty if $\kappa_\alpha = 0$.

Poverty evolution is simulated under three growth hypotheses: pro-poor, neutral or on the contrary less favorable to the poor than to the rest of the population (in accordance with the values of κ_α). To simulate the effect of a growth that changes the distribution of the incomes among the population, we assume that the Lorenz curve moves the following way:

$$L_{t+1}(P) = L_t(P) - \mu(P - L_t(P)) \qquad (21.5)$$

In Eq. (21.5) L (P) is the percentage of the income owned by the first p percents of the population, when individuals are classified according to an ascending order of income. When μ is positive, this relation implies that the Lorenz curve moves towards the bottom, otherwise inequality increases. This hypothesis is useful because it can be showed that μ is equal to the variation in percentage of the Gini coefficient. It only remains to link the evolution of the Gini index to that of the standard deviation of the income logarithm distribution σ which is made possible under the hypothesis that the distribution is log-normal. The setting of the probable values of κ_α will be made through a simulation with data from China.

21.5 Results of Simulations

Simulations are made according to several growth scenarios. We followed Chen [15] who carried out an empirical analysis in China. The first scenario (SIM1) postulates for the hypothesis that the considered variables are estimated with their observed levels. The second scenario (SIM2) is interested in the context of the induced effects by a possible rise in the public expenditures. The results of the two simulations are given in the following table.

In general no matter what scenario (SIM1 or SIM2) is adopted, the economic growth in China can be qualified as pro-poor because the two values of κ_α are all superior to one. In other words, the Chinese governments have carried out economic policies actions which open out onto the improvement of the well-being of the poor population. That conclusion is established if the simulation has been made in SIM1 with $\kappa_\alpha = 1.01$. Similarly in SIM2, it is found that a possible increase of the public expenditures is profitable to the lower-income groups with $\kappa_\alpha = 1.21$.

In a specific way, the different areas haven't uniformly benefited by the fruits of the economic growth. Those which remained on the fringe are essentially the region of the middle, the middle-west, the north-east, the north-west and the south–west. They have a ratio of the global poverty elasticity related to poverty growth elasticity with 0.02, 0.94, −0.26, −0.07 and 0.01 respectively through the first scenario. The winners and losers in SIM1 can be observed in the second column of Table 21.1.

The poor population of the north-east and the north–west in rural China has not really profited from the economic growth. A majority of them are kept busy by working in the fields which could not bring them strong incomes. These weak incomes can be explained by the rise in the cost of the inputs and other pesticides and also by the reduction of the purchase price to the producer of the main agricultural products like corn and wheat etc. The socio-economic infrastructures are very few there.

Even if the results allow us observe a decrease in the poverty index, they don't authorize us to declare that the middle-west, the northern and the western regions display a pro-poor growth. The elasticity ratio of the total poverty is still between zero and one. Therefore the growth isn't strictly pro-poor.

Therefore, for the rest of the regions in China, the results testify that the growth is pro-poor. This power of the authorities is particularly observed in the half-southern part of the country. Indeed, this big area gathers the most important part of the industrial and commercial activities, but above all, with the presence of the big ports. The south-east part of China is equipped with modern social and economic infrastructures which make it one of the giants in south-east Asia. The economic activities prosper there and the distributed incomes are high. The public interventions are therefore productive because they support the economic growth and are profitable to the poor.

By considering the scenario (SIM2) and by taking into account the induced effects through a possible rise in the public expenditures, the results are partly modified. The hypothesis of the increase in the public expenditures has been accepted in the light of the theory of endogenous growth. This increase is inspired by the average annual growth rate of the public expenditures in China.

With that second simulation, the analysis of the results shows a sensitive rise in the number of regions which exhibit a pro-poor growth. Indeed, from nine regions with the SIM1, we obtain eight regions with the SIM2. Likewise, a possible increase in the public expenditures will allow the reduction of the number of

Table 21.2 Distribution of regions

Region	SIM1	SIM2
The north	$1 > \kappa_\alpha > 0$	$1 > \kappa_\alpha > 0$
The north–east	$\kappa_\alpha < 0$	$1 > \kappa_\alpha > 0$
The north–west	$\kappa_\alpha < 0$	$\kappa_\alpha < 0$
The south	$\kappa_\alpha > 1$	$\kappa_\alpha > 1$
The south–west	$\kappa_\alpha > 1$	$\kappa_\alpha > 1$
The east	$\kappa_\alpha > 1$	$\kappa_\alpha > 1$
The middle	$\kappa_\alpha > 1$	$\kappa_\alpha > 1$
The middle-west	$1 > \kappa_\alpha > 0$	$\kappa_\alpha > 1$

regions where poverty is more noticeable in comparison with Table 21.2. Only the north-western part seems less sensitive to the simulations because its poverty elasticity is smaller than zero. It is respectively by −0.07 and −0.01 with the SIM1 and SIM2. This can be explained by the difficult access to that area. Some roads there are not asphalted and most economic and social infrastructures are insufficient.

21.6 Conclusion

Our analysis has sought to analyze the contribution of the public expenditures to pro-poor growth in China. The economic growth in China can be considered pro-poor no matter what simulation scenario is adopted. The economic policies implemented have resulted in an economic growth whose results were profitable to the poor. This study reveals that not all the regions have profited uniformly from the growth. Most of the northern regions has negative poverty elasticity or lower than one, contrary to those of the centre and the south. The simulation confirmed the hypothesis that an increase in public expenditures is beneficial because pro-poor growth is noticeable in an increasing number of regions. In comparison with those results, it is important to emphasize the necessity to fight against the deep regional disparities for a better social cohesion of populations. The governments, who are responsible for the implementation of economic policies aimed at creating public infrastructures, should privilege the north-eastern, the north-western and the south-western regions. The remedy of the basic economic and social structures in the above mentioned zones should deserve to be priority projects in development strategies.

Acknowledgments This work is supported by National Science Foundation of China (NSFC) under Grant 71262012.

References

1. Epaulard A (2003) Poverty increase and reduction in developing countries and countries in transition. Impacts Perspect 13(2):9–20
2. Ravallion M, Datt G (2002) Why has economic growth been more pro-poor in some states of India than others? J Dev Econ 68(3):381–400
3. Kakwani N, Pernia E (2000) What is pro-poor growth. Asian Dev Rev 16(4):1–16
4. Barro RJ (1991) Economic growth in a cross section of countries. Quart J Econ 106(2):407–433
5. Kraay A (2004) When is growth pro-poor? Evidence from a panel of countries. *The World Bank Policy Research Working Paper*
6. National Bureau of Statistics (NBS) (2012) China Development Report
7. Sahn DE, Stiefel DC (2003) Progress towards the millennium development goals in Africa. World Dev 3(1):221–241
8. Berg SV, Pollitt MG, Tsuji M (2002) Private initiatives in infrastructure: priorities? Incentives and performance. Edward Elgar Publishing Limited, pp 256–259
9. Gillis M, Perkins D, Roemer M, Snodgrass D (2001) Development economics. Boeck University, pp 784–788
10. Dévarajan S, Swaroop V, Zou H (1996) The composition of public expenditure and economic growth. J Monetary Econ 37(2):313–344
11. Gunter BG, Cohen MJ, Lofgren H (2005) Analyzing macro-poverty linkages: an overview. Dev Policy Rev 23(3):285–298
12. Gupta I, Mitra A (2004) Economic growth, health and poverty: an exploratory study for India. Dev Policy Rev 22(2):193–206
13. Ravallion M, Chen S (2003) Measuring pro-poor growth. Econ Lett 78(1):93–99
14. Pereira da Silva LA, Essama-Nssah B, Samaké I (2003) Linking aggregate macro-consistency models to household surveys: a poverty analysis macroeconomic simulator (PAMS). World Bank
15. Chen S, Ravallion M (2004) Welfare impacts of China's accession to the world trade organization. World Bank Econ Rev 18(1):29–57

Chapter 22
Robust Control for Air-Breathing Hypersonic Cruise Vehicles

Wenbiao Zhu

Abstract The integral ramjet engine has gained wide popularity as one of the versatile air-breathing engines for hypersonic flight vehicles. Due to the uniqueness of tightly integrated engine-airframe configuration, results in significant coupling between the structure, propulsion system and vehicle aerodynamics, the hypersonic flight vehicle modeling and control system design become very challenging. For air-breathing hypersonic cruise vehicles, the flight Mach number control is one of the important control system task. In this paper under the condition that the air-breathing hypersonic cruise vehicle dynamics is considered, the robust Mach number controller for an air-breathing hypersonic cruise vehicle is designed by using the H_∞ robust control method. The simulation and validation results indicate that the designed controller makes the Mach number control system have a very good performance, the designed controller meets the Mach number control system performance and robustness demands. The effectiveness of the presented methods is demonstrated by the simulation and validation results.

Keywords Robust control · Hypersonic flight vehicle · Air-breathing ramjet engine · Integrated engine-airframe configuration

22.1 Introduction

In order to keep a cruise flight state at high altitudes and Mach numbers, the integral ramjet engine has gained wide popularity as one of the versatile air-breathing engines for hypersonic flight vehicles. The integral ramjet-engine combines the

---

W. Zhu (✉)
Beijing Aerospace Automatic Control Institute, National Key Laboratory of Science and Technology on Aerospace Intelligent Control Technology, Beijing 100854, People Republic of China
e-mail: zwb528@ sohu.com

Z. Sun and Z. Deng (eds.), *Proceedings of 2013 Chinese Intelligent Automation Conference*, Lecture Notes in Electrical Engineering 255,
DOI: 10.1007/978-3-642-38460-8_22,

performance characteristics of both the hypersonic flight vehicle and ramjet engine. Due to the uniqueness of tightly integrated engine-airframe configuration, results in significant coupling between the structure, propulsion system and vehicle aerodynamics. Because these significant couplings, and the sensitivity to changes in flight condition and the difficulty in measuring and estimating the aerodynamic characteristics of the vehicle, the hypersonic flight vehicle modeling and control system design become very challenging [1–3]. Therefore, robust control has been the effective technique used for hypersonic flight control. The robust control techniques have been pursued to design flight control systems for many years [4–12]. In Ref. [4], investigates a control-oriented design approach along with the introduction of frequency-dependent weighting functions in the H_∞-control synthesis framework as design variables. The methodology is implemented for vibration-attenuation of a coupled and complex dynamics of a hypersonic vehicle. In Ref. [6], an H_∞ standard problem based on the acceleration sensitivity function is proposed as a basic scheme. The standard problem can be used to analyze or design control laws taking onto account new dynamic elements (actuators dynamics, navigation filter) or additional specification. The scheme is applied on Reusable Launch Vehicle (RLV) during atmospheric reentry to design control laws and to determine the worst-case along the reentry trajectory. In Ref. [7], The H-based and classical engineering approaches to flight control law design are compared regarding similarity between the derived controllers structure. The gain/phase margins requirements are satisfied in both cases. The closed-loop system robustness with the integrated μ-control law is higher than with the traditional control law. Flight regime scheduling via blending of two neighboring controllers against dynamic pressure doubles the control law order, however this approach allows a smaller number of design points in the flight envelope. In Ref. [11], describes an augmenting adaptive control law for attitude stabilization of the first stage of the Crew Launch Vehicle. Low bandwidth control action is maintained by using an $H\infty$ Norm Minimization approach for adaptive control. Simulation results show that the adaptive control law always improves the performance of the cases that are stable with the existing linear design. For the cases that are unstable with only the linear control law, the adaptive control algorithm is able to recover acceptable tracking performance in almost all of the cases. In Ref. [12], introduces the H∞ adaptive control architecture applied to the Generic Transport Model. The flight control system implemented on the Generic Transport Model is an inversion based design that uses the $H\infty$ adaptive control architecture to efficiently compensate for uncertainty using redundant actuators. Simulation results show the performance of the control algorithm in the presence of severe vehicle damage.

This paper investigates the robust Mach number controller design for an air-breathing hypersonic cruise vehicle under the condition that the air-breathing hypersonic cruise vehicle dynamics is considered. This paper is organized as follows. Section 22.2 discusses the mixed sensitivity optimization problem. The robust Mach number controller design and simulation results for an air-breathing hypersonic cruise vehicle are presented in Sect. 22.3. In Sect. 22.4 is the conclusion.

22.2 Mixed Sensitivity Optimization Problem

The H_∞ control design problem can be represented by the mixed sensitivity optimization problem. The mixed sensitivity problem considered in this paper is of the configuration shown in Fig. 22.1.

Where $G(s)$ is the plant, $W_1(s)$, $W_2(s)$ and $W_3(s)$ are weighting functions, $K(s)$ is the controller, r is the reference input, e is the error, u is the controller output, d is the disturbance input, n is the measuring noise, and y is the system output.

The sensitivity function S is defined as

$$S(s) = (I + G(s)K(s))^{-1} \tag{22.1}$$

The controller sensitivity function R is defined as

$$\begin{aligned} R(s) &= K(s)(I + G(s)K(s))^{-1} \\ &= K(s)S(s) \end{aligned} \tag{22.2}$$

The complementary sensitivity function T is defined as

$$\begin{aligned} T(s) &= G(s)K(s)(I + G(s)K(s))^{-1} \\ &= I - S(s) \end{aligned} \tag{22.3}$$

Using the three weighting functions and the aforementioned functions S, R, and T to form the following cost function:

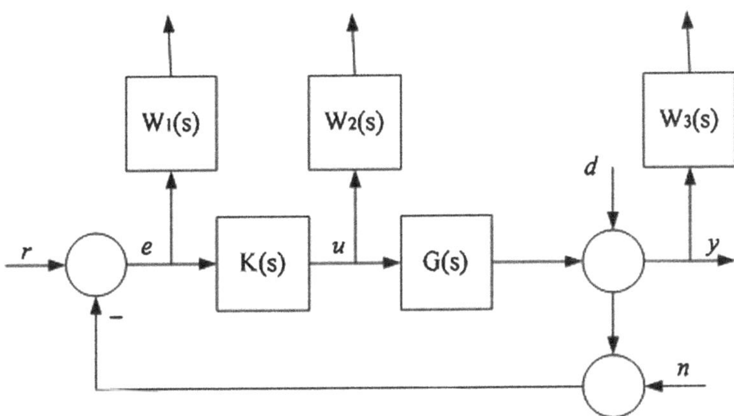

Fig. 22.1 The mixed sensitivity problem

$$T_{zw} = \begin{bmatrix} W_1 S \\ W_2 R \\ W_3 T \end{bmatrix} \tag{22.4}$$

Then the mixed sensitivity optimization problem can be described as, find the controller K, such that the resulting closed-loop system is stable and $\|T_{zw}(s)\|_\infty$ is minimized.

The optimization problem does not have a closed solution, therefore a solution needs to be accomplished by an iterative process. The steps followed to design the H_∞ controller are summarized as follows:

Step 1: design the weighting functions. Weighting functions are very important for the H_∞ controller design. Generally, in order to enhance the ability of rejecting disturbances, the weighting function $W_1(s)$ should be of the low pass filter performance, and the amplitude of $W_1(s)$ should be higher at the low frequencies for controlling the overshoot. While in order to get better dynamic performance, the weighting function $W_3(s)$ should be of the high pass filter performance, and the amplitude of $W_3(s)$ should be higher at the high frequencies, be lower at the low frequencies. For lower the controller order, the $W_2(s)$ generally is designed as a smaller positive constant.

Step 2: derive the augmented plant. Augment the plant G by including the weighting functions, and building the augmented plant P described by the following formula:

$$\begin{aligned} \dot{x} &= Ax + B_1 w + B_2 u \\ z &= C_1 x + D_{11} w + D_{12} u \\ y &= C_2 x + D_{21} w + D_{22} u \end{aligned} \tag{22.5}$$

Step 3: compute the H_∞ controller. By minimizing

$$\|T_{zw}(s)\|_\infty = \left\| \begin{matrix} W_1 S \\ W_2 R \\ W_3 T \end{matrix} \right\|_\infty,$$

compute the H_∞ controller.

22.3 Robust Mach Number Controller Design and Simulation

The longitudinal dynamics of the air-breathing hypersonic vehicle model can be described as follows

$$\begin{cases} m\frac{dV}{dt} = p\cos\alpha - X - mg\sin\theta + F_{xd} \\ mV\frac{d\theta}{dt} = p\sin\alpha + Y - mg\cos\theta + F_{yd} \\ I_z\frac{d\omega_z}{dt} = M_z + M_{zd} \\ \frac{d\vartheta}{dt} = \omega_z \\ \frac{dh}{dt} = V\sin\theta \\ \vartheta = \theta + \alpha \end{cases} \tag{22.6}$$

where p is the thrust, X is the drag force, Y is the lift force, M_z is the pitching moment, V is the velocity, h is the altitude, m is the mass, I_z is the moment of inertia, α is the angle of attack, θ is the flight-path angle, ϑ is the pitching angle, ω_z is the pitching angular rate, F_{xd}, F_{yd} and M_{zd} respectively are X direction disturbance force, Y direction disturbance force and pitching disturbance moment.

For air-breathing hypersonic cruise vehicles, the flight Mach number control is one of the important control system task. The velocity of the air-breathing hypersonic cruise vehicle considered here is decided by the thrust from the air-breathing ramjet engine. Regulating the mass flow rate of the fuel into the engine combustion chamber can control the thrust from the engine.

At cruise flight state, linearizes the model (formula 22.6), and under the condition that the air-breathing hypersonic cruise vehicle dynamics is considered, derives the transfer function G from the mass flow rate of the fuel into the engine combustion chamber to the Mach number, then follows the **Step** 1 described in Sect. 22.2, designs the weighting functions $W_1(s)$, $W_2(s)$ and $W_3(s)$. In this paper, for the studied air-breathing hypersonic cruise vehicle, the weighting functions $W_1(s)$, $W_2(s)$ and $W_3(s)$ are designed as follows

$$W_1(s) = \frac{0.32s^2 + 16s + 768}{s^2 + 2s + 1.2} \tag{22.7}$$

$$W_2(s) = 10^{-5} \tag{22.8}$$

$$W_3(s) = \frac{8s + 80}{s + 300} \tag{22.9}$$

Using the MATLAB Robust Control Toolbox, augment the plant G by including the above weighting functions $W_1(s)$, $W_2(s)$ and $W_3(s)$, and building the augmented P described by the formula (22.5). Then by minimizing

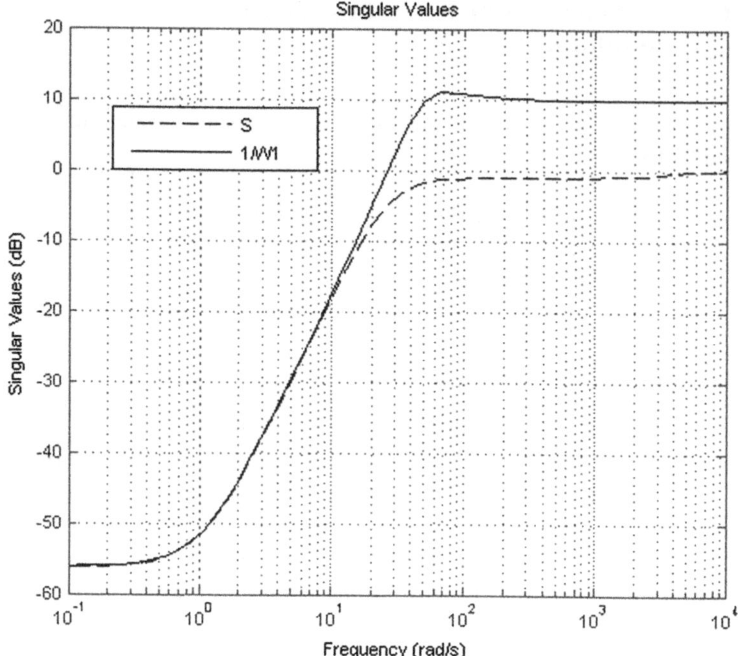

Fig. 22.2 The relationship between S and $W_1^{-1}(s)$

$$\|T_{zw}(s)\|_\infty = \left\| \begin{matrix} W_1 S \\ W_2 R \\ W_3 T \end{matrix} \right\|_\infty,$$

work out the H_∞ controller $K(s)$.

In order to validate the designed controller, Fig. 22.2 shows the singular values relationship between the sensitivity function S and $W_1^{-1}(s)$, Fig. 22.3 shows the singular values relationship between the complementary sensitivity function T and $W_3^{-1}(s)$.

As can be seen in Fig. 22.2, the singular values satisfy $\bar{\sigma}[S(jw)] < \bar{\sigma}[W_1^{-1}(jw)]$, that is, the condition $\|W_1 S\|_\infty < 1$ is satisfied, therefore the designed controller meets the Mach number control system performance demand.

Also in Fig. 22.3, the singular values satisfy $\bar{\sigma}[T(jw)] < \bar{\sigma}[W_3^{-1}(jw)]$, that is, the condition $\|W_3 T\|_\infty < 1$ is satisfied, therefore the designed controller meets the Mach number control system robustness demand.

The Mach number control closed system step response curve is plotted in Fig. 22.4. This curve shows that the closed system is stable, the regulating time is less than 0.2 s, and the stable state precision is higher, so the designed controller makes the Mach number control system have a very good performance.

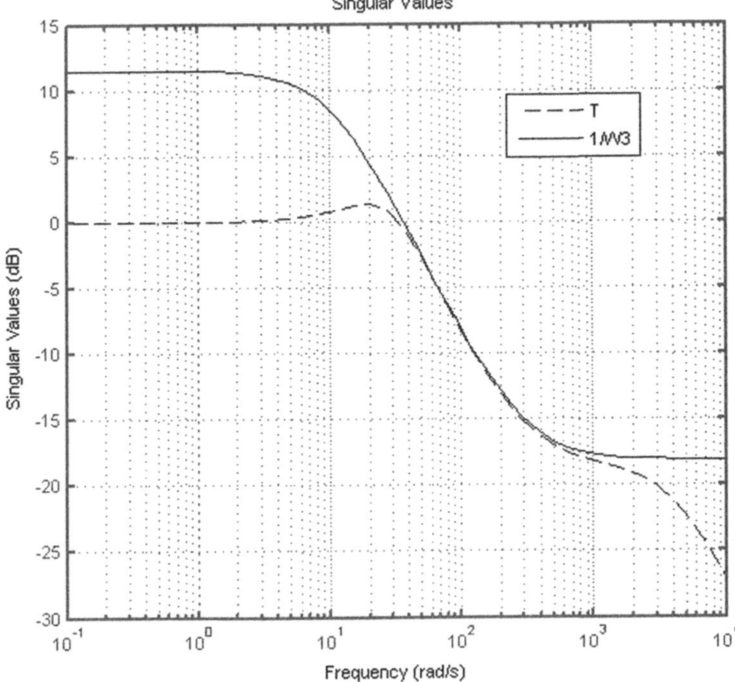

Fig. 22.3 The relationship between T and $W_3^{-1}(s)$

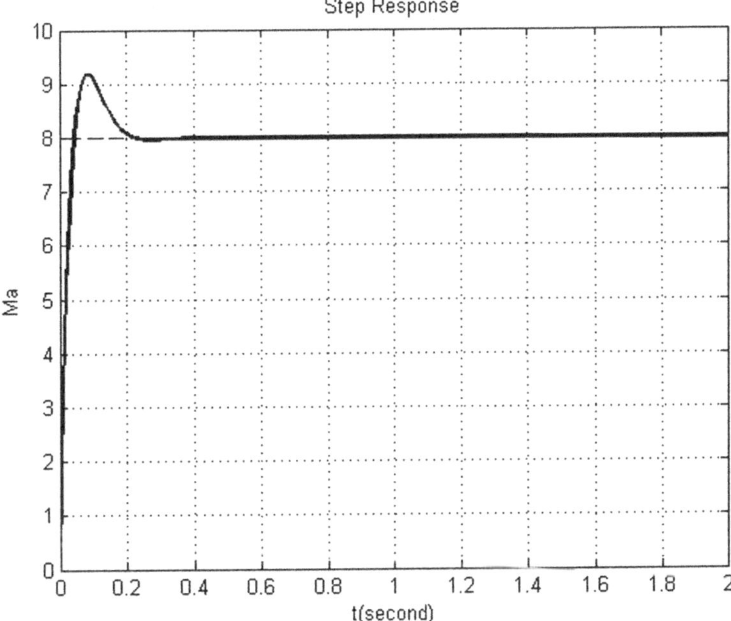

Fig. 22.4 The closed system step response

22.4 Conclusion

For air-breathing hypersonic cruise vehicles, the flight Mach number control is one of the important control system task. In this paper under the condition that the air-breathing hypersonic cruise vehicle dynamics is considered, the robust Mach number controller for an air-breathing hypersonic cruise vehicle is designed by using the H_{∞} robust control method. The simulation and validation results indicate that the designed controller makes the Mach number control system have a very good performance, that is, the closed system is stable, the regulating time is less than 0.2 s, and the stable state precision is higher. As can be seen in the simulation and validation results, the designed controller meets the Mach number control system performance and robustness demands. The effectiveness of the presented methods is demonstrated by the simulation and validation results. The robust controller design for integrating the Mach number control and the vehicle attitude control is the future work.

References

1. Chavez FR, Schmidt DK (1994) Analytical aeropropulsive/aeroelastic hypersonic-vehicle model with dynamic analysis. AIAA J Guidance Control Dyn 17(6):1308–1319
2. Schmidt DK (1992) Dynamics and control of hypersonic aeropropulsive/aeroelastic vehicles. AIAA Paper p 4326
3. Walton JT (1989) Performance sensitivity of hypersonic vehicles to change angle of attack and dynamic pressure. AIAA Paper 89–2463:1989
4. Bhat S, Viana FAC, Lind R, Haftka RT (2010) Control-oriented design using H_{∞} synthesis and multiple surrogates. AIAA Paper p 3089
5. Hitachi Y, Liu HHT (2009) H_{∞}-LTR technique applied to robust control of propulsion-controlled aircraft. AIAA Paper p 6176
6. Fezans N, Alazard D, Imbert N, Carpentier B (2007) H_{∞} control design for multivariable mechanical systems—Application to RLV reentry. AIAA Paper p 6349
7. Goman M, Sidoryuk M, Ustinov A (2005) Control law design for flexible aircraft: comparison of the $H\infty$-based and classical methods. AIAA Paper p 6265
8. Garcia GA, Keshmiri S, Colgren RD (2010) Advanced H-infinity trainer autopilot. AIAA Paper p 8363
9. Dorobantu A, Murch AM, Balas GJ (2012) H_{∞} robust control design for the NASA air STAR flight test vehicle. AIAA Paper p 1181
10. Marcos A (2010) Again scheduled H-infinity controller for a re-entry benchmark. AIAA Paper p 7567
11. Muse JA, Calise AJ (2010) H_{∞} norm minimization approach for adaptive control of the crew launch vehicle. AIAA Paper p 7566
12. Muse JA, Calise AJ (2010) H_{∞} adaptive flight control of the generic transport model. AIAA Paper p 3323

Chapter 23
Sensor Failure Detection and Diagnosis via Polynomial Chaos Theory—Part II: Digital Realization

Weilin Li, Xiaobin Zhang, Huimin Li and Wenli Yao

Abstract The reliability and performance of complex power systems with many sensors are largely dependent on the accuracy and reliability of the sensors. Thus, it is very important for monitoring and diagnosis the sensor faults. In this paper, the digital realization of polynomial chaos theory (PCT) based sensor failure detection and diagnosis (SFDD) algorithm considering parameter uncertainties has been detailed. A digital signal processor (DSP), TMS320F2812, is adopted to implement the proposed SFDD algorithm. Both hardware-in-the-loop testing and laboratory experimental testing have been done to verify the online practical application capability of the proposed algorithm. A three-stage synchronous generator with automatic voltage regulator (AVR) has been used as a case study. The control performance of AVR is guaranteed to be more reliable by the application of SFDD algorithm based on PCT. Experiment results show good consistency with the theory analysis.

Keywords Sensor failure detection and diagnosis · Digital realization · Polynomial chaos theory · Synchronous generator · Automatic voltage regulator

23.1 Introduction

It is necessary to provide the controllers and operators with accurate measurements of the system in industrial processes. The reliability and performance of complex power system with many sensors are largely dependent on the accuracy and

W. Li (✉) · X. Zhang · W. Yao
School of Automation, Northwestern Polytechnical University,
Xi'an 710072 Shaanxi, China
e-mail: li.weilin@hotmail.com

H. Li
EE Department, University of South Carolina, Columbia,
SC 90089, USA

Z. Sun and Z. Deng (eds.), *Proceedings of 2013 Chinese Intelligent Automation Conference*, Lecture Notes in Electrical Engineering 255,
DOI: 10.1007/978-3-642-38460-8_23, © Springer-Verlag Berlin Heidelberg 2013

reliability of the sensors. However, the sensors are not significantly more reliable than the systems being monitored, and the indication of an abnormal state may be the result of a sensor failure rather than a system failure. Thus, it is very important for monitoring and diagnosis the sensor fault.

Paper [1] proposes a new analytical approach for sensor failure detection and diagnosis (SFDD) under parameter uncertainty, based on the mathematics of polynomial chaos theory (PCT). The objective is to improve the self-healing capability of power systems under damage, to achieve a higher level system reliability and safety. The sensor failure diagnosis solutions are developed using polynomial chaos mathematical procedures to create a general technique for bounding the dynamic behavior of sensors. Bad data are handled with the resulting algorithms to reject them from control unit and to reconstruct the new data during sensor failure.

The developed algorithms have been implemented on a digital signal processor (DSP) in this paper, and validated through both hardware-in-the-loop (HIL) testing and real time online experimental testing. This makes a migration in the field of sensor failure diagnosis towards on-line operation by countering for parameter uncertainties. Furthermore, the proposed method can not only be used for the purpose of sensor failure diagnosis, but also for system monitoring.

A variable frequency synchronous generator in the application of more electric aircraft (MEA) power systems has been adopted as a case study. Automatic voltage regulator (AVR) is used to regulate the output voltage of the synchronous generator. In particular, the reliability of AVR is specifically focused in order to avoid catastrophic malfunctions that may be introduced by failed sensors to propagate into the whole system.

23.2 Brief Introduction of PCT Based SFDD Algorithm

Polynomial chaos is a method that uses a polynomial based stochastic space to represent the evolution of the uncertainty propagation into the system [2]. PCT concept was first introduced by Wiener in 1938 as "Homogeneous Chaos" [3]. The theory evolved into the Wiener–Askey polynomial chaos, which extended the theory to the entire Askey scheme of orthogonal polynomials [4]. The detailed process of the PCT based SFDD algorithm has already been detailed in [1], thus in this paper only a brief introduction is presented.

By choosing certain polynomial base ψ, uncertain parameter X can be expanded into PCT domain as shown in Eq. (23.1).

$$X(\theta) = \sum_{n=0}^{\infty} a_n(\theta)\Psi_n(\xi(\theta)) \tag{23.1}$$

where: X (θ) Random process or function under analysis in terms of θ, which represents the random event; a_n The coefficients of the expansion; ψ_n. The selected polynomial basis; ξ Random vector with a probability distribution function (PDF) according to selected ψ_n.

Considering a linear time-invariant dynamic system in state space form, i.e.,

$$\begin{aligned} \dot{x}(t) &= Ax(t) + Bu(t) \\ y(t) &= Cx(t) + Du(t) \end{aligned} \tag{23.2}$$

where $x(t) \in R^n$ system state vector; $u(t) \in R^r$ input vector; $y(t) \in R^m$ output vector; A, B, C system metrics with appropriate dimensions.

Given that the parameters in A and/or B are uncertain and that the uncertain parameters can be described by a known probability density function (PDF), then the system can be expanded using PCT. The PCT expanded system constitutes a new set of state equations that describe how the parameter uncertainty propagates to the output. The generic form of the system in PCT domain can be presented as in Eq. (23.3).

$$\begin{aligned} \dot{x}_{pct}(t) &= A_{pct}x_{pct}(t) + B_{pct}u_{pct}(t) \\ y_{pct}(t) &= C_{pct}x(t) + D_{pct}u_{pct}(t) \end{aligned} \tag{23.3}$$

where $x_{pct}(t) \in R^{n_{pct}}$ system state vector; $u_{pct}(t) \in R^{r_{pct}}$ input vector; $y_{pct}(t) \in R^{m_{pct}}$ output vector; $A_{pct}, B_{pct}, C_{pct}$ system metrics with appropriate dimensions.

If the above generated system model is observable, a closed-loop state observer can be designed. The observer based on the stochastic PCT model is mathematically defined as

$$\frac{d\hat{\mathbf{x}}_{\mathbf{pct}}(t)}{dt} = \mathbf{A}_{\mathbf{pct}}\hat{\mathbf{x}}_{\mathbf{pct}}(t) + \mathbf{B}_{\mathbf{upct}}\mathbf{u}(t) + \mathbf{B}_{\mathbf{wpct}}\mathbf{w}(t) + \mathbf{G}_{\mathbf{pct}}[\mathbf{m}(t) - \hat{\mathbf{m}}(t)] \tag{23.4}$$

where: $\hat{\mathbf{x}}_{\mathbf{pct}}(t)$: estimated uncertain states in PCT form; m(t):available measurements; $\hat{\mathbf{m}}(t)$: expected values of the estimated values of measured variables. A_{pct}, B_{upct}, B_{wpct}: system matrices with appropriate dimensions after expansion. G_{pct}: gain matrix.

As a result of the PCT approach, a reasonable interval, z_t, can be determined for each measured variable. The diagnosis of the measured data then becomes a decision making process based on the fact that measured value belonging or not to a continuously evolving interval.

The basic operation logic of SFDD algorithm could be summarized as follows:

If measurement is within the uncertain thresholds, Sensor is healthy;

Else, sensor has failed. Identifying and isolating the failed sensors, rebuilding the bad data.

Figure 23.1 illustrates the process of data reconstruction. In this work, by leveraging on the characteristics of PCT (detailed in 2) to propagate uncertainty in power systems, it enables, in principle, the propagation of power system uncertainties in real time, given a known PDF of the main uncertainty sources.

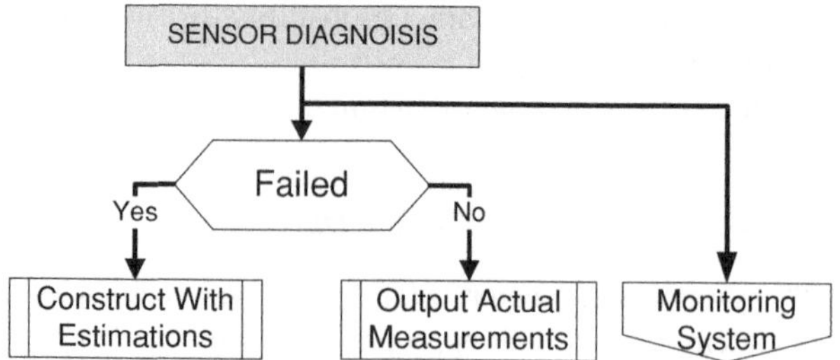

Fig. 23.1 Missing data reconstruction algorithm

23.3 Digital Realization

Configuration of the hardware platform for digital realization of the proposed algorithm is introduced in this section. Its main component is digital signal processor (DSP) TMS320F2812 (microprocessor). The CPU board for implementation of the proposed sensor failure diagnosis algorithm can be seen from Fig. 23.2, two DSPs have been applied, one is used for calculation of the sampled data and realization of SFDD algorithm, the other one is used for control algorithm application and communication.

The C code of the proposed SFDD algorithm is generated from MATLAB with MATLAB Coder. The generated C code can be used for both real-time and non-real-time applications. Thus in this paper, the proposed algorithm is first modeled and simulated in MATLAB. And the C code generated from MATLAB Coder is then compiled in Visual DSP environment until it is executable, and finally loaded into DSP01.

Fig. 23.2 CPU board for digital realization of the proposed algorithm

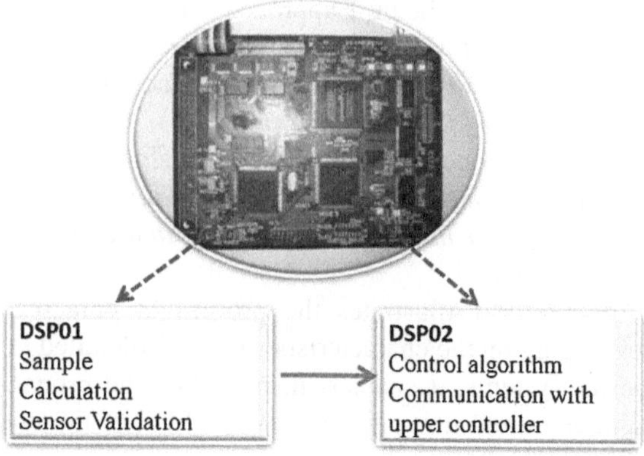

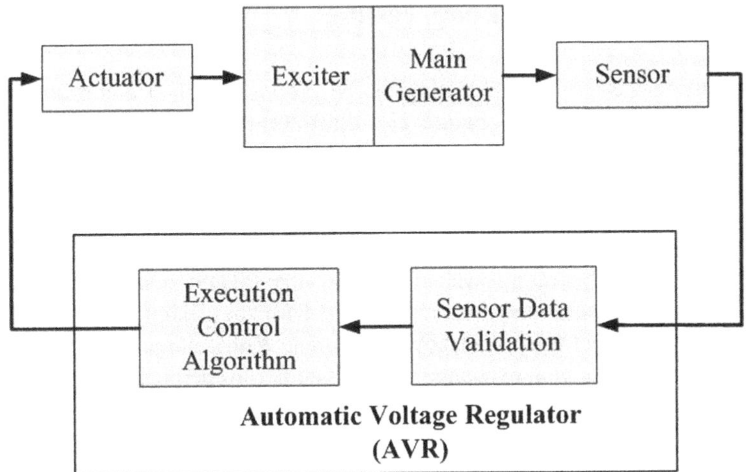

Fig. 23.3 Control structure with sensor data validation

23.4 Case Study

In this paper, a variable frequency synchronous generator used in the more electric aircraft (MEA) is adopted as a case study to verify practical application of the proposed SFDD algorithm. The rated output voltage of the generator is 115 V. And the rated frequency ranges from 360 to 800 Hz.

The generator in aircraft applications is a so called three-stage generator which contains two synchronous machines, a rectifier and a small permanent magnet generator (PMG), driven by the same shaft [5]. The mathematical equations of the three stage synchronous generator can be found from [6]. The performance of the controller of the generator depends on the availability of data from the sensors. However, the traditional control algorithm executes upon one assumption: all the local measurements from the sensors are reliable. In this situation, once the sensor has been failed, a catastrophic consequence on the whole system will occur. For example, the controller gives wrong signals according to the bad sensor data, even though the real system is still in good condition.

So, validity of the measurements is one of the important priors for a reliable operation of AVR [7]. In this work, the acquired data are diagnosed before they are fed into the control center. By this method, bad data is able to be rejected to avoid a catastrophic malfunction. By sensor diagnosis, the availability of measurements which is also a further strengthening on controller performance is achieved. With the application of the proposed SFDD technology based on PCT, the control system of the generator becomes like Fig. 23.3. The detailed process of SFDD algorithm for this particular synchronous generator is provided in [7]. Both hardware-in-the-loop testing and laboratory experimental verification will be provided in the following Sections.

23.5 Hardware-in-the-Loop Testing

A hardware-in-the-loop (HIL) testing of the proposed SFDD algorithm is done prior to practical applications. Figure 23.4 illustrates the laboratory structure of the HIL testing platform. The three-stage synchronous generator model is modeled and simulated in real time with real time digital simulator (RTDS). RTDS [8] is a platform which utilizes a combination of custom software and hardware to perform real time simulation of power systems, with a typical time-step of 50 μs.

The proprietary operating system used by the RTDS guarantees "hard real time" during all simulations. It is an ideal tool for the design, development and testing of power electronics and power systems. With a large capacity for both digital and analogue signal exchange (through numerous dedicated, high speed I/O

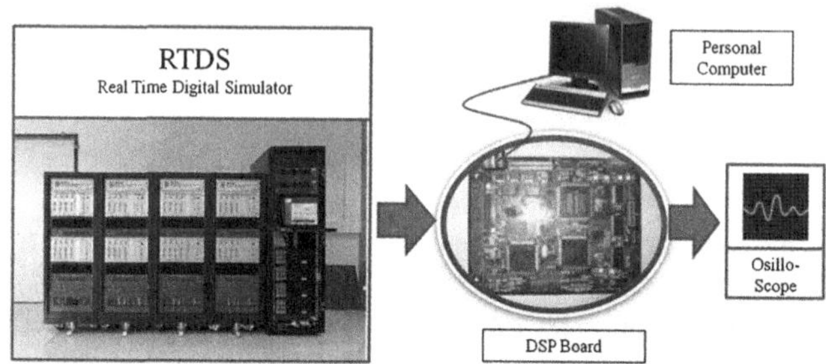

Fig. 23.4 Hardware-in-the-loop testing platform

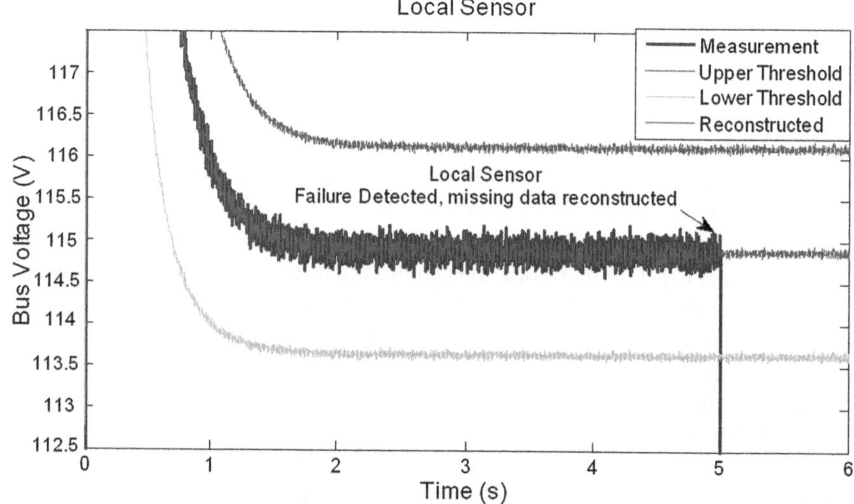

Fig. 23.5 Bus voltage sensor monitoring and diagnosis

Fig. 23.6 Experimental system

ports), physical protection and control devices are connected to the simulator to interact with the simulated systems.

The output voltage of the synchronous generator is sampled and calculated in DSP with PCT. As Fig. 23.5, both the upper and lower thresholds of the output voltage can be successfully built to bound the sensor behavior and the measured signal can be successfully reconstructed after the sensor failed at 5 s.

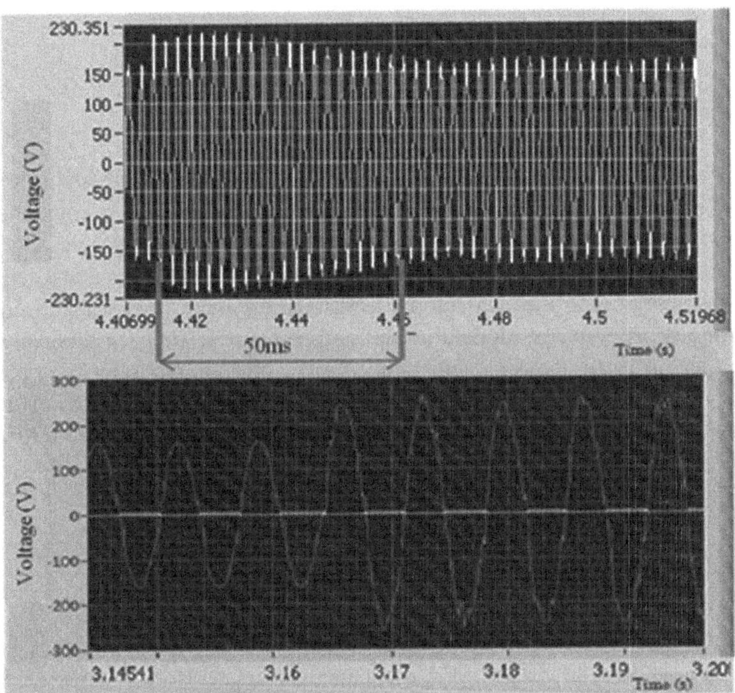

Fig. 23.7 Output voltage with (*up*) and without (*down*) considering sensor failure

23.6 Experimental Validation

Online testing has also been done to verify the proposed algorithm. And the experimental system structure is shown in Fig. 23.6. The controller performances of the synchronous generator with and without the proposed PCT based SFDD algorithm have been compared in this Section.

If no sensor failure validation is applied in the controller, once one of the voltage sensors of the three phases is failed, the measured output voltage of this phase will become zero which leads to the decrease of the averaged RMS value of the output voltage. The controller then will take actions based on the wrong measured data and make the output voltage increase although the real system is still in good condition. This problem is solved by sensor failure diagnosis method based on PCT in this paper. The failed sensor will be detected and the missed data will be reconstructed with the help of sensor diagnosis, which makes sure the system working in a steady state.

Figure 23.7 shows the results of the output voltages with and without considering sensor failure under rated resistive load. As we can see, it takes about 50 ms for the system to recover from a sudden sensor failure to steady state. If no sensor validation is applied, the output voltage will rise up to 172 V (rms). These experimental results show good consistancy with theritical analysis, which further validate the capability of the PCT based SFDD algorithm for online operation.

23.7 Conclusion

This paper illustrates the digital realization of PCT based SFDD algorithm considering parameter uncertainties. DSP TMS320F2812 is adopted to implement the proposed SFDD algorithm. Both hardware-in-the-loop testing and laboratory experimental testing have been done to verify the practical application capability of the proposed algorithm. An automatic voltage regulator for variable frequency synchronous generator is adopted as an appropriate case study in this paper. The control performance of AVR is guaranteed to be more reliable by the application of the proposed SFDD method based on PCT. Control performances of the regulator with and without considering sensor validation are investigated. Experiment results show good consistency to the theory analysis.

References

1. Li W, Zhang X, Li H, Yao W (2013) Sensor failure detection and diagnosis via polynomial chaos theory-Part I: theoretical background. Lecture notes in electrical engineering, pp 1–8
2. Smith AHC (2007) Robust and Optimal Control Using Polynomial Chaos Theory. PHD dissertation, University of South Carolina

3. Wiener XXX (1938) The homogeneous chaos. Amer J Math 60:897–936
4. Xiu D, Karniadakis GE (2002) The Wiener–Askey polynomial chaos for stochastic differential equations. SIAM J Sci Comput 24(2):619–644
5. Rosado S, Ma X, Han C, Wang F, Boroyevich D (2008) Model-based digital controller for a variable frequency synchronous generator with brushless exciter. IEEE Trans Energy Convers 23(1):42–52
6. Li W, Zhang X, Li H (2010) A novel digital automatic voltage regulator for synchronous generator. Proceedings of IEEE POWERCON, pp 1–6
7. Li W, Li H, Ni F, Zhang X, Monti A (2011) Digital automatic voltage regulator for synchronous generator considering sensor failure. Eur Trans Electr Power 22(8):1037–1052
8. Forsyth P, Maguire T, Kuffel R (2004) Real time digital simulation for control and protection system testing. Proc IEEE APESC 1:329–335

Chapter 24
Covariance Intersection Fusion Robust Steady-State Kalman Filter for Two-Sensor Systems with Time-Delayed Measurements

Wenjuan Qi, Peng Zhang, Wenqing Feng and Zili Deng

Abstract For two-sensor systems with time-delayed measurements and uncertain noise variances, this paper presents a measurements transformation approach which transforms the systems with time-delayed measurements into the equivalent systems without measurement delays. Further the local robust steady-state Kalman filter with conservative upper bounds of unknown noise variances is presented, and then the covariance intersection (CI) fusion robust steady-state Kalman filter is also presented. The robustness of these filters is proved based on the Lyapunov equation. It is proved that the robust accuracy of the CI fuser is higher than that of each local robust Kalman filter. A Monte-Carlo simulation example shows its correctness and effectiveness.

Keywords Multi-sensor information fusion · Covariance intersection fusion · Robust Kalman filter · Time-delayed measurements · Uncertain noise variances

24.1 Introduction

The multi-sensor information fusion has received great attentions and has been widely applied in many high-technology fields, such as tracking, signal proceeding, GPS position, robotics and so on. Usually, the standard systems without time-delayed observations are considered, but in many fields, such as the communication and control engineering, the systems with observations delays exist [1, 2].

The optimal Kalman filtering needs to exactly know system model and noise variances, the robust Kalman filters are designed to solve the filtering problems for

W. Qi · P. Zhang · W. Feng · Z. Deng (✉)
Department of Automation, Heilongjiang University, Electronic and Engineering College
130, XueFu Road 74, Harbin 150080, China
e-mail: dzl@hlju.edu.cn

Z. Sun and Z. Deng (eds.), *Proceedings of 2013 Chinese Intelligent Automation Conference*, Lecture Notes in Electrical Engineering 255,
DOI: 10.1007/978-3-642-38460-8_24,

uncertain systems. In recent years, several results have been derived for any admissible uncertainty of model parameters [3, 4] based on the Riccati equations.

Recently, the covariance intersection fusion (CI) method has been presented by Julier and Uhlman [5], which can handle the systems with unknown variances and cross-covariances.

In this paper, the two-sensor systems with uncertain noise variances and time-delayed measurements are considered. The local steady-state robust Kalman filter is presented and the covariance intersection (CI) fusion robust Kalman filter is proposed by the convex combination of the local robust Kalman filters. The robustness of the filters is proved based on the Lyapunov equation.

24.2 Measurement Transformation

Consider the two-sensor uncertain system with time-delayed measurements

$$x(t+1) = \Phi x(t) + \Gamma w(t) \tag{24.1}$$

$$z_i(t) = H_i x(t-k_i) + e_i(t), i = 1,2, \tag{24.2}$$

where t is the discrete time, $x(t) \in R^n$ is the state, $z_i(t) \in R^{m_i}$ is the measurement of the ith subsystem, $k_i \geq 0$ is the time-delay, $w(t) \in R^r, e_i(t) \in R^{m_i}$ are uncorrelated white noises with zeros mean and unknown actual variances $\bar{Q}$ and $\bar{R}_i$, respectively. Φ, Γ and H_i are known constant matrices. Assume that Q and R_i are conservative upper bounds of $\bar{Q}$ and $\bar{R}_i$, respectively, i.e.

$$\bar{Q} \leq Q, \bar{R}_i \leq R_i, i = 1,2 \tag{24.3}$$

where $A \leq B$ means that $B - A \geq 0$ is a semi-positive definite matrix. Assume that each subsystem is completely observable and completely controllable.

Introducing the new measurements $y_i(t)$ and the measurement noises $v_i(t)$

$$y_i(t) = z_i(t+k_i), v_i(t) = e_i(t+k_i), i = 1,2 \tag{24.4}$$

From (24.2), we have the observation equations without time-delayed

$$y_i(t) = H_i x(t) + v_i(t), i = 1,2 \tag{24.5}$$

where $v_i(t)$ also has the variances $\bar{R}_i$. From (24.4), we have

$$\hat{x}_i^z(t|t) = \hat{x}_i(t|t-k_i), i = 1,2 \tag{24.6}$$

where $\hat{x}_i^z(t|t)$ are the estimates of $x(t)$ based on $(z_i(t), z_i(t-1), \cdots)$ $\hat{x}_i(t|t-k_i)$ are the estimates of $x(t)$ based on $(y_i(t-k_i), y_i(t-k_i-1), \cdots)$.

Define the local steady-state cross-covariance as

$$P_{ij}^z = E\left[\tilde{x}_i^z(t|t)\tilde{x}_j^{zT}(t|t)\right], P_{ij}(k_i, k_j) = E\left[\tilde{x}_i(t|t-k_i)\tilde{x}_j^T(t|t-k_j)\right] \tag{24.7}$$

where $\tilde{x}_i^z(t|t) = x(t) - \hat{x}_i^z(t|t)$, $\tilde{x}_i(t|t-k_i) = x(t) - \hat{x}_i(t|t-k_i)$, from (24.6), we can get $P_{ij}^z = P_{ij}(k_i, k_j)$. When $i = j$, defining $P_i^z = P_{ii}^z$, $P_i(k_i) = P_{ii}(k_i, k_i)$, we have $P_i^z = P_i(k_i)$.

The problem is to find the local robust steady-state Kalman filter $x_i^z(t|t)$ and the CI fused robust steady-state Kalman filter $x_{CI}^z(t|t)$.

24.3 Local Robust Steady-State Kalman Filter

For two-sensor system (24.1) and (24.5), the local conservative steady-state Kalman one-step predictor with conservative variances Q and R_i are given by Sun and Deng [6], Kailath et al. [7]

$$\hat{x}_i(t+1|t) = \psi_{pi}\hat{x}_i(t|t-1) + K_{pi}y_i(t) \tag{24.8}$$

$$\psi_{pi} = \Phi - K_{pi}H_i, K_{pi} = \Phi\Sigma_i H_i^T\left(H_i\Sigma_i H_i^T + R_i\right)^{-1} \tag{24.9}$$

where Ψ_{pi} is a stable matrix and conservative one-step predictor error variance Σ_i satisfies the steady-state Riccati equation

$$\Sigma_i = \Phi\left[\Sigma_i - \Sigma_i H_i^T\left(H_i\Sigma_i H_i^T + R_i\right)^{-1} H_i\Sigma_i\right]\Phi^T + \Gamma Q\Gamma^T \tag{24.10}$$

From (24.8), it can be rewritten as the Layapunov equation

$$\Sigma_i = \psi_{pi}\Sigma_i\psi_{pi}^T + \Gamma Q\Gamma^T + K_{pi}R_iK_{pi}^T \tag{24.11}$$

Defining the actual steady-state one-step predictor error variance as

$$\bar{\Sigma}_i = \mathrm{E}\left[\tilde{x}_i(t+1|t)\tilde{x}_i^{\mathrm{T}}(t+1|t)\right], \tilde{x}_i(t+1|t) = x(t+1) - \hat{x}_i(t+1|t) \tag{24.12}$$

Theorem 1 *The Kalman one-step predictor* (24.8)–(24.11) *is robust for all admissible actual variances* $\bar{Q}$ *and* $\bar{R}_i$ *satisfying* $\bar{Q} \leq Q, \bar{R}_i \leq R_i$, *in the sense that*

$$\bar{\Sigma}_i \leq \Sigma_i \tag{24.13}$$

Proof From (24.1), we have $\hat{x}_i(t+1|t) = \Phi\hat{x}_i(t|t)$, applying (24.12) yields $\tilde{x}_i(t+1|t) = \Phi\tilde{x}_i(t|t) + \Gamma w(t)$, where $\tilde{x}_i(t|t) = \left[I_n - K_{fi}H\right]\tilde{x}_i(t|t-1) - K_{fi}v_i(t)$ and $K_{fi} = \Sigma_i H_i^T\left(H_i\Sigma_i H_i^T + R_i\right)^{-1}$, we have the actual prediction error formula

$$\tilde{x}_i(t+1|t) = \psi_{pi}\tilde{x}_i(t|t-1) + \Gamma w(t) - K_{pi}v_i(t) \tag{24.14}$$

According to (24.12), applying (24.14) yields the actual steady-state one-step predictor error variance as

$$\bar{\Sigma}_i = \psi_{pi}\bar{\Sigma}_i\psi_{pi}^T + \Gamma\bar{Q}\Gamma^T + K_{pi}\bar{R}_iK_{pi}^T \tag{24.15}$$

Defining $\Delta\Sigma_i = \Sigma_i - \bar{\Sigma}_i$, subtracting (24.15) from (24.11) yields the Lyapunov equation

$$\Delta\Sigma_i = \psi_{pi}\Delta\Sigma_i\psi_{pi}^T + \Gamma(Q - \bar{Q})\Gamma^T + K_{pi}(R_i - \bar{R}_i)K_{pi}^T \tag{24.16}$$

Applying (24.3), noting that ψ_{pi} is a stable matrix, and applying the property of the Lyapunov equation [1] yield that $\Delta\Sigma_i \geq 0$, i.e. $\bar{\Sigma}_i \leq \Sigma_i$. □

For (24.1) and (24.5), the steady-state multi-step Kalman predictors are given by [6, 7]

$$\hat{x}_i(t + k_i|t) = \Phi^{k_i-1}\hat{x}_i(t + 1|t), k_i \geq 2 \tag{24.17}$$

The local steady-state multi-step predictor error variances are given as

$$P_i(k_i) = \Phi^{k_i-1}\Sigma_i\left(\Phi^{k_i-1}\right)^T + \sum_{j=0}^{k_i-2}\Phi^j\Gamma Q\Gamma^T\left(\Phi^j\right)^T, k_i \geq 2 \tag{24.18}$$

Defining the actual steady-state multi-step predictor error variance as

$$\bar{P}_i(k_i) = \mathrm{E}\left[\tilde{x}_i(t + k_i|t)\tilde{x}_i^{\mathrm{T}}(t + k_i|t)\right], \tilde{x}_i(t + k_i|t) = x(t + k_i) - \hat{x}_i(t + k_i|t) \tag{24.19}$$

Theorem 2 The conservative Kalman multi-step predictor (24.17)–(24.18) is robust for all admissible actual variances $\bar{Q}$ and $\bar{R}_i$ satisfying $\bar{Q} \leq Q, \bar{R}_i \leq R_i$. i.e.

$$\bar{P}_i(k_i) \leq P_i(k_i) \tag{24.20}$$

Proof Iterating $N - 1$ steps for (24.1), we obtain the non-recursive formula as

$$x(t + k_i) = \Phi^{k_i-1}x(t + 1) + \sum_{j=0}^{k_i-2}\Phi^j\Gamma w(t - 1) \tag{24.21}$$

Substituting (24.17) and (24.21) into $\tilde{x}_i(t + k_i|t) = x(t + k_i) - \hat{x}_i(t + k_i|t)$, we have

$$\tilde{x}_i(t + k_i|t) = \Phi^{k_i-1}\tilde{x}(t + 1|t) + \sum_{j=0}^{k_i-2}\Phi^j\Gamma w(t - 1) \tag{24.22}$$

Substituting (24.22) into (24.19) yields the actual steady-state filtering error variance as

$$\bar{P}_i(k_i) = \Phi^{k_i-1}\bar{\Sigma}_i\left(\Phi^{k_i-1}\right)^T + \sum_{j=0}^{k_i-2}\Phi^j\Gamma\bar{Q}\Gamma^T\left(\Phi^j\right)^T, k_i \geq 2 \tag{24.23}$$

Defining $\Delta P_i(k_i) = P_i(k_i) - \bar{P}_i(k_i)$, subtracting (24.23) from (24.18) yields

$$\Delta P_i(k_i) = \Phi^{k_i-1}(\Sigma_i - \bar{\Sigma}_i)\left(\Phi^{k_i-1}\right)^T + \sum_{j=0}^{k_i-2}\Phi^j\Gamma(Q - \bar{Q})\Gamma^T\left(\Phi^j\right)^T \tag{24.24}$$

Applying (24.3) and (24.13) yields $\Delta P_i(k_i) \geq 0$, (24.20) holds. □

24.4 CI Fusion Robust Steady-State Kalman Filter

For two-sensor system (24.1) and (24.2), applying the CI fused algorithm [5], the CI fusion robust steady-state Kalman filters are given as

$$\hat{x}^z_{CI}(t|t) = P^z_{CI}\left(\omega\left(P^z_1\right)^{-1}\hat{x}^z_1(t|t) + (1-\omega)\left(P^z_2\right)^{-1}\hat{x}^z_2(t|t)\right) \tag{24.25}$$

$$P^z_{CI} = \left[\omega\left(P^z_1\right)^{-1} + (1-\omega)\left(P^z_1\right)^{-1}\right]^{-1} \tag{24.26}$$

Applying (24.25), (24.26), (24.6) and (24.7) yields

$$\hat{x}^z_{CI}(t|t) = P^z_{CI}\left(\omega P_1^{-1}(k_1)\hat{x}_1(t|t-k_1) + (1-\omega)P_2^{-1}(k_2)\hat{x}_2(t|t-k_2)\right) \tag{24.27}$$

$$P^z_{CI} = \left[\omega P_1^{-1}(k_1) + (1-\omega)P_2^{-1}(k_2)\right]^{-1} \tag{24.28}$$

with the constraint $\omega \geq 0$, when $k_i = 1$, we have $P_1(1) = \Sigma_1, P_2(1) = \Sigma_2$.

The weighting coefficient ω is obtained by minimizing the performance index

$$\min_{\omega} trP^z_{CI} = \min_{\omega\in[0,1]} tr\left\{\left[\omega P_1^{-1}(k_1) + (1-\omega)P_2^{-1}(k_2)\right]^{-1}\right\} \tag{24.29}$$

where the symbol tr denotes the trace of matrix. The optimal weights ω can be quickly obtained by the 0.618 method or the Fibinacci method.

Theorem 3 *The covariance intersection fused filter* (24.27) *and* (24.28) *has the actual error variance* $\bar{P}_{CI}$ *as*

$$\begin{aligned}\bar{P}^z_{CI} &= E\left[\tilde{x}^z_{CI}(t|t)\tilde{x}^{zT}_{CI}(t|t)\right]\\ &= P^z_{CI}\left[\omega^2 P_1^{-1}(k_1)\bar{P}_1(k_1)P_1^{-1}(k_1) + \omega(1-\omega)P_1^{-1}(k_1)\bar{P}_{12}(k_1,k_2)P_2^{-1}(k_2)\right.\\ &\quad \left.+\omega(1-\omega)P_2^{-1}(k_2)\bar{P}_{21}(k_2,k_1)P_1^{-1}(k_1) + (1-\omega)^2 P_2^{-1}(k_2)\bar{P}_2(k_2)P_2^{-1}(k_2)\right]P^z_{CI}\end{aligned} \tag{24.30}$$

where $\bar{P}_{12}(k_1,k_2) = E\left[\tilde{x}_1(t|t-k_1)\tilde{x}_2^T(t|t-k_2)\right]$ and

$$\begin{aligned}\bar{P}_{12}(k_1,k_2) =& \Phi^{k_1-1}\psi_{p1}^{k_2-k_1}\bar{\Sigma}_{12}\left(\Phi^{k_2-1}\right)^T\\ &+\sum_{r=k_1-1}^{k_2-2}\Phi^{k_1-1}\psi_{pi}^{k_1-r-1}\Gamma\bar{Q}\Gamma^T(\Phi^r)^T + \sum_{r=0}^{k_1-2}\Phi^r\Gamma\bar{Q}\Gamma(\Phi^r)^T, k_2\geq k_1\geq 2\end{aligned} \tag{24.31}$$

$$\begin{aligned}\bar{P}_{12}(k_1,k_2) =& \Phi^{k_1-1}\bar{\Sigma}_{12}\psi_{p2}^{k_1-k_2}\left(\Phi^{k_2-1}\right)^T\\ &+\sum_{r=k_2-1}^{k_1-2}\Phi^r\Gamma\bar{Q}\Gamma^T\psi_{p2}^{(k_2-r-1)T}\Psi^{(k_2-1)T} + \sum_{r-0}^{k_2-2}\Phi^r\Gamma\bar{Q}\Gamma(\Phi^r)^T, k_1\geq k_2\geq 2\end{aligned} \tag{24.32}$$

Especially

$$\bar{P}_{12}(1,1)=\bar{\Sigma}_{12},\bar{\Sigma}_{12}=\psi_{p1}\bar{\Sigma}_{12}\psi_{p2}^{T}+\Gamma\bar{Q}\Gamma^{T} \quad (24.33)$$

$$\bar{P}_{12}(k_1,1)=\Phi^{k_1-1}\bar{\Sigma}_{12}\psi_{p2}^{k_1-1}+\sum_{r=0}^{k_1-2}\Phi^{r}\Gamma\bar{Q}\Gamma^{T}\Psi_{p2}^{rT} \quad (24.34)$$

$$\bar{P}_{12}(1,k_2)=\psi_{p1}^{k_2-1}\bar{\Sigma}_{12}\left(\Phi^{k_2-1}\right)^{T}+\sum_{r=0}^{k_2-2}\psi_{pi}^{r}\Gamma\bar{Q}\Gamma^{T}(\Phi^{r})^{T} \quad (24.35)$$

Proof From (24.28), we have $x(t)=P_{CI}^{z}\left[\omega P_1^{-1}(k_1)+(1-\omega)P_2^{-1}(k_2)\right]x(t)$. Using (24.27), we easily obtain the CI actual fused filtering error

$$\tilde{x}_{CI}^{z}(t|t)=P_{CI}^{z}\left[\omega P_1^{-1}(k_1)\tilde{x}_1(t|t-k_1)+(1-\omega)P_2^{-1}(k_2)\tilde{x}_2(t|t-k_2)\right] \quad (24.36)$$

which yields (24.30). Equations (24.31)–(24.35) have been proved in Ref. [6]. □

Remark 1 Applying (24.20), Ref. [5] proved that the two-sensor CI fuser is robust for all admissible $\bar{Q}$ and $\bar{R}_i$ satisfying (24.3), i.e.

$$\bar{P}_{CI}^{z}\leq P_{CI}^{z} \quad (24.37)$$

24.5 Accuracy Analysis

Theorem 4 For the two-sensor system (24.1)–(24.2) with time-delayed measurements, the local steady-state robust Kalman filter and CI fuser have the accuracy relations

$$\bar{P}_i^{z}=\bar{P}_i(k_i),P_i^{z}=P_i(k_i) \quad (24.38)$$

$$tr\bar{P}_i^{z}\leq trP_i^{z},i=1,2 \quad (24.39)$$

$$tr\bar{P}_{CI}^{z}\leq trP_{CI}^{z}\leq trP_i^{z},i=1,2 \quad (24.40)$$

Proof From the robustness (24.20) the accuracy relation (24.39) holds. From (24.37), the first inequality of (24.40) holds. Applying (24.29), taking $\omega=1$ yields $trP_{CI}^{z}=trP_1^{z}$ and $\omega=0$ yields $trP_{CI}^{z}=trP_2^{z}$, Hence when $\omega\in[0,1]$, we have the accuracy relation $trP_{CI}^{z}\leq trP_i^{z},i=1,2$ □.

Remark 2 Inequalities (24.39) and (24.40) show that the robust accuracy of the CI fuser is higher than that of each local robust filter.

24.6 Simulation Example

Consider the two-sensor tracking system (24.1)–(24.2)with time-delayed measurements, where $\Phi = \begin{bmatrix} 1 & T_0 \\ 0 & 1 \end{bmatrix}, \Gamma = \begin{bmatrix} 0.5T_0^2 \\ T_0 \end{bmatrix}$, $H_1 = [1 \quad 0]$, $H_2 = I_2$, $T_0 = 0.25$ is the sampled period, $x(t) = [x_1(t), x_2(t)]^T$ is the state, $x_1(t)$ and $x_2(t)$ are the position and velocity of target at time tT_0. $w(t)$ and $v_i(t)$ are independent Gaussion white noises with zero mean and unknown variances Q and R_i respectively. In the simulation, we take $Q = 0.5$, $R_1 = 0.58$, $R_2 = diag(4, 0.25)$, $\bar{Q} = 0.45$, $\bar{R}_1 = 0.5$, $\bar{R}_2 = diag(3,\ 0.16)$, $k_1 = 1$, $k_2 = 2$

In order to give a geometric interpretation of the accuracy relations, the covariance ellipse is defined as the locus of points $\{x : x^T P^{-1} x = c\}$, where P is the variance matrix and c is a constant. Generally, we select $c = 1$. It has been proved in [8] that $P_1 \leq P_2$ is equivalent to that the covariance ellipse of P_1 is enclosed in that of P_2. The accuracy comparison of the covariance ellipses is shown in Fig. 24.1. From Fig. 24.1, we see that the ellipse of the actual variances $\bar{\Sigma}_1 or \bar{P}_2(2)$ is enclosed in that of $\Sigma_1 or P_2(2)$, respectively, which verify the consistent (24.13) and (24.20). The ellipse of actual CI fused variance $\bar{P}_{CI}$ is enclosed in that of P_{CI}, which verifies the robustness of (24.37).

In order to verify the above theoretical accuracy relations, taking $\rho = 200$ runs, the curves of the mean square errors (MSE) of local and fused Kalman filters are shown in Fig. 24.2, which verifies the accuracy relations (24.39), (24.40) and the accuracy relations in Table 24.1.

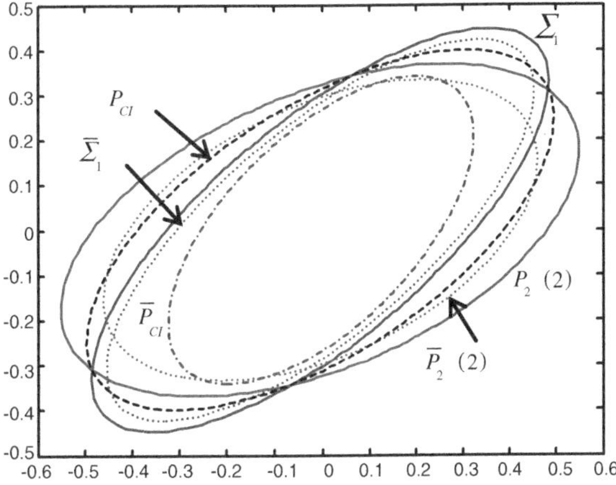

Fig. 24.1 The covariance *ellipses* of robust Kalman filters

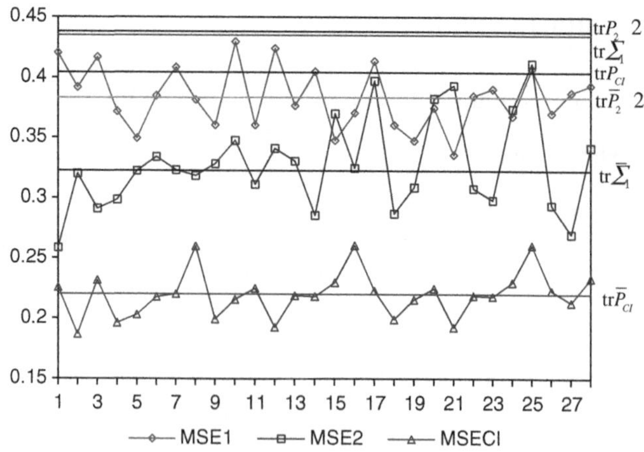

Fig. 24.2 The MSE *curves* of local and fused filters

Table 24.1 The accuracy comparison of local and fused filters

$tr\Sigma_1$	$trP_2(2)$	$tr\bar{\Sigma}_1$	$tr\bar{P}_2(2)$	trP_{CI}	$tr\bar{P}_{CI}$
0.4348	0.4377	0.3833	0.3225	0.4042	0.2202

24.7 Conclusion

For two-sensor systems with uncertain noise variances and time-delayed measurements, the local and CI robust fused robust steady-state Kalman filters have been presented, and their robustness was proved based on the Lyapunov equation. The robust accuracy of CI fuser is higher than that the robust accuracy of each local filter

Acknowledgments This work is supported by the Natural Science Foundation of China under grant NSFC-60874063, the 2012 Innovation and Scientific Research Foundation of graduate student of Heilongjiang Province under grant YJSCX2012-263HLJ, and the Support Program for Young Professionals in Regular Higher Education Institutions of Heilongjiang Province under grant 1251G012.

References

1. Kailath T, Sayed AH, Hassibi B (2000) Linear estimation. Prentice Hall, New York
2. Lu X, Zhang HS, Wang W (2005) Kalman filtering for multiple time-delay systems. Automatica 87(4):1455–1461
3. Zhu X, Soh YC, Xie L (2002) Design and analysis of discrete-time robust Kalman filters. Automatica 38:1069–1077

4. Theodor Y, Sharked U (1996) Roust discrete-time minimum-variance filtering. IEEE Trans Sig Process 44(2):181–189
5. Julier SJ, Uhlman JK (2009) General decentralized data fusion with covariance intersection. In: Liggins ME, Hall DL, Llinas J (eds) Handbook of multisensor data fusion, theory and practice. CRC Press, Boca Raton, pp 319–342
6. Sun X-J, Deng Z-L (2009) Information fusion wiener filter for the multisensory multichannel ARMA signals with time-delayed measurements. IET-Sig Process 3:403–415
7. Kailath T, Sayed AH, Hassibi B (2000) Linear estimation. Prentice Hall, New York
8. Deng Z, Zhang P, Qi W, Liu J, Gao Y (2012) Sequential covariance intersection fusion Kalman filter. Inf Sci 189:293–309

Chapter 25
Centralized Fusion Steady-State Robust Kalman Filter for Uncertain Multisensor Systems

Peng Zhang, Wenjuan Qi and Zili Deng

Abstract For multisensor systems with uncertain noise variances, a centralized fusion steady-state robust Kalman filter with the upper bound of noise variances is presented. Based on the Lyapunov equation, it is proved that the local and centralized fusion steady-state Kalman filter are robust, i.e., the variances of the actual filtering errors are less than conservative variances of the corresponding robust Kalman filters. Further, it is proved that the robust accuracy of the centralized fusion robust Kalman filter is higher than that of the local robust Kalman filter. A Monte-Carlo simulation example shows its correctness.

Keywords Centralized fusion · Steady-state Kalman filter · Uncertain systems · Robust

25.1 Introduction

Multisensor information fusion Kalman filtering has been applied to many fields, such as guidance, defence, robotics, tracking, GPS positioning [1, 2]. For Kalman filtering-based fusion, two basic fusion methods are centralized and distributed fusion methods, depending on whether raw data are used directly for fusion or not. Although the centralized fusion method may require a larger computational burden, it can give the globally optimal state estimation by directly combining the

P. Zhang
Department of Computer and Information Engineering, Harbin Deqiang College of Commerce, Harbin 150025, China
e-mail: zp52218@sina.com.cn

W. Qi · Z. Deng (✉)
Department of Automation, Heilongjiang University, Electronic and Engineering College, XueFu Road 74 130, Harbin 150080, China
e-mail: dzl@hlju.edu.cn

Z. Sun and Z. Deng (eds.), *Proceedings of 2013 Chinese Intelligent Automation Conference*, Lecture Notes in Electrical Engineering 255,
DOI: 10.1007/978-3-642-38460-8_25, © Springer-Verlag Berlin Heidelberg 2013

local measurement data. However, since the system model is usually an approximation to a physical situation in many applications, the model parameters and noise statistics are seldom exact. To solve the filtering problems for uncertain systems with model parameters, based on Riccati equation, several results have been derived which give an upper bound on the filtering error variances for all admissible uncertainty of model parameters [3–5].

In this paper, for the multisensor systems with uncertainty of noise variances, the local and centralized fusion robust steady-state Kalman filter are presented. Based on Lyapunov equation, their robustness is proved. They respectively give the upper bounds of actual filtering error variances.

25.2 Centralized Fused Steady-State Robust Kalman Filter

Consider the multisensor system with uncertainty of noise variances

$$x(t+1) = \Phi x(t) + \Gamma w(t) \tag{25.1}$$

$$y_i(t) = H_i x(t) + v_i(t),\ i = 1, \ldots, L \tag{25.2}$$

where t is the discrete time, $x(t) \in R^n$ is the state, $y_i(t) \in R^{m_i}$ is the measurement of the *ith* subsystem, $w(t) \in R^r$, $v_i(t) \in R^{m_i}$ are white noises with zero mean and unknown actual variances $\bar{Q}$ and $\bar{R}_i$, respectively. Φ, Γ and H_i are known constant matrices. Assume that Q and R_i are conservative upper bounds of $\bar{Q}$ and $\bar{R}_i$, respectively, i.e.,

$$\bar{Q} \le Q, \bar{R}_i \le R_i, i = 1, \ldots, L \tag{25.3}$$

and each subsystem is completely observable and completely controllable.

Combine all measurement equations as

$$y(t) = Hx(t) + v(t) \tag{25.4}$$

$$y(t) = \begin{bmatrix} y_1(t) & \ldots & y_L(t) \end{bmatrix}^{\mathrm{T}}, H = \begin{bmatrix} H_1 & \ldots & H_L \end{bmatrix}^{\mathrm{T}}, v(t) = \begin{bmatrix} v_1(t) & \ldots & v_L(t) \end{bmatrix}^{\mathrm{T}} \tag{25.5}$$

and $v(t)$ has the conservative upper variance matrix R as

$$R = \text{block} - \text{diag}[\mathrm{R}_1, \ldots, \mathrm{R}_\mathrm{L}] \tag{25.6}$$

Therefore from (25.3), the actual noise variance $\bar{R}$ has the property

$$\bar{R} \le R \tag{25.7}$$

Based on (25.1) and (25.4), the globally optimal centralized fused steady-state robust Kalman filter is given by [6, 7]

$$\hat{x}(t|t) = \Psi\hat{x}(t-1|t-1) + Ky(t) \tag{25.8}$$

where Ψ is a stable matrix, Ψ and K as following

$$\Psi = [I_n - KH]\Phi,\ K = \Sigma H^{\mathrm{T}}(H\Sigma H^{\mathrm{T}} + R)^{-1}, \tag{25.9}$$

the conservative prediction error variance Σ satisfies the Riccati equation

$$\Sigma = \Phi\left[\Sigma - \Sigma H^{\mathrm{T}}\left(H\Sigma H^{\mathrm{T}} + R\right)^{-1}H\Sigma\right]\Phi^{\mathrm{T}} + \Gamma Q\Gamma^{\mathrm{T}} \tag{25.10}$$

The fused upper bound filtering error variance P_0 is obtained by

$$P_0 = (I_n - KH)\Sigma \tag{25.11}$$

and from [8], it can be written as

$$P_0 = \Psi P_0\Psi^{\mathrm{T}} + (I_n - KH)\Gamma Q\Gamma^{\mathrm{T}}(I_n - KH)^{\mathrm{T}} + KRK^{\mathrm{T}} \tag{25.12}$$

Defining the actual fused filtering error variance $\bar{P}_0 = \mathrm{E}[\tilde{x}(t|t)\tilde{x}^{\mathrm{T}}(t|t)]$, substituting (25.1), (25.4), (25.8) and (25.9) to $\tilde{x}(t|t) = x(t) - \hat{x}(t|t)$, we can obtain

$$\tilde{x}(t|t) = \Psi\tilde{x}(t-1|t-1) + (I_n - KH)\Gamma w(t-1) - Kv(t) \tag{25.13}$$

Therefore, we can obtain the actual fused filtering error variance $\bar{P}_0$ as

$$\bar{P}_0 = \Psi\bar{P}_0\Psi^{\mathrm{T}} + (I_n - KH)\Gamma\bar{Q}\Gamma^{\mathrm{T}}(I_n - KH)^{\mathrm{T}} + K\bar{R}K^{\mathrm{T}} \tag{25.14}$$

Theorem 1 *For the uncertain multisensor systems* (25.1) *and* (25.2), *the centralized fused steady-state Kalman filter* (25.8)–(25.11) *is robust, i.e.,*

$$\bar{P}_0 \leq P_0 \tag{25.15}$$

Proof Defining $\Delta P_0 = P_0 - \bar{P}_0$, and subtracting (25.14) from (25.12) yields the Lyapunov equation

$$\Delta P_0 = \Psi\Delta P_0\Psi^{\mathrm{T}} + \Lambda \tag{25.16}$$

where

$$\Lambda = (I_n - KH)\Gamma(Q - \bar{Q})\Gamma^{\mathrm{T}}(I_n - KH)^{\mathrm{T}} + K(R - \bar{R})K^{\mathrm{T}} \tag{25.17}$$

From (25.3) and (25.7), we can obtain that $\Lambda \geq 0$, noting that Ψ is stable, applying the property of Lyapunov equation [6], we have $\Delta P_0 \geq 0$, i.e., $\bar{P}_0 \leq P_0$, the proof is completed.

Remark 1 For uncertain multisensor system (25.1) and (25.2), similar to the centralized fusion steady-state Kalman filter, the local robust steady-state Kalman filter with conservative upper variances Q and R_i given by (25.3) is given by [6, 7]

$$\hat{x}_i(t|t) = \Psi_i\hat{x}_i(t-1|t-1) + K_i y_i(t) \tag{25.18}$$

$$\Psi_i = [I_n - K_i H_i]\Phi, \ K_i = \Sigma_i H_i^{\mathrm{T}}\left(H_i\Sigma_i H_i^{\mathrm{T}} + R_i\right)^{-1} \tag{25.19}$$

$$\Sigma_i = \Phi\left[\Sigma_i - \Sigma_i H_i^{\mathrm{T}}\left(H_i\Sigma_i H_i^{\mathrm{T}} + R_i\right)^{-1} H_i\Sigma_i\right]\Phi^{\mathrm{T}} + \Gamma Q\Gamma^{\mathrm{T}} \tag{25.20}$$

$$P_i = (I_n - K_i H_i)\Sigma_i \tag{25.21}$$

or

$$P_i = \Psi_i P_i \Psi_i^{\mathrm{T}} + (I_n - K_i H_i)\Gamma Q\Gamma^{\mathrm{T}}(I_n - K_i H_i)^{\mathrm{T}} + K_i R_i K_i^{\mathrm{T}} \tag{25.22}$$

and the actual local filtering error variance is given by

$$\bar{P}_i = \Psi_i \bar{P}_i \Psi_i^{\mathrm{T}} + (I_n - K_i H_i)\Gamma \bar{Q}\Gamma^{\mathrm{T}}(I_n - K_i H_i)^{\mathrm{T}} + K_i \bar{R}_i K_i^{\mathrm{T}} \tag{25.23}$$

Similarly, the local steady-state Kalman filter is also robust, i.e.,

$$\bar{P}_i \leq P_i \tag{25.24}$$

25.3 The Accuracy Comparison Between Local and Centralized Fused Robust Kalman Filters

Theorem 2 *For the uncertain multisensor systems* (25.1) *and* (25.2), *accuracy relation between local and centralized fused robust Kalman filters is given asthe*

$$\bar{P}_0 \leq P_0 \leq P_i \tag{25.25}$$

Proof The first inequality $\bar{P}_0 \leq P_0$ has been proved in (25.15). The local and centralized optimal robust Kalman filter are time varying, and has the form in information filter [6, 8]

$$P_0^{-1}(t|t) = P_0^{-1}(t|t-1) + H^{\mathrm{T}}R^{-1}H, P_i^{-1}(t|t) = P_i^{-1}(t|t-1) + H_i^{\mathrm{T}}R_i^{-1}H_i \tag{25.26}$$

$$\begin{aligned} P_0(t|t-1) &= \Phi P_0(t-1|t-1)\Phi^{\mathrm{T}} + \Gamma Q\Gamma^{\mathrm{T}}, \\ P_i(t|t-1) &= \Phi P_i(t-1|t-1)\Phi^{\mathrm{T}} + \Gamma Q\Gamma^{\mathrm{T}} \end{aligned} \tag{25.27}$$

By the mathematical induction, when $t = 1$, from (25.27), we can obtain

$$P_0(1|0) = \Phi P_0(0|0)\Phi^{\mathrm{T}} + \Gamma Q\Gamma^{\mathrm{T}}, P_i(1|0) = \Phi P_i(0|0)\Phi^{\mathrm{T}} + \Gamma Q\Gamma^{\mathrm{T}} \tag{25.28}$$

Taking the same initial value $P_0(0|0) = P_i(0|0)$, we have

$$P_0(1|0) = P_i(1|0) \tag{25.29}$$

By (25.26) and (25.5), we can obtain

$$P_0^{-1}(1|1) = P_0^{-1}(1|0) + H^{\mathrm{T}}R^{-1}H, P_i^{-1}(1|1) = P_i^{-1}(1|0) + H_i^{\mathrm{T}}R_i^{-1}H_i \tag{25.30}$$

$$H^{\mathrm{T}}R^{-1}H = \sum_{i=1}^{L} H_i^{\mathrm{T}}R_i^{-1}H_i \geq H_i^{\mathrm{T}}R_i^{-1}H_i \tag{25.31}$$

From (25.30) and (25.31), we have $P_0^{-1}(1|1) \geq P_i^{-1}(1|1)$, so that

$$P_0(1|1) \leq P_i(1|1) \tag{25.32}$$

We now assume that the formula holds for $t = k$, that is

$$P_0(k|k) \leq P_i(k|k) \tag{25.33}$$

We must now verify that the formula holds for $t = k + 1$ as well. Using (25.27)

$$P_0(k+1|k) = \Phi P_0(k|k)\Phi^{\mathrm{T}} + \Gamma Q \Gamma^{\mathrm{T}}, P_i(k+1|k) = \Phi P_i(k|k)\Phi^{\mathrm{T}} + \Gamma Q \Gamma^{\mathrm{T}} \tag{25.34}$$

Then by (25.33), we have

$$P_0(k+1|k) \leq P_i(k+1|k) \tag{25.35}$$

So we can obtain that $P_0^{-1}(k+1|k) \geq P_i^{-1}(k+1|k)$, then by (25.26),

$$\begin{aligned} P_0^{-1}(k+1|k+1) &= P_0^{-1}(k+1|k) + H^{\mathrm{T}}R^{-1}H, \\ P_i^{-1}(k+1|k+1) &= P_i^{-1}(k+1|k) + H_i^{\mathrm{T}}R_i^{-1}H_i \end{aligned} \tag{25.36}$$

From (25.31), we have

$$P_0^{-1}(k+1|k+1) \geq P_i^{-1}(k+1|k+1) \tag{25.37}$$

i.e.,

$$P_0(k+1|k+1) \leq P_i(k+1|k+1) \tag{25.38}$$

Therefore, for the arbitrary t, we can obtain that

$$P_0(t|t) \leq P_i(t|t) \tag{25.39}$$

Taking limitation to (25.39), we have $P_0 \leq P_i$. The proof is completed.
Taking the trace of (25.25), we have the accuracy relation as

$$tr\bar{P}_0 \leq trP_0 \leq trP_i, i = 1, \ldots, L \tag{25.40}$$

Remark2 We define the robust accuracy of the centralized fusion steady-state Kalman filter is trP_0, the robust accuracy of local steady-state Kalman filter is trP_i. Equation (25.31) shows that the robust accuracy of centralized robust Kalman filter is higher than that of the local robust Kalman filter.

25.4 Simulation Example

Consider 3-sensor tracking system

$$x(t+1)=\begin{bmatrix}1 & T_0\\ 0 & 1\end{bmatrix}x(t)+\begin{bmatrix}0.5T_0^2\\ T_0\end{bmatrix}w(t) \quad (25.41)$$

$$y_i(t)=H_ix(t)+v_i(t),\ i=1,2,3 \quad (25.42)$$

$$H_1=[1\quad 0],\ H_2=I_2,\ H_3=[1\quad 0] \quad (25.43)$$

where T_0 is the sample period, $x(t)\in R^2$ is the state, $x(t)=[x_1(t),x_2(t)]^{\mathrm{T}}$, $x_1(t)$ and $x_2(t)$ are the position and velocity of target at time tT_0, $y_i(t)$ is the measurement for sensor i, $w(t)$ and $v_i(t)$ are white noise with zero mean and unknown variances $\bar{Q}$ and $\bar{R}_i$, respectively. Assume that Q and R_i are conservative estimates of $\bar{Q}$ and $\bar{R}_i$, respectively. In the simulation, we take $T_0=0.4$, $Q=2.2$, $R_1=1.2$, $R_2=\mathrm{diag}(4.9,3.6)$, $R_3=\mathrm{diag}(5,0.81)$, $\bar{Q}=2$, $\bar{R}_1=1$, $\bar{R}_2=\mathrm{diag}(4,0.25)$, $\bar{R}_3=\mathrm{diag}(4.2,0.64)$. The comparisons of the filtering error variance matrices between the local and centralized fused Kalman filters are shown in Tables (25.1), (25.2) and (25.3). From Tables (25.1) and (25.2), we can obtain that $P_i-\bar{P}_i\geq 0$, $P_0-\bar{P}_0\geq 0$, $P_i-P_0\geq 0, i=1,2,3$, i.e., these matrices are positive definite, and also shows that the relations (25.15), (25.24) and (25.25) hold.

In order to verify the above theoretical results for the accuracy relation, the 200 Monte-Carlo runs are performed. The mean-square error (MSE) curves of the local and centralized fused Kalman filters are shown in Fig. 25.1, where the straight lines denote trP_i and $tr\bar{P}_i$, $i=0,1,2,3$, the curves denote $MSE_i(t)$ values, and $MSE_i(t)$ value at time t is defined as the sampled average for $trP_i=tr\mathrm{E}[\tilde{x}_i(t|t)\tilde{x}_i^{\mathrm{T}}(t|t)]$ with $\tilde{x}_i(t|t)=x(t)-\hat{x}_i(t|t)$, i.e.,

Table 25.1 The accuracy comparison of P_i $i=0,1,2,3$

P_1	P_2	P_3	P_0
$\begin{bmatrix}0.5769 & 0.4683\\ 0.4683 & 0.9079\end{bmatrix}$	$\begin{bmatrix}0.4862 & 0.0845\\ 0.0845 & 0.2192\end{bmatrix}$	$\begin{bmatrix}0.7016 & 0.1826\\ 0.1826 & 0.3755\end{bmatrix}$	$\begin{bmatrix}0.1512 & 0.049\\ 0.049 & 0.1645\end{bmatrix}$

Table 25.2 The accuracy comparison of $\bar{P}_i$, $i=0,1,2,3$

$\bar{P}_1$	$\bar{P}_2$	$\bar{P}_3$	$\bar{P}_0$
$\begin{bmatrix}0.4884 & 0.4051\\ 0.4051 & 0.8054\end{bmatrix}$	$\begin{bmatrix}0.3703 & 0.058\\ 0.058 & 0.1658\end{bmatrix}$	$\begin{bmatrix}0.5751 & 0.1451\\ 0.1451 & 0.3133\end{bmatrix}$	$\begin{bmatrix}0.1192 & 0.0356\\ 0.0356 & 0.1272\end{bmatrix}$

Table 25.3 The accuracy comparison of trP_i and $tr\bar{P}_i$, $i=0,1,2,3$

$trP_1(tr\bar{P}_1)$	$trP_2(tr\bar{P}_2)$	$trP_3(tr\bar{P}_3)$	$trP_0(tr\bar{P}_0)$
1.4848(1.2938)	0.70538(0.53611)	1.0771(0.88841)	0.31566(0.24638)

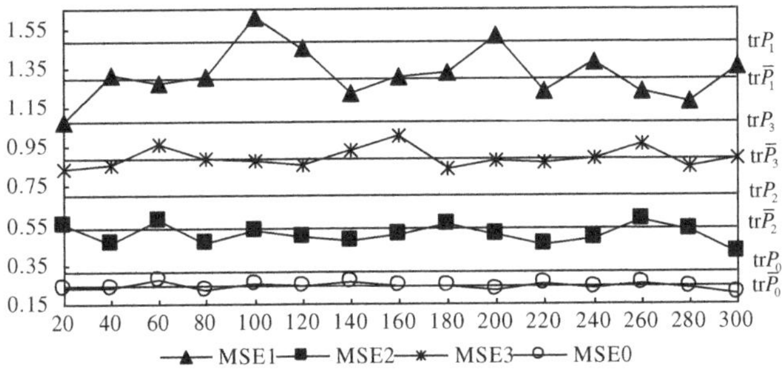

Fig. 25.1 The comparison of $MSE_i(t)$ and trP_i, $i = 0, 1, 2, 3$

$$MSE_i(t) = \frac{1}{N}\sum_{j=1}^{N} \tilde{x}_i^{(j)\mathrm{T}}(t|t)\tilde{x}_i^{(j)}(t|t), \ i = 0, 1, 2, 3 \tag{25.44}$$

where $N = 200$ is the run number, $\tilde{x}_i^{(j)}(t|t) = \hat{x}_i^{(j)}(t|t) - x^{(j)}(t)$, $\hat{x}_i^{(j)}(t|t)$ or $x^{(j)}(t)$ denote the jth realization (sample) of $\hat{x}_i(t|t)$ or $x(t)$, $t = 1, \ldots, 200$ denotes the time (step). From Fig. 25.1, we see that the values of the $MSE_i(t)$ are close to the corresponding $tr\bar{P}_i$ and the accuracy relation (25.40) holds.

25.5 Conclusion

For multisensor systems with uncertainty of noises variances, the local and centralized fused steady-state Kalman filter are presented. Based on the Lyapunov equation, it is proved that the local and centralized fused steady-state Kalman filters are robust, and we proved that the robust accuracy of centralized fusion steady-state Kalman filter is higher than that of the local steady-state Kalman filter.

Acknowledgments This work is supported by the Natural Science Foundation of China under grant NSFC-60874063, the 2012 Innovation and Scientific Research Foundation of graduate student of Heilongjiang Province under grant YJSCX2012-263HLJ, and the Support Program for Young Professionals in Regular Higher Education Institutions of Heilongjiang Province under grant 1251G012.

References

1. Shalom YB, Li XR, Kirubarajan T (2001) Estimation with applications to tracking and navigation. Wiley, New York
2. Sun SL, Deng ZL (2004) Multisensor optimal information Kalman filter. Automatica 40:1017–1023
3. Zhu X, Yeng CS, Xie LH (2002) Design and analysis of discrete-time robust Kalman filters. Automatica 38:1069–1077

4. Xie LH, Yeng CS, De Souza CE (1994) Robust Kalman filtering for uncertain discrete-time systems. IEEE Trans Autom Control 39:1310–1314
5. Theodor Y, Sharked U (1996) Roust discrete-time minimum-variance filtering. IEEE Trans Signal Process 44(2):181–189
6. Kailath T, Sayed AH, Hassibi B (2000) Linear estimation. Prentice Hall, New York
7. Jazwinski AH (1970) Stochastic processed and filtering theory. Academic Press, New York
8. Deng ZL (2012). Information fusion estimation theory and applications. Science Press

Part II
Intelligent Technology and Systems

Chapter 26
Agent Based Railway Network Rescheduling System

Li Wang, Yong Qin, Jie Xu and Limin Jia

Abstract An agent based train rescheduling system for the railway network is introduced. The system is formed by four layers that including implementation layer, section rescheduling layer, line rescheduling layer and network rescheduling layer from bottom to top of the system. Four kinds of rescheduling agents are involved in each layer based on its feature. The agents in the same layer or in the neighbor layers can communicate with each other. The agent based network train rescheduling system benefits to get a global optimization in the perspective of railway network. In the end a simulation on Beijing-Shanghai high speed railway is realized to prove the system effectiveness.

Keywords Agent · Train rescheduling · Railway · Network

26.1 Introduction

The infrastructure of high-speed railway is extensively developed in China for the past several years. The network topology structure and operation mode of the railway are changing profoundly. The target is to cover its major economic areas with a high speed railway network, which consists of four horizontal and four vertical lines, in the following several years. The network scale is much larger than any existing ones in the world. Train rescheduling is a focal point to improve the operation mode with the characteristics of high train speed, high train frequency and mixed train speed (HHM).

L. Wang · Y. Qin · J. Xu · L. Jia (✉)
State Key Laboratory of Rail Traffic Control and Safety, Beijing Jiaotong University, Beijing, China
e-mail: jialm@vip.sina.com

L. Wang
School of Traffic and Transportation, Beijing Jiaotong University, Beijing, China

Z. Sun and Z. Deng (eds.), *Proceedings of 2013 Chinese Intelligent Automation Conference*, Lecture Notes in Electrical Engineering 255,
DOI: 10.1007/978-3-642-38460-8_26,

In China train rescheduling is mostly manual developed by the operator with the support of computer in the existing line or the high speed railway. The operators dominate only one line or some section in one line with fixed three hours period depending on their experience. In future China will establish five comprehensive operation centre of the railway network. Each of them will dominate several lines that total length over thousands of kilometers. So it is very urgent to establish an agility and expansibility system to make full use of railway network capacity and improve the railway service level.

A four-layer architecture rescheduling system is proposed to reduce the complexity of the problem. Four kinds of rescheduling agents are involved in each layer according to its feature and the agents in the same layer or in the neighbor layers can communicate with each other. Each kind of the agents has its own rescheduling rules or knowledge, and all of the agents have the ability of study to improve the rules or knowledge. The agent based architecture adapts the development of railway network and is more conducive to the overall optimization.

Many papers studied train rescheduling systems and models. [1] presents an algorithm based on the global information to reduce the total travel time on the single-track railway and avoid the deadlock. [2–5] investigate the difference between fixed block and moving block, and build expert system to realize railway train rescheduling simulation. [6] research on subway train control mechanism and rescheduling method based on multi-agent. Although none of them come down to train rescheduling in the network, all of them provide good reference.

26.2 Railway Network Rescheduling System

In this section, we propose an agent based distributed train operation rescheduling system and model with the principle of hierarchical optimization of large system due to above characters of the railway operation problem shown in the first section. The system is formed by four layers that include implementation layer, section rescheduling layer, line rescheduling layer and network rescheduling layer. Implementation layer is formed by section device and station interlocking equipment and the top three layers is realized by three kinds of agents. The layers have the ability of communication with the neighbor layers. The lower provide process state information to the upper and the upper provide rescheduling strategies to the lower. Each kind of the agents has its own rescheduling rules or knowledge, and all of the agents have the ability of study to improve the rules or knowledge. Figure 26.1 illustrates the architecture of railway network rescheduling system (RNRS).

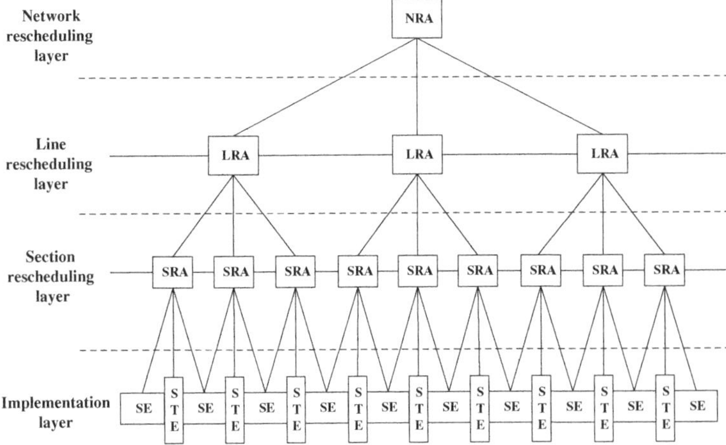

Fig. 26.1 Architecture of RNRS

26.2.1 Implementation Layer

Implementation layer is the bottom of the railway network rescheduling system, which is formed by a variety of equipment directly related to train control. This layer does not have the ability of decision-making and learning. It provides the information of the train running status and all equipment using states and executes the strategies of the upper layer to complete the train rescheduling.

26.2.2 Section Rescheduling Layer

Section rescheduling layer is formed by the section rescheduling agent, which is the bottom of the decision-making layers, to complete the section constraints. Figure 26.2 illustrates the structure of section rescheduling agent (SRA). SRA includes six parts that are communication module, coordination module, section rescheduling knowledge base, learning module, behavior decision module and man–machine interface, which provide the agent the ability of learning, collaboration and decision.

Communication module realizes the communication ability between SRA and line rescheduling agent (LRA). It also transfers the line control strategies and section rescheduling constraints to the knowledge base. Coordination module realizes the collaboration between the agents in this layer. Section rescheduling knowledge base stores the section and station rescheduling rules. It is the basic of SRA decision strategies. The neighbor agent and interlocking equipment provide the learning module of the train running status and equipment using state. The learning module will create new knowledge according to historical experience if

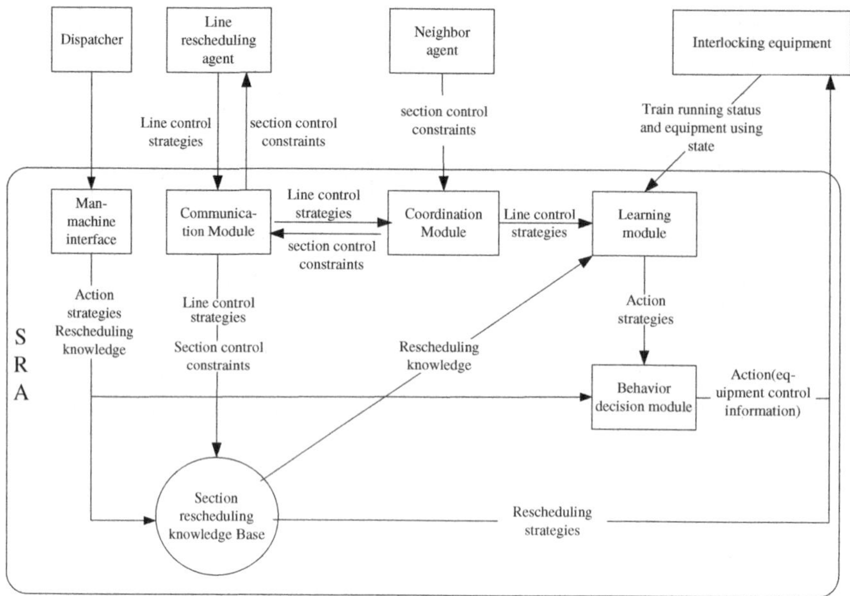

Fig. 26.2 Structure of SRA

there are not feasible strategies. The learning ability increases the whole system expansibility. Behavior decision module chooses the action to interlocking equipment according to the learning module strategies to complete the train rescheduling. Man–machine interface provide the possibility of dispatchers send the action strategies to the behavior decision module directly in emergency.

26.2.3 Line Rescheduling Layer

Line rescheduling layer is formed by the line rescheduling agents (LRA), which realize the control of train rescheduling of line and create the line control strategies for SRA. Figure 26.3 illustrates the structure of LRA. It is similar to SRA, which also includes six parts. The difference between them is the knowledge and control strategies. For LRA the main knowledge is line timetable rescheduling rules and its behavior action is sending the section control strategies to SRA.

26.2.4 Network Rescheduling Layer

Network rescheduling layer is formed by the section rescheduling agent, which is the top of RNRS, to optimize the network constraints. Figure 26.4 illustrates the

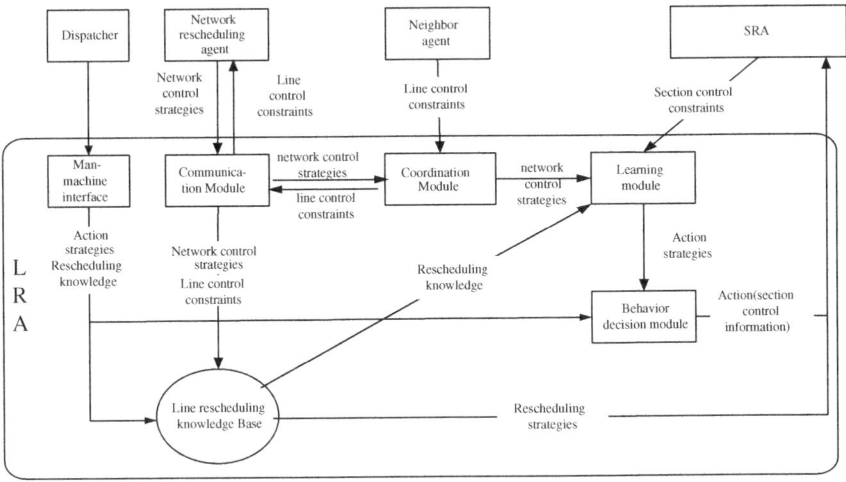

Fig. 26.3 Structure of LRA

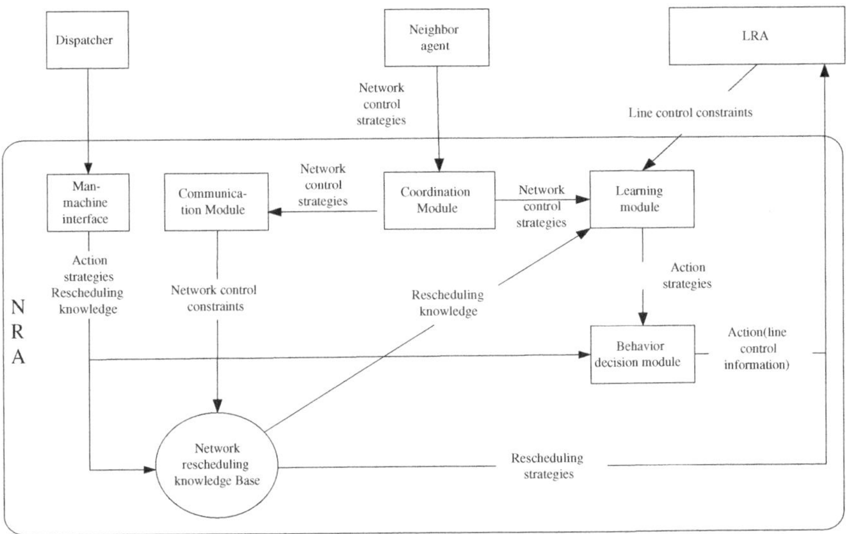

Fig. 26.4 Structure of NRA

structure of network rescheduling agent (NRA). NRA is similar to SRA, the difference between them is the knowledge and control strategies. For NRA the main knowledge is capacity calculation rules and rerouting strategies, and its behavior action is sending the line control strategies to LRA.

26.3 Realization of RNRS

This section mainly describes the knowledge rules for the three decision making layers that include section rescheduling layer, line rescheduling layer and network rescheduling layer, since implementation layer does not have the ability of decision-making and learning. The relevant notations and parameters are listed first in the following:

i	Train index;
j	Station index;
$k_{j,j+1}$	Section index, whose endpoint stations are j and $j+1$;
n_i	Grouping of train i;
$x_{i,j}$	Departure time of train i at station j;
$y_{i,j}$	Arrival time of train i at station j;
$x^*_{i,j}$	Scheduled departure time of train i in station j;
$y^*_{i,j}$	Scheduled arrival time of train i in station j;
$a_{i,k}$	Minimum running time of train i in section k;
$\tau^t_{i,j}$	Minimum parking time of time of train i in station j;
τ^s_j	Minimum headway in station j;
$T_{i,j}$	Minimum operating time of train i in station j;
P_j	Platform number in station j;
t	Current time;
$T_{e,j}$	Emergency type for the station j;
$T^s_{e,k}$	Emergency type for the section k;
$L^s_{e,k}$	Emergency grade for the section k;
$L_{e,j}$	Emergency grade for the station j;
b_i	If the train i reaches its final stop with a delay larger than the tolerance, $b_i = 1$, otherwise, $b_i = 0$.

26.3.1 Section Rescheduling Layer

This layer focuses on the section constraints. We call all the unnormal situations as emergency in the model. If some emergency happened that like some trains delay for long time, limited speed for heavy storm, communication failure, etc. So many constraints are listed below.

- Minimum running time constraint

$$a_{i,k} \geq f\left(T_{e,j}, T^s_{e,k}, L^s_{e,k}, L_{e,j}\right) \tag{26.1}$$

– Minimum parking time constraint

$$\tau^{t}_{i,j} \geq f\left(T_{e,j}, T^{s}_{e,k}, L^{s}_{e,k}, L_{e,j}\right) \tag{26.2}$$

– Minimum headway constraint

$$\tau^{s}_{j} \geq f\left(T_{e,j}, T^{s}_{e,k}, L^{s}_{e,k}, L_{e,j}\right) \tag{26.3}$$

– Minimum operating time in station constraint

$$T_{i,j} \geq f\left(T_{e,j}, T^{s}_{e,k}, L^{s}_{e,k}, L_{e,j}\right) \tag{26.4}$$

26.3.2 Line Rescheduling Layer

This layer focuses on the line timetable rescheduling optimization. In the model we take two objectives, one is the minimum total delay time and the other is minimum train number that delay beyond the tolerance.

$$\text{Minimize} \sum_{i=1}^{n} n_i \sum_{j=1}^{m} \left(\left(x_{i,j} - x^{*}_{i,j}\right)^2 + \left(y_{i,j} - y^{*}_{i,j}\right)^2\right). \tag{26.5}$$

$$\text{Minimize} \sum_{i=1}^{n} b_i \tag{26.6}$$

Line timetable rescheduling constraints are listed as below.

– Minimum running time interval constraint

$$y_{i,j+1} \geq x_{i,j} + a_{i,k} \tag{26.7}$$

– Train stop interval constraint

$$x_{i,j} - y_{i,j} \geq \max\left\{\tau^{t}_{i,j}, T_{i,j}\right\} \tag{26.8}$$

– Train headway constraint

$$x_{i+1,j} - x_{i,j} \geq \tau^{s}_{j} \tag{26.9}$$

Train $i + 1$ is departure at station j after train i.

– Train arrival headway constraint

$$y_{i+1,j} - y_{i,j} \geq \tau^{s}_{j} \tag{26.10}$$

Train $i + 1$ arrivals at station j after train i.

– Station capacity constraint

$$\sum_{i} \varphi(x_{i,j} - t)\varphi(t - y_{i,j}) \le f(P_j, n_i), \ \varphi(x) = \begin{cases} 1, x > 0 \\ 0, x \le 0 \end{cases} \tag{26.11}$$

26.3.3 Network Rescheduling Layer

This layer focuses on the network timetable rescheduling optimization. In the model we take the minimum rerouting cost as the objective. The rerouting strategies include cancelling trains, merging trains and making a detour,

$$\text{Min} \sum_{i \in Trn} \left(c_i^{cancel} q_i + c_i^{\det our} o_i + c_{ij}^{merge} \sum_{j \in Trn} m_{ij} \right) \tag{26.12}$$

Line timetable rescheduling constraints are listed as below.

– Canceled trains cannot make detour.

$$q_i + o_i \le 1, \ i \in Trn \tag{26.13}$$

– If a train is eligible to be merged, it can be merged to (with) at most one other train. The merged train cannot be canceled.

$$\sum_{i=j} m_{ij} = 0, \ i,j \in Trn \tag{26.14}$$

$$\sum_{j \in Trn} m_{ij} + q_i \le 1, \ i \in Trn \tag{26.15}$$

$$\sum_{i \in Trn} m_{ij} + q_j \le 1, \ j \in Trn \tag{26.16}$$

$$m_{ij} + m_{ji} \le 1, \ i,j \in Trn \tag{26.17}$$

– The detoured trains cannot exceed the residual capacity of alternative lines.

$$\sum_{i \in Trn} o_i \le \Delta C \tag{26.18}$$

– Constraint on the minimal number of trains being applied rerouting actions.

$$\sum_{i \in Trn} \sum_{j \in Trn} m_{ij} + \sum_{i \in Trn} q_i + \sum_{i \in Trn} o_i \ge \delta \tag{26.19}$$

Some additional parameters are listed as below.

c_i^{cancel}, c_i^{detour}, c_{ij}^{merge} : The cost of canceling train i, detouring train i and merging train i to train j, respectively. ΔC : The total residual capacity of alternative lines. δ : The minimal number of trains to be applied rerouting actions.

26.4 Case Study

The simulation is put on the busiest part of Jing-Hu high speed line, between Nanjing (Ning for short) and Shanghai (Hu for short). As shown in Fig. 26.5, besides Jing-Hu high speed line, there are two other lines connecting Hu and Ning, which are Hu-Ning inter-city high speed line ($\overline{NH}_{ic}$) and existing Jing-Hu line ($\overline{NH}_e$). There are seven stations on the Hu-Ning Part, thereby six sections. Its daily service starts at 6:30 am and ends at 11:30 pm. As currently planned, there are 60 trains (14 high speed trains and 38 quasi-high speed trains) from Jing-Hu line and 8 quasi-high speed trains from Riverside line that go from Ning to Hu by the Hu-Ning part [7]. The trains from Jing-Hu line are called self-line trains, while those from Riverside line are called cross-line trains. As in reality, self-line trains have higher priority than cross-line trains. As for self-line trains, the high speed trains have higher priority than the quasi-high speed trains.

First we take train G101 delays 20 min in the first section and train G103 delays 20 min in the second section. The minimum separation time on track possession in each station is set to 1 min and the departure headway is set to 3 min. All of other parameters are same to normal. So the section rescheduling layer provides the normal constraints, and this situation can be handled in the line rescheduling layer without the network rescheduling layer rescheduling. Figure 26.6 is the optimization result for delay situation. The gray lines mean the initial timetable and the blue lines mean the rescheduling timetable. The two delay trains involved other 4 trains delay. The total delay time is 76 min and the computer time is 40 s.

The simulation of heavy storm situation is presented and the height of snow on the track is more than 22 cm and less than 30 cm. In this situation SRA provides the limited speed 100 km/h for the impacted section and stations, and then LRA do the line rescheduling optimization. Figure 26.7 shows that three trains reaches its final stop late than 24 p.m. The total delay time is 6,476 min. Since it is unacceptable that

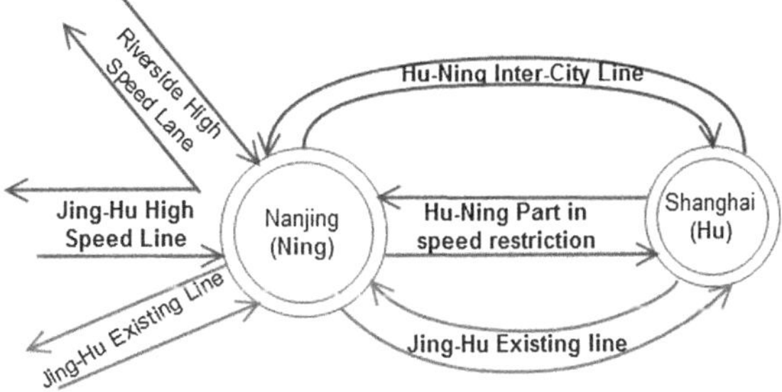

Fig. 26.5 Railway network around Nanjing

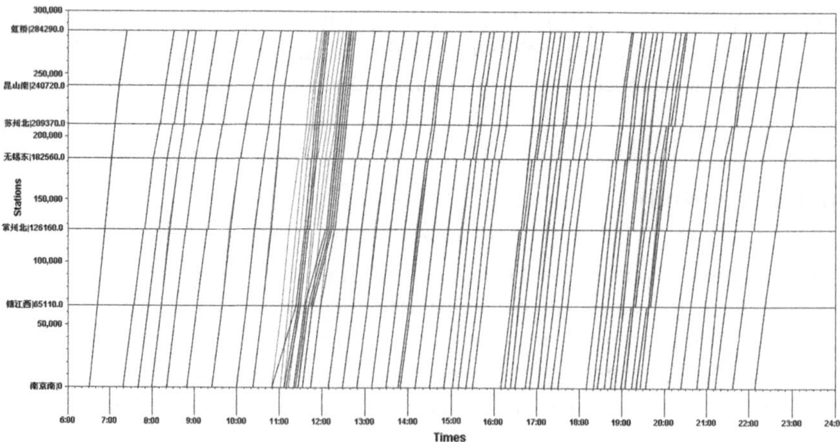

Fig. 26.6 Simulation result of delay situation

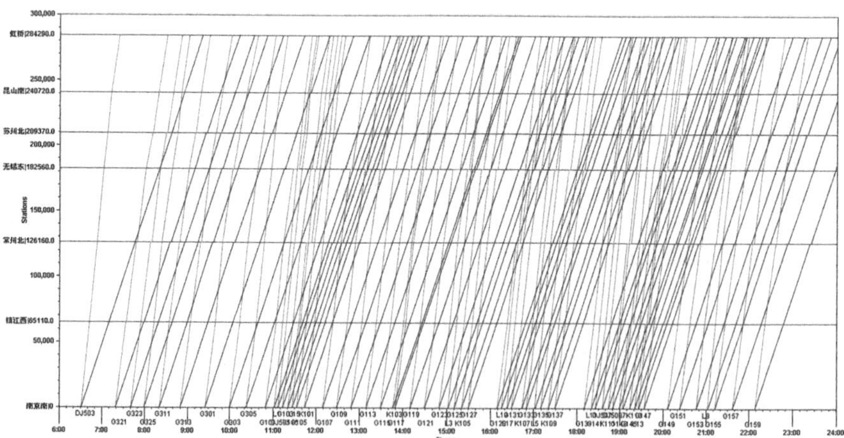

Fig. 26.7 First simulation in heavy storm

2 trains are late then 24 p.m. to LRA, LRA react the problem to NRA. NRA gets the residual capacity of $\overline{NH}_{ic}$ and $\overline{NH}_{e}$ is 2 and 3 respectively. Figure 26.8 shows the new rescheduling result with the NRA control strategies. Train K115, K105, K113 detour to $\overline{NH}_{e}$, train G157 and G159 detour to $\overline{NH}_{ic}$, train G143 is merged to G145. The total delay time is 5,863 min and there is no train late then 24 p.m.

From Fig. 26.8 we can see detouring trains strategy has a high priority, merging trains strategy has a medium priority and canceling train strategy has a low priority according to the costs respectively. The trains Ks (K105, K113 and K115) and Gs (G157 and G159) take a detour or are merged because that the Ks are cross-line trains that have lower c_i^{cancel} and c_i^{detour} than other trains. The Gs depart from Ning after 9 pm have a lower c_i^{cancel} and c_i^{detour} than other trains.

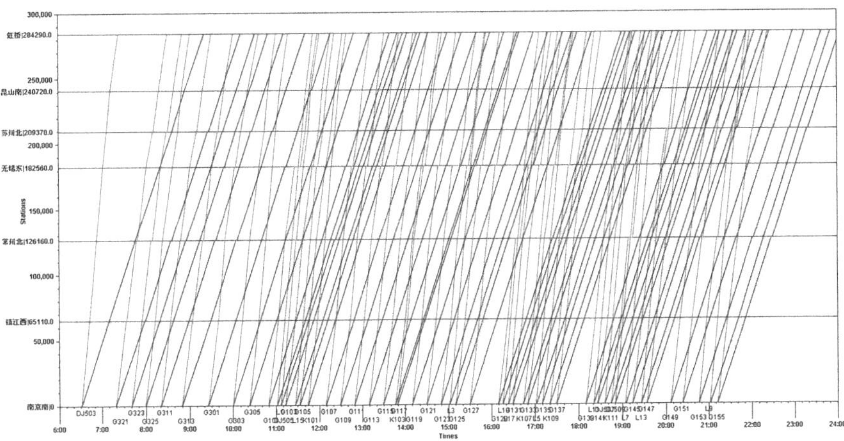

Fig. 26.8 New simulation in heavy storm

26.5 Conclusion

This paper presents an agent based train rescheduling system for the railway network. The system is formed by four layers that including implementation layer, section rescheduling layer, line rescheduling layer and network rescheduling layer from bottom to top of the system. Four kinds of rescheduling agents are involved in each layer according to its feature, and the agents in the same layer or in the neighbor layers can communicate with each other. The agent based network train rescheduling system benefits train operation mode in china to get a global optimization in the perspective of railway network. The case study is simulated on Beijing-Shanghai high speed railway which proved the system effectiveness.

Acknowledgments This work has been supported by China National Technique Supporting Plan Project (2009BAG12A10), Chinese State Key Laboratory of Rail Traffic Control and Safety Independent Project (Grant: RCS2010ZZ002, RCS2011ZZ004) and National Natural Science Project (Grant: 61074151).

References

1. Li F, Gao Z, Li K, Yang L (2008) Efficient scheduling of railway traffic based on global information of train. Transp Res Part B 42(2008):1008–1030
2. Zhang L, Li P, Jia L (2004) Study on the simulation for train operation adjustment under moving block. J Syst Simul 16(10):2257–2263 (in Chinese)
3. Meng X, Jia L, Qin Y (2010) Train timetable optimizing and rescheduling based on improved particle swarm algorithm. Transp Res Rec 2197(1):71–79
4. Joaquin R (2007) A constraint programming model for real-time train scheduling at junctions. Transp Res Part B 41(2):231–245

5. Yang L, Li K, Gao Z, Li X (2012) Optimizing trains movement on a railway network. Omega 40(5):619–633
6. Lu F, Song M, Tian G (2007) Method for subway operation adjustment based on multi-agent. China Railway Sci 28(1):123–126 (in Chinese)
7. Wang L, Jia L (2011) A two-layer optimization model for high-speed railway line planning. J Zhejiang Univ Sci A 12(1):902–912

Chapter 27
The BLDCM Control System Based on Fuzzy-PI Controller

Xue Lv and Hongsheng Li

Abstract Recently, lots of researches have been investigated for the BLDCM system. The purpose of this paper is to design a controller which can reduce the torque and speed fluctuation when load changes suddenly. The advantage of traditional PI controller is that there is no static error, but the dynamic performance is not good especially the load changes suddenly. The advantage of Fuzzy controller is the dynamic performance is good when the load changes suddenly, but there exists static error. Compared to traditional PI and Fuzzy controller, the new algorithm based on Fuzzy-PI can improve the dynamic performance of the system. Also, the speed and torque fluctuations are reduced at the same time. Simulation results show that the validity of the proposed method. The new control algorithm can be used in BLDC control system. Simulation results can provide reference for the actual BLDC control system.

Keywords Fuzzy-PI control · Simulink · BLDCM

27.1 Introduction

BLDCM have been widely used due to its high power density, efficiency, high torque-to-inertia ratio and reliable operation. In case of the control of robot arms and tracking applications with lower stiffness, the optimal BLDCM with small size and its robust control scheme are needed.

Up to now, 90 % of the controllers are PI and PID controllers. This is because of its simplicity and ease of design. The PID control is still the most used control in variable speed operations of BLDCM. However, the PI control method is a

X. Lv (✉) · H. Li
School of Mechanical and Electronic Engineering, Wuhan University of Technology, No 122 LuoShi Road, Wuhan, Hongshan, China
e-mail: lvxueer200@163.com

Z. Sun and Z. Deng (eds.), *Proceedings of 2013 Chinese Intelligent Automation Conference*, Lecture Notes in Electrical Engineering 255,
DOI: 10.1007/978-3-642-38460-8_27,

linear control method and do not work well in some discrete-time system [6]. Therefore, when the operating condition changes such as disturbances and load changes suddenly, the retuning process of control gains is necessary [7]. Controllers using artificial intelligent tools, such as fuzzy logic can be applied to overcome this problem. The paper mainly focuses on designing the new fuzzy PI algorithm in order to control speed and torque. The purpose is to decrease the speed and torque fluctuations of the control system when load changes suddenly.

In this paper, the mathematics model of the BLDCM is presented in Sect. 27.2. The Fuzzy PI controller design is provided in Sect. 27.3. Simulation results on analysis of the algorithms are made in Sect. 27.4. Finally, concluding remarks are presented in Sect. 27.5.

27.2 The Mathematics Model of the BLDCM

The model of BLDCM is similar to permanent magnet DC motor. The characteristic equations of BLDCM can be expressed by the following equations. The voltage equation can be written as.

$$\begin{bmatrix} U_a \\ U_b \\ U_c \end{bmatrix} = \begin{bmatrix} R_a & 0 & 0 \\ 0 & R_b & 0 \\ 0 & 0 & R_c \end{bmatrix} \begin{bmatrix} I_a \\ I_b \\ I_c \end{bmatrix} + \begin{bmatrix} L_a & L_{ab} & L_{ac} \\ L_{ba} & L_b & L_{bc} \\ L_{ca} & L_{cb} & L_c \end{bmatrix} p \begin{bmatrix} I_a \\ I_b \\ I_c \end{bmatrix} + \begin{bmatrix} E_a \\ E_b \\ E_c \end{bmatrix} \tag{27.1}$$

where I_a, I_b, I_c, U_a, U_b, U_c are stator current and voltage. R_a, R_b, R_c are equivalent stator resistance. L_{ab}, L_{bc}, L_{ca} are equivalent mutual inductance. L_a, L_b, L_c are equivalent self inductance. E_a, E_b, E_c are back electromotive force, and p represents differential operators.

In this paper, we assume $R_a = R_b = R_c = R$, $L_a = L_b = L_c = L$, $L_{ab} = L_{bc} = L_{ca} = M$, also, $I_a + I_b + I_c = 0$. The (27.1) can be rewritten as:

$$\begin{bmatrix} U_a \\ U_b \\ U_c \end{bmatrix} = \begin{bmatrix} R_a & 0 & 0 \\ 0 & R_b & 0 \\ 0 & 0 & R_c \end{bmatrix} \begin{bmatrix} I_a \\ I_b \\ I_c \end{bmatrix} + \begin{bmatrix} L-M & 0 & 0 \\ 0 & L-M & 0 \\ 0 & 0 & L-M \end{bmatrix} p \begin{bmatrix} I_a \\ I_b \\ I_c \end{bmatrix} + \begin{bmatrix} E_a \\ E_b \\ E_c \end{bmatrix} \tag{27.2}$$

The mechanical equation of the BLDCM can be described as

$$T_e - T_L - B_1 w = J\frac{dw}{dt} \tag{27.3}$$

At the same time, the electromagnetic torque equation is

$$T_e = (E_a I_a + E_b I_b + E_c I_c)/w \tag{27.4}$$

where B_1 is the friction coefficient, J is the inertia coefficient and w is the mechanical speed of the rotor and T_L is the load torque. The speed equation can also be given as.

$$\frac{d\theta}{dt} = \omega \tag{27.5}$$

A great many various control strategies can be used for BLDCM system. In this paper, a method of self-adapting parameters for PI controller based on the fuzzy self-tuning is proposed. Simulation results illustrate that the validity of the proposed method.

27.3 Fuzzy-PI Controller Design

27.3.1 The Structure of Fuzzy Self-Tuning PI Controller

The Fuzzy-PI controller takes conventional PI as the foundation, which uses the theory of fuzzy reason and variable discourse of universe to on-line regulate the parameters of PI automatically [8]. The structure of fuzzy self-tuning PI controller is shown in Fig. 27.1.

The error and error changing rate are used as the input variables in control system, and the output variables are the parameters of PI control, those are ΔKp and ΔKi. Here, E denotes the system error, EC denotes the system error changing rate. The fuzzy sets of E, EC, ΔKp, ΔKi is {NB, NM, NS, 0, PS, PM, PB}, the region of E, EC are $\{-3, -2, -1, 0, 1, 2, 3\}$, and the region of ΔKp, ΔKi, are $\{-3, -2, -1, 0, 1, 2, 3\}$ respectively. Where NB, NM, NS, 0, PS, PM and PB are linguistic values. They respectively represent "negative big", "negative medium", "negative small", "0", "positive small", "positive medium" "positive big" [1].

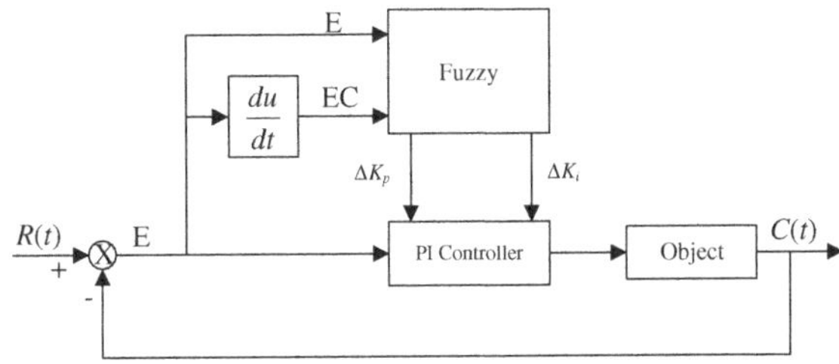

Fig. 27.1 The structure of fuzzy- PI controller

Table 27.1 Fuzzy rule for ΔKp

e ec	NB	NM	NS	ZO	PS	PM	PB
NB	PB	PB	PM	PM	PS	ZO	ZO
NM	PB	PB	PM	PS	PS	ZO	NS
NS	PM	PM	PM	PS	ZO	NS	NS
ZO	PM	PM	PS	ZO	NS	NM	NM
PS	PS	PS	ZO	NS	NS	NM	NM
PM	PS	ZO	NS	NM	NM	NM	NB
PB	ZO	ZO	NM	NM	NM	NB	NB

27.3.2 Forming and Reasoning of Fuzzy Control Rules

The essential part of the fuzzy logic controller is a set of linguistic rules. In many cases it is easy to translate an expert's experience into such rules [2]. Any number of rules can be created to define the actions of the fuzzy controller. In this paper, the fuzzy control rules design is based on the medical robot can approach the target quick and stable.

The application of conventional fuzzy conditions and fuzzy relations "If E is A and EC is B then ΔKp is C, ΔKi is D," [3] can establish fuzzy rules. The finally determined fuzzy rules are shown in the following table (Tables 27.1, 27.2).

After constructing the table of fuzzy control rule $(\Delta Kp, \Delta Ki)$ we can make adaptive correction by the following method [4, 5]:

$$\begin{aligned} Kp &= Kp' + \Delta Kp \\ Ki &= Ki' + \Delta Ki \end{aligned} \tag{27.6}$$

Based on the above analysis, according to the principle of self-tuning parameters of PI, the model of Fuzzy-PI is presented in Fig. 27.2.

Table 27.2 Fuzzy rule for ΔKi

e ec	NB	NM	NS	ZO	PS	PM	PB
NB	NB	NB	NM	NM	NS	ZO	ZO
NM	NB	NB	NM	NS	NS	ZO	ZO
NS	NB	NM	NS	NS	ZO	PS	PS
ZO	NM	NM	NS	ZO	PS	PM	PM
PS	NM	NS	ZO	PS	PS	PM	PB
PM	ZO	ZO	PS	PS	PM	PB	PB
PB	ZO	ZO	PS	PM	PM	PB	PB

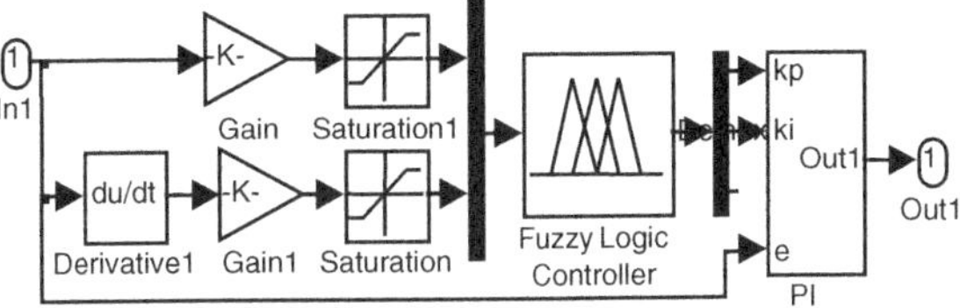

Fig. 27.2 The mathematic model of fuzzy-PI controller

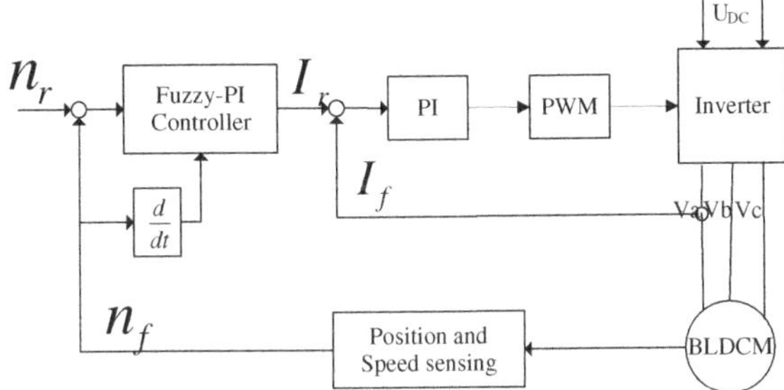

Fig. 27.3 Structure diagram of the system

Table 27.3 BLDCM motor specification

P	1	J	$0.0006\ \mathrm{Kg \cdot m^2}$
R	1Ω	B	$0.0002\ \mathrm{N \cdot m \cdot s/rad}$
L	$0.012H$	$P0$	380W
M	$-0.07H$	U_{DC}	220V

27.4 Simulation Results

A control system based on Fuzzy-PI control for BLDCM is simulated in Matlab. The control block diagram of the whole system is shown in the following Fig. 27.3.

The parameters of the BLDCM are shown in Table 27.3. The reference rotor speed is set to 140 rad/s, load torque is $1\ \mathrm{N \cdot m}$ and $2\ \mathrm{N \cdot m}$ when $t = 0.2\mathrm{s}$,. The simulation time is set to 0.5s.

Simulation results are as follows. Figure 27.4 shows the waveforms of the stator current based on PID controller. Figure 27.5 shows the waveforms of the stator current based on the Fuzzy-PI controller. Simulation results show the effectiveness of modeling in this paper. There is about 5 A current ripple in the PID controller and 1A current ripple in the Fuzzy- PI controller.

Figures 27.6 and 27.7 show the waveforms of the speed response. Speed response curve based on the PID controller is illustrated in the Fig. 27.6. Speed response curve based on Fuzzy-PI controller is illustrated in the Fig. 27.7. There exists about 8 rad/s overshoot in the PID Controller and 0.5 rad/s overshoot in the Fuzzy-PI controller.

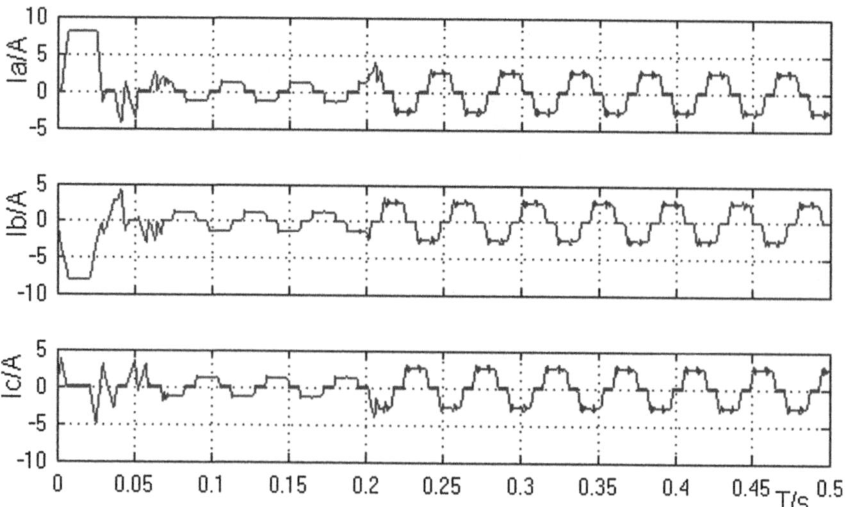

Fig. 27.4 Current response based on PID controller

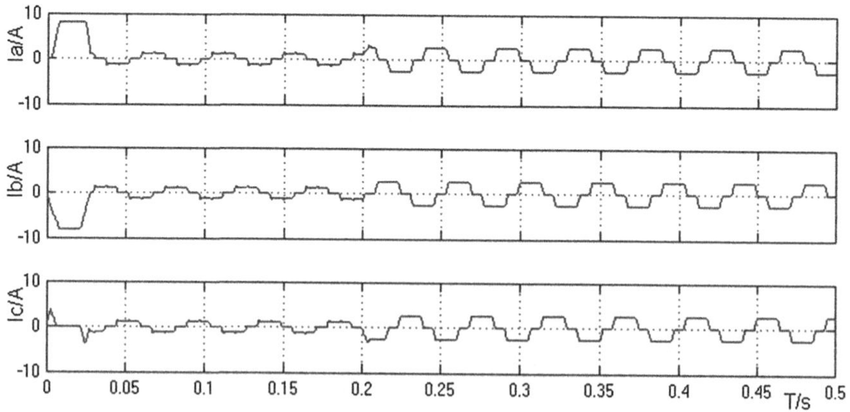

Fig. 27.5 Current response based on fuzzy-PI controller

Figure 27.8 shows the torque waveform based on the PID controller. Torque response curve based on the Fuzzy-PI controller is presented in the Fig. 27.9. The maximum Torque ripple in the PID controller is about −2.5 N m and the maximum Torque ripple in the Fuzzy-PI controller is about −0.4 N m.

From the Fig. 27.4, 27.5, 27.6, 27.7, 27.8, 27.9. We can get the result that the current speed and Torque performance of system is improved in the Fuzzy-PI controller compared to PID controller. At the same time, the torque ripple is reduced abruptly. Simulation results show that the proposed method is effective for the BLDCM system.

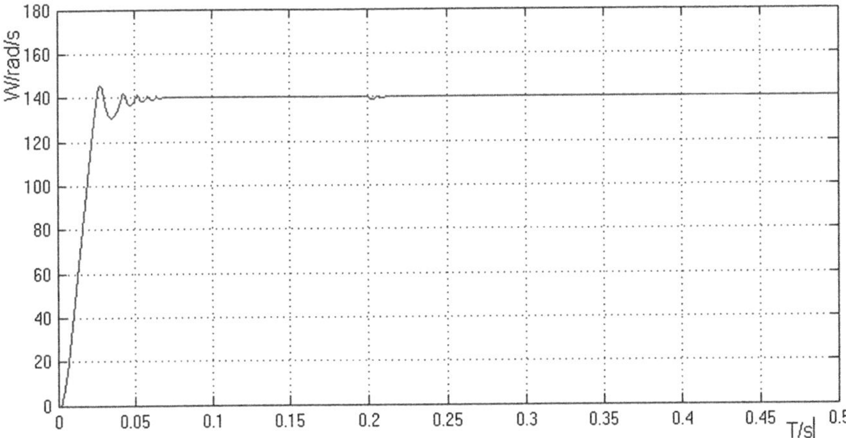

Fig. 27.6 Speed response based on PID controller

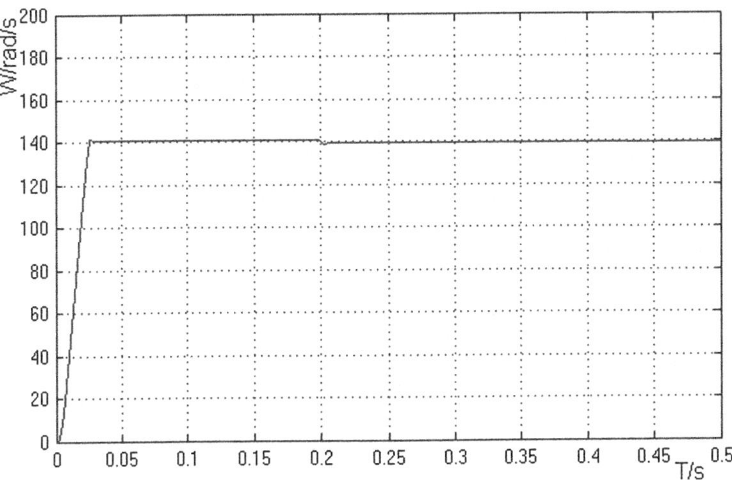

Fig. 27.7 Speed response based on fuzzy-PI controller

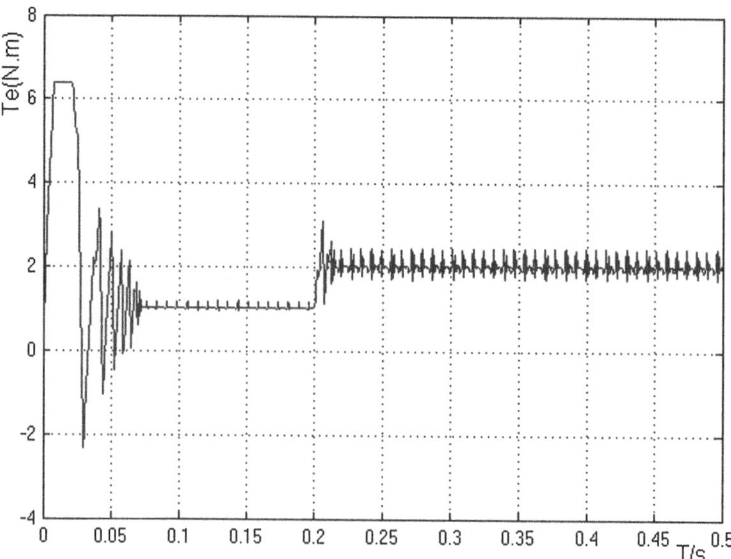

Fig. 27.8 Torque response based on PID controller

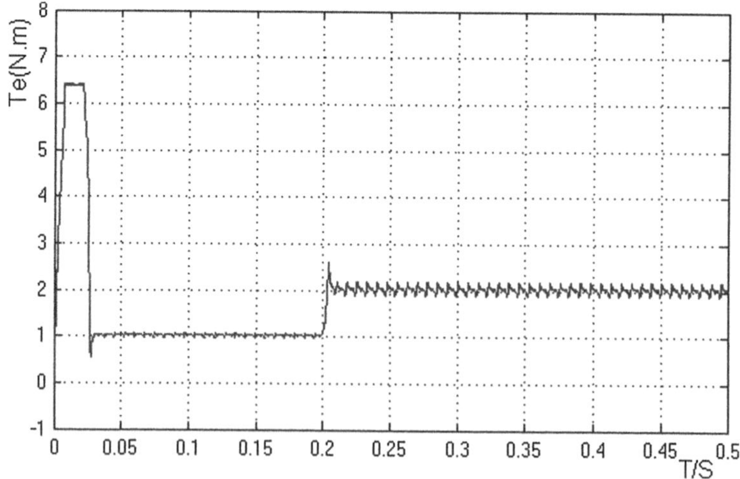

Fig. 27.9 Torque response based on fuzzy-PI controller

27.5 Conclusion

This paper proposes a Fuzzy-PI controller for BLDCM system. The Fuzzy-PI controller combines the advantage of Fuzzy controller and PI controller. From the simulation results, the new algorithm based on Fuzzy-PI can improve the dynamic performance of the system compared to traditional algorithms. The speed and torque fluctuations are reduced obviously from the added chart. It is very useful for the actual BLDCM control system.

References

1. Li H (1997) The essence of fuzzy control and a kind of fine fuzzy control. Control Theory Appl 14(6):015
2. Hori Y (1993) Robust and adaptive control of a servomotor using low precision shaft encoder. In: Proceedings IEEE IECON'93, pp 73–78
3. Wu X, Sanjib KP, Xu J (2007) Effects of pulse-width modulation schemes on the performance of three-phase voltage source converter. IECON 2026–2031
4. Ogasawara S, Akagi H (1991) An approach to position sensorless drive for brushless dc motors. IEEE Trans Ind Appl 27(5):928–933
5. Moore KL, Bhattacharyya SP, Dahleh M (1993) Capabilities and limitations of multirate control schemes. Automatica 29(41):941–995
6. Jacobina CB, Silva RC, Lima AMM, Ribeiro RLA (1995) Vector and scalar control of a four switch three phase inverter. Proc IEEE Ind Appl Conf 3:2422–2429
7. Er MJ, Anderson BDO (1995) Design of reduced-order multirate output linear functional observer-based compensator. Automatica 31(2):237–242
8. Johnson JP, Ehsai M, Guzelgunler Y (1999) Review of sensorless methods for brushless. Proc IEEE Ind Appl Conf 1:143–150

[illegible] compared to traditional algorithms. The speed and torque [illegible] from the [illegible] it is [illegible] the [illegible] system.

References

[illegible]

Chapter 28
Research and Comparison of CUDA GPU Programming in MATLAB and Mathematica

Xiongwei Liu, Lizhi Cheng and Qun Zhou

Abstract This paper discusses and analyzes the reasons of fast development of GPU computing, CUDA-enabled GPU programming model, direct matrix multiplication algorithms based on CPU and CUDA-enabled GPU. In three different environments of VC++2008, Matlab 2009b and Mathematica 8, for matrices of different dimensions, we compare execution efficiency by means of runtime. For research and study, GPU Computing in mathematica software systems provides a good choice because of their advantages of low cost, ease of use, high efficiency.

Keywords GPU · CUDA · Mathematica · MATLAB · Matrix multiplication

28.1 Introduction

It is well-known that upgrading CPU frequency will result in high temperature, power consumption and other technical bottleneck. The speed of increase of CPU frequency is no longer consistent with Moore's law. Two major CPU manufacturers Intel and AMD in the world improved CPU performance by increasing the number of cores in a CPU chip. However, for technical reasons, a CPU chip only contains a low number of cores. Up to now, the number of cores on a desktop CPU on the market is no more than 16. During this period, GPU with large arithmetic

X. Liu (✉) · L. Cheng
Department of Mathematics and System Science, College of Science, National University of Defense Technology, Changsha, China
e-mail: liuxiongwei@yahoo.com.cn

Q. Zhou
School of Information Science and Engineering, Hunan International Economics University, Changsha, China
e-mail: shwzhouqun@163.com

Z. Sun and Z. Deng (eds.), *Proceedings of 2013 Chinese Intelligent Automation Conference*, Lecture Notes in Electrical Engineering 255,
DOI: 10.1007/978-3-642-38460-8_28,

units obtained increasingly people's attention in the area of general computing. The development in recent years has shown that the floating-point processing performance of a high-end GPU was ten times higher than that of a high-end desktop CPU, its memory bandwidth was also about five times. Furthermore, while GPU provided the same computing power, its cost and power consumption were smaller than those of CPU-based systems [1].

However, for general programmer, it was not easy to take advantage of GPU to carry out general computing previously. It could only be achieved through a standard graphics interface (API), such as OpenGL or DirectX. Therefore, if users wanted to use GPU to do common computing tasks, they had to learn special graphics programming language [2]. This made the efficiency of development with GPU and its application very low. To this end, people wish to have convenient, fast GPU programming interface and wish to solve their problems with the aid of the high performance computing power of GPU in a familiar programming environment. GPU programming based on CUDA in MATLAB and Mathemaitca were designed to address these issues.

28.2 CUDA Programming

The programmable graphics processing unit (GPU) makes those problems whose data are fit for parallel computing are processed faster and more efficiently because of its strong computing power and multicore processor with very high memory bandwidth. Therefore, programmers always pay close attention to more quickly and easily develop its ability by programming. In order to meet this need, NVIDIA GPU manufacturers introduced a new universal parallel computing architecture, namely CUDA, in 2006. The architecture leveraged the advantages of CPU and GPU respectively. It was able to solve many complex computational problems more effectively with the aid of NVIDIA GPU Computing engines and significantly improved computing performance [3].

CUDA is a complete GPGPU solution and provide a direct access interface to the hardware. It is not like traditional ways which has to rely on graphics API interfaces to access GPU and not only contains the CUDA instruction set architecture (ISA) and parallel computing engine in GPU inside, but also gives developers a high similar development environment for C programming [4]. In this way, developers can benefit from the familiar, similar to the C programming language to write programs for CUDA architecture, which allow programs to run with high-performance on GPU supported by CUDA. Of course, it also provides programming interfaces to other programming environments, such as Fortran and C++.

In addition, CUDA architecture provides technical development libraries. Developers are able to set up their own applications quickly and easily on the basis of development libraries (such as universal math library: CUFFT and CUBLAS). The codes based on CUDA are divided into two types in the actual implementation of a program: the host code that runs on the CPU (usually serial code), the device

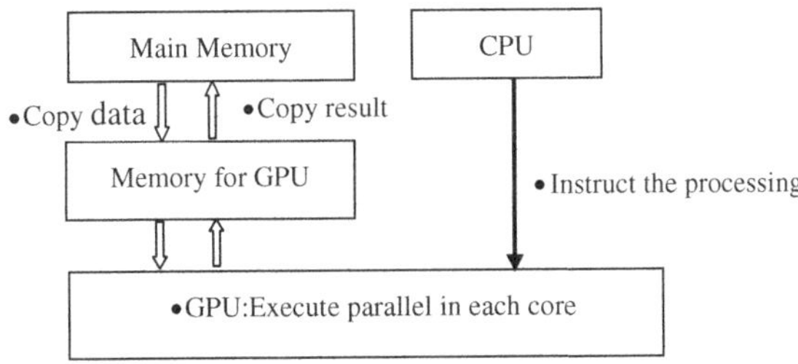

Fig. 28.1 Processing flow on CUDA

code that runs on the GPU (usually parallel code) [3]. Processing flow on CUDA is shown in Fig. 28.1 [4].

28.3 CUDA Programming in Mathematica

Mathematica 8 introduced a package of GPU programming. It provides a convenient way to use CUDA through CUDALink. It can automatically solve those issues which have to be considered by CUDA programmer [5]. It not only provides some internal functions (such as image operation functions, linear algebra operations, etc.) to directly call GPU to improve operation efficiency [6], and also can integrate some program with development tools. It allows users to load their own CUDA function into Mathematica to improve the efficiency of GPU general computing programming. These allow users to easily create compound algorithms which mix CPU code with GPU code to improve Mathematica calculation efficiency.

GPU programming is very easy for Users in Mathematica. The process of GPU programming with Mathematica is shown in Fig. 28.2. Users need not worry about many steps of CUDA GPU programming. The only thing they have to consider is write CUDA kernels [5].

28.4 CUDA Programming in MATLAB

In order to improve computing efficiency, MATLAB provided support for NVIDIA CUDA GPU programming by using the parallel computing toolbox in the latest version, which provided users with a new way to accelerate scientific

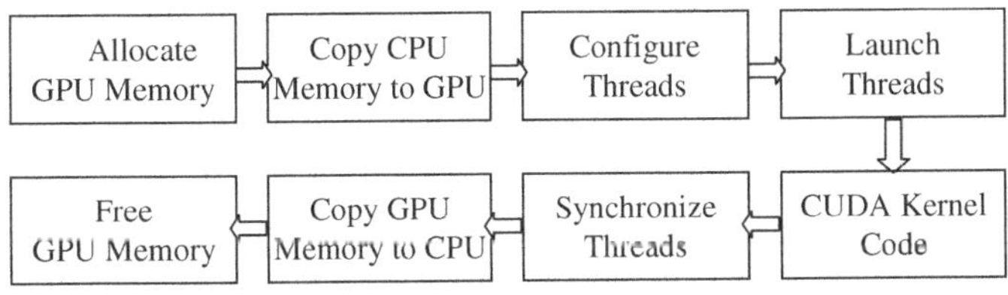

Fig. 28.2 GPU programming using mathematica

computing. MATLAB users need not learn CUDA GPU programming or make significant changes to their applications, and does not need to know the underlying programming with GPU, then can use the NVIDIA CUDA-enabled GPU and CUDA library to easily achieve acceleration of parallel code.

Implementation of GPU Computing and data processing in MATLAB are mainly achieved through the GPUArray object and arrayfun methods. MATLAB user can not only pass workspace data of MATLAB to GPU memory, but also can define data directly on GPU for processing. MATLAB user can use functions which can be run directly on GPU, or call user-defined functions on the GPU to improve the efficiency of operations and programming.

In order to achieve maximum flexibility and ease of GPU programming, MATLAB provides different programming models to better meet the developer's preferences. In addition, due to GPU programming was integrated in Parallel Computing Toolbox, therefore, all applications with parallelism, regardless of its is located on GPU or on CPU, can be accelerated by using rare code. This makes tasks and data with parallelism in MATLAB easily be extended to more hardware platform, and eventually be extended to clusters [7].

28.5 CUDA Programming of Matrix Multiplication

Matrix multiplication is often used in scientific computing and is the most basic, time-consuming arithmetic operations. It has a wide range of applications in many areas. Studying and testing various implementations of matrix multiplication to find fast and effective algorithm has been of continuing concern in scientific computing field.

Let $A = \left(a_{ij}\right)_{m\times p}$, $B = \left(b_{ij}\right)_{p\times n}$, matrix multiplication is defined as

$$c_{ij} = a_{i1}b_{1j} + a_{i2}b_{2j} + \cdots + a_{ik}b_{kj} = \sum_{k=1}^{p} a_{ik}b_{kj} \tag{28.1}$$

where $i = 1, 2, \cdots, m$, $j = 1, 2, \cdots, n$. Only when the number of columns of the matrix on the left is equal to the number of rows of the matrix on the right, the left matrix can be multiplied by the right. Matrix multiplication has a very good parallelism. It can be carried out in blocks. Let $\mathbf{A} = \left(A_{ij}\right)_{s\times t}$, $\mathbf{B} = \left(B_{ij}\right)_{t\times r}$, then $\mathbf{C} = \left(C_{ij}\right)_{s\times r}$, where

$$C_{ij} = \sum_{k=1}^{t} A_{ik}B_{kj} \quad (i = 1, \cdots, s;\ j = 1, \cdots, r). \tag{28.2}$$

Literature [8] described three methods on CPU and four methods on GPU based on CUDA architecture and compared their performance for the implementation of matrix multiplication. Literature [9] showed a specific implementation idea of an efficient algorithm of matrix multiplication on GPU. Experiments showed that its speed of calculation was comparable with the speed of matrix multiplication with CUBLAS library.

28.6 Experimental Results

We compared the runtime of the basic matrix multiplication algorithm based on CUDA GPU programming in VC++2008, MATLAB2009 and Mathematica 8. The matrix multiplication was performed on CPU, on GPU and on CPU+GPU. The computer system configuration as follows: Windows 7(32-bit), 3.5 G available memory, Pentium(R) Dual-Core CPU E5400 2.7 GHz, NVIDIA GeForce GTX 550 Ti(1G). The matrices were generated automatically by the software system. The values of all elements of two matrices were 1.5. Two matrices were loaded into main memory before they were processed. The time only counted calculation time, regardless of the time which the data transfer between memories consumed. And the time was given by statistical functions which were included in programming environment.

In Table 28.1, CPU time was the computing time for algorithm which operated directly on elements of matrices by for-loop in Visual C++ 2008. GPU time was the computing time for algorithm which operated on blocks of matrices, the block size was 256. In Table 28.2, CPU time (function) was the time for matrix multiplication carried out by A*B in MATLAB, CPU time (direct) was the computing time for algorithm which operated directly on elements of matrices by for-loop, GPU time was the time for matrix multiplication carried out by using the GPU-Array object directly and GPU operation function. In Table 28.3, CPU time (function) was the time for matrix multiplication carried out by Dot which was internal function in Mathematica, CPU time (direct) was same to that in Table 28.2, GPU time was the time for matrix multiplication carried out by CUDADot which was internal GPU operation function in Mathematica. The values of CPU time (direct) in Table 28.3 were not presented when the dimension were 1024 or 2048 because the operations were very time-consuming. GPU computing time in MATLAB did not include the time for passing data into the GPU memory and taking the data back from the GPU memory, but it was included in Table 28.3.

Experimental results showed that when the dimensions of matrices were less than 128, matrix multiplication on CPU commonly took less time than that on GPU, and when the dimensions were greater than 128, and larger dimensions,

Table 28.1 Test results for matrices of different dimensions in Visual C++2008

Dimension	64	128	256	512	1024	2048
CPU time	0.000	0.016	0.109	0.967	44.601	358.145
GPU time	0.031	0.032	0.032	0.047	0.109	0.359

Table 28.2 Test results for matrices of different dimensions in MATLAB 2009b

Dimension	64	128	256	512	1024	2048
CPU time(function)	0.0003	0.0007	0.005	0.030	0.214	1.605
CPU time(direct)	0.003	0.0257	0.1922	2.5845	64.2386	533.502
GPU time	0.0013	0.0016	0.004	0.011	0.053	0.367

Table 28.3 Test results for matrices of different dimensions in mathematica 8

Dimension	64	128	256	512	1024	2048
CPU time(function)	7.28e–17	0.00	0.016	0.078	0.422	2.745
CPU time(direct)	1.186	9.188	72.961	583.131	–	–
GPU time	0.016	0.016	0.047	0.219	0.904	3.557

GPU computing advantages were obvious. For example, in VC++2008, when the dimension of the matrices was 256, CPU time was 3.4 times more than GPU time. When the dimension was 512, reached to 20.57 times, when the dimension was 1024, 2048, it was expanded to about 409.18 times and 997.62 times. In MATLAB and Mathematica, matrix multiplication based on for-loop were not comparable, the time consumed was not only much more than the time consumed by the internal functions, but more than CPU time in Visual C++2008. In Mathematica, when the dimension of the matrices reached above 1024, the speed was very slow, while in MATLAB it was acceptable. Compared with matrix multiplication function in MATLAB, the acceleration of GPU computing in Matlab was obvious without consideration of data transfer between memories. When the dimension of matrices was less than 512, the acceleration was not obvious. When the dimension of matrices was 1024 or 2048, the time consumed by function carried out on CPU was about $4 \sim 5$ times more than the time consumed by GPU. Since GPU time in Mathemaitca included the time of data input and output, matrix multiplication carried out by using function Dot was faster than by using direct GPU operation function CUDADot. GPU Computing took considerably longer time than CPU calculation, and smaller dimension, gap clear. But with the increase of dimension, the gap continued to shrink.

28.7 Conclusion

Analysis and experimental results showed that in Visual C++ and MATLAB GPU computing had distinct advantages. Compared with Visual C++ and Mathematica, GPU programming based on CUDA in MATLAB was more convenient, fast, efficient, and MATLAB had the advantage in computing speed on the smaller dimensions of matrices. With the increase of the dimensions of matrices, this difference was not evident, and when the dimension was very large, Visual C++ should be more efficient. In Mathematica, compared with GPU programming in Visual C++, the efficiency of GPU programming based on CUDA was higher, but efficiency of implementation was relatively low. This was mainly because the acceleration of computing with help of GPU depended largely on data transfers overhead between the host memory and GPU devices. In General, for compute-intensive applications in parallel, when data transfer was a small amount, using the GPU computing would be able to achieve a faster program execution. Therefore, before using the GPU to perform calculations, we need to do more to consider data traffic optimization, so as to reduce the data transfer and achieve greater efficiency in the implementation.

References

1. Zhang S, Chu Y, Zhao K, Zhang Y (2009) High performance calculation of GPU CUDA. China Water Conservancy and Hydropower Press, Beijing in Chinese
2. Sanders J, Kandrot E (2010) CUDA by example: an introduction to general-purpose GPU programming. Pearson Education, Inc
3. NVIDIA GPU computing documentation. http://developer.nvidia.com
4. Wikipedia, CUDA. http://en.wikipedia.org/wiki/CUDA
5. Wolfram White Paper: CUDA programming within mathematica. http://www.wolfram.com/products/mathematica-cuda-free-white-paper.html
6. GPU computing. http://reference.wolfram.com/mathematica/guide/GPUComputing.html
7. Zucchelli G (2012) Using MATLAB to easily enjoy the power of the GPU. EDN China 3:57–58
8. Liu J, Guo L (2011) Comparison and analysis of matrix multiplications on GPU and CPU. Comput Eng Appli 47(19):9–11 (in Chinese)
9. Liang JJ, Kaixin R, Guo L, Liu Y (2011) Design and implementation of matrix multiplication on GPU. Comput Syst Appli 20(1):178–181 (in Chinese)

Chapter 29
Researching on the Placement of Data Replicas in the System of HDFS Cloud Storage Cluster

Guangbin Bao, Chaojia Yu, Hong Zhao and Yangyang Luan

Abstract The default model and the dynamic model of data placement based on the need of fault tolerance in HDFS cloud storage cluster system were built. On this basis, the $Rack_i$ selection algorithm based on the weight to process the problems of data transmission was designed and the DN_{ik} selection algorithm based on the gray evaluation criteria of entropy weight to evaluate each DataNode in the $Rack_i$ comprehensively were proposed, that come to a DataNode with the best Map/Reduce performance eventually. Analysis and experiments showed that the $Rack_i$ selection algorithm and DN_{ik} selection algorithm can provide the data replicas pre-written into HDFS reasonable DataNodes.

Keywords HDFS · Cloud storage · The placement of data replicas · Map/Reduce

29.1 Introduction

Cloud storage is the cloud computing system which provides the storage and management of data to the cloud computing. By using distributed file system, grid technology and application clustering technologies etc., the cloud storage provides

G. Bao (✉) · C. Yu · H. Zhao · Y. Luan
School of Computer and Communication, LanZhou University of Technology, LanZhou 730050, China
e-mail: 280321576@qq.com

C. Yu
e-mail: yuchaojia@163.com

H. Zhao
e-mail: zhaoh@lut.cn

Y. Luan
e-mail: luanyangyang01@126.com

Z. Sun and Z. Deng (eds.), *Proceedings of 2013 Chinese Intelligent Automation Conference*, Lecture Notes in Electrical Engineering 255,
DOI: 10.1007/978-3-642-38460-8_29, © Springer-Verlag Berlin Heidelberg 2013

business computing and data storage to the users in a variety of different types of equipment which application software set can work in a collaborative mode in cloud storage system. The characteristics of cloud storage cluster are as follows: (1) The price advantage: comparing with the same scale data backed up in other systems, the application of cloud storage data center is much lower than the cost of devices purchased which are dedicated as data storage; (2) The complete advantage: cloud storage only backups data to improve the high availability of data, so users do not have to worry about the loss of control of their data; (3) The complete storage services: intelligent storage technology and the technology of data backup management. At present, the existing Cloud storage products are Google GFS, Amazon S3, IBM Blue Cloud, Yahoo HDFS etc. HDFS is one of an open source cloud storage products of the Apache Software Foundation. Because of HDFS can run on cheap hardware equipment, it is used in business development and academic research by Facebook, Amazon, Last. FM Company, Cornell University, Carnegie Mellon University and Palo Alto Research Center etc. Therefore, the paper studied the placement strategy of data replicas in cloud storage cluster which was based on the HDFS cluster architecture.

29.2 Related Work

There are some relative mature placement strategies of data replicas for reference about other distributed environments. The early study related to the placement of data replicas considered mainly the system performance in data grid system, such as literature [1]. With the development of electronic manufacturing, appearing the research on high availability of data copies: According to the dynamic characteristics of data grid, Literature [2] put forward three models: the P-center, the P-median, the Multi-objective. The three models were aiming at three factors: network delay, the user request, hybrid of network delay and the user request. Experiments showed that the Multi-objective model is superior to the P-center and the P-median model; Literature [3] proposed a creation strategy that the inter-domain data copies derived from the domain data copies. The strategy balanced the network load of users' access by deriving data copies from domain and drove the expansion of data copies within inter-domain by the frequency of data replicas accessed to reduce the consumption of system bandwidth and customers' access delay; According to the characteristics of European data grid, such as the storage ability of DataNode, hierarchical topology structure of grid system, source data stored in the top-level node of grid system etc., Literature [4] and Literature [5] proposed the rapid spread of replication strategy, the best client replication strategy, waterfall replica creation strategy and caching falls replica creation strategy etc.

29.3 The Placement Strategy of Data Replicas in HDFS Cloud Cluster

The balancer constantly moves data blocks until the balancing of cluster load, when load exceeds the given threshold in the HDFS cluster. It means that the utilization rate of each DataNode and the utilization rate of cluster are very close, but does not exceed the cluster threshold. The adjustment of cluster load is heavily dependent on the balancer in HDFS, but the balancer has the shortcomings about hysteresis quality and passive trigger etc. According to the rack node $Rack_i$ selection algorithm and DataNode DN_{ik} selection algorithm in the paper, the placement strategy fully considers cluster load balance, and actively places data replicas into optimal nodes when data copies are written into the cluster. The optimal node is the lightest load DataNode. It can make the cluster avoid the potential risk of load imbalance caused by the data copies placement of random rack node selection and DataNode selection, and can reduce the occupancy of network resources caused by subsequent adjustments of balancer.

29.3.1 The Default Placement Strategy of Data Copy

To enhance the high availability of the HDFS cloud storage cluster system, the strategy will create a default main data copy DB_1 and two default data copies DB_2 and DB_3 for each data block. DB_1 and DB_2 are saved in the different DataNodes of the same rack, and DB_3 is saved in the rack node different from main data copy DB_1 located in.

The flow diagram about the default placement strategy of data copy was shown in Fig. 29.1.

29.3.2 The Dynamic Placement Strategy of Data Copy

The dynamic placement strategy of data copies is executed by system according to the following situation: the adaptive model SLA based on cost-driven, the model of single-node load state and the model of cluster system reliability, so that, the system can be more flexible to create the number of copy greater than 3 data copies for a data block.

When the strategy creates the fourth data replica for a data block, the replica numbered DB_3 is defined as the second main data copy DB_3^* of the data block. The placement follow-up for the data block to create other data replicas was shown in Fig. 29.2.

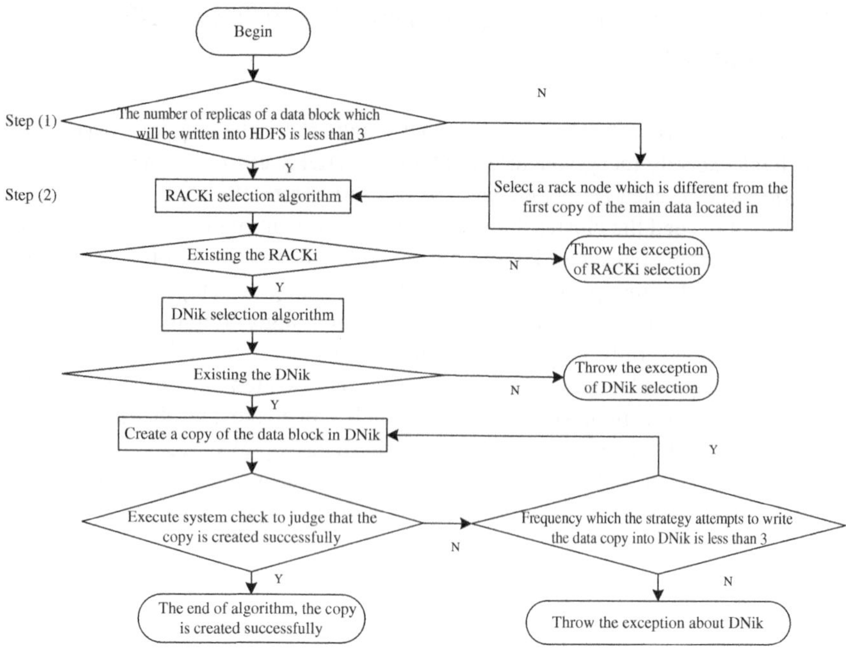

Fig. 29.1 The flow diagram of default placement strategy of data copy

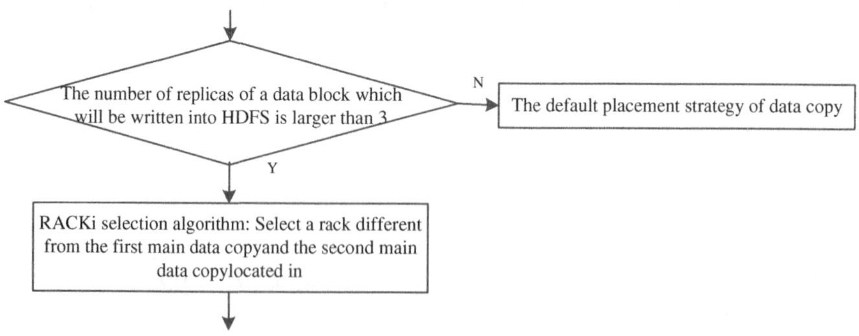

Fig. 29.2 The flow diagram of dynamic placement strategy of data copy

1. First, the strategy judges that the number of copies which data block will be pre-written in HDFS is larger or less than 3, if larger than 3, then turn to (2), or turn to step (1) of the default placement strategy of data copy in Sect. 3.1.
2. Select a rack different from the first main data copy DB_1 and the second main data copy $DB_3^*(DB_3)$ located in, then turn to step (2) of the default placement strategy of data copy in Sect. 3.1.

29.4 Rack Node $Rack_i$ Selection Algorithm

When some copies of a data block will be created, the $Rack_i$ selection algorithm will choose a suitable rack to improve system fault tolerance, and consider both load balancing and communication costs in this section to realize the problem of placement of data copies be transformed into the problem of weighting the factors between communication costs d_i [6] and current storage condition r_i of rack.

Definition 1 The shortest distance between a client and all rack nodes can be expressed as a *MAX_RACK_NUM* dimensional vector $CtoR = [d_1, d_2, \ldots, d_i, \ldots]$ in HDFS cloud storage cluster. d_i is the shortest network distance between the client and the rack numbered $Rack_i$.

Definition 2 Direct distance between adjacent communication nodes is expressed as e_{ml} in HDFS cloud storage cluster (the direct network distance between communication node m and communication node l), then $d_i = \min\left(\sum_{e_{ml}\in(Client\rightarrow RACK_i)} d_{(e_{ml})}\right)$. $Client \rightarrow RACK_i$ indicates the reachable path set constituted by intermediate communication links between *Client* and $Rack_i$. When the distance between a client and a rack node equals 0, it means that the client is in the rack.

Definition 3 Current redundant capacity of each rack node $Rack_i$ storage can be expressed as a *MAX_RACK_NUM* dimensional vector $Sr = [r_1, r_2, \ldots, r_i, \ldots]$ in HDFS cloud storage cluster. $r_i = \frac{u_i}{t_i}$ (u_i is $Rack_i$'s capacity which has been used, t_i is the total capacity of $Rack_i$).

The condition of a data replica DB_jwritten into a rack node $Rack_i$ of HDFS cloud cluster can be expressed as sum weighted by network communication and rack node capacity:

$$f(d_i, r_i) = \delta \cdot d_i + (1 - \delta) \cdot r_i \quad \delta \in [0, 1] \tag{29.1}$$

δ is the value of the two states weighted. Rack node selection mainly considers the communication cost between clients and racks, and the current capacity status of rack.

In order to express the topology distances between the client and all rack nodes, a *MAX_RACK_NUM* dimensional vector $F = [f(d_1, r_1), f(d_2, r_2), \ldots\ldots, f(d_i, r_i), \ldots]$ which corresponds to the rack nodes in HDFS was introduced. The cost about the distance between a client and each rack node in HDFS cloud storage cluster can be expressed as:

$$F = \delta \cdot CtoR + (1 - \delta) \cdot Sr \tag{29.2}$$

When the number *MAX_RACK NUM*of rack is given, the straight distance e_{ml} between any two adjacent communication nodes is available and the current redundant capacity of each rack node can be calculated in HDFS cloud cluster, the problem of how to determine the set $\{Rack_i\}$ of rack node constituted by the shortest topological distances between the client and the rack nodes to balance cluster load and save network bandwidth can be converted to solving the minimum set about Eq. (29.2).

The $Rack_i$ is calculated as follows:

$$f(d(e_{ml}), r(u_i, t_i)) = \delta \cdot \min\left(\sum_{e_{ml} \in (Client \to RACK_i)} d_{(e_{ml})}\right) + (1 - \delta) \cdot \frac{u_i}{t_i} \quad (29.3)$$

Selecting the minimum values from F to constitute the set $\{Rack_i\}$, then the cluster system votes for the elements in$\{Rack_i\}$. Voting rules are: The cluster system sorts the racks in accordance with the number of frequency accessed by the client, the higher frequency the higher the number of votes, and then elects the rack which owns the highest frequency number as the rack where the data copy will be written in.

29.5 DataNode DN_{ik} Selection Algorithm

This section will introduce the grey relation evaluation [7, 8] on the basis of Sect. 3, and then builds the evaluation system of current status of DataNode DN_{ik} in rack node $Rack_i$ based on entropy weight [9, 10].

According to the above requirements, DN_{ik} performance evaluation can be divided into as shown in Fig. 29.3:

Definition 4 $X_{ik} = \begin{bmatrix} x_{11}x_{12} \cdots x_{15} \\ \cdots \\ x_{k1}x_{k2} \cdots x_{k5} \\ \cdots \end{bmatrix}$ is the evaluation matrix of DataNodes in rack node numbered i. $x_{kh}(h = 1, 2, \cdots, 5)$ is the hth indicator weight of the DataNode numbered k in the rack node.

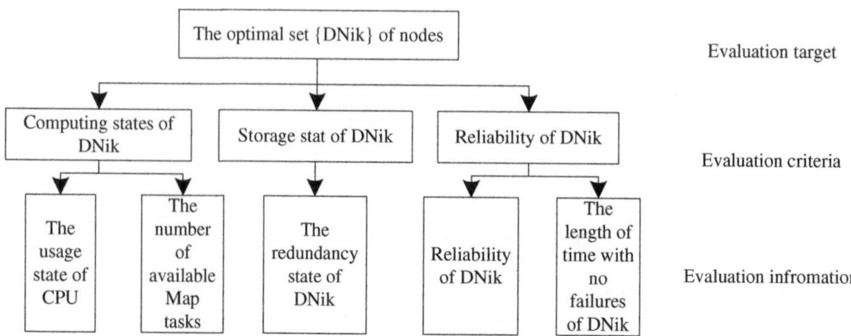

Fig. 29.3 The three evaluation system of DN_{ik}'s performance

Definition 5: The calculation steps of evaluation indicators of DN_{ik} based on entropy are as follows:

Step 1: Calculate the hth indicator's entropy of X_{ik}

$$e_h = -\frac{1}{\ln MAX\_DN\_NUM} \sum_{k=1}^{MAX\_DN\_NUM} p_{kh} \ln(p_{kh}) \quad \left(p_{kh} = \frac{x_{kh}}{\sum_{k=1}^{MAX\_DN\_NUM} x_{kh}} \right) \tag{29.4}$$

Step 2: Calculate the hth indicator's entropy weight of X_{ik}

$$w_h = \frac{1 - e_h}{\sum_{h=1}^{5} 1 - e_h} \tag{29.5}$$

The core idea of the gray relation is to analyze the degree of correlation of the geometry of the sequence set of curves, then to judge it whether close contact. The distance between any two curves is greater, while the degree of correlation between the two corresponding curves is smaller, whereas the degree of correlation is greater.

The calculation steps of gray relational grade are as follows:

Step 1: According to the evaluation matrix defined in Definition 4 to determine the reference vector and comparing vectors:

Reference vector

$$x = (\max_{1 \le k \le MAX\_DN\_NUM} x_{k1}, \max_{1 \le k \le MAX\_DN\_NUM} x_{k2}, \ldots, \max_{1 \le k \le MAX\_DN\_NUM} x_{k5}) \tag{29.6}$$

Comparing vectors

$$x_k = (x_{k1}, x_{k2}, \ldots, x_{k5}) \qquad (k = 1, 2, \ldots, MAX\_DN\_NUM) \tag{29.7}$$

Step 2: Normalize the reference vectors and the comparing vectors to 1, and process the vectors dimensionlessly. The processing result is:

$$x^*(h) = \frac{x(h)}{\sum_{h=1}^{5} x(h)} \qquad x_k^*(h) = \frac{x_k(h)}{\sum_{h=1}^{5} x_k(h)} \tag{29.8}$$

Step 3: Calculate the weight of the gray relational grade:

$$r_k(h) = \frac{\min_k \min_h \Delta k(h) + \xi \max_k \max_h \Delta k(h)}{\Delta k(h) + \xi \max_k \max_h \Delta k(h)} \quad \xi \in [0, 1] \tag{29.9}$$

$$\Delta k(h) = |x^*(h) - x_k^*(h)|$$

Step 4: Calculate the relational grade about x^* and x_k^*:

$$r_k = \sum_{h=1}^{5} e_h \cdot r_k(h) \quad (29.10)$$

Calculate and compare the current DN_{ik}'s r_k in $Rack_i$ to analyze and evaluate the comprehensive state of DN_{ik}, the larger r_k and the better comprehensive state of DataNode DN_{ik}, then add the DataNode into the set of $\{DN_{ik}\}$. At last, the cluster system votes for the elements in $\{DN_{ik}\}$ using the same voting rules in Sect. 4, and then elects the DataNode DN_{ik} which owns the most votes as the DataNode where the data copy will be written in.

29.6 The Experimental Results and Analysis

In order to facilitate comparison performance conveniently, 50000 data blocks were distributed into HDFS cloud cluster by the three placement strategy: the random strategy, $\delta = 0$ and $\delta = 1$ of $Rack_i$ selection algorithm. Figure 29.5 shows

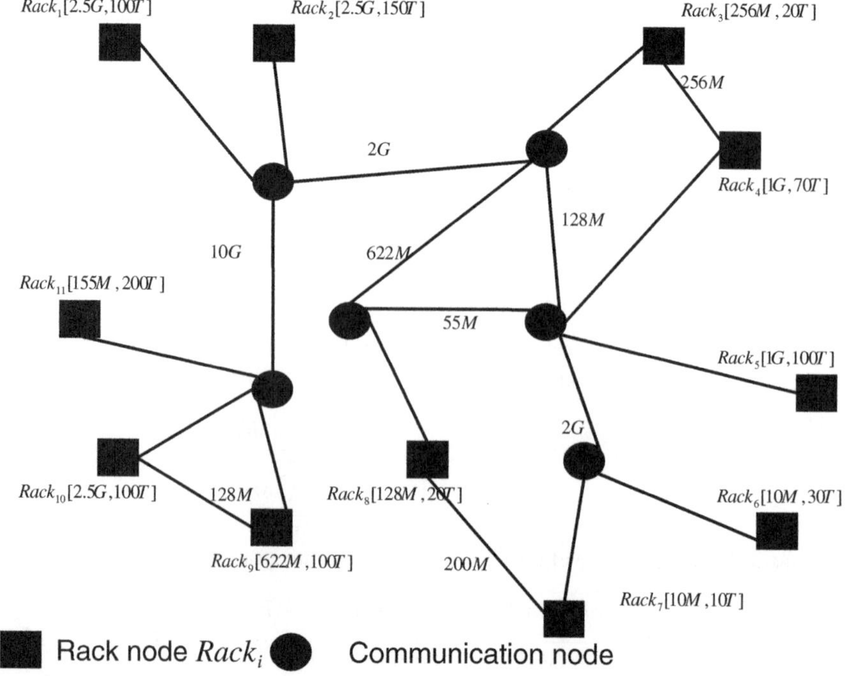

Fig. 29.4 The topological structure of rack nodes in HDFS

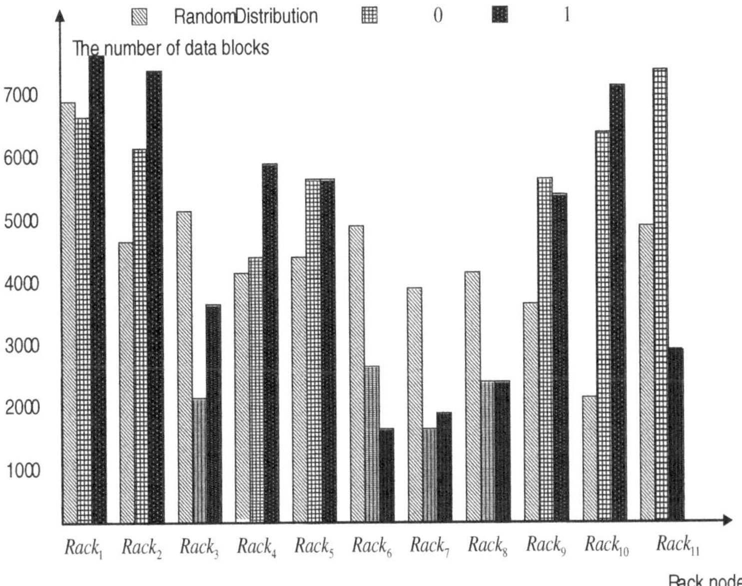

Fig. 29.5 The distribution of the data blocks in HDFS

the distribution of the data blocks obtained by the use of simulation as shown in Fig. 29.4. According to the Fig. 29.5, δ can be adjusted by the algorithm according to the need, weighting the network bandwidth and the storage state of rack node in cluster system, to achieve more excellent performance of the system's services and to provide customers with cloud storage services.

Now, The DN_{ik} selection algorithm will be analyzed. Five equipment were tested at times t_1, t_2 and t_3 in order to evaluate the comprehensive states of data nodes in a rack node, shown as follows:

$$X_{t_1} = \begin{bmatrix} 0.70 & 0.82 & 0.78 & 0.88 & 0.65 \\ 0.67 & 0.73 & 0.57 & 0.58 & 0.66 \\ 0.81 & 0.34 & 0.63 & 0.74 & 0.75 \\ 0.75 & 0.56 & 0.87 & 0.72 & 0.65 \\ 0.56 & 0.75 & 0.61 & 0.58 & 0.88 \end{bmatrix}$$

$$X_{t_2} = \begin{bmatrix} 0.61 & 0.72 & 0.75 & 0.64 & 0.58 \\ 0.88 & 0.91 & 0.77 & 0.74 & 0.69 \\ 0.55 & 0.34 & 0.60 & 0.78 & 0.68 \\ 0.83 & 0.74 & 0.90 & 0.77 & 0.71 \\ 0.64 & 0.72 & 0.87 & 0.55 & 0.69 \end{bmatrix}$$

$$X_{t_3} = \begin{bmatrix} 0.64 & 0.58 & 0.74 & 0.67 & 0.55 \\ 0.79 & 0.73 & 0.81 & 0.76 & 0.70 \\ 0.57 & 0.54 & 0.59 & 0.76 & 0.80 \\ 0.81 & 0.69 & 0.85 & 0.71 & 0.60 \\ 0.57 & 0.79 & 0.74 & 0.60 & 0.71 \end{bmatrix}$$

According to the X_{t_1}, X_{t_2} and X_{t_3} above, the corresponding gray relational grade can be calculated. The results were shown as follow:

$$X_{t_1} \rightarrow \begin{vmatrix} DN_{i1} & DN_{i2} & DN_{i3} & DN_{i4} & DN_{i5} \\ 0.935 & 0.667 & 0.641 & 0.734 & 0.694 \end{vmatrix} \quad X_{t_2} \rightarrow \begin{vmatrix} DN_{i1} & DN_{i2} & DN_{i3} & DN_{i4} & DN_{i5} \\ 0.657 & 0.831 & 0.531 & 0.831 & 0.594 \end{vmatrix}$$

$$X_{t_3} \rightarrow \begin{vmatrix} DN_{i1} & DN_{i2} & DN_{i3} & DN_{i4} & DN_{i5} \\ 0.667 & 0.794 & 0.631 & 0.671 & 0.763 \end{vmatrix}$$

According to the results above, DN_{i1} could be the best DataNode at time t_1 to place a data block; DN_{i2} and DN_{i4} could be the best DataNodes at time t_2, and can be as the more best DataNode according to the voting rules; DN_{i2} could be the best DataNode at time t_3 to place a data block. The experimental results showed that the DataNodes gained from the DN_{ik} selection algorithm based on the gray evaluation criteria of entropy weigh conformed to the actual performance of the five equipment at three times t_1, t_2 and t_3.

29.7 Conclusion

First of all, the paper built the default placement strategy of data replicas and the dynamic placement strategy of data replicas according to the different management of data replicas between HDFS cloud storage cluster and other distribution environments; On this basis, the paper designed the $Rack_i$ selection algorithm based on weight to process data transmission problems; At last, the paper proposed the gray evaluation, and built the current evaluation criteria of DataNode based on entropy weight located in $Rack_i$.

Acknowledgments This work is financially supported by the national natural science foundation of China under Grant NO. 61262016 and the natural science foundation of GanSu province under Grant NO. 1010RJZA050 & 14-0220.

References

1. Korupolu M, Plaxton G, Rajaraman R (2001) Placement algorithms for hierarchical cooperative caching. In: Proceedings of the 10th annual symposium on discrete algorithms vol 38 No 1 Jan pp 260–302

2. Rashedur RM, Barker K, Alhajj R (2008) Replica placement strategies in data grid. J Grid Comput 6:103–123
3. Buyya R (2002) Economic-based distributed resource management and scheduling for grid computer. Monash University, Melbourne
4. Ranganathan K, Foster I (2001) Identifying dynamic replication strategies for a high-performance data grid. In: Proceeding of the second international workshop on grid computing Denver, Nov pp 75–86
5. Ranganathan K, Foster I (2001) Design and evaluation of dynamic replication strategies for a high-perfomance data grid. International conference on computing in high energy and nuclear physics, Beijing, Sep
6. Xiong Fu, Wang R, Song D (2010) Heuristic replica placement algorithm in data grids. Syst Eng Electron , 32(7):1513–1516 (in Chinese)
7. Julong D (1996) Basic methods of grey system. Huazhong University of Science and Technology Press, Wuhan (in Chinese)
8. Yannian Z, Zhu Yi' (2011) The grey correlation evaluation algorithm of comprehensive performance about embedded computer based on the combination weight. J Northwestern Polytechnical Univ 29(1):12–16 (in Chinese)
9. Yi XQ (2010) Application of entropy weight coefficient method to the evaluation of the profitability of listed firms. 2010 international conference on computer application and system modeling 22–24 pp 108–110
10. Xiong R, Junzhou L, Song A, Jin J-H (2011) QoS preference-aware replica selection strategy in cloud computing. J Commun 32(7):93–103

Chapter 30
Optimization and Implementation of the Sobel Edge Detection on Davinci Platform

Wancai Li, Zekun Liu and Zhiwei Tang

Abstract Sobel is one of the best edge detection algorithms. However, it often loses some textural information when complex image is processed. This paper mainly discusses how to improve the shortcomings of Sobel and optimize the filtering, templates, and noise algorithms for edge detection. Beside the optimization, the paper presents the implementation of edge detection on Davinci platform. The algorithm code has been optimized which based on the TMS320DM6446 dual-core DSP. Multiplication operations have been split into the shift and add operations through integration algorithm engine which use the DSP's pipelining operations. After the test experiment on Davinci platform, it can work efficiently. This method can not only provide the possibility of the real-time image processing, but also eliminate the impact of noise, and enhance the detection accuracy.

Keywords Davinci DSP · Sobel · Image processing · Edge detection

30.1 Introduction

With the rapid development of image processing, its application field is increasingly, such as automobiles, computers, mobile phones and network. However, image processing hardware does not seem to be so desirable [1]. TI Davinci platform which developed specifically to handle the high-speed image processing can provide the possibility of the real-time image processing. The edge detection which is one of the

W. Li · Z. Liu (✉) · Z. Tang
The Third Research Institute of Public Security of P.R.C,
Shanghai 201204, China
e-mail: briskliu@163.com

W. Li
e-mail: liwancai1004@hotmail.com

Z. Sun and Z. Deng (eds.), *Proceedings of 2013 Chinese Intelligent Automation Conference*, Lecture Notes in Electrical Engineering 255,
DOI: 10.1007/978-3-642-38460-8_30,

important features in the image is the foundation of the computer vision and pattern recognition [2]. Sobel is one of the best edge detection algorithms during the classic edge detection algorithms. It can be easy to be implemented and could get the better edges [3]. This paper mainly discusses how to improve the shortcomings of Sobel and optimize the filtering, templates, and noise algorithms for edge detection. Beside the optimization, the paper presents the implementation of edge detection on Davinci platform. The algorithm code has been optimized which based on the TMS320DM6446 dual-core DSP. Multiplication operations have been split into the shift and add operations through integration algorithm engine which use the DSP's pipelining operations. After the test experiment on Davinci platform, it can work efficiently. This method can not only provide the possibility of the real-time image processing, but also eliminate the impact of noise, and enhance the detection accuracy [4].

30.2 System Architecture

The system has an internal module and peripheral module. The internal module composed by the ARM subsystem, DSP subsystem and the video processing subsystem. ARM system composed by ARM926EJS core, 16 kB I-cache, 8 kB D-cache, 16 kB RAM, 8 kB ROM. DSP subsystem composed by C64x+DSP core, 64 kB L2 RAM, 32 kB L1 cache, 80 kB L1 Data cache. Video process sub system composed by VPFE and VPBE. DM6446 peripherals interface include memory interface, serial interface, system interface. Memory interface include DDR2 interface, Async EMIF/NAND interface, ATA interface; serial port. The DM6446 system architecture is shown in Fig. 30.1

DM6446's ARM core is the main control component. ARM is responsible for configuration and control the DSP subsystem, VPSS subsystem, and most of the peripherals and external memory.

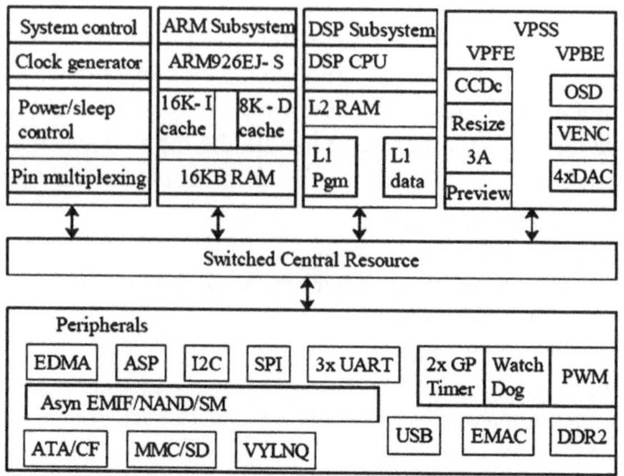

Fig. 30.1 System architecture

30.3 Sobel Edge Detection

The Sobel image edge detection algorithm is designed by using two directions templates convolution [5]. These two directions template is vertical and level edge detection. The basic principle of this algorithm is base on the variation of the brightness on the image edges. If the gray scale variation exceeds an appropriate threshold value TH, this pixel is the edge point. Simplicity and fast is the advantage of Sobel algorithm. However, it can only detect the horizontal direction and the vertical direction of the edge. For more complex image edge detection, it cannot be used. Meanwhile, the Sobel algorithm base on that, where the gradation value is greater than or equal to the threshold value, the pixel points are edge points. This may cause misjudgment, because noise point's value is also including [6].

30.3.1 Median Filtering

Image edges and the noise are performed as a high frequency component. Therefore, it's necessary to use the filtering process to reduce the impact of noise [7]. The median filter is a nonlinear signal processing method. This is commonly used to protect the edge information. The weighted median filter includes the following steps. Firstly, sort for each window and takes the appropriate ratio; secondly, curve fitting and the slope of the curve fitting characterized by the image feature of this window; then, re-weighted according to the characteristics of the various parts of the image [8].

30.3.2 Increasing the Direction Template

In addition to the horizontal and vertical directions, there are other directions, such as 135° and 45° [9]. As shown in Fig. 30.2, in order to increase the accuracy of edge detection, we increase the direction template from 2 to 4.

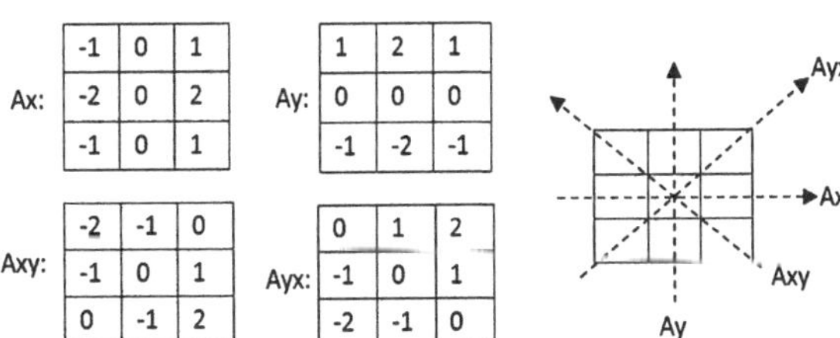

Fig. 30.2 The edge detection of four direction template

30.3.3 The Edge Positioning and Noise

Usually the edge of the image is continuous and smooth, and the edge having a direction and magnitude, but the noise is random. Along with any edge point can always find another edge point, and the difference of the gray-scale and direction are similar [10]. But noise is different; Along with any edge point can not find another edge point. Based on this principle, the noise points and edge points can be distinguished. According to (30.1), for an image f (x, y), using the above four directions template, Sobel operator calculated for each pixel in the image to obtain the maximum value. The pixel direction is indicated by the template corresponding to the maximum value. For any $i = 0,1,-1$; $j = 0,1,-1$ are established, it is judged point (x, y) as noise point.

$$|F(x, y) - F(x + i, y + j)| > TH \tag{30.1}$$

30.4 The Implementation of the Sobel Edge Detection on Davinci Platform

This system uses the TI's TMS320DM6446 evaluation development board DVEVM. Development system kernel is Montavista Linux.DM6446 is dual-core architecture, so there is a communication problem between the two cores. The communication between the DM6446 ARM core and DSP core is achieved mainly through mutual interrupt. ARM and DSP shared many peripherals resources, such as EDMA, the Timer0/1, Power Sleep Controller, ASP, and EMIFA, EMIFB the Data. Their internal memory, such as the ARM internal RAM and DSP L2/L1P/L1D RAM, these are share resources of two CPU cores. The evaluation board hardware platform is divided into five parts, they are video capture, data storage, image processing, result display and power management. Video capture part uses the analog PAL camera with the high-precision video A/D converter. The entire system process consists of three parts which are image acquisition, edge processing, and output display.

The collected video camera signal is digitized by the video encoder SAA7115. The SAA7115 is configured through the I2C bus. The video is pre-processing in the SAA7115, and then output to the processing unit DM6446. After the SAA7105 D/A conversion, the results are displayed on the monitor.

The optimization of Sobel has improved the performance, however, it increase the computation. Such as the increased direction template, each pixel which has 2 convolution operations has increased to 4 times. We improve the algorithm design based on the Davinci hardware structural. On the TMS320DM6446 dual-core DSP, multiplication operation can be split into the shift and add operations through integration algorithm engine [11]. The entire software program writes and debug in

Fig. 30.3 Original image

Fig. 30.4 Before optimization

the TI's Davinci software development environment. We generate C code in Linux compiler environment, and optimize C code in TI's code composer studio.

30.5 Result

The test video is PAL format and the frame rate is 25 frames per second. Figures 30.3, 30.4 and 30.5 are the results of the edge detection. According to the test, Davinci platform which developed specifically to handle the high-speed image processing can provide the possibility of the real-time image processing. Optimized edge detection operators can eliminate the impact of noise, and give a clear outline of the edge. The algorithm can not only suppress most of the noise in the image, and also enhance the detection accuracy.

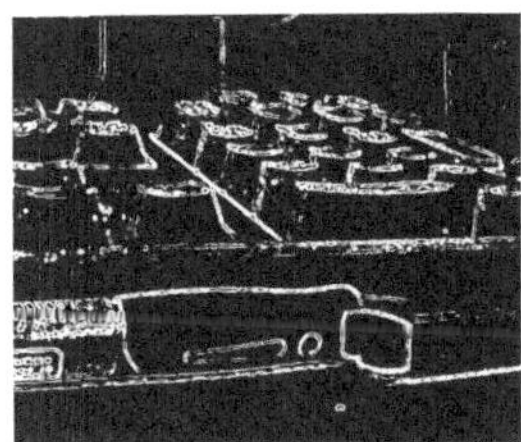

Fig. 30.5 After optimization

30.6 Conclusion

This paper presents the optimization and implementation of the Sobel edge detection on Davinci platform. We optimize the Sobel algorithm on filtering, templates, and noise. According to the dual-core DSP's instruction and software pipelining operations, we implement the algorithm efficiently. It not only improves the accuracy of image detection but also achieve the real-time requirements. Optimized edge detection operators can eliminate the impact of noise. It will be widely used in target tracking, non-contact detection, autopilot, video surveillance applications.

References

1. Bhandarkar SM, Zhan Y, Poter WD (1994) An edge detection technique using genetic algorithm-based optimization. Elsevier Science Ltd 0031-3203(94)E0028-J
2. Harris C, Buxton B (1996) Evolving edge detectors with genetic programming. Stanford, California, 1088-4750,0-262-61127-9, pp 309–314
3. Zhou L, Chen M (2005) Image's edge detection of stores of pets based on grey relational system. J Huazhong Univ Sci Technol (Nat Sci Ed) 33(4):61–63
4. Zheng Y, Rao J, Wu L (2010) Analyzing and comparing several edge detection methods of digital image. Mech Eng Autom
5. Saenthon A, Kaitwanidvilai S (2010) Development of new edge detection filter based on genetic algorithm: an application to a soldering joint inspection. Int J Adv Manuf Technol 46:1009–1019
6. Saskoro AS, Masuda R, Kabutomori M, Suga Y (2009) An application of genetic algorithm for edge detection of molten pool in fixed pipe welding. Int J Adv Manuf Technol 45:1104–1112
7. Hsich C-T, Juan Y-S, Hung K-M (2005) Multiple license plate detection for complex background. Proc Int Conf AINA 2:389–392
8. Gonzalez RC, Woods RE (2004) Digital image processing using MATLAB. Prentice Hall, Englewood Cliffs
9. Pratt WK (2001) Digital image processing. A Wiley-Interscience Publication
10. Xie F, Zhao D (2008) Visual C++ digital image processing. Electronic Industry Press, Beijing
11. Wang Q, Li Z, Li Q, Sun J, Fu J (2003) Edge detection algorithm for imaging ladar. Chin Opt Lett 1(5):272–274

Chapter 31
Modeling for Penicillin Fermentation Process Based on Weighted LS-SVM and Pensim

Weili Xiong, Xiao Wang, Qian Zhang and Baoguo Xu

Abstract In order to solve the problem when monitoring and controlling penicillin fermentation processes, So an intelligent modeling method based on Quantum Particle Swarm Optimization (QPSO) algorithm and Weighted Least Squares Support Vector Machines (WLS-SVM) is presented, which can overcome the noise of sample data, the high non-linear. Applied the method in penicillin fermentation processes and compared with the Pensim simulation platform data, it obviously found that the WLS-SVM is superior to the unweighted LS-SVM modeling method that has a better estimation accuracy and robustness.

Keywords Weighted SVM · Quantum particle swarm algorithm · Penicillin · Modeling

31.1 Introduction

In fermentation process, biomass concentration is one of the most important key indicators of the process performance. Due to the non-linear and time dependencies of fermentation processes, the lack of biosensor along with the serious correlation of all parameters, the classical system theory can hardly describe these processes very well. This problem has led to the development of a range of 'soft sensors'. These sensors utilize mathematical models (ranging from structured to data-based) and algorithms, together with available online information e.g. temperature, pH value, dissolved oxygen tension, relative pressure and agitator-rotated speed, to estimate the key bioprocess parameters. There are a wide number of

W. Xiong (✉) · X. Wang · Q. Zhang · B. Xu
Key Laboratory of Advanced Process Control for Light Industry (Ministry of Education), Automation Institute of Jiangnan University, Wuxi, Jiangsu, China
e-mail: greenpre@163.com

Z. Sun and Z. Deng (eds.), *Proceedings of 2013 Chinese Intelligent Automation Conference*, Lecture Notes in Electrical Engineering 255,
DOI: 10.1007/978-3-642-38460-8_31,

techniques available to formulate such models, each varying in complexity and ease of development.

Some literatures adopted artificial neural networks (ANN) to model the fermentation processes [1, 2], however, ANN based on empirical risk minimization theory needs a large number of data samples and training epochs to minimize the training error and this often lead to the over-fitting problem that decreases the model generalization capability. The SVM proposed by Vapnik uses the structural risk minimization principle instead of empirical risk minimization principle [3, 4]. SVM overcomes the intrinsic limitation of ANN and is powerful for modeling nonlinear systems and be applied in many classification problems successfully [5]. According to the practice, Least Squares Support Vector Machines (LS-SVM) [6] presented by Suykens J.A.K can overcome the disadvantage of slow training velocity in the large-scale problem, which transforms the quadratic optimization problem into that of solving linear equation set. However, the transform makes the solution lose sparsity and robustness, WLS-SVM presented in reference, can overcome this two shortcomings in a certain extent [7].

Therefore, an intelligent modeling method based on QPSO algorithm and WLS-SVM is presented in this paper, which to overcome the disadvantage of low predicted accuracy caused by lacking the priori knowledge, the disadvantage of noise when collected on-line and finite learning samples. With the data based on Pensim simulation platform use the QPSO algorithm train the WLS-SVM model, which not only improve the prediction accuracy, but also holds the solution's sparsity and robustness.

31.2 Weighted Least Squares Support Vector Machine

Based on the standard SVM, Suykens proposes the least square support vector machine, which solves in a set of linear equations instead of quadratic programming, and adds error quadratic terms in objective function. So it can gain solutions by solving linear equations and improve the computing speed. The LS-SVM doesn't need ε by comparison with SVM, which avoids deciding ε to model for fermentation processes.

Given a training data set of n points $\{(x_i, y_i),\ i = 1, 2\ldots n\}$, with input data $x_i \in R^n$ and output data $y_i \in R$, one considers the following optimization problem in primal weight space:

$$min\ J(w, \xi) = \frac{1}{2}\|w\|^2 + \frac{1}{2}C\sum_{i=1}^{n}\xi_i^2 \tag{31.1}$$

such that

$$y_i = w^T\varphi(x_i) + b + \xi_i \quad i = 1, 2\ldots n \tag{31.2}$$

With $\varphi(x)$ a function which maps the input space into a so-called higher dimensional (possibly infinite dimensional) feature space, weight vector $w \in R^n$ in primal weight space, error variables $\xi_i \in R$ and bias term b. Therefore, one computes the model in the dual space instead of the primal space. One defines the Lagrangian:

$$L(w, b, \xi, \alpha) = \frac{1}{2} w^T w + \frac{1}{2} C \sum_{i=1}^{n} \xi_i^2 - \sum_{i=1}^{n} \alpha_i \left(w^T \varphi(x_i) - y_i + b + \xi_i \right) \tag{31.3}$$

with Lagrange multipliers $\alpha_i \in R$, $i = 1, 2 \ldots n$ (called support values), and can easily obtains the solution:

$$\begin{bmatrix} 0 & 1 & \cdots & \\ 1 & K(x_1, x_1) + \frac{1}{C} & \cdots & K(x_1, x_n) \\ \vdots & \vdots & \ddots & \vdots \\ 1 & K(x_n, x_1) & \cdots & K(x_n, x_n) + \frac{1}{C} \end{bmatrix} \begin{bmatrix} b \\ \alpha_1 \\ \vdots \\ \alpha_n \end{bmatrix} = \begin{bmatrix} 1 \\ y_1 \\ \vdots \\ y_n \end{bmatrix} \tag{31.4}$$

The resulting LS-SVM model for function estimation becomes:

$$f(x) = \sum_{i=1}^{n} \alpha_i K(x, x_i) + b. \tag{31.5}$$

In order to obtain a robust estimate based upon the previous LS-SVM solution, in a subsequent step, one can weight the squared error ξ_i^2 by weighting factor v_i. This leads to the optimization problem:

$$min\ J(w^*, \xi^*) = \frac{1}{2} \|w^*\|^2 + \frac{1}{2} C \sum_{i=1}^{n} v_i \xi_i^{*2} \tag{31.6}$$

such that $y_i = w^{*T} \varphi(x_i) + b^* + \xi_i^* \quad i = 1, 2 \ldots n$

The Lagrangian becomes:

$$\begin{aligned} L(w^*, b^*, \xi^*, \alpha^*) = & \frac{1}{2} w^{*T} w^* + \frac{1}{2} C \sum_{i=1}^{n} v_i \xi_i^{*2} \\ & - \sum_{i=1}^{n} \alpha_i^* \left(w^{*T} \varphi(x_i) - y_i + b^* + \xi_i^* \right) \end{aligned} \tag{31.7}$$

From the conditions for optimality, one obtains the KKT system:

$$\begin{bmatrix} 0 & 1_n^T \\ 1_n & \Omega + V_n \end{bmatrix} \begin{bmatrix} b^* \\ \alpha^* \end{bmatrix} = \begin{bmatrix} 0 \\ Y \end{bmatrix} \tag{31.8}$$

where the diagonal matrix V_n and Ω are given by

$$V_n = diag\left\{ \frac{1}{Cv_1} \quad \cdots \quad \frac{1}{Cv_n} \right\}, \Omega = \{ K(x, x_i), k, i = 1, \ldots n \}$$

The choice of the weights v_i is determined based upon the error variables $e_i = \frac{\alpha_i}{C}$ from the (unweighted) LS-SVM case. Robust estimates are obtained then by taking

$$v_i = \begin{cases} 1 & |e_i/\hat{s}| \leq c_1 \\ \frac{c_2 - |e_i/\hat{s}|}{c_2 - c_1} & c_1 \leq |e_i/\hat{s}| \leq c_2 \\ 10^{-4} & other \end{cases} \tag{31.9}$$

where $\hat{s}$ is a robust estimate of the standard deviation of the LS-SVM error variables e_i:

$$\hat{s} = \frac{IQR}{2 \times 0.6745} \tag{31.10}$$

The interquartile range IQR is the difference between the 75th percentile and 25th percentile. In the estimate $\hat{s}$ we take into account how much the estimated error distribution deviates from a Gaussian distribution. The constants c_1, c_2 are typically chosen as $c_1 = 2.5$ and $c_2 = 3$. The kernel function has a great effect to the system's generalization capability. We choose the radial basis function $K(x, x_i) = exp\left(\frac{\|x - x_i\|^2}{2\sigma^2}\right)$ in this paper.

31.3 WL-SVM Algorithm Based on QPSO

31.3.1 Quantum Particle Swarm Algorithm

In PSO algorithm, the trajectory of the particle is determined by position vector and velocity vector of particle, but it is no meaning in quantum theory, because we can't determine simultaneously the particle's next position according to the uncertainty principle [8, 9]. Researching the convergence behavior of particle with quantum mechanics, Sun has presented a new algorithm model-QPSO based on PSO algorithm [10, 11]. For the property of particle in aggregation state is different absolutely, the particle can search the optimum solution in the space of feasible solution. So the QPSO algorithm is superior to all presented PSO algorithms in search capability.

In QPSO algorithm, the particles motion is as following iterative equation:

$$\begin{gathered} mbest = \frac{1}{M}\sum_{i=1}^{M} p_i = \left(\frac{1}{M}\sum_{i=1}^{M} p_{i1}, \ldots, \frac{1}{M}\sum_{i=1}^{M} p_{id}\right) \\ P = f * p_i + (1 - f) \times p_g \quad (f = rand) \\ \begin{cases} x(t+1) = P - \beta \times |mbest - x(t)| \times ln\left(\frac{1}{u}\right) & u > 0.5 \\ x(t+1) = P + \beta \times |mbest - x(t)| \times ln\left(\frac{1}{u}\right) & u \leq 0.5 \end{cases} \end{gathered} \tag{31.11}$$

The global point, called mean best (*mbest*) of the particles is defined as the mean of all particles' *pbest*. To keep the convergent property of algorithm, every particle must be convergent to one point $P(P = (p_1, p_2, \ldots, p_d))$.

In the above equation, p_i is the local optimal point, p_g is the global optimal point, β is a design parameter, called contraction expansion coefficient; u is a random number distributed in range [0,1], δ is a control parameter, the convergent rate of algorithm can be controlled by adjusting the parameter δ.

$$\beta = \delta + (1 - \delta) \times (t_{max} - t)/t_{max} \tag{31.12}$$

where, $t_{\max}$ is the maximum iterative epoch, t is the current number of iterations.

31.3.2 The Main Steps of the WLS-SVM Algorithm

A soft sensing model based on WLS-SVM is a black-box model, which based only on input–output measurements of an industrial process. In the modeling procedure, the relationship between the input and output of the plant can be emphasized while the sophisticated inner structure is ignored. Procedures of soft sensing modeling based on WLS-SVM whose parameters optimized by QPSO are summarized as following.

Step 1 Give training data $\{x_k, y_k\}$, $k = 1, \ldots n$, get the LS-SVM model by using QPSO algorithm optimizing the optimum regularization parameter C and the kernel parameter σ

Step 2 Calculate $e_i = \frac{\alpha_i}{C}$

Step 3 Calculate the robust estimation $\hat{s}$ from the e_i distribution

Step 4 Determine the weights v_i based on e_i and $\hat{s}$

Step 5 Solve the Eq. (31.8), then obtain α^* and b^*, finally obtain the function model as follows $y(x) = \sum_{i=1}^{n} \alpha_i^* K(x, x_i) + b^*$.

Step 6 To show the superiority of the soft sensor model based on WLS-SVM, more work is done in the following experiments. Minimum absolute error (MIAE), maximum absolute error (MAE), mean absolute difference (MAD) and mean squared error (MSE) are generally employed to evaluate the performance of a soft sensor model. They can be defined as following:

$$\begin{cases} MIAE = min|y_i - y_i^*| \\ MAE = max|y_i - y_i^*| \\ MAD = \frac{1}{n}\sum_{i=1}^{n} |y_i - y_i^*| \\ MSE = \frac{1}{n}\sum_{i=1}^{n} |y_i - y_i^*|^2 \end{cases} \tag{31.13}$$

31.4 Simulation

31.4.1 Simulation of One-Dimensional Function

In order to verify that the above WLS-SVM algorithm is accuracy and validity, the one dimensional and two dimensional functions are used as evaluation of the prediction power. The one dimensional function is as follows:

$$y = sin(x) + \zeta \quad -\pi \leq x \leq \pi$$

where ζ is Gaussian noise that average value is equal to 0 and the variance is equal to 0.01. 150 samples are generated by using $sin x$ function (as before) with Gaussian noise ($\sigma = 0.01$). The 100 points are trained in the WLS-SVM and LS-SVM combined with QPSO for selecting optimized parameters. The rest 50 points are used to test the prediction errors.

In Fig. 31.1 prediction results of $sin x$ function by the LS-SVM and WLS-SVM regressions respectively, whose parameters are automatically tuned using the QPSO and the Fig. 31.2 shown the error curve of one-dimensional function.

To show the superiority of the soft sensor model based on WLS-SVM, more work is done in the following experiments, MIAE, MAE, MAD and MSE are calculated in Table 31.1.

31.4.2 Simulation of Two-Dimensional Function

The two-dimensional function as follows:

$$z = sin x \cdot cos y + \zeta \quad -\pi \leq x, y \leq \pi$$

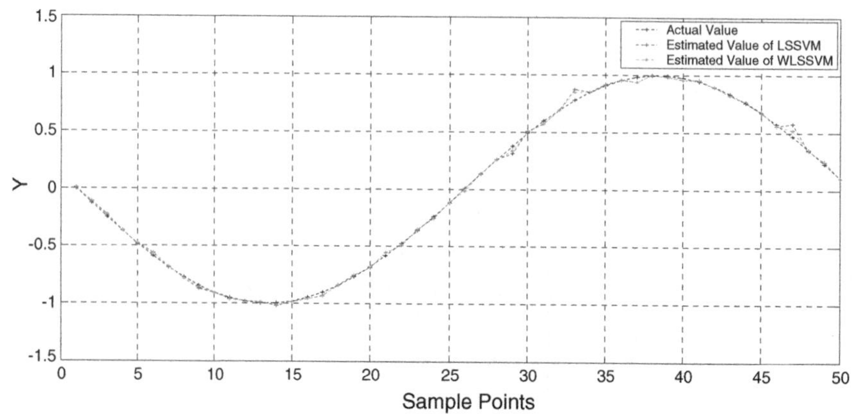

Fig. 31.1 Simulation result of one-dimensional function in WLS-SVM and LS-SVM

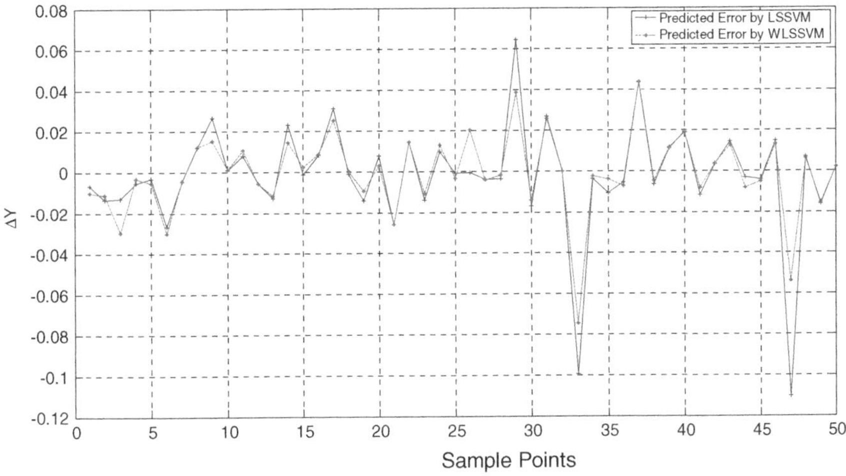

Fig. 31.2 The error curve of one-dimensional function

where ζ is Gaussian noise that average value is equal to 0 and the variance is equal to 0.01. The training sample set contains 100 data which are generated within the definition randomly and the testing sample set contains 100 data which are generated within the definition by equidistance.

The comparison pre-estimated result about the two-dimensional function of two methods is shown in the Fig. 31.3, the Fig. 31.4 shown the error surface of two-dimensional function and the Table 31.2 given the performance comparisons between WLS-SVM and LS-SVM.

From the Table 31.2, it clearly shows that the MSE of modeling based on WLS-SVM is little better and the MIAE, MAE and MAD are smaller than LS-SVM.

31.5 WLS-SVM Model Based on Pensim Simulation Platform

In order to further test the performance of the proposed algorithm, applying this algorithm to penicillin fermentation process. It is a very complicated biochemistry process that possesses the characteristics of high non-linear, time dependencies

Table 31.1 Performance comparisons of one-dimensional function between WLS-SVM and LS-SVM

Method	C	σ	MIAE	MAE	MAD	MSE
LS-SVM	653.9	1.2461	1.243e-4	0.0078	0.0024	8.653e-06
WLS-SVM	653.9	1.2461	0.114e-4	0.0053	0.0024	8.320e-06

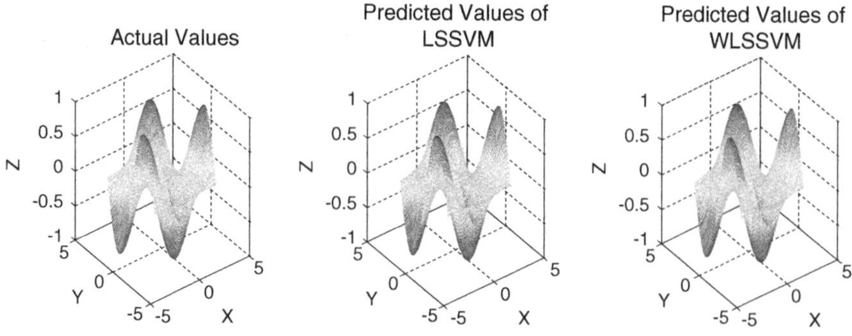

Fig. 31.3 Simulation result of two-dimensional function in WLS-SVM and LS-SVM

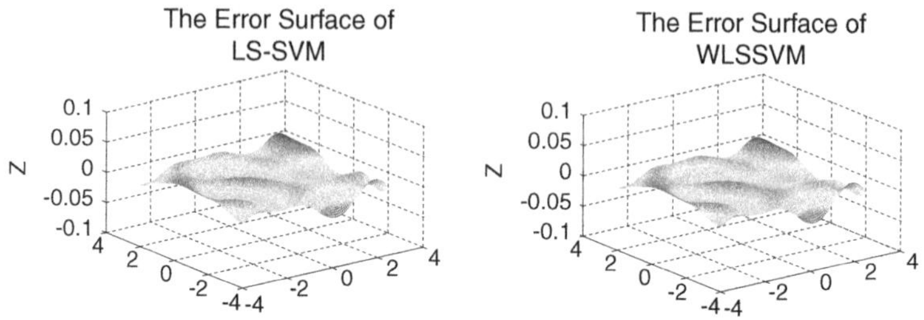

Fig. 31.4 The error surface of two-dimensional function

Table 31.2 Performance comparisons of two-dimensional function between WLS-SVM and LS-SVM

Method	C	σ	MIAE	MAE	MAD	MSE
LS-SVM	3956.1	1.7393	8.303e-5	0.0175	0.0044	4.060e-5
WLS-SVM	3956.1	1.7393	3.240e-5	0.0170	0.0041	3.907e-5

and uncertainty along with the serious relevancy of all parameters. Due to the nonlinear and time dependencies of fermentation processes, the lack of biosensor along with the serious relevancy of all parameters, the classic system theory can hardly describe these processes very well. The training sample set with noises which affect accuracy of the model, therefore the algorithm in this paper is raised to reduce the effect of noise to improve the precision.

In this study, the data of penicillin fermentation process is generated by Pensim v2.0, simulation software developed by Undey et al. in Illinois Institute of Technology [12] based on the mechanistic model of Bajpai and Reuss [13]. Process flow sheet is shown in Fig. 31.5. It considers a variety of physical and biomass factors which will impact the quantity of bacteria and penicillin product [14]. Related studies have shown the practicality and effectiveness of the simulation platform. Select model input variables which will impact the penicillin

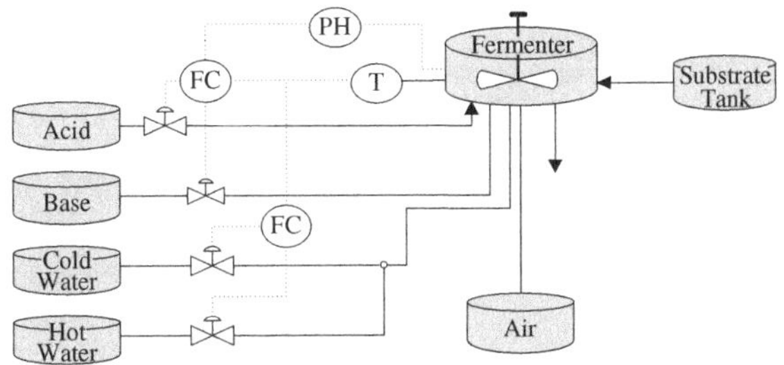

Fig. 31.5 Penicillin fermentation process

fermentation process and can be on-line measured: Aeration Rate $(L \cdot h^{-1})$, Agitator Power (W), Substrate Feed Rate $(L \cdot h^{-1})$, Dissolved Oxygen Concentration (%), Culture Volume (L), Carbon Dioxide Concentrations $(mmol \cdot L^{-1})$, pH, Temperature (K). Select Penicillin Concentration as the model output variables.

So it is very difficult to find a precise mathematic model for the penicillin fermentation process. There are many factors that affect the yield of the penicillin fermentation [13]. These factors mainly include the fermentation temperature, pH value, the substrate concentration, the dissolved oxygen concentration, the biomass concentration, the grown rate of biomass and the hyphal shape.

By analyzing the real data collected from Pensim simulation platform, we adopted some factors that greatly affecting titer as the input variables. These variables are the cell concentration, CO_2 concentration, substrate concentration, the air flux, agitation power, generate heat and the cold water flows respectively.

The data of penicillin fermentation process is generated by Pensim in the normal operating conditions. The 100 training samples which are selected randomly are used to build models and 100 validation data whose sample interval time is 1.25 h are used to choose the best model between WLS-SVM and LS-SVM. Figure 31.6 shows the simulation result of penicillin fermentation process in WLS-SVM and LS-SVM, and the error curve is shown in Fig. 31.7.

Figures 31.6 and 31.7 shows that the accuracy of WLS-SVM model is higher than LS-SVM model. The concentration of penicillin can be predicted more accurately. In order to verify the performance of WLS-SVM based on QPSO more comprehensive, compare the MIAE, MAE, MAD, MSE of WLS-SVM modeling method based on QPSO with LS-SVM modeling method, the results shown in Table 31.3.

From Table 31.3, we can easily find out that the modeling method based on WLS-SVM can estimate the key parameters with a high precision of penicillin fermentation process. Actually, WLS-SVM can generate MSE value 7.609×10^{-7} and MAD value 5.908×10^{-4}, while LS-SVM yield MSE value 8.077×10^{-7} and MAD value 6.067×10^{-4}. We also can see that other two indicator of the four criterions of the proposed WLS-SVM soft sensor model is less than the LS-SVM soft sensor model.

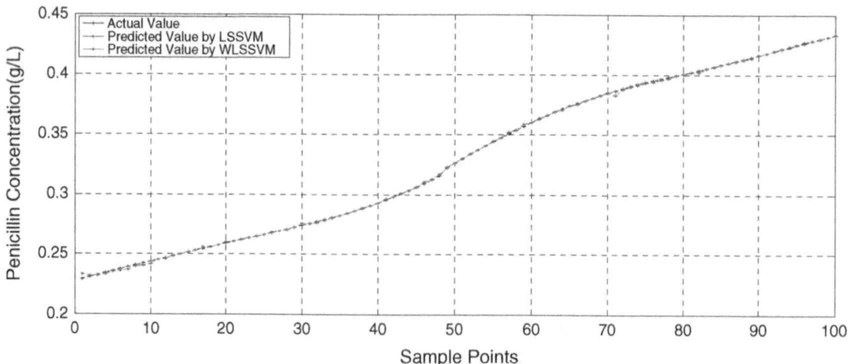

Fig. 31.6 Simulation result of penicillin fermentation process in WLS-SVM and LS-SVM

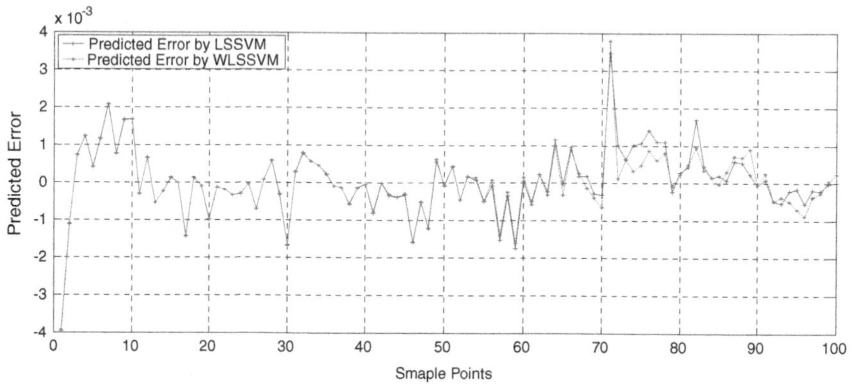

Fig. 31.7 The error surface of penicillin fermentation process

Table 31.3 Performance comparisons of penicillin fermentation process between WLS-SVM and LS-SVM

Method	C	σ	MIAE	MAE	MAD	MSE
LS-SVM	9713.6	1.0178	2.853e-6	0.0040	6.067e-4	8.077e-7
WLS-SVM	9713.6	1.0178	1.151e-6	0.0039	5.908e-4	7.609e-7

31.6 Conclusion

Penicillin fermentation process is a very complicated biochemistry process that possesses the characteristics of high non-linear, time dependencies and uncertainty. To realize the optimization control, the pre-estimate model should be built, but in case of outliers and noise with non-Gaussian distribution the performance of the LS-SVM is affected. In order to reduce the impact, a method WLS-SVM algorithm based on QPSO is used to modeling the penicillin fermentation process in this paper. By omitting a relative and small amount of the least meaningful data points (this corresponds to setting these α_i values to zero), one can re-estimate the LS-SVM. A large number of simulation experiments show that the WLS-SVM algorithm based on QPSO is helpful to improve the prediction accuracy and robustness.

Acknowledgments This work was supported by the National Natural Science Foundation of China (21206053,21276111); General Financial Grant from China Postdoctoral Science Foundation (2012M511678); A Project Funded by the Priority Academic Program Development of Jiangsu Higher Education Institutions.

References

1. Becker T, Enders T, Delgado A (2002) Dynamic neural networks as a tool for the online optimization of industrial fermentation. Bioprocess Biosyst Eng 24(2):347–354
2. Narendra KS, Parthasarathy K (1990) Identification and control of dynamic systems using neural network. IEEE Trans Neural Network 1(1):4–27
3. Vapnik VN (1995) The nature of statistical learning theory. Springer, New York
4. Vapnik VN (1999) An overview of statistical learning theory. IEEE Trans Neural Network 10(5):988–999
5. Cristianini N, Shawe-Taylor J (2000) An introduction to support vector machines and other kernel-based learning methods. Cambridge University Press, Cambridge
6. Suykens JAK, Vandewalle J (1999) Least squares support vector machine classifiers. Neural Process Lett 9(3):293–300
7. Suykens JAK, De Brabanter J, Lukas L et al (2002) Weighted least squares support vector machines robustness and sparse approximation. Neurocomputing 48(2):85–105
8. Kennedy J, Eberhart RC (1995) Particle swarm optimization. IEEE international conference on neural network, pp 1942–1948
9. Shi Y, Eberhart R (1998) A modified particle swarm optimizater. In: Proceedings of the IEEE conference on evolutionary computation, pp 69–73
10. Sun J, Xu WB, Feng B (2005) Adaptive parameter control for quantum-behaved particle swarm optimization on individual level. In: Proceedings of 2005 IEEE international conference on systems, man and cybernetics, vol 4, pp 3049–3054
11. Sun J, Xu WB, Feng B (2004) A global search strategy of quantum-behaved particle swarm optimization. In: Proceedings of 2004 IEEE conference on cybernetics and intelligent systems, vol 1, Singapore, pp 111–116
12. Birol G, Undey C, Cinar A (2002) A modular simulation package for fed-batch fermentation: penicillin production. Comput Chem Eng 26(11):1553–1565
13. Bajpai RK, Reuss M (1980) A mechanistic model for penicillin production. J Chem Technol Biotechnol 30(1):332–344
14. Lee JM, Yoo CK, Lee B (2004) Enhanced process monitoring of fed-batch penicillin cultivation using time-varying and multivariate statistical analysis. J Biotechnol 110(2):119–136

Chapter 32
Testing-Oriented Simulator for Autonomous Underwater Vehicles

Jinhua Wang and Yongzhong Ma

Abstract Over the past decade, there has been a growing interest in utilizing formation control and path planning in autonomous underwater vehicles (AUVs) designs. In this paper we present a novel method to create an AUV simulator using the Hardware in the Loop Simulation (HILS) and Virtual Reality (VR). The developed setup offers an alternative to difficult, costly, and possibly hazardous real-time testing and validation of formation control or path planning algorithms for autonomous underwater vehicles. The hardware of the platform is provided with data flow, followed by detailed descriptions of the AUV sensor models that are employed. An example of fuzzy algorithm development obstacle avoidance system is also described. Experimental results are presented showing the feasibility of methods. Finally, the suggestion of the hardware-in-loop simulator is given.

Keywords AUVs · Virtual reality · Hardware in the loop simulation · Fuzzy algorithm testing

32.1 Introduction

AUVs have become a hot research topic in the last decade worldwide. They have become a main tool for surveying below the sea in scientific, military and commercial applications because of the significant improvement in their performance. To successfully implement new technologies in the field, a number of sub-functions have to be tested and verified in advance. They could be formation control, path planning and communication functions as well as robust control for AUVs, including an emergency architecture for survival. Since it would be very expensive and time consuming to conduct all these tests at sea, researchers and engineers

J. Wang (✉) · Y. Ma
Chinese People's Armed Police Force Engineering University, Xi'an, China
e-mail: lljhw10@sina.com

Z. Sun and Z. Deng (eds.), *Proceedings of 2013 Chinese Intelligent Automation Conference*, Lecture Notes in Electrical Engineering 255,
DOI: 10.1007/978-3-642-38460-8_32,

engaged in the operation and development of underwater vehicles need easier test schemes and faster means.

To study the algorithms of formation control and path planning, the virtual-environment-based testbed are usually useful to save development cost and time. The virtual reality is developed following the physics laws and the simulation codes are developed through software like OpenGL and Matlab/Simulink_VRML. Examples of virtual environment involving autonomous underwater vehicles can be found in Brutzman [1] and Denis Gracanin [2]. In Bouxsein and An [3], Harris and Recce [4], Raezkowsky [5], the robot sensor simulation were developed based on computer simulation. However, navigation controller software of AUVs must run only to special hardware because it is the real-time embedded software and thus scheduling and operating testing of control algorithms is usually not available in VR system.

This paper has present a simulatior for an AUV based on hardware in the loop simulation and virtual reality. The platform includes AUV mathematical models, a VR system and a navigation and control processor from a real AUV. The simulator is a testbed that satisfies the various needs of the experimental tests required for the development of control and navigational architectures or software algorithms for underwater flying vehicles. Moreover, this simulator has real hardware modules from an AUV, which significantly improves the accuracy of simulation. Besides, such an in-lab validation saves time and development cost and thus is preferable during navigation controller development course. To carry out a number of tests while avoiding many difficulties in field trials, the first version of this simulator was designed to develop an obstacle avoidance system (OAS). The results of application show that the simulator presented in this paper has better practicability and prospect in project. And it is a useful tool for developing AUV behaviors.

32.2 Hardware of the Test Platform

32.2.1 Introduction of a New Long-Range AUV

A new long-range AUV is a kind of test platform which could carry out the module combination abided by the mission requirements to achieve a variety of complex functions. It could not only cruise like the traditional underwater vehicle in use of rudder and contrary-turning propeller, but also could hover, climb or swing in the front and back vertical thruster and lateral thruster. The AUV consists of 10 sections of the basic functions, including: collision avoidance sonar segment, carrying section, former vertical thruster segment, front lateral thruster segment, instruments and battery cabin, navigation controller section, power battery cabin, the back lateral thruster segment, back vertical thruster and the rear section, its shape and structural layout is shown in Fig. 32.1.

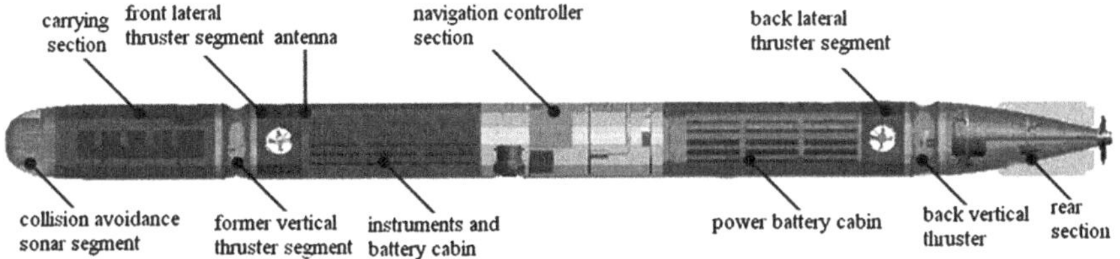

Fig. 32.1 Sketch map of AUV structure

32.2.2 Configuration

Intelligent control system of the underwater vehicle is made of real objects, and that the perception system and enforcement body are in use of the virtual reality technology. The configuration of the HILS system is shown in Fig. 32.2, which includes actual AUV navigation and control embedded processor, AUV hydrodynamic model computer, graphics workstation, network service and COM port. Each computer node is connected to the network server by means of optical fiber, which can meet the demand of real-time test. AUV navigation and control embedded processors connect with graphics workstation via the serial port.

In the Fig. 32.2, the AUV's mathematical model of computer (A) is a personal computer installed WINDOWS XP (Service pack 2). Its function is to calculate dynamics and kinematic differential equations to obtain the position and attitude information of the AUV, and then transmits it to the workstation node via the network.

Graphics workstation (IBM 9228 IntelliStation Z pro) installed Windows XP professional X64 Edition, build ocean environment, display dynamically formation cooperation, establish three dimensional solid model (such as: propeller), and

Fig. 32.2 Configurations of the platform

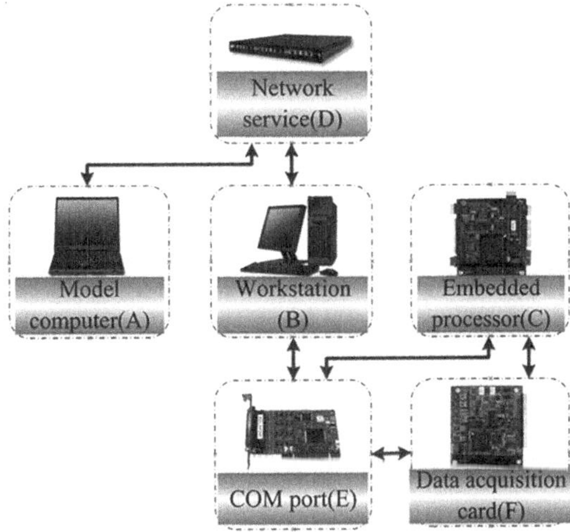

receive solution data of (A) to drive AUV solid model. Then, extract the position and attitude information in the visual simulation system as a substitute for various sensors data in the AUV, and transfer the simulation data to the processor (C) through the serial port (E) and data acquisition card (F).

Based on the μc/os-II embedded operating system and PC104 board, the navigation and control system (C) is constructed.

The function of a real-time network server (D) is to deliver data between A and B. All computer systems are connected to a star via network service (SCRAMNet GT), which characters 100Mbps Ethernet exchange and UDP/IP network protocol.

32.2.3 Flow Diagram

In Fig. 32.3, ① is the calculated results of AUV mathematical model, which include longitude, latitude, course angle, pitching angle, roll angle linear velocity and angular velocity in the local coordinate system. To take into account the influence of earth curvature on underwater vehicles, the coordinate systems are converted among geodetic coordinate system, geocentric coordinate system and north-east-down (NED) coordinate system in the ②. In order to drive the AUV model, the NED coordinates is converted to the coordinates of virtual environment in ③. ④ is the emulational data of sensor models, such as the velocity from the Doppler sonar, the depth from bathometer and the pose from the strap down inertial navigation system. The above data is transmitted through serial interface according to the actual transmission format and frequency in order to simulate the

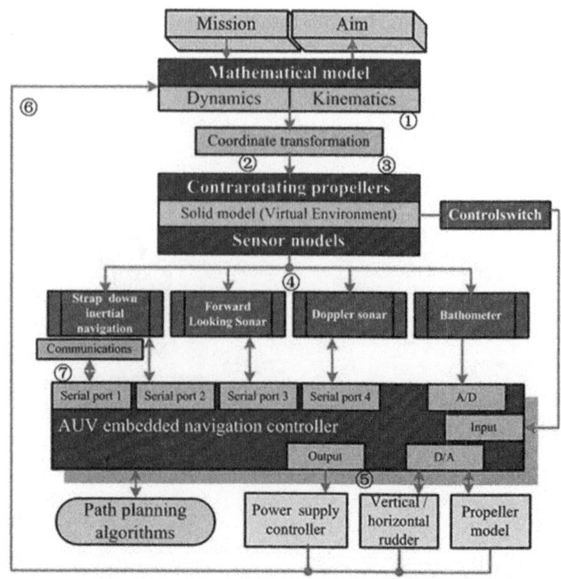

Fig. 32.3 Information flow chart

transmission characteristic of the real sensors, and emerge the course of sensor data sent to the embedded navigation controller. The AUV provided with multi-sensor, gather the different sensor data according to the disparate of speed. Thus, the platform adopts the PCI data acquisition card to handle the sensor data efficiently. ⑤ is the control instruction of embedded navigation controller sent to steering gear, propeller and power supply controller. ⑥ is the rotate speed, rudder angle and shut down sent to embedded navigation controller. The xPC sampling analog signals, such as the voltage of propeller revolutions, horizontal rudder and vertical rudder, from embedded navigation controller through A/D interface. And change into corresponding physical quantity according to the scale. ⑦ is the TXD or RXD which the AUV send the depot ship.

32.3 Simulation Experiment

In this section, the performance of the testbed platform is testified with its availability based on the fuzzy control algorithm.

The obstacle-avoidance algorithm divides the detection region of the AUV Forward Looking Sonar into three parts: front, left front and right front. And take the shortest distance between the craft in each region and the obstacle as the input, then get three inputs $[dl, dc, dr]$, which include the distance and the direction information between AUV and the obstacle. Besides, take the azimuth angle *tr* between the craft and the target as another input, these four inputs could determine the forward direction of AUV. As the direction of travel is decided by the yaw angle, so we prescribe the yaw angle *Sa* as the input which is shown in Fig. 32.4.

The values of *dl*, *dc*, *dr* could be got from the data generated from the Forward Looking Sonar simulator. The algorithm stipulate that there is no obstacle detected when values of *dl*, *dc*, *dr* are all greater than 200 m. Provided that the sonar simulator detects the obstacles, the distance between AUV and obstacle is calculated and the shortest distance in each domain act as the input values of *dl*, *dc*, *dr*.

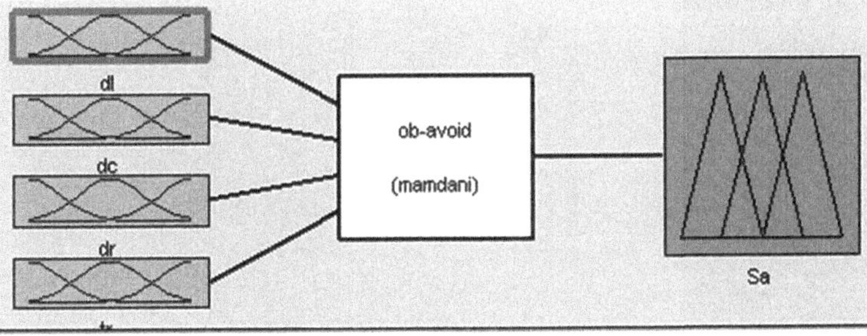

Fig. 32.4 Fuzzy reasoning of the obstacle avoidance

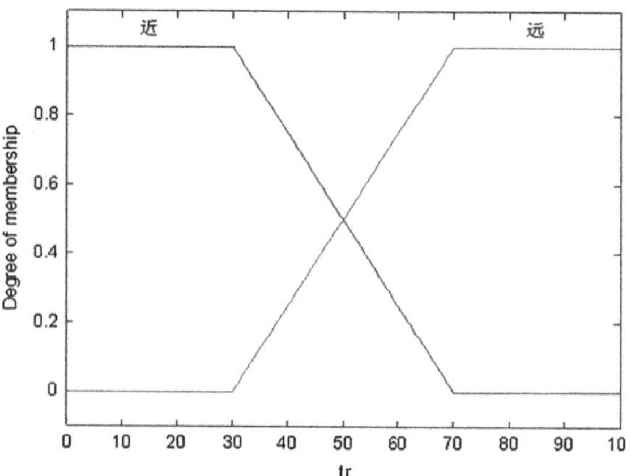

Fig. 32.5 Membership function of the distance between AUV and obstacle

The fuzzy linguistic variables of *dl*, *dc*, *dr* are Near, Far, And its domain $X = [0, 200]$. The membership function is shown in Fig. 32.5.

The fuzzy linguistic variables of direction angle *tr* are LB(left big), LS(left small), ZE(zero), RS(right small), RB(right big), and its domain $Y = [-180, 180]$. Its membership function is shown in Fig. 32.6.

The fuzzy linguistic variables of the output *Sa* are TLB (turning left to big), TLS (turning left to small), TZE (turning to zero), TRS (turning right to small), TRB (turning right to big), and its domain $Z = [-30, 30]$. The membership function is shown in Fig 32.7.

When the target is located on the right side of the AUV, *tr* is positive, otherwise negative. When the AUV turns right, *Sa* is positive; and AUV turns left, *Sa* is negative. The basic principles of obstacle avoidance are as follows: when the obstacles are detected in the underwater vehicle's left (right) and dead ahead, the AUV turns right (left) immediately.

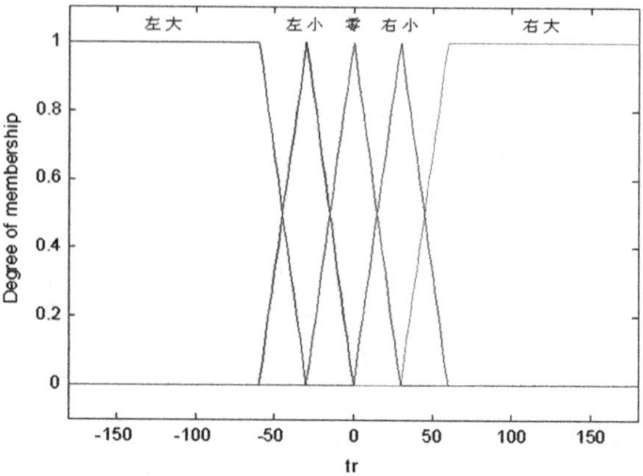

Fig. 32.6 Membership function of the azimuth between AUV and targets

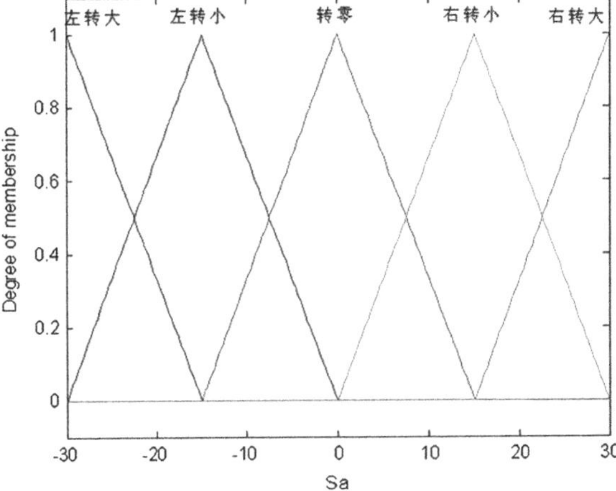

Fig. 32.7 Membership function of the yaw angle

The obstacle avoidance trajectories are shown in Fig. 32.8. In Fig. 32.8, the line segments on the bow of AUV are the simulative sound wave engendered by sonar simulator. The red line segments mean no obstacle in front of AUV, and the green line segments indicate the obstacle perceived. Yellow triangle marking stands for the AUV trajectory and its position and direction angle are from the experimental data. The results of application show that obstacle avoidance behavior is correct and credible with high image fidelity and good real-time. The methods of simulation were feasible. And this testbed can work for the development of AUV hardware and software.

AUV velocity vectors (v_x, v_y, v_z), rudder angles (horizontal rudder, vertical rudder), positional information (latitude, longitude and depths) and attitude information (pitch angle, heading angle, roll angle) is shown in Fig. 32.9.

Fig. 32.8 The path diagram of the virtual robot to avoid cylindrical obstacle

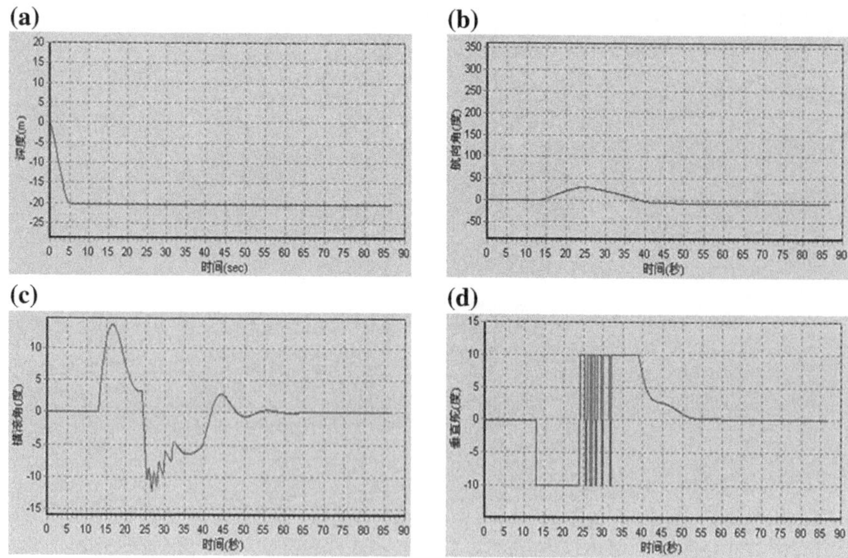

Fig. 32.9 Parameter records of AUV cylindrical-obstacle avoidance. **a** Depth. **b** Course angle. **c** Roll angle. **d** Vertical rudder

32.4 Conclusion

The main contribution in this paper is to develop an innovative hardware-in-the-loop simulator to simulate the long distance AUV for the purpose of validation of formation control or path planning algorithms. The simulator includes the actual embedded navigation controller from the AUV, and the sensor simulation model. A graphic workstation is applied to draw the virtual environment, including the ocean surroundings and the three-dimensional solid models. The fuzzy control algorithm is presented showing the feasibility of simulator.

It is of note that we have only given the designs of the test-bed platform of the AUVs here. By using the same idea, however, we are optimistic that the HILS presented here can be extended to Unmanned Underwater Vehicles (UUVs) systems. In addition, the models designs of the UUVs can be easily discussed via similar way.

References

1. Brutzman DP (1994) A virtual world for an autonomous underwater vehicle. Naval Postgraduate School, California
2. Denis Gracanin, Kimon P. Valavanis. (1998). Virtual Environment Testbed for Autonomous Underwater Vehicles. J Control Eng Pract 6:653–660

3. Bouxsein P, An E (2007) A SONAR simulation used to develop an obstacle avoidance system. OCEANS 2006–Asia Pac 16(19):1–7
4. Kenneth D. Harris, Michael Recce. (1998). Experimental Modeling of Time-of-Flight Sonar. Robot Auton Syst 24:33–42
5. Raezkowsky J (1989) Simulation of cameras in robot application. IEEE JCGA 9(1):16–25

Chapter 33
Virtual Reality-Based Forward Looking Sonar Simulation

Jinhua Wang, Renjun Zhan and Xulin Liu

Abstract Forward Looking Sonar simulators are useful tools for developing Obstacle Avoidance system of Autonomous underwater vehicle (AUV). A simulating the single and multi-beam forward looking sonar are established through the process simulation model. A new type beam model for sensing the environment is contrived based on collision detection technique of Virtual Reality. And it implements accurate realization of sonar resolution, which is able to keep the system real-time. The results of experiment show that the methods of simulation are feasible and the simulation is close to the AUV course of movement.

Keywords Visual simulation · Forward looking sonar · Collision detection

33.1 Introduction

In the development process of AUV obstacle-avoidance system, the Forward Looking Sonar plays a pivotal role in the craft's perception of the underwater environment, and it is widely applied in the obstacle avoidance, path planning, and target tracking and formation coordination. To develop an underwater obstacle-avoidance system in use of the real Forward Looking Sonar requires a lot of experiments under all kinds of the atrocious environment, which has the disadvantages of long research, costs and risks.

Many researchers have carried out the simulation research on the underwater vehicle sonar. Kenneth and Michael [1] have built sonar time-of-flight experimental model; Bo-Chang Chen and Jung-hua Chou have designed sonar echo

J. Wang (✉) · R. Zhan
Chinese people's Armed Police Force Engineering University,
Xi'an, China
e-mail: lljhw10@sina.com

X. Liu
Xi'an Institute of Modern Control Technology, Xi'an, China

Z. Sun and Z. Deng (eds.), *Proceedings of 2013 Chinese Intelligent Automation Conference*, Lecture Notes in Electrical Engineering 255,
DOI: 10.1007/978-3-642-38460-8_33,

model [2, 3] have simulated sonar echo feedback based on the ray-tracing algorithm. From the above studies, we can see that the current sonar simulation generally use the time-of-flight or the ray-tracing algorithm to accurately simulate the sonar. However, the abovementioned methods have the disadvantages such as the complex mathematical modeling, and complicated computation, which has a strong impact on the real-time of system. Besides, the obstacle model in the above methods is the simple geometrical configuration and it could not be the real representation of the complex seabed; therefore, it is difficult to apply the above sonar simulation model in the practical engineering. In order to effectively detect the complex terrain and motion entities (such as torpedoes, submarines or surface warships) and simulate the working process of the obstacle-avoidance system, Forward Looking Sonar Simulator based on the collision detection technique of virtual reality and related algorithms are presented, which could solve these problems such as the complex virtual environment perception and sonar sector scanning simulation etc. The modeling methods proposed in this paper accord fully with the sensor modeling pattern defined by Huck [4]. We can establish all kinds of virtual seafloor terrains based on real data (trench, sea walls and undersea mountains etc.), and carry out the terrain perception to provide the obstacle information for obstacle-avoidance system. In addition, three-dimensional display could make us clearly observe the all experiments and facilitates the obstacle-avoidance system improvement in a rapid way.

33.2 Forward Looking Sonar Simulator Description

Owing to the power, size and processing unit of AUVs, two major types of Forward Looking Sonar are used: the single-beam scanning sonar and multi-beam sonar system; as there is no essential difference between the single-beam sonar and the multi-beam sonar, multi-beam sonar system is composed of multiple transducer array modules, so it could send and receive multiple beams in working and forms a sector sound transmission area, through which acquire data of the underwater environment. Based on the above principles, two modes of the sonar simulator are designed, including the single-beam and multi-beam scanning. The single-beam sonar modeling parameters is obtained from the 881A sonar produced by IMAGENEX as shown in Table 33.1. The multi-beam sonar modeling data is from the Dolphin 6201 sonar produced by the British Marine Electronics Company, and the values are shown in Table 33.2.

Table 33.1 The single-beam sonar parameters

Parameter	Value
Acoustic width (υ)	2.4°
Horizontal angle (η)	360°
Stepping angle (μ)	0.9°

Table 33.2 The multi-beam sonar parameters

Parameter	Value
Vertical angle (τ)	15°
Horizontal angle (η)	90°
Included angle of transducer (μ)	1.5°

33.2.1 Simulation Method of Sound Wave

According to the theory that the acoustic waves through the short-distance can be taken as the ray, using the collision detection segment (VR technique) as a substitute for the sound wave propagation to perceive the virtual environment, is brought forward in the paper. At the same time, a new type of bamboo-shaped detector to reduce the computation and work in the actual resolution of the sonar is devised, as shown in Fig. 33.1.

According to the Fig. 33.1, the collision detection is designed to be bamboo-shaped segment, that is to take the beam transmitted by sonar transducer as the lines, the length of the first detection bamboo segment is the minimum detection distance (its value is 5 m, which means that the smallest resolution is 5 m), the bamboo-shaped segment length increases by 5 m successively, the last cumulative segment length is 200 m (that is the detecting radius of the sonar). Three bamboo-shaped detectors in this figure simultaneously place in the horizontal plane perpendicular to AUVs, which could simulate the acoustic beam transmitted by the transducer channel (called to be a collision detection channel) of Forward Looking Sonar in certain direction, and its angle is the vertical angle and its range can be adjusted. Each detection segment could sense the virtual obstacle. When there is the intersection between the segment and the obstacle, there will be the triggering signal given.

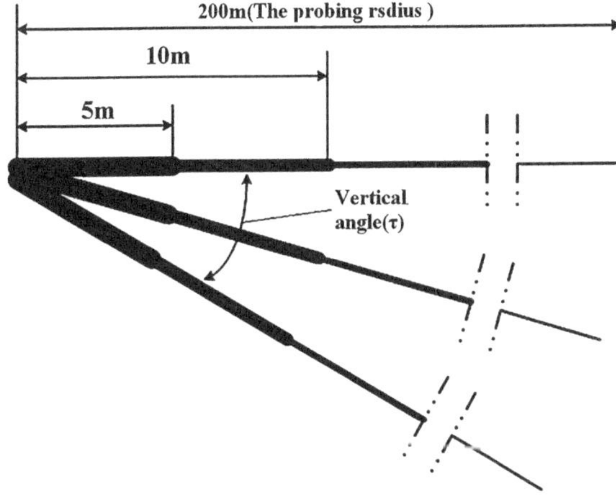

Fig. 33.1 Sketch map of collision detection segments

1. The Bamboo-Shaped Collision Detection Segments

In order to facilitate the management, three-dimensional array $seg[K][J][I]$ is used to define all bamboo-shaped collision detection segment, and I stands for the number of the horizontal beams of sonar, while the values of the single-beam sonar is different from that of the multi-beam sonar, J for the number of vertical beams of sonar, set $J = 3$, K for the number of collision detection segment needed to achieve the sonar detecting radius, the calculating formula of I, K are as follows:

$$I = \eta/\mu \tag{33.1}$$

$$K = \omega/\sigma \tag{33.2}$$

where η stands for the horizontal angle of Forward Looking Sonar, μ for the horizontal angle between two adjacent transducers of multi-beam Forward Looking Sonar or the stepping angle of single-beam Forward Looking Sonar scanning, ω for the detecting radius of Forward Looking Sonar, σ for the smallest resolution of the length of the collision detection segment. The value of ω to be 200 m and σ to be 5 m; the others are showed in Tables 33.1 and 33.2.

2. Location of the collision detection segments

After the definition of a bamboo-like collision detection segments, its location needs to be set which includes the initial location of bamboo-shaped collision detection segments and relative positions among them. The structure of the detection channel of single-beam Forward Looking Sonar is the same as that of the multi-beam sonar (Fig. 33.1), while they only differ in the angles. The multi-beam Forward Looking Sonar is illustrated by serving as an example in this paper. Figure 33.2 shows the sound propagation characteristics of the multi-beam Forward Looking Sonar.

It is essential to accurately place each collision detection segment (showed in Fig. 33.3). Supposed that the Forward Looking Sonar is installed in the bow of AUV, the starting points of all collision detection segments are in the bow of

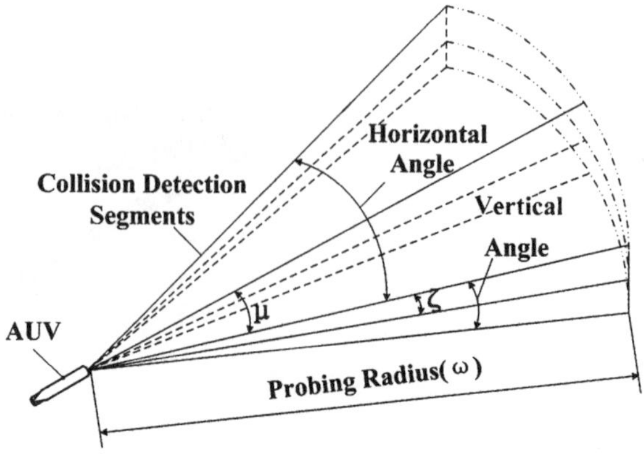

Fig. 33.2 Sonar beam pattern

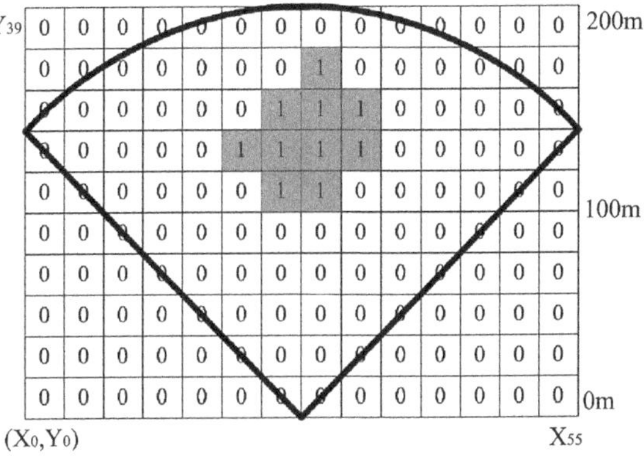

Fig. 33.3 2D grid recording data

AUV. The relative position among collision detection segments can be respectively written as,

$$\begin{cases} x[k][j][i] = \cos\left(\frac{1}{2}\tau - \mu \times j\right)\cos\left(\frac{1}{2}\eta - \xi \times i\right) \\ y[k][j][i] = \cos\left(\frac{1}{2}\tau - \mu \times j\right)\sin\left(\frac{1}{2}\eta - \xi \times i\right) \\ z[k][j][i] = \sin\left(\frac{1}{2}\tau - \mu \times j\right) \end{cases} \tag{33.3}$$

$$len[k] = 5 + 5k \tag{33.4}$$

where $x[k][j][i]$, $y[k][j][i]$ and $z[k][j][i]$ are the coordinates of the unit vector of collision detection segment $seg[k][j][i]$ in the detecting coordinates system of the Forward Looking Sonar, $len[k]$ is the length of $seg[k][j][i]$, τ stands for the vertical angle of the Forward Looking Sonar, η for the horizontal angle of Forward Looking Sonar, μ for the horizontal angle between two adjacent transducers of multi-beam Forward Looking Sonar or the stepping angle of single-beam Forward Looking Sonar, ξ for the vertical spatial angle between two adjacent collision detection (Fig. 33.2). i, j, k is the signs for the collision detection segment corresponding to the elements of seg, where $i \in \{0, 1, \ldots I - 1\}, j \in \{0, 1, 2\}$, and $k \in \{0, 1, \ldots, K - 1\}$.

3. Simplified Collision Detection Algorithm

First is to define the obstacle as the detection target of the collision detection segments, that is to identify the virtual targets which can be recognized by collision detector(for example, undersea terrain, ships etc.) and set the intersection test method and output the collision detection results. As the collision detection algorithm is too complicated and taking up too much resource that worsens the real-time of system; in order to improve the implementation efficiency, a simplified collision detection algorithm and the test results are presented in the paper.

The specific approach is as follows: to cancel the coordinates of intersection, normal vector of intersected primitive and the model name of intersection. When the intersection occurs, the triggering signal "0" or "1" of collision detection channel is only recorded in the Detecting result, "0" indicates that there is no obstacle in the current collision detection channel; "1" denotes that there is obstacle encountered by the collision detection channel.

33.2.2 Scanning Simulation and Data Record

1. Forward-looking sonar data recording format

In this paper, 2D grid recording data (the method proposed by the Carnegie Mellon University) is used to express the local environment information detected by sonar detector. The reason is that the intelligent terminal of AUV is with limited capability to process the information and the grid data is with little calculation, therefore, it could meet the requirement of obstacle avoidance system. Schematic diagram of the grid method is shown in Fig. 33.3. 2D rectangular grid is played in the Local Coordinates System to describe the obstacle information, and each rectangular grid has a cumulative CV value (CV value is the possibility of the existence of obstacle after the sensor information fusion). In addition, the grid size has a direct impact on the performance of path planning algorithm. The authors make the following stipulations: each bamboo segment represents that the area can be effectively detected is 5×5 m^2, which will achieve the [56] $\times$ [40] grid matrix to completely describe the virtual environment detected by sonar, and the value of CV is "1" when bamboo segments are triggered, otherwise CV is "0". Specific steps are as follows: simplify the algorithm and results of collision detection. As the sonar emulator starts working, its detecting result only records the triggering information of the collision bamboo segment, and write (data) on the 2D coordinate grid.

2. Scanning Simulation and Data Recording of Single-beam Forward Looking Sonar

To the single-beam sonar simulator as an example, a collision detection channel is used to simulate the acoustic beam sent by the transducer, and activates or shuts adjacent collision detection channels in the proper sequence. The steps are as follows

Step a: set the collision detection channel and the initial state; The *I* collision detection channels are distributed over the bow of AUV at intervals of the stepping angle, as shown in Fig. 33.4. Set the collision detection channel in the 90° to be the current collision detection channel, and turn off all collision detection channels.

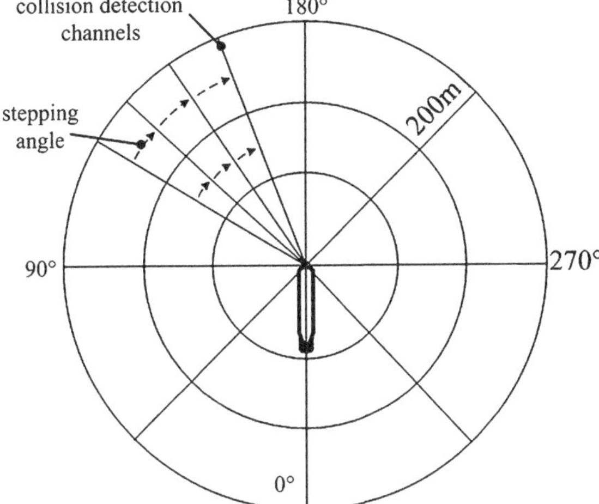

Fig. 33.4 Position of single beam channels

Step b: activate the current collision detection channel; In this step, activate the current collision detection channel to perceive the external environment, while the others are a state of closedown (Fig. 33.5a).

Step c: whether the activation time of the current channel is equal to the stepping time (f) of the single-beam sonar? When the conditions are not satisfied, continue to be in step c. Otherwise, go to step d;

The value of f is calculated by

$$f = T/I \tag{33.5}$$

where T is the scanning period of the single-beam Forward Looking Sonar.

Step d: write the triggering signal of the channels on the 2D coordinate grid; 2D grid data of the single-beam sonar is generated as follows: a two-dimensional array $[I] \times [K]$ is used to store the triggering states of I collision detection channels; each column of the array corresponds to the triggering state of channels. Figure 33.6 is

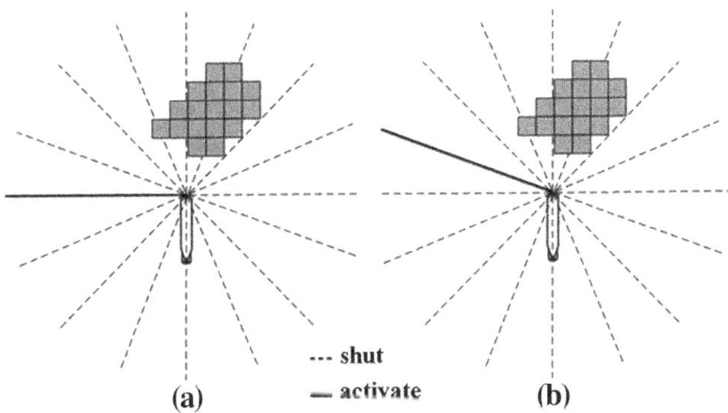

Fig. 33.5 Course of current activated channel

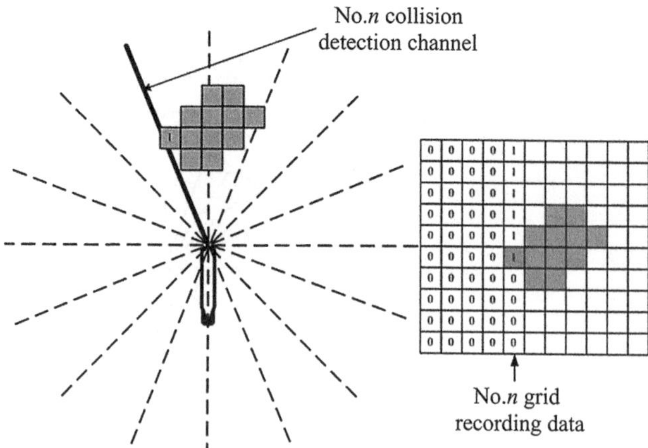

Fig. 33.6 Data logging of single beam sonar

simplified diagram of No.n collision detection channel, the dashed area denotes the obstacle, testing the triggering state of the collision detection and writing the CV on the 2D coordinate grid, channels intersected with the terrain is thought AUV can't get across, "1" is used to makes up a whole in accordance. The triggering information of the channels is "0000111111", and save the simulation sonar data to No.n column of the two-dimensional array.

Step e: close the current collision detection channel. In this step, turn off the current channel after f (Fig. 33.5b), and judge whether the current collision detection channel is the last one? If it is, record a period of sonar array into the database package according to the actual format of sonar. Otherwise, set the next channel adjacent to be the current channel and return to the step b characterized by cycles.

The multi-beam sonar simulator's approach is omitted in the paper.

33.3 Experiments

An 11,000 m × 8,500 m virtual terrain is called from the seabed model database. Based on contour line data, this model is developed by the polymesh rectangular grid algorithm, including the sea walls, trenches and ridges. The simulation result of the Forward Looking Sonar is shown in Fig. 33.7. The collision detection channel appeared green stands for the triggering, that means the obstacle is detected (seafloor terrain, etc.), while the red denotes that it is not triggered, which shows that there is no any obstacle detected.

The simulator stores the data based on the sampling frequency of the sonar. Besides, it is able to record the simulation data into the text file (.txt). The real-time simulation data of multi-beam sonar is shown in Fig. 33.8 (the data generated

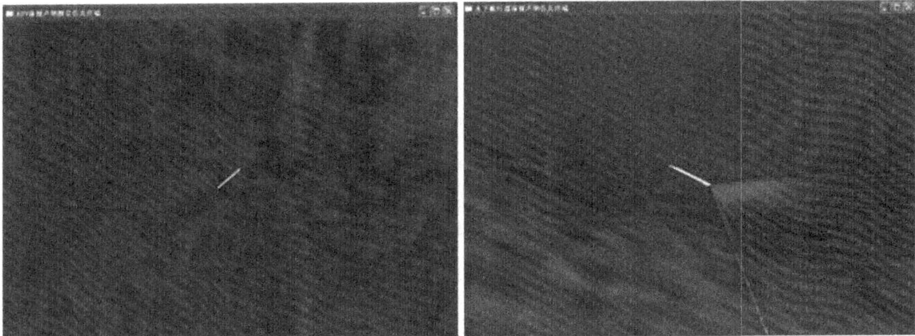

Fig. 33.7 Simulated images of forward looking sonar (*left* single-beam and *right* multi-beam)

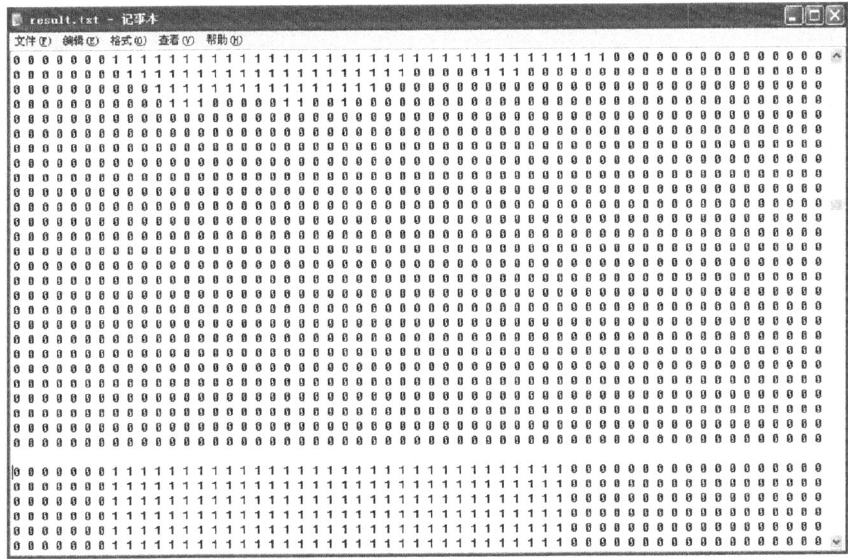

Fig. 33.8 Simulation data of multi-beam sonar (two successive periods)

by the single-beam simulator is identical with the multi-beam simulator). According to the actual format and frequency, the system sends simulation data to the sonar microprocessor via serial port. And the purpose is to better simulate the transmission characteristics of the sensor and reproduce the process of sonar data transmitting to the navigation system of AUVs.

33.4 Conclusion

The main contribution in this paper is to develop an innovative Forward Looking Sonar simulator. A new type beam model for sensing the virtual environment is contrived based on collision detection technique in Virtual Reality. And it implements accurate realization of sonar resolution, which is able to keep the system real-time. It is of note that we have only given the designs of the two-dimensional plane of the sonar here. By using the same idea, however, we are optimistic that the simulator presented here can be extended to three-dimensional space [5]. In addition, the models designs of the sonar can be easily discussed via similar way.

References

1. Kenneth DH, Michael R (1998) Experimental modeling of time-of-flight sonar. Robot Auton Syst 24:33–42
2. Chen B-C, Chou J-H (2008) A jump-U model of echo pattern for a sonar ranging module. Appl Acoust 69:1299–1307 (in Chinese)
3. Philip B, Edgar A (2007) A SONAR simulation used to develop an obstacle avoidance system. OCEANS'06—Asia Pac 16(19):1–7
4. Raezkowsky J (1989) Simulation of cameras in robot applieation. IEEE JCGA 9(1):16–25
5. Auran PG, Silven O (1996) Underwater sonar range sensing and 3D image. Control Eng Pract 4(3):393–400

Chapter 34
A Machine Vision System for Bearing Greasing Procedure

Hao Shen, Chengfei Zhu, Shuxiao Li and Hongxing Chang

Abstract Bearing is widely used in many machines, which can reduce the friction between connected components. However, many manufacturers still use human or machine methods to inspect the bearing production, which is inefficient, costly and unreliable, especially for the miniature bearing. In this paper, we propose a machine vision system for bearing greasing procedure. The proposed system uses image processing technology to process digital image captured by a camera and can locate the bearing cage quickly and accurately. Firstly, the bearing is separated from the whole image using the RANSAC least square circle fitting method. Secondly, to facilitate the process algorithm, the bearing area is transformed into a rectangle image. Next, some novel projection methods are involved. Finally, a center map is calculated to get the final greasing location. Experimental results show that the proposed machine vision system has high accuracy and efficiency, and can fully meet the online production requirement.

Keywords: Machine vision · Bearings · Vision system · Bearing greasing

34.1 Introduction

In the industry of machinery, bearings are important components that connect different machine parts to reduce frictions. They have been widely used in many machines, such as air conditioners, cars, and many other rotating machines. The quality of bearings can directly influence the performance of many machines, and may even cause serious disasters.

H. Shen (✉) · C. Zhu · S. Li · H. Chang
Institute of Automation, Chinese Academy of Sciences,
95 Zhongguancun East Road, Beijing 100190, China
e-mail: hao.shen@ia.ac.cn

Z. Sun and Z. Deng (eds.), *Proceedings of 2013 Chinese Intelligent Automation Conference*, Lecture Notes in Electrical Engineering 255,
DOI: 10.1007/978-3-642-38460-8_34, © Springer-Verlag Berlin Heidelberg 2013

Bearings are usually mass-produced with high precision demand, and a lot of inspection measures have been adopted in the production process to ensure the quality of bearings. The inspection measures can be classified into three steps: material inspection, assembling inspection and final goods inspection. The material inspection is mainly focused on the dimension and surface inspection of the receiving materials, such as inner rings, outer rings and balls. The assembling inspection is used to inspect the defects that are caused by assembling process, including surface inspection and vibration test. The final goods inspection is mainly focused on the surface defects, and gives a full inspection before packing. The greasing procedure is a vital procedure in assembly production, which occurred after the cage assembling. The purpose of greasing procedure is to add the oil into the gap of cage. Currently, the inspection of bearings in manufacture mainly depends on skilled human inspectors with the help of intensive lights. The manual activity of inspection is subjective and highly depending on the experience of human inspectors, which can't provide the guarantee of quality. For the greasing procedure, some electric methods, such as fiber location methods, have been involved for the detection of the cage gap position. However, these methods usually have high demand of localization, which is difficult to meet. Besides, for miniature bearings, these methods can't be used as a result of the dimension limitation. Therefore, automatically locating the position of bearing cage becomes an important issue for the greasing procedure, and computer vision can play a crucial role in it.

Machine vision, as a non-contacted technology, can simulate the function of human vision, and has many advantages compared with other technology. More and more researchers have changed their focus on machine vision. In the past few decades, great progresses have been made and much work has been done in industry production inspections [1]. Chiou et al. [2] proposed a machine vision system for the inspection of PU-packing. Sun et al. [3] designed a multi-view system to inspect electric contacts. For bearing inspection, Wu et al. [4] proposed a vision system for the inspection of tapered roller bearing. Deng et al. [5] and Shen et al. [6] designed the vision system for the inspection of bearing surfaces. Wang et al. [7] proposed a bearing characters recognition system for the inspection of characters defects. Telljohann [8] designed a diameter inspection system for rivets, which is used to rivet the bearing and the shafts during the production of float bearings. However, there are still few methods for the bearing assembly production.

In this paper, we develop a machine vision system for the bearing cage localization in greasing procedure. The proposed algorithm is simple but effective. The proposed system is capable of localizing the bearing cage quickly and accurately.

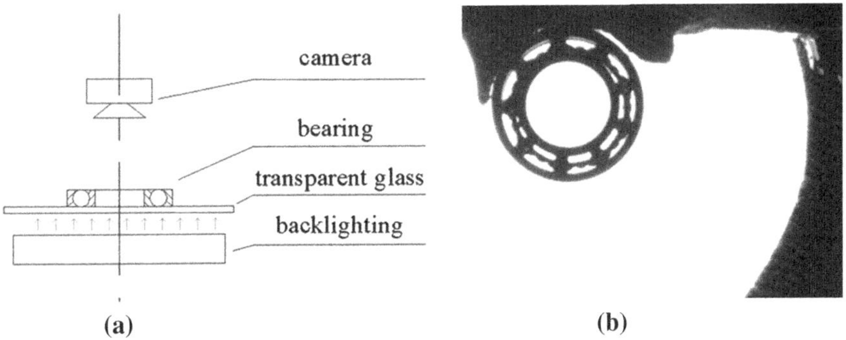

Fig. 34.1 Lighting and image acquisition system. **a** System illustration. **b** The captured image

34.2 Lighting and Image Acquisition System

The proposed system, as shown in Fig. 34.1a, mainly consists of LED backlighting, CCD camera, computer and mechanical components. The quality of the obtained image is a key factor in machine vision, which can greatly simplify the processing algorithm, and ensure the accuracy of processing results. In our system, a backlighting device is used to provide uniform illumination and avoid the environmental noises. Fig. 34.1b presents the captured image with proper lighting. After the image acquisition, the image will be processed by the inspection algorithm, and the returned cage position can be used to control the mechanical components to accomplish the greasing procedure.

34.3 Inspection Algorithm

The purpose of the proposed system is to get the location of bearing cage. And the position result will be used to control mechanical components to accomplish the greasing process. As shown in Fig. 34.2, the proposed inspection algorithm consists of six parts: image preprocessing, bearing segmentation, P2C transformation, vertical projection, distribution map generation and result return.

In order to remove noises in image, we involve a preprocessing procedure after the image is captured from the camera. Here, a 3*3 median filter is used to smooth the image.

34.3.1 Bearing Segmentation

As we can see in Fig. 34.1b, thanks to the backlighting illuminator, the background of the bearing is quite simple. Also, as the bearing is located by mechanical

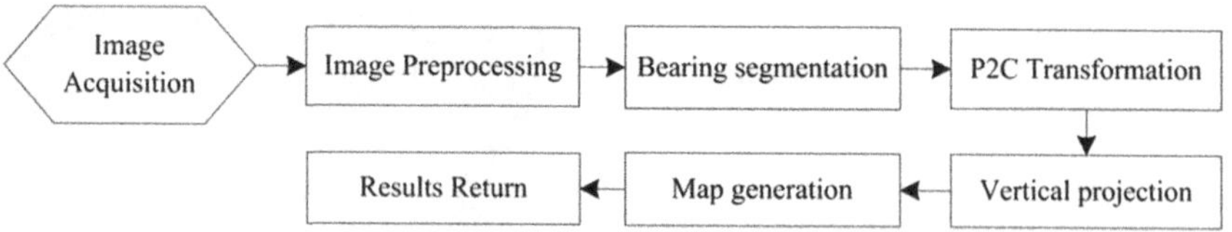

Fig. 34.2 Main procedure of the proposed algorithm

devices, the position is fixed. So we can get the approximate location of the bearing centre directly. Then the edge points of inner ring and outer ring are sampled, prepared for the circle fitting. The method of edge points sampling is illustrated in Fig. 34.3. From the initial centre to an outward direction, the first foreground point is sampled as an inner ring edge point, whereas the sample direction for the outer ring is just on the contrary.

As the outer ring edge is not complete, and there are some noises existed among the edge points, we propose a robust RANSAC [9] least square circle fitting method to estimate the accurate parameters of the inner and outer rings' shape:

Step 1. Randomly choose 6 edge points.
Step 2. Use least square method to estimate the parameters of circle.
Step 3. Check all the edge points, count the number of points belonging to the estimate circle
Step 4. If the number of inliers is larger than the threshold, return the circle parameter. Else go to step 1, continue to do step1 to step 4.

To facilitate the inspection algorithm, the ring-shaped image of bearing is transformed into a rectangle image using Polar to Cartesian (P2C) coordinate transformation method [2]. As shown in Fig. 34.4, R is the outer radius of the expanded bearing area; R_0 is the inner radius of the expanded bearing area; W is the width of the resulting rectangle image; F and D denote the origin image and destination image. To obtain the result point $D(w,r)$ corresponding to the source image point $F(x,y)$, the following equation is used.

$$\begin{cases} x = C_x + (r + R_0) \times \cos(\theta) \\ y = C_y - (r + R_0) \times \sin(\theta) \end{cases} \quad (34.1)$$

where (C_x, C_y) is the center of bearing and $\theta = 2\pi w / W$.

Fig. 34.3 Edge points sampling method

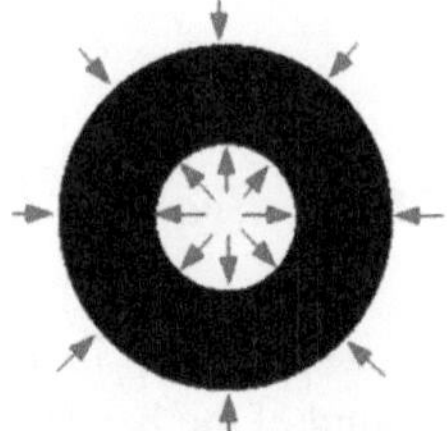

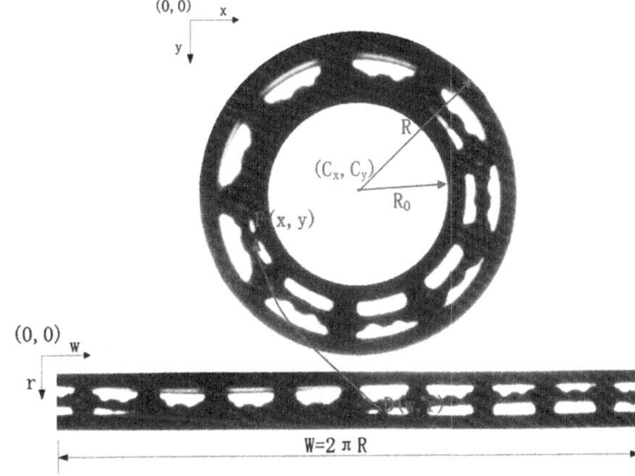

Fig. 34.4 P2C transformation

Besides, in order to keep a good quality of the result image, the bilinear interpolation method is used to get a proper pixel value. The final segmented rectangle image is shown in Fig. 34.5a.

34.3.2 Vertical Projection

After the rectangle image is obtained, we utilize the projection method to get the distribution of bearing cage. Firstly, the rectangle image is projected on the vertical direction, which can be formulated as Formulation 2; then the maxima suppression is applied; after that the threshold method is involved to get the binarized distribution and the threshold value is taken as *M/6*, where *M* is the max value of the suppressed distribution; finally the projection is realigned from the beginning of block to simplify the later processing. The processing results are shown in Fig. 34.5.

$$projection_j = \left(\sum_{i=1}^{H} F(i,j) \right) / H \qquad (34.2)$$

34.3.3 Map Generation

After the aligned binary projection is obtained, we generate a vote map to find the optimized gap position of the bearing cage. Firstly, the centre of each gap is exploited. Then, the centre position is projected evenly on the other gaps, based on the number of them, and each projection is calculated as follows.

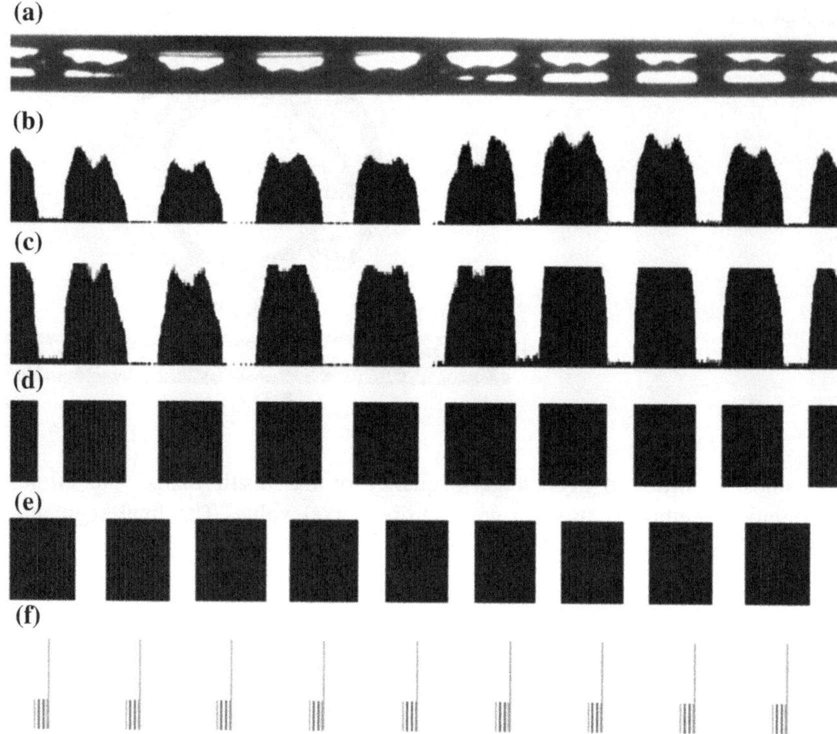

Fig. 34.5 Algorithm processing results. **a** Rectangle image. **b** Vertical projection image. **c** Maxima suppression result. **d** Binary projection image. **e** Aligned binary projection image. **f** Vote map

```
For each gap i
  Calculate the centre of the gap Ci;
  StartPos = Ci-(i-1)*SPACE;
  For j = 0:num

           voteMap[j*SPACE+StartPos] ++;

  end

end
```

here, SPACE is the average space between balls, which can be obtained by *SPACE* = *W*/*num*, *W* is the stretched width of bearing, *num* is the number of balls.

The final vote map is shown in Fig. 34.5f. Finally after the mean filter process of the vote map, the max vote position will be chosen as the returned position of the bearing cage.

34.4 Results and Conclusions

The proposed inspection system is implemented on a personal computer with Pentium® dual-Core 2.50 GHz CPU and 2G RAM. The inspection algorithm is programmed with the language of C++. The average cost for the proposed system is 16 ms. All the test images are captured in a bearing manufactory. The proposed machine vision system can correctly locate all the cages, and help to control the mechanical components to accomplish the greasing procedure. Some of the inspection results are shown in Fig. 34.6. The red lines are the detected center of cage gaps. We can see that, although the first two images are under bad camera condition, the inspection results are not affected at all.

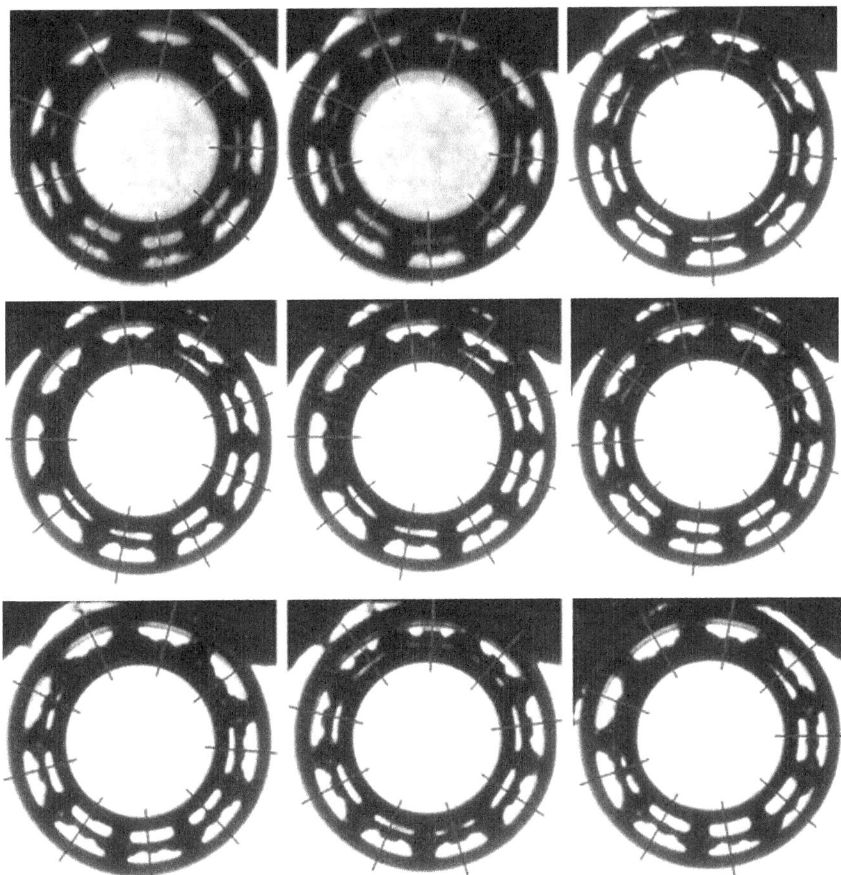

Fig. 34.6 Processing results

The experimental results show that the proposed system can control the mechanical components to fulfill the greasing procedure automatically and effectively. The speed and accuracy could satisfy the requirements of real-time production, meanwhile, improve the quality of the bearing production.

Acknowledgments This work is partly supported by National Natural Science Foundation of China (No.61005028, No.61175032 and No.61101222) and by The CAS Special Grant for Postgraduate Research, Innovation and Practice (No.2010.17).

References

1. Malamas EN, Petrakis EGM, Zervakis M, Petit L, Legat J-D (2003) A survey on industrial vision systems, applications, and tools. Image Vis Comput 21:171–188
2. Chiou Y-C, Li W-C (2009) Flaw detection of cylindrical surfaces in PU-packing by using machine vision technique. Measurement 42:989–1000
3. Sun T-H, Tseng C-C, Chen M-S (2010) Electric contacts inspection using machine vision. Image Vis Comput 28:890–901
4. Wu Q, Lou X, Zeng Z, He T (2010) Defects inspecting system for tapered roller bearings based on machine vision. In: International conference on electrical and control engineering
5. Deng S, Cai WW, Xu QY, Liang B (2010) Defect detection of bearing surfaces based on machine vision technique. In: International conference on computer application and system modeling
6. Shen H, Li S, Gu D, Chang H (2012) Bearing defect inspection based on machine vision. Measurement 45:719–733
7. Hengdi W, Yang Z, Sier D, Erdong S, Yong W (2011) Bearing characters recognition system based on LabVIEW. International conference on consumer electronics, communications and networks
8. Telljohann A (2006) Introduction to building a machine vision inspection. In: Hornberg A (ed) Handbook of machine vision. Wiley-VCH Verlag GmbH & Co KGaA, Weinheim, pp 35–71
9. Fischler M, Bolles R (1981) Random sample consensus: a paradigm for model fitting with application to image analysis and automated cartography. Commun ACM 24:381–395

Chapter 35
Active Control of Sound Transmission Through Double Plate Structures Using Volume Velocity Sensor

Qibo Mao

Abstract Based on coupling structural–acoustic modal model, using PZT actuators and volume velocity sensor as actuator/sensor, the analytical simulations are presented for the actively controlled the sound transmission through double plate structure. Firstly, the results show the potential for using volume velocity sensors to improve sound transmission loss through double plate structures. Secondly, the effects of physical parameters of actuator/sensor and double plate structure on control performances are discussed. And some useful conclusions are obtained, for example, if volume velocity sensor is applied to radiating plate, sound transmission loss will improve significantly, no matter where the actuators (i.e. PZT actuators on incident or radiating plate) are located.

Keywords Double plate structure · PZT actuator · Volume velocity sensor

35.1 Introduction

Double wall structures provide good noise insulation from the exterior of an aircraft fuselage to the interior noise field. However, the acoustic performance of the double wall structure deteriorates rapidly at low frequency around the mass-air-mass resonance, where it can be even became worse than that of a single plate [1]. There has been a great deal of analytical and experimental research on active control of sound transmission through double wall structures, and significant progress was obtained in both the active noise control (ANC) system [2, 3] and the active structural acoustic control (ASAC) system [4, 5]. In general, previous work on the active control of sound transmission through the double wall structures

Q. Mao (✉)
School of Aircraft Engineering, Nanchang HangKong University,
696 South Fenghe Avenue, Nanchang CN-330063, China
e-mail: Qibo_mao@yahoo.com

Z. Sun and Z. Deng (eds.), *Proceedings of 2013 Chinese Intelligent Automation Conference*, Lecture Notes in Electrical Engineering 255,
DOI: 10.1007/978-3-642-38460-8_35,

employs microphones or accelerometers as sensors. If microphones are used as the pressure sensors, microphones in the far field in general perform better than those located within the cavity. However, positioning performance microphone in the far field is often impractical to implement. In most recent research, microphones located in the cavity were used as discrete pressure sensors so that the control system can be made more compact and the acoustical path between the exciting source and sensor is less sensitive to environmental changes and to time delays in the control loop.

The aim of this paper is to illustrate the potential of improvement of sound transmission loss of double wall structures using volume velocity sensor. The paper presents results of simulation of a controlled double wall structure and discusses control performances with PZT actuators applied to one of the panes.

35.2 System Modelling for Active Control

A model is demonstrated in Fig. 35.1 to describe the mechanical behaviour of a double plate structure. Two plates (plane, parallel, same finite size with length L_x and width L_y), denoted by incident plate and radiating plate, are located on a rigid framework and baffled in an infinite rigid wall. The incident plate is set to be $z = 0$ and the radiating plate at $z = L_z$. The radiated acoustic field of the double plate structure is assumed as an acoustic free field.

Assume that the double plate structure is excited by the random indent acoustic wave (diffuse field), and there are K_i control PZT actuators on incident plate and K_r control PZT actuators on radiating plate. The acoustical field of the cavity can be described using a homogeneous wave equation [6]:

$$\nabla^2 p - \frac{1}{c_o^2}\frac{\partial^2 p}{\partial t^2} = -\rho_o \sum_{k=1}^{K_p} \frac{\partial Q_{con,k}}{\partial t} \tag{35.1}$$

with boundary conditions

$$\frac{\partial p}{\partial \vec{n}} = \begin{cases} \rho_o \frac{\partial^2 w^i}{\partial t^2} & \text{on incident plate } (z = 0) \\ -\rho_o \frac{\partial^2 w^r}{\partial t^2} & \text{on radiating plate } (z = L_z) \\ 0 & \text{otherwise} \end{cases}$$

where ρ_o and c_o are the density and sound-speed of the air, respectively. p is the sound pressure in cavity, w^i and w^r are displacements of incident plate and radiating plate, respectively.

Assume that the stiffness and mass of PZT actuators added on plates are negligible. The vibration of the incident and radiating plate are governed by the following well-known equation [7]

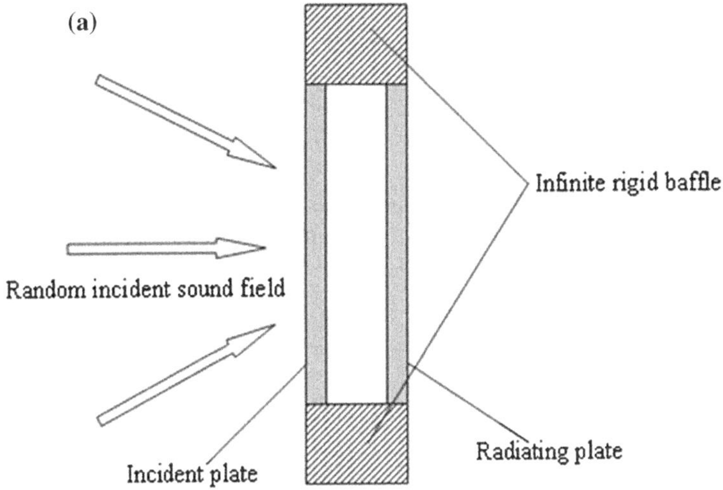

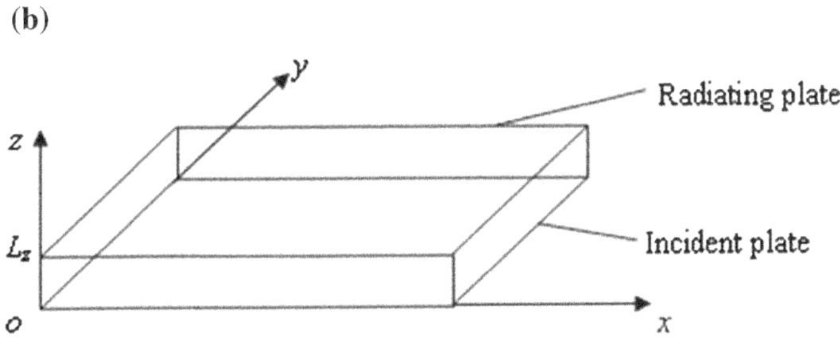

Fig. 35.1 **a** Double plate structure. **b** The coordinates for double plate system

$$D^i\nabla^4 w^i + m_s^i \frac{\partial^2 w^i}{\partial t^2} = p^{in} - p(z=0) + \sum_{k=1}^{K_i} C_0^i\left(\nabla^2\chi_k\right)\frac{d_{31}}{h^{pzt}} V_{con,k}^i \tag{35.2}$$

$$D^r\nabla^4 w^r + m_s^r \frac{\partial^2 w^r}{\partial t^2} = p(z=L_z) - \sum_{k=1}^{K_r} C_0^r\left(\nabla^2\chi_k\right)\frac{d_{31}}{h^{pzt}} V_{con,k}^r \tag{35.3}$$

With

$$D^i = \frac{(h^i)^3 E^i}{12\left(1-(v^i)^2\right)},\ D^r = \frac{(h^r)^3 E^r}{12\left(1-(v^r)^2\right)},\ m_s^i = \rho_s^i h^i,\ m_s^r = \rho_s^r h^r \tag{35.4}$$

where $\nabla^2 = \frac{\partial^2}{\partial x^2} + \frac{\partial^2}{\partial y^2}$, superscript i and r denote the incident plate and radiating plate, respectively. p^{in} is the excited sound pressure on the incident plate. D, m_s, h, E, v and ρ_s are the bending stiffness, the mass per unit area, the thickness, the

Young's modulus, the Poisson's ratio, and the density of the plate, respectively. C_0^i and C_0^r are constants that are the functions of the plates and piezoelectric actuator properties and geometry [7]. χ_k is a unit function, where the kth PZT actuator exits and is zero otherwise. V_k is the kth PZT actuator control input voltage. h^{pzt}, E^{pzt}, v^{pzt} and d_{31} are the density, Young's modulus, Poisson's ratio, and piezoelectric strain constant, respectively.

The cavity pressure p and the displacement distribution of the plates can be represented by summation of N and M uncoupling modes, respectively.

$$p(x,y,z) = \sum_{n=0}^{N-1} \Omega_n(x,y,z)P_n = \mathbf{\Omega}^T\mathbf{P} \tag{35.5}$$

$$w^i(x,y) = \sum_{m=1}^{M} \Phi_m^i(x,y)\eta_m^i = \left(\mathbf{\Phi}^i\right)^T \boldsymbol{\eta}^i \tag{35.6}$$

$$w^r(x,y) = \sum_{m=1}^{M} \Phi_m^r(x,y)\eta_m^r = \left(\mathbf{\Phi}^r\right)^T \boldsymbol{\eta}^r \tag{35.7}$$

where $\boldsymbol{\Omega}_n(x,\ y,\ z)$ and $\boldsymbol{P}_n$ are acoustic mode shape and corresponding modal pressure, respectively. $n = (n_x,\ n_y,\ n_z)$ denote the index of uncoupled acoustic modes in x, y and z axis, respectively. $\Phi_m(x,\ y)$ and $\boldsymbol{\eta}_m$ are structural mode shape and corresponding modal displacement, respectively. $m = (m_x,\ m_y)$ denote the index of structural mode in the x and y axis, respectively.

Substituting Eqs. (35.5–35.7) into Eqs. (35.2–35.4), and using the orthogonal properties of the mode shape functions. The complete set of equations for the double plate system can be expressed in modal coordinates and in matrix form as

$$\mathbf{P} = \mathbf{Y}^p\left(\mathbf{L}^i\dot{\eta}^i - \mathbf{L}^r\dot{\eta}^r\right) \tag{35.8}$$

$$\dot{\eta}^i = \mathbf{Y}^i\left(\mathbf{P}^{ext} + \sum_{k=1}^{K_i} \varepsilon_k^i V_{con,k}^i - \left(\mathbf{L}^i\right)^T\mathbf{P}\right) \tag{35.9}$$

$$\dot{\eta}^r = \mathbf{Y}^r\left(-\sum_{k=1}^{K_r} \varepsilon_k^r V_{con,k}^r + \left(\mathbf{L}^r\right)^T\mathbf{P}\right) \tag{35.10}$$

where $\dot{\eta}^i$ and $\dot{\eta}^r$ are the modal velocity vector of the incident plate and radiating plate. $\boldsymbol{Y}^p$, $\boldsymbol{Y}^i$ and $\boldsymbol{Y}^r$ are diagonal matrix.

Substituting Eqs. (35.9) and (35.10) into Eq. (35.8), we get

$$\begin{aligned}\mathbf{P} &= \left[\mathbf{I} + \mathbf{Y}_p\left(\mathbf{L}_i\mathbf{Y}_i\mathbf{L}_i^T + \mathbf{L}_r\mathbf{Y}_r\mathbf{L}_r^T\right)\right]^{-1}\Big[\mathbf{Y}_p\mathbf{L}_i\mathbf{Y}_i \cdot p^{ext} \\ &\quad + \mathbf{Y}_p\mathbf{L}_i\mathbf{Y}_i \cdot \sum_{k=1}^{K_i} \varepsilon_k^i V_{con,k}^i + \mathbf{Y}_p\mathbf{L}_r\mathbf{Y}_r \sum_{k=1}^{K_r} \varepsilon_k^r V_{con,k}^r\Big] \\ &= \mathbf{P}_{pri} + \mathbf{P}_c^{incident} + \mathbf{P}_c^{radiating} = \mathbf{P}_{pri} + \mathbf{P}_{c\_unit}\mathbf{f}_c\end{aligned} \tag{35.11}$$

where $\boldsymbol{P}_{pri}$ is the sound pressure's modal amplitude due to the primary source (random incident wave). $\mathbf{P}_c^{incident}$ and $\mathbf{P}_c^{radiating}$ are the sound pressure's modal amplitude due to the control sources on the incident plate and the radiating plate, respectively. $\mathbf{P}_{c\_unit} = \begin{bmatrix} \mathbf{P}_{c\_unit}^{incident} & \mathbf{P}_{c\_unit}^{radiating} \end{bmatrix}$, is the modal amplitude matrix due to the unit control sources. $\boldsymbol{f}_c = \begin{bmatrix} \mathbf{V}_{con}^{i} & \mathbf{V}_{con}^{r} \end{bmatrix}^T$, is the complex amplitude of control sources.

Combining Eqs. (35.8–35.11), we can get the fully coupled structural–acoustic responses for the double plate structure.

35.3 Cancellation Volume Velocity

Since the stated objective is the reduction of the sound transmission of the double plate structures, the appropriately designed sensor should be able to mainly detect the strongly radiating vibration distributions. It has been widely accepted that PVDF sensors can be designed based on the volume displacement/velocity. At low frequencies, the volume velocity accounts for most of the radiated sound power. The strategy of cancellation the volume velocity strategy has been shown to be very efficient for reducing sound radiation in the low frequency range.

The structural volume velocity V_{vol} is defined as the integral of the velocity over the surface of the plate,

$$V_{vol} = \int_0^{L_x} \int_0^{L_y} v(x,y)dydx \tag{35.12}$$

By means of additive theory, the velocity vector in Eq. (35.12) can be written as

$$\boldsymbol{v} = \boldsymbol{v}_{pri} + \boldsymbol{v}_c = \boldsymbol{v}_{pri} + \boldsymbol{v}_{c\_unit}\boldsymbol{f}_c \tag{35.13}$$

where $\boldsymbol{v}_{pri}$and $\boldsymbol{v}_{unit}^c$ are the velocity due to the primary source and unit control source, respectively. $\boldsymbol{f}_c$ is the complex amplitude of the control source.

Substituting Eq. (35.13) into Eq. (35.12), yields

$$V_{vol} = \int_0^{L_x} \int_0^{L_y} v_{pri}(x,y)dydx + \int_0^{L_x} \int_0^{L_y} v_{c\_unit}(x,y)dydx \cdot f_c \tag{35.14}$$

We can calculate the optimal control force strength f_c by setting volume velocity V_{vol} to zero.

$$f_c^{opt} = -\frac{\int_0^{L_x} \int_0^{L_y} v^p(x,y)dydx}{\int_0^{L_x} \int_0^{L_y} v_{unit}^c(x,y)dydx} \tag{35.15}$$

The sound transmission loss $TL = 10\log_{10}(W_l/W_r)$ is defined as the sound power incident on the incident plate divided by the sound power radiated by the radiating plate. Many of the results quoted in the following sections will be expressed in terms of the frequency averaged transmission loss, defined as

$$TL_{avg} = 10\log_{10}\left(\frac{W_{i,avg}}{W_{r,avg}}\right) \text{ with } W_{avg} = \frac{1}{N}\sum_{n=1}^{N} W_{\omega_n} \tag{35.16}$$

where $[\omega_1, \omega_N]$ is the frequency range of interest.

35.4 Numerical Results

The numerical results calculated from the double plate structure model developed in the previous section are presented. Assume that the boundary conditions for both plates are simply supported. The incident and radiating plate are made of aluminum of dimensions 500 × 300 mm, the thickness of the cavity between the double plate is 60 mm. Here, three types of double plate structures are discussed in detail, namely, DP-3/1, DP-2/2, DP-1/3 structures. The first and second numeral of each name denotes the thickness of incident plate, radiating plate (unit: mm), respectively.

The incident plate is excited by a random incident acoustic wave of 1 Pa amplitude, to make sure that all the structural modes are well excited. Four symmetrical symmetrical PZT actuators applied on either plate are used as actuators. Here, only the single-input single-output (SISO) system is calculated, i.e. four symmetrical PZT actuators are constrained to operate with the same voltage to achieve a single input system. For the sake of brevity, in the following sections, the symmetrical PZT actuators on the incident plate and radiating plate will be shortened to radiating-PZT and incident-PZT, respectively.

For simplicity of analysis, we treat the frequency range of interest from 20 to 800 Hz in all cases. The improvements of frequency averaged sound transmission loss TL_{avg} are listed in Tables 35.1 and 35.2 for different types of double plate structures. Notice that the control performance for the minimization of sound power is also listed in Tables 35.1 and 35.2. This is because the minimization sound power method represents the maximum achievable attenuation according to the actuator location, although this method only has theoretical meanings, because the sound power is difficult to obtain for a real control system.

Table 35.1 Frequency averaged transmission loss TL_{avg} improvement using volume velocity sensor with radiating-PZT (dB)

	DP-3/1	DP-2/2	DP-1/3
Sound power	27.5	31.8	33.9
Incident volume velocity	−26.0	−19.7	−7.3
Radiating volume velocity	27.3	31.3	33.5

Table 35.2 Frequency averaged transmission loss TL_{avg} improvement using volume velocity sensor with incident-PZT (dB)

	DP-3/1	DP-2/2	DP-1/3
Minimization sound power	34.0	27.6	33.4
Cancellation incident volume velocity	33.2	20.1	29.3
Cancellation radiating volume velocity	33.8	26.4	27.2

From Tables 35.1 and 35.2, it can be found that cancellation of the volume velocity on the incident plate will result in very low improvement or even large reduction of frequency averaged transmission loss, except that incident-PVDF/incident-PZT collocated configuration can obtain as similarly significant improvement as minimization of sound power of the radiating plate.

However, cancellation of volume velocity of radiating plate can achieve almost the same transmission loss improvement as minimization of sound power of radiating plate in many cases. This is because at low frequencies, the volume velocity accounts for the majority of the sound power [1, 7, 8]. It is important to note that the PZT actuators can be applied to either plate without effect on control performance, when a volume velocity sensor is applied to radiating plate.

35.5 Conclusion

In this paper, a fully coupling structural–acoustic modal model is developed for the active control of the sound transmission through double plate structures using volume velocity sensor. The simulation results show the potential for using piezoelectric actuator/sensors to improve the sound transmission loss through double plate structures. If a volume velocity sensor on the radiating plate is used as a sensor, sound transmission loss can achieve almost the same improvement as the minimization of sound power method. Furthermore, location of the actuators has little influence on control performance.

Acknowledgments This work was sponsored by the National Natural Science Foundation of China (no. 51265037); Scientific Research Foundation for the Returned Overseas Chinese Scholars, State Education Ministry; Technology Foundation of Jiangxi Province (no. KJLD12075).

References

1. Mao Q, Pietrzko S (2006) Control of sound transmission through double wall partitions using optimally tuned Helmholtz resonators. Acta Acust Unit Acust 91(4):723–731
2. Jakob A, Möser M (2003) Active control of double-glazed windows. Part 1: feedforward control. Appl Acoust 64:163–182

3. Kaiser OE, Pietrzko S, Morari M (2003) Feedback control of sound transmission through a double glazed window. J Sound Vib 263:775–795
4. Carneal JP, Fuller CR (2004) An analytical and experimental investigation of active structural acoustic control of noise transmission through double panel systems. J Sound Vibr 272:749–771
5. Pietrzko S, Mao Q (2008) New results in active and passive control of sound transmission through double wall structures. Aerosp Sci Technol 12(1):42–53
6. Nelson PA, Elliott SJ (1992) Active control of sound. Academic Press, London
7. Fuller CR, Elliott SJ, Nelson PA (1997) Active control of vibration. Academic Press, London
8. Mao Q, Pietrzko S (2010) Experimental study for control of sound transmission through double glazed window using optimally tuned Helmholtz resonators. Appl Acoust 71(1):32–38

Chapter 36
Research on Prognostics of Dynamically Tuned Gyroscope Storage Life

Yuxiong Pan, Qingdong Li and Zhang Ren

Abstract The storage life of dynamically tuned gyroscope (DTG) widely used in the field of guidance, navigation and control (GNC) is an important index for its military application. Aiming at this issue, the step-down-stress accelerated life testing (SDS-ALT) is implemented to test the DTG to get statistical and degradation data for its characteristics of long life and high reliability. A three-step analysis method is developed to establish Arrhenius accelerated model by using statistical data. Then, degradation data at higher stress can be converted to those at normal stress by data conversion formula. A new individual-based prognostic algorithm, named the path classification and estimation (PACE) model, which is entirely based on the converted degradation data, has been developed to predict remaining storage life estimation of DTG. Using the degradation data of SDS-ALT, the remaining storage life of DTG is evaluated. The evaluated results show that the model has higher prediction accuracy and smaller variance especially for small sample condition.

Keywords Dynamically tuned gyroscope · Remaining storage life · Degradation data · PACE algorithm · Small sample

36.1 Introduction

Dynamic tuned gyroscope (DTG, also known as flexible gyroscope), which uses a flexible supporting to suspend gyroscope rotor that is separated from drive motor, is a new gyroscope of two degrees of freedom. In recent years, because of its advantages, such as simple structure, small size, light weight, high reliability, long

Y. Pan (✉) · Q. Li · Z. Ren
Science and Technology on Aircraft Control, Laboratory, Beihang University, Beijing, China
e-mail: twhpyx@gmail.com

Z. Sun and Z. Deng (eds.), *Proceedings of 2013 Chinese Intelligent Automation Conference*, Lecture Notes in Electrical Engineering 255,
DOI: 10.1007/978-3-642-38460-8_36,

life, etc., DTG is widely used in the fields of GNC [1]. Prognostics is a term given to equipment life prediction technique and may be thought of as the holy grail of condition based maintenance [2]. The ability to monitor and control complex systems has been of particular interest for decades with a multitude of successful applications. However, the ability to identify system degradation and predict remaining storage life has proved to be difficult.

Accelerated life testing (ALT), especially SDS-ALT, can shorten the test time and save cost in order to assess DTG's storage life and reliability index, for it has characteristics of long life and high reliability [3].

Perhaps the most effective method in terms of high-precision prediction of system degradation is the application of physics-of-failure to structural degradation and structural health monitoring systems. Physics-of-failure methods focus on issues such as material deformation, fracture, fatigue, and material loss. One example of this type is the General Path Model (GPM). GPM, which was first proposed by Lu and Meeker [4], considers physics-of-failure of components to particularly determine a degradation parameter to form a prognostic prediction. Garvey and Hines put forward an algorithm called PACE which is based on GPM [5]. Chen and Song employed PACE model based method to solve the problem of prediction accuracy for small sample pneumatic cylinders' prognosis [6].

36.2 Establishment of Accelerated Storage Life Model

36.2.1 Basic Assumptions

Assumptions of the SDS-ALT are proposed as follows [7]:

1. Storage life of product obeys Weibull distribution in Eq. (36.1) at any different stresses

$$F_i(t) = 1 - \exp\left\{-\left(\frac{t}{\eta_i}\right)^{m_i}\right\} \tag{36.1}$$

2. The remaining storage life of Product is only dependent on the cumulative ineffective portion and current level of stress, and has nothing to do with the accumulation mode.
3. Product performance degradation process is monotone, namely the performance degradation is irreversible.
4. Product characteristic life and the stress level obey the Arrhenius accelerated model.

$$\ln \eta_i = a + b/S_i \tag{36.2}$$

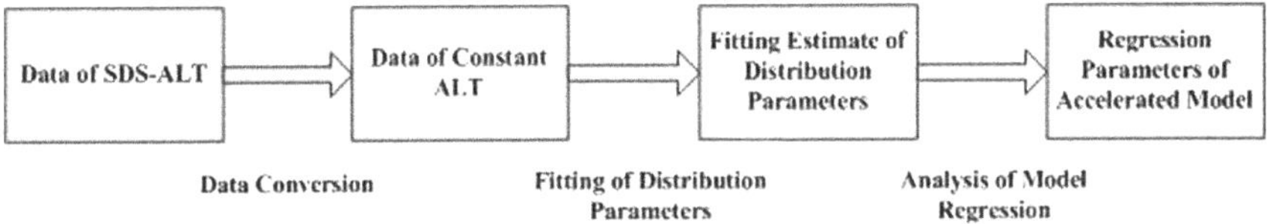

Fig. 36.1 Three-step analysis method

36.2.2 Three-Step Analysis Method

Compared with two-step analysis method of constant stress accelerated life testing, as shown in Fig. 36.1, statistical analysis method of SDS-ALT is established with the name of three-step analysis method. The basic process is as follows:

1. The complete failure samples at each step are obtained through the data conversion process, and the conversion from data of SDS to those of constant stress is implemented.
2. Distribution of equivalent life data at each step is fitted and distribution parameters are estimated.
3. Parameters of accelerated life model are estimated by regression analysis of life characteristic parameter estimation and stress levels.

Data conversion analysis (step 1) is the key of three-step analysis method, and the next two steps actually constitute two-step analysis method for constant stress testing which are widely used [8], so data conversion is deeply discussed in this paper.

36.3 PACE Model Based on Kernel Regression

GPM is founded on the concept that a degradation signal collected from an individual product will follow a general path until it reaches a failure threshold. PACE model modifies the GPM as a classification problem for a product's degradation path classified according to a series of exemplary paths. The results of the classification are used to estimate the remaining storage life of the product. The requirement of the existence of a failure threshold is removed, thereby, enabling PACE to be applied to systems, where a distinct failure threshold is not likely to occur.

As its name suggests, the PACE model is fundamentally composed of two operations: (1) classify a current degradation path as belonging to one or more of previously collected exemplary degradation paths (path classification) and (2) use the resulting memberships to estimate the remaining storage life (estimation).

Kernel regression (KR) is an empirical and nonparametric modeling technique that uses exemplary observations to make predictions. For a query observation of

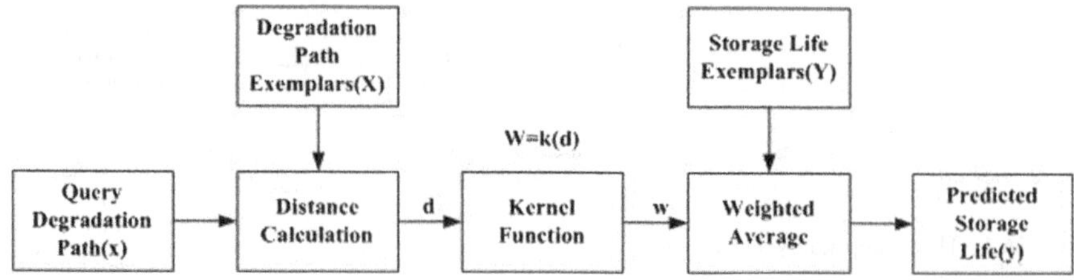

Fig. 36.2 Process diagram for kernel regression prediction algorithm

the inputs, the KR estimation process can be structured into three steps. First, the distance of the query from each of the input exemplars is calculated. Next, the distances are supplied as inputs to a kernel function, which converts the distances to weights (similarities). Finally, the weights are used to estimate the output by calculating a weighted average of the output examples. Taking DTG as an example, these steps are depicted in Fig. 36.2.

The three steps of KR prediction process are discussed below:

1. Distance Calculation
   A common distance function is the Euclidean distance, which is also known as the L2-norm. The Euclidean distance for the *i*th exemplar path and a query is given by:

$$d(X_i, x) = \sqrt{(X_i - x)^2} = |X_i - x| \tag{36.3}$$

2. Similarity Quantification
   To transform the distance into a similarity or weight, a kernel function is used. In general, a kernel function should have large values for small distances and small values for large distances. In other words, when a query point is nearly identical to a reference point its distance should be small and, therefore, that particular reference point should receive a large weight and vice versa.

$$w_i = \frac{1}{\sqrt{2\pi h^2}} e^{-d_i^2/2h^2} \tag{36.4}$$

The result is a vector of $n$weights:

$$w = \begin{bmatrix} w_1 & w_2 & w_3 & w_4 \end{bmatrix} \tag{36.5}$$

3. Output Estimation
   Notice that in KR the weighted sum of the outputs is normalized by the sum of weights defined as follows:

$$\hat{y}(x) = \frac{1}{a} \sum_{i=1}^{n} (w_i Y_i) \tag{36.6}$$

where $a = \sum_{i=1}^{n} w_i$, the KR Based PACE Model combine kernel regression and PACE model together to create an algorithm which can perform lifetime prediction successfully.

36.4 Results of Remaining Storage Life Prediction

36.4.1 SDS-ALT of DTG and Data Conversion

SDS-ALT of some type DTG is implemented to obtain degradation data by taking temperature as accelerated stress. Applied stress levels of the accelerated storage life testing are 80, 66.7, 53.4, 40 °C, and storage temperature is 23 °C. Total small samples are five, and failure numbers at each stress are 2, 1, 1, 1 respectively. Time to failure and distribution parameter estimates at each stress through data conversion process are shown in Table 36.1.

Relationship between temperature and storage life satisfy Arrhenius model, accelerated life model of the type DTG is as follows:

$$\ln \eta = -2.9501 + 2517.7/S \tag{36.7}$$

When $S = S_0 = 23^{\circ}\mathrm{C},\ \eta_0 = 296.15$

36.4.2 Prediction of Remaining Storage Life

Drift rate coefficient $D(Y)_y$ related to a_y and along y-axis is selected as dominant parameter through data analysis and comparison. According to Eq. (36.5), test time can be converted from higher stresses to those of storage stress.

Shown as Fig. 36.3, data1 to data5 designates the growth curve of drift rate coefficient of DTG1 to DTG5, respectively. The curves are obtained by polynomial regression according to raw data. This is for easy handling in mathematical calculation.

Table 36.1 11SDS-ALT storage data and characteristic parameters of DTG

Stress level (°C)	Failure data (days)	Characteristic life (days)	Shape parameter
80	20.40	63.6917	1.6544
66.7	20	77.1046	1.7637
53.4	40	143.5002	1.8419
40	90	147.9489	1.8341

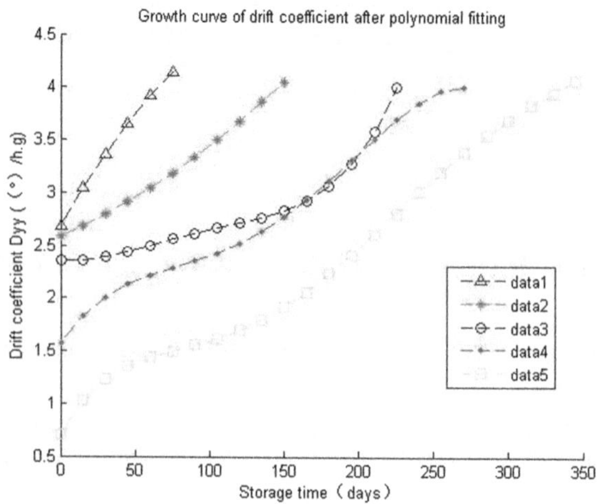

Fig. 36.3 Degradation path after polynomial fitting

The query path has identifier of 4, which will be classified to other four degradation paths, creating corresponding memberships. Shown as Fig. 36.4, data1, data2, data3 and data5 designate the remaining storage life of DTG1, DTG2, DTG3, DTG5, respectively. The remaining storage life of DTG4 can be dynamically predicted, as drift coefficient values are continuously input. In the first 60 days, DTG1, 2, 3, 5 have the effect on DTG4, and then as DTG1, 2, 3 gradually failed, especially after 230 days, the line of DTG4 is gradually close to that of DTG5. And finally the remaining storage life of DTG4 is 30 days, while the real value is 24 days.

Two performance measures are considered to dominate the prognosis task: average bias and average dispersion [9]. Then average bias and average dispersion:

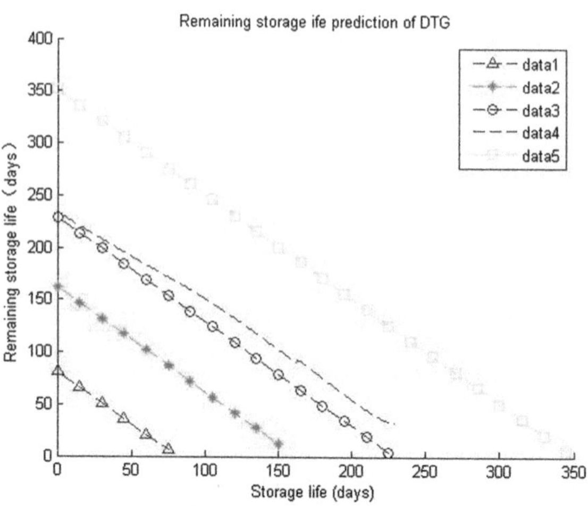

Fig. 36.4 Estimation of remaining storage life prediction

$$\begin{cases} \bar{E} = \frac{1}{N}\sum_{i=1}^{N} E_i \\ \sigma^2 = \frac{1}{N}\sum_{i=1}^{N} E_i^2 \end{cases} \quad (36.8)$$

where $E_i = t_{pf}(i) - t_{af}(i)$ (predicted time − actual time). The average bias and dispersion of PACE model based prognostic algorithm are $\overline{E} = 20$ and $\sigma^2 = 400$, while the average bias of statistical model based prognostic algorithm are $\overline{E} = 25$ and $\sigma^2 = 625$. It is observed that the average bias and dispersion of PACE model based prognostic algorithm are smaller and have better precision.

36.5 Conclusions

Aiming at the problem of storage life prediction of DTG, SDS-ALT is conducted to collect accelerated storage test time which is used to statistically analyze by three-step analysis method, and degradation data is also gathered to predict the remaining storage life of DTG by using PACE method. The following conclusions can be drawn by analysis and verification results.

1. Parameters of the Arrhenius accelerated life model can be estimated under the condition of small samples.
2. The major advantage of the PACE compared with the other prognostic algorithms is that precise prediction results can be obtained even though a failure threshold is unknown.
3. The PACE model is flexible enough to innovation by dynamically changing parameters to realize dynamic remaining life prediction.

Acknowledgments This work is supported under National Natural Science Foundation of China, under grant 61101004.

References

1. Zhou B (2002) Design and manufacture of dynamically tuned gyroscope. Southeast University Press, Nanjing (in Chinese)
2. Coble JB, Hines JW (2008) Prognostics algorithm categorization with PHM challenge application. In: Proceedings of the international conference of prognostics and health management, PHM 2008, Denver, pp 1–11
3. Zhang CH, Chen X, Wen X (2005) Step-down-stress accelerated life testing statistical analysis. Acta Armamentarii 26(5):666–669 (in Chinese)
4. Lu C, Meeker WQ (1993) Using degradation measures to estimate a time-to-failure distribution. Technometrics 35(2):161–173

5. Garvey DR (2007) An integrated fuzzy inference based monitoring. Dissertation, University of Tennessee, Knoxville
6. Chen J, Song CW (2012) Path classification and estimation model based prognosis of pneumatic cylinder lifetime. Chin J Mech Eng 25(2):392–397
7. Ma J, Chao DH (2010) Accelerated storage life evaluation of FOG based on drift brownian movement. J Chin Inertial Technol 18(6):756–760 (in Chinese)
8. Zhang CH, Chen X, Li Y (2002) Analysis for constant-stress accelerated life testing data under Weibull life distribution. J Nat Univ Defense Technol 24(2):81–84 (in Chinese)
9. Roemer MJ, Jim DS (2005) Validation and verification of prognostics and health management technologies, aerospace conference, pp 3941–3947

Chapter 37
Automatic Testing Device for Gas Cylinder Based on LabVIEW

Fei Chen, Yingjuan Yue, Wenxia Sun and Yaque Jing

Abstract In this paper, a testing device for gas cylinder, which is composed of PC, data acquisition card on USB bus technology and the application software compiled on the labVIEW platform, is designed on the basis of virtual instrument technology. The assembly and disassembly of the cylinder plug, weighing, temperature measuring, being irrigated upright, and hydraulic pressure testing are all realized. This device has been put into use with highly practical effect.

Keywords Virtual instrument (VI) · Automatic testing device · Gas cylinder · LabVIEW

37.1 Introduction

Gas cylinder in a most universally used mobile pressure container worldwide in social life, industry, national defense and so on. As one of the seven special equipments, gas cylinder should be tested every three or five years in accordance with Gas Cylinder Safety Supervision Rules by State Quality Inspection Administration. In the testing, the cylinder plug disassembled and tightened, the cylinder weighed, watered in or out frequently, all of which bring work density [1]. A testing device is designed based on virtual instrument in the light of gas cylinder with maximum pressure of 35 MPa.

F. Chen (✉) · Y. Yue · W. Sun · Y. Jing
Xi'an Research Institution of High Technology, Xi'an 710025 Shanxi, China
e-mail: chenf_1234@yahoo.com.cn

Z. Sun and Z. Deng (eds.), *Proceedings of 2013 Chinese Intelligent Automation Conference*, Lecture Notes in Electrical Engineering 255,
DOI: 10.1007/978-3-642-38460-8_37, © Springer-Verlag Berlin Heidelberg 2013

37.2 Virtual Instrument and Development Platform

Virtual instrument is a computer equipment system, in which a computer is the core, users design the definitions, virtual panel is offered, and the test function is realized by test software [1–3]. This instrument, by making use of the display function of a computer monitor, simulates the control panel in traditional instrument so that the test result is output in various ways. At the same time, by making use of the software function of a computer, the operation, analysis and processing of the signal data are dealt with. And signal acquisition, measurement and adjustment would be handled by the equipment of I/O interface. Two characteristics of virtual instrument are as follows [4]:

For one thing, virtual is the panel of such instrument, which is quite easy to design. For another, the measurement function of virtual instrument is carried out by software programming, i.e. a variety of measurement functions are realized by the composition of software modules which have different functions.

The National Instrument of the United States (NI) introduces LabVIEW—a development platform of typical virtual instrument. In the field of instrument control, measurement and test, industrial monitoring, it has the following characteristics [5]:

1. Graphical programming, easy to learn and use, flexible to operate.
   Icon is the basic programming unit. Different icons represent different function modules. The process of programming is that of numbers of icons lined. These lines show the data transmitted by all function modules. The connection between data flow controls the execution order and allows synchronous operation for many a few data channels. This programming is similar to the process of human thinking, which is easier for engineering personnel to adopt.
2. Being hierarchical and modular.
   LabVIEW follows and develops the structured and modular design, which makes virtual instrument hierarchical and modular. In other words, any virtual instrument can be regarded as the top program, and as other virtual instruments, or its subroutine. Thus, users can decompose a complicated application task into systematical and multi-level sub-tasks. Then with a sub-virtual instrument arranged for each sub-task, these sub-virtual instruments are composed, modified, crossed or combined with the help of block diagram principle. And the top virtual instrument ultimately built would be a collection of sub-virtual instruments with all application functions.
3. Powerful hardware drivers and graphical display.
   LabVIEW provides hundreds of source drivers. It can develop its own hardware driver library through Dynamic Link Library (DLL), by which its scope and functions of hardware are expanded. Besides, LabVIEW shows powerful graphics and image processing functions, which is applied in system test, control and simulation.
4. Rich mathematical analysis and signal processing module.
   LabVIEW offers rich computing nodes in mathematics, including formula

node, estimation, calculus, linear algebra, curve fitting, mathematical statistics, optimization method, roots and numerical node. It also offers numerous signal processing nodes, including signal generation, time domain processing, frequency domain processing, signal measurement, digital filtering and window function etc., all of which can meet the needs of signal measurement and signal processing.

37.3 The Composition and Working Principle of the Testing Device

37.3.1 The Composition of the Testing Device

This device is mainly composed by mechanical work platform, hydraulic torque wrenches, hydraulic erection mechanism, pressure test pumps, sensors, signal conditioning circuit, data acquisition card and computer. Figure 37.1 is the diagram of the device structure. Around the core of a computer installed with LabVIEW development platform, the data acquisition card of domestic 12 and single 16 input simulations, dual 8 input and 16 input and output is adopted. The digital to analog conversion time is 100 us.

Corresponding to pressure, weight and temperature respectively, the device has four channel analog signal inputs, which reveals 0-5VDC or 4-20mADC standard signal after transmission. 1-5VDC voltage signal is to be achieved when 4-20mADC current signal flows through a 250 Ω precision resistor. All the voltage

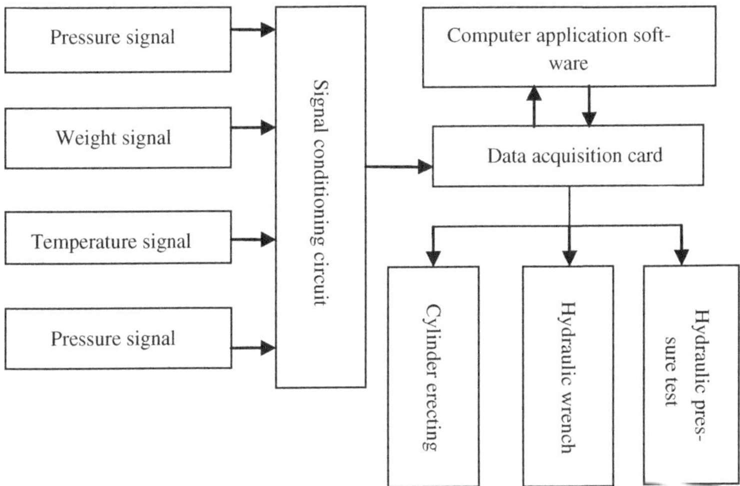

Fig. 37.1 Diagram of the device structure

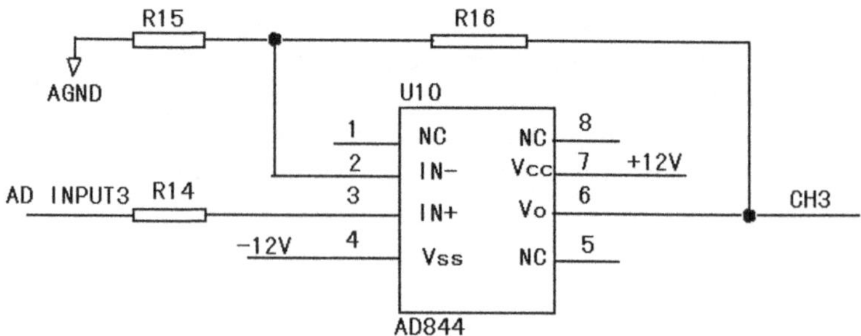

Fig. 37.2 AD844 amplifying circuit

signals, after amplified twice by an amplifying circuit of AD844 [6], enter the data acquisition card. The input impedance is raised and the output impedance is reduced by the amplifying circuit, which is shown in Fig. 37.2.

37.3.2 Main Function of the Testing Device

Multiple projects in gas cylinders' periodic inspection are completed by this testing device through sensors collecting signals, through data acquisition card and hardware circuit driving execution component.

1. Cylinder weighing: by monitoring the change of the cylinder weight, when emptied and when full of water, the actual volume can be worked out. The device is symmetrically accompanied by four JHBT strain type weighing sensors, with the range of 0–1,200 kg, the precision of 0.1 % FS, working temperature of −30–70 °C, three-line output signal of 0-5VDC, load resistance of 350 Ω, and 12VDC power supply.
2. Cylinder valve assembly and disassembly: hydraulic torque wrench, whose maximum torque is 5,000 N.m, is applied to deal with the cylinder valve assembly and disassembly in the testing. The device sends control signals to stepping motor driver by the control software through data acquisition card and other hardware. This would drive the stepping motor to adjust the maximum output pressure of the hydraulic pump, which leads to a precise output torque by hydraulic wrench.
3. Temperature measurement: the platinum thermal resistance temperature sensor of ALT-RA is adopted in the device to monitor the temperature in the water pressure test.
4. Hydraulic erecting: the air in the cylinder must be emptied in the water pressure test. An air cylinder can be tilted to a certain angle so that all the air in it could escape from the upper outlet when it is filled with water. The cylinder erecting is realized by the output switch quantity of data acquisition card which controls hydraulic mechanism.

5. Pressure test: the pressure test pump is started by data acquisition card, while having a real-time measurement of the pressure pump. When reaching the test pressure required, turn off the pressure test pump and its valve to the holding pressure state.

37.3.3 Software Design of the Testing Device

The program, which can meet the needs of users, is easily designed by utilizing its rich library and powerful interface function in the development platform of LabVIEW.

37.3.3.1 Application Diagram

In LabVIEW, a block diagram, similar to the structure of general flow chart, is the foundation of procedure operation [7]. The device software applies structured and modular design to put each function into effect by the corresponding SubVI. In each module, pressure regulation, control of hydraulic wrench, cylinder weighing, cylinder erecting, real-time measurement and display of hydraulic wrench torque, pressure test pump control etc. are fulfilled by the block diagram. Figure 37.3 is the block diagram of cylinder weighing module weight VI.

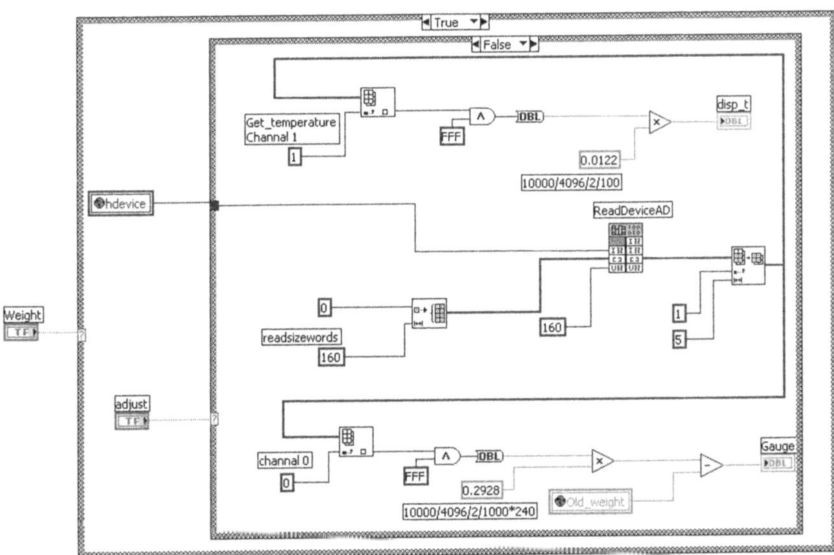

Fig. 37.3 Block diagram of cylinder weighing module weight VI

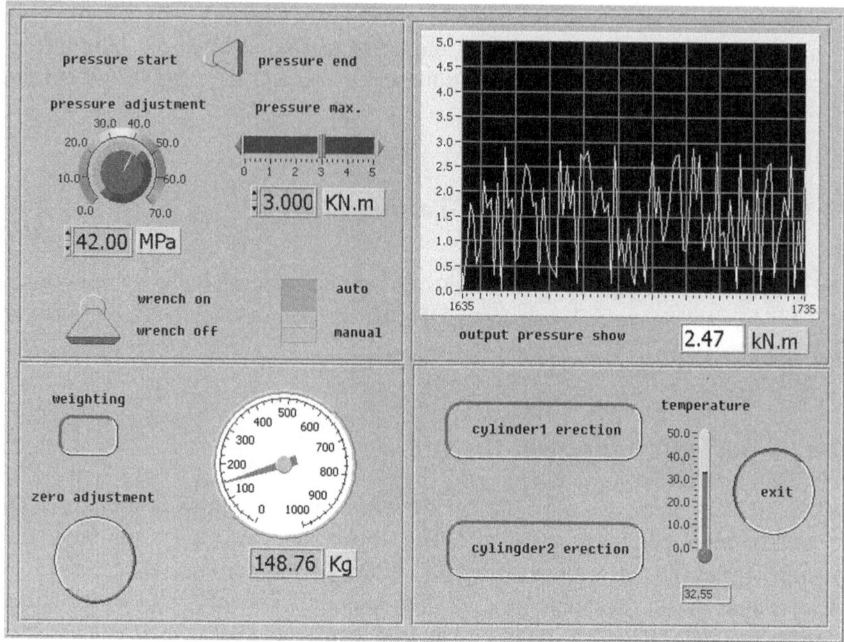

Fig. 37.4 Operating panel of detection system

37.3.3.2 User Interface (Front Panel)

Front panel [8] is the interface on which virtual instrument is operated by users, and on which not only the virtual interface is offered by front panel, but also various instruments can be called and program flow can be controlled by operating various controls. Figure 37.4 is the operating panel of the testing device.

37.4 Conclusion

The testing device has been put into use. According to the function and actual detection program of gas cylinder, this is a gas cylinder automatic testing device, which is good in quality and stable in operation in testing fast hold, weighing, hydraulic pump pressure regulation, hydraulic erecting irrigation and drainage, the assembly and disassembly of cylinder plug, temperature measurement, and hydraulic pressure test.

References

1. Zhuang R, Wu X (2012) LabVIEW-based control of stepping motors. Mod Electron Tech 4(4):202–204 (in Chinese)
2. Singh C, Poddar K (2008) Implementation of a LabVIEW-based automated wind tunnel instrumentation system. National Wind Tunnel Facility Indian Institute of Technology Kanpur (INDIA), 978-1-4244-2746-8/08/$25.00@2008 IEEE
3. Jimenez FJ, Frutos JD (2005) Virtual instrument for measurement, processing data, and visualization of vibration patterns of piezoelectric devices. Computer Standards and Interfaces
4. Zhang J, Wu J, Han Z (2009) Testing system based on LabVIEW. In: Proceedings of the 2009 international conference on advances in construction machinery and vehicle engineering, pp 214–216
5. Chen H, Zou Y (2011) Development of a LabVIEW-based interactive digital power control development system. In: Proceedings of the 10th PCIM Asia conference, pp 183–1686
6. Chen L (2012) Room temperature control system simulation based on NI. Comput Digital Eng 40(2):146–148 (in Chinese)
7. Cai C, Wang F, Dai H (2008) The DAQ system of the vehicle test based on the virtual instruments technology. In: Proceedings of 2008 international seminar on future biomedical information engineering, pp 202–205
8. He Y, Yue L (2011) Dynamic pressure analysis data management system based on LabVIEW. Mech Eng Autom (2):22–24 (in Chinese)

References

[illegible]

Chapter 38
Study on the FBG and ZigBee Technologies in Telemetry System of Flow Velocity

Bing Han, Dongjie Tan, Xingtao Zhou, Liangliang Li and Shaopeng Yu

Abstract In order to overcome the disadvantage of low accuracy for traditional monitoring technologies of flow velocity, and solve the problems in real-time data acquisition and remote transmission for the existing hydrological observation, we design a new telemetry system of flow velocity based on fiber grating technology and wireless network technology, which can realize high sensitivity and automatic monitoring of flow velocity for multiple measure points in a large range of areas. The telemetry system has been put into practical application in a monitoring project and the measurement results indicate that this system is of such advantages as real-time capability, high accuracy, unattended operation, and can well meet the present requirements of hydrological observation and measurement.

Keywords FBG · Flow velocity · Telemetry system · ZigBee · Wireless network

38.1 Introduction

The flow velocity is an important parameter in the industrial production process, energy metering, environmental monitoring, hydraulic model and river engineering model test, and the flow velocity sensor is an indispensable device to the process of fluid detection [1]. As the earliest and most extensive application of the flow velocity sensor, the technology of mechanical rotor types are relatively mature, but they have some disadvantages such as the measurement with large error and low measuring precision due to the limitation conditions of mechanical structure themselves. Some developed recently measuring instruments of flow velocity, such as the ultrasonic current meter, electromagnetic velocimeter and

B. Han (✉) · D. Tan · X. Zhou · L. Li · S. Yu
Pipeline R&D Center, PetroChina Pipeline Company,
No.51, Jinguang Road, Langfang, Hebei, China
e-mail: kjhanbing@petrochina.com.cn

Z. Sun and Z. Deng (eds.), *Proceedings of 2013 Chinese Intelligent Automation Conference*, Lecture Notes in Electrical Engineering 255,
DOI: 10.1007/978-3-642-38460-8_38, © Springer-Verlag Berlin Heidelberg 2013

Acoustic Doppler Velocimetry, which have a higher accuracy and are convenient to use, but they are costly and susceptible to electromagnetic interference; and it is difficult to build a large-scale sensor network for the above technologies due to their short signal transmission distance. These make the existing monitoring systems of flow velocity have some limitations in application space and geographical environment, and it is difficult to achieve real-time measurement of flow velocity for multiple measure points in large-scale regional.

Fiber Bragg grating (FBG) is a type of new optical fiber passive device, which is formed by creating periodical distribution of refraction index in fiber core based on the photosensitive characteristics of fiber material, and it has the advantages of anti-electromagnetic interference, waterproof, wide dynamic range, high sensitivity, convenient to build up network and easy to realize distributed measurement [2]. It is reported that there are lots of quantities can be measured by using FBG sensing technology, such as temperature, strain, pressure, stress, acceleration, vibration and so on [3–5]. With the development of optical fiber technology, there are more and more flow velocity sensors based on a variety of optics principles. In this article, we have designed a set of automatic device of flow velocity monitoring based on FBG sensing technology, which has the characteristics of high measurement accuracy, stability and reliability. In order to ensure the real-time, validity and persistence of data acquisition during the monitoring process, we have built a wireless sensor network based on ZigBee technology to centralize the flow velocity information of multiple measure points to the monitoring stations in the field, and realized the remote monitoring of flow velocity by communication with the remote monitoring terminal through network center node of ZigBee with GPRS.

38.2 Structure and Principle of FBG Flowmeter

38.2.1 Principle of FBG Sensing

FBG is the simplest and most widely used fiber grating, it is an optical fiber which the refractive index varies periodically, and the modulation depth of refractive index and grating period are generally constant. The changes in temperature or strain can cause the changes in period and refractive index of FBG, and then lead to the changes in reflection spectrum and transmission spectrum of FBG. We can obtain the variation values of temperature and strain by detecting reflection spectrum and transmission spectrum of FBG. When the ambient temperature, strain, or other physical quantities change, the grating period Λ or fiber core refractive index n_{eff} will change, and the reflection spectrum and transmission spectrum of FBG will change as well, which will cause the center wavelength of the fiber grating to occur a displacement $\Delta\lambda$, as shown in Fig. 38.1.

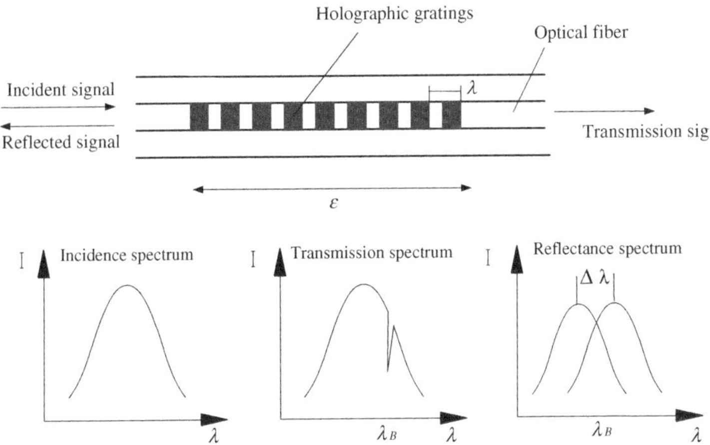

Fig. 38.1 Schematic diagram of FBG sensing principle

Based on coupled mode theory, a guided mode which is propagated in uniform FBG may be coupled to another one propagated in the opposite direction to form a narrow-band reflection, and the peak reflection wavelength λ_B can be written

$$\lambda_B = 2n_{eff}\Lambda \tag{38.1}$$

where, λ_B is the Bragg wavelength, n_{eff} is the effective refractive index of fiber transmission mode, Λ is the grating pitch. We can get the wavelength shift of FBG, λ_B, which is caused by strain ε and temperature difference ΔT by differential transformation with respect to Eq. (38.1), as follows

$$\frac{\Delta\lambda_B}{\lambda_B} = (\alpha_f + \xi)\Delta T + (1 - P_e)\varepsilon \tag{38.2}$$

where, α_f, ζ and P_e are thermal expansion coefficient, thermo-optical coefficient and elasto-optical coefficient of fiber optic materials.

38.2.2 Design of FBG Flowmeter

The FBG flowmeter mainly includes two major parts, the venturi tube and pressure sensing device based on fiber grating, they are connected to each other with two conduits, as is shown in Fig. 38.2. On the basis of principle of fluid mechanics, when flux density is constant, the Bernoulli equation of incompressible fluid can be written as follow [6]

$$P_1 + \frac{1}{2}\rho v_1^2 = P_2 + \frac{1}{2}\rho v_2^2 \tag{38.3}$$

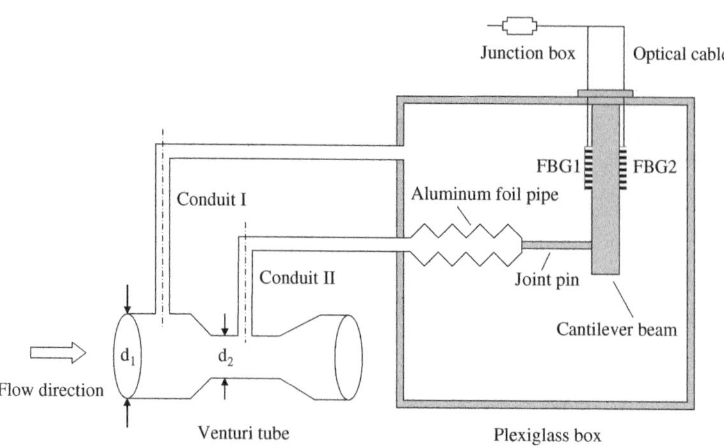

Fig. 38.2 Schematic diagram of FBG flowmeter

Based on fluid continuity equation [7], we have

$$\pi\left(\frac{1}{2}d_1\right)^2 v_1 = \pi\left(\frac{1}{2}d_2\right)^2 v_2 \tag{38.4}$$

By combining Eqs. (38.3) and (38.4), the pressure difference between the cross-section of conduit I and conduit II can be deduced, as follow

$$\Delta P = P_1 - P_2 = \frac{\rho}{2}\left[\left(\frac{d_1}{d_2}\right)^4 - 1\right]v_1^2 = \frac{\rho}{2}\left[\left(\frac{d_1}{d_2}\right)^4 - 1\right]v^2 \tag{38.5}$$

where, d_1 and d_2 are inner diameters of venture tube; v_1, v_2, and P_1, P_2, are flow velocity and pressure on longitudinal section of conduit I and conduit II; ρ is fluid density; $v = v_1$. The flow velocity of fluid has a positive correlation with pressure difference between conduit I and conduit II, and ΔP can be measured with the pressure sensing device based on fiber grating whose main body is a plexiglass box with an aluminum foil pipe and a cantilever beam inside it, they are connected by a joint pin; two fiber grating sensors are adhered to the beam with longitudinal symmetrical structure and accessed into the junction box with optical cable.

Because the FBG1 and FBG2 are adhered to beam with longitudinal symmetrical structure, both the two Bragg wavelength shifts of fiber grating caused by temperature change are the same, while both of them caused by the strain change are equal in magnitude but opposite in direction [5]. We have

$$\frac{\Delta\lambda_{B1}}{\lambda_B} = \left(\alpha_f + \xi\right)\Delta T + (1 - P_e)\varepsilon \tag{38.6}$$

$$\frac{\Delta\lambda_{B2}}{\lambda_B} = \left(\alpha_f + \xi\right)\Delta T - (1 - P_e)\varepsilon \tag{38.7}$$

Combining Eqs. (38.6) and (38.7), we have

$$\frac{\Delta\lambda_B}{\lambda_B} = \frac{\Delta\lambda_{B1} - \Delta\lambda_{B2}}{\lambda_B} = 2(1 - P_e)\varepsilon \tag{38.8}$$

Based on material mechanics theory, the axial strain of cantilever beam ε meets the following formulas

$$\begin{cases} \varepsilon = \frac{Mh}{2EI_y} \\ \frac{M}{I_y} = \frac{12Fl}{bh^3} \end{cases} \tag{38.9}$$

where E, h, l and b are the elasticity modulus, thickness, length and width of beam, I_y is the section moment of inertia of beam, M is the bending moment of beam, F is the pressure force applied to beam. Because the aluminum foil pipe is stretchable only along axial direction and its elasticity coefficient is very small, the pressure force F can be expressed as follow

$$F = \Delta PSg = (P_1 - P_2)\pi\left(\frac{d}{2}\right)^2 g \tag{38.10}$$

where ΔP is pressure difference inside and outside of aluminum pipe, S is cross-section area of pipe, g is acceleration of gravity, d is pipe diameter. The pressure difference ΔP is applied to the beam through joint pin, which results in the deformation of beam, and causes the wavelength shift of FBG sensors. By combining Eqs. (38.5), (38.8–38.10), we have

$$\frac{\Delta\lambda_B}{\lambda_B} = \frac{3(1 - P_e)\pi l d^2 \rho g}{2Ebh^2}\left[\left(\frac{d_1}{d_2}\right)^4 - 1\right]v^2 \tag{38.11}$$

As is known from Eq. (38.11), the variation of wavelength $\Delta\lambda_B$ has a linear relation with the square of flow velocity v, so we can achieve the measurement of flow velocity by using the FBG flowmeter.

38.3 Design of Telemetry System

38.3.1 System Composition

The telemetry system is composed of the following three components:

1. ZigBee wireless sensor network. It consists of multiple sensor nodes equipped with the FBG flowmeter in the monitoring area, and they are connected to a network coordinator in a star topology structure [8]. The flow velocity data

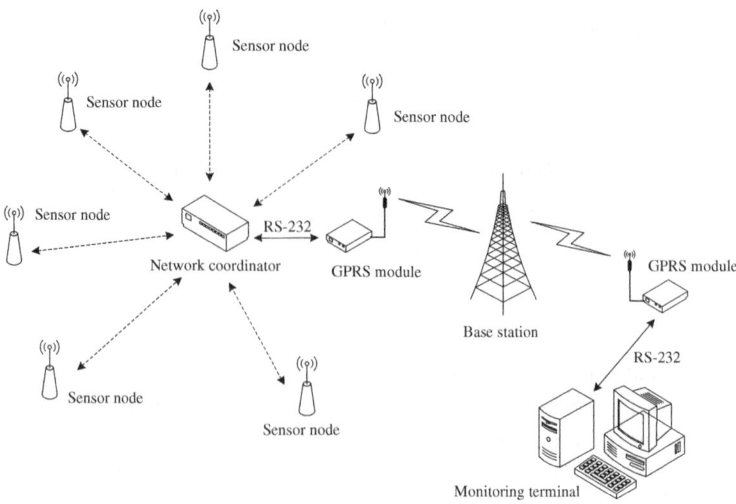

Fig. 38.3 Schematic diagram of telemetry system

acquired by each sensor node is sent wirelessly to the network coordinator, and the coordinator is connected to a GPRS module via RS-232 port.

2. GPRS network. It will be transmitted to the remote monitoring terminal that the monitoring data are aggregated in the network coordinator via GPRS network.
3. Monitoring terminal. The GPRS module of upper computer is connected to the center server via RS-232 port, which is responsible for the real-time data storage, analysis and processing, and it will give an alarm signal when the flow velocity reaches a critical value (Fig. 38.3).

38.3.2 Hardware Design of the System

ZigBee wireless sensor network consists of a number of sensor nodes and a network coordinator [9]. The sensor nodes are composed of FBG flowmeter, MCU module and radio frequency (RF) module, which is responsible for real-time data acquisition and transmission of flow velocity in the monitoring area. After receiving the monitoring data, the FBG flowmeter will convert them to digital signals and send them to the MCU module [10], and the MCU module is responsible for data processing and transmission. All the sensor nodes are powered by solar batteries. The network coordinator includes RF module, MCU module and GPRS module, which is responsible for data gathering and processing of ZigBee network, data communication with lower computer and executing some instructions. The network coordinator is supplied by an external power source. The hardware architecture of this telemetry system is shown in Fig. 38.4.

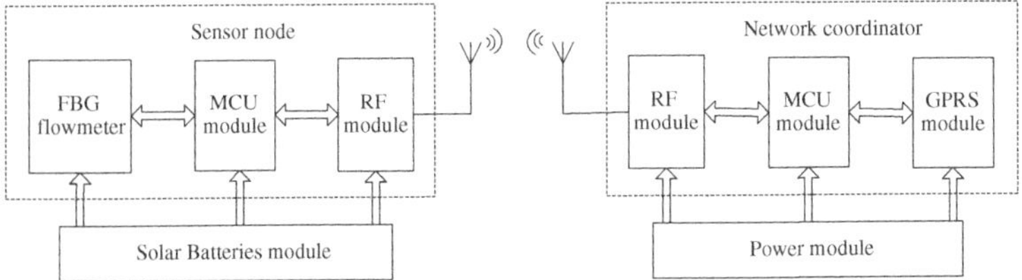

Fig. 38.4 Hardware structure diagram of the telemetry system

38.4 Practical Example

In order to test the telemetry system, field experiments were carried out in a river in southwest area of China, which is a typical mountainous seasonal stream, the flood hydrograph is characterized by rapid rise and recession, and the change of peak discharge during the low water period and flood season is obvious. Figure 38.5 shows the monitoring results of flow velocity during 3 days of persistent rain. As can be seen from the figure, the variation tendency of flow velocity is basically consistent with the rainfall distribution based on meteorological station.

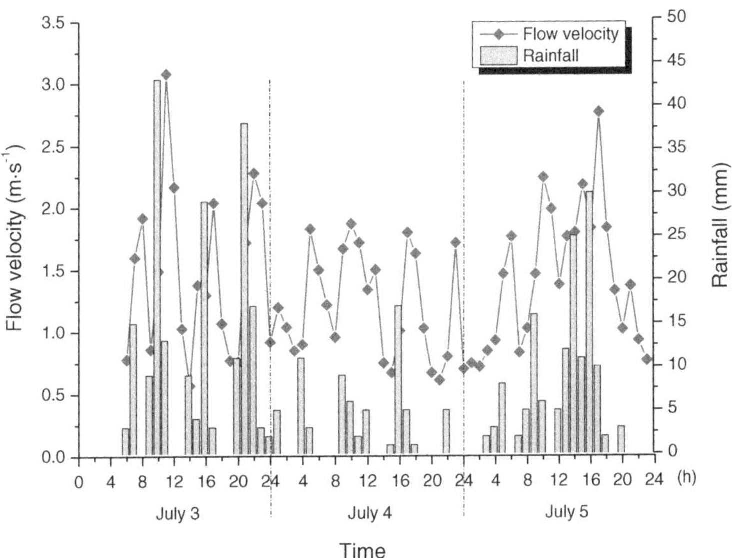

Fig. 38.5 Monitoring results of flow velocity of the river during three days of persistent rain

38.5 Conclusion

This article has designed a new FBG flowmeter by applying fiber Bragg grating sensing technology into hydrologic observation, which has the advantages of high accuracy, anti-electromagnetic interference, corrosion resistance etc., and the influence of temperature on measurement results could be eliminated by using the dual-fiber pattern. In order to achieve the real-time and effective performances of data collection, all the measure points in field are linked to each other to make up monitoring network based on ZigBee technology, and the communication between center node of ZigBee wireless network and remote monitoring terminal has been realized by GPRS. The field experiments have proved that the telemetry system of flow velocity has the characteristics of fast networking, low power consumption, high reliability and unattended operation and can be applied well in practice.

References

1. Sheng SZ, Xu YT, Yuan HJ (2002) New development in the technology of flow measurement over the last decade. Mech Eng 24(5):1–14 (in Chinese)
2. Zhao Y (2007) Optical fiber gratings and sensing technology. Tsinghua University Press, Beijing, pp 145–179 (in Chinese)
3. Rao YJ, Webb DJ, Jackson DA et al (1997) In fiber Bragg grating temperature sensor system for medical application. J Lightwave Technol 15(5):779–785
4. Kersey AD, Davis MA, Patrick HJ (1997) Fiber grating sensors. Light wave Technol 15(8):1442–1463
5. Chung SH, Kim JH, Yu BA (2001) A fiber bragg grating sensor demodulation technique using a polarization maintaining fiber loop mirror. IEEE Photo Tech. Lett 13(12):1343–1345
6. Yang SL, Shen J, Li TZ (2009) Flow velocity sensor based on double fiber Bragg gratings. Semicond Optoelectron 30(5):759–762 (in Chinese)
7. Chen JJ, Zhang WG, Tu QC et al (2006) High-sensitivity flow velocity sensor based on fiberg grating. Acta Optica Sinica 26(8):1136–1139 (in Chinese)
8. Sun LM, Li JZ, Chen Y (2005) Wireless sensor network. Tsinghua University Press, Beijing, pp 185–191 (in Chinese)
9. Kinney P. ZigBee Technology: wireless control that simply works. [DB/OL]. (2004-08-30). http://www.hometoys.com/htinews/oet03/articl2.es/kinney/zigbee.htm
10. Kintner MM, Conant R (2005) Opportunities of wireless sensors and controls for building operation. Energ Eng J 102(5):27–48

Chapter 39
Statics of Supporting Leg for a Water Strider Robot

Licheng Wu, Yu Yang, Guosheng Yang and Xinkai Gui

Abstract Biologically inspired by water strider, water strider robot uses surface tension force produced by supporting leg to stay and walk on water surface. The robot's loading capacity can be known through statics analysis. A static model of the supporting leg is created and the maximum surface tension condition, that is the surface-breaking condition, is analyzed. Then the calculation methods of supporting force and maximum allowed onto-water depth are proposed. For several materials, the curves between surface tension and contact angle, which are based on the proposed model and method, are shown using the Matlab program. The theoretical calculation results of the supporting force and the maximum allowed onto-water depth are compared with experimental data. Then the validity of the proposed models and methods is verified.

Keywords Water strider robot · Statics · Surface tension

39.1 Introduction

Water strider robot was first introduced in 2003. A robot named Water Strider was reported in [1], and this robot has six legs composed of 0.2 mm gauge stainless steel wire. Its structure is simple and has limited locomotion capacity, but as the first robot in the world that can be stay and move on water surface, it gets sharp focus. Its theoretical research and experimental results was published in Nature and along with a commentary article [2]. Then a robot named Water Walker was reported in [3], which has 8.2-inch-long steel wire legs with hydrophobic Teflon coated. A new prototype of motor-driven floating robot proposed in [4] has

L. Wu (✉) · Y. Yang · G. Yang · X. Gui
School of Information Engineering, Minzu University of China,
Beijing 100081, China
e-mail: wulichenggg@hotmail.com

Z. Sun and Z. Deng (eds.), *Proceedings of 2013 Chinese Intelligent Automation Conference*, Lecture Notes in Electrical Engineering 255,
DOI: 10.1007/978-3-642-38460-8_39,

12 supporting legs to improve the robot's loading capability. Based on Water Dancer in [5], Water DancerII is proposed in this paper.

Simulation results of the curve between supporting force and contact angle was given in [6], but the exact supporting force was not calculated [7] gave only the experimental results of supporting force, lacking the overall theoretical analysis, and the calculation data were not elaborate.

39.2 Numerical Model of Supporting Legs

Analyze and calculate the surface tension force of supporting leg, and thus calculate the lift force acting on robot. The numerical model of supporting leg and its interface with water need to be established first.

39.2.1 Description of the Problem

As a bionic robot of water strider, the robot's supporting leg is usually long and thin cylinder, with the material is stainless steel wire or carbon fiber. It is assumed that the supporting leg is long and thin rigid cylinder, θ_c is the contact angle with water. The supporting legs push down the surface a certain depth. The cylinder's symmetry axis maintains horizontal. As the leg's length is much larger than the diameter, it's assumed there are no complex end-effects at the two ends of the cylinder, the problem then can be taken as a 2-D problem, shown in Fig. 39.1.

When an object is placed on water without breaking the surface, the force acting on this object consists of surface tension force F_b and buoyancy force F_s [8]. The lift force Fcan be described as follows:

$$F = F_b + F_s \tag{39.1}$$

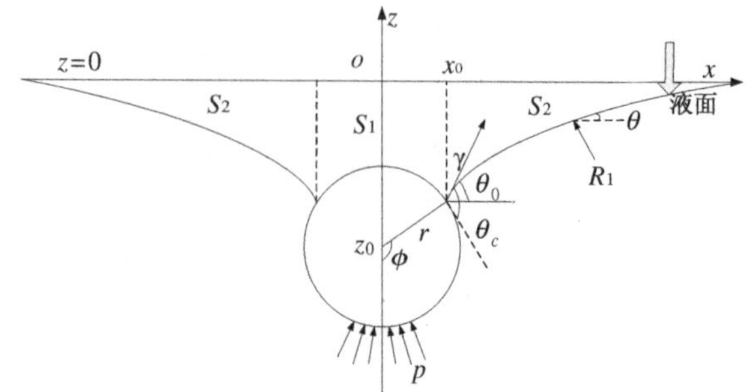

Fig. 39.1 Model of supporting leg

where F_b is deduced by integrating hydrostatic pressure p over the cylinder in contact with water. F_s is the vertical component of surface tension force γ along the water–air interface. The equations are as followed:

$$F_b = \int_0^{\phi_0} p\cos\phi \cdot r d\phi = \rho g S_1 = \rho g\left(-2z_0 r\sin\phi + r^2\sin\phi\cos\phi\right) \quad (39.2)$$

$$F_s = 2\gamma\sin\theta_0 = \rho g S_2 \quad (39.3)$$

where r is the radius of the cylinder. z_0 is the leg's onto-water depth. Definitions of ϕ and θ_0 are shown in Fig. 39.1. ϕ_0 is the value of ϕ at the junction point where water, air and cylinder all meet together. ρ is the density of water. g is the gravitational acceleration. S_1 and S_2 are the area of water displaced, S_1 is the portion above cylinder-water interface, S_2 is the portion above air–water interface. They are shown in Fig. 39.1.

According to Young–Laplace equation [9], the hydrostatic pressure at a certain point on water–air interface is:

$$p = \gamma \cdot (1/R_1 + 1/R_2) \quad (39.4)$$

where γ is the surface tension coefficient of water at temperature 20° $(\gamma = 0.072\text{N/m})$. R_1 and R_2 are the curvature radius of contact surface at the given point. Based on our assumption, it can be considered that $R_2 = \infty$. Then Eq. (39.4) can be simplified as:

$$\rho g h(x) = -\gamma \cdot \frac{1}{R_1} = -\gamma \cdot \frac{d^2}{dx^2}h(x) \Big/ \left(1 + (\frac{d}{dx}h(x))^2\right)^{\frac{3}{2}} \quad (39.5)$$

where $z = h(x)$ is the curve equation of water–air interface, there is a minus sign '-' because $z = h(x) < 0$. The boundary conditions are:

$$h'(x)|_{x=x_0} = \tan(\theta_0) \quad (39.6)$$

$$h(\infty) = 0 \quad (39.7)$$

where x_0 is the x axis value at the point where water, air and cylinder all meet together, therefore exist the following relationship:

$$x_0 = r\sin\phi \quad (39.8)$$

$$\phi = \pi + \theta_0 - \theta_c \quad (39.9)$$

If ϕ is known, x_0 and θ_0 can be got from Eqs. (39.8) and (39.9), and the surface tension force can be derived from Eq. (39.3). The profile of water–air interface $h(x)$ can be solved from(39.5 to 39.7). $z_0 = h(x_0)$, then the buoyancy force can be calculated from Eq. (39.2). so if ϕ is known, the lift force can be got.

39.2.2 The Maximum Surface Tension Force

It is known from Fig. 39.1 that when the leg pushes onto water surface deeper and deeper, the solid–liquid-gas junction point move up along the leg accordingly. When the two junction point on leg's two sides meet at leg's top (at this time $\phi = 180^\circ$), the two water surface meet together, and the leg was submerged, that is, the water surface was broken. While the pushing depth increases, the pressure force from the leg on water surface increases consequently. When this pressure force reaches its limitation, that is, when supporting force equals to its maximum value, water surface may be broken even if the two water surface has not met. At this time, ϕ reaches a critical value. When water surface is broken the exact value of ϕ is related to the material's contact angle. The specific analysis is shown in Fig. 39.2.

With the supporting leg pressing onto the originally horizontal water surface deeper and deeper, θ_0 increases constantly from zero, and ϕ increases accordingly from Eq. (39.9). As shown in Fig. 39.2. At this time, the total lift force F also achieves its maximum (the assumption's validity is verified in Table 39.2).

Based on the above analysis, the ϕ value when water surface is broken is:

$$\phi_{\max} = \begin{cases} 180^\circ & \theta_c < 90^\circ \\ 270^\circ - \theta_c & \theta_c \geq 90^\circ \end{cases} \tag{39.10}$$

ϕ can take value between 0 and $\phi_{\max}$. According to Eq. (39.9), for material that $\theta_c < 90^\circ$, the range of θ_0 is from 0 to θ_c. For material that $\theta_c \geq 90^\circ$, $\theta_0 = 90^\circ$ is possible. The maximum surface tension force of supporting leg is [from Eq.(39.3)]:

$$F_{s\max} = \begin{cases} 2\gamma \sin\theta_c & \theta_c < 90^\circ, \theta_0 = \theta_c \\ 2\gamma & \theta_c \geq 90^\circ, \theta_0 = 90^\circ \end{cases} \tag{39.11}$$

Equation (39.11) implies that for material that $\theta_c \geq 90^\circ$, the maximum surface tension force is 2γ no matter how much is the exact value of θ_c. However, the material with bigger θ_c has better performance, as Eq. (39.10) shows that bigger θ_c produce smaller $\phi_{\max}$. Smaller ϕ means smaller area in contact with water, which can reduce the resistance and then improve the robot's stability.

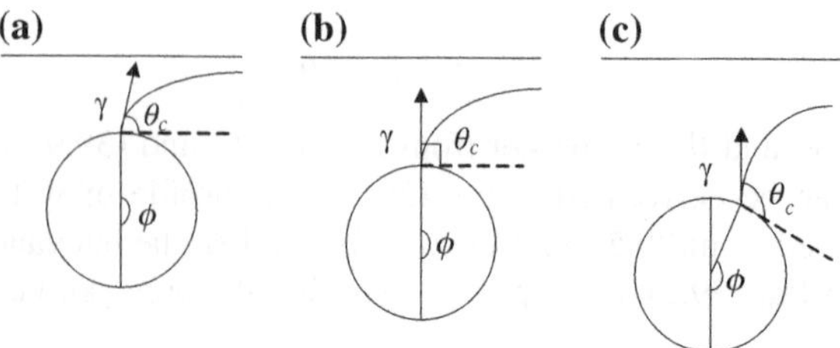

Fig. 39.2 Limit position of surface just before broken **a** $\theta_c < 90^\circ$ **b** $\theta_c = 90^\circ$ **c** $\theta_c > 90^\circ$

39.2.3 Interface Profile Equation and Maximum Onto-Water Depth Solving

Theoretically, the equation of water–air interface profile $z = h(x)$ can be obtained by integrating Eq. (39.5). This paper solves the problem by using its inverse function form $x = f(z)$. The following is derived from Eq. (39.5):

$$\rho g z = -\gamma \cdot \frac{1}{R_1} = -\gamma \cdot \frac{d^2}{dz^2} f(z) \Big/ \left(1 + (\frac{d}{dz} f(z))^2\right)^{\frac{3}{2}} \tag{39.12}$$

Integrate Eq. (39.12) from 0 to z, it can be got that:

$$\frac{\rho g z^2}{2\gamma} = 1 - \frac{df(z)}{dz} \Big/ \sqrt{1 + \left(\frac{df(z)}{dz}\right)^2} = 1 - \cos\theta \tag{39.13}$$

where θ is the slope angle of water surface. Defining $a = \sqrt{2\gamma/\rho g}$ then:

$$z = -a\sqrt{1 - \cos\theta} \tag{39.14}$$

$$z_0 = -a\sqrt{1 - \cos\theta_0} \tag{39.15}$$

Equation (39.13) can be simplified as: $\frac{df(z)}{dz} = \frac{z^2 - a^2}{z\sqrt{2a^2 - z^2}}$ integrate it once again:

$$x = \frac{a}{\sqrt{2}} \cosh^{-1}\left(\frac{\sqrt{2}a}{-z}\right) - \sqrt{2a^2 - z^2} + c \tag{39.16}$$

where c is the integration constant. c can be solved from boundary conditions (39.8) and (39.15), as follows:

$$c = r\sin\phi - \frac{a}{\sqrt{2}} \cosh^{-1}\left(\frac{\sqrt{2}a}{-z_0}\right) + \sqrt{2}a\sqrt{1 - \frac{z_0^2}{2a^2}} \tag{39.17}$$

Substitute Eq. (39.9) into Eq. (39.15), the relationship between z_0 and ϕ is:

$$z_0 = -a\sqrt{1 - \cos(\phi + \theta_c - \pi)} \tag{39.18}$$

Thus, if z_0 or ϕ is known, water–air interface profile equation can be derived from (39.16 to 39.18). From Eq. (39.10), the range of ϕ is $[0,\ \phi_{\max}]$, therefore the leg's maximum depth onto water without breaking surface can be got by Eq. (39.18) :

$$z_{0\max} = \begin{cases} -a\sqrt{1 - \cos\theta_c} & \theta_c < 90^\circ \\ -a = \sqrt{2\gamma/\rho g} & \theta_c \geq 90^\circ \end{cases} \tag{39.19}$$

39.3 Simulation and Analysis

When water is 20, the parameters are: $\gamma = 0.072\mathrm{N/m}, g = 9.81\mathrm{m/s^2}$, $\rho = 9.98 \times 10^2\mathrm{kg/m^3}$. Assuming supporting leg is long and straight cylinder with radium 0.2 mm

39.3.1 Water–Air Interface Simulation

For different materials that have various contact angles, Fig. 39.3 shows the surface profile just before water surface is broken, at this time, ϕ is very close to $\phi_{\max}$. It's shows that when $\theta_c > 90°$, the water–air interface profile of different material looks almost identical. This simulation result verifies the numerical solution of water–air interface profile and the validity of our analysis.

39.3.2 Calculation of Supporting Force and Allowed Onto-Water Depth

Given supporting leg's material is hydrophilic stainless steel wire with radium 0.2 mm and contact angle 60. To increase surface tension force, the steel wire is

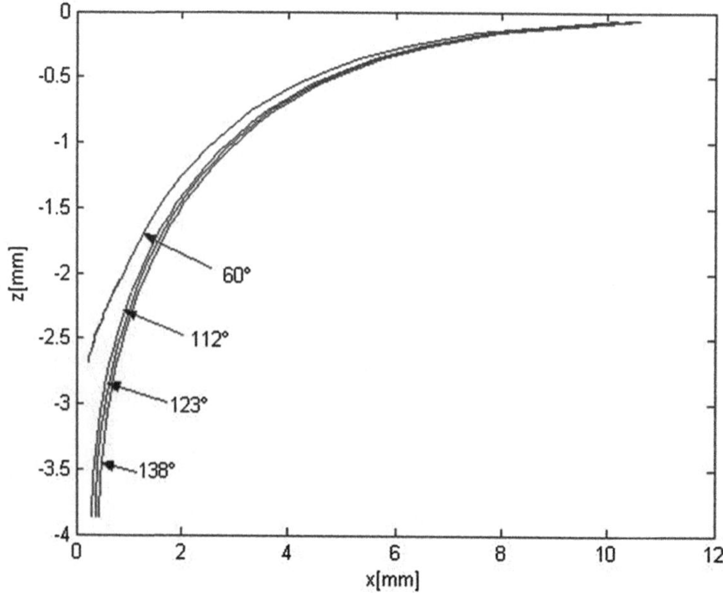

Fig. 39.3 Air-water surface profile of various materials just before surface is broken

Table 39.1 Water strider robot supporting force of the wire leg (L = 10 mm)

Leg material	$\theta_c/°$	D/mm	Z_{0max}/mm	F_s/mN	F_b/mN	F/mN
Stainless steel wire	60	0.2	−2.7	1.18	0.09	1.27
Hydrophobic steel wire	≈120	0.2	−3.86	1.44	0.04	1.48

Table 39.2 The lift force of supporting leg with Teflon coated material when $\phi > \phi_{max}$ (L = 10 mm)

$\theta_c/°$	$\phi/°$	F_s/mN	F_b/mN	F/mN	Z_0/mm
90	158	1.44	0.06	1.5	3.86
90.1	158.1	1.4399	0.0562	1.496	3.867
92	160	1.439	0.053	1.492	3.995

paint-sprayed or wax-coated to make contact angle exceeds 90. For the above two leg material, supporting force and maximum allowed onto-water depth and other parameters are shown in Table 39.1.

Song et al. [6] propose a good hydrophobic Teflon coated material. It is taken as one of our experiment samples. The parameters are: $\theta_c = 112°$, $D = 0.33$mm. Table 39.2 shows the calculation results of total lift force. When F_s exceed maximum value $(\theta_0 > 90°)$. The results suggest that if increase continuously leg's onto-water depth after supporting force reaches its maximum, the supporting force and buoyancy force would get smaller rather than bigger. This means the leg will break water surface and sink, thus the assumption in 2.2 is verified.

39.4 Conclusion

A detailed statics analysis on water strider robot is made based on Young–Laplace equation. The governing equation of interface between supporting leg and water surface is established. The maximum surface tension force conditions, that is, the surface breaking conditions is analyzed, therefore the calculation method of leg's supporting force and maximum allowed onto-water depth is proposed. For different material, the relationship between surface tension force and contact angle is simulated using Matlab, and the numerical validation of surface breaking conditions is performed too. This method and result can serve as rules for leg design of water strider robot.

References

1. Hu DL, Chan B, Bush JWM (2003) The hydrodynamics of water strider locomotion. Nature 424(7):663–666
2. Dickson Michael (2003) How to walk on water. Nature 424(7):621–622

3. Suhr SH, Song YS, Lee SJ et al (2005) Biologically Inspired Miniature Water Strider Robot. In: Proceedings of the robotics: science and systems I. Boston
4. Song YS, Sitti M (2007) STRIDE: a highly maneuverable and non-tethered water strider robot. In: 2007 IEEE international conference on robotics and automation. Italy, 980–984
5. Wu L, Ding L, Guo D et al (2008) A bionic water spider robot moving on the water: China, ZL 200610112601.7 (in Chinese)
6. Song YS, Suhr SH, Sitti M (2006) Modeling of the supporting legs for designing biomi metic water strider robots. In: Proceedings of the 2006 IEEE international conference on robotics and automation, Orlando
7. Suzuki K, Takanobu H, Noya K et al (2007) Water strider robots with microfabricated hydrophobic legs. In: 2007 IEEE/RSJ international conference on intelligent robots and systems, San Diego, 590–595
8. Keller JB (1998) Surface tension force on a partly submerged body. Phys Fluids 10(11):3009–3010
9. Adamson AW, Gast AP (1997) Physical chemistry of surfaces, 6th edn. Wiley-Interscience, New York

Chapter 40
Orientation Control for Mobile Robot with Two Trailers

Jin Cheng, Yong Zhang and Zhonghua Wang

Abstract The control problem for a mobile robot with two trailers is addressed in this paper. A nonlinear smooth control law for tracking of the orientations of the trailers with asymptotic stability in backward motion is firstly proposed. The stability of the closed loop feedback system is proved with theory about cascade system and input-to-state stability and also Lypunov technique. The effectiveness of the proposed control law and the controller are illustrated by numeric simulation results.

Keywords Mobile robot with two trailers · Asymptotic stability · Cascade system · Input-to-state stable

40.1 Introduction

As a noholonomic system, motion control for mobile robot with trailers has given rise to plenty of work in recent years. However, motion control for the tractor-trailer mobile robot is much more challenging due to the nonholonomic constraints existing on the motion of transverse direction and additional holonomic constraints introduced by the joint axis [1]. In addition, the subsystem of the trailers is open-loop unstable when the robot moves backward, which results in the so-called jack-knife effects on the relative angle between the tractor and the trailer and also between the two trailers [2].

A lot of related work has been done for the control problem of mobile robot with one trailer [3–6]. Brockett has proven that there exits no simple continuous state feedback control law for the stabilization of such noholonomic systems about

J. Cheng (✉) · Y. Zhang · Z. Wang
School of Electrical Engineering, University of Jinan, No. 106, Road Jiwei, Jinan 250022, China
e-mail: cse_chengj@ujn.edu.cn

Z. Sun and Z. Deng (eds.), *Proceedings of 2013 Chinese Intelligent Automation Conference*, Lecture Notes in Electrical Engineering 255,
DOI: 10.1007/978-3-642-38460-8_40,

a given configuration [7]. Whereas, related work about motion control of the mobile robot with two trailers is few at present. In this paper, stable control laws for the orientation angle when the mobile robot with two trailers moves backward along rectilinear trace is proposed. The robot can be reversed with asymptotic stability at given orientation to track rectilinear trajectory.

The paper is organized as follows. Section 40.2 gives model of the system kinematics of the tractor-trailer mobile robot with two trailers and describes the control problem. Section 40.3 deals with the control problems of the robot in backward motion and proposes the feedback control laws for tracking desired orientation angle. Numeric simulation results are presented in Sect. 40.4 to illustrate the effectiveness of proposed control laws. Finally, Sect. 40.5 gives the conclusion.

40.2 Kinematics of Mobile Robot with Two Trailers

The tractor-trailer mobile robot considered in this paper is driven by a car-like tractor towing two passive trailers as shown in Fig. 40.1.

The first trailer is hooked up at the middle point of the rear wheels of the tractor and the second trailer is hooked up at the middle point of the rear wheels of the first trailer. The tractor is rear-wheel droved with linear speed v. The configuration of the system can be described by a vector $(x_1, y_1, \theta_1, \theta_2, \theta_3)$, which x_1 and y_1 denote the position coordinates of the tractor's in Cartesian coordinates, θ_1 and θ_2 measure the orientation of the tractor and the trailer with respect to the x axis, φ is the steering angle. L_1, L_2 and L_3 are length parameters of the tractor and trailers.

The kinematics of the mobile robot with two trailers are described by (40.1), where $\gamma_1 = \theta_1 - \theta_2$ and $\gamma_2 = \theta_2 - \theta_3$ are respectively the intersection angles between the two adjacent axis of the tractor and the two trailers.

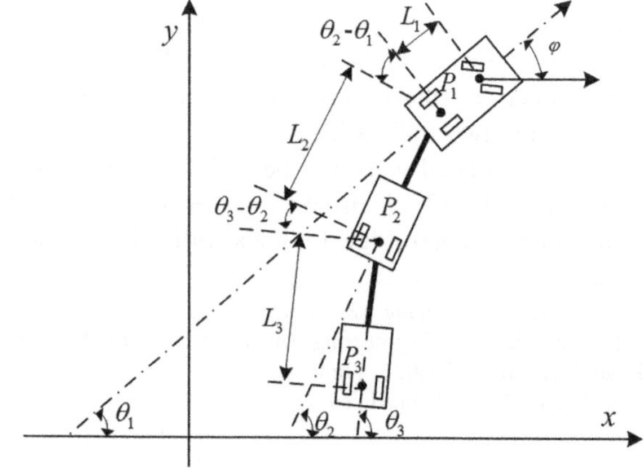

Fig. 40.1 Geometry of mobile robot with two trailers

$$
\begin{aligned}
\dot{x}_1 &= v\cos\theta_1 \\
\dot{y}_1 &= v\sin\theta_1 \\
\dot{\theta}_1 &= \frac{v\tan\varphi}{L_1} \\
\dot{\theta}_2 &= \frac{v}{L_2}\sin\gamma_1 \\
\dot{\theta}_3 &= \frac{v}{L_3}\cos\gamma_1\sin\gamma_2
\end{aligned}
\tag{40.1}
$$

40.3 Design of the Orientation Tracking Control Law

According the kinematics of the mobile robot with two trailers, the system can only be controlled under the limitation $|\gamma_i| < \pi/2$ $(i = 1,\ 2)$. Assume the desired constant orientation angle θ_{3d} for the second trailer satisfies with $|\theta_{3d} - \theta_3(0)| \leq \pi/2$, and if there exits a asymptotic control law that make the orientation angle θ_3 converges to θ_{3d}, then it yields $|\theta_{3d} - \theta_3(t)| \leq \pi/2,\quad \forall t \geq 0$. Therefore, assuming that the desired orientation angle θ_{3d} satisfies the condition $|\theta_{3d} - \theta_3(0)| \leq \pi/2$, we have the following proposition.

Proposition *Consider the system* (40.1) *with control input* $v = -1$. *Assume* $\dot{\theta}_{3d} = 0$ *and* $|\theta_{3d} - \theta_3(0)| \leq \pi/2$, *there exits a feedback control law for the control input* φ, *which the orientation angle of the second trailer converges to* θ_{3d} *with locally asymptotic stability.*

Proof Define

$$\sin\gamma_{2d} = -\sin(\theta_{3d} - \theta_3) \tag{40.2}$$

and let

$$s_1 = \sin\gamma_{2d} - \sin\gamma_2 \tag{40.3}$$

The Lyapunov function is then designed with

$$V_1 = \frac{1}{2}s_1^2 \tag{40.4}$$

Computing the time derivative of V_1 along the trajectories of the system (40.1) yields

$$
\begin{aligned}
\dot{V}_1 = &-v(\sin\gamma_{2d} - \sin\gamma_2)\cos\gamma_1\cos\gamma_2 \cdot \\
&\left(\frac{1}{L_2}\tan\gamma_1 - \frac{1}{L_3}\sin\gamma_2 - \frac{1}{L_3}\tan\gamma_2\cos(\theta_{3d} - \theta_3)\right)
\end{aligned}
\tag{40.5}
$$

Define

$$s_2 = \frac{1}{L_2}\tan\gamma_1 - \frac{1}{L_3}\sin\gamma_2 - \frac{1}{L_3}\tan\gamma_2\cos(\theta_{3d} - \theta_3) + k_1 s_1 \tag{40.6}$$

with $k_1 > 0$, and the Lyapunov function

$$V_2 = V_1 + \frac{1}{2}s_2^2 \tag{40.7}$$

Differentiate V_2 with respect to time along system (40.1), one obtains

$$\dot{V}_2 = kv(\sin\gamma_{2d} - \sin\gamma_2)^2\cos\gamma_1\cos\gamma_2 + s_2(\dot{s}_2 - v\cos\gamma_1\cos\gamma_2(\sin\gamma_{2d} - \sin\gamma_2)) \tag{40.8}$$

Let

$$\dot{s}_2 = v\cos\gamma_1\cos\gamma_2(\sin\gamma_{2d} - \sin\gamma_2) - k_2 s_2 \tag{40.9}$$

with $k_2 > 0$. The solution of (9) is calculated with

$$\varphi = -a\tan\left(\frac{L_1 \cdot \tilde{\eta}(\gamma_1, \gamma_2, \theta_3, \theta_{3d})}{vL_2L_3^2(1 + \tan^2\gamma_1)}\right), \tag{40.10}$$

where

$$\begin{aligned}
\tilde{\eta}(\gamma_1, \gamma_2, \theta_3, \theta_{3d}) = {} & vL_3^2\sin\gamma_1 + vL_3^2\sin\gamma_1\tan^2\gamma_1 + vL_2L_3\sin\gamma_1\cos\gamma_2 \\
& + vL_2L_3\sin\gamma_1\cos(\theta_{3d} - \theta_3) + vL_2L_3\sin\gamma_1\cos(\theta_{3d} - \theta_3)\tan^2\gamma_2 \\
& + vk_1L_2L_3^2\sin\gamma_1\cos\gamma_2 - vL_2^2\cos\gamma_1\sin\gamma_2\cos\gamma_2 \\
& - vL_2^2\cos\gamma_1\sin\gamma_2\cos(\theta_{3d} - \theta_3) - vL_2^2\cos\gamma_1\sin\gamma_2\cos(\theta_{3d} - \theta_3)\tan^2\gamma_2 \\
& + vL_2^2\cos\gamma_1\sin\gamma_2\sin(\theta_{3d} - \theta_3)\tan\gamma_2 - vk_1L_2^2L_3\cos\gamma_1\sin\gamma_2\cos\gamma_2 \\
& + vL_2^2L_3^3\cos\gamma_1\cos\gamma_2\sin\gamma_{2d} - vL_2^2L_3^3\cos\gamma_1\cos\gamma_2\sin\gamma_2 - k_2L_2L_3^2\tan\gamma_1 \\
& + k_2L_2^2L_3\sin\gamma_2 + k_2L_2^2L_3\tan\gamma_2\cos(\theta_{3d} - \theta_3) \\
& - k_1k_2L_2^2L_3^2\sin\gamma_{2d} + k_1k_2L_2^2L_3^2\sin\gamma_2
\end{aligned}$$

Substitute (40.9) into (40.8), it yields

$$\dot{V}_2 = v\cos\gamma_1\cos\gamma_2 k_1 s_1^2 - k_2 s_2^2$$

For $|\gamma_i| < \pi/2\ (i = 1, 2)$ and $v < 0$, one has $\dot{V}_2 \leq 0$.

Define $s = [s_1, s_2]^T$, according to Barhashin-Krasovskii *Theorem* [8], the equilibrium $s = 0$ is locally asymptotically stable. Also, one has

$$\lambda_1\|s\|^2 \leq V_2 \leq \lambda_2\|s\|^2 \tag{40.11}$$

with $0<\lambda_1 \leq \frac{1}{2} \leq \lambda_2$, and

$$\dot{V}_2 \leq -\lambda_3 \|s\|^2 \tag{40.12}$$

with $\lambda_3 \geq \max\{k_1, k_2\}$. As a conclusion of (40.11) and (40.12), the equilibrium $s = 0$ is also exponentially stable, i.e., when $s \to 0$, with (40.3), one has $\gamma_2 \to \gamma_{2d}$. The next is to prove that s_3 also converges to zero with asymptotical stability when $s_1 \to 0$. Define $s_3 = \sin(\theta_{3d} - \theta_3)$. Computing the time derivative of s_3 with respect to time yields

$$\dot{s}_3 = -\frac{1}{L_3} v \cos\gamma_1 \cos(\theta_{3d} - \theta_3) \sin\gamma_2 \tag{40.13}$$

From (40.2) and (40.3), one has

$$\sin\gamma_2 = -\sin(\theta_{3d} - \theta_3) - s_1 \tag{40.14}$$

Substituting $\sin\gamma_2$ with (40.14), rewrite (40.13) with

$$\dot{s}_3 = -\frac{1}{L_3} v \cos\gamma_1 \cos(\theta_{3d} - \theta_3)(-s_3 - s_1) \tag{40.15}$$

Denote the derivative of s_3 and s_1 as

$$\dot{s}_3 = f_1(t, s_1, s_3) \tag{40.16}$$

$$\dot{s}_1 = f_2(t, s_1) \tag{40.17}$$

where f_1 and f_2 are piecewise continuous in t and locally Lipschitz in $\delta = [s_1, s_3]^T$. Subsystem (40.16) and (40.17) form a cascade system, which can be analyzed with input-to-output stability. With the proposed feedback control law (40.10), it has been proved above that the subsystem (40.17) at the equilibrium points $s_1 = 0$ is uniformly asymptotically stable.

Define a Lyapunov function

$$V_3 = \frac{1}{2} s_3^2 \tag{40.18}$$

With assumption $|\gamma_i| \leq \frac{\pi}{2}, \quad i = 1, 2$ and $|\theta_{3d} - \theta_3| \leq \frac{\pi}{2}$ and $v = -1$, one has

$$\dot{V}_3 = \frac{1}{L_3} v \cos\gamma_1 \cos(\theta_{3d} - \theta_3) s_3^2 \leq -\frac{1}{L_3} s_3^2 \tag{40.19}$$

So, the subsystem $\dot{s}_3 = f_1(t, 0, s_3)$ is asymptotically stable and also exponentially stable at equilibrium point $s_3 = 0$.

Choosing s_1 as the input of system (40.16), the system (40.16) is input-to-state stable [8]. Then the origin (s_1, s_3) of the cascade system composed by (40.16) and (40.17) is uniformly asymptotically stable, i.e., s_3 converges to zero with asymptotical stability when $s_1 \to 0$.

As a conclusion, when $(s_1, s_3) \to \mathbf{0}$, it yields $\theta_3 \to \theta_{3d}$. The orientation angle of the second trailer locally converges to θ_{3d} with asymptotically stability.

40.4 Simulation Results

Numeric simulations were performed to illustrate the effectiveness of the proposed control laws. The parameters of the mobile robot with two trailers are set with $L_1 = 1$, $L_2 = 2$, $L_3 = 2$. The initial configuration of the robot is set with $[x_1, y_1, \theta_1, \theta_2, \theta_3] = [0, 0, \pi, \pi, \pi]$ and the robot moves backward with speed $v = -1 \ \ textm/s$. The parameters in control law (40.10) are chosen with $k_1 = k_2 = 1$ and the desired angle for the trailer is set with $\theta_{3d} = \frac{7}{6}\pi$.

Simulation results in Figs. 40.2 and 40.3 shows that the robot is reversed along a rectilinear trajectory while tracking the desired orientation angle θ_{3d} successfully.

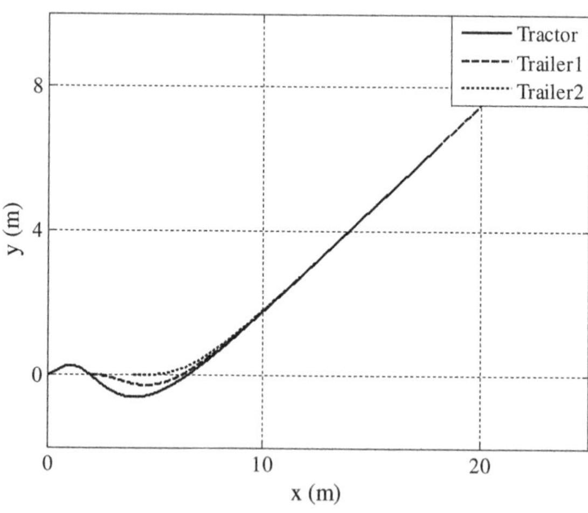

Fig. 40.2 Traces of the tractor and the two trailers

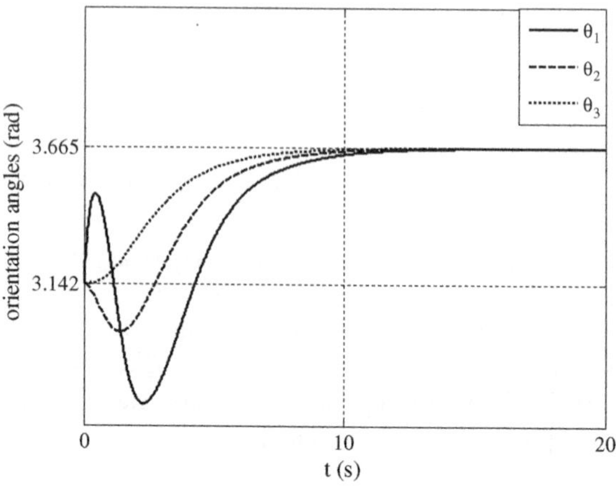

Fig. 40.3 Response of orientation angles of $\theta_i (i = 1, 2, 3)$

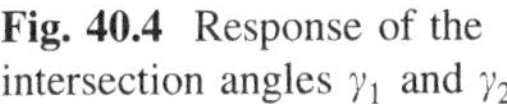

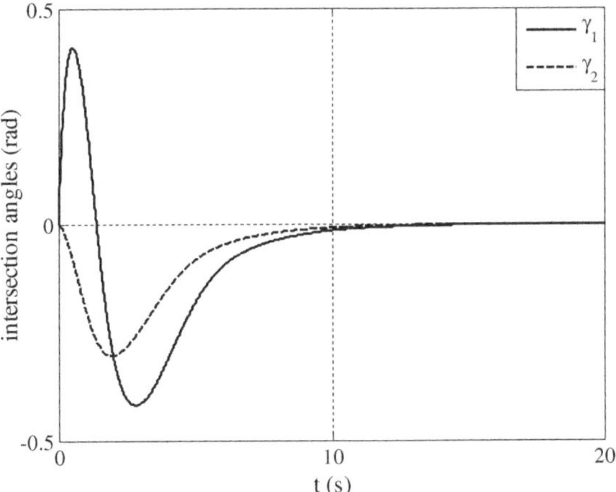

Fig. 40.4 Response of the intersection angles γ_1 and γ_2

Result in Fig. 40.4 also confirms the effectiveness of the proposed control law (40.10) which the intersection angles $\gamma_1 \to 0$ and $\gamma_2 \to 0$ with time t increases.

40.5 Conclusion

The control problem of reversing the mobile robot with two trailers has been addressed in this paper. With the proposed feedback control law, the robot with two trailers can be reversed with asymptotic stability on tracking desired orientation angle along rectilinear trajectory. According to Dubins theorem, an optimal curve with minimal length is composed of circular arcs and/or line segments. The control laws proposed in this paper will contribute to the future work about robot motion planning problem.

Acknowledgments This work was supported by the National Natural Science Foundation of China (No. 61203335) and partly supported by the National Natural Science Foundation of China (No. 61074021).

References

1. Dong Y et al (2005) Kinematics and constraint analysis of tractor-trailer mobile robot. In: machine learning and cybernetics in 2005 proceedings of international conference. IEEE, Guangzhou, China, pp 1380–1384
2. Martinez JL et al (2008) Steering limitations for a vehicle pulling passive trailers. IEEE Trans Control Syst Technol 16(4):809–818
3. Lee J et al (2001) A passive multiple trailer system for indoor service robots. In: Intelligent robots and systems in 2001 proceeding of IEEE/RSJ international conference, vol 2. Societies and Associations, Maui, Hawaii, USA, pp 827–832

4. Romano MD et al (2002) A experimental stabilization of tractor and tractor-trailer like vehicles. In: Intelligent control in 2002 proceeding of IEEE international symposium. IEEE, Vancouver, Canada, pp 188–193
5. Liu Z et al (2008) Path planning for tractor-trailer mobile robot system based on equivalent size. In: Intelligent control and automation in proceeding of the 7th World congress. IEEE, Chongqing, China, pp 5744–5749
6. Park M et al (2004) Control of a mobile robot with passive multiple trailers. In: Robotics and automation in 2004 proceeding of IEEE international conference, vol 5. IEEE, New Orleans, Louisiana, USA pp 4369–4374
7. Brockett RW (1983) Asymptotic stability and feedback stabilization. In: Progress in math. Of proceeding of conference, vol 27. Birkhauser, Boston, pp 181–208
8. Khalil HK (1996) Nonlinear systems. Prentice Hall, America

Chapter 41
The Effects of Leg Configurations on Trotting Quadruped Robot

Bin Li, Yibin Li, Xuewen Rong, Jian Meng and Hui Chai

Abstract This paper studied the configuration modes between the legs and the body frame of the mammalian quadruped robot. Based on the quadruped robot models with four different leg configuration approaches, the optimal leg configuration mode of the quadruped robot with trotting gait is verified by means of ADAMS and MATLAB co-simulation for improving the locomotion stability and decreasing the slip of quadruped robot, which provide a theoretical basis for the design of leg configuration on real quadruped robot Scalf.

Keywords Quadruped robot · Leg configuration · Stability

41.1 Introduction

Legged robots have potential to move on uneven terrain than tracked and wheeled robots. Among legged robots, the quadruped robots have good mobility and stability of locomotion. Current research on the quadruped robot has made great progress, but the ability of the robot is still lagging far behind the bionic objects. Most of them only are in a laboratory simulation stage and can not be used in practice. In order to increase the practical utility and transportation power and improve the environmental adaptability of complex terrain of the quadruped robot, the robot requires highly dynamics, fast speed and more payload capability [1].

In the development of quadruped robots, the driving mode has great influence on the performance of robots. In recent years, with the requirements of highly

B. Li
School of Science, Shandong Polytechnic University, Jinan 250353, China

B. Li (✉) · Y. Li · X. Rong · J. Meng · H. Chai
School of Control Science and Engineering, Shandong University, Jinan 250061, China
e-mail: ribbenlee@126.com

Z. Sun and Z. Deng (eds.), *Proceedings of 2013 Chinese Intelligent Automation Conference*, Lecture Notes in Electrical Engineering 255,
DOI: 10.1007/978-3-642-38460-8_41,

dynamics and large load capacity for quadruped robots in military and civil fields, as well as the progress of the hydraulic drive technology, hydraulic is gaining renewed interest in the field of robotics [2]. Especially in recent years, a new upsurge of studying for mammalian bionic, hydraulically actuated quadruped robots has emerged [3].

The construction of high performance quadruped robot with highly dynamics, fast speed and more payload capability is a dominant research direction and development trend. Therefore, the research of scientific problems and intrinsic mechanism on quadruped robot and the understanding of robot basic characteristics are urgently needed. The quadruped robots have many degrees of freedom (DOF), which require more careful design of leg configurations for improving the performance of robots. Several researchers have investigated the relation between leg configurations and performance of robots based on simple two-link model [4]. Besides the research focusing on leg configuration itself, some researchers have noted the importance of passive legs on the influence of stability of the quadruped trotting robot [5].

In this paper, we focused on the configuration approaches between the legs and the body frame of quadruped robot to improve the locomotion stability and decrease the slip during the dynamic walking with trotting gait. Firstly, the leg joints structure of the quadruped robot is designed. Then, based on the gait planning method and the kinematics equations, the optimal configuration method between the legs and the body of quadruped robot is verified by means of ADAMS and MATLAB co-simulation, which provide a theoretical basis for improving the performance of quadruped robot.

41.2 Bionic Leg Structure of the Quadruped Robot

Legged robots are classified into mammalian robots and reptilian robots based on the mechanical structures as shown in Figs. 41.1 and 41.2. Compared with the reptilian robots, the basically vertical leg orientation of mammalian robots may be regarded as weight carrying adaptation [6]. These robots can carry much more payload in an efficient way. The mammalian robots with joint actuators have good

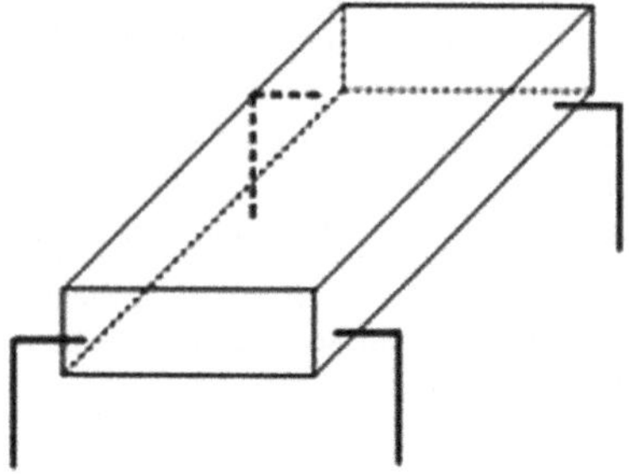

Fig. 41.1 The reptilian quadruped robot

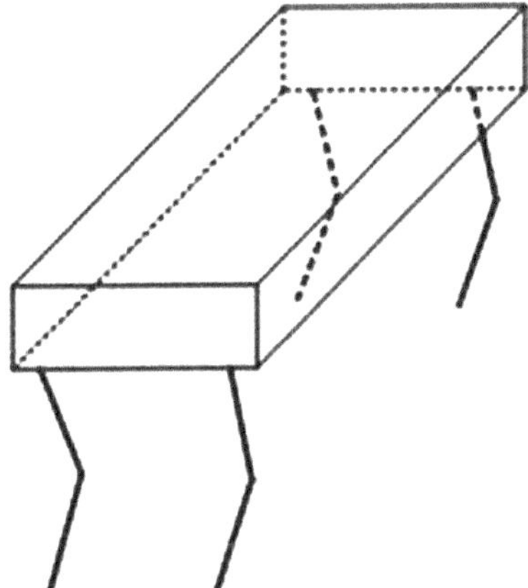

Fig. 41.2 The mammalian quadruped robot

walking speed and transportation power. Consequently, the research of bionic robots based on mammal animals becomes an important development direction in robotics field and the leg structure of the virtual model of robot is designed based on the bionic structure of horse/mule of the quadruped mammal animals [7].

41.3 Leg Configurations of the Quadruped Robot

Except for the design of leg structure of robots, the connection mode between the legs and the body frame of the quadruped robot has much influence on the stability performance. These leg configurations of the quadruped robot are illustrated in Fig. 41.3, which are forward elbow and backward knee model (Posture I), forward knee and backward elbow model (Posture II), knees model (Posture III) and elbows model (Posture IV), respectively. The models of the quadruped robot with

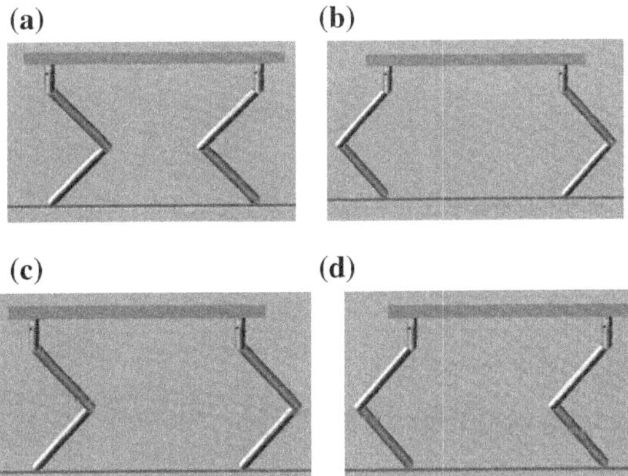

Fig. 41.3 Different leg configurations of the quadruped robot (motion is from *left* to *right*) **a** Posture I. **b** Posture II. **c** Posture III. **d** Posture IV

different leg configurations are set up by means of UG/Modelling and the motion direction of the robot is from left to right.

The simulation model of the quadruped robot is consists of four legs and a body frame. Each leg has three DOF and comprises one roll joint, two pitch joints as shown in Fig. 41.4.

The size of the body frame is 1,000 mm × 500 mm × 50 mm (length, width and height, respectively). And the variables of the robot model are shown in Table 41.1.

In the trotting gait planning of the virtual quadruped robot, the relative velocity and acceleration between the foot and the ground are zero both at the time of departure and landing for decreasing the impact force to the ground and reducing the control complexity of the robot [8]. Based on above-mentioned assumption, the foot gait planning can be described as:

$$x(t) = S \times \left(-\frac{16}{T^3}t^3 + \frac{12}{T^2}t^2 - \frac{1}{T}t - \frac{1}{4} \right) \tag{41.1}$$

$$z(t) = H \times \left(-\frac{128}{T^3}t^3 + \frac{48}{T^2}t^2 \right) \tag{41.2}$$

Based on the gait planning and the kinematics of the quadruped robot with three DOF solutions [9], the walking gait sequence can be generated. For simplicity sake, there is no stability control strategy in the gait planning for the robot. Therefore, in order to prevent the robot falling down, the maximum foot height of the robot is 30 mm, the stride length is 60 mm and the stride period is 1 s

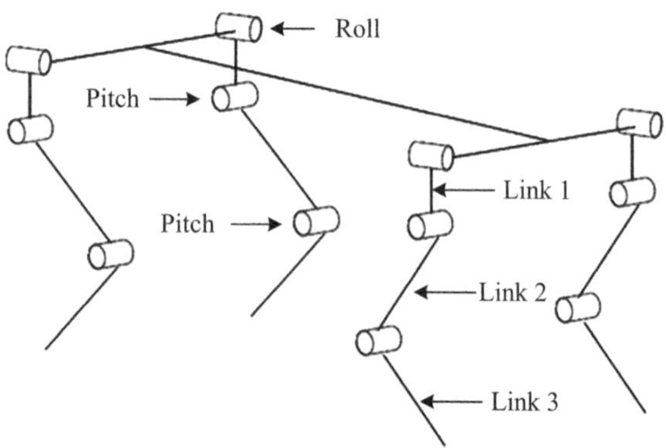

Fig. 41.4 The physical configuration of simulation model of the quadruped robot

Table 41.1 Variables of the simulation model of the quadruped robot

Links (kg)	Parameters	Unit size (mm)
Body (150)	Length × width × height	1,000 × 500 × 50
Link 1 (1)	Length × diameter	80 × 40
Link 2 (3)	Length × diameter	322 × 40
Link 3 (3)	Length × diameter	326 × 40

respectively. In addition, for easiness of simulation, the forward trotting gait is demonstrated in the walking time of 5 s. In the trotting gait of the quadruped robot, the first half cycle is used to initialize the quadruped gait. After the end of the half gait cycle, the quadruped robot enters a normal gait mode.

41.4 Simulations and Results Analysis

Numerical simulations are implemented to verify the performance of the quadruped robot with different leg configurations between the legs and the body frame. This paper has focused on determining the optimal leg configuration for the quadruped robot based on the locomotion stability and impact force between the leg and the ground.

Under the same experimental conditions, the ADAMS and MATLAB co-simulation is carried out to the virtual models of quadruped robot with four different leg configurations. Firstly, the gait planning trajectory is generated by using the software MATALB based on the inverse kinematics and the trotting gait planning approach of the quadruped robot, then, the driving of the joints of the quadruped robot with the output of MATLAB in ADAMS environment.

41.4.1 The Locomotion Stability Analysis of Quadruped Robot

Robot stability must be carefully controlled during the process of gait planning. However, there is no necessary and sufficient dynamic stability defined for higher dynamic walking of quadruped robot. Therefore, for simplicity and practical simulation limitation, the measure of the locomotion stability of the virtual model of quadruped robot is the pitching and rolling angles in this paper.

In Figs. 41.5 and 41.6, the motion curve of the body of the virtual robot on the pitching direction and the rolling direction are compared respectively. From Fig. 41.5, it can be noted that the fluctuation of the pitching angles of Posture III and Posture I is basically identical, which is better in locomotion stability than the Posture II and Posture IV. In terms of the motion curve of the rolling angles in Fig. 41.6, the Posture I is the optimal leg configuration mode in the four different leg configurations. Moreover, because of the equivalent compensation of the Posture III of the robot and the hind leg dragging problem in high speed dynamic walking [10], the Posture I with forward elbow and backward knee is the optimal leg configuration mode for the locomotion stability of the quadruped robot.

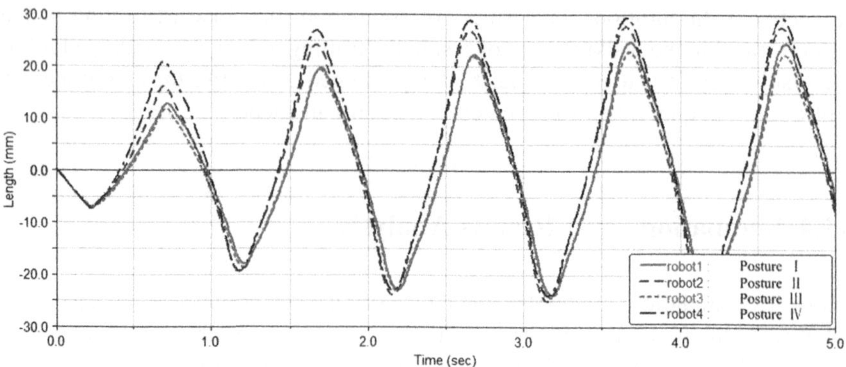

Fig. 41.5 The motion *curve* of the body for the robot on the pitching direction

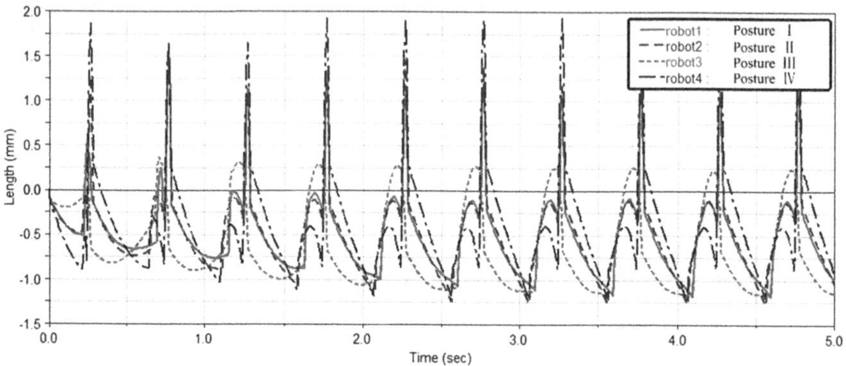

Fig. 41.6 The motion *curve* of the body for the robot on the rolling direction

41.4.2 The Impact Force Analysis of Quadruped Robot

In general, a slip occurs when the angle of the impact force is outside the friction cone of the quadruped robot. Figure 41.7 gives the forward walking curve of the body of the quadruped robot with different leg configurations. In an ideal condition without slipping, the walking distance of the robot with trotting gait is 585 mm. However, in the pure kinematics control condition, the hind-leg dragging problem of the robot occurs due to the acceleration of the body causing the body backward and the actual walking distance is less than the ideal condition. Meanwhile, the forward offset of the center of gravity of Posture III compensates for the hind-leg dragging problem, while the Posture IV aggravates the hind-leg dragging problem.

As is shown in Fig. 41.7, the Posture I of the quadruped robot has the longest forward walking distance. It is shown that the slip between the foot and the ground for Posture I is relatively less than the other leg configurations and this leg configuration mode is the optimal one for the locomotion stability of the quadruped robot.

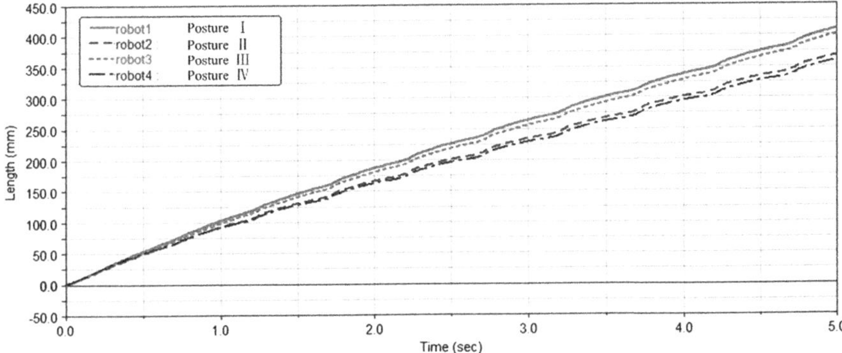

Fig. 41.7 The forward walking distance of the body of the robots with different leg configurations

The impact force between the foot and the ground is illustrated by using the acceleration of the leg of the link 3. Figures 41.8, 41.9, 41.10 and 41.11 show the acceleration curve of the RH (right hind) and RF (right front) legs on the vertical direction for the four different leg configurations of quadruped robot. It is worth noting that the vertical acceleration of the Posture I of the quadruped robot is less than the others, which is demonstrated that the motion of the Posture I has little impact force between the foot and the ground and good stability performance.

Based on the kinematics and the co-simulation of ADAMS and MATLAB, comprehensive analysis on effects of the fluctuation of the body pitching and rolling angles, the slip and the impact force between the foot and the ground, it is obvious that the forward elbow and backward knee model (Posture I) of the quadruped robot is the optimal configuration model. And the leg configuration approach of the quadruped robot Scalf using the Posture I in SUCRO (Shandong University, Center for Robotics) has being developed and as shown in Fig. 41.12.

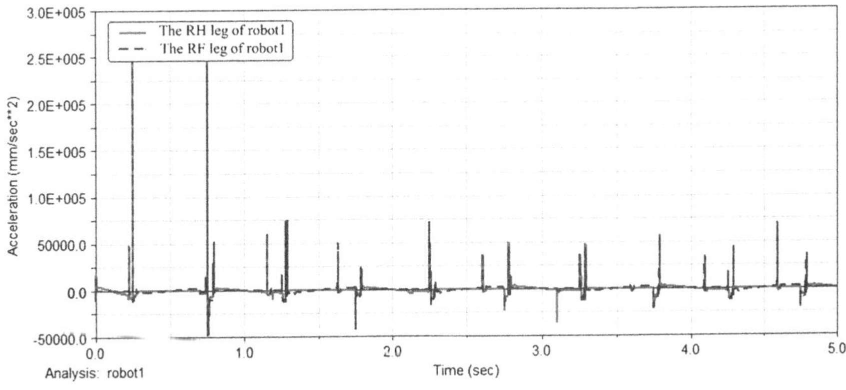

Fig. 41.8 The acceleration *curve* of the RH and RF legs for the Posture I

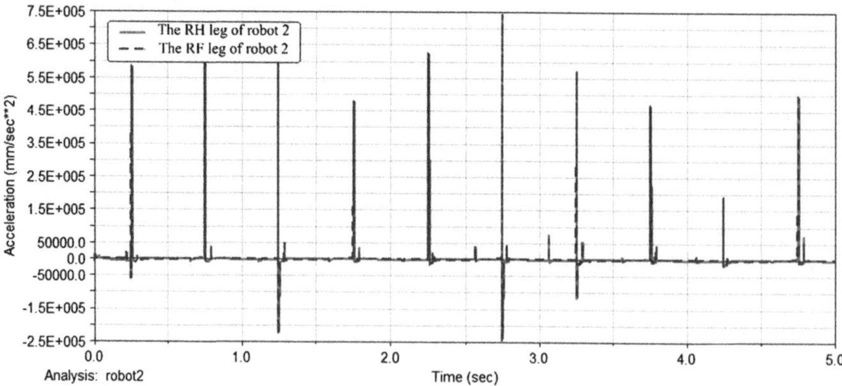

Fig. 41.9 The acceleration *curve* of the RH and RF legs for the Posture II

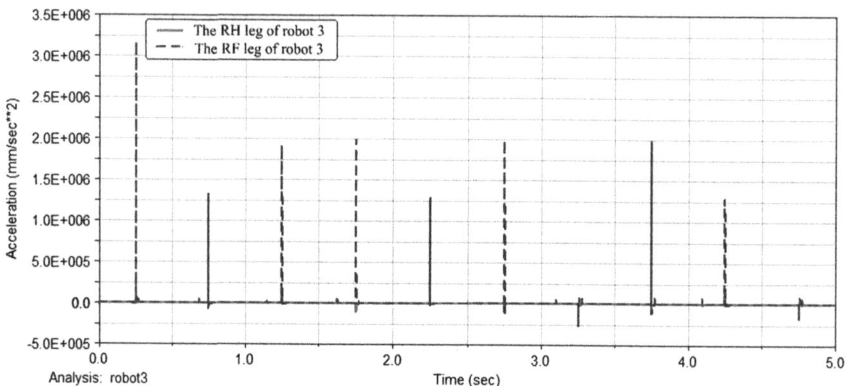

Fig. 41.10 The acceleration *curve* of the RH and RF legs for the Posture III

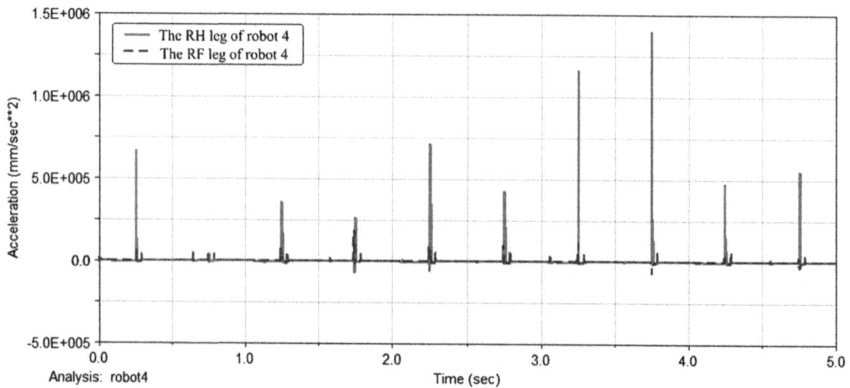

Fig. 41.11 The acceleration *curve* of the RH and RF legs for the Posture IV

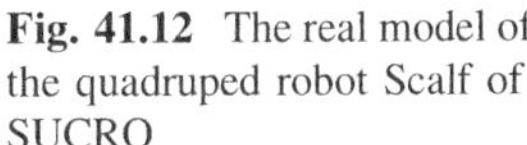

Fig. 41.12 The real model of the quadruped robot Scalf of SUCRO

41.5 Conclusion

In this paper, based on the forward and inverse kinematics solutions and the trotting gait planning approach, the virtual model of the quadruped robot is constructed using ADAMS software. Then, we thoroughly investigated different leg configuration modes for the quadruped robot and the optimal configuration of quadruped robot is verified by means of ADAMS and MATLAB co-simulation, which provide a theoretical basis for the design of leg configuration between the legs and the body frame on real quadruped robot.

Acknowledgments This paper is supported by the State Key Program of National Natural Science of China (61233014) and the Independent Innovation Foundation of Shandong University (IIFSDU) under grant 2012TS214.

References

1. Buehler M (2002) Dynamic locomotion with one, four and six-legged robots. J Robot Soc Jpn 20(3):15–20
2. Semini C (2010) HyQ- design and development of a hydraulically actuated quadruped robot. Ph.D dissertation, Italian Institute of Technology and University of Genoa
3. Li YB, Li B, Ruan JH, Rong XW (2011) Research of mammal bionic quadruped robots: a review. In: Proceedings of 5th international conference on robotics, automation and mechatronics (RAM), Qingdao, pp 166–171
4. Jones MS, Hurst JW (2012) Effects of leg configuration on running and walking robots. In: Proceedings of the 5th international conference on climbing and walking robots and the support technologies for mobile machines, Baltimore, pp 519–526

5. Meek S, Kim J, Anderson M (2008) Stability of a trotting quadruped robot with passive, underactuated legs. The IEEE international conference on robotics and automation, Pasadena, 347–351
6. Todd DJ (1985) Walking machines: an introduction to legged robots. Kogan Page, London
7. Boston Dynamics, BigDog Overview.pdf [EB/OL] (2010) http://www.bostondynamics.com/img/BigDog_Overview.pdf
8. Rong XW, Li YB, Ruan JH, Li B (2012) Design and simulation for a hydraulic actuated quadruped robot. J Mech Sci Technol 26(4):1171–1177
9. Li B, Li YB, Rong XW, Meng J (2011) Trotting gait planning and implementation for a little quadruped robot. Lecture Notes in Electrical Engineering, 97 LNEE, vol 1. Springer, Berlin Heidelberg, pp 195–202
10. Zhang XL, Zeng XY, Zheng HJ (2011) Resolution of the hind leg dragging problem of a quadruped robot in high-speed dynamic walking. High Technol Lett 21(4):404–410 (in Chinese)

Chapter 42
Accuracy Improvement of Ship's Inertial System by Deflections of the Vertical Based Gravity Gradiometer

Jihang Jin, Shengquan Li and Guobin Chang

Abstract The error of high-accuracy Inertial Navigation System (INS) due to the gravity disturbance can't afford to be neglected. As gradiometer aided INS has been used as a resultful method for improving accuracy of underwater navigation, a novel method compensating the vertical deflections by gradiometer is introduced. In this paper, the architecture of compensation deflections of vertical by gradiometer is firstly analyzed. Then Taking pure INS as an example, an error dynamics equation of INS, including the influence of gravity disturbance, is given and the position error due to deflections of vertical and corresponding characteristics of error propagation are analyzed. Finally, simulation is done on deflections of vertical database, and form the simulation result we can see that the horizontal error of INS due to the gravity disturbance on the sailing course can reach as large as one kilometer.

Keywords Gradiometer · Inertial system · Compensation · Aided navigation

42.1 Introduction

Gravity uncertainties are an prominent source of error in all inertial navigation systems and are particularly important in modern, high quality equipment. One promising approach for mitigating gravity induced navigation errors is to measure the earth's gravity gradient field with a gravity gradiometer and use these measurements to reduce the gravity uncertainty and its errors. Recent advances in gradiometer technology indicate that real-time gradiometer compensation of inertial systems for deflections of the vertical and gravity anomalies may be possible [1, 2].

J. Jin (✉) · S. Li · G. Chang
Institute of Hydrographic Surveying and Charting, No. 40 YouYi Road, HeXi, Tianjin City, China
e-mail: jihjin@126.com

Z. Sun and Z. Deng (eds.), *Proceedings of 2013 Chinese Intelligent Automation Conference*, Lecture Notes in Electrical Engineering 255,
DOI: 10.1007/978-3-642-38460-8_42,

A gravity gradiometer development program has been underway for some time at some companies, as Bell Aerospace and Lockheed-Martin Federal System Inc [3]. An instrument of another type, FTG has been development [4]. The gravity measured systems GSS developed by Bell Aerospace is prominent [5], which is used aided inertial system to support real-time vertical deflection, estimate and improve Schuler errors and platform errors in inertial systems.

In principle, real-time deflections and anomalies can be found by continuous integration of measured gravity gradients. However, straightforward integration of the gradients transforms gradiometer errors into growing errors in the recovered gravity vector. From a systems analysis viewpoint, mechanization of the gradiometer is essentially immaterial and the instrument can be considered a black box that measures an certain physical quantity to within a certain accuracy [6, 7].

In this paper, the gravity gradiometer is only treated a way to support the real-time vertical deflection data, the key is analyses errors induced the gravity disturbance. As we all know, vehicle's inertial navigation errors will increase with time going on [8], a causation is we use reference gravity model in inertial system, neglect the gravity disturbance. The error state equation of inertial system including normal gravity model and gravity disturbance is introduced and later, the computer program for solving the error equations is described, followed by the results.

42.2 Principle Overview

In an inertial navigation system, an analytic (smooth) reference shape is universally chosen which closely approximates the geoid [9, 10]. The reference shape chosen is an ellipsoid. The gravity field of the Earth and its effect upon navigation systems can be defined in terms of differences between the geoid and reference ellipsoid (Fig. 42.1) [11]. The "gravity anomaly", Δg, is defined as the difference between the magnitude of the true gravity vector at a point on the geoid, and the magnitude of the reference gravity vector at a corresponding point on the ellipsoid. The "deflection of the vertical" is defined as the angular difference between the true vertical (perpendicular to the geoid) and the reference vertical (perpendicular to the reference ellipsoid). This angular difference is broken up into two components, a north–south component of ξ and an east–west component of η. The uncertainty of gravity, $\delta\bar{g}$ is defined to be the difference between the true gravity vector g, and the reference gravity vector $\bar{g}_0$ at the same point. This vector is normally expressed in north, east, down coordinates as [12].

$$\delta\bar{g} = \begin{bmatrix} -g_0\xi \\ -g_0\eta \\ \Delta g \end{bmatrix} \tag{42.1}$$

where g_0 is the nominal (reference) value of gravity.

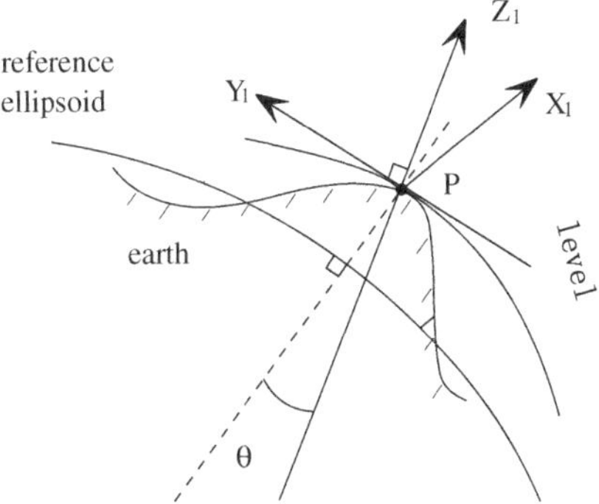

Fig. 42.1 Deflections of the vertical

As we all know, the gravity gradient tensor correlates to gravity disturbance, it is definition that gravity disturbance vector T_i and gravity gradient tensor T_{ij} in navigation frame, here [13]

$$T_i = \begin{bmatrix} \frac{\partial T}{\partial x} & \frac{\partial T}{\partial y} & \frac{\partial T}{\partial z} \end{bmatrix}^T = \begin{bmatrix} -g_0\xi \\ -g_0\eta \\ \Delta g \end{bmatrix} \tag{42.2}$$

$$T_{ij} = \begin{bmatrix} T_{xx} & T_{xy} & T_{xz} \\ T_{yx} & T_{yy} & T_{yz} \\ T_{zx} & T_{zy} & T_{zz} \end{bmatrix} = \begin{bmatrix} \frac{\partial^2 T}{\partial x^2} & \frac{\partial^2 T}{\partial x \partial y} & \frac{\partial^2 T}{\partial x \partial z} \\ \frac{\partial^2 T}{\partial y \partial x} & \frac{\partial^2 T}{\partial y^2} & \frac{\partial^2 T}{\partial y \partial z} \\ \frac{\partial^2 T}{\partial z \partial x} & \frac{\partial^2 T}{\partial z \partial y} & \frac{\partial^2 T}{\partial z^2} \end{bmatrix} \tag{42.3}$$

The gravity disturbance of three direction is output by straightforward integration of the gravity gradiometer measured (T_{xx}, T_{yy}, T_{zz}), but it transforms gradiometer errors into gravity vector [14]. In this paper, we get real vertical deflection data to substitute measured data supported by gradiometer (Fig. 42.2). It is convenience to study the propagation of the errors induced by gravity disturbance.

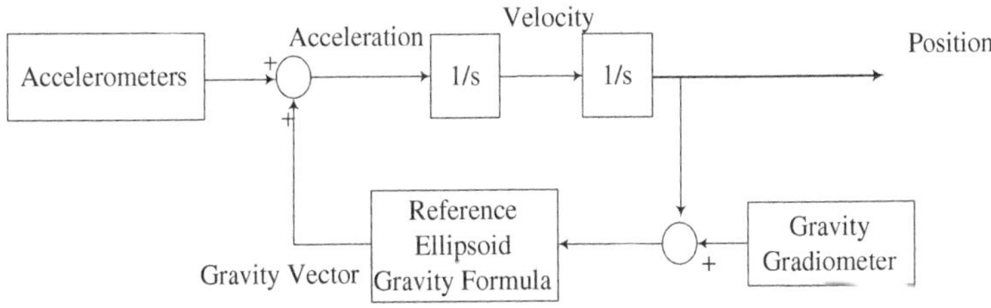

Fig. 42.2 Gradiometer aided INS

42.3 Navigation Error Equations

Compensation algorithm in this paper takes into account the global (log-wavelength) characteristics of the earth's gravity field and the gradiometer measurements of the local (short-wavelength) gravity field. We take Bi-Quadratic Polynomial Interpolation on the vertical deflection, and it is facility to observe the value on the track.

According the Newton second law, the navigation equation in local-level is

$$f = \dot{V} + (2\Omega_{ie} + \omega_e) \times V - g \tag{42.4}$$

where $(2\Omega_{ie} + \omega_e) \times V$ is Coriolis acceleration, $\dot{V}, V$ is acceleration vector and velocity vector of vehicle, f is the output of the accelerometer, g is gravity vector. Equations (42.1) and (42.4) yield the navigation error propagation equation

$$\Delta\dot{\mathbf{X}}(t) = \mathbf{F}(t) \cdot \Delta\mathbf{X}(t) + \mathbf{G}(k) \cdot \mathbf{W}(t) \tag{42.5}$$

where $\Delta X = \left[\delta V_E,\ \delta V_N,\ \delta V_U,\ \delta\phi,\ \delta\lambda,\ \delta h,\ \varphi_x,\ \varphi_y,\ \varphi_z\right]^T$ is the error vector of INS, $W(t) = [g_E(t)\quad g_N(t)\quad 0\quad 0\quad 0\quad 0\quad 0]^T$ is gravity disturbance vector, $\mathrm{G}(k)$ is the system control matrix, $\mathrm{F}(t)$ is the system matrix; $\delta V_E,\ \delta V_N$ and δV_U denote the velocity error of measurement instrument in east, north and down, $\delta\phi,\ \delta\lambda, \delta h$ are latitude error, longitude error and height error, $\varphi_x,\ \varphi_y,\ \varphi_z$ are roll error, pitch error and heading error respectively. See [8] for the exact expression of these variables. Note that the state navigation error equation (42.5) is continuous, and a discretization transform should be taken to it by the step of $t = kT_0 \quad (k = 1, 2, 3\ldots)$, the result is

$$X[(k+1)T_0] = A(kT_0)X(kT_0) + W_d(kT_0) \tag{42.6}$$

where $A(kT_0) = e^{F(kT_0)T_0}$ is the state transform matrix from $X(kT_0)$ to $X[(k+1)T_0]$, $W_d(kT_0)$ is power of the error source. Then the INS position errors can be estimated on navigation error equation by iterative calculation.

42.4 Simulation

In this section we apply the real vertical data which is calculated by the geoid data downloaded from NOAA web to the aided navigation to estimate the INS errors. Simulation I is done on the area in $N21^\circ \sim N24^\circ$, $E120^\circ \sim E123^\circ$, the initial position of vehicle is $(21.5^\circ, 120.9^\circ)$with none initial errors. The assumptions for simulations are as Table 42.1 is shown.

Before measurements, we take the initial state vector $\mathbf{X}(0) = zero(9, 1)$.

Figure 42.3 shows the 32-hour aided navigation results. The estimated error due to north–south component is a line and north–south gravity component a dashed line. When the gravity component grows with linearity, the position errors also

Table 42.1 Assumptions for simulations

	I	II	III
Vehicle velocity (nm/h)	5	4.5	13.6
Vehicle heading (deg)	6	91	91
Acceleration (nm/h^2)	East: 0.035 North: −0.035	None	None
Additional error sources	None	None	None

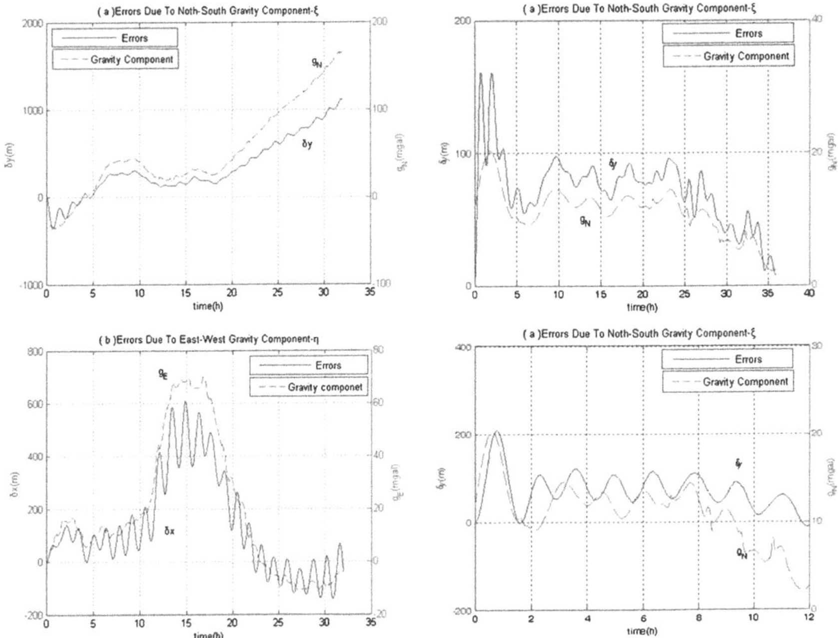

Fig. 42.3 Errors due to deflections of the vertical in simulation I (*left*), II and III (*right*) by different velocity

grow fast, and it is bound that the deflection of vertical grows gently. Here in Table 42.2, The max INS horizontal error caused by north–south component ξ reach to 1112 m.

Simulation II and III are done on the area in N20° ~ 24°, E201° ~ 205°, the characters of simulation is described on Table 42.1. The initial position of vehicle is $(23.5°, 201.3°)$ and two simulations are done on the same track, only the velocity is different between two simulations. we can see when the velocity gets 13.6 nm/h, the gravity component changing frequency is closed to Schuler frequency, and the INS errors then grows discretely.

Table 42.2 Error due to deflections of the vertical

	$\|\Delta g_E\|$ (mgal)	$\|\delta x\|$ (m)	$\|\Delta g_N\|$ (mgal)	$\|\delta y\|$ (m)
Max value	70	608	170	**1112**

42.5 Conclusion

Gravity disturbance is an important error source of high quality inertial system, especially in the area the vertical deflection varies greatly, the errors due to gravity disturbance are not to be neglected.

The influence of gravity disturbance is especially significant, if the INS is non-damped or no other external aided information, the distributional forms of the gravity disturbance on the track are a dominant cause, it is a feasible method to trail off errors due to gravity disturbance. Additionally, cutting off the time of vehicle running and changing the vehicle' velocity is another method to bate the increase errors, it can also weaken the increase of system errors with time on.

A method to compensate the gravity disturbance is to put it in navigation calculation, it need to prestore gravity data. This method has some limitless and it is restricted by time and area. We can also estimate the position error and revise INS based real gravity data supported by gravity database on track, it is also can be supported by gradiometer measured, it can improve position precision commendably.

References

1. GuoY G, Zhong B, Bian SF (2003) The determination of Earth gravity field and the matched navigation in gravity field. Hydrogr Surv Charting 23(5):61–64 (in Chinese)
2. Affleck CA, Jircitano A (1990) Passive gravity gradiometer navigation system. In: Position location and navigation symposium, 1990. Record. The 1990′s-A decade of excellence in the navigation sciences. IEEE PLANS'90. IEEE, 60–66
3. DeWall J et al (2001) Ship augmented gravity enhancement (SAGE). In: Proceedings of IEEE/ION PLANS 2006, 36–43
4. Hammond S, Murphy C (2003) Air-FTG™: bell Geospace's airborne gravity gradiometer–a descripton and case study. ASEG Preview 105:24–26
5. Martin L (2007) Advances and challenges in the development and deployment of gravity gradiometer systems. EGM 2007 international workshop
6. Grubin C (1975) Accuracy improvement in a gravity gradiometer-aided cruise inertial navigator subjected to deflections of the vertical. In: American Institute of Aeronautics and Astronautics, Guidance and Control Conference, Boston, Mass, 1975
7. Heller W, Jordan S (1975) Mechanization and error equations for two new gradiometer-aided inertial navigation system configurations. In: AIAA, Guidance and control conference
8. Jekeli C (2000) Inertial navigation systems with geodetic applications. de Gruyter, Berlin/New York
9. Hsu DY (1998) An accurate and efficient approximation to the normal gravity. In: Position location and navigation symposium, IEEE 1998, 38–44

10. Moryl J et al (1998) The universal gravity module for enhanced submarine navigation. In: Position location and navigation symposium, IEEE 1998, 324–331.
11. Grejner-Brzezinskaa DA et al (2001) On improved gravity modeling supporting direct dereferencing. IEEE
12. Mandour I, El-Dakiky M (1988) Inertial navigation system synthesis approach and gravity-induced error sensitivity. Aerosp Electron Syst, IEEE Trans 24:40–50
13. Xu ZY et al (2007) Situation and development of marine gravity aided navigation system. Prog Geoph 22:104–111 (in Chinese)
14. Eissfeller B, Spietz P (1989) Shaping filter design for the anomalous gravity field by means of spectral factorization. Manuscripta Geodaetica 14:183–192

Chapter 43 Probability Distribution of the Monthly Passenger Flow from Chongqing to Yichang and Monte Carlo Simulation

Peng Wan, Shoujiang Zhao, Jian Li and Ni Zhan

Abstract Based on the data of each section's monthly passenger flow from 2006 to 2009, this paper studied the probability distribution of passenger flow from Chongqing to Yichang by use of mathematical analysis methods. By the chi-square testing, the monthly passenger flow of Chongqing and Three Gorges are accord with the normal distribution, while Yichang's is accord with the exponential distribution. The monthly passenger flow from Chongqing to Yichang (the sum of the monthly passenger flow of Chongqing, Three Gorges and Yichang) is close to normal distribution which is approved by monte carlo method.

Keywords Passenger flow · Monte carlo · Chi square test · Exponential distribution · Normal distribution

43.1 Introduction

With the completion of the three gorges dam, fleet of ten thousand tons can reach Chongqing directly, the transportation industry of Three Gorges has a great development and transport capacity also increases heavily. Meanwhile, splendid scene appears in the Three Gorges [1], which cause the tourism boom and the increase of passenger volume. However, traffic accidents have occurred frequently as a result of the complicated waters. In order to strengthen water safety management and improve the safety management level of ferry crossing and ferryboat [2], based on the data of each section's monthly passenger flow from 2006 to 2009 counted by Yangtze River shipping administration, this paper dealed with the

P. Wan (✉) · S. Zhao · J. Li · N. Zhan
School of Science, China Three Gorges University,
8 University Avenue, Yichang City, China
e-mail: shjzhao@yahoo.com.cn

Z. Sun and Z. Deng (eds.), *Proceedings of 2013 Chinese Intelligent Automation Conference*, Lecture Notes in Electrical Engineering 255,
DOI: 10.1007/978-3-642-38460-8_43,

Table 43.1 Statistics table for the monthly passenger volume data of chongqing

Group(ten thousand)	100–130	130–160	160–190	190–220	220–250	250–280
Group mark	115	145	175	205	235	265
Frequency number	4	12	3	2	1	1
Frequency	0.1739	0.5217	0.1304	0.0870	0.0435	0.0435

passenger flow condition of this water area and obtained the distribution function of the monthly passenger flow from Chongqing to Yichang, by which theoretical basis was provided for related administration.

43.2 Date Processing

With the sample of each section's monthly passenger flow data from 2006 to 2009 counted by the Yangtze River shipping administration, this paper ranges the possible value of passenger flow of Chongqing, Yichang and Three Gorges into groups, each group's mid-value is used as group mark, then we count the frequency, the result is in the following Tables 43.1, 43.2 and 43.3:

The frequency histograms of monthly passenger flow of Chongqing, Yichang and Three Gorges are as follows (Figs. 43.1, 43.2, 43.3).

43.3 Analysis of the Probability Distribution

From the above frequency histograms, it is seen that they probably correspond the exponential distribution or the normal distribution. We first assume that they correspond with a certain distribution, then, we use chi-square test to verify whether they correspond with the assumed distribution. The density function of exponential distribution is:

Table 43.2 Statistics table for the monthly passenger volume data of yichang

Group(ten thousand)	37–48	48–59	59–70	70–92	92–125
Group mark	42.5	53.5	64.5	81	108.5
Frequency number	16	4	1	1	1
Frequency	0.6957	0.1739	0.0435	0.0435	0.0435

Table 43.3 Statistics table for the monthly passenger volume data of three gorges

Group(ten thousand)	24–33	33–42	42–51	51–69	69–96
Group mark	28.5	37.5	46.5	60	82.5
Frequency number	2	8	8	4	1
Frequency	0.0870	0.3478	0.3478	0.0652	0.0435

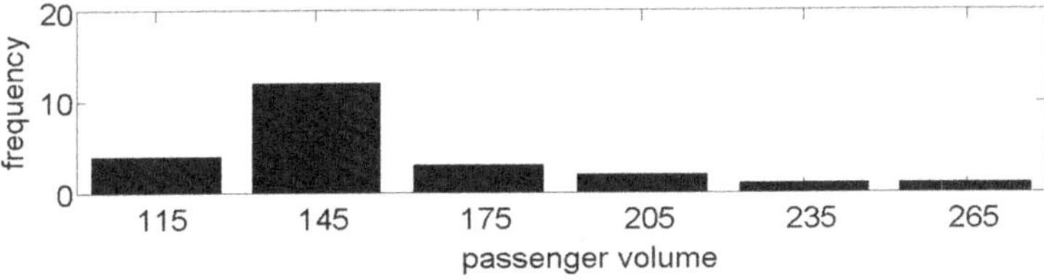

Fig. 43.1 The frequency histograms of monthly passenger flow of chongqing

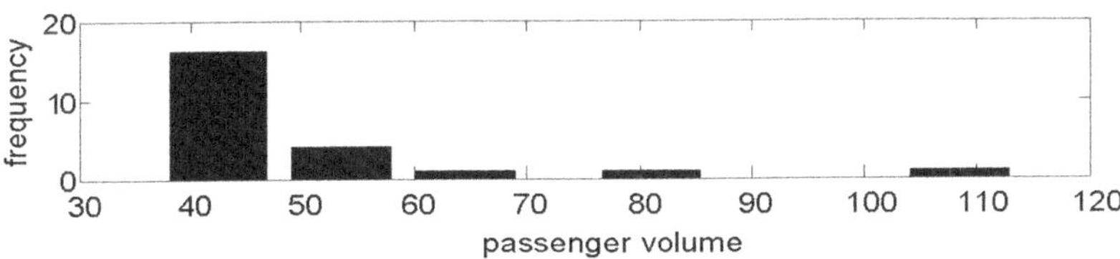

Fig. 43.2 The frequency histograms of monthly passenger flow of yichang

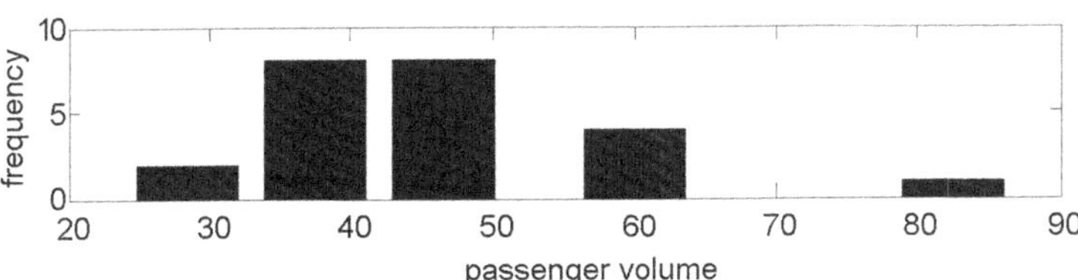

Fig. 43.3 The frequency histograms of monthly passenger flow of three gorges

$$f(x) = \begin{cases} \lambda e^{-\lambda x}, & x \geq 0 \\ 0, & x < 0 \end{cases} \tag{43.1}$$

The density function of normal distribution is:

$$f(x) = \frac{1}{\sqrt{2\pi}\sigma} e^{-\frac{(x-\mu)^2}{2\sigma^2}} \tag{43.2}$$

where μ means average value and σ^2 means variance.

In Statistics, there are many test methods for distribution functions, while the chi-square test is the most commonly used [3–5]. The formula for this test is

$$\chi^2 = \sum_{k=1}^{r} \frac{(n_k - np_k)^2}{np_k}$$

Table 43.4 The chi-square test of the probability distribution for monthly passenger flow of chongqing

Group (ten thousand)	Group mark (x_i)	Measured frequency (n_k)	Theoretical frequency (np_k)	$n_k - np_k$	$\frac{(n_k-np_k)^2}{np_k}$
100–130	115	4	3.82	0.18	0.00080
130–160	145	12	6.77	5.23	4.0326
160–190	175	3	6.51	−3.51	1.8916
190–220	205	2	3.39	−1.39	0.5721
220–250	235	1	0.96	0.04	0.0018
250–280	265	1	0.15	0.085	4.9805
Subtotal		23			11.4866

where r is for group number, n_k is for the measured frequency, np_k is for the theoretical frequency; n is for the total sample.

In the chi-square distribution, parameters are only related to the significant level α and the degree of freedom R, while the parameters of R is related to the following variables: $R = r - s - 1$, where s is the number of estimated parameters for the statistical distribution functions being used, r is for packet number. According to Pearson theorem, for the significant level, if the statistic χ^2 is less than $\chi^2_{R,1-\alpha}$, then we accept hypothesis test, otherwise refuse.

For Chongqing's monthly passenger flow, we wish to see if the data deviate from a normal distribution $N(\mu, \sigma^2)$, so we get best estimate of μ and σ are as follows: $\mu = 158.0413$, $\sigma = 37.3508$ by maximum likelihood method. To calculate the theoretical probability of each interval, we get the following Table 43.4:

If α takes the value of 0.005, according to the table, we get:

$$11.4866 = \chi^2 < \chi^2_{3,1-0.005} = 12.838$$

Thus we accept H_0 : Chongqing's passenger flow obey: $N(158.0413, 37.3508^2)$

For monthly passenger flow of Yichang, we suspect the data obey the following exponential distribution:

$$f(x) = \begin{cases} 0.02e^{-0.02(x-37)}, & x \geq 37 \\ 0, & x < 37 \end{cases} \tag{43.3}$$

The observed value of chi-square can be calculated as follows (Table 43.5): when the test is performed at $\alpha = 0.05$, we get:

$$\chi^2 = 3.6673 < 7.815 = \chi^2_{3,0.95}$$

Thus we accept H_0 : Yichang's passenger flow obey exponential distribution.

For the monthly passenger flow of three gorges, we obtain (Table 43.6)

If α takes the value of 0.05, according to the table, we get:

$$\chi^2 = 3.1723 < 7.815 = \chi^2_{3,0.95}$$

Table 43.5 The chi-square test of the probability distribution for monthly passenger flow of Yichang

Interval (ten thousand)	Group mark (x_i)	Observed frequency (n_k)	Expected frequency (np_k)	$n_k - np_k$	$\frac{(n_k-np_k)^2}{np_k}$
37–48	42.5	16	13.19	2.81	0.5998
48–59	53.5	4	5.63	−1.63	0.4700
59–70	64.5	1	2.40	−1.40	0.8169
70–92	81	1	1.46	−0.46	0.1454
92–125	108.5	1	0.30	0.70	1.6352
Subtotal		23			3.6673

Table 43.6 The chi-square test of the probability distribution for monthly passenger flow of three gorges

Interval (ten thousand)	Group mark (x_i)	Observed frequency (n_k)	Expected frequency (np_k)	$n_k - np_k$	$\frac{(n_k-np_k)^2}{np_k}$
24–33	28.5	2	2.52	−0.52	0.1057
33–42	37.5	8	5.38	2.62	1.2781
42–51	46.5	8	6.72	1.28	0.2459
51–69	60	4	6.98	−2.98	1.2754
69–96	82.5	1	0.60	0.40	0.2673
Subtotal		23			3.1723

Then we accept H_0 : Three Gorges' passenger flow obey normal distribution $N(45.7147, 11.9908^2)$.

43.4 Generating Random Numbers Simulation

Based on the density function of probability distribution of this three areas' monthly passenger flow, this study generate 1 million random numbers with the monte carlo method [6–8], and count the passenger volume of Chongqing to Yichang, namely, the sum of this three random numbers. After that, we divide them into groups to get the distribution table of the passenger flow from Chongqing to Yichang.

The simulation results are shown in the table below (Table 43.7):

Fitting the upper data with the Normal distribution whose average value $\mu_0 = 216.6753$ and variance $\sigma_0 = 41.2929$, we acquire the results as the following graph (Fig. 43.4):

Table 43.7 The simulation results

Groups	Frequency number	Frequency
100–125	12166	0.012169
125–150	39297	0.038887
150–175	104234	0.104682
175–200	189680	0.190076
200–225	238520	0.238075
225–250	208465	0.208631
250–275	128055	0.127643
275–300	56240	0.056480
300–325	18027	0.017982
325–350	5316	0.005375

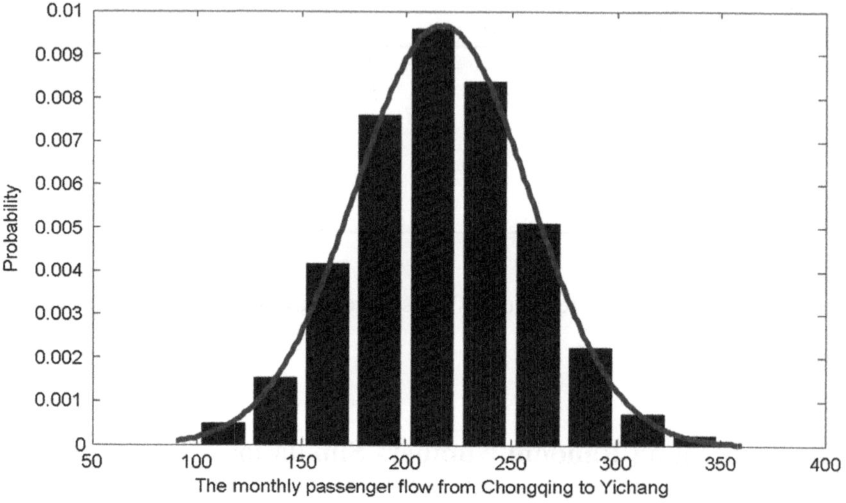

Fig. 43.4 Probability graph of monthly passenger flow from chongqing to yichang

43.5 Conclusion

In this paper, we deal with the data counted by the Yangtze River shipping administration, and study the monthly passenger flow from Chongqing to Yichang using the Mathematical Statistics. The results show that the passenger flow from Chongqing to Yichang is close to normal distribution, which provide theoretical basis for further study in shipping problems.

Acknowledgments This research supported by the National Natural Science Foundation of China (No. 11226202).

References

1. Li X (2011) Analysis on monthly passenger flow from Chongqing to Yichang section based on monte carlo method. J Wuhan Inst Shipbuild Technol 10(4):14–16 (in Chinese)
2. Yu J (2006) The characteristics of the probability distribution of xijiang waterway vessel flow. J Transp Eng 6:88–93 (in Chinese)
3. Zhang W, Liao P (2004) Analysis on characteristics of ship arrival at Shiqiao lock. Hydro-Sci Eng 26(1):70–73 (in Chinese)
4. Zhang W, Liao P, Wu L (2004) Main parameters of waterway lock capacity. J Traffic Transp Eng 4(3):108–110 (in Chinese)
5. Carroll J, Bronzini M (1973) Waterway transportation simulation models: Development and application. Waterw Resour Res 9(1):51–63
6. Wang H, Guo XC (2000) The book of the traffic engineering. Southeast University Press, Nanjing (in Chinese)
7. Wei ZS (2006) Probability theory and mathematical statistics. Higher Education Press, Beijing (in Chinese)
8. Chen H (2007) Matlab and its application and guidelines in science and engineering course. Xidian University Press, Xian (in Chinese)

Chapter 44
A Design of Bus Automatic Broadcast Station System Based on RFID

Yuping Su and Weixin Yang

Abstract In view of current broadcast station problems existed in the bus, bus automatic voice broadcast station system based on RFID is designed in this paper. The system can realize function of automatic broadcast station by combining single chip microcomputer technology with digital speech and wireless communication technology. The overall scheme, hardware and software design of the system are given in the paper, the system can lower the driver's labor intensity, increase bus security, it has lower cost and higher practical value. The system is suitable for the development of intelligent traffic and the construction of intelligent city.

Keywords Automatic voice broadcast · RFID · Digital speech · Single chip microcomputer

44.1 Introduction

Bus automatic voice broadcast station plays an important role in bus business, which affects the bus service quality directly. At present, there are two main ways in the field of bus automatic broadcast station, one is the traditional voice broadcast station which is manual controlled by the driver, the other is the use of global satellite positioning system in broadcast station system. In the traditional manual broadcast station system, station identification, voice playback depend on

Y. Su (✉) · W. Yang
College of Electrical and Engineering, Northwest University for Nationalities, Lanzhou, China
e-mail: syp_abc@163.com

W. Yang
e-mail: zhywx369@163.com

Z. Sun and Z. Deng (eds.), *Proceedings of 2013 Chinese Intelligent Automation Conference*, Lecture Notes in Electrical Engineering 255,
DOI: 10.1007/978-3-642-38460-8_44,

the driver's observation and judgment, it not only increase the driver's operation difficulty, but also there are some potential hazards. GPS has been put into use in part of city in the United States, the kind of product is also been researched and developed in domestic market. Though the technology can relieve driver operation on station identification and broadcast because of its strong function and good stability, its investment price is very expensive [1], the small and medium-sized city can not afford it. So the system can't suit the current development of bus broadcast station system.

Radio frequency identification technology is a new technology which is developed rapidly in recent years, it uses radio frequency to recognize non-contact object automatically [2, 3]. The system adopts RFID technology to realize the automatic voice broadcast station in the paper, it not only improves the bus safety coefficient but also meet the requirement for low cost, it could be applied to the current urban buses.

44.2 Hardware Structure of the System

44.2.1 System Hardware Circuit Design

The whole broadcast station system is composed of bus stop board and bus carrier [4]. Bus stop board is made up of SCM and wireless launcher. Bus carrier mainly consists of the SCM controller, wireless receive machine, a voice chip and the power amplifier circuit, system diagram is shown in Fig 44.1.

Design ideas: install a particular Launcher in bus stop board, the Launcher send signals continuously, when the bus moves at a certain distance from bus stop board, the receive machine in the bus can receive signals that come from the Launcher, and send the coding information to SCM, then SCM recognize it according to former information storaged in the database send corresponding voice message address to voice chip. Then voice chip broadcast site's voice information that is pre-recorded in voice memory module. And voice is amplified by an amplifying circuit in order to drive a loudspeaker, thus automatic station broadcast is completed. At the same time, the LCD display can also display the name of destination station.

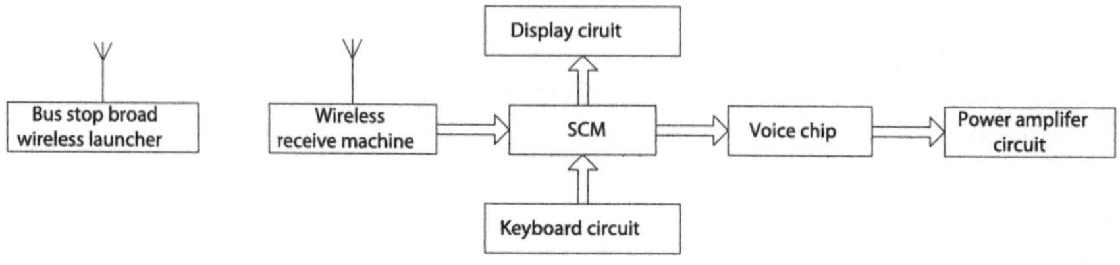

Fig. 44.1 System diagram

44.2.2 Unit Circuit Design

44.2.2.1 Transmitting Circuit

Wireless launcher in bus stop board regard PT2262 as the core of the remote control transmitter module, transmitting module can be installed in advertising light box with different codes in every station, send out specific RFID identification signal regularly. PT2262 has a maximum of 12 bit three state address bus, can provide 5314413(3^{12}) address coding [5], avoiding address conflict maximally.

44.2.2.2 Receiving Circuit

Wireless receive machine is made up of the remote control receiver module which regard PT2272 as core, broadcast station system and Wireless receive machine are integrated together, installing in the bus, when the bus is close to the station. Wireless receiver machine will receive a RFID signal that is transmitted by transmitting module and decode it, send the decoded digital signal to the SCM.

44.2.2.3 Main Control Module

Control core of the design is STC89C52RC which is one of primary components of the system. It control the emission of signal in transmitting circuit and take charge of speech signal recording, signal storage and playback, respond to disruptions that caused by press key in receiving ciruit. P0 port of STC89C52RC is linked to ISD4004 voice system, P1 port is linked to PT2272 decode module, P2 port is connected with the liquid crystal display, P3 port is connected with the keyboard.

44.2.2.4 Voice Recording-Playing Module

ISD4004 is a new kind of voice chip that is manufactured by ISD company in the United States. Recording time is about 8–16 min, the sound quality is good. Its working voltage is 3 V [6]. We can convert 5 V into 3 V by transformation circuit.

The ISD4004 chip SS (select signal line), MOSI (serial input), SCLK (clock input) are connected to the P0.0, P0.1, P0.2 port of SCM separately. ISD4004 and SCM interface circuit and the peripheral circuit is as shown in Fig 44.2. MIK is microphone and is used to record the speech, ordinary live recording can be completed by it.

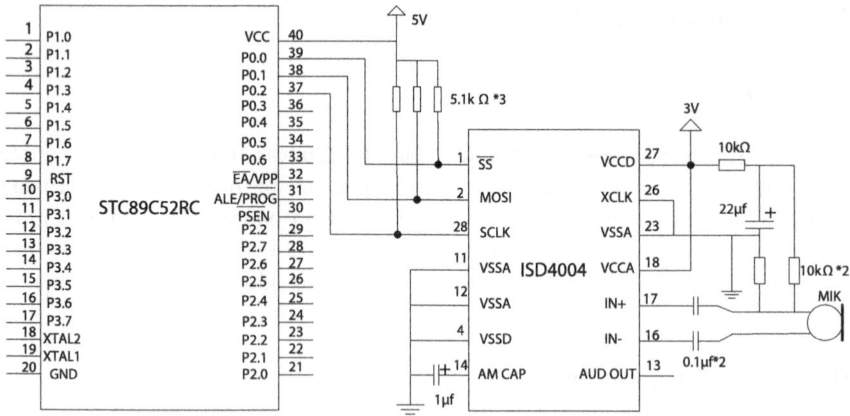

Fig. 44.2 Hard circuit of voice recording-playing module

44.2.2.5 Power Amplifier Circuit

Pin audout in ISD4004 can be connected to the loudspeaker directly [7], speech signal can be output directly through the speaker, but driving power is small relatively though such connection, taking into account the larger volume demand in the bus, low-frequency power amplification chip LM386 as the loudspeaker driving circuit can be connected between the 1SD4004 chip and the loudspeaker. Power amplifier circuit is shown in Fig 44.3. 10 μf bypass capacitor is linked between 1 pin and 8 pin, amplification factor can be Multiplied 200 times in the amplification circuit [8, 9].

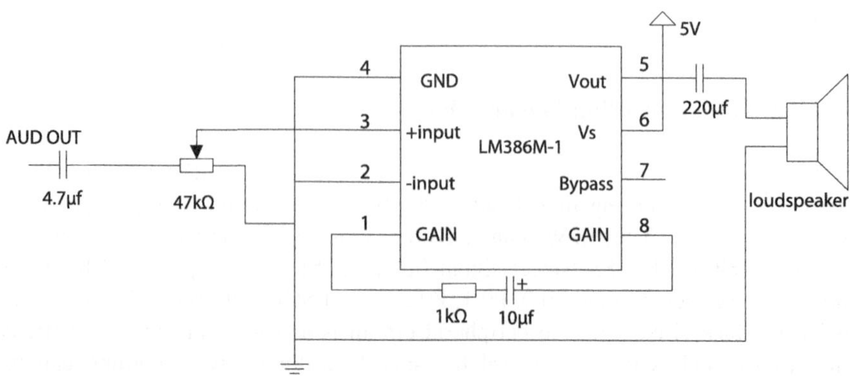

Fig. 44.3 Power amplifier circuit

44.2.2.6 Keyboard and Display Module

The bus broadcast system also retains a manual control function, when the station information broadcast is abnormal, the bus driver can adjust station information through the operation of keys on the keyboard. Display module adopts 128*64 liquid crystal display that is used to realize synchronous display of related sites when broadcast speech signals.

44.3 System Software Design

The software design program divides mainly into five parts, the system main program, transmitting program, receiving program, press key program and display program. The following part describe the main program, transmitting program, receiving program in detail.

44.3.1 System Main Program

SCM detect the emission signal constantly, once the signal is detected receiving circuit board reads data according to the received data coding, judges the information at the site, and then sends the voice message address to the voice chip, playing the prerecorded voice information. The main program complete the initialization of the system in no-signal situation. The main program flow chart is shown in Fig. 44.4

44.3.2 Transmitting Program

The transmitting part mainly produces a periodic coding signal. Coding signal includes the start mark, data coding and end mark, because the wireless transmit-receive module only receive or transmit half a byte of data every time. A frame data is at least 16 bits message format, needs four times to complete transformation. In the actual application, we can transmit signal discontinuous every three seconds or five seconds. Therefore the program also need to have a second or five seconds delay. The system transmitting program diagram is illustrated in Fig. 44.5.

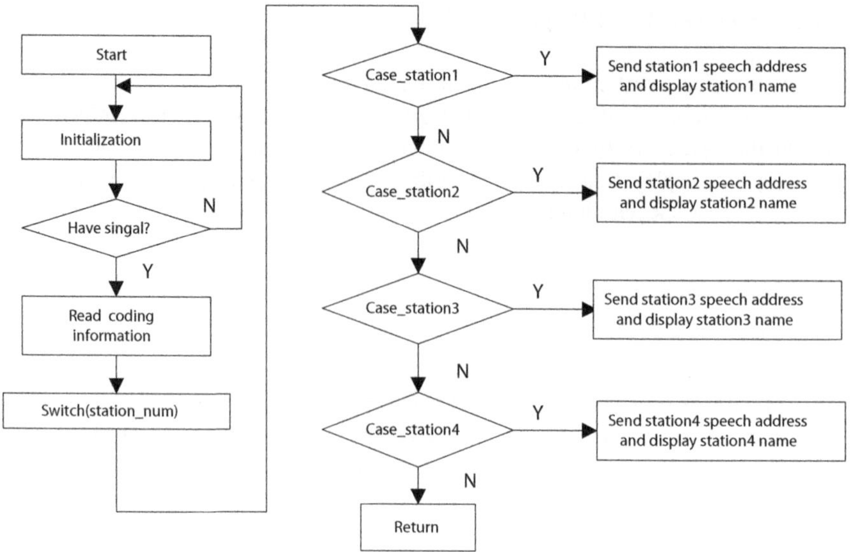

Fig. 44.4 System main program diagram

Fig. 44.5 Transmitting program diagram

44.3.3 Receiving Program

VT port of receiving plate is connected to a port of the SCM in hardware circuit. When the RF signal come, receiving board VT level signal changes from low to high, the signal is sent to SCM, the SCM reads information through the signal. Specific program flow chart is shown in Fig. 44.6.

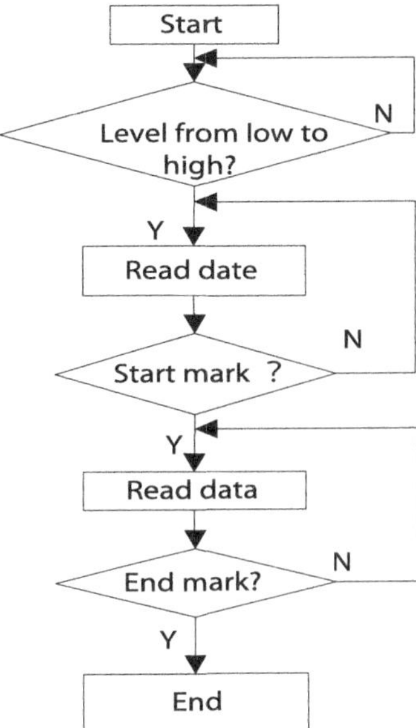

Fig. 44.6 Receiving program flow chart

44.4 Conclusion

The system adopts a novel idea, achieve the bus automatic station report by using MCU and RFID. System operates well by experiment test. Owing to economy and safety of the broadcast station system, market prospect is good. The system has only realized the bus station automatic identification, broadcast and display, in the next stage we will use wireless network or wired network, transmit bus stop data to the public management department or traffic control department, thus the bus running state can be real-time monitored. When alarm condition occurs in the bus, the monitoring center can also send a message to the driver in time so as to realize bidirectional data communication.

References

1. Lin L, Wang LC, Lu Q (2006) Intelligent bus monitor and control system based on GPS and GPRS. J Instrum Apparatuses 27(6):561–563 (in Chinese)
2. Yang L (2011) Intelligent public traffic broadcast station system based on RFID. J Changchun Univ Sci Technol 6(3):191–192 (in Chinese)
3. Huicong LI (2010) Automated bus stop announcement system based on active RFID technology. Mod Electron Technol 23:155–157 (in Chinese)
4. Lu Y, Yao H (2005) The automatic recognition and broadcasting system in bus station based on RFID. Electr Technol Appl 13:113–115 (in Chinese)

5. Ye X (2011) Design and realization of automatic report station system in the bus based on MCU. Inf Technol 4:40–43 (in Chinese)
6. Zhou G, Xu MQ (2007) City bus auto broadcast station system design and its application. Equip Manuf Technol 4:109–111 (in Chinese)
7. Guo T (2009) New concept 51 single chip microcomputer C language tutorial. Start, improve, develop. Electronics Industry Press, Beijing (in Chinese)
8. Qiu Y, Zhang G (2008) Family voice alarm system design based on ISD4004. Single Chip Microcomput Embed Syst Appl 1:47–50 (in Chinese)
9. Zhou B, Fen W, Hu J (2008) Bus automatic broadcast station system design. J Sichuan Coll Sci Eng (Nat Sci Ed) 6:79–81 (in Chinese)

Chapter 45
A Strategy for Improving Interference Suppression of Airborne Array Radar Under Clutter Environment

Hao Jiang, Wen Zhai, Nini Rao, Bo Zhou and Chaoyang Qiu

Abstract This paper proposed a strategy for improving interference suppression performance of airborne array radar under clutter environment. This strategy detects the range cells containing clutters and then eliminates them. After the estimation of weight coefficients in adaptive interference cancellation algorithm using the left range cells without clutters, the eliminated range cells are put back on their original positions and carry out the interference cancellation. The simulation results show that the proposed strategy improves the interference cancellation performance and ensures the freedom degrees required by adaptive interference cancellation system under clutter environment Furthermore, the resolution that distinguishes different interferences from different angles is also improved. This proposed strategy is simple and has low computational complexity. So it has the practical application prospects in airborne array radar.

Keywords Airborne array radar · Adaptive interference cancellation · Clutter

45.1 Introduction

The adaptive interference cancellation technique is commonly applied to suppress the active interferences in airborne array radars. Up to now, a lot of efforts have been made for improving the interference suppression ability of airborne array radar, especially under clutter environment. For example, Haimovich [1, 2] proposed a

H. Jiang · W. Zhai · N. Rao (✉) · B. Zhou
University of Electronic Science and Technology of China,
Chengdu 610054, China
e-mail: raonn@uestc.edu.cn

C. Qiu
Institute of Leihua Electronic Technology, Wuxi 214063, China

Z. Sun and Z. Deng (eds.), *Proceedings of 2013 Chinese Intelligent Automation Conference*, Lecture Notes in Electrical Engineering 255,
DOI: 10.1007/978-3-642-38460-8_45, © Springer-Verlag Berlin Heidelberg 2013

characteristic canceller to suppress the active interferences. Principal component (PC) rule [3, 4] used the feature-based vector of clutter subspace to form the weight vector for filtering the clutters in the main branching. Based on the structure of the generalized sidelobe canceller (GSC), Goldstein and Reed [5, 6] proposed the cross-spectral method (CSM), a dimensionality reduction method. However, the operation of these dimensionality reduction methods are so excessive. To solve this problem, Goldstein and Scharf [7, 8] further proposed the improved Wiener filter, in which the computational complexity is significantly reduced. However, it requires the clutters obey the evenly distribution. This is obviously inconsistent with the real clutters that airborne radar faces. Due to some special limitations in the airborne radars, such as the size and weight, most of the existing methods are hard to be realized in airborne array radar.

This paper proposed a strategy for improving the performance of adaptive interference cancellation system in airborne array radar under clutter environment. The simulation experiments confirm the effectiveness and feasibility of proposed method. In addition, this method is simple, has low computational complexity and a good practical application prospects.

45.2 Signal Models

45.2.1 The Receiving Signal Model

Assuming that the antenna of an airborne radar consists of N_s array elements, the number of pulses per coherent processing time is N_t and the *n-th* element and the *k*-th snapshot of receiving data is denoted by x_{nk}, the array data vector of the *n-th* element is $\mathbf{x}(n) = [x_{n1}, \ldots, x_{nNt}]^T$. Then, a $N_s N_t \times 1$ column vector $\mathbf{x}$ is formed using $x(n), n = 1, 2, \ldots, N_s$, as below:

$$x = [x^T(1), \ldots, x^T(N_s)]^T$$

According to the components of the receiving signal, $\mathbf{x}$ can also be expressed as:

$$\mathrm{x} = \mathrm{S} + \mathrm{C} + \mathrm{J} + \mathrm{N} \tag{45.1}$$

where S is the target steering vector: $\mathrm{S} = \mathrm{S_s} \otimes \mathrm{S_t}$, in which $S_s = \left[1, e^{jfs}, \ldots, e^{j(Ns-1)fs}\right]$, $S_t = \left[1, e^{jft}, \ldots, e^{j(Nt-1)ft}\right]$, f_s and f_t represent the space and time normalized frequencies respectively; C is the clutter component; J is the interference component; N is the noise component.

45.2.2 Clutter Model

The clutters of radar are usually divided into the ground, sea and weather clutters. Different types of clutters have different probability distribution properties. For the ground clutter, the probability distribution of its amplitude can be described by the model of Rayleigh distribution, lognormal distribution or Weibull distribution; its power spectrum is the Gaussian spectrum or cubic spectrum while the commonly used type is the Gaussian spectrum. For the sea clutter, it can be described by the Gauss spectrum model whose amplitude obeys the lognormal distribution or K distribution. For the weather clutter, it can be indicated by the Gaussian spectrum model whose amplitude obeys Rayleigh distribution. When an airborne radar works in the downward-looking state, the major clutter it faces is the ground/sea clutter. Thus, this paper respectively simulates the ground and sea clutters using the Gaussian spectrums whose amplitude obeys the Rayleigh distribution in (45.2) and the K-distribution in (3).

$$p(x_1) = \begin{cases} \frac{x_1}{\sigma^2}\exp\left(-\frac{x_1^2}{2\sigma^2}\right), & x_1 \geq 0 \\ 0, & x_1 < 0 \end{cases} \tag{45.2}$$

where x_1 denotes the amplitude envelope of clutter with Rayleigh distribution in the receiving signal; σ is the standard deviation of Rayleigh distribution.

$$f(x_2;\alpha,v) = \frac{2}{\alpha\Gamma(v)}\left(\frac{x_2}{2\alpha}\right)^v K_{v-1}\left(\frac{x_2}{\alpha}\right), (x_2 > 0, v > 0) \tag{45.3}$$

where x_2 denotes the envelope amplitude of clutter with K-distribution in the receiving signal; v is the shape parameter, α is the scaling function; $\Gamma(\cdot)$ is the Gamma function; $K_v(\cdot)$ is the second class modified Bessel function; σ^2 is the average power of clutter. The relationship among σ^2,v and α is given as below:

$$\alpha^2 = \frac{\sigma^2}{2v} \tag{45.4}$$

For most of radar clutters, the range of the shape parameter is $0.1 < v < \infty$. For a smaller v, such as $v \to 0.1$, the clutter has a long tail. Under $v \to \infty$, the K-distribution is close to the Rayleigh distribution.

45.3 Strategy

The distribution trajectories of clutters in the angle–Doppler space for airborne radar are dependent on the speed of the carrier aircraft, the array element spacing and the radar pulse repetition frequency (PRF). However, these parameters above are known for the radar signal processor. So we can firstly estimate location information of the clutter ridge in angle–Doppler plane using the information above. Then, we use the information of clutter slash ridge to determine whether a range cell contains the clutter. Next, we knock out those range cells with clutters. Finally, we calculate the interference nulling using the left range cells without clutters based on the adaptive interference cancellation algorithm for interference suppression. The details of this strategy are as follows:

1. Randomly take five points on clutter ridge line for calculating the sample power spectrum in the range cell.
2. Judge whether the sample power spectrum calculated in step 1 is greater than the threshold of clutter-to-noise ratio (CNR) or not;
3. If the result in step 2 is true, this range cell is judged to contain the clutter and is removed. If the result is false, this range cell is retained.
4. The sample data in the left range cells are applied to calculate the weight coefficients of adaptive interference cancellation algorithm.
5. Put the removed range cells back on their original positions and then perform the cancelation in all range cells using the weight coefficients in step 4.

This proposed strategy can avoid the problem that the interference nulling can not be established since of the increasing of clutter freedom degree and can effectively improve the anti-jamming performance of airborne radar with adaptive interference cancellation system.

45.4 Results

In the simulation experiments, we assume that $N_s = 10$, $N_t = 12$, the total Number of range cells is $L = 26$, and the number of range cells with clutters is $L_C = 6$. The Rayleigh distribution in (45.2) and K-distribution clutter models in (45.3) are respectively used to generate two simulated clutter data sets. The first set of data containing Rayleigh distribution represents the range cells with the ground clutters. The second set of data with K-distribution represents the range cells with the sea clutters. We randomly assign the range cells containing clutters. The desired signal direction is 0°. There are two interference signals in the receiving signal, which are respectively the frequency modulation (FM) interference with azimuth of 45° and amplitude modulation (AM) interference with azimuth of −15°.

We used the sensitivity, specificity and accuracy to evaluate the effectiveness of the proposed strategy. These indexes are defined as follows:

$$\text{sensitivity} = \text{TP}/\text{AE} \tag{45.5}$$

$$\text{specificity} = \text{TP}/(\text{TP} + \text{FP}) \tag{45.6}$$

where the TP is the number of range cells containing clutters correctly detected, i.e., the true positive; the FP is the number of range cells containing clutters wrongly detected, i.e., the false positive; the AE is the number of known range cells containing clutters. Since the specificity and sensitivity are usually contradictory, we define the mean of the specificity and sensitivity as the accuracy to comprehensively measure the performance of the proposed strategy.

45.4.1 The Clutter Ridge Information

The side looking uniformly spaced linear array is taken as an example in the simulation. For other types with the different clutter ridge locus, the processing ideas and methods are similar with this example. In our simulation, we take SNR = 0 dB, CNR = 40 dB and JNR = 40 dB.

Figures 45.1 and 45.2 respectively show the results of power spectral analysis from the space-time sampling points on the first range cell in the first set of data and on the 10th range cell in the second set of data. The clutter oblique ridge can be obviously seen from Figs. 45.1 and 45.2. The results illustrate that there is the clutters on two ranging cells.

Figure 45.3 shows the results of power spectral analysis from the space-time sampling points on the thirty-first range cell in the first set of data. The clutter oblique ridge information obviously disappears in Fig. 45.3. The results show that there is no clutter in this range cell.

45.4.2 Detection Results for the Range Cells Containing Clutters

Table 45.1 shows the average detection results for 100 experiments. It can be seen from Table 45.1 that the detection accuracy of this strategy reach 100 %. Thus, it is very effective when detecting the range cells containing clutters.

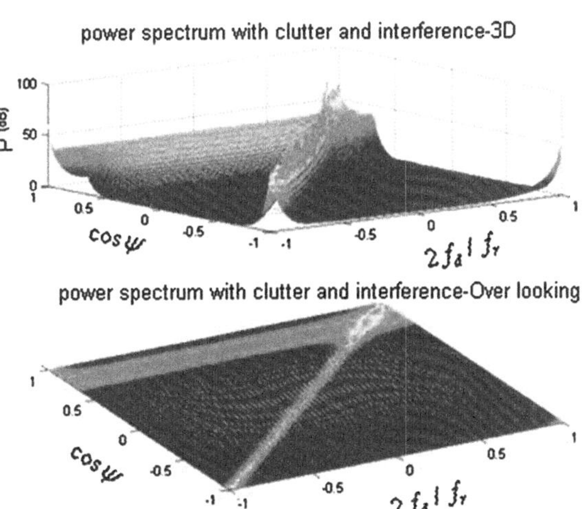

Fig. 45.1 The clutter ridge information for the case with Rayleigh distribution

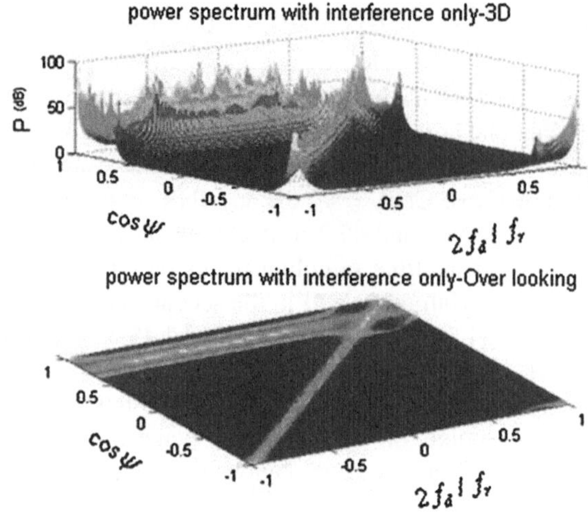

Fig. 45.2 The clutter ridge information for the case with K-distribution

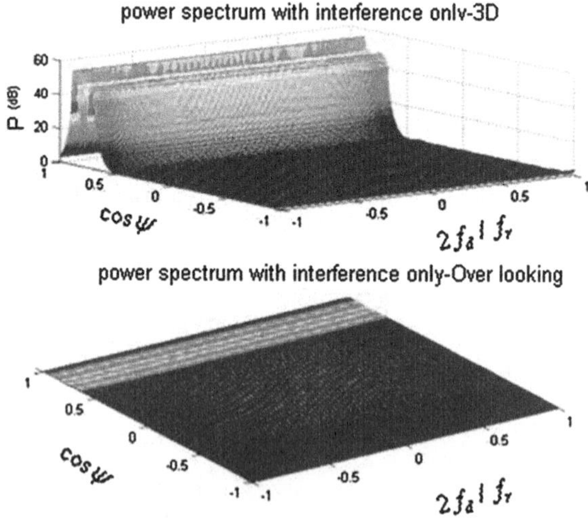

Fig. 45.3 The clutter ridge information for the case without clutter

45.4.3 The Results of Interference Cancellation

Figure 45.4 shows the beampatterns formed in spatial domain using the adaptive interference cancellation algorithm before and after the strategy was used. Without the strategy, the clutters cause the beampattern to be distorted and the main beam can not be effectively formed in the expected direction although the nulling can be formed in the direction of the interferences in the beampattern. With this proposed strategy, we can see from Fig. 45.4 that not only the deep nulls can be formed in the directions of the interference arriving, but also the main beam is well established in the desired direction. The results show that this strategy can effectively

Table 45.1 Detection results for the range cells containing clutters

Types of clutter	TP	FP	Sensitivity	Specificity	Accuracy(%)
Rayleigh distribution	6	0	1	1	100
K-distribution	6	0	1	1	100

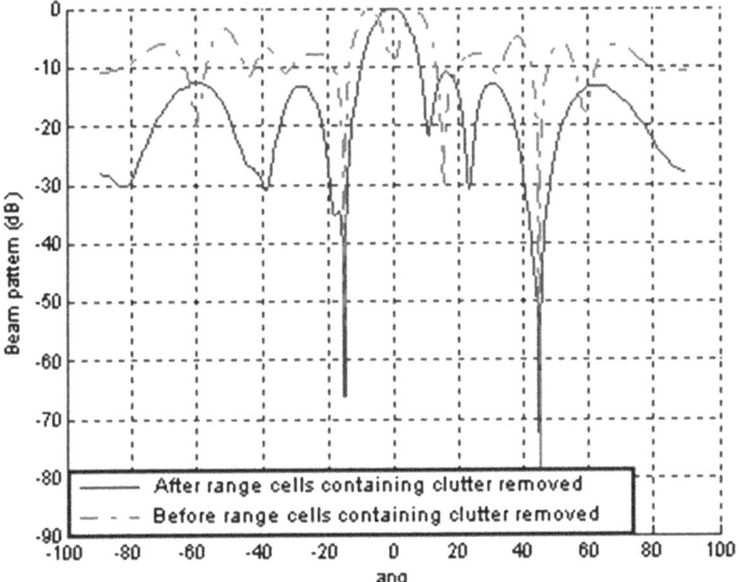

Fig. 45.4 The beampatterns before and after the strategy was used

improve the performance of adaptive interference cancellation and increase the airborne radar capability of anti-jamming in spatial domain.

Figure 45.5 shows the changes of normalized nulling depth caused by the changes of difference between two interference angles after performing our strategy, where those nulling depth less than −60 dB are truncated to −60 dB for observation conveniently. This result reveals that the sensitivity of adaptive interference cancellation system for different interference angles when performing the proposed strategy, that is, the interference resolution. It can be seen from Fig. 45.5 that the normalized nulling depth shows a horizontal line when the angle difference is greater than 5°. It illustrates that the adaptive interference cancellation system can distinguish two different interferences and robustly realizes the interference cancellation with the nulling depth up to -60 dB under the resolution condition above. However, when the angle difference of two interference is 0°, that is, two interferences come from the same direction, the adaptive interference cancellation system can also form the nulls with depth of −60 dB in two interference directions of arrival, but can not distinguish two interferences.

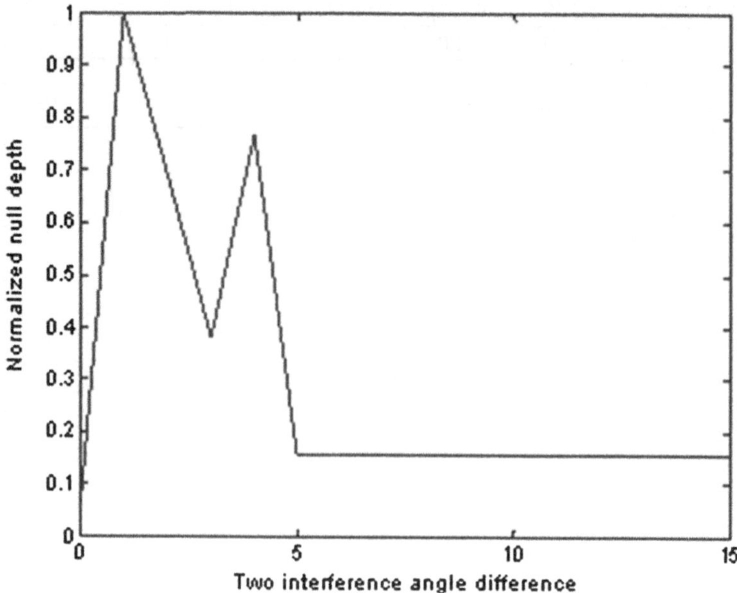

Fig. 45.5 The changes of normalized nulling depth caused by changes of difference between two interference angles

45.5 Conclusion

After performing the proposed strategy under clutter environment, the adaptive interference cancellation system used in airborne radar can provide the degree of freedom required by the system and also improve the performance of the adaptive interference cancellation. Further, the system can well and robustly distinguish the different interferences from different angles. In addition, this strategy is very simple and easy to be realized. Therefore, it has practical application prospects in airborne phased array radar.

Acknowledgments This work was supported in part by a grant from Aeronautical Science Foundation of China (Grant No. 20112080014).

References

1. Alexander MH, Yeheskel B-N (1991) An eigenanalysis interference canceler. Signal Process, IEEE Trans 39(1):76–84
2. Alexander MH, Yeheskel B-N (1988) Interference cancellation by eigenanalysis methods. Proceedings 1988 digital signal processing workshop

3. Peter AZ, Goldstein JS, Joseph RG et al (1998) Comparison of reduced-rank signal processing techniques. Signals, systems & computers, 1998. Conference record of the Thirty-Second Asilomar conference on 1998, vol 1, pp 421–425
4. Christopher DP, Alexander MH, Tareq FA et al (2000) Reduced-rank STAP performance analysis. Aerosp Electr Syst, IEEE Trans 36(2):664–676
5. Goldstein JS, Irving SR (2002) Reduced-rank adaptive filtering. Signal Process, IEEE Trans 45(2):492–496
6. Goldstein JS, Irving SR (2002) Subspace selection for partially adaptive sensor array processing. Aerosp Electr Syst, IEEE Trans 33(2):539–544
7. Goldstein JS, Irving SR, Louis LS (1998) A multistage representation of the Wiener filter based on orthogonal projections. Inf Theory, IEEE Trans 44(7):2943–2959
8. Louis LS, Thomas JK (1998) Wiener filters in canonical coordinates for transform coding, filtering, and quantizing. Signal Process, IEEE Trans 46(3):647–654

Chapter 46
Magnetometer Calibration Scheme for Quadrotors with On-Board Magnetic Field of Multiple DC Motors

Haiwei Liu, Ming Liu, Yunjian Ge and Feng Shuang

Abstract This paper presents a 3D calibration approach of magnetometer with a new fitting error function. Unlike some iterative and numerical approaches, the method has closed form solution with great simplicity and therefore can be easily implemented by microcontroller-based on-board pilots of miniature unmanned aerial vehicles (UAV) with a potential of in-flight calibration. The experimental validation on a quadrotor in our laboratory shows its efficiency profoundly owing to its close analytical solutions. The approach was further applied to evaluate the impact of the on-board magnetic field generated by the permanent-magnets of DC motors of the quadrotor at different thrust levels, and the result shows that the relation between UAV's thrust command and motors-induced magnetic bias can be simply fitted by a linear function, which can be used for precise heading calculation during flight.

Keywords UAV · Quadrotor · Magnetometer · Calibration

46.1 Introduction

Quadrotors are a kind of miniature UAVs with broad applications. Strap-down Inertial Navigation Systems (INS) is a substantial part of such systems in the need of location awareness. Comparing with those precise, but very expensive north-seeking apparatus applied in the commercial aircraft and UAVs, the MEMS

H. Liu
Department of Automation, University of Science and Technology of China, Hefei 230027, China

H. Liu · M. Liu (✉) · Y. Ge · F. Shuang
Institute of Intelligent Machines, Hefei 230031, China
e-mail: ming.liu@iim.ac.cn

Z. Sun and Z. Deng (eds.), *Proceedings of 2013 Chinese Intelligent Automation Conference*, Lecture Notes in Electrical Engineering 255,
DOI: 10.1007/978-3-642-38460-8_46,

magnetometer becomes a realistic choice for the small scale UAVs due to their special features such as low power consumption, low cost, small size and easy installation as well as very limited load capacity of small UAVs. However, since these low cost INSs suffer from measurement noise, DC drifts, great sensitivity to temperature changes and vibration etc., great concern must be paid to the calibration and alignment of such sensors to guarantee the good performance of UAVs. This concern becomes even serious for the quadrotor UAVs driven by servo motors, since the magnetic field generated by the motors will make big impact on the earth field to be measured.

There are mainly two kinds of calibration approaches with many variant versions. The first one, called "swinging" method, is based on the perturbation of the basic heading equation [1]. It offers a direct estimation in the heading angle domain, but a more precise heading reference is needed during the calibration. The second is the non-linear two-step calibration (also known as ellipsoid fitting) method [2], which fits the raw data from magnetometer in the sensor frame with an ellipsoid. The drawback is that the ellipsoid fitting requires complicated optimal and iterative algorithms, which is not suitable for real-time in-flight calibration by UAVs' on-board MCU (micro-controller unit).

Our research objective is to develop autonomous quadrotors UAV systems, which can not only fly according to their own planning under the guidance of GPS, but also perceive and overcame the uncertainty or error of the sensor data. Therefore, precise attitude estimation and heading control are basic and fundamental. Especially for the miniature quadrotors driven by multiple brushless DC motors, the magnetic field generated by the permanent-magnets in motor assembling will affect the earth magnetic field to be measured because they are usually mounted quite close to sensor board.

In this paper, after a profound study on the existing approaches, which could only provide numerical solutions by iterations, we first introduce a 3D calibration algorithm with closed-form analytical solution, based on a fitting function originally proposed for hybrid camera calibration [3]. The algorithm is based on a precise magnetic field model consisting of two different modeling sources, the device-related ones and measurement-related ones [4]. Subject to the important issue of DC motors' magnetic field, the influence of the field on magnetometer was then explored. The results of experiment conducted on our quadrotor will be given which shows that this algorithm can obviously promote the speed of calibration and can provide a benchmarking relation between on-board magnetic field bias and motors with sufficient precision.

This paper is organized as follows. In Sect. 46.2, a unified error model of magnetometer and prior calibration approaches were reviewed and its simplified form with respect to quadrotors was presented. Section 46.3 derived analytical solutions to the new sphere fitting error function. In Sect. 46.4, we compared the numerical result with the earlier one and examined the impact of rotating motors on the measurement of earth magnetic field with our approach experimentally. Conclusions and future works were given in Sect. 46.5.

46.2 Theoretical Background

46.2.1 Unified Error Model

Magnetometer is applied to measure the earth's magnetic field, for the purpose of determining the heading reference. There are a number of kinds of sources of error to corrupt the output of a magnetometer. A rather complete mathematical modeling is given by Gebre-Egziabher et al. [2], Foster and Elkaim [4] with five sources in two categories: null shift errors C_{zb}, scale factor errors C_{sf} and misalignment errors C_m, which are caused by the device itself; and hard iron errors $\delta \overrightarrow{B^b}$ and soft iron errors C_{si}, which are inherent from magnetic fields,

$$\widehat{\overrightarrow{B^b}} = C_m C_{sf} C_{si} (\overrightarrow{B^b} + \delta \overrightarrow{B^b}) + C_{zb} \tag{46.1}$$

The null shift of a sensor C_{zb} (dc offset zero bias) is a 3×1 constant vector without any relation of the magnetic vector whether being measured or not. The scale error of a sensor C_{sf} is a side effect of varying sensitive between different sensors. The misalignment error rises from the non-orthogonally between the three axes, which can be regulated with the 3×3 projective matrix C_m. When the particular sensor is selected, these three error sources are actually settled and normally only influenced by the internal conditions, such as the measurement integrated circuit or such factors as temperature drift.

The hard iron errors $\delta \overrightarrow{B^b}$ are generally constants and unwanted magnetic fields observed by a magnetometer and fixed on the sensor coordinate frame. Mathematically, this source is indistinguishable with the null shift error. Relatively, the soft iron errors C_{si} are due to the materials' responding to exposure to an external field. In general, this source is often assumed to be miniature, linear and attached to the sensor frame so that a square matrix close to identity matrix is sufficient.

46.2.2 Prior Art of Calibration

The first calibration method of modern solid-state magnetometers is the so-called "swinging" method [1], starting from the heading equation perturbation in angle domain. In calibration, rotate a leveled magnetometer around its vertical axis and compare the precise auxiliary heading reference angles with the computed angles for error estimation. Its major drawbacks are that it requires a more precise heading reference, which means costly testing equipment and that it only can be calibrated on the ground prior to flight.

Gebre-Egziabher proposed a 2D calibration method [2] based on the concept of ellipsoid fitting in the beginning of this century, which has been widely applied by industry since then. Its main idea is that the locus of error-free measurements from two perpendicularly mounted single-axis magnetometers on a horizontal plane is a circle centering at the origin. The errors mentioned above shift from circle to an eclipse both in the shape and center location. Later literature extended this approach to 3-D case [4] and took the misalignment between the coordinate systems of the platform and the magnetometer into account [5].

46.3 3D Calibration Approach for Quadrotors

46.3.1 Simplified Error Model

For our apparatus and IMU, the error model is not necessarily so complex and can be simplified for in-flight calibration of the trial-axis magnetometer with certain precision limited by its resolution.

The self-examination mode [6] of the magnetometer can be used to calibrate for device-related error sources, i.e., its zero bias, scale factors and misalignment errors, with an 1,100 mGa internal standard reference magnetic field generated under this test mode. Also, the misalignment between the frame of the quadrotor body and that of sensor, suggested in [5], is not considered here. With the sufficient internal benchmarking, the initial magnetometer outputs are denoted as $\overrightarrow{\widetilde{B^b}}$ and a simplified model applied is

$$\overrightarrow{\widetilde{B^b}} = \overrightarrow{B^b} + \delta \overrightarrow{B^b} \tag{46.2}$$

Typically, a test showed that the magnetic intense of the permanent-magnet of our DC motor could be up to 15,000 mGa, almost 30 times weaker than the earth magnetic field intense. Therefore, this influence counts for most of the bias $\delta \overrightarrow{B^b}$ in (46.2). It also showed that the influence of motors is fortunately attenuated rapidly as the distances between the motors and the sensor increase; otherwise the magnetometer can't be applied in quadrotor at all.

For a complete 3D case here, we make two assumptions: First, the magnitude of the earth's magnetic field vector is supposed to be invariant during calibration, i.e., $|\overrightarrow{B^b}|$ is constant. (The variation of the earth's magnetic field vector at different location of the earth is well known and well modeled [7]). Vector $\overrightarrow{B^b}$ can point to any direction, depending on how we rotate the sensor in the navigation coordinate frame. Secondly, the influence of motors is attached to the frame of sensor. Therefore, $\delta \overrightarrow{B^b}$ should be a constant vector in sensor's frame. With these two assumptions, the endpoint of measured vector $\overrightarrow{\widetilde{B^b}}$ should be a sphere in error-free case. Thus, the calibration of motors' influence on magnetic measurement turns out to be a problem of fitting the data points to a perfect sphere in the frame of magnetometer, if we rotate the quadrotor body mounting with sensors in every possible direction to sample the magnetometer data points.

It should be noticed that the real $\overrightarrow{B^b}$ includes not only the earth magnetic field, but also any possible environmental magnetic field that doesn't attach on the sensor frame, even those vary with location or time. At the same time, the bias $\delta \overrightarrow{B^b}$ includes all influences that attached in the platform. So, it should be treated cautiously especially if the data has large variation.

46.3.2 Fitting Error Estimation

Define the i-th output sample of magnetometer as $n_i = [x_i\, y_i\, z_i]^T$ and the bias and radius to be estimated as $n_c = [x_c\, y_c\, z_c]^T$ and R respectively, the i-th distance fitting error of the sphere is given by $d_i = \sqrt{(n_i - n_c)^T(n_i - n_c)} - R = \|n_i - n_c\| - R$. Given the sample set $\{n_i | i = 1, 2, 3, \ldots N\}$, the calibration objective is to find the "best" estimates of n_c and R to minimize cost function

$$J = \min \sum_{i=1}^{N} d_i^2 \tag{46.3}$$

However the estimation based on this fitting function (called RE scheme) shows strong nonlinearity caused by its square root operation and has no closed form solution. It has to be solved by means of numerical iteration optimal approaches [8], even though it has a direct geometric meaning, i.e., the Euclidian distance from the data point to the surface of the sphere.

On the contrary, we consider another fitting error function $d_i^o = (n_i - n_c)^T(n_i - n_c) - R^2$ and the cost function

$$J^o = \min \sum_{i=1}^{N} (d_i^o)^2 \tag{46.4}$$

This function and its corresponding algorithm were proposed in [3] originally for calibrating the hybrid camera system for dynamic motion tracking. The fitting algorithm is derived by letting its partial derivatives with respective to n_c and R to be zeros and then solve the estimates of n_c and R, as shown in the following. First get $\partial J^o / \partial(R^o) = 0$ we can solve

$$R^2 = \frac{1}{N} \sum_{i=1}^{N} (n_i - n_c)^T (n_i - n_c) \tag{46.5}$$

Further, from $\partial J^o/\partial n_c = 0$ and taking (46.5) into account we have that

$$2An_c = B \tag{46.6}$$

where A and B are 3 by 3 matrixes given by

$$A = \begin{pmatrix} \overline{X^2} - \overline{X}^2 & \overline{XY} - \overline{X}\cdot\overline{Y} & \overline{XZ} - \overline{X}\cdot\overline{Z} \\ \overline{XY} - \overline{X}\cdot\overline{Y} & \overline{Y^2} - \overline{Y}^2 & \overline{YZ} - \overline{Y}\cdot\overline{Z} \\ \overline{XZ} - \overline{X}\cdot\overline{Z} & \overline{YZ} - \overline{Y}\cdot\overline{Z} & \overline{Z^2} - \overline{Z}^2 \end{pmatrix},$$

$$B = \begin{pmatrix} (\overline{X^3} + \overline{XY^2} + \overline{XZ^2}) - \overline{X}\cdot(\overline{X^2} + \overline{Y^2} + \overline{Z^2}) \\ (\overline{Y^3} + \overline{YX^2} + \overline{YZ^2}) - \overline{Y}\cdot(\overline{X^2} + \overline{Y^2} + \overline{Z^2}) \\ (\overline{Z^3} + \overline{ZY^2} + \overline{ZX^2}) - \overline{Z}\cdot(\overline{X^2} + \overline{Y^2} + \overline{Z^2}) \end{pmatrix}$$

where the bar means average, such as $\overline{X} = \frac{1}{N}\sum_{i=1}^{N} x_i$. The estimates are then given by

$$n_c = \frac{1}{2}A^{-1}B \tag{46.7}$$

and (46.5). Theoretically, as far as we rotate the sensor with sufficient orientations in full 3D space, the non-singularity of A should be guaranteed which ensures the existence of the solutions. We regarded this approach as R2E.

46.4 Experimental Results

In this section, we present experiment results based on the theory given in the previous sections. The experiments were conducted on our X configuration quadrotor with carbon fiber framework shown in Fig. 46.1.

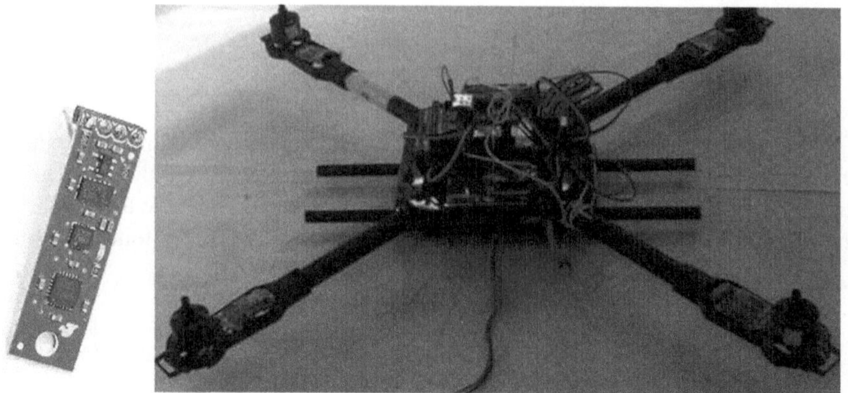

Fig. 46.1 The quadrotor UAV and the 9 DOF IMU board (*left*) applied in test

This quadrotor has been equipped with a 9 DOF IMU, a range finder, a barometer, a GPS, a data log unit, and Zigbee wireless communication with ground station. The mounted units ensured us to fly the quadrotor, and to acquire and log the all flight data for analysis. The IMU consists of a tri-axial AMR magnetometer HMC5883L by Honeywell with a resolution of 5 mGa, a tri-axial accelerometer ADXL345 by Analog Devices and a tri-axial gyroscope PS-ITG-3200 by InvenSense. Four DC brushless motors contain permanent-magnet parts and can drive more than ten amps of current at starting time. The distance between the IMU and the motors is about 25 cm.

46.4.1 Calibration Results and Approach Comparison

The raw magnetometer sample data set 1, obtained during a calibration of 200 s with a sampling frequency of 50 Hz, is plotted in Fig. 46.2. In Fig. 46.2, there are 3 regions of interest (ROI) that last for more than 25 s each when the sensor was rotated around *X*, *Y*, and *Z* axis, respectively. The other data points show the transition phase between ROIs, sampled when the sensor was rotated in random directions. We firstly validate the efficiency of the fitting approach RE based on fitting function (46.3). We then figure out the minimal number of data points to operate a precious calibration using approach R2E based on cost function (46.4).

The typical fitting spheres with two approaches are shown in Fig. 46.3 and parameter estimates are compared in Tables 46.1 and 46.2. The RE was solved by GlobalSearch optimal function in Matlab [9]. The tables indicate that both approaches have almost identical converging speed in terms of the numbers of the

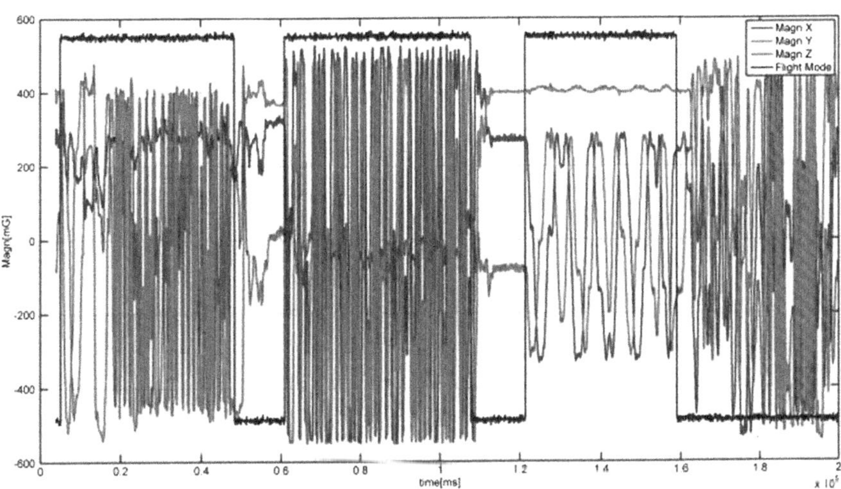

Fig. 46.2 Dataset 1. The flight mode signal serves to mark the ROI with high value

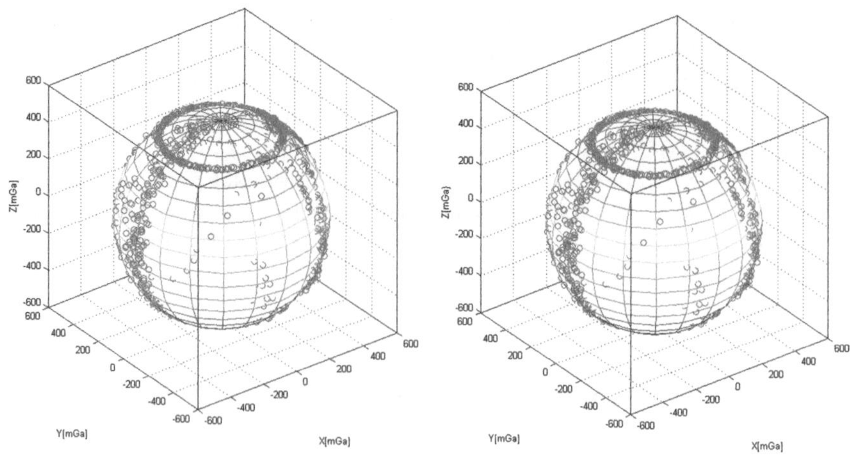

Fig. 46.3 Typical fitting results with "RE" (*left*) and "R2E" (*right*) estimations

sample supplied as well as estimate values. But, as shown by their last columns, two approaches show significant difference in their computing cost time. Calculating on the same dual core 2.7 GHz PC with 4G RAM, the R2E approach is nearly 400 times faster than RE approach. It indicates that the R2E is significantly superior to RE especially in terms of computing efficient, which is very encouraging for the small scale quadrotor UAVs with limited on board computing power.

Further, the test results showed that a rotation around each of three orthometric axes for a period of 10 s, respectively, was sufficient to fit the relatively accurate sphere, since the n_c, n_c, n_c, R estimates rapidly converged to their final values in about 30 s and then fluctuated inside a bounded ranges less than 3 mGa. This value is fairly small comparing with the stationary SD (6 mGa) of the low-grade magnetometer.

46.4.2 Evaluation on Influence of Motors' on Magnetic Field

To further investigate the dynamic influence of the magnetic field generated by the DC servo motor on the magnetometer calibration, we conducted calibration at different rotation speed of motors. As the motor speed cannot be measured directly we used motor speed command, the pulse width (PW) signal to the motor speed controller, as an approximation. The influence at different speed level is characterized by a relation between sensor bias $\delta \overrightarrow{B^b}$ and motor speed. At each speed level we assume magnetic bias $\delta \overrightarrow{B^b}$ is a constant, but it varies as the motor speed changes.

Table 46.1 Fitting results with RE estimation. The label TimePeriod of 5, 10, 15, 20, 25 means that the data selected for fitting is from the front 5, 10, 15, 20, 25 s of each ROI

TimePeriod (s)	5	10	15	20	25	200
Xc (mGa)	−19.33	−18.45	−20.22	−19.99	−19.18	−16.48
Yc (mGa)	−9.24	−15.04	−18.70	−18.08	−17.61	−15.90
Zc (mGa)	−30.29	−30.86	−29.51	−30.25	−30.77	−31.40
R (mGa)	529.34	526.52	526.02	526.31	526.56	523.52
SD (mGa)	7.26	7.63	9.38	9.10	9.11	9.50
Cost (s)	2.313	2.243	3.434	3.083	2.602	4.006

Table 46.2 Fitting results with R2E estimation. TimePeriod means the same as Table 46.1

TimePeriod (s)	5	10	15	20	25	200
Xc (mGa)	−19.25	−18.38	−20.01	−19.80	−19.00	−16.43
Yc (mGa)	−9.30	−14.98	−18.47	−17.89	−17.46	−15.78
Zc (mGa)	−30.28	−30.85	−29.58	−30.31	−30.82	−31.42
R (mGa)	529.37	526.57	526.08	526.36	526.62	523.60
SD (mGa)[a]	7.26	7.63	9.38	9.10	9.11	9.50
Cost (s)	0.005	0.006	0.005	0.005	0.005	0.006

[a] The SD here defined the same as RE estimation for comparison, i.e. $SD = \sqrt{\frac{1}{N}\sum_{i=1}^{N}\left(\sqrt{(Xi-Xc)^2+(Yi-Yc)^2+(Zi-Zc)^2}-R\right)^2}$

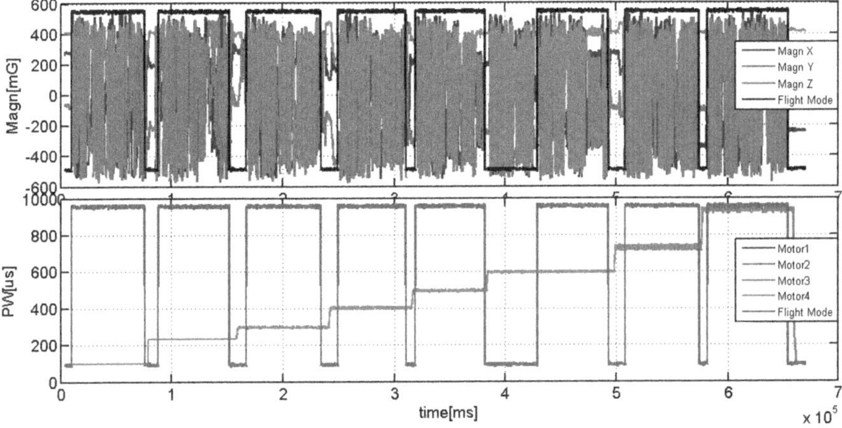

Fig. 46.4 Dataset 2. Flight mode s marked the stable periods of PW. The motor command had a pulse width modulation period of 2,000 us and triggered the rotation at 100 us

We sampled the raw magnetometer output, as shown in Fig. 46.4, at a frequency of 50 Hz with the motor command PW ascending step by step and maintaining constant at each step for more than 60 s (shown by blue plot).

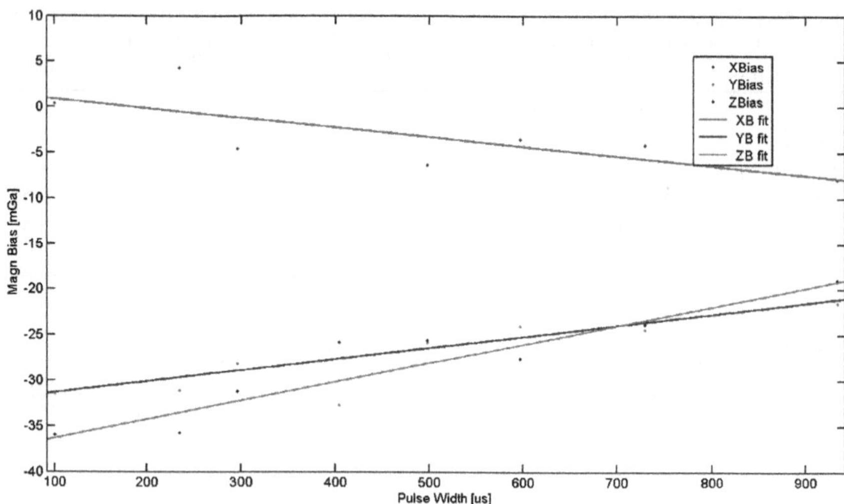

Fig. 46.5 Fitting results of motors' influence on magnetometer

Table 46.3 Linear fitting parameters of the magnetic bias owing to motors' rotation

	Slope (mGa/us)	Intercept (mGa)	STD (mGa)
X bias	−0.01045 (−0.02148, 0.0005751)	1.95 (−4.18, 8.08)	2.62
Y bias	0.01234 (0.008596, 0.01609)	−32.56 (−34.65, −30.48)	0.89
Z bias	0.02063 (0.01441, 0.02684)	−38.39 (−41.84, −34.93)	1.47
I bias	−0.02249 (−0.02846, −0.01651)	50.19 (46.87, 53.51)	1.42

The estimated biases $\delta \overrightarrow{B^b}$ at different motor speed are indicated by the dots shown in Fig. 46.5, which shows a nonlinear relationship between speed and bias coordinates. For simplicity, a linear fitting is applied with a confidence coefficient of 95 % after excluding the 4th point an obvious outlier. The fitting parameters and their bounds are given in Table 46.3. They show that the trends of biases of *X, Y, Z* axis are $0 \rightarrow -8$, $31 \rightarrow -21$, $-35 \rightarrow -18$ mGa separately, and the maximal fitting SD is smaller than 3 mGa, which is acceptable comparing with the magnetometer resolution (5 mGa). This approach is useful for the in-flight calibration as the motor speeds may have rather large variation when a UAV is flying.

46.5 Conclusions and Future Works

The paper presented a fast calibration method of magnetometer on quadrotors, based on an error fitting function originally proposed for camera calibration. Its major advantages over the traditional numerical approaches are that it has a closed form solution, is 400 time faster and is simpler enough to be executed by on-board micro-computers of small scale UAVs. The influence of the magnetic field of permanent-magnets of quadrotors' DC motors on the magnetometer calibration was investigated. The relationship between motors' speed and the earth magnetic field was also studied, which can be approximated by linear function.

Future works will be the verification of this calibration approach on the auto-pilot of our UAV, for autonomous flight under the guidance of GPS.

Acknowledgments This work is supported by NSFC (No. 61201076). The authors like to thank Xuekun Zhuang at Institute of Intelligent Machines, Hefei for his valuable support on realizing the numerical global search algorithm in Matlab.

References

1. Bowditch N (1995) The American practical navigator. Defense Mapping Agency, Hydrographic/Topographic Center, Bethesda, Maryland, USA
2. Gebre-Egziabher D, Elkaim GH, Powell JD, Parkinson BW (2001) A non-linear, two-step estimation algorithm for calibrating solid-state strapdown magnetometers. In: Presented at the 8th International St. Petersburg conference on navigation systems, St. Petersburg, Russia
3. Jennings A, Black J, Allen C (2011) Empirically based modeling of tape spring hinge deployment for space structures. In: AIAA SPACE 2011 conference & exposition, AIAA 2011-7249, Long Beach, California
4. Foster CC, Elkaim GH (2008) Extension of a two-step calibration methodology to include nonorthogonal sensor axes. IEEE Trans Aerosp Electron Syst, IEEE AES 44(3):1070–1078
5. Li X, Li Z (2012) A new calibration method for tri-axial field sensors in strap-down navigation systems. Meas Sci Technol 23(10):105105
6. http://www51.honeywell.com/aero/common/documents/myaerospacecatalog-documents/Defense_Brochures-documents/HMC5883L_3-Axis_Digital_Compass_IC.pdf
7. http://www.ngdc.noaa.gov/geomag-web/#igrfwmm
8. Vasconcelos JF, Elkaim G, Silvestre C (2011) Geometric approach to strapdown magnetometer calibration in sensor frame. IEEE Trans Aerosp Electron Syst, IEEE AES 47(2):1293–1306
9. Ugray Z, Lasdon L, Plummer J, Glover F, Kelly J, Martí R (2007) Scatter search and local NLP solvers: a multistart framework for global optimization. INFORMS J Comput 19(3):328–340

Chapter 47
TSVD Regularization with Ill-Conditioning Diagnosis in GNSS Multipath Estimation

Qingming Gui, Ke Chen and Yongwei Gu

Abstract Multipath is one of the main error sources of the differential applications of global positioning systems (GPS). The maximum likelihood estimation of multipath is simplified with grid method using the idea of deploying estimation points. There is ill-conditioning in the simplified model, decreasing the numerical stability of the maximum likelihood estimation grid maximum likelihood estimation in some conditions. The truncated singular value decomposition (TSVD) regularization with ill-conditioning diagnosis is introduced to mitigate the effects of ill-conditioning. The truncation level is selected based on ill-conditioning diagnosis by comparing the values of the diagnosis statistics of the estimation under different truncation level. TSVD regulation with the truncation level selected in ill-conditioning diagnosis approach can resist the ill-conditioning effectively and the estimation is numerical stable. The multipath effect can be effectively mitigated by the method proposed in this paper.

Keywords Multipath estimation · TSVD regulation · Ill-conditioning diagnosis

47.1 Introduction

Multipath is a main error source in modern applications of differential global positioning system (GPS) [1]. Multipath estimation is a receiver-internal methodology estimating multipath parameters with statistical methods. In multicorrelator

Q. Gui (✉) · Y. Gu
Institute of Science, Infomation Engineering University, 62 Kexue Revenue, Zhengzhou 450001, Henan, China
e-mail: guiqm@public.zz.ha.cn

K. Chen
The Unit 68029 of PLA, 79-1 Jiaojiawan South Road, Lanzhou 730020, Gansu, China
e-mail: ckhtl23@gmail.com

Z. Sun and Z. Deng (eds.), *Proceedings of 2013 Chinese Intelligent Automation Conference*, Lecture Notes in Electrical Engineering 255,
DOI: 10.1007/978-3-642-38460-8_47,

techniques the correlation between received base-band signal and local code signal is sampled with a bank of correlators and the parameters are estimated with some statistic methods. Usually the estimated multipath signals is subtracted from the received compound signal, then receiver tracking loop is used to lock the delay of refined signal. One patented technique is NovAtel's multipath estimating delay lock loop [2], and the computational burden of it is very high. In some simplified methods (e.g. [3]), only one signal multipath component is considered. A grid approach is proposed in [4], which deploys estimation points uniformly and estimates multipath parameters with maximum likelihood estimation and Kalman filtering, but the method is disturbed by noise severely. Some grid search methods are proposed to detect multipath components and estimate the parameters, e.g. [5], but the computational burden is high as well.

The grid method in [4] is not stable in computation and is vulnerable. In [4], a small bias is added in the diagonal of the normal matrix as an offset to make the numerical calculation stable in multipath estimation problem. In fact, this is equivalent to ordinary ridge estimation [6], which is a kind of biased estimation overcoming the ill-conditioning. However, the causation and influences of ill-conditioning is not studied, the rationality of the method is not discussed. In [7], it is revealed that ill-conditioning decreases the stability of the MLE, making the MLE disturbed by noise severely and TSVD regularization is used to overcome the ill-conditioning of the normal matrix.

In this paper, TSVD regularization with ill-conditioning diagnosis is introduced to improve the method proposed in [7]. The paper is organized as follows. Firstly, the grid maximum likelihood estimation (GMLE) of multipath and the ill-conditioning in the GMLE is reviewed. Secondly, the TSVD regularization method is introduced to resist the ill-conditioning, and the method choosing the truncation level with ill-conditioning diagnosis is proposed. Finally, demonstrate the validity of the method proposed in this paper through simulations. The paper concludes that under some conditions, the performances of the methods proposed in this paper can resist the ill-conditioning and mitigate multipath efficiently.

47.2 GMLE of Multipath Based on Multicorrelator

47.2.1 Multicorrelator Receiver Structure

Considering the C/A code of GPS, In multipath conditions, the received base band signal can be expressed by Gadallah [4] and Gadallah et al. [5]

$$z_{\mathrm{I}}(t) + jz_{\mathrm{Q}}(t) = \sum_{i=0}^{N-1} A_0(t)a_i(t)x(t - \tau_i(t))e^{\Delta\varphi_i(t)} + n(t) \tag{47.1}$$

where A_0 is the amplitude of the line-of-sight (LOS) signal, $x(t)$ includes the pseudorandom noise (PRN) code and the digital modulation, N is the number of signal components including the LOS signal and multipath signals, $a_i(t)$, $\tau_i(t)$ and $\Delta\varphi_i(t)$ are the time-varying attenuation coefficient, the propagation delay and the remaining phase of the ith signal component respectively, and $n(t) = n_I(t) + jn_Q(t)$ is the noise. It follows that $a_0(t) = 1$, $a_i(t) > 0$ and $\tau_i(t) - \tau_0(t) > 0$ for $i = 1, \ldots, N$, where $i = 0$ refers to the LOS signal, $\Delta\varphi_i(t) = \varphi_i(t) - \hat{\varphi}_0(t)$, $\varphi_i(t)$ is the carrier phase of the ith signal component and $\hat{\varphi}_0(t)$ is the local carrier phase estimation. Let $\alpha_i(t)$ denote the delay coefficient such that $\tau_i(t) = \tau_0(t) + \alpha_i(t)T_C$, where T_C is the chip duration.

Through the bank of correlators, the received signal is correlated with a series of delayed replica with delays coefficients $\boldsymbol{\delta} = (\delta_1, \delta_2, \ldots, \delta_n)$, and the components of $\boldsymbol{\delta}$ are referred to as sample points. Usually, the sample points could be deployed uniformly, for which it means that the spacing of two adjacent sample points $\Delta = \delta_{i+1} - \delta_i$ is a constant for $i = 1, \ldots, n-1$. For the jth correlator, taking $a_i(t)$, $\tau_i(t)$, $\alpha(i)$ and $\varphi_i(t)$ as constants in the coherent integration interval, the received signal is correlated with delayed local replica of PRN code $c(t - \hat{\tau}_0 - \delta_j T_C)$ and the in-phase and quadrature output can be expressed in matrix form as

$$\mathbf{R}_I = \mathbf{HX} + \mathbf{n}'_I \tag{47.2}$$

$$\mathbf{R}_Q = \mathbf{HY} + \mathbf{n}'_Q \tag{47.3}$$

where $\mathbf{R}_I$, $\mathbf{X}$, $\mathbf{n}'_I$, $\mathbf{R}_Q$, $\mathbf{Y}$, $\mathbf{n}'_Q$ are column vectors and $\mathbf{H}$ is a matrix with $[\mathbf{R}_I]_j = R_{I,j}$, $[\mathbf{R}_Q]_j = R_{Q,j}$, $[\mathbf{n}'_I]_j = n'_{I,j}$, $\left[\mathbf{n}'_Q\right]_j = n'_{Q,j}$, $[\mathbf{X}]_i = x_i$, $[\mathbf{Y}]_i = y_i$, $[\mathbf{H}]_{j,i} = R_C((\delta_j - \alpha_i)T_C)$, for $i = 0, \ldots, N$ and $j = 1, \ldots, n$, and $x_i = a_i A_0 \cos(\varphi_i - \varphi_0)$, $y_i = a_i A_0 \sin(\varphi_i - \varphi_0)$, $R_C(\bullet)$ is the auto-correlation of PRN code, T is the coherent integration time.

Considering that $n(t)$ is additive white Gaussian noise with power density $N_0/2$, the noise samples, $\mathbf{n}'_I$ and $\mathbf{n}'_Q$ are uncorrelated and Gaussian distributed with zero mean, both of which can be characterized by the covariance matrix $\sigma_0^2\boldsymbol{\Sigma}$ [6], where $\sigma_0^2 = N_0/T$ and $[\boldsymbol{\Sigma}]_{j,j'} = R_C((\delta_j - \delta_{j'})T_C)$, $j, j' = 1, \ldots, n$.

47.2.2 Grid Maximum Likelihood Multipath Estimation

In (47.2) and(47.3), N, $\mathbf{X}$ and $\mathbf{Y}$ are all unknown parameters. Consider the parameter estimation in (47.2). The maximum likelihood estimation (MLE) of N, $\boldsymbol{\alpha}$ and $\mathbf{X}$ is

$$(N, \hat{\boldsymbol{\alpha}}, \hat{\mathbf{X}})_{ML} = \arg\max_{N,\boldsymbol{\alpha},\mathbf{X}} L(N, \boldsymbol{\alpha}, \mathbf{X}|\mathbf{R}_I) \tag{47.4}$$

where the likelihood function equals to the probability density function of Gaussian distribution $\mathrm{N}(\mathbf{HX}, \sigma_0^2\boldsymbol{\Sigma})$.

It is quite difficult to solute (47.4), which is a non-linear equation. Following the ideal the grid method proposed in [4] and [5], the multipath is estimated in a grid approach. Assuming that $N = N_{\mathrm{E}}$ and $\boldsymbol{\alpha}$ are fixed, $\boldsymbol{\alpha} = (\alpha_0, \alpha_1, \ldots, \alpha_{N_{\mathrm{E}}})$ is referred to as estimation points and $\mathbf{H}_{\mathrm{E}}$ denotes the corresponding design matrix $\mathbf{H}$. Usually the estimation points are uniformly deployed in the interval $0 < \alpha < 1.5$. Under this assumption, the grid maximum likelihood estimation (GMLE) of $\mathbf{X}^{\alpha}$ can be expressed as

$$\hat{\mathbf{X}}^{\boldsymbol{\alpha}}_{\mathrm{ML}} = \arg\max_{\mathbf{X}^{\alpha}} L(\mathbf{X}^{\alpha} | \mathbf{R}_{\mathrm{I}}) \tag{47.5}$$

the solution of which is

$$\hat{\mathbf{X}}^{\boldsymbol{\alpha}}_{\mathrm{ML}} = (\mathbf{H}_{\mathrm{E}}^{\mathrm{T}} \boldsymbol{\Sigma}^{-1} \mathbf{H}_{\mathrm{E}})^{-1} \mathbf{H}_{\mathrm{E}}^{\mathrm{T}} \boldsymbol{\Sigma}^{-1} \mathbf{R}_{\mathrm{I}} \tag{47.6}$$

The estimation procedure of $\mathbf{X}^{\alpha}$ is to fit the real correlation function using a series of transformed autocorrelation functions $\{\hat{x}^{\alpha_i} R_{\mathrm{C}}(t - \alpha_i) | i = 1, \ldots, N_{\mathrm{E}}\}$ of the C/A code.

Similarly, the GMLE of $\mathbf{Y}^{\alpha}$ can be given by

$$\hat{\mathbf{Y}}^{\boldsymbol{\alpha}}_{\mathrm{ML}} = (\mathbf{H}_{\mathrm{E}}^{\mathrm{T}} \boldsymbol{\Sigma}^{-1} \mathbf{H}_{\mathrm{E}})^{-1} \mathbf{H}_{\mathrm{E}}^{\mathrm{T}} \boldsymbol{\Sigma}^{-1} \mathbf{R}_{\mathrm{Q}} \tag{47.7}$$

The multipath estimation problem is an inverse problem because it is to reconstruct the signals from the samples of the correlation curve. This kind of problems might be ill-conditioned sometimes [7, 8]. Considering (47.6), it can be seen that the GMLE of multipath is equivalent to correlated least squares estimation and $\mathbf{N} = \mathbf{H}_{\mathrm{E}}^{\mathrm{T}} \boldsymbol{\Sigma}^{-1} \mathbf{H}_{\mathrm{E}}$ is referred to as normal matrix. The numerical stability of the MLE decreases if there exists ill-conditioning in normal matrix. The ill-conditioning can be considered by examining the condition number of the normal matrix [9, 10]. In [7], the condition numbers of the normal matrix are listed under several configurations of sample points and estimation points and it is revealed that the normal matrix is ill-conditioned severely, decreasing the numerical stability of the GMLE in most configurations. It is necessary to introduce other estimation methods to mitigate the effect of ill-conditioning.

47.3 TSVD Multipath Estimation with Ill-Conditioning Diagnosis

47.3.1 TSVD Multipath Estimation

To overcome the ill-conditioning in inverse problems, regularization approaches is usually efficient methods [11]. The truncated singular value decomposition regularization is one of the regularization approaches [12, 13], which is simple and easy to implement.

Considering (47.8), The Cholesky decomposition of $\boldsymbol{\Sigma}^{-1}$ is

$$\boldsymbol{\Sigma}^{-1} = \mathbf{W}\mathbf{W}^{\mathrm{T}} \tag{47.8}$$

where $\mathbf{W}$ is a lower triangular matrix.

Letting $\mathbf{L} = \mathbf{W}^{\mathrm{T}}\mathbf{R}_{\mathrm{I}}$, $\mathbf{A} = \mathbf{W}^{\mathrm{T}}\mathbf{H}_{\mathrm{E}}$, the GMLE of $\mathbf{X}^{\boldsymbol{\alpha}}$ is

$$\hat{\mathbf{X}}^{\boldsymbol{\alpha}}_{\mathrm{ML}} = (\mathbf{A}^{\mathrm{T}}\mathbf{A})^{-1}\mathbf{A}^{\mathrm{T}}\mathbf{L} \tag{47.9}$$

The singular value decomposition (SVD) of $\mathbf{A}$ is

$$\mathbf{A} = \mathbf{V}\mathbf{D}\mathbf{U}^{\mathrm{T}} = \sum_{r=1}^{N_{\mathrm{E}}} \sigma_r \mathbf{u}_r \mathbf{v}_r^{\mathrm{T}} \tag{47.10}$$

where $\mathbf{D} = \mathrm{diag}(\sigma_1, \ldots, \sigma_{N_{\mathrm{E}}})$, $\sigma_1 \geq \cdots \geq \sigma_{N_{\mathrm{E}}} > 0$ are the singular values of $\mathbf{A}$, $\mathbf{V}_{N_{\mathrm{E}} \times N_{\mathrm{E}}} = (\mathbf{v}_1, \ldots, \mathbf{v}_{N_{\mathrm{E}}})$ is orthogonal matrix and $\mathbf{U}_{n \times N_{\mathrm{E}}} = (\mathbf{u}_1, \ldots, \mathbf{u}_{N_{\mathrm{E}}})$ is orthogonal in columns.

The MLE of multipath is equivalent to

$$\hat{\mathbf{X}}^{\boldsymbol{\alpha}}_{\mathrm{ML}} = \mathbf{V}\mathbf{D}^{-1}\mathbf{U}^{\mathrm{T}}\mathbf{L} = \sum_{r=1}^{N_{\mathrm{E}}} \frac{\mathbf{u}_r^{\mathrm{T}}\mathbf{L}}{\sigma_r}\mathbf{v}_r \tag{47.11}$$

To overcome the ill-conditioning, (47.11) is truncated, omitting the components corresponding to tiny singular values of (47.11), TSVD estimation of (47.11) is defined as

$$\hat{\mathbf{X}}^{\boldsymbol{\alpha}}_{\mathrm{TSVD}} = \sum_{r=1}^{K} \frac{\mathbf{u}_r^{\mathrm{T}}\mathbf{L}}{\sigma_r}\mathbf{v}_r \tag{47.12}$$

where $K \leq N_{\mathrm{E}}$ is the truncation level. Specially, if $K = N_{\mathrm{E}}$, the TSVD estimation is equivalent to the GMLE.

The variance of the TSVD estimation of the ith parameter $\hat{x}^{\alpha_i}$ is

$$\mathrm{Var}(\hat{x}^{\alpha_i}_{\mathrm{TSVD}}) = \sigma_0^2 \sum_{r=1}^{K} \left(\frac{[\mathbf{V}]_{ir}}{\sigma_r}\right)^2 \tag{47.13}$$

from which it can be seen that the variance of TSVD estimation is smaller when the truncation level K is smaller. But on the other hand, the expectation of TSVD estimation is

$$\mathrm{E}(\hat{\mathbf{X}}^{\boldsymbol{\alpha}}_{\mathrm{TSVD}}) = \mathbf{V}_K\mathbf{V}_K^{\mathrm{T}}\mathrm{E}(\hat{\mathbf{X}}^{\boldsymbol{\alpha}}_{\mathrm{ML}}) \tag{47.14}$$

where $\mathbf{V}_K = (\mathbf{v}_1, \ldots, \mathbf{v}_K)$. It can be seen that the larger the truncation level K is, the smaller bias the TSVD estimation has.

The above two aspects are contradictive, and the properties of the TSVD estimation depends on the selection of the truncation level. The truncation level is simply selected to omit the singular values smaller than a threshold. But in this kind of methods the noise level is not considered. Methods choosing the truncation level considering both the singular values and the noises is appropriate.

47.3.2 Choosing the Truncation Level with Ill-Conditioning Diagnosis

One method choosing the truncation level considering both the singular values is based on ill-conditioning diagnosis.

Let $\mathbf{U}_K = (\mathbf{u}_1, \ldots, \mathbf{u}_K)$, $\mathbf{V}_K = (\mathbf{v}_1, \ldots, \mathbf{v}_K)$ and $\mathbf{D}_K = \mathrm{diag}(\sigma_1, \ldots, \sigma_K)$ while K is the truncation level. The TSVD estimation is equivalent to

$$\hat{\mathbf{X}}^{\alpha}_{\mathrm{TSVD}}(K) = \mathbf{V}_K \mathbf{D}_K^{-1} \mathbf{U}_K^{\mathrm{T}} \mathbf{L} \tag{47.15}$$

Letting $\hat{x}^{\alpha_i}_{\mathrm{TSVD}}(K)$ denote the estimation of the ith parameter if the truncation level is K, we define a statistic to descript the performance of $\hat{x}^{\alpha_i}_{\mathrm{TSVD}}(K)$ as follows [14]

$$F_i(K) = \frac{(N_{\mathrm{E}} - n)(\hat{x}^{\alpha_i}_{\mathrm{TSVD}}(K) - x_0^{\alpha_i})^2 (\mathbf{b}_i^{\mathrm{T}} \mathbf{M}_i \mathbf{b}_i)}{\mathbf{V}(K)^{\mathrm{T}} \mathbf{V}(K)} \tag{47.16}$$

where $x_0^{\alpha_i}$ is a comparing background, usually chosen as 0; $\mathbf{M}_i = \mathbf{I} - \mathbf{B}_{(i)} (\mathbf{B}_{(i)}^{\mathrm{T}} \mathbf{B}_{(i)})^{-1} \mathbf{B}_{(i)}^{\mathrm{T}} \mathbf{P}$, $\mathbf{B} = \mathbf{U}_k \mathbf{D}_k \mathbf{V}_k^{\mathrm{T}}$, $\mathbf{B}_{(i)}$ is the matrix $\mathbf{B}$ after deleting the ith column, $\mathbf{b}_i$ is the ith column of $\mathbf{B}$, $\mathbf{V}(K) = \mathbf{A}\hat{\mathbf{X}}^{\alpha}_{\mathrm{TSVD}} - \mathbf{L}$ is the residual of the TSVD solution. The diagnosis statistic between the N_{E} components of $\hat{\mathbf{X}}^{\alpha}_{\mathrm{TSVD}}$ is expressed as

$$\mathrm{dif}_{\mathrm{LS}}(K) = \frac{\max\limits_{i=1,\ldots,N_{\mathrm{E}}} \{F_i\}}{\min\limits_{i=1,\ldots,N_{\mathrm{E}}} \{F_i\}} \tag{47.17}$$

The TSVD estimation is satisfactory when diagnosis statistic (47.17) is small. In TSVD regularization with ill-conditioning diagnosis, the truncation level K is chosen as the following method: calculate the TSVD estimation under every truncation level, then choose the K which minimizes the diagnosis statistic (47.17).

47.4 Simulation Results

A channel of GPS receiver is simulated. The PRN number is 12 and the IF frequency $f_{\mathrm{IF}} = 4.309$ MHz. The coherent integration time is 0.02 s, i.e., one chip duration of the navigation message code. The carrier-noise-ration (CNR) of the direct signal is 38 dB-Hz. Simulate 2,000 epochs.

Calculate the GMLE and the TSVD estimation of multipath and subtract multipath components estimated with the GMLE and the TSVD estimation to refine the signals. The refined signal is tracked with the ordinary E-L discriminator. In the two estimation approaches, the spacing of sample points is 0.05 T_C and the spacing of estimation points is 0.1 T_C. The truncation level of the TSVD regularization is chosen in two approaches: (1) $K = 7$. The truncation level is chosen following the rule that the singular values smaller than the 0.5 % of the largest one are regarded as tiny and omitted. (2) Truncation level is chosen based on ill-conditioning diagnosis proposed in this paper. Use ordinary E-L discriminator with the correlator spacing 1 duration and narrow correlator with the correlator spacing 0.1 code chip to track with the compound signal for comparison. Calculate the tracking errors of the above five approaches.

For the above five approaches, ε_i denotes error of signal tracking in delay lock loop (DLL) of the ith epoch. Compare two indices of the performances of the three

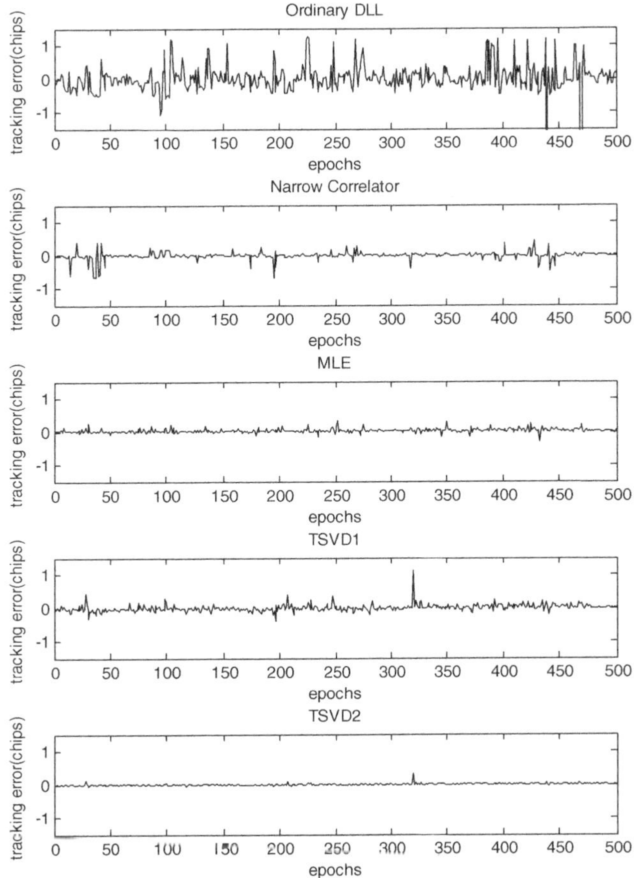

Fig. 47.1 Tracking errors of different approaches

Table 47.1 Comparisons of tracking errors

	Normal DLL	Narrow correlator	GMLE	TSVD(1)	TSVD(2)
Quadratic mean error	0.3727	0.1155	0.0568	0.0961	0.0210
Maximum error	1.8102	0.6836	0.3140	1.1197	0.3280

approaches: the quadratic mean error $\mathrm{Q} = \sqrt{\sum_{i=1}^{S} \varepsilon_i^2 \Big/ S}$ and the maximum error $\mathrm{M} = \max\limits_{1 \le i \le S} \{|\varepsilon_i|\}$ in the S epochs, where $S = 2{,}000$.

The tracking errors are illustrated in Fig. 47.1 and the quadratic mean error and the maximum error are listed in Table 47.1.

It can be shown from Fig. 47.1 and Table 47.1 that, because of the strong interference due to noise, the GMLE is unable to effectively mitigate the multipath effect. The TSVD approach with the truncation level omitting the singular value smaller than a threshold cannot mitigate multipath due to the irrationality of the arbitrary truncation level selection. The TSVD approach based on ill-conditioning diagnosis can still effectively mitigate the multipath effect, locking the signal delay much better the other four approaches.

47.5 Conclusion

The grid maximum likelihood of multipath is simplified using the idea of deploying estimation points uniformly, while in the simplified model there is severe ill-conditioning, which damages the numerical stability of the estimation, resulting in large locking errors. The TSVD regularization with ill-conditioning diagnosis is introduced to mitigate the effects of ill-conditioning. The truncation level is selected based on ill-conditioning diagnosis by comparing the values of the diagnosis statistics of the estimation under different truncation level. Simulations show that under some noise level, the truncation level selected in ill-conditioning diagnosis approach is much better than that in the threshold approach. This method can resist the ill-conditioning and the estimation is stable. The multipath effect can be effectively mitigated by the method proposed in this paper.

Acknowledgments The authors acknowledge the financial support by the Natural Science Foundation of China (40974009, 40474007).

References

1. Kaplan ED, Hegarty CJ (2006) Understanding GPS: principles and applications, 2nd edn. Artech House, Boston
2. van Nee RDJ (1992) The multipath estimating delay lock loop. In: IEEE 2nd International symposium on spread spectrum techniques and applications, Yokohama

3. Fantino M, Dovis F, Wang J (2005) Quality monitoring for multipath affected GPS signals. J Global Positioning Syst 4:151–159
4. Gadallah EA (1998) Global positioning system (GPS) receiver design for multipaths mitigation. Ph. D Thesis, Air Force Institute of Technology Air University, Ohio
5. Gadallah EA, Pachter M, DeVilbiss SL (1998) Design of GPS receiver code and carrier tracking loops for multipath mitigation. In: Proceedings of ION GPS-98, Nashville
6. Blanco-Delgado N, Nunes FD, Xavier JM (2008) A geometrical approach for maximum likelihood estimation of multipath. In: Proceedings of ION GNSS 2008, Savannah
7. Chen K, Gui QM (2011) TSVD regularization in GPS multicorrelator multipath estimation. In: 2011 2nd IEEE International conference on artificial intelligence, management science and electronic commerce, Zhengzhou
8. Hoerl AE, Kennard RW (1970) Ridge regression: biased estimation for non-orthogonal problems. Technometrics 12:55–82
9. Belsley DA, Kuh E, Welsch RE (1980) Regression diagnostics: identifying influential data and sources of collinearity. Wiley, New Jersey
10. Stewart GW, Sun JG (1990) Matrix perturbation theory. Academic Press, New York
11. Tenorio L (2001) Statistical regularization of inverse problems. SIAM Rev 43:347–366
12. Hansen PC (1990) Truncated singular value decomposition solutions to discrete ill-posed problems with ill-determined numerical rank. SIAM J Sci Stat Comput 11:503–518
13. Xu P (1998) Truncated SVD methods for discrete linear ill-posed problems. Geophys J Int 135:505–514
14. Gu YW, Gui QM, Han SH (2010) Regularization based on signal-to-noise index and its application in GPS rapid positioning. Scientia Sinica (Physica, Mechanica & Astronomica) 40:663–668 (in Chinese)

Chapter 48
Power Line Detection Based on Region Growing and Ridge-Based Line Detector

Xiwen Yao, Lei Guo and Tianyun Zhao

Abstract Automatic power line detection is one of the most important and challenging tasks for helicopter's low flight to avoid the power-line-strike accident. In this paper, we propose a novel method to detect power lines in the infrared image. Firstly, region growing is used to get the line support regions. Then a ridge-based line detector is applied to extract the accurate power lines points in the support regions followed by a linking step combined with chain codes analysis to refine the power lines. The experimental results against two state-of-the-art line detection algorithms demonstrate the performance of the proposed algorithm.

Keywords Power line detection · Region growing · Ridge-based line detector

48.1 Introduction

In the helicopter's low altitude search, rescue and reconnaissance operations, power lines impose a considerable hazard, especially in low visible and high cluttered environments. Automatic power line detection is one of the most important and challenging tasks for helicopter to avoid the power-line-strike accident, which has attract more attention in the past few years. So far, many algorithms for power line detection in various imagery sources have been developed.

The Hough transform is a well known line detection method, which was initially employed to detect candidate lines followed by various approaches to

X. Yao (✉) · L. Guo · T. Zhao
School of Automation, Northwestern Polytechnical University,
Xi'an, Shaanxi 710129, China
e-mail: yaoxiwen517@gmail.com

Z. Sun and Z. Deng (eds.), *Proceedings of 2013 Chinese Intelligent Automation Conference*, Lecture Notes in Electrical Engineering 255,
DOI: 10.1007/978-3-642-38460-8_48,

differentiate the cable lines from the noise lines [1, 2], Ma [2] trained a Support Vector Machine based on the Bragg pattern features to perform the classification, while summarized characteristics of power lines are used to discriminate power lines from other linear objects in [3]. However, the methods based on Hough transform are sensitive to noise and detect more false positive lines.

Candamo [4] proposed an edge-based power line detection, applying Canny edge detector combined with morphological filters to create a feature map, and then fit the line model on the feature map. For thick line in the image, this method will detect two parallel lines.

Line Segment Detector (LSD) is a technique that uses Burns' [5] combining pixels having the same gradient orientation to find good candidates for line segments and then validates the candidates using the Helmholtz principle of Desolneux's [6] line validation method. Although LSD produces good results for most types of images, it fails especially in images where the background contains a lot of white noise [7], and cannot get the accurate position of the lines.

Line can also be estimated by extracting the centre line or ridge, Steger [8] proposed a ridge-based line detector which can detect centre line with sub-pixel precision. Since this approach extracts lines as the maxima of the magnitude of the second directional derivatives, it can only detect salient lines [9], which makes it unsuitable for power line detection in the noisy infrared images.

To detect the power line from the infrared images, we propose a novel method based on the region growing and ridge-based line detector. Firstly, we group connected pixels that share the same gradient angle up to a certain tolerance to form the line support regions as described in the algorithm of Burns et al., and then in the line support regions, a ridge-based line detector is used to get the accurate power lines points followed by a linking step combined with a line validation method based on the chain code analysis to differentiate the power lines and the lines formed by the noise points.

48.2 The Proposed Method

48.2.1 Line Support Region

48.2.1.1 Image Gradient

The image gradient reflects a directional change of the intensity between adjacent pixels in an image, which provides two pieces of information that the magnitude of the gradient tells us how quickly the image is changing, while the direction of the gradient tells us the direction in which the image is changing most rapidly.

Burns [5] has shown that the mask appears to be the best choice to estimate gradient magnitude and orientation. Thus, here we define the gradient of the image I at pixel (x, y) as

$$G(x,y) = [(I(x,\ y+1) + I(x+1,\ y+1)) - (I(x,\ y) + I(x+1,\ y)] * i + [(I(x+1,\ y) + I(x+1,\ y+1)) - (I(x,\ y) + I(x,\ y+1))] * j \tag{48.1}$$

where $I(x,y)$ is the intensity value of image I at pixel (x,y).

And the local gradient magnitude and orientation are computed respectively by

$$|G(x,y)| = \sqrt{G_i(x,y)^2 + G_j(x,y)^2} \tag{48.2}$$

$$\theta = \arctan \frac{G_i(x,y)}{G_j(x,y)} \tag{48.3}$$

where $G_i(x,y)$ and $G_j(x,y)$ are the gradient magnitudes in the direction i and j respectively. Figure 48.1 shows the gradient of the infrared image we have computed using the 2×2 mask.

48.2.1.2 Region Growing

To get the line support regions, after we have got the magnitude and orientation of the image gradient, the next step is grouping the connected pixels that share the same gradient orientation.

The purpose of this step is to get the regions that may contain the candidate line segments, so we apply a simple scheme to partition the gradient directions as described in [5]. 360° range of gradient directions is quantized into sixteen intervals, of which each interval is 22.5°.

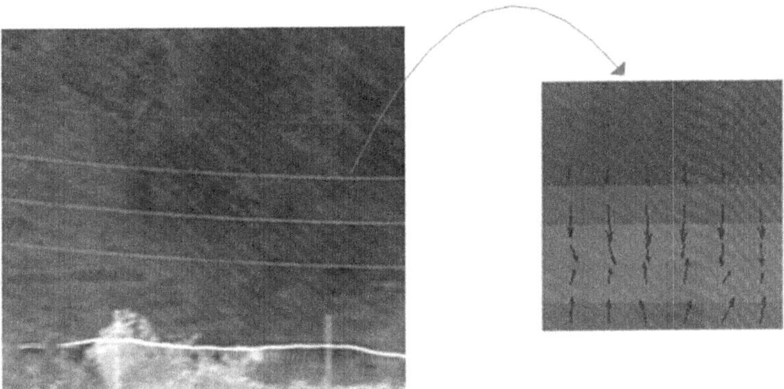

Fig. 48.1 Image gradient

Similar to [6], the first selected pixel to start the region growing is according to the magnitude of gradient. So pixels with larger magnitude will be first tested based on the fact that larger magnitude will show better to forecast the candidate line segments.

Once the pixel is selected, its neighbor pixels are tested. If the difference between the tested pixel's gradient orientation and the original pixel's does not exceed the threshold value that is 22.5° as for our partition scheme, the tested pixel is added to the region and the region gradient orientation is updated to the mean orientation of the region's whole pixels. The process will not end until there are no new pixels to be added.

Figure 48.3d illustrates the region growing results. The results show that region growing can get the line support region even from the noisy infrared image, which facilitates the subsequent processing and improve the performance of our proposed method in some degree.

48.2.2 Ridge-Based Line Points Detector

Ridge-based detector extracts the ridges as line points for which the second derivate is local maximal along the direction perpendicular to the line. The direction that we denote as n can be estimated by calculating the eigenvalues and eigenvectors of the Hessian matrix

$$H(x,y) = \begin{pmatrix} r_{xx} & r_{xy} \\ r_{xy} & r_{yy} \end{pmatrix} \tag{48.4}$$

where $r_{xx}\, r_{xy}\, r_{yy}$ is the partial derivative of the image estimated by convolving the image with discrete two dimensional Gaussian partial derivative kernels.

$$G_{i,j} = \frac{1}{2\pi\sigma^2} e^{-\frac{\left(i-\frac{n+1}{2}\right)^2+\left(j-\frac{n+1}{2}\right)^2}{2\sigma^2}} \tag{48.5}$$

where σ is the standard deviation which depends on the line's width, Steger [8] proved that σ should be set $w/\sqrt{3}$ for good performance in which w is the line's width. The kernel size n of Gaussian is set to 3σ without losing discernible accuracy.

The calculation of Hessian matrix can be done in a numerically stable and efficient way by using one Jacobi rotation to annihilate the r_{xy} term. Let the eigenvector correspond to the eigenvalue of maximum absolute value, i.e., the direction perpendicular to the line, will be given by (n_x, n_y) with $||(n_x, n_y)||_2 = 1$ [8].

And the point (P_x, P_y) is given by

$$(P_x, P_y) = \left(-\frac{r_x n_x^2 + r_x n_x n_y}{r_{xx} n_x^2 + 2r_{xy} n_x n_y + r_{yy} n_y^2}, -\frac{r_y n_y^2 + r_y n_x n_y}{r_{xx} n_x^2 + 2r_{xy} n_x n_y + r_{yy} n_y^2}\right) \in ([-0.5, 0.5], [-0.5, 0.5]) \quad (48.6)$$

48.2.3 Linking with Chain Codes

After the points have been detected, Steger [8] then links the points to find the curvilinear structures by the proposed method. To our specific purpose that is detecting power lines from infrared images, power line validation operation needs to be performed to remove the false positive lines. Inspired by the Steger's linking algorithm, we proposed a refined linking algorithm that introduces a line validation method based on the chain codes to get the final power line results.

Chain codes are one of the digital curves representations based on 4-connectivity or 8-connectivity of the segments. As for 8-connectivity scheme, the direction of each segments is encoded by using a numbering scheme $\{i|i = 0, 1, 2, \ldots, 7\}$ denoting an angle of $45° \times i$ counter-clockwise from the positive x-axis, as shown in Fig. 48.2 [10]. The chain codes can be viewed as a connected neighborhood operation which can be easily added in the linking step as a validation method to discriminate power lines from other linear objects.

Similar to Steger [8], we select the pixel with maximum second derivative to start the linking step, the appropriate neighbor to the current line will be added in and meanwhile the chain codes related to the line will be constructed. If in the line support region, there are no new points to be added in, the obtained chain codes will be examined. For an approximately straight power line segment within a certain range, only a single direction or two alternating direction will be the main

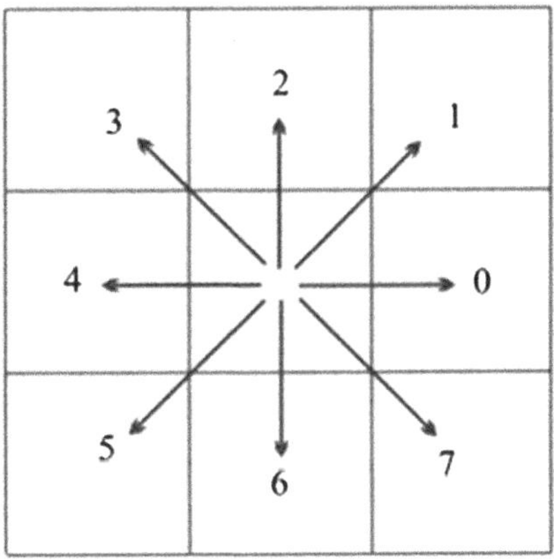

Fig. 48.2 Chain codes in eight directions

state of the chain codes. Based on the simplicity truth, we can easily remove the unrelated linked curves obviously formed by the noise.

48.3 Experiments and Discussions

A thorough evaluation of the proposed power line detection method is presented in this section. The experiments are performed on a set of image captured from 3 to 5 micron refrigeration remote infrared camera. Figure 48.3 shows the results of the proposed power line detection compared with LSD [6] and Steger's ridge-based method [8].

From Fig. 48.3 we can clearly see that the proposed power line detection can get better results than LSD and Steger's ridge-based method. Steger's ridge based method extracts lines as the maxima of the magnitude of the second directional derivatives, so it can detect only salient lines which can be seen from Fig. 48.3b. Although LSD gets a lot of false segments compared with Steger's method and our proposed method, this method can find the support regions even around the thin blurred lines as shown in Fig. 48.3d. Our algorithm combines the advantages of the above two methods, we detect power lines using the ridge-based detector in the support regions obtained by LSD and validate the detected lines based on the chain codes. The validation step can easily remove other linear objects like wire tower and lines formed by the noise points which can be seen from the middle row of Fig. 48.3.

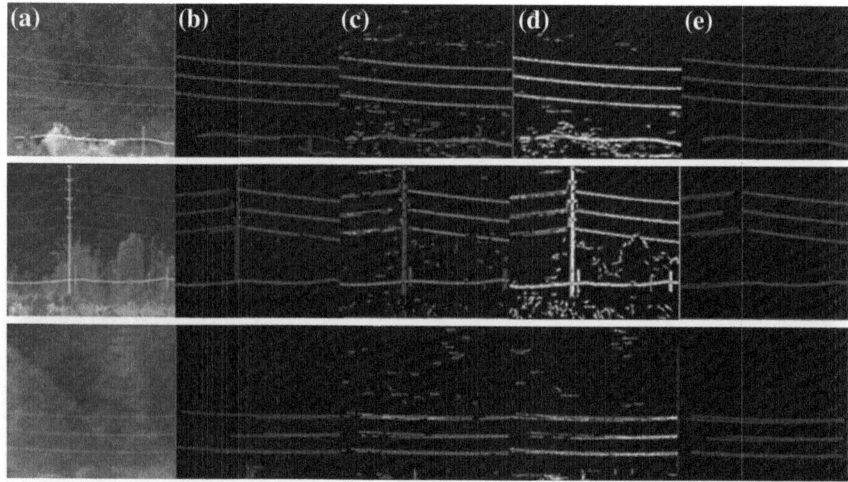

Fig. 48.3 (**a**) are the original images. The Steger's results are shown in (**b**) with LSD in (**c**). (**d**) are the line support regions and (**e**) are our algorithm's results

48.4 Conclusion

This paper presented a novel method based on region growing and ridge-based line detector to detect power lines in the infrared image. There are three steps involved in the proposed method. First, adjacent pixels that share the same gradient orientation are grouped to form the line support regions and then a ridge-based line detector is used to extract the accurate power lines points in the support regions. Finally, in the linking process we apply a line validation method based on the chain code analysis to remove the lines formed by the noise and other linear structures. Experiments results show that the proposed method can reduce false positive lines and extract the accurate power lines from infrared images and outperforms Steger's method and LSD.

Acknowledgments The work is supported by graduate starting seed fund of Northwestern Polytechnical University (NO. Z2012116).

References

1. Zhang J, Liu L, Wang B, Chen X, Wang Q, Zheng T, Jinan S (2011) High speed automatic power line detection and tracking for a UAV-based inspection. Am J Eng Technol Res 11(12):266–268
2. Ma Q, Goshi DS, Shih YC, Sun MT (2011) An algorithm for power line detection and warning based on a millimeter-wave radar video. IEEE Trans Image Process 20(12):3534–3543
3. Li Z, Liu Y, Hayward R, Zhang J, Cai J (2008) Knowledge-based power line detection for UAV surveillance and inspection systems. In: 2008 23rd International conference on image and vision computing New Zealand (IVCNZ), pp 1–6
4. Candamo J, Goldgof D (2008) Wire detection in low-altitude, urban, and low-quality video frames. In: 2008 19th International conference on pattern recognition (ICPR), pp 1–4
5. Burns JB, Hanson AR, Riseman EM (1986) Extracting straight lines. IEEE Trans Pattern Anal Mach Intell 8(4):425–455
6. Von Gioi RG, Jakubowicz J, Morel JM, Randall G (2010) LSD: a fast line segment detector with a false detection control. IEEE Trans Pattern Anal Mach Intell 32(4):722–732
7. Akinlar C, Topal C (2011) Edlines: real-time line segment detection by Edge Drawing (ed). In: 2011 18th IEEE international conference on image processing (ICIP), pp 2837–2840
8. Steger C (1998) An unbiased detector of curvilinear structures. IEEE Trans Pattern Anal Mach Intell 20(2):113–125
9. Liu L, Zhang D, You J (2007) Detecting wide lines using isotropic nonlinear filtering. IEEE Trans Image Process 16(6):1584–1595
10. Kui Y, Žalik B (2005) An efficient chain code with Huffman coding. Pattern Recogn 38(4):553–557

Chapter 49
The Implementation of the HTTP-Based Network Storage Queue Service

Bo Yu, Limin Jia, Guoqiang Cai and Honghui Dong

Abstract Network storage queue is essential for train network information collection. We developed an application of network storage on the train network queue. Based on the Hypertext Transfer Protocol, this queue can ensure security of network operation and enable effective information transfer. This queue provides an efficient way of data transmission from train network to B/S system. Network storage queue includes a real-time database Redis and an expandable external interface service layer. The benchmark shows that compared with traditional database storage, the network queue with higher performance, can ensure the safety of the train data transmission.

Keywords Redis · HTTP · Network storage · Queue service

49.1 Introduction

The development of train reliability, train control and network security has been in an important position. In a typical train control system, there exists an endless stream of data transmitted to the diagnostic server. Train diagnostic server should equip with a large capacity, high-speed storage system to ensure its normal operation. The emergence of Redis provides us with such a choice. Redis is a key-value storage system. It supports the storage of value types include string (string), list (linked list), set (collection) and zset (ordered collection). These data types are supported to push/pop, add/remove and take the intersection, these operations is atomic [1, 2]. On this basis, redis support a variety of ways of sortation. In order to ensure efficiency of data storage, data is cached in memory, and periodically updated data is written to disk or to modify the operation to write additional log files. Hypertext Transfer Association

B. Yu (✉) · L. Jia · G. Cai · H. Dong
Traffic Safety Engineering, Beijing Jiaotong University, Beijing 100044, China
e-mail: yusinuo@163.com

Z. Sun and Z. Deng (eds.), *Proceedings of 2013 Chinese Intelligent Automation Conference*, Lecture Notes in Electrical Engineering 255,
DOI: 10.1007/978-3-642-38460-8_49,

is a simple, fast, flexible, connectionless, stateless network transport protocol, and has good compatibility, easy to use computer language processing, suitable for use on trains in transit monitoring field [3, 4]. This article describes the Hypertext Transfer Protocol-based network storage queue which support a variety of methods with train data cache encrypted operation.

49.2 System Architecture Design and Implementation

The database server running Redis database services, mainly using disk array memory database persistence [5], uses the TCP protocol to communicate the application server. As shown on Fig 49.1.

TCP services provided by the main features are

1. Connection oriented transmission;
2. The end-to-end communication;
3. High reliability, ensuring the correctness of data transmission, does not appear lost or out of order;
4. Full duplex transmission;
5. The byte stream mode, i.e., in bytes transmitted byte sequence;
6. Urgent data transmission function.

Thus encapsulated a collection of method is need for required communication operation as TcpConnect, TcpRead, TcpResolve, TcpServer, TcpKeepAlive and so on.

49.3 Underlying Service Design and Implementation

As a service, it should support user interaction. According to the monitoring environment, train sensor via the RJ45 interface transfer data to the network storage queue, and train diagnostic hosts get data from the store queue for analysis, this needs an expansion of a series of operating interface to the underlying service.

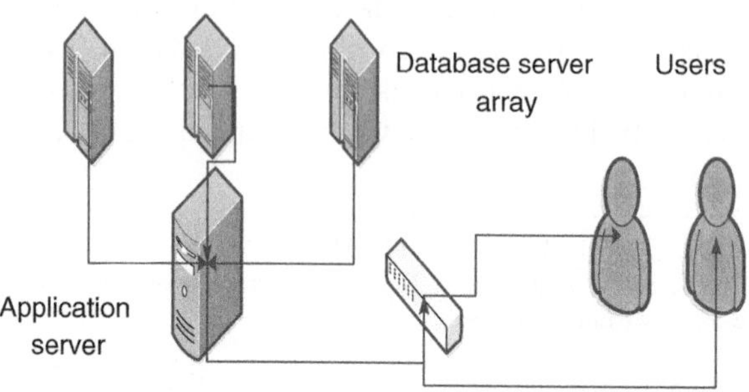

Fig. 49.1 Structure of the system design

49.3.1 Mode of Operation Set

It uses a typical list structure to form a storage queue, the input parameters for the queue name, operation method, the content of the data. The URL parameter of "opt" specifies the type of operation [6].

1. The queue operation of inputting data
   Option "put", now need to input queue name is the name field, into the queue operation need to enter authkey key field, whether the backup queue field backup, typical operation command to http://domain:portnum/?Name=queue_name&opt=put&data=xxxxx&authkey=xxx&backup=1, data for the input data content. The return value is the implementation of information operation. The successful implementation of httpmqs_success is returned, it returns httpmqs_failed.
2. Queue operation of getting data
   Option " get", then input queue name but no input data content, typical operation command to http://domain:portnum/?Name=queue_name&opt=get&authkey=xxx. The success of operation returns to obtain numerical operation failed queue, returns httpmqs_failed. If the operation succeeds then delete the head of the queue element.
3. Queue operation of observing queue state
   Option "status", a typical operation command is http://domain:protnum/?Name=queue_name&opt=status&authkey=XXX, then returns the input queue state information. Return the string as JSON format such as: {"name":"queue_name","length":6,"charset":"UTF-8"}
4. Queue operation of saving to disk
   Option "Sava", typical operation command to http://domain:protnum/&opt=save&authkey=XXX, then the database data persistence to the hard disk, data loss prevention.

49.3.2 Algorithm of Service Realization

We encapsulate a method for all operations, while the operations are not the same, it judges the function parameter, and enter the corresponding processing unit for processing, so we define a function as followed: void (struct httpsqs_handler evhttp_request * req, void * ARG), evhttp_request is a structure of evhttp, which includes the HTTP request information.

1. Message preprocessing
   When obtaining an operation request, (1) analysis of the HTTP request additional parameter values, char * decode_uri = strdup ((char) evhttp_request_uri (req)), (2) using the evkeyvalq structure definition of a linked list, linked list includes a key and a value, the other is a structure body pointer, the additional parameter value input to the list of struct evkeyvalq httpsqs_http_query;

evhttp_parse_query (decode_uri, &httpsqs_http_query), (3) release parameter value free (decode_uri).

2. The message processing
According to option processing unit selection, first get option of "opt" const char * httpsqs_input_opt = evhttp_find_header (&httpsqs_http_query,"opt") and then do option comparison, according to the options on the database server for operation.
3. Parameter setting of http header
According to the processing conditions, use int evhttp_add_header (evkeyvalq * char * struct, const, const char HTTP headers *) to fill [7].

49.4 Realization of the Network Storage Queue

The network storage queue introduced is part of the whole subway train safety monitoring system, but it is necessary for the subway train monitoring system, it ensures the whole system operating stable and secured [8]. HTTP network storage queue run under the Linux platform, using C language development, as a service, select the system automatically start running.

The store queue software with modular design, the whole project consists of 3 modulecs, (1) the software process module, is mainly responsible for the process of opening, the configuration file initialization process, signal setting, (2) http event handling module, because the store queue to run on HTTP, this module is mainly responsible for the response to HTTP on request information and call the interaction module, finally returning for results, (3) interact with the Redis server module, the module is mainly responsible for the Redis database operation and Redis database response information. The store queue operation processes as shown on Fig 49.2.

49.5 Application of Network Storage and Test

This service origin from National 863 project, the network storage queue service has been tested and put into use. Testing environment of Core2processor, 128G SSD, operated system is Redhat 5, the network is local area network 100 M.

49.5.1 Application of HTTP-Based Network Storage Queue Service

In real time train detection system vehicle sensor network collect data and then send it into the network storage queue, the http-based network storage queue service caches it, and at the same time establishing fail queue and retry queue, when the operation fails, data into the retry queue, waiting to try again, in retry

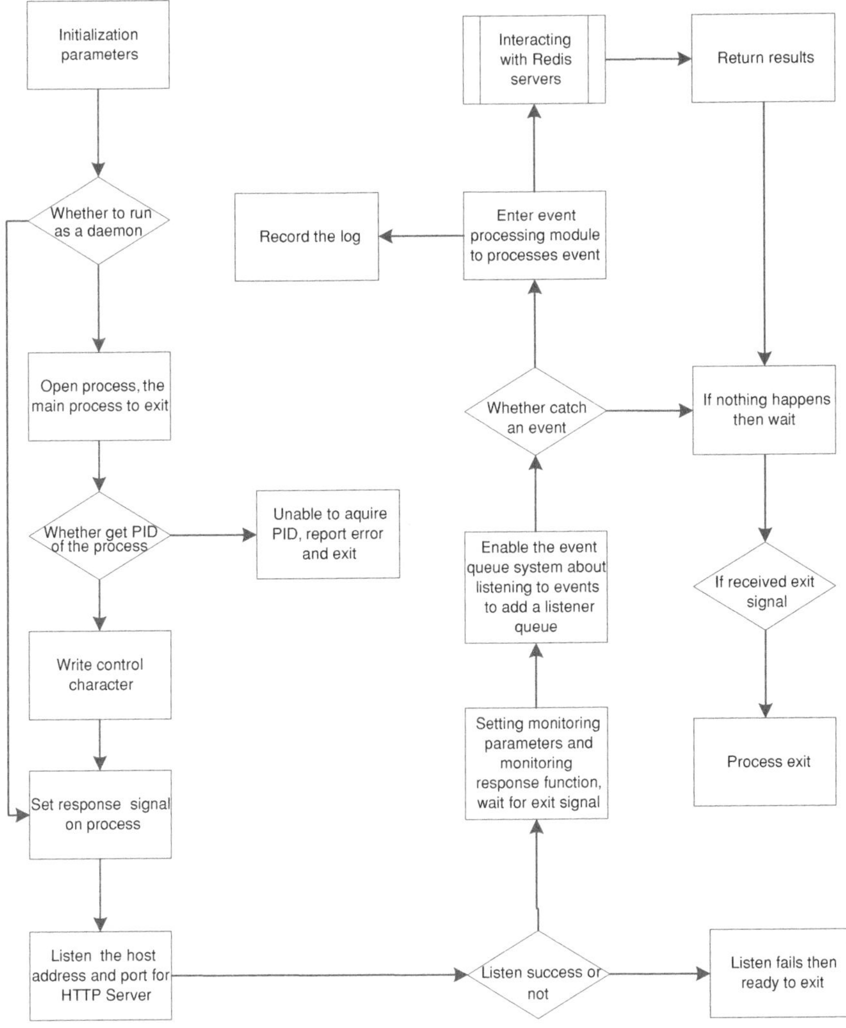

Fig. 49.2 Schematic diagram of the software life cycle

queue to retry, a failed times as the config defined, the operation will be put into manual processing process. The data stream shows on Fig. 49.3.

49.5.2 Network Storage Queue Performance Benchmark

Using the Apache AB command pressure test on 30 threads, insert 1,024 bytes into the queue, with http keep-alive, the result on average is about 7983.57 operations

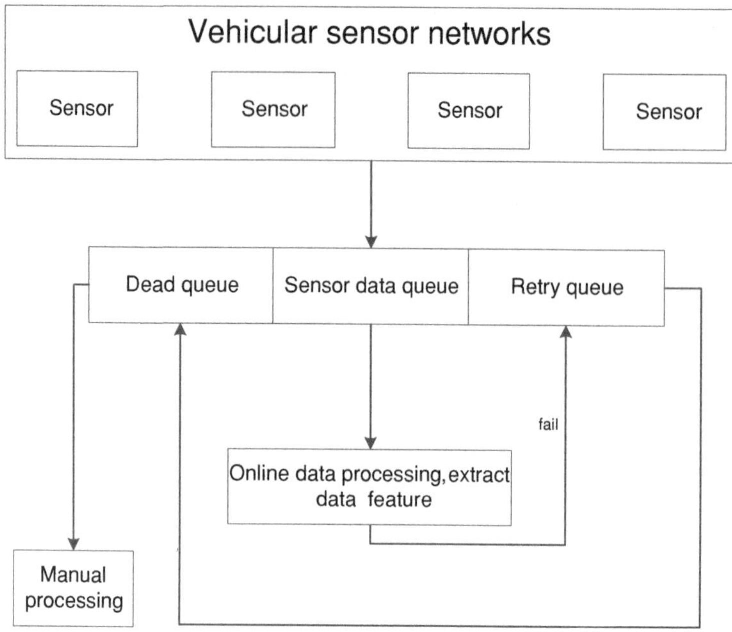

Fig. 49.3 Train production environment typical application architecture diagram

per second. Compared with Mysql of less than 1,000 operations per second, this http message queue is much more powerful. The real test result:

[root@localhost yatiomqs] # ab -k -c 10 -n 100000 "http://127.0.0.1:1680/?name=test&opt=put&data=....."

This is ApacheBench, Version 2.3 <$Revision: 655654 $>

Copyright 1996 Adam Twiss, Zeus Technology Ltd, http://www.zeustech.net/

Licensed to The Apache Software Foundation, http://www.apache.org/

Benchmarking 127.0.0.1 (be patient)

Finished 100,000 requests

Server Hostname: 127.0.0.1

Server Port: 1680

DocumentPath: /?name = test&opt = put&data=".....”

Document Length: 134 bytes

Concurrency Level: 10

Time taken for tests: 12.526 s

Complete requests: 100,000

Failed requests: 0

Write errors: 0

Non-2xx responses: 100,000

Keep-Alive requests: 0

Total transferred: 2,64,00,000 bytes

HTML transferred: 1,34,00,000 bytes

Requests per second: 7983.57 [#/sec] (mean)
Time per request: 1.253 [ms] (mean)
Time per request: 0.125 [ms] (mean, across all concurrent requests)
Transfer rate: 2058.26 [Kbytes/sec] received

49.6 The Ending

This paper proposes the use of a HTTP protocol to establish network storage queue implementation method, and gives the relevant algorithm and method of realization. Theoretical analysis and simulation results show that the model of HTTP protocol and real-time database Redis combination can better play the role of real time database, has wide range of application. And the separation of service layer and the business layer enables that if a new real time database is developed and it is more suitable for use to use, we can replace the service layer for other very convenient. Further research will be done according to the actual application of improving the architecture, mostly on the expansion of business layer function.

Acknowledgments Supported by The National High Technology Research and Development Program of China (Grant No. 2011AA110505), Beijing Jiaotong University Project (RCS) RCS2008ZZ004.

References

1. Liao G, Liu X, Xu B (2012) Design and realization of sales promotion display system based on J2EE + Redis + Greasemonkey. Softw Guide 96-98 (In Chinese)
2. Jiang H, Yang Q (2011) A keyword-based query solution for native XML database. International conference on internet technology and applications, iTAP 2011
3. Cohen E, Kaplan H, Oldham J (1999) Managing TCP connections under persistent HTTP. Comput Netw 31(11):1709–1723
4. Feng XX, Peng Y, Zhao YL (2011) The analysis of a botnet based on HTTP protocol. Adv Mater Res 575–579
5. Zhang X, Wei Y (2010) Research of functionally graded materials database. 2nd international workshop on education technology and computer science, pp 655–658
6. Davern P, Nashid N, Zahran A, Sreenan CJ (2011) Towards an automated client-side framework for evaluating HTTP/TCP performance. International symposium on performance evaluation of computer and telecommunication systems, SPECTS 2011, pp 205–212
7. Natarajan P, Baker F, Amer P (2009) Multiple TCP connections improve HTTP throughput–Myth or fact? IEEE 28th international performance computing and communications conference, IPCCC 2009, pp 41–48
8. Neumann D, Bodenstein C, Rana OF, Krishnaswamy R (2011) STACEE: enhancing storage clouds using edge devices. 8th international conference on autonomic computing, ICAC 2011, pp 19–26

Chapter 50
The Visual Internet of Things System Based on Depth Camera

Xucong Zhang, Xiaoyun Wang and Yingmin Jia

Abstract The Visual Internet of Things is an important part of information technology. It is proposed to strength the system with atomic visual label by taking visual camera as the sensor. Unfortunately, the traditional color camera is greatly influenced by the condition of illumination, and suffers from the low detection accuracy. To solve that problem, we build a new Visual Internet of Things with depth camera. The new system takes advantage of the illumination invariant of depth information and rich texture of color information to label the objects in the scene. We use Kinect as the sensor to get the color and depth information of the scene, modify the traditional computer vision technology for the combinatorial information to label target object, and return the result to user interface. We set up the hardware platform and the real application validates the robust and high precision of the system.

Keywords Vision internet of things · Kinect · Depth camera · Detection

50.1 Introduction

The Internet of Things (IoT) leads the trend of the next information technology and creates the communication between "things". Most of the Internet of things systems utilize the RFID or other non-contact wireless technology as their sensor and achieved successes in the past. However, the RFID label has to be attached on every object for recognition, which cannot be implemented in some situations. Besides, the cost of RFID labels should under consideration when there are huge amounts of objects. The Visual Internet of Things (VIoT) is proposed to provide a

X. Zhang (✉) · X. Wang · Y. Jia
The Seventh Research Division and the Department of Systems and Control, Beijing University of Aeronautics and Astronautics, F907, Xinzhu Building, NO 37 Xueyuan Road, Haidian District, Beijing, China
e-mail: yongshengsiling@163.com

Z. Sun and Z. Deng (eds.), *Proceedings of 2013 Chinese Intelligent Automation Conference*, Lecture Notes in Electrical Engineering 255,
DOI: 10.1007/978-3-642-38460-8_50, © Springer-Verlag Berlin Heidelberg 2013

visual method to access the object labels. With the help of visual cameras, the VIoT can get the object location via image information of the scene, and attach the visual label to the object, then return the label to the information network.

The color camera used by VIoT based on the passive light source, can be greatly influenced by the change of illumination condition, and may result in the serious performance decline. The average precision of the best object detection [1] based on the color image is still in a low stage according to the Pascal VOC challenge. So the object detection based on the color image can not match the requests of the real application in the VIoT.

To overcome the shortcoming of color image provided by current visual camera, we proposed to take depth camera as the sensor of VIoT. The depth camera can generate the depth image of the scene, in which every pixel indicates the distance between the point and camera. The depth camera offers a new dimension of the scene and will change the detection strategy profoundly.

The depth camera we used is the Microsoft Kinect sensor [2]. The Kinect obtain the depth information of the scene by the light code technology with active light source. Besides, the Kinect equipped with another color camera, so we can obtain the depth image and color image of the scene at the same time. In the process of building the new VIoT, we restrict our attention to the monocular case, where a serial image processing is to analyze the image and detect the target from the depth and color image. Once we get the location of the target, the visual labels will be attached on it and return the information to the system, thus we achieve the function of integrate VIoT.

50.2 System Architecture of VIoT

According to the mainstream VIT, our VIoT include three parts: perception, information processing and application. The Fig. 50.1 shows the whole architecture of our VIoT, and the following is detailed introduction.

50.3 Perception

The perception part of traditional VIT consists of RFID or other wireless sensor, while we use the Kinect as the sensor of our novel VIoT. The Kinect is illustrated in Fig. 50.2

The Kinect sensor can real-time get the color image and the depth image of the scene at the same time. Each image is VGA (480×640), and the color image is three channels while the depth image is only one channel. Every pixel of the depth image indicates the distance between the point of scene and camera with millimeter precision.

First, we make the calibration between color and depth image, because the initial data provided by Kinect is not calibrated. Second, we find that the initial

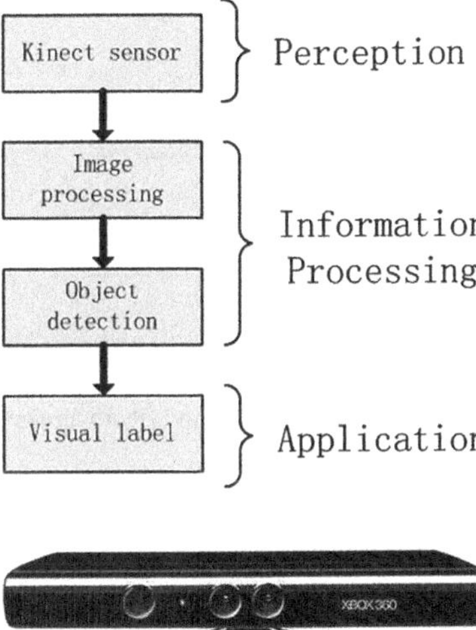

Fig. 50.1 System architecture of VIoT

Fig. 50.2 Kinect

depth data provided by Kinect is very noisy and incomplete. Noisy refers to variations between 2 and 4 different discrete depth levels. And the incomplete data which means the 0 value points of the depth data, will appears on the specular surface, edge of object, black region etc. We smooth the depth data by taking the mean over 9 depth frames for the noisy region. For incomplete regions we use a modified median filter: for every zero value point, we collect the depth value with a 5×5 pixel window and calculate the median of non-zero value as the new value of this point. After the two steps, we get the depth image ready for the next processing. On the other hand, the color image just smoothed by a single general Gaussian filtration. The Fig. 50.3 shows the color picture and depth picture after the steps above.

50.4 Information Processing

The procedure of information processing is illustrated in Fig. 50.4

We separate the data from Kinect to be color image and depth image. The color image is just smoothed by the methods introduced before. For the depth image, the depth information is obtained by the active light source, so it is independent of illumination condition and shadow. We take full advantage of it to use the depth image to build the background subtraction model. For the color image and depth image is calibrated, the foreground region of them have the same location in the image, so we can get the foreground region of the color image by contrasting with depth image. We fuse the color and depth information carefully to achieve both the rich feature and illumination variant. Then the modified HOG feature is extracted on the fused data. On the last step of information processing, we will implement

Fig. 50.3 Color image and depth image

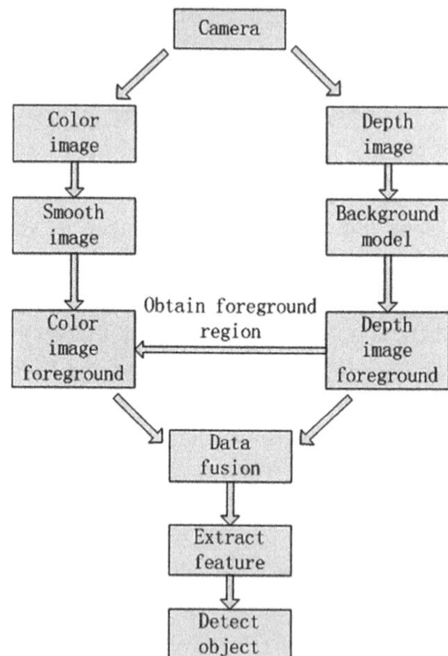

Fig. 50.4 Procedure of information processing

the object detection with a trained model, which learned from the labeled samples with support vector machine (SVM) [3]. In the following, we will introduce the important parts of the information processing.

50.4.1 Background Subtraction Model

The background subtraction model has been researched for decades, the recent stat-of-art methods are the Gaussian mixture model Maddalena [4], Zivkovic [5] and Barnich [6] according to the [7]. But all the methods applied for the color image will be great influenced by the illumination condition and shadow. The flash and other external interference will also be fatal for those methods. For the illumination invariant of the depth image, it will be the perfect source data for the background subtraction model. Although there are still much noise in the depth image from Kinect sensor, but that kind of noise will be eliminated easily. For the depth information of the image will changes only if there is really something moving in the scene, so we utilized the simply Gaussian model for the background, which can be described as a Gaussian function:

$$f(x) = \frac{1}{\sqrt{2\pi}\sigma} e^{-\frac{(x-\mu)^2}{2\sigma^2}} \quad (50.1)$$

where the x means the time sequence of pixel.

We compare the top background subtraction models with ours, and the performance is same but our model is faster and resource is economized.

When object moving in or out the scene, the background subtraction model will give the depth foreground region of the object, and we compare the region with color image to get the color foreground of the object. The procedure is illustrated in Fig. 50.5

50.4.2 Data Fusion and Feature Extraction

Since the pioneer of Kinect application, how to take advantage of the depth data had to be a hot topic again. The Microsoft researchers proposed "Depth image features" [8] for the Xbox application, which archived great success in the human pose recognition, but the method need large train dataset. The [9] evaluated the application HOG feature on depth images for recognition, and then extracted the

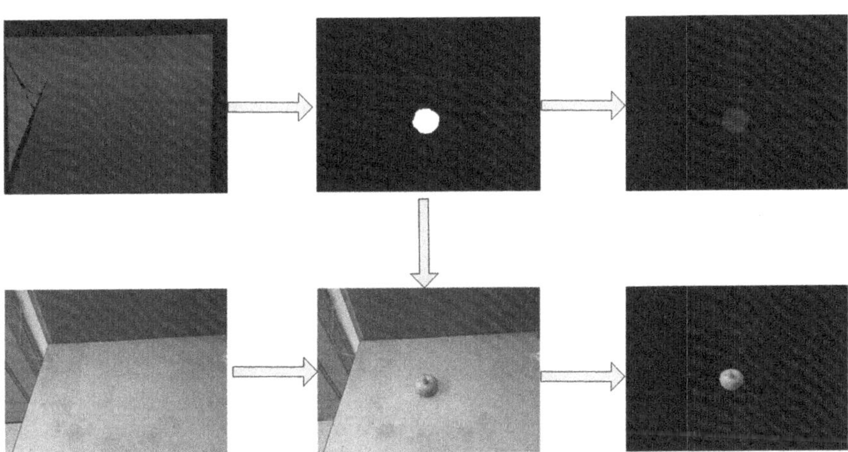

Fig. 50.5 Background subtraction

better 3D point cloud descriptor viewpoint feature histogram (VFH) in the point cloud, further combined it with the DPM [10] based on color image. Another attempt is [11], which proposed a simple "Average" descriptor for the depth data.

In the above works, the descriptors for the depth data are based on the single depth channel or weak connection with color information. The traditional color descriptors can't explore all the potential of depth, independent depth feature is not the perfect way either.

In this paper, we combine the color information and depth information to be a fused image, which include four channels: three for the RGB of color image and one for the depth image. We choose the HOG [12] as the feature of the fusion image, for its geometrical and optical transformation invariance. The feature is extracted for each channel of every point and cascaded to a vector. Thus the feature consists of color feature and depth feature. For the scale invariance, we resize the fusion image for HOG extracting and cascade all features to be a long vector as the description of the point.

50.4.3 Object Detection

We train the object model on the positive and negative samples, which can be collect off-line. Besides, our VIoT allow the on-line training, which means the user can select the object samples and random negative samples in the scene for training the model. We use the LibSVM [13] to train the model.

In detection, although the background subtraction model gives the foreground, we still should give the window of image to the model and get a confidence score when more than one objects in the scene. The traditional way to get the windows is sliding a fixed scale window in pyramid image. It is time-consuming. Inspired by [14], we utilize the image segmentation to get the independent parts of the foreground based on the fusion image. We build the frequency histogram for every channel of every part in foreground. If the difference between neighbor histograms is larger than threshold, the neighbor parts will merge to be one along with their frequency histogram. The segmentation result is illustrated in Fig. 50.6.

With the help of image segmentation, we take the independent parts to object model and get confident scores, which determine whether the part is targeted.

50.5 Application

After obtaining the object location, the information communication and feedback become an easy task. We program the user interface with C++ to achieve function of VIoT, including operation of Kinect sensor, information processing, collecting samples in the scene, parameter setting and object detection. The main interface of the system is illustrated in Fig. 50.7.

Fig. 50.6 Image segmentation based on fusion image

Fig. 50.7 Main interface of VIoT

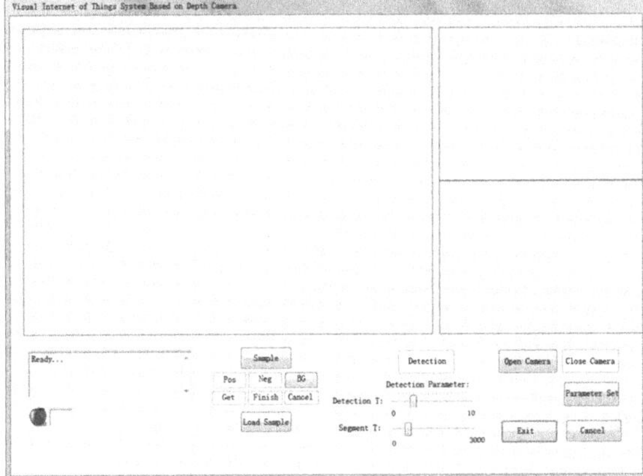

The system will get the color image and depth image of the scene, and we can select the object we want to train, as show in Fig. 50.8, where the left picture is the color image, the down-right picture is the depth image, and the top-right picture is the color foreground.

Fig. 50.8 Function of VIoT

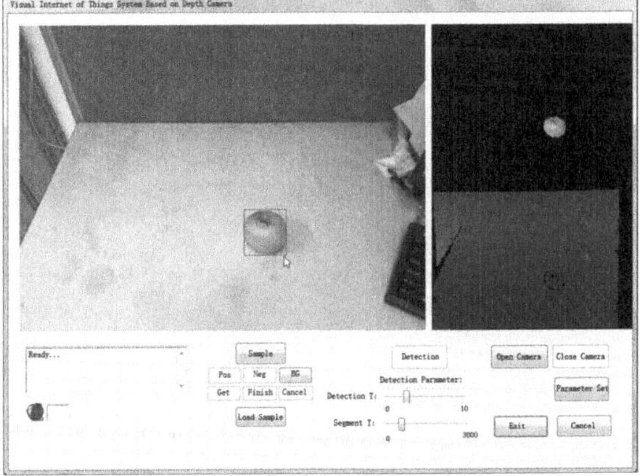

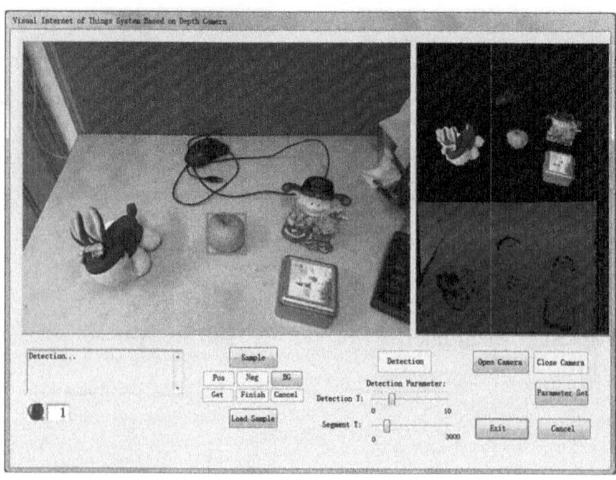

Fig. 50.9 Object detection

In the detection, the system will detect the target object in the scene and label it with yellow rectangle, as shown in the Fig. 50.9. The number of target object is on the down-left of the interface.

50.6 Conclusion

In this paper, the depth camera is first applied in the Visual Internet of Things. The proposed VIoT can get the color and depth information of the scene, and take full advantage of them to get the target object location in the scene. Besides, the fusion of color and depth has been discussed and the feature extraction method has been modified for the novel fusion image. The real application validates the robust and high precision of the system.

References

1. Van de Sande KEA, Uijlings JRR, Gevers T, Smeulders AWM (2011) Segmentation as selective search for object recognition. Computer Vision (ICCV), 2011 IEEE International Conference
2. Microsoft Corp. http://www.xbox.com/kinect
3. Suykens JAK (2001) Support vector machines: a nonlinear modelling and control perspective. Eur J Control 2001(7):311–327
4. Maddalena L, Petrosino A (2008) A self-organizing approach to background subtraction for visual surveillance applications. IEEE Trans Image Process 17(7):1168–1177
5. Zivkovic Z, van der Heijden F (2006) Efficient adaptive density estimation per image pixel for the task of background subtraction. Pattern Recogn Lett 27:773–780
6. Barnich O, Van Droogenbroeck M (2009) Vibe: a powerful random technique to estimate the background in video sequences. In: IEEE International Conference on Acoustics, Speech and Signal Processing, pp 945–948

7. Brutzer S, Hoferlin B, Heidemann G (2011) Evaluation of background subtraction techniques for video surveillance, computer vision and pattern recognition (CVPR), 2011 IEEE conference on 2011
8. Shotton J, Fitzgibbon A, Cook M, Sharp T, Finocchio M, Moore R, Kipman A, Blake A (2011) Real-time human pose recognition in parts from single depth images. In Computer Vision and Pattern Recognition (CVPR), 2011 IEEE Conference on, pp 1297–1304
9. Susanto W, Rohrbach M, Schiele B (2012) 3D object detection with multiple kinects. In computer vision–ECCV 2012. Workshops and demonstrations, pp 93–102. Springer, NY
10. Felzenszwalb PF, Girshick RB, McAllester D, Ramanan D (2010) Object detection with discriminatively trained part-based models. Pattern analysis and machine intelligence, IEEE transactions on, 32(9):1627–1645
11. Jourdheuil L, Allezard N, Chateau T, Chesnais T (2012) Heterogeneous adaboost with real-time constraints-application to the detection of pedestrians by stereovision. In VISAPP (1), pp 539–546. SciTePress, Berlin
12. Dalal N, Triggs B (2005) Histograms of oriented gradients for human detection. Computer vision and pattern recognition, 2005. IEEE computer society conference on, vol 1, pp 886–893
13. Chang C-C, Lin C-J (2011) LIBSVM: a library for support vector machines. ACM Trans Intell Syst Technol 2(27):1–27
14. Felzenszwalb PF, Huttenlocher DP (2004) Efficient graph-based image segmentation. Int J Comput Vision 2(59):167–181

Chapter 51
Design and Implementation of Fire-Alarming System for Indoor Environment Based on Wireless Sensor Networks

Xiuwen Fu, Wenfeng Li and Lin Yang

Abstract Wireless sensor network (WSN) as an emerging networking technology with strong capability to locally and collaboratively sense and interpret data from environment, has been widely implemented in various emergency situations including firefighting. In this paper, we propose a fire-alarming system called fire-alarming system for indoor environment (FASIE) which innovatively integrates wireless fire-alarming network with handheld fire-rescuing support system. Unlike traditional firefighting networks with implementation of WSN, FASIE is capable of providing full service to firefighting activities including fire-alarming, fire-rescuing and firefighter orientation. Aiming to better illustrate FASIE, we first give an overall description of FASIE and presents the architecture of the system. And then, the hardware and system performance are introduced.

Keywords Wireless Sensor Networks (WSN) · Fire-alarming system for indoor environment · Handheld fire-rescuing support system

51.1 Introduction

As new fabrication and integration technologies reduce the cost and size of wireless micro-sensors, we are witnessing another revolution that facilitates observation and control of our life and physical world, just as networking technologies have done for the ways that individuals and organizations exchange information [1]. Networks of interconnected tiny sensors deeply embedded into the physical environment enable us to observe and interact with the environment and human beings in real time in a fidelity that was previously unobtainable.

X. Fu (✉) · W. Li · L. Yang
Department of Logistics Engineering, Wuhan University of Technology, 1040#, Heping Road, 430063 Wuhan, People Rebuplic of China
e-mail: XiuwenFu@whut.edu.cn

Z. Sun and Z. Deng (eds.), *Proceedings of 2013 Chinese Intelligent Automation Conference*, Lecture Notes in Electrical Engineering 255,
DOI: 10.1007/978-3-642-38460-8_51,

Firefighting application as one of the most influential aspects to public safety activity, has also gained significant benefits from emergence of WSN and its further development [2]. In this paper, we present a wireless fire-alarming system fire-alarming system for indoor environment (FASIE) whose main contribution is to provide comprehensive technical support for fire-alarming and rescuing by introducing wireless fire-alarming network and handheld fire-rescuing support system.

The rest of this paper is organized as follows. Section 51.2 presents the state of the art of the existing fire-alarming systems. Section 51.3 introduces FASIE from the perspectives of system components and functions. Section 51.4 presents the layered architecture of FASIE. Section 51.5 presents the key hardware involved in FASIE. Section 51.6 analyses the performance of FASIE. Finally, the conclusion and future work are presented in Sect. 51.7.

51.2 Related Works

Due to the enormous application potential represented by WSN in the domain of firefighting, many firefighting systems are developed based on WSN. But in most cases, the only purpose of the network consists in acquiring environmental data and these data are gathered and displayed in a base station, stored in a database or sent to a remote location [3, 4]. Here, several representative proposals differing from traditional paradigm are listed as follows:

- *CFFDRS* [5] is a danger rating system for forest fire which has been widely implemented in the USA and New Zealand. The major component of this system is the Fire Weather Index (FWI) and this index is evaluated through fire critical weather elements such as temperature, relative humidity, wind speed and precipitation. The information acquirement of CFFDRS relies on the thousands of wireless sensing nodes deployed in the forest area. The monitoring network in CFFDRS collects multiple kinds of information (e.g., temperature, wind speed and humidity). The base station of the monitoring network delivers the environment information directly to remote server.
- *SCIER* [6] is an integrated system working for detecting, monitoring and predicting natural hazards. The target applications of SCIER are urban and rural areas. The unique component in SCIER is the Local Alerting Control Unit, which controls a Wireless sensor network (WSN) and is responsible for the early detection of potential fire, the fire location estimation and the subsequent alerting functions. SCIER develops two different kinds of wireless sensing nodes for urban areas and rural areas respectively: citizen owner sensors and publicly owned sensors. The authority, sensors and installation of citizen owner sensors and publicly owned sensors are different according to their application scenarios.

- *EIDOS* [7] is a comprehensive system including alarm alerting and firefighting rescuing. The main component of the system is a network composed of thousands of sensor nodes, deployed in the field by means of an Unmanned Aerial Vehicle (UAV). The other key element in the system is the handheld device that firefighters carry. The network is responsible for monitoring the surveillance zone and transmitting environmental information to a remote server. The handheld device is integrated with a light-weight browser which is capable of a visiting remote database or a data center.

To sum up, despite the fact that the systems mentioned above has experienced significant improvements compared with the traditional cases, most of them can only focus single implementation area like specialized fire-alarming network or fire-rescuing support network, leading to lack of abilities to undertake more complicated tasks. To some extents, they do waste many of their resources. Therefore, the goal of FASIE is to offer more comprehensive services to fire-fighting activities including fire-monitoring, fire-rescuing and firefighter orientation.

51.3 System Introductions

Figure 51.1 demonstrates the application architecture of FASIE system which is consisted by two components: the wireless fire-alarming network and the handheld fire-rescuing support system. The wireless fire-alarming network is composed of hundreds of wireless sensing nodes which are deployed in the buildings and other application scenarios. Due to wireless sensing nodes with specialized application in fire-fighting scenario, each node is encapsulated into a fireproof packaging. Unlike WSN for forest monitoring, the major application of FASIE mainly focuses on the monitoring of indoor environment. Thus, the node is equipped with two power-supply modules (battery module and fixed power module). When fire happens, wireless fire-alarming network is able to detect fire instantly and deliver alarming message to base station simultaneously. In FASIE, the base station is a small-sized high-performance embedded system with two communication interfaces (i.e. Wi-Fi and 3G), which can be deployed conveniently. When receiving alarm information, the base station will relay alarming information to surveillance platform through Wi-Fi/3G. And the surveillance platform will make relevant decisions like crowd evacuation and calling service of fire-rescuing corresponding to the information received. Handheld fire-rescuing support system as a key element in FASIE is the unique feature which can distinguish this system from other existing support systems for firefighting. The handheld fire-rescuing system is composed of two devices: a fireproofing PDA and a wireless sensing node with Bluetooth interface. The portable wireless sensing node is able to share information with PDA through Bluetooth interface. In this way, when firefighters enter into the fire zone, the portable wireless sensing nodes is able to receive location

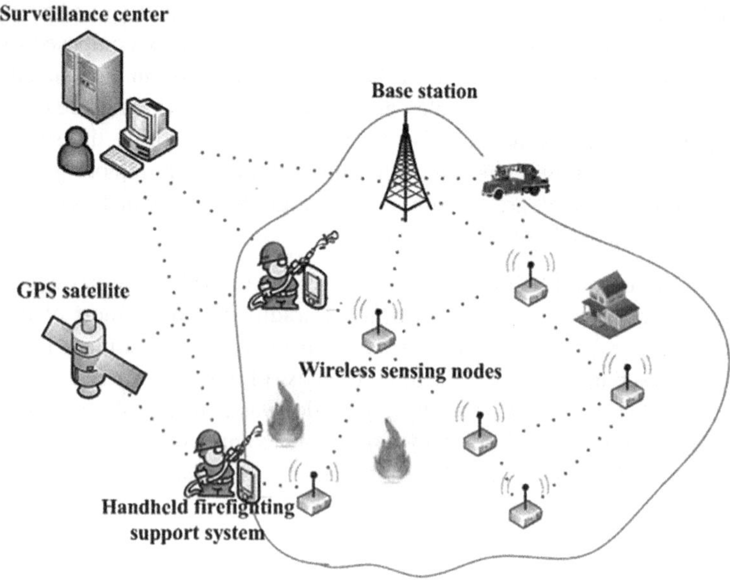

Fig. 51.1 Application architecture of FASIE

information (e.g. RSSI and LQI) through dynamic self-organization. By means of the location algorithm for indoor environment, the PDA is able to access the exact location of the firefighters and inform supervision platform via 3G network. Besides indoor location service, the functions of handheld fire-rescuing system also incorporate outdoor location service and monitoring of health status. With the support of GPS chips inside PDA, the PDA is able to access the accurate position which can provide tremendous help for dispatcher of firefighting equipments and crew. Due to the portable wireless sensing node integrating multiple sensors, the wireless sensing node is able to sense the surrounding environment and monitor the physical parameters of the firefighters. By means of 3G network, the surveillance center is able to supervise the health state of the firefighters and their surrounding situation. When firefighters face danger, the supervision platform is able to organize the rescuing activities immediately. Moreover, as the handheld fire-alarming system is integrated with a light-weight browser, firefighters are capable to visit remote database through web services.

51.4 Layered Architecture

In order to enhance versatility, flexibility and extensibility of the system, the design of FASIE adopts layered architecture following open system interconnection reference model (OSI) [8] and is divided into six layers. From top to bottom, layers are application layer, forwarding layer, assembly layer, Link management layer, device layer and sensing layer. The related details of the layers are described as follows (Fig. 51.2).

1. *Sensing layer* as the lowest level of the system is responsible for the environment sensing and information sampling. In general, sensing layer collects environment information and relay the information to device layer through data interfaces (e.g. SPI and I^2C). According to various devices, the data relayed from sensing layer is also different. For wireless sensor nodes, the sensing layer mainly incorporates smoke and temperature information. For handheld devices, besides the environment parameters mentioned above, the sensing layer also includes the physical parameters of the fire fighters (e.g. ECG and blood pressure). The selection of sensors and data interface and configuration of sampling rate are some key technologies worthy noticing in sensing layer.
2. *Device layer* is the basis of the system and all of the functions facing clients must be realized through this layer. The responsibility of the device layer is to provide hardware support for wireless communication (e.g. Wi-Fi, Zigbee and Bluetooth) between various devices. In general, the device layer mainly include multiple kinds of gateway equipments whose core technology is to achieve information-sharing between different types of equipments with various data interfaces.
3. *Link management layer* plays the most crucial role in self-organization of the network and is generally conceived as the major technical bottleneck to impede the advancement of WSN research. The responsibilities of the management link layer mainly include neighbor discovering, authentication and link estimation and self-organization. In FASIE, when a new device like wireless fire-alarming node joins the network, the link management detects this device and submits the related information to the forwarding layer. It's worthy noticing that when a fire fighter with handheld device enters into the detecting zone, the link management also should be able capture the moving track of the fire fighter.
4. *Forwarding layer* is the layer responsible for data delivering and relying. More specifically, if link management layer is used to build channel, the forwarding layer is to solve problems such as which channel should be selected and which path the data should follow. As far as FASIE, the role of forwarding layer playing in the system is to determine the target node that next hoop should forward into and achieve effective cache management by building message queue.
5. *Assembly layer* can be considered as the bond to connect high level of application with the transmitting devices of the network, of which functions usually includes encryption, protocol transform and name service. Due to the high sensitivity of energy-consuming in WSN, the data delivered from one device to another device need to be recoded, aiming to shorten the length of message and thus guarantee the information security and energy-saving. When an encoded message reaches the device, the assembly layer is also demanded to decode the message and guarantee the integrity of the information. FASIE involving several network media. Different network media demand different network protocols. Therefore, the assemble layer is required to transform protocols in order to keep information flowing among heterogonous networks.

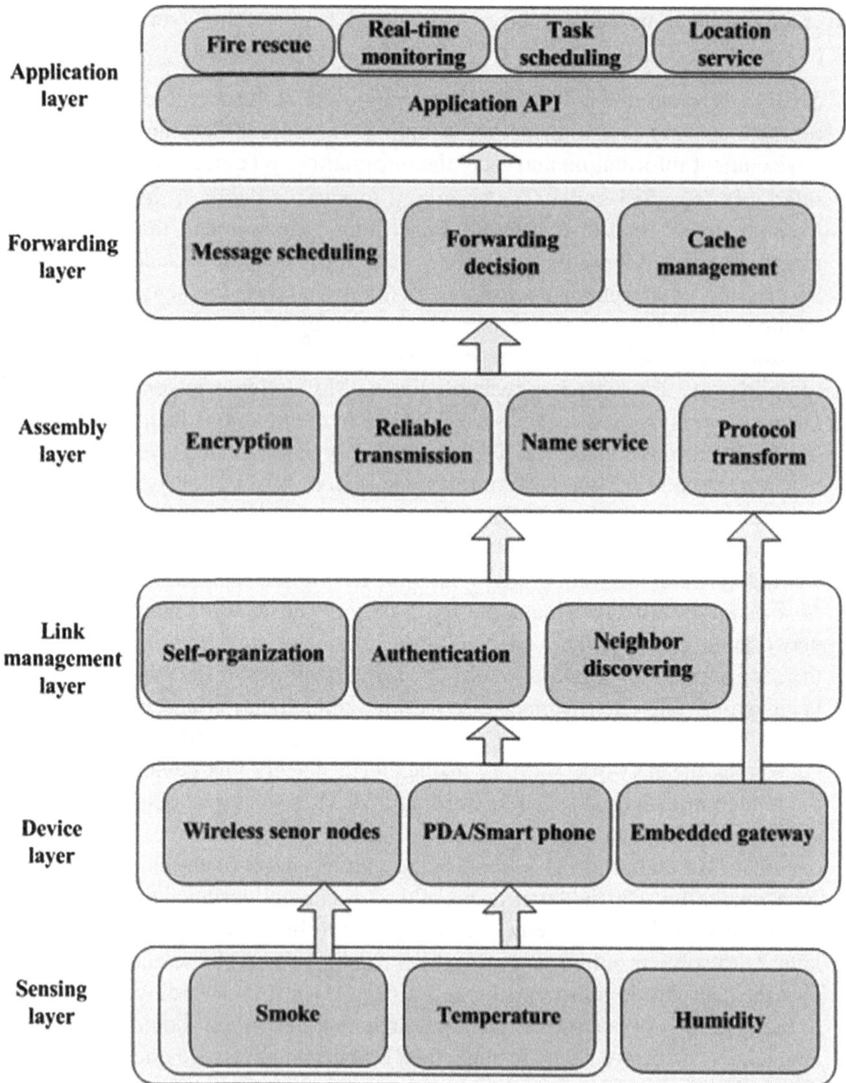

Fig. 51.2 Layered architecture of FASIE

6. *Application layer* as the highest level of the system is responsible to provide services to clients. The performance of the application layer closely concerns the users experience and thus influences the future application market. The extensible and flexible API is the essential component in the application layers, which is able to help clients implement diversified functions according to their own requirements.

51.5 Device Design

51.5.1 Wireless Sensing Node

For wireless sensing nodes in FASIE, we chose Atmel ATmega128 8-bit processor as the controller and CC2530 made by Chipcon as RF chip. In order to guarantee the accuracy of the sensing results, two types of sensors (e.g. smoke senor and thermometer) are adopted to detect fire happening. We chose MQ-2 resistance-type sensor as smoke sensor and select DS18B20 digital temperature sensor for temperature sampling (Fig. 51.3).

- *CC2530* is a true system-on-chip (SoC) solution for IEEE 802.15.4, Zigbee and RF4CE applications. It enables robust network nodes to be built with very low total bill-of-material costs. CC2530 combines the excellent performance of a leading RF transceiver with an industry-standard enhanced 8051 MCU, in-system programmable flash memory, 8-KB RAM, and many other powerful features. CC2530 has various operating modes, making it highly suited for systems where ultralow power consumption is required. Short transition times between operating modes further ensure low energy consumption [9].
- *MQ-2* smoke sensor as resistance-type sensor is used for measuring the combustible gas in the air. The required supply voltage for proper functioning is 5–15 V. The voltage supply of MQ-2 in sensing nodes is 6 V provided by Atmega128 controlling board.
- *DS18B20:* DS18B20 is a high-accuracy thermometer for sensing temperature in the air. It has an operating temperature range of −55 to +125 °C and is accurate to ± 0.5 °C over the range of −10 °C to +85 °C. The power Supply range is 3.0–5.5 V. For the wireless sensing node used in FASIE, the power supply relies on the Atmega128 controlling board and the value of voltage supply is 5 V [10].

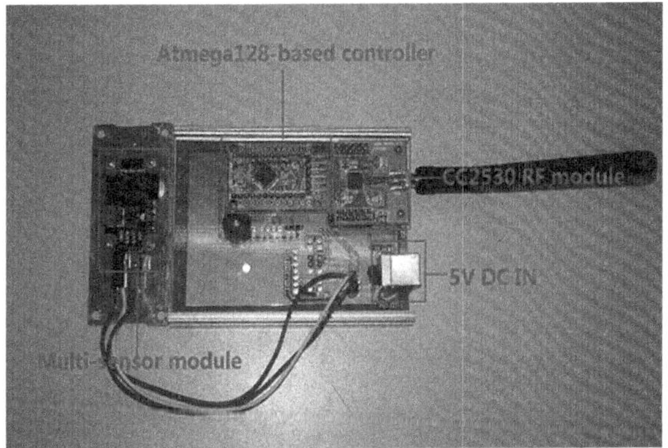

Fig. 51.3 Hardware architecture of wireless sensing node

51.5.2 Embedded Gateway

The embedded gateway is based on the X20ii board whose core controller is S5PV210 from Samsung. The embedded gateway is equipped with a wireless IEEE801.11 b/g interface and a fast Ethernet interface, enabling clients to choose different network interface with respect to the practical network environment. In addition, the embedded gateway also offers 2 RS232 serial ports and 4 USB ports (Fig. 51.4).

51.5.3 Handheld Fire-Rescuing Support System

As shown in Fig. 51.5, the handheld firefighting system is composed of two major components: a high-performance PDA and a portable firefighting sensing node. The function of the PDA is to receive, display and relay data from the portable firefighting sensing node, while the duty of portable sensing node mainly focuses on collecting environment information and receiving location data from wireless fire-alarming network deployed at a fire scene. Here, in FASIE we choose U880 from ZTE as the hardware platform of PDA. Since U880 is a powerful mobile telecommunication device which supports Bluetooth and GPS location, it is well qualified to shoulder the PDA task in FASIE. The portable fire-rescuing support system is consisted of four modules: ATmega128-based controller, CC2530 RF chip, Bluetooth module and multi-sensor module. Since the design and installation of Amtega128 controller and CC2530 RF chip in portable wireless sensing node are similar with the wireless sensing nodes described in Sect. 5.1. The Bluetooth module we chose in FASIE is HC-05 Bluetooth module made by CSR, which supports AT instruction set and master–slave model switching. In FASIE, the voltage supply of Bluetooth module is provided by Atmega128 controller and its data transmission is relied on TTL serial.

Fig. 51.4 Hardware architecture of embedded gateway

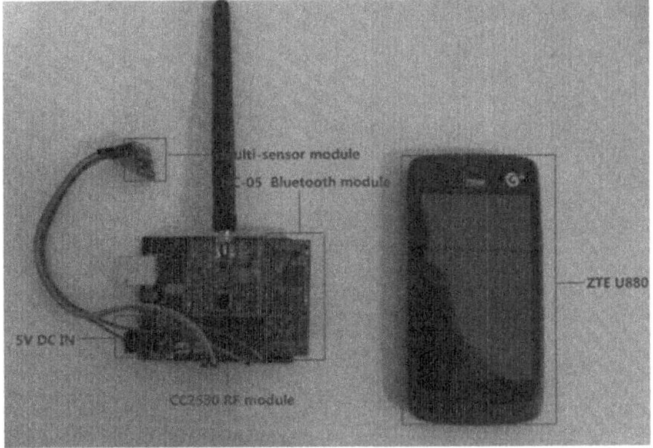

Fig. 51.5 Hardware architecture of handheld fire-rescuing support system

51.6 Experimental Evaluations

In order to further illustrate the performance of FASIE, in this section we choose the battery life of one single sensing node and the average time delay of the entire network as performance parameters. Among the longevity testing of the wireless sensing node, a fully changed 9 V battery was put into the sensing node of which sampling frequency was configured into 2 per second. In terms of the overall performance of the entire network, the experimental scene is located at a 2,000 km^2 train station where FASIE system is already deployed and operating normally. Aiming to conform to the state compulsory regulation on installation of fire-alarming node, the surveillance zone of a single sensing node is limited within 20 m^2. Additionally, we used a CSMA MAC protocol which would back-up in the presence of interference and further avoid the appearance of packets-losing.

51.6.1 Battery Life

Figure 51.6 demonstrates the battery performance of the wireless sensing nodes in FAISE during our deployment. At the beginning, the battery is fully charged, of which voltage value is 5 V. When the voltage value of the battery is below 4.1 V, the operation of the node becomes unstable. Thus, here we can consider the period when the voltage of battery is from 5 to 4.1 V as the survival period of the node. As shown in the graph, the voltage value of the battery decrease smoothly along with time and it will cost more than 2 months to reach the node-failure threshold value, which is well beyond our stated goal of 6 weeks. It's worth noticing that at the initial experiment, the node can only last maximum of 1 week. The reason responsible for this is that we only set the sleep mode on controller processor to idle but without concerning of the power consumption of the RF chip. Actually, according to parameters mentioned in [11], more than 70 % of energy is consumed during the stage of the data transmission. Due to this, the further improvement of

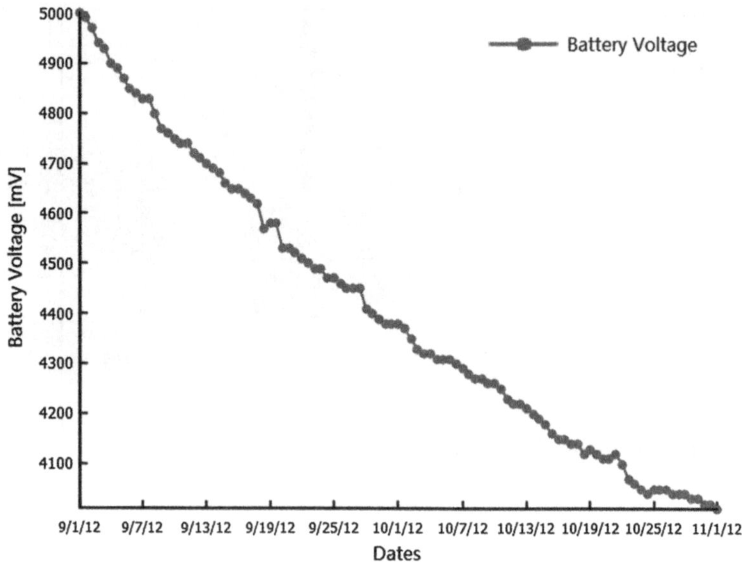

Fig. 51.6 Battery life

the nodes is implemented, in which the RF chips can only be woken up when the threshold of the smoke and temperature value is achieved in the fire scene.

51.6.2 Time Delay

Figure 51.7 demonstrates the time delay of the entire network. Here, we adopt a new parameter to measure the time delay that occurs during the process of data delivering. Assume that the node belonging to the network creates a pre-fixed packet attached with a timestamp T_{x_1} which is then transmitted to the base station. As soon as the base station receives a packet, a CONFIRM packet will be created and sent back instantly. When the node acquires the CONFIRM packet from base station, the time T_{x_2} when CONFIRM packet is received, will be recorded and the time delay of the node $\Delta T = T_{x_2} - T_{x_1}$ can be easily accessed and then the average time delay of the entire network can be acquired easily by repeating the same process. From the graph, when the number of the nodes in the network is < 30, the impacts of the network extending on overall time delay is not so obvious, while the network size increases to 100, the increasing tendency become much more aggressive. The reason responsible for this is that when network scale is within 20, sensing nodes communicate with base station mainly via single hop and the average time delay is 9 ms, while with the extending scale of the network, the average hop between base station and sensing nodes increases rapidly, leading to the dramatic increase of the average time delay and the time delay surges to

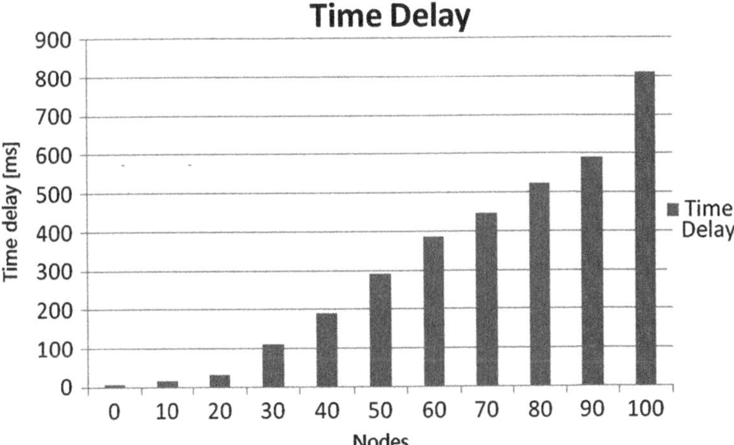

Fig. 51.7 Time delay

814 ms. Considering in most cases, the scale of the network is within 100. Thus, we can reasonably consider the time delay of the network FASIE performs well enough to deliver fire alarm to base station rapidly.

51.7 Conclusions and Future Work

In this paper, a comprehensive fire-alarming system called FASIE based on wireless sensor networks is designed and discussed on network architecture, hardware design. Several experiments have been implemented to evaluate the performance of the system. By integrating wireless fire-alarming network with handheld fire-rescuing support system, FASIE is capable of providing full service to firefighting activities including fire-alarming, fire-rescuing and firefighter location. Since one significant characteristic of the implementation on firefighting activities is the high dynamics of network resulting from firefighters or failure of nodes, a fault tolerance routing protocol is one of the most crucial standards to measure the performance of the firefighting support system, which will be a major focus in the future work.

Acknowledgments This work has been financially supported by National Science & Technology R&D Program (2012BAJ05B07), Fundamental Research Funds for the Central Universities (2012-JL-08) and International Cooperation Project funded by Hubei province (2011BFA012).

References

1. Mottola L, Picco GP (2010) Programming wireless sensor networks: fundamental concepts and state of the art. ACM Computing Surveys
2. Fortino G, Guerrieri A, O'Hare GMP et al (2012) A flexible building management framework based on wireless sensor and actuator networks. J Netw Comput Appl 35(6):1934–1952
3. Garcia EM, Bermudez A, Casado R, Quiles FJ (2007) Collaborative data processing for forest fire fighting. 4th European conference on wireless sensor networks
4. Son EHB, Her Y, Kim J (2006) A design and implementation of forest-fires surveillance system based on wireless sensor networks for South Korea Mountains. J Comput Sci Netw Secur 6(9):124–130
5. Kucuk G, Kosucu B, Yavas A, Baydere S (2008) FireSense: forest fire prediction and detection system using wireless sensor networks. 4th IEEE/ACM international conference on distributed computing in sensor systems (DCOSS'08), Santorini Island, Greece
6. Zervas E, Sekkas O, Hadjiefthymiades S, Anagnostopoulos C (2007) Fire detection in the urban rural interface through fusion techniques. 1st international workshop on mobile Ad hoc and sensor systems for global and homeland security (MASS-GHS 2007), Pisa, Italy
7. Garcia EM, Serna MA, Bermudez A, Casado R (2008) Simulating a WSN-based wildfire fighting support system. IEEE international workshop on modelling, analysis and simulation of sensor networks (MASSN'08), Sydney
8. Bachir A, Dohler M, Watteyne T, Leung KK (2010) MAC essentials for wireless sensor networks. IEEE commun surveys and tutorials, vol 12(2), second quarter 2010
9. CC2530 Datasheet, available at http://www.ti.com
10. DS18B20Datasheet,available at http://datasheets.maximintegrated.com/en/ds/DS18B20.pdf
11. Srivastava M (2002) Sensor node platform and energy issues. In: Proceeding of the 8th ACM international conference on mobile computing and networking (MobiCom'02)

Chapter 52
Sound Source Localization Strategy Based on Mobile Robot

Qinqi Xu and Peng Yang

Abstract Combining the movement of the mobile robot is proposed to locate the sound source. A tetrahedron array of four microphones is used as the robot's ears, and the azimuth and elevation of the sound source is calculated based on generalized cross-correlation algorithm. After measuring the azimuth and elevation of the sound source, robot moves a distance, and measure the azimuth and elevation again. Finally, calculate the distance of the sound source using the data obtained through the two measurements. The theoretical analysis and experiment results show that the robot is capable of localizing sound source in a three-dimensional space. The method is efficient and practical enough for real-time operation.

Keywords Mobile robot · Auditory · Sound source localization · Tetrahedron array

52.1 Introduction

With the continuous development and improvement of digital signal processing technology and acoustic electric technology, as well as the level of humanoid robot improving in artificial intelligence, auditory system, as an important component of the human senses, has become an important study in robot research. Acoustic detection is the sound wave information obtained by using acoustic principle and electronic devices to achieve the target identification and location technologies, have very broad application prospects in the military and civilian, such as smart bombs on the target positioning system and tracking is completed by passive acoustic detection system [1].

Q. Xu (✉) · P. Yang
School of Control Science and Engineering, Hebei University of Technology,
No 8 GuangRong Road, Tianjin, China
e-mail: shiyuse1@126.com

Z. Sun and Z. Deng (eds.), *Proceedings of 2013 Chinese Intelligent Automation Conference*, Lecture Notes in Electrical Engineering 255,
DOI: 10.1007/978-3-642-38460-8_52,

Robot auditory system is an important means of improving the robot intelligent and robot interaction with the environment. Auditory localization accuracy is much lower than the visual positioning accuracy, but hearing has some better characteristics than other senses. Auditory system is omni-directional. Microphone array can accept the sound from any direction in space [2]. Sound can bypass obstacles. Robot rely on hearing can work in the dark or low light environment to locate a sound source. These rely on vision can not be realized.

At present, such research is in its infancy. The sound source position information includes azimuth, elevation and distance, but most of the existing robot sound source positioning device can only locate azimuth and elevation, can not accurately positioning the distance. For example, a rectangular array of eight microphones is proposed by J. M. Valin et al. to locate the sound source azimuth and elevation [3]. The use of the artificial head and artificial ears is proposed by F. Keyrouz et al. to simultaneously locate two sound source azimuth and elevation [4]. In china, Xiaoning Liu et al. proposed a isosceles triangle array of three microphones to locate the sound source azimuth [5]. A five microphone array is proposed for robot auditory localization by Xiaoling Lü et al. Sound source position in space is calculated by the planar array of four microphones, another microphone is used as an auxiliary unit to judge the source in front or rear of the robot [6]. This method can achieve the measurement of the azimuth, elevation and distance, but distance error is too big, and can not be applied to the actual sound source localization. In response to this situation described above, combining the movement of the mobile robot is proposed to locate the sound source in this paper.

52.2 Robot System

The object of this study is the intelligent mobile robot which is composed of mobile robot body [7], sonar sensor and auditory sensor as shown in Fig. 52.1. The sonar positions in this pioneer 3 sonar arrays are fixed: one on each side, and six facing outward at 20° intervals. Together, fore and aft sonar arrays provide 360° of nearly seamless sensing for the platform. The auditory sensor is the tetrahedral array of four microphones, which is placed on the deck of the robot.

52.3 Sound Source Localization System

52.3.1 Sound Source Localization Method

There are many sound source localization algorithms, but we select the location method based on time delay, taking into account the real-time implementation of the system [8]. The method is divided into two steps. The first step is to estimate

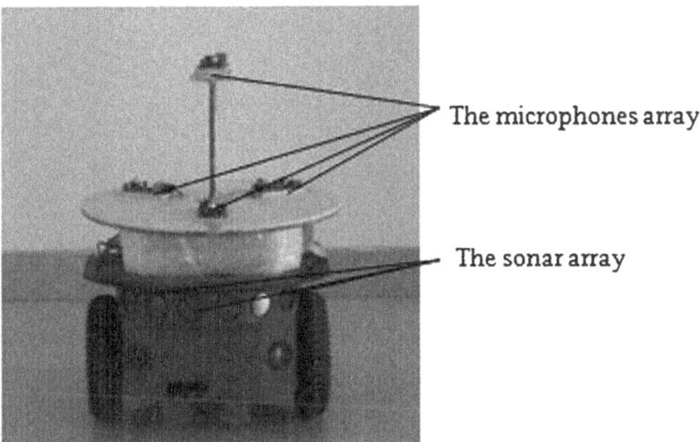

Fig. 52.1 Robot system

the relative delay of each microphone. The second step is to estimated sound source location based on these multiple time difference of arrival.

Suppose the discrete event mode for the signal two microphone received is $x_1(t) = a_1 s(t) + n_1(t),\ x_2(t) = a_2 s(t - \tau_{12}) + n_2(t)$. α_i is the attenuation coefficient for sound source signal, $s(t)$ is the target sound source signal, $x_i(t)$ is the signal collected by microphone, $n_i(t)$ is the additional noise source, $\tau_{1,2}$ is time delay of the s two signal collected by the two microphones [9, 10].

By fourier transform, the sound signal $x_i(t),\ i = 1, 2$ is changed from the time domain signal into the frequency domain signal $X_i(\omega)$. Cross-power spectrum function is $G_{X_1X_2}(\omega) = X_1(\omega)X_2{}^*(\omega)$, we know smooth coherence transform weighting function is $\psi_{12}(\omega) = 1/\sqrt{G_{X_1X_1}(\omega)G_{X_2X_2}(\omega)}$, $G_{X_2X_2}(\omega)$ is autocorrelation spectrum. So cross-correlation function is $R_{x_1x_2}(\tau) = \int_0^{\pi} \psi_{12}(\omega)G_{12}(\omega)e^{j\omega\tau}d\omega$. At last, the abscissa of the cross-correlation function peak is the time delay.

52.3.2 The Realization of Positioning Azimuth and Elevation

Microphone array we used is tetrahedral, shown an in Fig. 52.2. S1, S2, S3, S4 are the four microphones, O is the origin of the coordinate system, which is also the center of the underside plane of the regular tetrahedron. Q is the target sound source, coordinate is Q(x,y,z). The distance from coordinate origin to the destination point OQ is r, and the angle with the Z-axis is β. OQ is the projection for OQ on the underside plane of the regular tetrahedron, and the angle with the X-axis is α, with Z-axis is β.

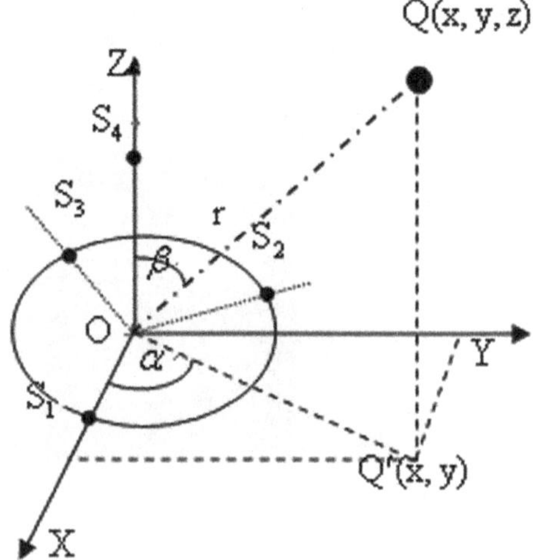

Fig. 52.2 Tetrahedral array model

Assume the distance from S1 to the origin O is a, then the four microphones coordinate are $S_1 = (a, 0, 0)$, $S_2 = (-a/2, \sqrt{3}a/2, 0)$, $S_3 = (-a/2, -\sqrt{3}a/2, 0)$, $S_4 = (0, 0, \sqrt{2}a)$.

We use c represents the speed of sound, r1, r2, r3, r4 represents the distance from the target sound source to the four microphones. Use τ_{4i} (i = 1,2,3) to represent the time difference between microphone S_i (i = 1,2,3) and S_4. Use $d_{4i}(\mathrm{i} = 1, 2, 3)$ to represent the distance difference between microphone $S_i(\mathrm{i} = 1, 2, 3)$ and S_4. So

$$d_{4i} = PS_i - PS_4 = \tau_{4i} \cdot c, \ (\mathrm{i} = 1,\ 2,\ 3) \tag{52.1}$$

Comprehensive reference to all sorts of Planar array and Three-dimensional array formula derivation principle, and by the geometric relationship of the model, you can obtain the following equation [7, 8]:

Azimuth formula:

$$\alpha \approx \arctan \sqrt{3} \frac{\tau_{43} - \tau_{42}}{\tau_{42} + \tau_{43} - 2 \cdot \tau_{41}} \tag{52.2}$$

Elevation angle formula:

$$\beta \approx arc\cot \frac{t_{42} + t_{43} + t_{41}}{2\sqrt{2}\sqrt{t_{42}^2 + t_{43}^2 + t_{41}^2 - t_{42}t_{41} - t_{43}t_{41} - t_{42}t_{43}}} \tag{52.3}$$

And the following distance formula which will be used in the next section,

$$\mathrm{r} = OQ'/\sin(\beta) \tag{52.4}$$

52.3.3 The Realization of Positioning Distance

The tetrahedral array of four microphones is placed on the deck of the robot, so that when the mobile robot move and in situ rotation, the microphone array will be driven to move and rotate. The microphone array 0° azimuth is consistent with the front of the robot, and the azimuth angle range is −180°–+180°.

Here we introduce the general process of calculating the distance. First, measure the azimuth and elevation of the sound source, then the robot moves a distance, and measure the azimuth and elevation again. Finally, calculate the distance of the sound source using the data obtained through the two measurements. The measurement process is clearly shown in Fig. 52.3. In Fig. 52.3, Numeral 2 represents the microphone array, numeral 3 represents the sound source, the black arrow represents the microphone array 0° azimuth and it is also the front of the robot, O1 is the center point of the bottom surface of the tetrahedron microphone array before the robot moves a distance, O2 is the center point of the bottom surface of the tetrahedron microphone array after the robot moves a distance, Z is the projection for the sound source on the underside plane of the regular tetrahedron microphone array. Black arrows represent the microphone array azimuth 0° direction, which is consistent with the front of the robot. A is the angle between the line segment O1Z and the black arrow at the point O1. B is the angle between the line segment O2Z and the black arrow at the point O2. The distance between the point O1 and the point Z is d1. The distance between the point O2 and the point Z is d2. The distance robot move is L. X is the angle between the line segment O1 Z and the line segment O1O2, δ is the angle between the line segment O2Z and the line segment O2O1.

Specific implementation steps are as follows:

Step 1: Measure the azimuth A and elevation E of the sound source before the robot moves a distance.

At the moment, the robot is located at the point O1, as shown in Fig. 52.3. The black arrow at the point O1 is the front of the robot. We use A to represent the azimuth of the sound source, and use E to represent elevation of the sound source. E is not shown in Figure 52.3, because Fig. 52.3 is a plan view.

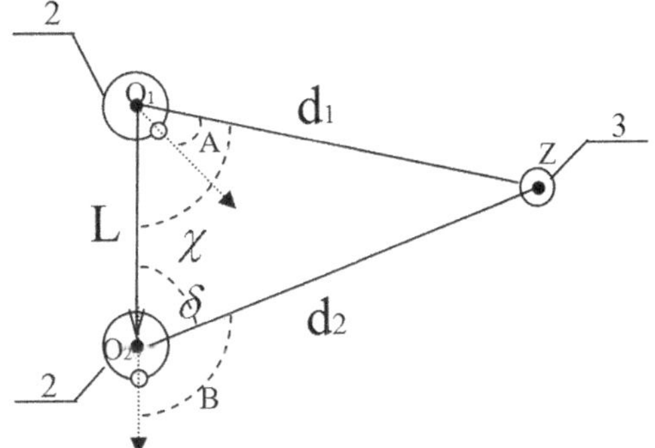

Fig. 52.3 Distance calculation model

Step 2: The robot Turn clockwise, and move a distance.
The robot turn clockwise through $\chi - A°$, and move a distance L. At the moment, the robot is located at the point O2, The black arrow at the point O2 is the front of the robot.

Step 3: Measure the azimuth B and elevation F of the sound source after the robot moves a distance.
At the moment, the robot is located at the point O2, as shown in Fig. 52.3. The black arrow at the point O2 is the front of the robot. We use B to represent the azimuth of the sound source, and use F to represent elevation of the sound source. F is not shown in Fig. 52.3, because Fig. 52.3 is a plan view.

Step 4: Calculate the distance of the sound source
We can obtain the formula from Fig. 52.3:

$$d_2 = L * \sin(\chi) / \sin(180 - \chi - (180 - B)) \tag{52.5}$$

So we obtain the distance formula 52.6 in accordance with Eqs. 52.4 and 52.5.

$$r = d_2 / \sin(F) \tag{52.6}$$

52.4 Experiment

Experiment was carried out in ordinary laboratory. The background noise, include computer fan noise and the sound incoming from the windows. We use speaker as the sound source. Experimental test environment is shown in Fig. 52.4. The test procedure is:

Step 1: Measure the azimuth A and elevation E of the sound source before the robot moves a distance.

Fig. 52.4 Experimental test environment

Table 52.1 Experimental data

The data before the robot moves a distance					The data after the robot moves a distance					
Experimental azimuth	Actual azimuth	Experimental elevation	Actual elevation	Actual distance	Experimental azimuth	Actual azimuth	Experimental elevation	Actual elevation	Experimental distance	Actual distance
60.12	60	87.25	88	1.0	134.84	135	88.84	88	1.42	1.41
29.38	30	89.08	88	1.5	123.42	124	87.21	89	1.82	1.80
14.84	15	88.14	88	2.0	116.79	117	88.39	89	2.22	2.24
9.55	10	88.45	88	2.5	112.29	112	89.06	89	2.64	2.69
2.23	2	87.62	88	3.0	109.02	108	88.10	89	3.07	3.16

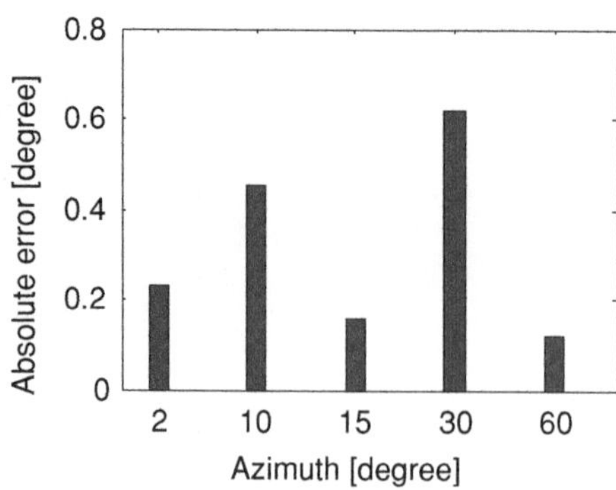

Fig. 52.5 Azimuth angle error

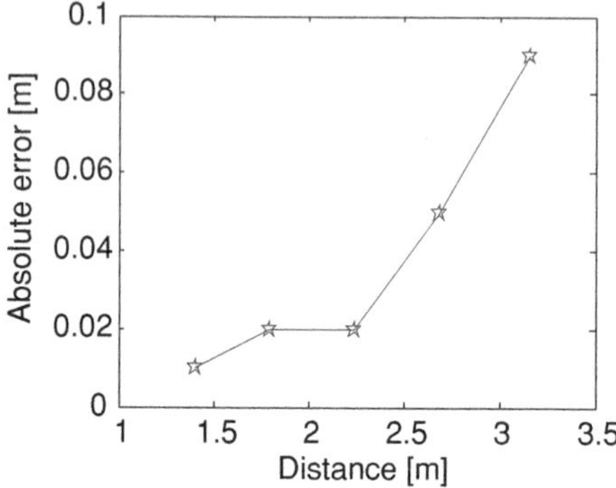

Fig. 52.6 Distance error

Step 2: The robot Turn clockwise, and move a distance.
The robot turn clockwise through $90 - A°$, and move one meter. Here the value of χ is 90°, the value of L is 1 m.

Step 3: Measure the azimuth B and elevation F of the sound source after the robot moves a distance.

Step 4: Calculate the distance of the sound source.

The above-described steps is repeated 5 times, and five groups of experimental data is obtained as shown in Table 52.1.

Figure 52.5 is the azimuth angle error figure before the robot moves a distance. We can see that the azimuth angle error is different with different azimuth angles and is less than 0.7°. Here are the azimuth angle data of 0°–60°. The microphone array is azimuth angle symmetry. For example, 0°–60° and 60°–120° are symmetrical. So based on the principle of symmetry, we can know that the azimuth angle error of the array is less 0.7°.

Azimuth angle error is mainly affected by noise and reverberation, and also is related to the accuracy of the microphone installation position. In addition, acquisition card rate that represents the accuracy of the time resolution is also the key factors which determine the azimuth accuracy.

Figure 52.6 is the distance error figure. We can see that the distance error is less than 0.1 m when sound source is within 3 m, and the maximum error percentage is 0.09/3.16 = 2.85 %, less than 3 %. This result is quite satisfactory. The distance error increases as the distance increases, and the trend is nonlinear.

Distance error is mainly affected by the azimuth angle error and the robot moves errors, and also is related to the accuracy of the robot rotation. In addition, acquisition card rate also affect the accuracy of the distance by influencing the azimuth accuracy.

52.5 Conclusion

The mobile robot sound source localization system based on combining the movement of the mobile robot is designed by this paper and has a higher positioning accuracy compared to other existing robot auditory system. The system is tested within the indoor environment, and the result show that the robot is capable of localizing sound source in a three-dimensional space. The method is efficient and practical enough for real-time operation.

Acknowledgments This work was supported by Hebei Province Science and Technology Research and Development Guide Plan (F2010000137). And the Research Institute of Prosthetics & Orthotics of the Ministry of Civil Affairs of P. R. China provided great aids on the experiments.

References

1. Lin Z, Xu B (2008) Sound source localization based on microphone array. Electro-Acoust Technol 5:19–24 (In Chinese)
2. Yamamoto S, Nakadai K, Tsujino H, Yokoyama T, Okuno H (2004) Improvement of robot audition by interfacing sound source separation and automatic speech recognition with missing feature theory. In: Proceedings IEEE international conference on robotics and automation, pp 1517–1523
3. Valin JM, Michaud F (2003) Robust sound source localization using a microphone array on a mobile robot. In: Proceedings IEEE/RSJ, pp 23–30
4. Keyrouz F, Maier W, Diepold K (2006) A novel humanoid binaural 3D sound localization and separation algorithm, IEEE-RAS international conference on humanoid robot. Genova, Italy, pp 296–301
5. Liu H, Li Z (2009) Sound source positioning system based on the mobile robot. J Sichuan Ordnance 4:77–79 (In Chinese)
6. Lü Xiaoling, Zhang Minglu (2010) Sound source localization strategy based on robot hearing. Chines J Sens Actuators 23(4):518–521 (In Chinese)
7. Yang P, Xu Q, Zu L (2009) Auditory system design based on mobile robot, 2009 Second international conference on intelligent networks and intelligent systems, ICINIS, pp 265–268

8. Weiwei C, Zhigang C, Jiangqiang W (2007) Time delay estimation techniques in source location. J Data Acquisition Process 22(1):90–99 (In Chinese)
9. Knapp CH, Carter GC (1976) The generalized correlation method for estimation of time delay. IEEE Trans Acoust Speech Signal Process 24(4):320–327
10. Chen JD, Benesty J, Huang YT (2005) Performance of GCC and AMDF based time delay estimation in practical reverberant environments. EURASIP J Appl Sig Process(1):25–36

Chapter 53
The Design of a Novel Artificial Label for Robot Navigation

Hao Wu, Guohui Tian, Peng Duan and Sen Sang

Abstract To assistant the localization and navigation of the robot, a novel artificial label based on QR Code technology is designed. This new label is made up of two sections: mode part and information part. Mode part is used to identify the artificial label and locate the robot. Information part is used to not only provide rich description information of the environment, but also provide effective navigation information to the robot. An algorithm based on shape and color is presented to recognize the label from far distance. The localization algorithm using vanishing line principle is designed to localize the artificial label quickly and accurately in the camera coordinates of the robot. The experiments demonstrate that the new artificial label has far recognition distance, large-scale recognition angle, good anti-disturbance ability high decoding efficiency and stability even in the complex indoor environment.

Keywords Artificial label · QR Code · Recognition · Camera

53.1 Introduction

It's a mainstream method that landmarks are used to offer the information of navigation for the robot nowadays. Landmarks are divided into two categories, natural and artificial landmark. Natural landmark means the natural object which already exists in the environment and can be used to assist localization and navigation. Artificial landmark are specially designed by people for robot localization and navigation. Comparing to the natural landmark, artificial landmark is more

H. Wu · G. Tian (✉) · P. Duan · S. Sang
School of Control Science and Engineering, Qianfoshan Branch,
Shandong University, Jinan 250061, China
e-mail: g.h.tian@sdu.edu.cn

Z. Sun and Z. Deng (eds.), *Proceedings of 2013 Chinese Intelligent Automation Conference*, Lecture Notes in Electrical Engineering 255,
DOI: 10.1007/978-3-642-38460-8_53,

informative and easier detected. Therefore, using artificial landmark for robot localization and navigation has been widely studied [1–4].

Zitova and Flusser [5] designed the artificial landmark with two concentric circles, however, its application is limited because of the lack of robustness and real time. Scharstein and Briggs [6] built a kind of self-similar artificial landmark by the scale similarity mode. But the reliability was decreased because of the use of interpolation method while the landmark was extracted. The landmark designed by Bing-yu [7] was based on the fractal and wavelet technique. However, this landmark still belonged to one- dimensional landmark mode. Zheng et al. [8] designed a two-dimensional landmark to locate the robot, but it couldn't afford plenty of semantic information.

A new artificial landmark based on QR Code technique is introduced in this paper. QR code technology is used to solve the complexity and limitations of semantic recognition and scene understanding. A fast and stable arithmetic of the recognition and orientation of the landmark is also designed, which can be used to navigate and locate the robot.

53.2 The Design and Recognition of QR Code Based Artificial Label

53.2.1 The Design of the Artificial Label

There are some shortages of current existed artificial labels [5–7]: the structure of code is single (one-dimensional mode), it is susceptible to noise pollution, it can't be recognized when the label is sheltered, it cannot be recognized in arbitrary angle, bar code stores limited information, etc. As the indoor environment is complex and the vision of cameras is limited, the robot often observes artificial labels of insufficient numbers and incomplete image. These problems reduce the accuracy of reading. So QR code based artificial label is designed in this article to solve above problem. QR code studied by Japanese Denso Corporation is the most representative two-dimensional barcode at present. Ultrahigh speed and all-round (360°) identifying is the main features of QR code. Moreover QR code has the merits of a large capacity, high reliability, high performance of keeping secret and anti-counterfeiting, etc. [9, 10].

Figure 53.1 shows the artificial object label which is used in this paper. The artificial object label's periphery is named as mode part which includes three chromatic rings (the three circle rings in Fig. 53.1 are red, other color also can be used) and blue rectangular; and the insider is named as information part which includes QR code. The mode part and information part are both two-dimensional structure, thus it is really a two-dimensional artificial object label.

The chromatic ring is applied to realize the label's fast recognition and localization in the complex environment (depending upon its color and concentric rings).

Fig. 53.1 Planar self-similar object mark based on QRcode

Its width ratio is 4:2:1. As long as the robot finds three concentric arcs, it can identify successfully. Therefore greater shaded proportion is allowed. The blue rectangular frame is used to adjust the robot's orientation.

53.2.2 The Recognition of the Artificial Label

When the robot obtains an image, first of all, it carries on the two-dimensional artificial label's recognition. If an artificial label is discovered, the robot will read the QR code information from the artificial label.

The video provided by camera is represented in RGB space. In order to reduce the influence of light, we transform the RGB space to YCbCr space which can separate the illumination from hue. The red regions can be detected and extracted after color space conversion. Gaussian model is used to color detection because of the easy parameter calculation and the high detection success rate. Several experiments are done to obtain the parameters of Gaussian model: means and variance.

$$m = \begin{bmatrix} 100.3866 & 229.2088 \end{bmatrix}, \quad C = \begin{bmatrix} 24.1167 & 0.977 \\ 0.977 & 12.7957 \end{bmatrix}. \tag{53.1}$$

The similarity degree comparing with the target is calculated to transform the image to YCbCr space:

$$p(CbCr) = \exp\left[-0.5(x-m)^T C^{-1}(x-m)\right], \tag{53.2}$$

where $\overline{Cb}$ and $\overline{Cr}$ are the means of Cb and $Cr, x = (CbCr)^T$, $m = E(x) = \left(\overline{Cb}, \overline{Cr}\right), C$ is the variance matrix.

$$C = E\left[(x-m)(x-m)^T\right] = \begin{bmatrix} \sigma_{CrCr} & \sigma_{CrCb} \\ \sigma_{CbCr} & \sigma_{CbCb} \end{bmatrix} \tag{53.3}$$

Then, the image is transformed to a binary image. In the binary image, the landmark should belong to the white region.

53.3 Robot Localization Based on QR Code Artificial Label

53.3.1 The Extraction of the Outside Contour's Corners

The largest blue outline is extracted by GAUSS transformation, and then the corner information also can be extracted. Firstly, an imminent method is used to get rough corner coordinates. The four points are A_1, B_1, C_1 and D_1 in sequence. Then least squares method is adopted to fit four straight lines between two adjacent contour points which are L_1, L_2, L_3 and L_4. The intersections are A_2, B_2, C_2, D_2. The average distance from all points of each straight line to fitted line are d_1, d_2, d_3, d_4. Then the formula (53.4) is calculated [11, 12]:

$$d_i \leq 1.0\, pixel, \qquad (i = 1, 2, 3, 4) \tag{53.4}$$

If the result of the formula (53.4) is satisfied, L_i is credible to adopt the current value. Otherwise, it means that excessive noise affects the linear fitting and L_i is not credible. L_i is divided into two sections from the center of it to fit straight lines L_{i1} and L_{i2}, as well as d_{i1} and d_{i2} are calculated respectively. Formula (53.4) is applied to judge the results of d_{i1} and d_{i2}. If the judging result is true, the straight line which satisfies the qualification is used to substitute L_i, and the calculation process is terminated. If the judging result is false, the dividing is done continually until getting the right line or exceeding the maximal dividing numbers.

53.3.2 Calculating Surface Gradient According to the Vanishing Line

The intersection of perspective projection (or extension line) for parallel line segments in the image plane is the vanishing point [13]. All the vanishing points of the parallel line on the same plane are collinear to form the vanishing line.

In the camera coordinate system, the plane is made by four detected corners (that is the plane of the artificial label) which can be expressed as:

$$Z = PX + QY + D \tag{53.5}$$

In formation (53.5) there are three unknown parameters P, Q, D.

Vanishing line equation can be calculated based on the perspective projection principle.

$$f_y \cdot P \cdot x + f_x \cdot Q \cdot y - f_y \cdot c_x \cdot P - f_x \cdot c_y \cdot Q - f_x \cdot f_y = 0 \tag{53.6}$$

The intersection of AD & BC extension lines and the intersection of AB & CD extension lines are used to calculate P and Q. The summation of the four sides' length is used to approximately calculate the parameter D.

Finally, the obtained P, Q, D values are placed into (53.6), so that the three-dimensional coordinates of artificial label's four corners in the camera coordinate system are calculated, and the artificial label positioning is achieved.

Suppose the detected corner points are $A(x_A, y_A), B(x_B, y_B)$, $C(x_C, y_C)$, $D(x_D, y_D)$, which are described anticlockwise at the beginning of left-up point. $E{:}(x_E, y_E)$ is the intersection of the extended lines of AD and BC, $F{:}(x_F, y_F)$ is the intersection of the extended lines of AB 和 CD. Then E and F can be calculated by the four points of A, B, C, D.

$$x_E = -\frac{\begin{vmatrix} x_A y_D - y_A x_D & x_D - x_A \\ x_B y_C - y_B x_C & x_C - x_D \end{vmatrix}}{\begin{vmatrix} y_A - y_D & x_D - x_A \\ y_B - y_C & x_C - x_B \end{vmatrix}} \quad y_E = -\frac{\begin{vmatrix} y_A - y_D & x_A y_D - y_A x_D \\ y_B - y_C & x_B y_C - y_B x_C \end{vmatrix}}{\begin{vmatrix} y_A - y_D & x_D - x_A \\ y_B - y_C & x_C - x_B \end{vmatrix}} \tag{53.7}$$

$$x_F = -\frac{\begin{vmatrix} x_A y_B - y_A x_B & x_B - x_A \\ x_C y_D - y_C x_D & x_D - x_C \end{vmatrix}}{\begin{vmatrix} y_A - y_B & x_B - x_A \\ y_C - y_D & x_D - x_C \end{vmatrix}} \quad y_F = -\frac{\begin{vmatrix} y_A - y_B & x_A y_B - y_A x_B \\ y_C - y_D & x_C y_D - y_C x_D \end{vmatrix}}{\begin{vmatrix} y_A - y_B & x_B - x_A \\ y_C - y_D & x_D - x_C \end{vmatrix}} \tag{53.8}$$

The formula (53.7–53.8) are took in the formula (53.6), P and Q are got.

$$P = \frac{\begin{vmatrix} 1 & y_E - c_y \\ 1 & y_F - c_y \end{vmatrix}}{\begin{vmatrix} x_E - c_x & y_E - c_y \\ x_F - c_x & y_F - c_y \end{vmatrix}} \cdot f_x \quad Q = \frac{\begin{vmatrix} x_E - c_x & 1 \\ x_F - c_x & 1 \end{vmatrix}}{\begin{vmatrix} x_E - c_x & y_E - c_y \\ x_F - c_x & y_F - c_y \end{vmatrix}} \cdot f_y \tag{53.9}$$

53.3.3 Adjusting the Pose of the Robot

A vertical line M from the center of artificial label to the plane of the artificial label is made to realize $MP \perp BC$. A plane $OP_1P_3P_4$ which includes the camera optical center O is made perpendicular to MP, and the point of intersection is P_3. OP_3 is used as the diagonal for rectangle $OP_1P_3P_4$, then P_1 is the projection point from the camera optical center O to the plane of artificial label. $l_1 = \left|\overrightarrow{OP_1}\right|$, $l_2 = \left|\overrightarrow{P_1P_3}\right|$ and $l_3 = \left|\overrightarrow{MP_3}\right|$ represent respectively the horizontal distance of the label, vertical distance and vertical distance from the cameras optical center to the center of artificial label.

Firstly the robot's position is determined at the left or right of the label according to the relative position between P_1 and P_3. The complementary angle of the robot current direction and the normal direction of artificial labels, $\alpha = \angle P_2OP_4$, is calculated. The length of line segment $l_2 = \left|\overrightarrow{P_1P_3}\right|$ is also calculated. Then the robot rotates an angle α, moves forward with l_1, turns in the opposite direction with 90°, and lifts up the panhead with $\beta = \arctan(l_3/l_1)$. During above steps, the robot now locates at the best position to read QR code clearly.

53.4 Experiments and Analysis

To ensure the long-range identification, in the lab, the peripheral circle radius of the artificial object label is set to 10 cm, and its color is red. The QR code and the blue square within the circle can be as large as possible. A label is affixed on a wall. After extracting red color and fitting ellipse, the identifying result is shown at the left of Fig. 53.2. The extracted blue square is shown at the middle and right in Fig. 53.2. As the recognition result of the outer design is close to circle, it indicates that the shooting direction of the robot camera is consistent with the normal direction of the plane pasting artificial object label.

But in most cases, the direction shoot by robot camera are not consistent with the normal direction of the plane pasting the artificial label. When the distance between the artificial label and the robot is less than 1 m, label images extracted at different angle of view are shown in Fig. 53.3a. The corresponding results of the extracted squares are shown in Fig. 53.3b. The angle of view orderly is 0°, 30°, 45°, 60°.

Fig. 53.2 The identifying result of the artificial object label

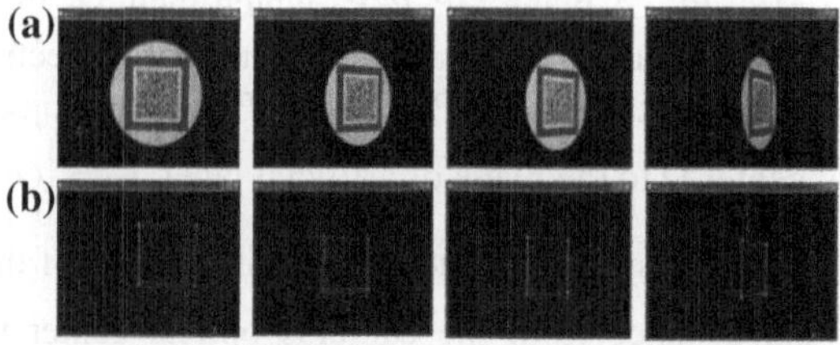

Fig. 53.3 The identifying results at different angle of view, (**a**) the extraction results of the QR code label by the robot at different angles of view, (**b**) the extraction results of the blue square by the robot at different angles of view

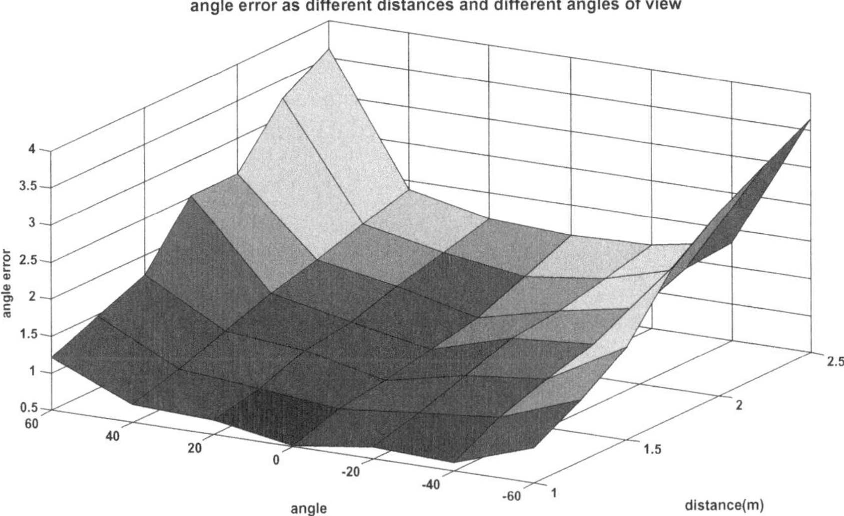

Fig. 53.4 The error graphs of distances and the angles as the robot locates different positions

In the effective range of [−60°, 60°] for artificial label recognition, the artificial label is recognized respectively at the distance of 1, 1.5, 2, and 2.5 m from the robot to the artificial label, at the angle of $\theta = 0°, \pm 20°, \pm 40°, \pm 60°$.

As shown in Fig. 53.4, when the angle is within ±40° and the distance is less than 2.5 m, the accuracy of reading artificial label can meet the requirements of the robot localization.

Fig. 53.5 The experiment result of disturbance with the same color objects

When the camera of the robot catches other objects which color is similar with the ring of the artificial object label, the artificial label can be extracted accurately by the above algorithm. As shown in Fig. 53.5, the artificial label is pasted on the door of the refrigerator. When two other items with red color appear in the range of the camera eyeshot, the program still identified the artificial label successfully.

53.5 Conclusion

To assistant the localization and navigation of the robot, a novel artificial label based on QR Code technology is designed. This new label is made up of two sections: mode part and information part. An algorithm based on shape and color is presented to recognize the label from far distance. The localization algorithm using vanishing line principle is designed to localize the artificial label quickly and accurately in the camera coordinates of the robot. The experiments demonstrate that the new artificial label has far recognition distance, large-scale recognition angle, good anti-disturbance ability high decoding efficiency and stability even in the complex indoor environment.

Acknowledgments Research supported by National Nature Science Foundation under Grant 61203330, 61104009 and 61075092, and Shandong Nature Science Foundation under Grant ZR2012FM031, ZR2011FM011 and ZR2010FM007, Independent Innovation Foundation of Shan dong university 2011JC017 and 2012TS078, Shandong Postdoctoral Innovation Foundation 201203058.

References

1. Lerner R, Rivlin E, Shimshoni I (2007) Landmark selection for task-oriented navigation. IEEE Trans Rob 23(3):494–505
2. Jang G, Kim S, Lee WH, Kweon IS (2002) Color landmark based self-localization for indoor mobile robots. In: Proceedings of the 2002 IEEE international conference on robotics and automation, Washington, USA 2002(1):1037–1042
3. Li G, Jiang Z (2010) Artificial landmark design for mobile robot localization and navigation. J Beijing Univ Civ Eng Archit 26(4):49–53 (in Chinese)
4. Taylor CJ, Kriegman DJ (1998) Vision-based motion planning and exploration algorithms for mobile robots. IEEE Trans Robot Autom 14(3):417–426
5. Zitova B, Flusser J (1999) Landmark recognition using invariant features. Pattern Recogn Lett 20(5):541–547
6. Scharstein D, Briggs AJ (2001) Real-time recognition of self-similar landmarks. Image Vis Comput 19(11):763–772
7. Bing-yu S (2003) Research on the problem concerning visual artificial landmark based navigation of autonomous mobile robot, Doctoral Dissertation. Dongbei University, ShenYang (in Chinese)
8. Zheng R, Yuan K, Li Y (2008) A 2D code for mobile robot localization and navigation in indoor environments. Chin High Technol Lett 18(4):269–374 (in Chinese)

9. Liu Y, Liu M (2006) Automatic recognition algorithm of quick response code based on embedded system. In: Proceedings of the sixth international conference on intelligent systems design and applications (ISDA'06), pp 81–86
10. Rouillard J (2008) Contextual QR Codes. The third international multi-conference on computing in the global information technology. pp 50–55
11. Harris C, Stephens M (1988) A combined corner and edge detector. In: Proceedings of the 4th Alvey vision conference. pp 147–151
12. Hough PVC (1959) Machine Analysis of Bubble Chamber Pictures. In: Proceedings of the international conference on high energy accelerators and instrumentation. pp 554–556
13. Chen Y, Sun Q (1989) The calculation of surface direction. J Univ Sci Technol China 18 (6):35–39 (in Chinese)

Chapter 54
Information System Design for Ship Surveillance

Zhi Zhao, Kefeng Ji, Xiangwei Xing and Huanxin Zou

Abstract Ship surveillance technology has been developed rapidly especially for the increasing request for maritime security and safety. In order to mitigate the maritime issues we plan to design a Ship Surveillance System (SSS), which will integrate data sources including space-borne SAR, shore-based AIS and space-based AIS. It will be inevitable to integrate other data sources such as HF-radar, optical satellite imagery, hyper-spectral and IR imagery in the future, so we adopt service-oriented architecture (SOA) concept to enhance the system's flexibility and extensibility. This paper firstly presents the SOA-based design of ship surveillance system on the basis of research on ship surveillance using space-borne SAR and AIS. Then the main functional modules are analyzed profoundly, and the security strategy is also proposed.

Keywords Ship surveillance · Information system · Service-oriented architecture

54.1 Introduction

Ship surveillance has always been focused on by many countries for its wide applications in surveillance of exclusive economic zone, environmental monitoring and anti-piracy operations, etc. Especially it is significant from the strategic view. Space-borne Synthetic Aperture Radar (SAR) is optimal for its high resolution and all weather working capabilities for ship surveillance compared to optical satellite. But the classification and identification of ship targets in SAR imagery is still immature. Automatic Identification System (AIS) plays an important role in maritime surveillance, and it becomes more efficient with rapid

Z. Zhao (✉) · K. Ji · X. Xing · H. Zou
School of Electronic Science and Engineering, National University of Defense Technology, Sanyi Avenue, Kaifu District, Changsha, Hunan, China
e-mail: zhaozhi_nudt@yahoo.com

Z. Sun and Z. Deng (eds.), *Proceedings of 2013 Chinese Intelligent Automation Conference*, Lecture Notes in Electrical Engineering 255,
DOI: 10.1007/978-3-642-38460-8_54,

development of space-based AIS which could cover larger ocean area. Using AIS as an auxiliary tool is beneficial to the interpretation technology for SAR imagery. But not all ships equip or operate with AIS equipment, so the combination of SAR and AIS for ship surveillance could overcome each limitation and improve the quality of detection and identification results.

On the further consideration, information fusion of multiple data sources is necessary for ship surveillance and has attracted more attention. The ship surveillance system should have good self-improvement and extensible features. So we attempt to develop the system using SOA, which provides a design framework with a view to realizing rapid and low-cost system development and to improving total system-quality. It builds applications out of software services, and uses defined protocols to make sure that services could "talk" to each other [1]. The core SOA lies on the concept of services, but SOA architecture is not only about services; it is a relationship of three kinds of participants: the service provider, the service discovery agency, and the service requestor (user). Publish, bind and find are three main functions which the middleware must provide [2]. SOA is characterized by loose coupling, coarse-grained service and standard interface.

Though State-of-the-art methods for ship detection and identification are numerous, there is not a single one that could satisfy all the applications. A SOA-based ship surveillance system which could serve different users for different applications requires flexible combination of several functional modules. It contains various algorithm modules to satisfy various conditions. SAR data acquired from different spatial resolutions, different modes and even different sensors need different algorithms to gain better results. Additionally, the process for ship surveillance in different applications also varies. For example, traffic monitoring may not need to identify every ship, and detection is enough to avoid accidents. But the rescue and track applications not only need to detect, but also need to identify and track so as to make rescue plans. So how to choose or combine different algorithms for special application is important, which needs flexibility. New algorithms or functions could also be added to improve the quality of service, which needs extensibility. Meanwhile, system security has also attracted users' attention.

This paper emphasizes on designing an applied ship surveillance system (SSS).

Firstly, the theory of ship surveillance using space-borne SAR and AIS is investigated. Secondly, the SOA-based design of ship surveillance system is presented. Thirdly, the main functional modules are analyzed profoundly, and the security strategy is also discussed.

54.2 Information Fusion of Space-Borne SAR and AIS

Ship identification and track rely on information fusion of space-borne SAR and AIS. Information fusion includes data level fusion, feature level fusion and decision level fusion. Information fusion relies on data characters. Coherent speckle and Doppler displacement for moving targets exist in SAR image. AIS

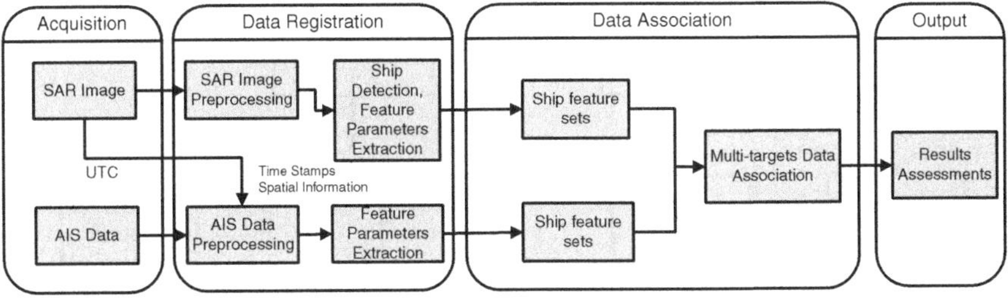

Fig. 54.1 The feature level fusion of space-borne SAR and AIS Data

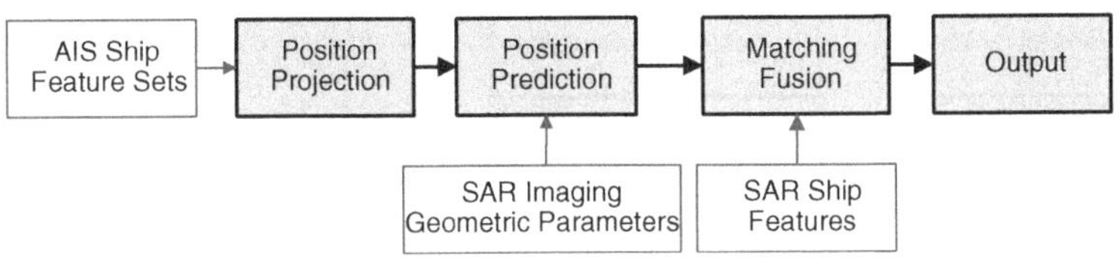

Fig. 54.2 Multi-targets data association based on single position feature

report contains static information, dynamic information and voyage information, etc. The dynamic messages are of particular interest to large-area ocean surveillance, as they are reported often and contain a ship's identity and position [3]. Because the format of space-borne SAR data differs from the one of AIS data, so the data level fusion is not available. After processing of SAR image the feature parameters of ship targets could be achieved. Meanwhile, these feature parameters can also be extracted from AIS reports. So the feature level fusion is available (see Fig. 54.1), which indicates data registration and association of targets' features [4]. Data acquisition adopts UTC to keep synchronous. Data registration prepares for data association and provides input data. SAR image pre-processing includes geometrical registration and RCS reconstructed filtering. Then detecting ships and extracting the feature parameters. AIS data pre-processing includes validation and space-time filter to improve the efficiency. Traditional data Association method based on single feature (position) [5] is shown in Fig. 54.2. To improve the accuracy, multi-targets data association based on multiple features is proposed (see Fig. 54.3), which adopts Bayes decision theory, Dempter-Shafer theory and Fuzzy Reasoning, etc.

54.3 Architecture for SOA-Based Ship Surveillance System

The layered architecture for ship surveillance system is illustrated in Fig. 54.4. The data layer which stores service data and control data interacts with data access layer directly and receives the instructions from data access layer. It is

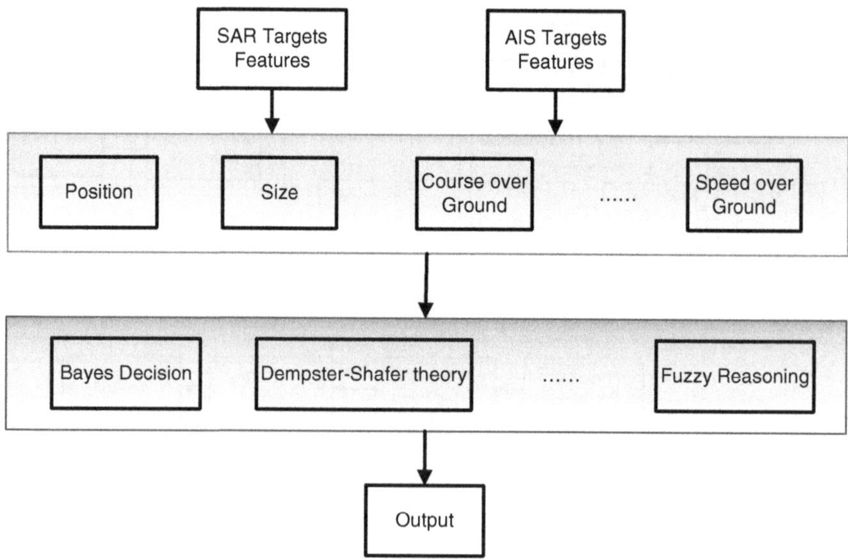

Fig. 54.3 Multi-targets data association based on multiple features

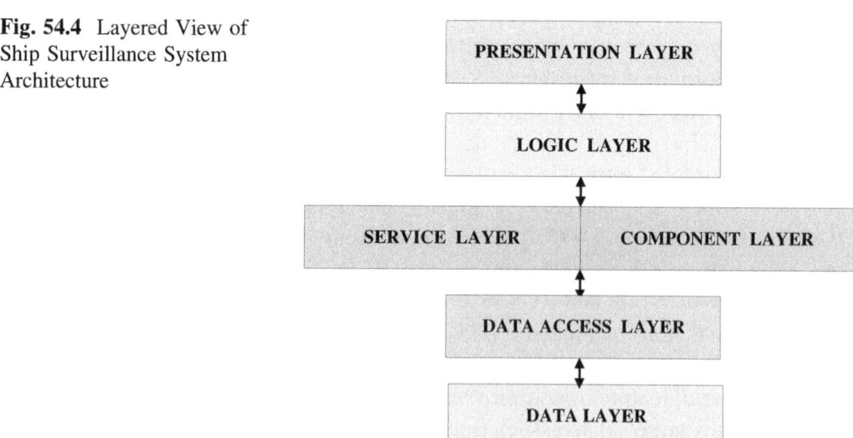

Fig. 54.4 Layered View of Ship Surveillance System Architecture

implemented by using Oracle database. Data access layer operates with the database according to services. Service layer contains all necessary services for ship surveillance applications. They are deposited and identified using WSDL protocol. Logic layer both performs the response from users and completes the service orchestration task to compose the services according the idiographic application [2, 6]. Component layer is designed to reduce cost and to enhance the reusing efficiency. Presentation layer interacts with users directly. It provides different service interfaces for various applications and the necessary parameters

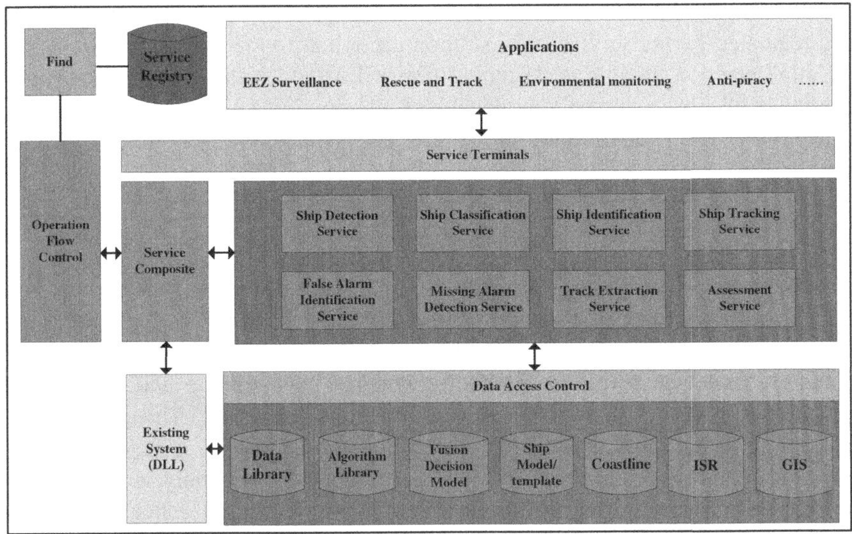

Fig. 54.5 SOA-Based Ship Surveillance System Architecture

transferred to logic layer. When the services are implemented, the results are returned automatically.

The proposed architecture [2] of SOA-based ship surveillance system is illustrated in Fig. 54.5. The databases include: SAR image database, AIS report database, SAR image ship targets interpretation algorithms, AIS data preprocessing and space-time filtering algorithms, fusion decision model, ship templates and models, sea clutter model library, coastline database, GIS and ISR, etc. After receiving the request from service client and the response from service registry, the corresponding services are composed according to different strategies for various applications. For example, when environmental monitoring application is required and a polluter is needed to track, the services including ship detection, identification, tracking and track extraction are employed. Then the temporary results or real-time track process will be presented using friendly interfaces.

54.4 Module Design and System Security

54.4.1 Module Design

All the modules are designed using .Net framework. Service management module is designed to discover and register the services so that all the services can be managed in the inventory. Through service management users can apply different applications for ship surveillance. To different applications the flow is different and regulated by the service orchestration according to the strategy.

Ship detection module is designed using CFAR (Constant False Alarm Rate) method in which sea clutter is modeled by K distribution considered to be the most suitable [7]. The K-distribution probability density function is expressed as:

$$p(x) = \frac{2}{x\Gamma(v)\Gamma(n)}\left(\frac{nvx}{\mu}\right)^{\frac{n+v}{2}} \times K_{v-n}\left(2\sqrt{\frac{nvx}{\mu}}\right) \quad x > 0 \tag{54.1}$$

where v is shape parameter, n is ENL, $\Gamma(\cdot)$ is Gamma function, μ is mean value, K_{v-n} is the revised Bessel function of the second type with rank number $v-n$ respectively.

False Alarm identification module is designed to eliminate undesired things which exist in images such as floater, sea clutter and islands, etc. The algorithms to discriminate the false alarms include the methods based on feature choice, knowledge and resolution, etc. Missing alarms detection is mainly to detect small ships based on detection in transferring field or improved sea clutter modeling.

Ship discrimination module is designed to discriminate ships based on SAR image interpretation technology and information fusion of SAR and AIS data. If ships detected in SAR image also reported by AIS, then the results could be got directly from AIS information. If ships exist in SAR image but not reported by AIS, the discrimination should be relied on SAR image interpretation.

Ship tracking module is designed to track interested targets. The algorithm is based on information fusion of SAR and AIS because AIS has the advantages in continuous observation. Also it relies on the data extraction and mining technology to search the space-time information.

54.4.2 Security Strategy

The security credit of ship surveillance system is sensitive to users especially when applications need to be kept secret. So the security has to be considered for the overall design from user access to database access. Firstly, data encryption [8] is concerned. Data transmission uses secure FTP, which needs to install encryption and decryption module for every client so that receiving encrypted data could be decrypted and authenticated. Data storage includes raw data storage and processing results. Access control, including examination and restriction of users' qualification, is used for data storage and extraction. Additionally, processing results produced in applications should also be stored in security for a longer period so that they could be extracted at any moment. Operating system encryption should also be adopted to protect data against physical data security risks. Secret key management is also considered. Secondly, identification technology is adopted. Public key Infrastructure is the major technology. RSA algorithm is adopted to resist password attacks. On consideration of comprehensive authorization principle, least authorization principle should also be concerned. The ship surveillance system is designed for multiple applications which need different accessing services, so layered authentication or restricted authentication is taken into account to restrict the non-authorized access to other databases or services. Thirdly, user access and operation record need to be stored to track insecure sources or illegal actions.

54.5 Conclusion

With the increasing demand for ocean surveillance, ship surveillance technology based on space-borne SAR and AIS has attracted more attention. This paper presents a SOA-based ship surveillance system based on this technology which could be extended to multi-source data fusion system. On the further consideration, the SOA-based architecture is more efficient. Further researches include both intelligent decision promotion and system architecture optimization.

References

1. Zhou L (2010) SOA based image retrieval approach. IEEE third international conference on information and computing, ICIC 2010, pp 319–322
2. Ashish S, Himanshu A, Ashim RS (2011) Designing a SOA based model. ACM SIGSOFT Softw Eng Notes 36(5):1–6
3. Eriksen T, Skauen A, Narheim B, Helleren O, Olsen O, Olsen R (2010) Tracking ship traffic with space-based AIS: experience gained in first months of operations. 2010 International Waterside Security Conference (WSS). Carrara, pp 1–4
4. Sun JX (2008) Modern pattern recognition (Second edn). Higher Education Press, Beijing, pp 454–458 (in Chinese)
5. Paris WV, Ryan AE, John W (2007) Ship signatures in RADARSAT-1 ScanSAR narrow B imagery analysis with AISLive data. DRDC, Ottawa, pp 6–22
6. Zheng Z, Huang D, Zhang J, He S, Liu Z (2010) A SOA-based decision support geographic information system for storm disaster assessment. The 18th international conference on geoinformation. Beijing, pp 1–5
7. Zhao Z, Ji KF, Xing XW, Zou HX (2012) Research on ship surveillance by integration of space-borne SAR and AIS. Chinese high resolution observation to earth academic seminar-satellite remote sensing and applications. Beijing 4(62):1–8 (in Chinese)
8. Ross E, Arifin B, Brodsky Y (2011) An information system for ship detection and identification. IEEE international geoscience and remote sensing symposium (IGARSS). Vancouver BC, pp 2081–2084

Chapter 55
Modeling of an Electric Vehicle for Drivability Improvements

Manli Dou, Gang Wu, Chun Shi and Xiaoguang Liu

Abstract A control-oriented drivability model for en electric vehicle is described. The developed model is capable of predicting longitudinal vehicle responses that affect drivability. The model is implemented in Simulink and validated by CRUISE, and the results demonstrate sufficient accuracy for the developed model. The model is useful for design, improvement and calibration of control algorithms and strategies.

Keywords Electric vehicle · Modeling · Drivability

55.1 Introduction

Electric vehicle (EV) technologies are currently being developed globally by a wide variety of manufacturers in response to existing, proposed and potential future legislation in various countries around the world. The EVs have several advantages over vehicles with internal combustion engines, such as energy efficiency and environmentally friendliness, and is seen to the right way to solve the environment and energy problems.

During development and calibration phases of EV control strategy, it is of crucial importance to assess and optimize the vehicle drivability. Accurate modeling of the vehicle propulsion system is therefore necessary to evaluate performance in early stages of the vehicle development process. The objective of this paper is to present such a control-oriented EV drivability model that allows for the prediction of vehicle responses in longitudinal responses [1].

M. Dou (✉) · G. Wu · C. Shi · X. Liu
Institute of Industrial Automation of Department of Automation,
University of Science and Technology of China, HeFei, China
e-mail: dmlzkd@mail.ustc.edu.cn

Z. Sun and Z. Deng (eds.), *Proceedings of 2013 Chinese Intelligent Automation Conference*, Lecture Notes in Electrical Engineering 255,
DOI: 10.1007/978-3-642-38460-8_55,

This paper is organized as follows: Sect. 55.2 presents the architecture of the experimental vehicle. Section 55.3 describes the details of the EV model. Section 55.4 provides comparisons of the developed model and the result of CRUISE. Conclusions are made in Sect. 55.5.

55.2 Experimental Vehicle

The experimental vehicle is a pure electric bus of Anhui Ankai Automobile Co., Ltd. Its drive system is mainly constituted by a traction electric machine, propeller shaft, universal joint, final drive and drive shaft. Figure 55.1 shows the vehicle configuration.

The traction electric machine (EM) is an asynchronous AC induction machine. The main parameters of the experimental vehicle are shown in Table 55.1.

55.3 Vehicle Modeling

A dynamic model of the test vehicle is developed to facilitate the evaluation of vehicle drivability. The model is implemented in the MATLAB/Simulink environment using a variable-step solver that is suited for stiff dynamic systems [2].

As the model is made for drivability, only longitudinal vehicle dynamics are taken into consideration. The propeller shaft is assumed to be rigid, while the rear and front half-shafts are assumed to be elastic. The impacts of environmental factors such as temperature, pressure and humidity are not taken into consideration.

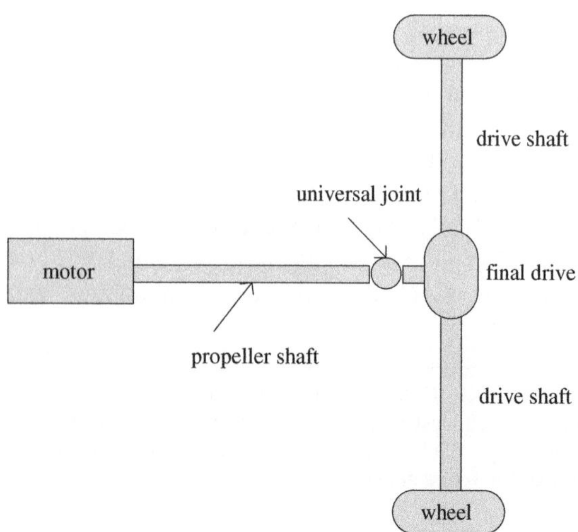

Fig. 55.1 Experimental vehicle configuration

Table 55.1 Main parameters

Parameter	Value
Frontal area	6 m^2
Drag coefficient	0.6
Load on front axle	5,096 kg
Load on rear axle	9,403 kg
Vehicle weight	14,500 kg
Radius of the wheel	0.465 m
Wheel inertia	9 kg × m^2
Efficiency of the transmission system	0.97
Transmission inertia	0.03 kg × m^2
Transmission ratio	5.63
Max motor torque	2,400 N
Max motor speed	2,520 rpm
Max motor power	240 Kw
Motor inertia	4 kg × m^2
Motor efficiency	0.88

Drivetrain losses are represented by lumped efficiency. The backlash impact is ignored.

Mathematical models of the vehicle subsystems and components are described as following.

55.3.1 High Voltage Battery

The battery affects the vehicle's drivability because the state-of-charge (SOC) has influence on control strategy. In order to predict this effect, a simplified battery model is used to estimate the battery SOC.

The power supplied by the battery, P_{batt}, is given as:

$$P_{batt} = P_{acc} + \begin{cases} T_{em} w_{em} \frac{1}{\eta_{em}}, T_{em} \geq 0 \\ T_{em} w_{em} \eta_{em}, T_{em} < 0 \end{cases} \tag{55.1}$$

where the P_{acc} is the power of the accessories, T_{em} is the EM output torque, w_{em} is the angular velocity of the motor rotor shaft, η_{em} is the efficiency of the electric motor.

Using the power supplied by the battery P_{batt}, the open-circuit voltage of the battery V_{oc}, and the internal resistance of the battery Z_{batt}, the battery current I_{batt} and voltage V_{oc} can be computed as [3]:

$$\begin{aligned} I_{batt} &= \frac{V_{oc} - \sqrt{V_{oc}^2 - 4Z_{batt}P_{batt}}}{2Z_{batt}} \\ V_{oc} &= V_{batt} - Z_{batt}I_{batt} \end{aligned} \tag{55.2}$$

The battery SOC can be obtained by the current integration:

$$SOC(t) = SOC_{init} + \frac{1}{C_{batt}} \int_0^t I_{batt} dt \tag{55.3}$$

where the C_{batt} is the battery energy capacity in [A.s].

55.3.2 Electric Motor and Propeller Shaft

The dynamics of the electric motor and propeller shaft can be described using a lumped model with inertias reflected to the electric motor (EM) output shaft [3] as:

$$\dot{w}_{em} = \frac{1}{J_{em} + J_{mech}} \left(T_{em} - T_{em,fr} - \frac{2}{\varsigma_{rd}} T_{shaft} \right) \tag{55.4}$$

where w_{em} is the angular velocity of the motor rotor shaft, J_{em} is the EM rotor inertia, J_{mech} is the mechanical inertia at the EM rotor shaft, T_{em} is the EM output torque, $T_{em,fr}$ is the friction torque, T_{shaft} is the reaction torque of a single half shaft on the rear axle, ς_{rd} is the rear main reducer speed ratio. The $T_{em,fr}$ is modeled as a constant value.

As the power supplied by the motor is limited by the motor max power P_{em}, so the T_{em} is limited by the max power, then the current power exceeds the max power, T_{em} has to be reduced by the equation:

$$T_{em} = 9550 \frac{P_{em}}{n_{em}} \tag{55.5}$$

where n_{em} is the speed(unit rpm) of the motor.

55.3.3 Universal Joint

Universal joint connects the propeller shaft and drive shaft. When the angle between the input shaft (propeller shaft) and output shaft (drive shaft) is not zero, the rotating speed of the output shaft is not the same to the rotating speed of the input shaft, even the input drive shaft axle rotates at a constant speed. Because the vehicle model is assumed drives on a straight road, so the speed of input and output shaft is the same. Under normal circumstances, the transmission efficiency of universal joint is 97 % ~ 99 % [4].

55.3.4 Drive Shaft

The half shafts are modeled as elastic rods with a damping coefficient. The torque applied to the shaft are proportional to the speed difference and twist angle between the shaft's terminals.

The torque of the driveshaft can be obtain as [5]:

$$T_{shaft} = k_{shaft} \int \left(w_{diff} - w_{wheel}\right) dt + c_{shaft}\left(w_{diff} - w_{wheel}\right) \tag{55.6}$$

where the k_{shaft} [N · m/rad] is the torsional stiffness, c_{shaft} [N · m/(rad/s)] is the shaft damping coefficient.

55.3.5 Brakes

The vehicle meets the driver's braking commands using a combination of conventional pneumatic braking and regenerative braking. Regenerative braking torques is realized by the electrical motor, pneumatic braking torques is modeled using an empirically determined relationship between the brake pedal position and the pneumatic brake pressure. So the brake torques on the front and rear wheels can obtain as [6]:

$$T_{br} = R_{br} p(\alpha_{br,pedal}) A_{br} \mu_{br} \tag{55.7}$$

where the T_{br} is the brake torque, R_{br} is the effective friction radius, A_{br} is the brake piston surface, μ_{br} is the friction coefficient, $p(\alpha_{br,pedal})$ is the function that maps the brake position $\alpha_{br,pedal}$ to the brake air pressure.

55.3.6 Tire Model

The frictional properties of the road surface are assumed to be uniformly acting on all tires.

The tire model proposed by Pacejka [7] is used to represent the longitudinal tire dynamics. This model is widely used in professional vehicle dynamics simulations; it uses the semi-empirical "Magic Formula" to compute the tractive forces generated by the tires:

$$F_x = D \cdot \sin(\mathrm{C} \cdot \mathrm{atan}(\mathrm{B} \cdot \kappa - \mathrm{E}(\mathrm{B} \cdot \kappa - \mathrm{atan}(\mathrm{B} \cdot \kappa)))) \tag{55.8}$$

where B is stiffness factor, C is shape factor, D is peak factor, E is curvature factor, κ is transient tire slip, F_x is tractive forces of the tire. B, C, D are obtained from the manufacturer's tire data. $D = \mu F_z$, μ is the coefficient of friction between the tires and the road surface, F_z is vertical forces acting on the front and rear wheels.

The vertical forces acting on the front and rear wheels ($F_{z,f}$ and $F_{z,r}$) change with the vehicle acceleration. Assume the vertical forces acting on each front (rear) wheel is the same. Then they can be calculated from the following equations [8].

$$\begin{aligned} 2F_{x,f}L &= M_{veh}gl_r - M_{veh}\dot{v}_{veh}H \\ 2F_{x,r}L &= M_{veh}gl_f + M_{veh}\dot{v}_{veh}H \end{aligned} \tag{55.9}$$

where the L is the distance between the front axle and rear axle, l_r is the distance between the rear axle to the center of mass, l_f is the distance between the front axle to the center of mass, H is the distance between the center of mass to the grand, and M_{veh} is the mass of the vehicle.

The wheel slip is $\kappa = \frac{r_{wheel}w_{wheel} - v_{veh}}{\max(v_{veh}, r_{wheel}w_{wheel})}$, where r_{wheel} is the radius of the wheel, w_{wheel} is the wheel's angular velocity, v_{veh} is the vehicle's longitudinal velocity.

The wheel's angular velocity is calculated from a torque balance at the wheels:

$$\dot{w}_{wheel} = \frac{1}{J_{wheel}}(T_{shaft} - T_{brake} - r_{wheel}F_x - r_{wheel}F_{roll,fr}) \tag{55.10}$$

where J_{wheel} is the wheel inertia and $F_{roll,fr}$ is the tire rolling force.

$F_{roll,fr}$ is modeled as a predominantly static function that also has a slight linear dependence on vehicle speed at constant tire pressure:

$$F_{roll,fr} = F_z \cos\gamma(c_1 + c_2 v_{veh}^2) \tag{55.11}$$

where c_1 and c_2 are rolling coefficients and γ is the road inclination.

55.3.7 Vehicle

Using Newton's second law in the longitudinal direction, the following equation is obtained:

$$\dot{v}_{veh} = \frac{1}{M_{veh}}(2F_{x,f} + 2F_{x,r} - \frac{1}{2}\rho_{air}C_dAv_{veh}^2 - M_{veh}g\sin(\gamma)) \tag{55.12}$$

where ρ_{air} is the air density, C_d is the vehicle's drag coefficient, A is the vehicle's frontal area, γ is the climbing angle.

In summary, the total model developed in Simulink is shown in Fig. 55.2.

55.4 Validation

The developed model is validated by AVL CRUISE, which is a simulation package that supports common tasks in a vehicle system and driveline analysis throughout all development phases, from concept planning to launch and beyond [9, 10].

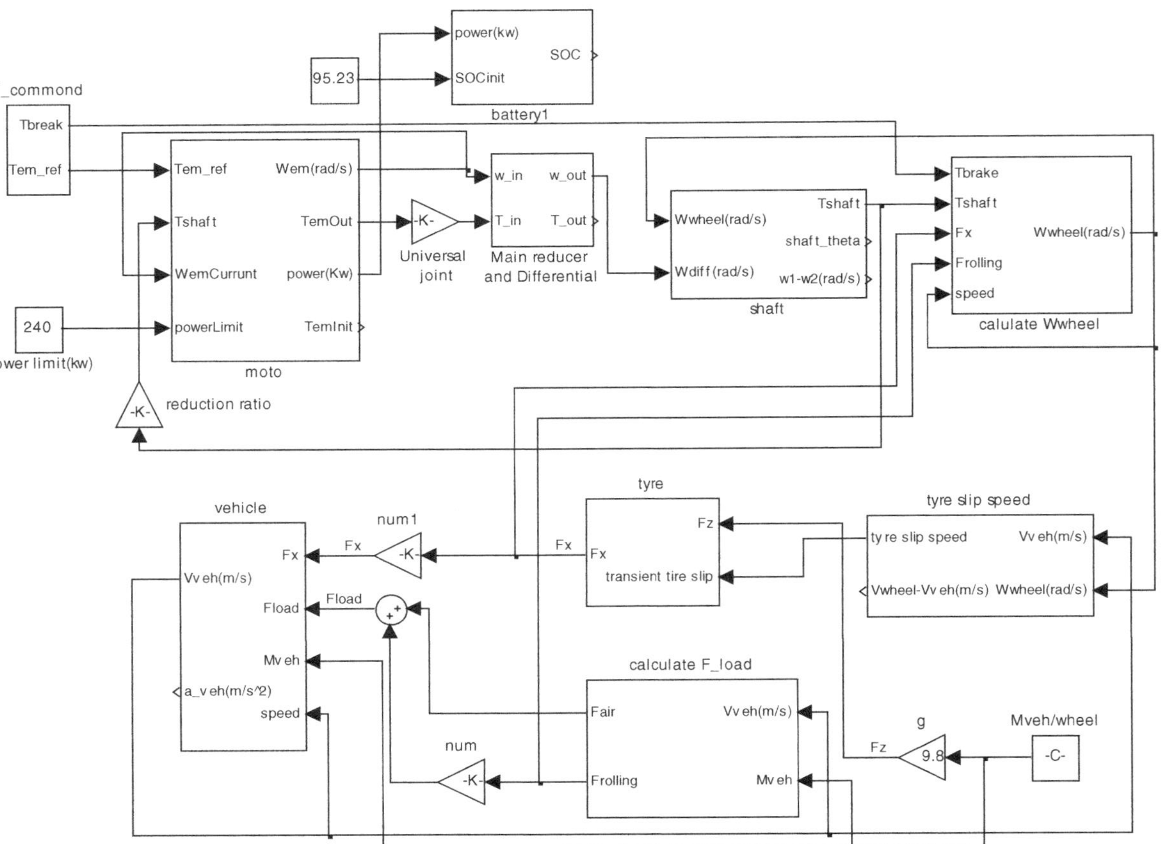

Fig. 55.2 Model block diagram

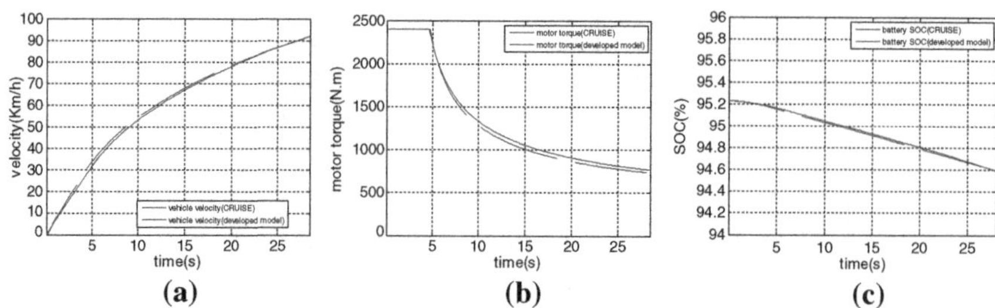

Fig. 55.3 (**a**) vehicle speed comparison (**b**) motor torque comparison (**c**) battery SOC comparison

A precise pure electric bus model is created in CRUISE, and is used for validate the developed model. The driving performance for acceleration is used to compare the CRUISE model and developed model. In the acceleration performance test, the acceleration pedal is set to max value, so the torque command sent to the motor is always the max state, and the vehicle speed will always increase. The test lasts 28 s, and the vehicle speed, motor torque acquired and battery SOC value from the two models is in Fig. 55.3a, b and c.

55.5 Conclusion

This paper demonstrates the development and validation of pure electrical vehicle model that is suited for the evaluation of vehicle drivability. The model is limited to describing longitudinal dynamic response. The primary application of the model is to assist in developing algorithms that can improve drivability. Future work will focus on the control algorithms and their implementation in the test vehicle.

Acknowledgments The authors would like to thank Anhui Ankai Automobile Co., Ltd, and National Electric Vehicle System Integration Engineering Research Center. This work was supported by Key Technologies R&D Program of Anhui Province (No.1201a0201003, The development and application of control system of the electric bus drive motor).

References

1. Koprubasi K (2008) Modeling and control of a hybrid-electric vehicle for drivability and fuel economy improvements [D]. The Ohio State University
2. Koprubasi K, Rizzoni G (2012) Development and experimental validation of a low-frequency dynamic model for a Hybrid Electric Vehicle [J]. Int J Powertrains 1(3)
3. Minh VT, Rashid AA (2012) Modeling and model predictive control for hybrid electric vehicles. Int J Automot Technol 13(3):477–485
4. Mills A, Hooke's R (2007) universal joint' and its application to sundials and the sundial-clock [J]. Notes Rec—Royal Soc London 61(2):219–236

5. Bayar K (2011) Development of a vehicle stability control strategy for a hybrid electric vehicle equipped with axle motors [D]. The Ohio State University
6. Gao Y, Ehsani M (2001) Electronic braking system of EV And HEV—integration of regenerative braking, automatic braking force control and abs [J]. SAE Technical Paper 2001-01-2478. doi:10.4271/2001-01-2478
7. Pacejka HB (1987) Tyre modeling for use in vehicle dynamics studies[J]. Soc Automot Eng
8. Zhisheng Xu (2011) Automobile Theories. China Machine Press, China (in Chinese)
9. Yang X (2011) The optimizing matching for transmission system based on cruise [D]. Chongqing University, Chongqing, China
10. Sangtarash F (2009) Effect of different regenerative braking strategies on braking performance and fuel economy in a hybrid electric bus employing cruise vehicle simulation [J]. SAE Int J Fuels Lubr 1(1 828-837). doi: 10.4271/2008-01-1561

Chapter 56
Simulation of the Dynamic Environment of Carrier Aircraft Approach and Landing

Xianjian Chen, Gang Liu and Guanxin Hong

Abstract Simulation of marine environment is very important for the system of carrier aircraft approach and landing. The effect of environment on landing of carrier aircraft has been discussed. It designs an environment scenario generator from dynamic characteristics of environment which as a simulation node accesses to the federal system of carrier aircraft landing. It can increase the reusability and operability of the environmental data. The realization methods of fog, cloud and illumination which affect the visibility of the environment of carrier-based aircraft landing have been discussed from the basic algorithm. And it provides a basis for the visualization of the environmental data.

Keywords Marine environment · Dynamic characteristics · Environment scenario generator · Realization methods

56.1 Introduction

The technology of carrier-based aircraft approach and landing has been hot and difficult today.

X. Chen (✉) · G. Hong
School of Aeronautic Science and Engineering, Beihang University, Beijing 100191, China
e-mail: xianjian97@sina.com.cn

G. Hong
e-mail: honggx@buaa.edu.cn

X. Chen
State Key Laboratory of Virtual Reality Technology and Systems, Beihang University, Beijing 100191, China

G. Liu
National Laboratory for Aeronautics and Astronautics, Beihang University, Beijing 100191, China
e-mail: lg@buaa.edu.cn

Z. Sun and Z. Deng (eds.), *Proceedings of 2013 Chinese Intelligent Automation Conference*, Lecture Notes in Electrical Engineering 255,
DOI: 10.1007/978-3-642-38460-8_56, © Springer-Verlag Berlin Heidelberg 2013

In recent years, along with the rising and development of the simulation technology it provides a new way to the study of carrier aircraft landing.

The process of carrier-based aircraft approach and landing is influenced by environmental factors, so environmental simulation is an extremely important part for the whole system simulation of carrier aircraft approach and landing. The United States Defense Modeling and Simulation Office (DMSO) released of an overall plan of modeling and Simulation early in 1995, which chose the authority description and expression of the natural environment as one of main goals [1]. And that promoted the follow-up series achievements.

China has tried to apply the HLA standard to establish environment server, and launched a series of design work [2].

This article summaries the impact of dynamic environment on the process of carrier-based aircraft approach and landing from dynamic characteristics of environment. It studies of the visual simulation method of dynamic environment and the environment scenario.

56.2 Impact of Environment on Carrier Aircraft Landing

Carrier is a steel fortress cruising at sea. And the main environment is the ocean. The marine environment consists of two parts atmosphere and ocean. The effect of the sea on carrier aircraft landing lies in the attitude motion caused by sea waves. This may lead to the movement of ideal landing point and may also cause the danger of the collision between the carrier and aircraft. Meanwhile it affects the efficacy of the landing aid system. If carrier moves too violently exceeded the stability limit of FLOLS system, it will lead the FLOLS system invalid. So it will affect the manual landing.

The effect on the carrier-based aircraft landing of the atmosphere is mainly reflected in two aspects. First, because the visibility affects the visual distance of carrier-based aircraft pilots and the scope of Optical Landing System, the approach procedure of the carrier-based aircraft is influenced. Second, due to the presence of the wind field, the body movement of the carrier-based aircraft is disturbed so that the landing is influenced. According to the difference of the visibility, the environment is divided into three kinds, and there are different landing procedures and different guidance modes of the carried-based aircraft in different environments [3]. That is because the scopes of the guidance system of carrier-based aircraft landing in different environments are different. Table 56.1 summarizes the impact of the environment on the guidance system.

But, weather at sea is more changeable. And the environment of carrier-based aircraft landing may not always keep in an unchanged state. So the effect of environmental cross should be considered. For example, an aircraft approaches and lands according to the environment II. When it flies at 5 nautical miles from

Table 56.1 Scopes of the guidance system in the three kinds of environment

	Weather conditions		Photoelectric guidance system		Visual optical guide system	
			Multi-sensor photoelectric guidance	Laser holographic image-guided system	ICOLS	FLOLS
The effective distance of the system (Nautical mile)	Daytime	I	10	4	15	4
		II	6	4	10	4
		III	3	4	3	0.75
	Night	I	10	4	15	4
		II	6	4	10	4
		III	3	4	3	0.75

the carrier, which is the approach decision point, if the pilot can't see the carrier, the aircraft must to be guided to the missed approach route to approach and land according to Environment II. That is to say the environment has changed from Environment II to III. In order to better research of effect of environmental cross on carrier aircraft approach and landing, an environment scenario generator which can direct all kinds of environment of carrier aircraft approach and landing should be built.

56.3 Design of the Environment Scenario Generator of Carrier Aircraft Approach and Landing

56.3.1 Environmental Database

Currently there are two ways to get environmental data. The first is to establish an environmental database by collecting the environmental data. The advantage of this method is that the data is real and reliable. But due to the huge amount of data, storage is extremely resource-intensive, and to the constraints of geographical conditions and acquisition devices limitations collecting environmental data is not comprehensive. The other way to get environmental data is mathematical modeling techniques. Based on the principles of physics, such as the laws of fluid dynamics and atmospheric motion the corresponding kinematic equations are established to numerically simulate the environment by solving these equations. In the issue of the design of environment scenario generator of carrier aircraft approach and landing, we consider the advantages and disadvantages of the two ways, and choose a combination of both to build environmental underlying database.

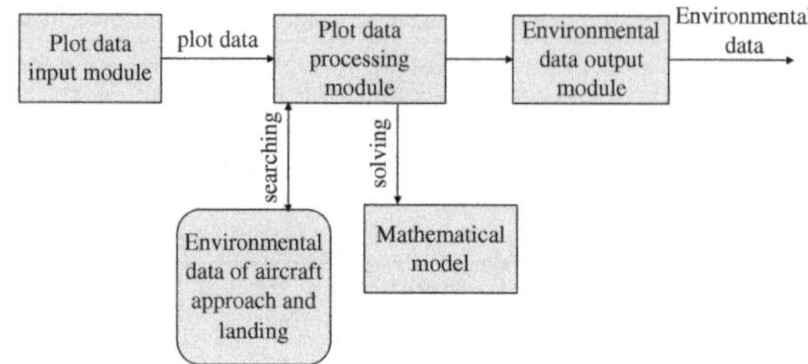

Fig. 56.1 Environment scenario generator of carrier aircraft approach and landing

56.3.2 Module Design

The environment scenario generator mainly consists of five modules as shown in Fig. 56.1. They are Data input interface, Drama data processing module, Environmental Data Output, Environmental database and Environment Mathematical Model. Drama data processing module comprises three functions:

Plot data parsing function: Receiving the scenario data and parse them, to calculate the data required by the environment script.

Search function: search environment data library of carrier-based aircraft approach and landing, and judge whether the required data exists in the environment library, thereby to make the call.

Solver activating function: when environmental database can't satisfy the need of some of the plots, it needs to activate the environment mathematical model solver module to solve the environmental data required, as a supplement to existing environment database.

56.3.3 Structure of Environmental Data of Aircraft Approach and Landing

Because the landing environment is mainly composed by the ocean and the sky, and these state of the environment is changing with the place and time. We can use four-tuple to describe the environment [4]:

$$\mathrm{E} = <\mathrm{O,A,L}, t>$$

O, A, L, t represent the ocean, atmosphere, location and time. Environment of carrier aircraft approach and landing is actually composed by a sequence of discrete parameters. And each tuple can be subdivided as Table 56.2 shows.

Table 56.2 Environment data classification

Atmosphere (A)				Ocean (O)
Cloud	Thickness	Air	Temperature	Sea state
			Pressure	
	Bottom height		Humidity	
			Density	
	Top height	Wind	Wind speed	Wave height
	Type	Fog	Visibility	Flow velocity
	Coverage		Coverage	Flow direction

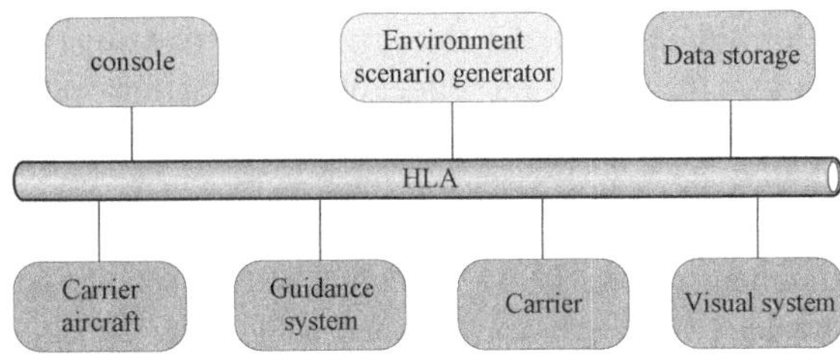

Fig. 56.2 Components of carrier aircraft approach and landing simulation system

56.3.4 Environmental Data Applications

We adopt the form of a data server, and access it to the federal system of carrier aircraft landing as a simulation node. Then visualize these environmental data through visual system node. In this way simulation program can be separated from the underlying database, so that the environmental data has better reusability. As shown in Fig. 56.2

We can initialize the environment of carrier aircraft approach and landing by the console. Simulation system calls environmental data according to the tasks required and then transmit the data to the visual system. Thus the environmental data will be visualized.

56.4 Rendering Algorithms and Implementation of the Environment

The environment of carrier-based aircraft carrier landing is classified based on visibility. So this article mainly considers the rendering method of factors affecting visibility. The main factors affect the landing visibility in the marine environment are fog, light, cloud, which are discussed below.

56.4.1 Fog

The principle of realization of the fog effect is that fog color and object color superimposed to the different weights. We study of algorithm of fog effect in OpenGL [5], and take RGBA mode as an example. The final color of the object can be written as:

$$C = fC_i + (1 - f)C_f$$

C_i is value of the color of the fragment. C_f is the value specified by the fog color. f is one of the most critical parameters, called fog effect factor. It can be seen when $f = 0$, the object is completely fogged, and its color is the background color. When $f = 1$, the object isn't fogged at all. There is no fog color mixed in the color of the object. Fog effect blending factor f in OpenGL is calculated according to the following three equations, and intercepted to [0, 1]. Table 56.3 summarizes three fog effect models and their scopes of use.

In the formula, z is the eye coordinate distance between the viewpoint and fragment center. It can be controlled by specifying the coordinates of fog coordinates of each vertex. The *density* is value of fog density. *start* and *end* are the beginning and end of the measure for the fog. *Density*, *start*, *end* can be defined by a particular function, the default are 1, 0, 1.

During the approach process the carrier aircraft is far away from the aircraft carrier. So we use the linear model fog effect. Vega Prime has packaged the fog effect algorithm [6]. We can directly regulate opaque value to visualize the distance affected by the fog.

In Environment II, the approach decision point is 5 nautical miles from carrier. It's about 9,000 m. If the pilot can't see the carrier, it shows that the environment has deteriorated. By Vega Prime we set the observation point at 9,000 m behind the carrier and 300 m above the carrier. With opaque changed, the effect is as Fig. 56.3 shows.

As we can see when the opaque value is 9,000 we could hardly find the carrier. Therefore we define the opaque 9,000 as the boundaries for Environment II between Environmental III.

Table 56.3 Three models of the fog effect factor

Fog effect factor	GL_EXP	GL_EXP2	GL_LINEAR
f	$e^{-(density \cdot z)}$	$e^{-(density \cdot z)^2}$	$\frac{end - z}{end - start}$
Range of application	Near viewpoint	Far from viewpoint	Farther from viewpoint
Concentration	Uniform	Uneven	Vary linearly

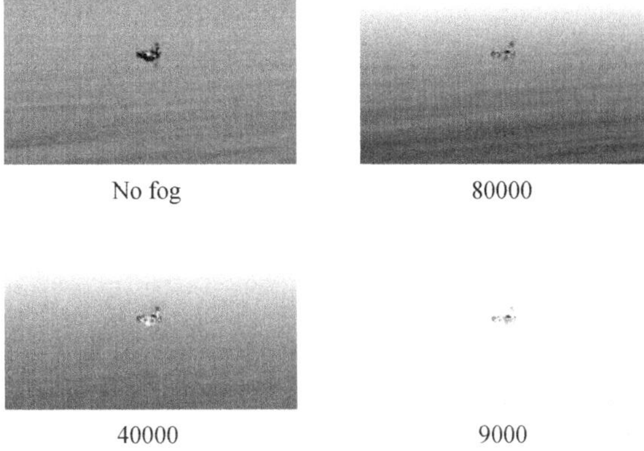

Fig. 56.3 The corresponding effect of different visibility

56.4.2 Cloud

Cloud is also an important factor to affect the vision of aircraft approach and landing. When the carrier aircraft located above the clouds, the sea vision will be obscured by the clouds. When carrier-based aircraft is in the clouds, the pilot's visibility will become smaller. In environmental database we use six data to describe the clouds: thickness, height of the bottom, height of the top, middle visibility, type and coverage.

The three-dimensional graphics engine generally packages the method of cloud rendering. We can directly control these parameters to simulate the clouds.

Cloud type is determined by RGB values. We can change the color values to achieve the dark clouds or white clouds.

56.4.3 Light

Day and night are the important boundaries conditions of the environmental of carrier-based aircraft carrier approach and landing. The biggest difference lies in the light. The light is the reason for all things to be seen. During the day due to the presence of the sun the light is sufficient. In the case of sufficient visibility, the pilot can easily find aircraft carrier. At night because there is not enough light, the pilot is hard to see the carrier. That will give pilot the psychological burden to increase landing difficulty.

In China, the sun goes down at about 18:30 in summer, and about 17:00 in winter. After the sunset brightness will last about one hour. If carrier-based aircraft approaches and lands during this period of time, it may occur environmental cross. Therefore, the processing of the light is very important in the simulation. OpenGL adopts simple light model, which the surface reflected light is defined as the material emission, scaled global ambient light and lights from source three parts superposed.

$$I = I_{emit} + K_a \cdot I_{global} + \sum_{lightsources} \frac{1}{K_c + K_l d + K_q d^2} \cdot C_{spotlight}(K_a I_a + K_d K_d \cos\theta + K_s I_s \cos^n a)$$

The first item of the formula is the light from the object itself. The second item is global light. It refers that the light scattering from the surrounding environment (such as indoor walls) incidents to the surface then reflects to our eyes. As the carrier-based aircraft landing in the open sea, and therefore the ambient light items can be ignored in the daytime. For the third item of the formula, the sun is infinity. So the light intensity isn't affected by distance. It belongs to directional light source. So attenuation factor $1/(K_c + K_l d + K_q d^2) = 1$ For condensing effect items, the sun does not belong to the spotlight, so $C_{spotlight}$ takes 1.

In order to increase friction, steel material of the carrier's deck is generally coated with a non-slip coating. The specular reflectance items $K_s I_s \cos^n a$ can be approximated as 0. In summary, in the daytime aircraft carrier illumination model can simplify as follow:

$$I = I_{emit} + (K_a I_a + K_d I_d \cos\theta)$$

When the sun was just downhill, due to atmospheric refraction and scattering effects, there is some time bright in the sky. So we can give certain intensity of ambient light, and ignore the sun light. The illumination model becomes as follow:

$$I = I_{emit} + K_a \cdot I_{global}$$

When there was the moon at night, the light model is $I = I_{emit} + (K_a I_a + K_d I_d \cos\theta)$ and is $I = I_{emit}$ without moon.

The effect of rendering is shown as Figs. 56.4, 56.5, and 56.6.

Fig. 56.4 Ocean ambient light at 18:00

Fig. 56.5 Ocean ambient light at 18:30

Fig. 56.6 Ocean ambient light at 19:30

56.5 Conclusion

The impact of the environment on approach and landing of carrier-based aircraft is mainly reflected in the operating distance of guidance system and in process of approach and landing. This article designs an environment scenario generator which meets the requirement of the carrier-based aircraft landing simulation system refer to international advanced environmental modeling techniques and separates the environment databases and visual link, and each as a separate node access to carrier-based aircraft simulation system. It can increase the reusability and operability of the environmental data. Finally, based on OpenGL the industry standard simulation software, this article studies the render algorithm about fog, cloud and illumination which influence the visibility of the environment, and achieves a smooth transition of environmental cross.

References

1. U.S. Department of Defense (1995) Defense modeling and simulation office, modeling and simulation master plan [R]
2. Cai J, Xu L (2010) Application research of atmospheric environmental simulation based on HLA. Equip Environ Eng 7(3):66–70 (in Chinese)
3. Chen X, Liu G, Hong G (2011) Selection and analysis of key points and factors in carrier aircraft landing process. Flight Dyn 29:20–24 (in Chinese)
4. Pang G (2003) The research and implementation on virtual natural environment server based on HLA. J Syst Simul 15(1):41–44 (in Chinese)
5. Shreiner D (2007) OpenGL programming guide, 6th edn. Addison-wesley, Boston
6. Multigen-Paradigm (2002) Vega prime options guide, version 2.2

Chapter 57
Script Based Spacecraft Fault Automatic Rapid Disposal Method Research and Application

Jun Zheng, Dan Luo and Benjin Li

Abstract In order to achieve maximize efficiency of on-orbit satellite application, realize independent fast spacecrafts fault emergency disposal, and minimum the fault effect, this paper presents a new method of script based spacecraft fault automatic rapid disposal. In the method, planning and scheduling platform is the core and satellite fault disposal script is one means. We build a complete script set for spacecraft fault disposal to realize the spacecraft fault detection, automatic completion of disposal. The method successfully applied into one on-orbit satellite operation and management and obtains good effect.

Keywords Spacecraft · Fault disposal · Satellite script · Planning and scheduling

57.1 Introduction

With the development of Chinese aerospace, different orbit types such as LEO, MEO, GEO and IGSO, different functions of the spacecrafts are increasing. According to incomplete statistics, China has scores of existing spacecrafts. According to China the twelfth five-year space development plan, in the end of the twelfth five-year, China will reach 100 spacecrafts in orbit. The spacecrafts will be in our country economy, marine, detection, defense and other fields and play an important role. However, the spacecrafts would be affected by the space environment in the life and arise a variety of different types of faults. These faults directly affect the benefits of the satellite, and cause the satellites life shortening or even losing life [1]. Therefore the in-orbit spacecrafts for fault detected timely and proper disposal call for more stringent requirements. At present according to the

J. Zheng (✉) · D. Luo · B. Li
Beijing Space Information Relay Transmit Technology Research Center, 5117, Beijing 100094, China
e-mail: zheng_raindrop@163.com

Z. Sun and Z. Deng (eds.), *Proceedings of 2013 Chinese Intelligent Automation Conference*, Lecture Notes in Electrical Engineering 255,
DOI: 10.1007/978-3-642-38460-8_57,

related literature material and papers can be shown, aerospace university, scientific research institute or the unit of operation of spacecraft control management give a high importance to spacecraft faults detection. They adopted a variety of methods for automatic fault diagnosis and recognition based on knowledge, neural network, and faults model and so on [2–7]. However, the fault response and disposal is mainly dependent on manual processing [8]. Even if there is a plan of the fault handling, we only take artificial judgment, artificial disposal methods. Especially for the disposal of time particularly demanding abnormalities, once the time delay, will affect the normal operation of the satellite. Based on the insufficient satellite fault disposal, we put forward the fast, timely and accurate demand for spacecrafts fault disposal.

57.2 Spacecraft Fault Automatic Disposal Method Based on Script

In order to achieve spacecrafts fault timely and correct disposal after the faults appear, we put forward an autonomic spacecrafts fault dispose method based on the script.

57.2.1 Autonomic Fault Disposal Mechanism

The mechanism is to quickly complete spacecrafts fault disposal and sets planning and scheduling platform as the core based on the fault disposal scripts [6]. The mechanism of satellite fault dispose logical relationship is shown in Fig. 57.1.

In Fig. 57.1, the main function of spacecrafts fault diagnosis system input data is the satellite telemetry data and bases on the fault model diagnosis method of fault recognition. Found that when the fault is detected the spacecrafts fault diagnosis system immediately is posterior to transmit abnormality information to planning and scheduling platform.

Planning and scheduling platform mainly completes normally satellite operations plans and anomalies of the disposal. Normal scheduling operation control unit is operative to normally operation events with planning process. The satellite fault disposal unit is based on the exception information sent by planning and scheduling platform to realize the abnormalities associated with existing satellite fault disposal scripts matching.

Fault disposal of the script is a pseudo instruction language performed by computer machine. The commands sets of the script is include instruction sending, process of judgment and effect evaluation, as shown in Fig. 57.2. Each fault countermeasure only corresponds to a disposal script. The main satellite system

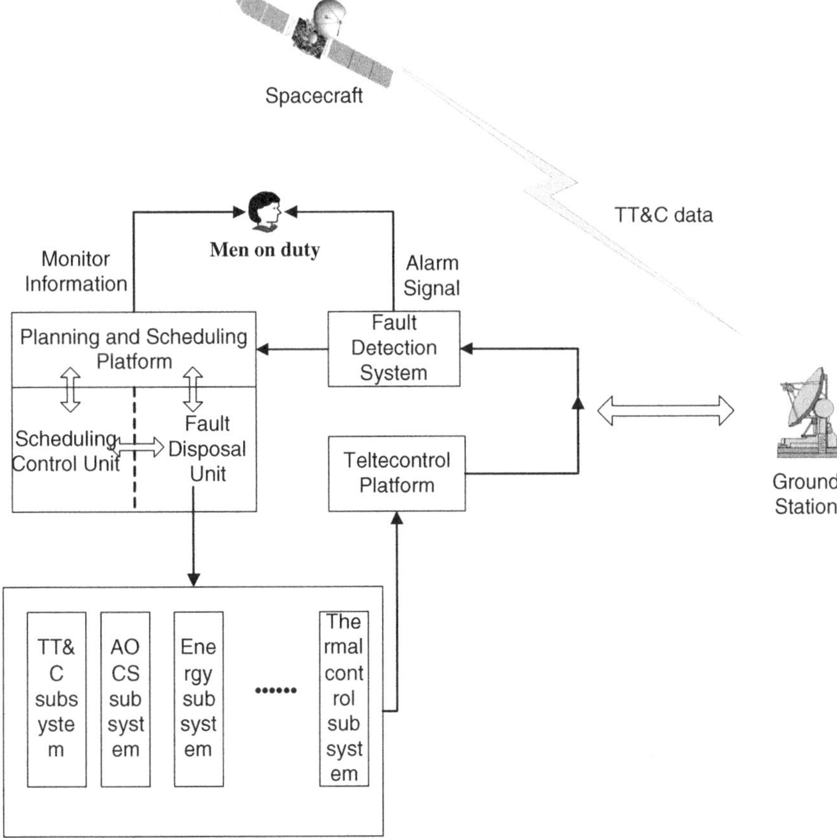

Fig. 57.1 The mechanism of satellite fault disposal logical relationship

fault disposal scripts are categorized and storage which facilitates the rapid and accurate script scheduling.

57.2.2 Disposal Process

The main process is shown in Fig. 57.3. The main steps are as followed.

Step 1: Real time fault diagnosis system for spacecrafts downlink telemetry data to analyze and judge. When one or several telemetry data beyond out of scope, the system sends alarm signal to men on duty, at the same time, sends abnormal satellite telemetry information to planning and scheduling platform;

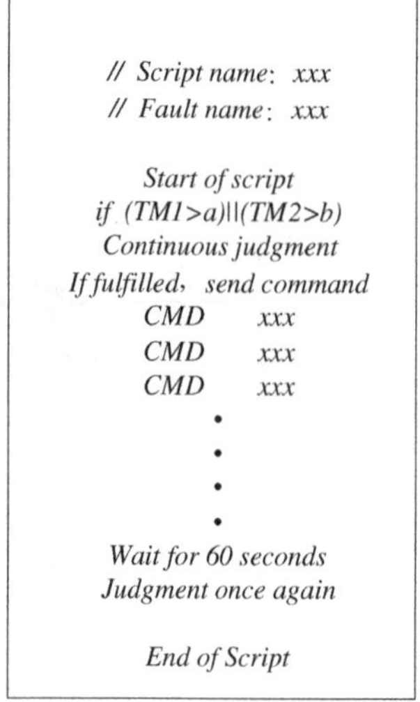

Fig. 57.2 Script content structure

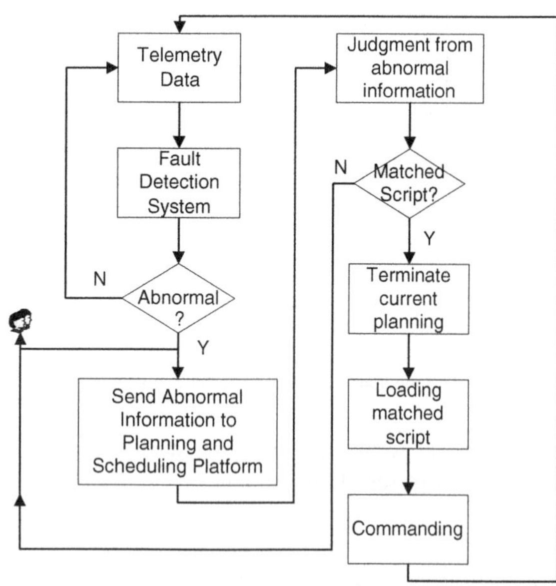

Fig. 57.3 Satellite fault disposal flow chat

Step 2: After planning and scheduling platform receiving the abnormal information from satellite fault disposal unit, the satellite fault disposal unit analyzes the abnormal information. When the abnormal need to dispose, disposal unit sends a terminate command to the normal operation control unit, and sends scheduling command to fault disposal scripts set for choosing one proper script.

Step 3: The fault disposal script set receives scheduling command and chooses one correct script. The matched script loads and executes by the spacecraft telecontrol platform.

Step 4: The instruction through the ground station equipment sent to the satellite, and the fault diagnosis system based on telemetry data to determine whether the fault disappear or not.

57.3 Engineering Application Case

As a result of the spacecraft fault disposal method is fast, high accuracy and low risk; we applied this method to a satellite control operation and management.

57.3.1 Fault Principle

The satellite has two infrared earth sensor (IRES) for measuring the rolling and pitching axis attitude. Each infrared earth sensor has four probes, as shown in Fig. 57.4. The four probe field constantly scan the earth's boundary. The IRES is sensitive infrared signal changes to determine the rolling and pitching axis attitude angle. The IRES three working probes can meet the work requirements. Once the sun into one of four probes within the field of view of infrared earth sensor, it can interfere with the IRES normal work. It may affect the satellite attitude measurement data. If attitude measurement datum goes wrong, the satellite rocks widely, even the satellite looses the attitude datum and satellite working mode changes into the cruising mode. If the satellite works in cursing mode, the satellite cannot perform the task in more than 10 h. Because of interference time from the sun to the earth sensor is fixed, so in general satellite automatically finishes its protection from interference probe according to the time.

57.3.2 Disposal Script Design

But if the satellite failed to automatically finish its interference probe protection, the satellite attitude determination will seriously be affected and cause the satellite to lose attitude reference. In order to prevent the occurrence of the abnormality, we designed the fault disposal script. The script design process is shown in Table 57.1.

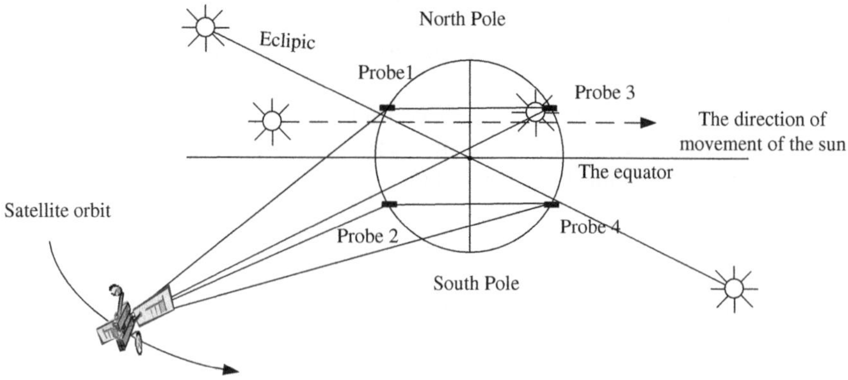

Fig. 57.4 The earth sensor principle of operation

Table 57.1 The earth sensor probe fault disposal script content

Order number	Telecontrol Command	Illustration
1	Start of script	
2	Wait for 30 s	
3	To judge whether the probe 1 or probe 2 whether in the state of "protection".(continuous judgment 7 frames telemetry data in 5 s): if not in the state of "protection", sending the telecontrol instruction; if in the state of "protection", continuous judging after wait XX seconds	
4	CMD xxx	Disposal probe 1 or probe 2 fault
5	CMD xxx	
6	CMD xxx	
7	Wait for xx seconds	
8	CMD xxx	Disposal probe 1 or probe 2 fault
9	CMD xxx	
End of script		

57.3.3 Engineering Application Effect

At 16:33 on September 15, 2011, the satellite fault diagnosis software suddenly reported that probe 1 not in the state of "protection". According to satellite fault disposal flow, the system then from satellite fault disposal unit obtained fault disposal of script and the script implemented in the telecontrol platform after loading. It only took 60 s to successfully complete the fault disposal. From the Figs. 57.5 and 57.6, can be seen that the disposal of abnormal did not produce any effects of satellite attitude.

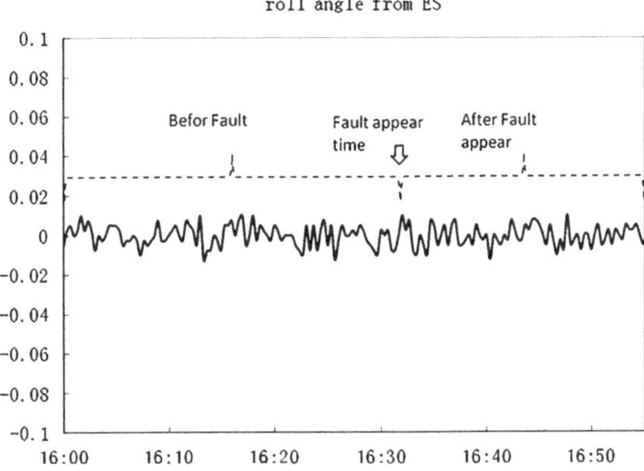

Fig. 57.5 Roll angel variation before and after the fault disposal

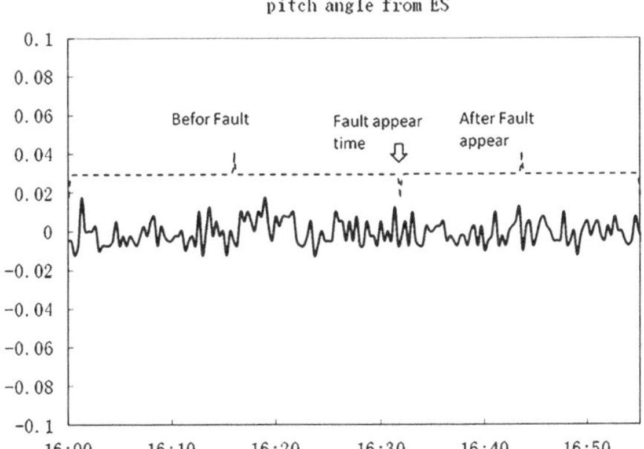

Fig. 57.6 Pitch angel variation before and after the fault disposal

57.4 Conclusion

This paper puts forward a new method of script based spacecraft fault automatic rapid disposal. In the method, planning and scheduling platform is the core and satellite fault disposal script is one means. The method successfully applied into one on-orbit satellite operation and management and achieve maximize efficiency of the satellite application.

References

1. Zhang S, Shi J, Wang JL (2010) Satellite on-board failure statistics and analysis. Spacecraft Eng 19(4):41–44 (in Chinese)
2. Zhang XG, Ren Z (2009) A new modeling method of satellite fault diagnosis based on the qualitative/quantitative hybrid model. Aerosp Control Appl 35(5):38–42 (in Chinese)

3. Qin W, Guo YF (2010) A failure waring system based on historical telemetry data for satellite on orbit. Spacecraft Eng 19(6):40–45 (in Chinese)
4. WU CM, (2010) Design of satellite fault expert diagnosis system based on knowledge. computer measure and control 18(4):743–745 (in Chinese)
5. Guo CJ, Teng YL (2011) Analysis and research of RAIM algorithms for identifying simultaneous two faulty satellites. Appl Res Comput 28(3):924–927 (in Chinese)
6. Wang ZH, Shen Y, Zhang XL (2012) Satellite gyroscopes fault diagnosis based on numerical differentiation scheme. J Astronaut 33(9):1262–1267
7. Chen HM, Jia YS, Mu YS (2012) Application of remote monitoring and real-time fault diagnosis technology. Inf Tech 15(10):116–120 (in Chinese)
8. Walt T, Lou H, Christopher R, Jay K, James R, Michael G, Roy S (2009) Autonomous and autonomic systems: with applications to NASA intelligent spacecraft operations and exploration systems. Springer, New York

Chapter 58
Synthetical Ramp Shift Strategy on Electric Vehicle with AMT

Zhifu Wang, Fulin Zhang and Yang Zhou

Abstract The performance of the drive motor system was better with AMT system. A synthetical ramp shift strategy that identified the riding condition based on rotational speed of the drive motor before and after the shifting process has been provided. The ramp shift test on one the electric truck with AMT was done, and the test results showed that by dropping gear continuously or discontinuously, the proper gear could be found eventually when the vehicle was running on ramp, though the original gear might be improper.

Keywords Electric vehicle · AMT · Ramp shifting strategy

58.1 Introduction

Electric Vehicles have been used widely all over the world. The driving system mostly consists of the motor system and the transmission system (mainly the automatic transmission) to balance the power demand and the driving characteristics of the electric vehicles at present. Now automatic transmission includes continuously variable transmission (CVT), dual clutch transmission (DCT), automatic transmission (AT) and Automated Mechanical Transmission (AMT). Limited by the torque, CVT and DCT are mainly used in passenger vehicles. AT and AMT are applied in heavy commercial vehicles, however, compared to AT, AMT is more efficient, smaller mass, more cost-effective and is ideally used in all kinds of electric vehicles including electric buses [1]. Ramp driving, especially driving up, is a typical dynamic condition of vehicles and improper shifting choice would cause too much shifting action to increase part wearing of transmission

Z. Wang (✉) · F. Zhang · Y. Zhou
National Engineering Laboratory for Electric Vehicles, Beijing Institute of Technology, Beijing, China
e-mail: wangzhifu@bit.edu.cn

Z. Sun and Z. Deng (eds.), *Proceedings of 2013 Chinese Intelligent Automation Conference*, Lecture Notes in Electrical Engineering 255,
DOI: 10.1007/978-3-642-38460-8_58,

system and reduce the comfort and dynamic performance of vehicles [2]. To be worse, it would make vehicles slipping down and cause some dangerous situation. So, it is important to study the control strategy of the ramp situation.

58.2 Classification and Features of Vehicle Ramp Running Condition

When the electric vehicle is running, the output of the power system will change with road conditions and these changes depend on the operation of drivers, including accelerating pedals, brake pedal, steering system, gear lever and so on. But the controller of automatic transmission voluntarily chooses gears on vehicles which have automatic transmission according to related parameters. When the vehicle is running on ramp, especially driving up, the resistance is different from general driving situation. In some cases, the ramp resistance is the main factor of the vehicle resistance. At the same time, when vehicles are driving up, the passing ability becomes the main task of the vehicles. At the same time the acceleration and economy would less important. When Manual shift vehicles are running on different ramps, the operating characteristics and shift strategies of drivers are different [3].

It follows that focusing on different ramps, the operating characteristics and shift strategies of the driver greatly different. It's almost impossible for automatic transmission vehicles to recognize these ramps and adopt relevant shift strategies by existing sensor. Even if fuzzy reasoning method and neural network method are adopted, it's also difficult to get accurate results and right shifting position.

58.3 Recognition of Ramp Running Condition

AMT shifting process on electric vehicle with AMT generally includes: pick-idle-shift. During the three stages, in addition to judge related gears on the basis of the current vehicle state and pilot operation, AMT control system sends relative command to the drive motor system to realize shifting. The corresponding work of the drive motor system contains: zero torque (free rotation state)—mediate speed (adjust the speed of the drive motor system to suitable number)—output torque. In a word, the drive motor system always outputs zero torque during the shifting process. Influenced by rolling, wind and ramp resistance, the vehicle gradually slows down [4]. At the same time, the drive motor system has accurate speed identification to recognize running condition of the vehicle slope by motor speed changes around the ramp shift gear driving.

The forces acted on the vehicles are shown in Fig. 58.1 [5]. Total traction, which affects interface between the drive wheel tire and the road, makes vehicles

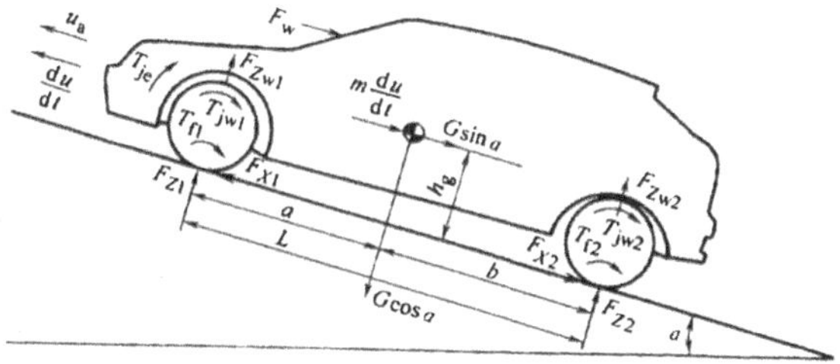

Fig. 58.1 The force of the vehicle

move. The power produced by the torque of power device promotes wheel through the transmission of transmission device. When the vehicle is moving, vehicles is limited by resistance. The resistance force usually contains tire rolling, air, climbing and acceleration resistance. Therefore, the total resistance of car is as follows:

$$\Sigma F_t = F_r + F_w + F_g + F_j \tag{58.1}$$

Where: ΣF_t-the total resistance force, F_r-the rolling resistance force, F_w-the air resistance force, F_g-climbing resistance force, F_j-acceleration resistance force.

Supposing beginning to pick block, transmission gear change rate is i_1, motor speed is ω_1. After the shift process, gear change rate is i_2, motor speed is ω_2, and the whole shift time is t_0. The slow speed rate of vehicle reducer is i_0, vehicle wheel radius is r, as a result, the vehicle reducing speed during shift is as follows:

$$\alpha = dv/dt = 0.3777 * r * (\omega_2/i_2 - \omega_1/i_1)/t_0/i_0 \tag{58.2}$$

Because power interrupts during shift, the reducing speed shown in formula (58.2) is affected by driving resistance, so when vehicles are driving up, driving resistance is as follows:

$$\Sigma F_t = \mathrm{M} \times \alpha = \mathrm{M} \times dv/dt = \mathrm{M} \times 0.3777 * r * (\omega_2/i_2 - \omega_1/i_1)/t_0/i_0 \tag{58.3}$$

where: M—total quality of vehicle.

58.4 Synthetical Control Shift Strategy and Gear Decision of Ramp Running Condition

Electric vehicle structure consists of the drive motor system, AMT system and other important parts which communicate and exchange information by CAN BUS. With the development of power technology and automatic control technology, benefiting from perfect technology, drive motor control is much better than engine control. When vehicles are driving, they are moving normally without further gear judgements if speed change is smaller around shift; if speed change is greater around shift, the driver will launch uphill shift control strategy [6].

When drive motor are operating at different rotating speed, we calculate maximum torque differently. Therefore, after calculating driving resistance, we also calculate driving force of drive system according to rotating speed of drive motor in order to limit the range of the transmission change and shift effectively [78].

When rotating speed of drive motor is less than the rated speed of the motor, the present gear range is as follows:

$$i \geq \sum F_t / \left(T_{max}/r/i_g\right) \quad (\omega_2 < \omega_n) \tag{58.4}$$

where: T_{max}-maximum torque of drive motor, ω_n-drive motor data speed.

While the rotating speed of the drive motor is more than or equal to the rated speed of the motor the present gear range is as follows:

$$i \geq \left(\sum F_t \times \omega_2\right) / \left(P_{max}/r/i_g/9549\right) \quad (\omega_2 \geq \omega_n) \tag{58.5}$$

where: P_{max}—maximum power of drive motor.

After uphill shift control system is activated, driving system calculates transmission gear range under the current driving condition by equation four or five and sends message to AMT system. If the gear suits the current chosen gear, driving will start; if not suitable, AMT system will adjust gear referring to the information. The calculation process is shown in Fig. 58.2.

The drive motor system parameter of light electric sanitation truck is shown in Table 58.1. And the electric vehicle parameters are shown in Table 58.2.

Figures 58.3 and 58.4 show rated speed and torque of drive motor system, efficiency, peak speed and torque, efficiency curve. In conclusion, drive motor system exports maximum torque below 3500 rpm when suitable gear promotes vehicle driving uphill.

According to above shift strategy, when vehicles are running in the ramp, after AMT system shifts for the first time according to the vehicle state and pilot operation, ramp shift strategy calculates corresponding slope. If the gear can overcome driving resistance, drive motor system will identify gear of AMT system. Greater slope will cause greater driving resistance, and then the drive motor system will immediately send changed gear information to AMT system. AMT system will no longer shift in sequence, but directly become appropriate gear by combining vehicle state, pilot operation and drive motor system.

Considering a vehicle demanding a certain backup power, once drive motor system is more than rated speed, it will keep constant power and torque will decrease with rising speed. If the driver shifts at rated speed point and drive torque equals resistance torque, the system will constantly regulate, therefore, 3200 rpm is chosen as climbing shift point, which certain backup power will exist in drive system.

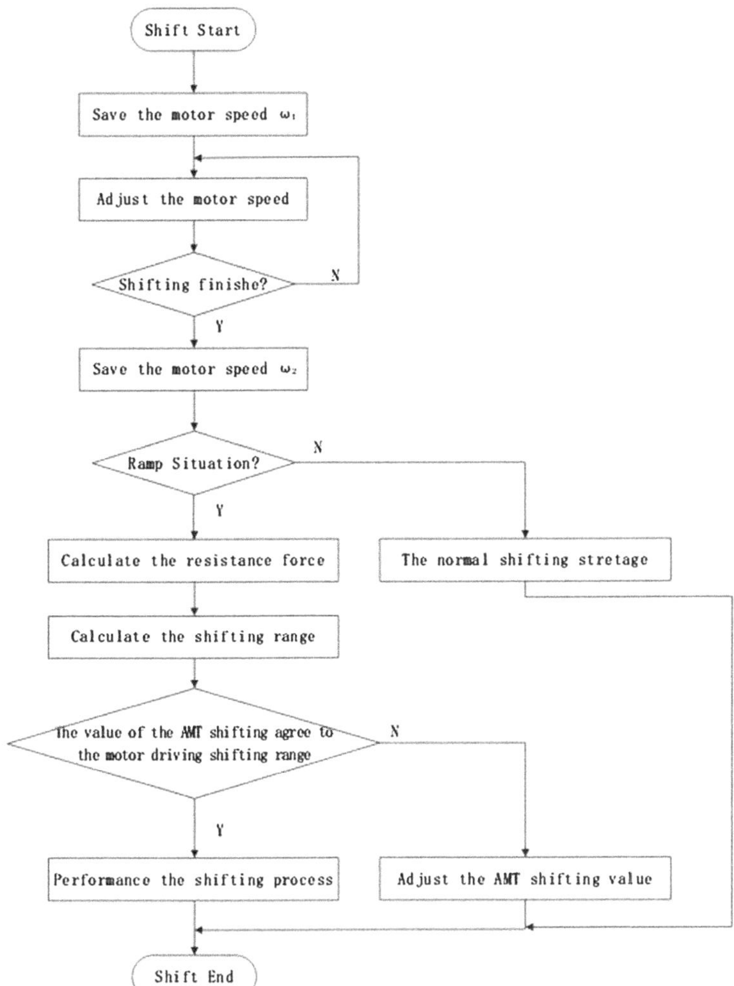

Fig. 58.2 The shifting calculation process

Table 58.1 Drive motor system parameters

Parameter	Value	Parameter	Value
Rated power/kw	60	Rated speed/(r/min)	3000
Peak power/kw	110	Highest rotating speed/(r/min)	6000
Rated/Nm	165	Peak torque/Nm	300

Table 58.2 Vehicle related parameters

Parameter	Value	Parameter		Value
Total quality/kg	8000		i1	5.594
Tire radius/m	0.381	Transmission change rate	i2	2.814
Frontal area	4.5267		i3	1.66
Air resistance ratio	0.5		i4	1
Main decreases in the ratio	6.833		i5	0.794

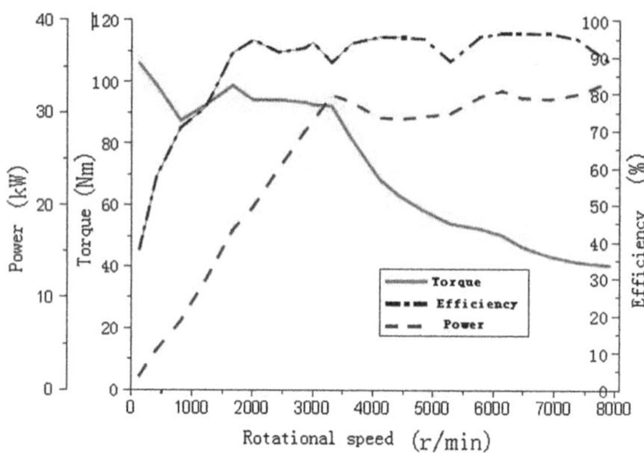

Fig. 58.3 Date curve of the motor

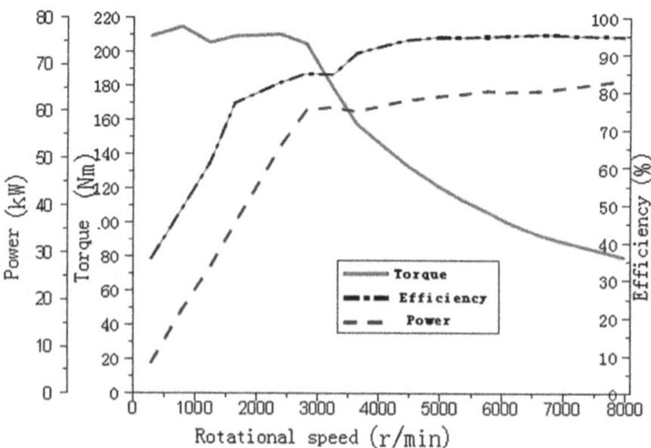

Fig. 58.4 Peak curve of the motor

58.5 Ramp Shift Test and Results Analysis

Figures 58.5 and 58.6 show the continuous drop block and block discontinuously process in different slopes.

Two conditions are based on the fifth block gear and 40 km/h for driving by means of 4 and 20 % slope respectively. When the vehicle goes onto 4 % slope, it would slow down under the slope resistance. At this time, AMT system sends shift signal and asks vehicle to switch to the fourth gear keep constant speed through

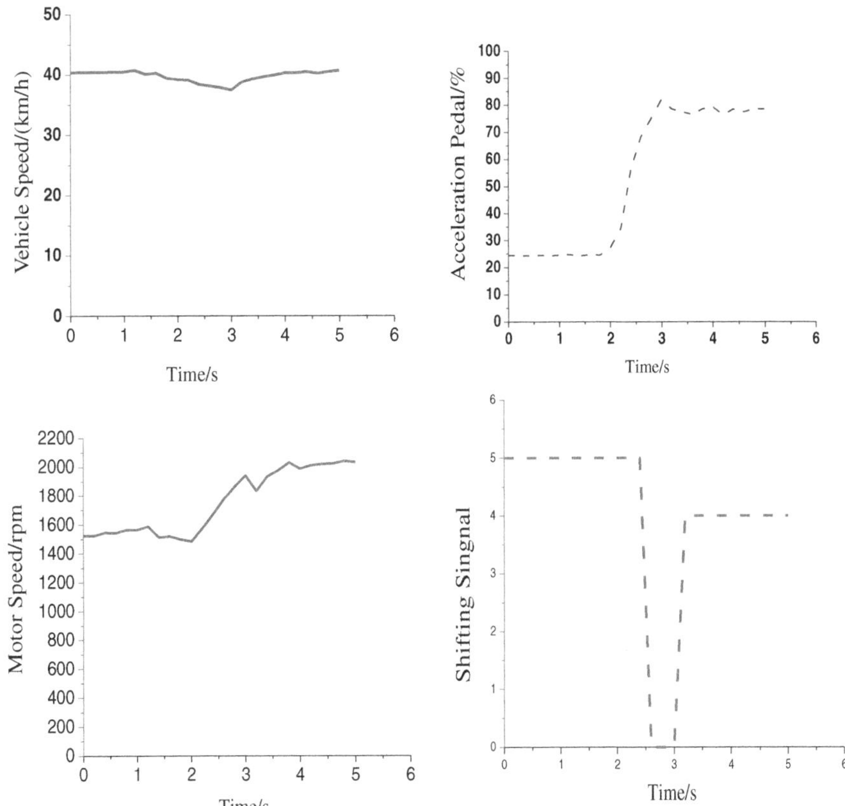

Fig. 58.5 The working curve about the 4 % slop

vehicle speed. Before and after shift of the drive system, the speeds are respectively $v_{a1} = 41$ km/h and $v_{a2} = 38$ km/h, and the speed difference is smaller, so usual shift strategy will take effect. So the vehicle goes along with the 4th shift at the 4 % slope which shown in Fig. 58.5. When the vehicles goes onto 20 % slope, they will decrease rapidly from 42.1 km/h to 31 km/h. At this time, AMT system sends shift signal and judges that the vehicle should be switched to the third block to maintain moving, however, the synthetical control shift strategy system should be taken because the speed difference is big. This time driving resistance calculated is about 16193 N, thus speed range calculated by driving system should be $i \geq 3.1$. As a result, AMT system continues to adjust shift strategy and the speed drops at the same time. Finally the vehicle goes onto 20 % slope with the first shift at the speed $v_{a1} = 6.5$ km/h which shown in Fig. 58.6.

In conclusion, the vehicle drops gear continuously and blocks discontinuously when driving on the ramp, which conforms to ramp shift strategy. Ramp shift test indicates that the slope judgment method is better to judge the ramp running

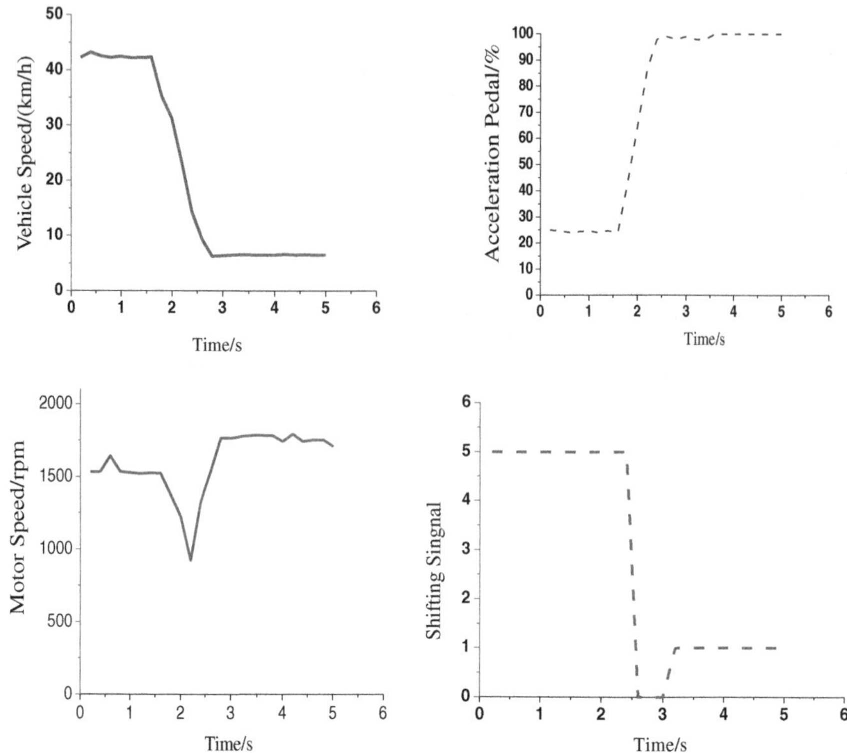

Fig. 58.6 The working curve about the 20 % slop

condition according to the drive motor system after shift and the vehicle can choose suitable gear by dropping gear continuously or discontinuously block when driving up.

58.6 Conclusion

Focused on the difficulty of recognizing the ramp running condition, ramp shift strategy is established in the paper. The shift control strategy is decided by the change of the motor speed and calculate the slope value. Therefore, the vehicle is able to choose gear by dropping gear continuously or discontinuously in the course of driving. The experiment shows that it would not be the proper gear at the first time. However the proper shift could be chosen by dropping gear continuously or discontinuously with the control strategy. This method achieves ramp shift control by using existing sensor, which is not only simple and effective, but also simplifies the hardware of control system.

Acknowledgments This research was supported by the National Natural Science Foundation(51105032), the National Laboratory for Electric Vehicles(2013NELEV004) and the Research Fund of Key Laboratory of Automotive Engineering of Sichuan Province(SZJJ2012-037). We thank Dr. Zhang Chengning(Beijing Institute of Technology, China) for the test vehicle.

References

1. Xi J, Wang L, Fu W et al (2010) Shifting control technology on automatic mechanical transmission of pure electric buses. Trans Beijing Inst Techn 30(1):43–45 (in Chinese)
2. Cheng Z, Cheng L, Fengchun S (2008) Dynamic modeling and analysis of AMT Shifting process for EV-BUS. In: IEEE vehicle power and propulsion conference (VPPC), Sep 3–5, Harbin, China
3. He Z, Bai H, Zhang P et al (2007) Ramp shift strategy and test research on AMT vehicle. Trans Chin Soc Agric Mach 38(2):13–16 (in Chinese)
4. Haj-Fraj A, Pferffer F (2001) Optimal control of gear shift operations in automatic transmissions. J Franklin Inst 338:371–390
5. YU Z (2000) Theory of automobile. China Machine Press (in Chinese)
6. Ding J, Wang L (2008) A research on the synthetical shifting gear schedule of the AMT Automobiles. Trans Shenyang Ligong Univ 27(4):87–90 (in Chinese)
7. Guo Y, Wang Z (2011) Synthetical ramp shift strategy and test research on electric vehicle with AMT. J Beijing Inst of Tech 20(2):93–96
8. He Z, Liang X, Han Z et al (2004) Dynamic model of shifting process and analysis of the main factors affecting shift quality for AMT Vehicle. J Ordnance Eng Coll (12) (in Chinese)

Chapter 59
The Study of Taxi Drivers' Fatigue Relieving Ways

Wang Hong, Fuwang Wang, Zuoqiu Qi and Tianwei Shi

Abstract The driving state of taxi drivers relate to passengers and the driver's life safety. So the study of taxi drivers' fatigue driving is of great significance, especially detection fatigue and fatigue ease. The taxi drivers usually have a short time rest after a long time driving. The rest mode can be divided into two kinds. One is relaxing in a taxi, another is outside a taxi. We have studied the two different relaxing ways in this paper. We have gathered the EEG signals of eight taxi drivers. Then we extracted δ, θ, α and β rhythm from drivers' EEG signals using wavelet packet decomposition. We judged the effective ways to alleviate fatigue through comparing the EEG fatigue state index F. The results show that relaxing outside a taxi is more effective than in a taxi after a long time driving.

Keywords Fatigue ease · EEG · Wavelet packet decomposition · EEG fatigue state index F

59.1 Introduction

In recent 10 years, the number of automobile has increased dramatically in China. However, traffic accidents have increased at the same time. There are lots of factors leading to traffic accidents. The driver fatigue driving is the main reasons that lead to traffic accidents [1, 2]. Therefore, accurately detecting driver fatigue and relief in time is very important. The study of detecting fatigue driving

W. Hong (✉) · F. Wang · Z. Qi · T. Shi
School of Mechanical Engineering and Automation, Northeastern University,
No.11, Lane 3, WenHua Road, HePing District, Shenyang, Liaoning 110819, China
e-mail: hongwang@mail.neu.edu.cn

F. Wang
e-mail: wfwmly@yahoo.com.cn

Z. Sun and Z. Deng (eds.), *Proceedings of 2013 Chinese Intelligent Automation Conference*, Lecture Notes in Electrical Engineering 255,
DOI: 10.1007/978-3-642-38460-8_59,

condition generally divided into subjective and objective aspects [3, 4]. It's hard to ensure the reliability of results, because the subjective research was affected by the subjective judgment ability of the drivers and researchers. Therefore, the researchers who are at home and abroad mainly study on fatigue driving from the perspective of an objective scientific. The objective research is mainly testing the physiological signals' transformation rules of drivers' electroencephalography (EEG), myoelectricity (EMG) and electrocardiograph (ECG). In recent years, there are remarkable achievements in objective research.

In this paper, we detected drivers' fatigue state from an objective perspective. And we studied the way of fatigue ease after a long time driving. We collected taxi drivers' EEG signals using EEG acquisition equipment called Emotiv. We judged the drivers' fatigue degree through analyzing EEG features of the central and occiput parts of a brain. The results showed that relaxing outside a taxi was more effective than in a taxi after a long time driving.

59.2 Experiment

59.2.1 Subjects and Procedure

The subjects were eight taxi drivers (male; mean age 32.75 years; range 26–40 years). They needed to be healthy and without sleep diseases. Drivers were required not to drink alcohol and coffee or eat drugs in 48 h before the experiment. The experiment was done from two o 'clock to three o 'clock in the afternoon. We gathered four types of the EEG signal (before and after rest) of a driver when he was in his car and out of his car. The rest time given to the drivers was 15 min. And each signal acquisition lasted 3 min. We collected taxi drivers' EEG signals Using EEG acquisition equipment called Emotiv in this experiment. Electrodes were placed according to the extended international 10–20 system. The signals of four channels (F3, F4, O1 and O2) were chosen in this experiment.

59.2.2 Data Pre-Processing

In wavelet decomposition, the wavelet transform frequency resolution will descend with the ascending of the signal frequency. However, wavelet packet decomposition can overcome this defect, and provide more sophisticated analysis to signals [5, 6]. It can also choose the appropriate band to match with signal spectrum according to the characteristics of signal analyzed, which can reflect the essential characteristics of signals [7, 8]. Therefore, we used wavelet packet decomposition to decomposed EEG signals in this paper. We gave 4 layer decomposition to band to extract $\delta(0 \sim 4\ \mathrm{Hz})$, $\theta(4 \sim 8\ \mathrm{Hz})$, $\alpha(8 \sim 12\ \mathrm{Hz})$ and

β(12 ~ 32 Hz) rhythms. Figure 59.1 is wavelet packet decomposition tree diagram.

Wavelet packet decomposition formula is shown as follows:

The *f(t)* means source signals. We got 2i band in the i points class after wavelet packet decomposition. The source signal f (t) can be expressed as:

$$f(t)=\sum_{j=0}^{2^i-1} f_{i,j}(t_j)=f_{i,0}(t_0)+f_{i,1}(t_1)+\cdots+f_{i,2^i-1}(t_{2^i-1}) \tag{59.1}$$

The j = 0, 1, 2,...,2^i-1, $f_{i,j}$ (t_j) is the reconstruction of signals in the i layer node (i,j) when we use wavelet packet decomposition. According to Parseval theorem [9] and formula (59.1) we can calculate and gain the energy spectrum of the signal *f (t)* after wavelet packet decomposition.

$$E_{i,j}(t_j)=\int_{\Gamma}|f_{i,j}(t_j)|^2 dt=\sum_{k=1}^{m}|x_{j,k}|^2 \tag{59.2}$$

The $E_{i,j}(t_j)$ is band energy that f(t) was decomposed to node (i, j) by using wavelet packet decomposition.

Here $x_{j,\ k}$(j = 0,1,2,...,2^i-1; k = 1,2,...,m) is the discrete points amplitude of the reconstruct signal $f_{i,j}(t_j)$. And m is signal sampling points. In this paper, we got low frequency sub-band of EEG signals by decomposing EEG signals to the fourth layer after resampling. The result was shown in Table 59.1.

Delta wave (0 ~ 4 Hz), theta wave (4 ~ 8 Hz) and alpha wave (8 ~ 12 Hz) were gotten by reconstructing sub-band s(4,0), sub-band s(4,1) and sub-band s(4,2) using wavelet packet respectively. And alpha wave (12 ~ 32 Hz) was gotten by reconstructing sub-band s(4, 3), s(4, 4), s(4, 5), s(4, 6) and s(4, 7) using wavelet packet. In this paper, we chose EEG signals of O1channel in two different rest ways to analyze EEG signals' characteristic. The result was shown in Figs. 59.2, 59.3.

The δ, θ, α and β rhythms were extracted from drivers' EEG signals using wavelet packet decomposition. However it can't distinguish the difference between before and after rest. So it need to make further analysis.

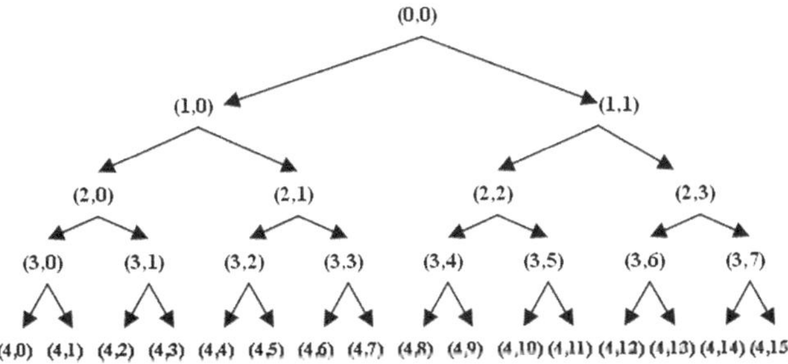

Fig. 59.1 Wavelet packet decomposition tree diagram

Table 59.1 The sub-bands of wavelet packet decomposition

Sub-band	s(4, 0)	s(4, 1)	s(4, 2)	s(4, 3)	s(4, 4)	s(4, 5)	s(4, 6)	s(4, 7)
Band range/Hz	0 ~ 4	4 ~ 8	8 ~ 12	12 ~ 16	16 ~ 20	20 ~ 24	24 ~ 28	28 ~ 32

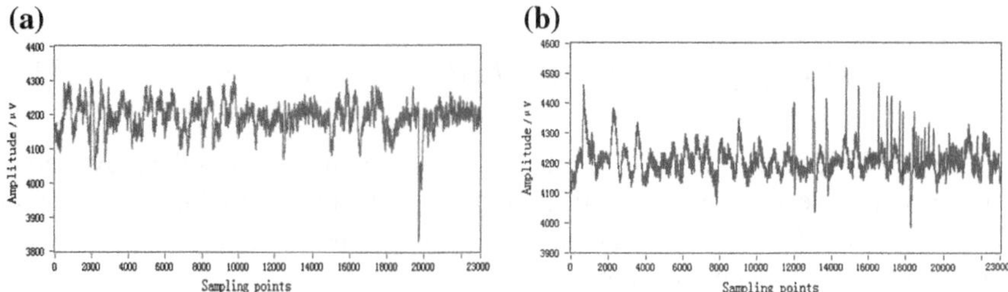

Fig. 59.2 The comparison of driver's EEG signals

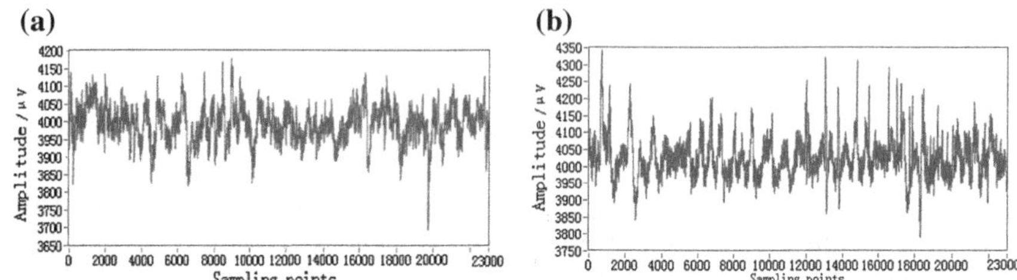

Fig. 59.3 The comparison of driver's EEG signals

59.3 Quantitative EEG Analyses

The EEG signals were gathered in different time periods in the experiment. And there may be some changes of mental state for drivers in different period. So we chose to calculate relative power spectrum of drivers' EEG signals in different frequency band to analyze the fatigue driving EEG features in order to remove the influences mentioned in this section.

The relative power spectrum of EEG rhythm is refers to the proportion of each rhythm's power spectrum in total power spectrum. The formula shows as (59.3):

$$P_i = \frac{E_i}{E_\alpha + E_\beta + E_\theta + E_\delta} \tag{59.3}$$

P_i is defined as the relative power spectrum of i rhythm ($i = \delta, \theta, \alpha, \beta$).

The average relative power spectrum were used to analyse EEG signals in order to reduce the fluctuation influence of each rhythm relative power spectrum. Researches had shown that EEG signals' slow wave gradually ascended and fast wave gradually descended meanwhile when adults' mental state change from the normal state to fatigue state [10, 11]. Therefore, we can use the energy ratio of the fast and slow wave to estimate the fatigue state index F. The formula shows as (59.4):

$$F = \frac{E_\beta + E_\alpha}{E_\theta + E_\delta} \tag{59.4}$$

E_i is defined as the energy of i rhythm (i = δ, θ, α, β).

$$F = \frac{\overline{P}_\alpha + \overline{P}_\beta}{\overline{P}_\delta + \overline{P}_\theta} \tag{59.5}$$

We calculated the EEG characteristic fatigue index F of drivers when taxi drivers chose different types of rest. And we analyzed the EEG signals of channels (O1, O2, F3, F4). The result was shown in Fig. 59.4.

Obviously, we could conclude that the growth rate of the EEG characteristic fatigue index F which drivers rest outside a car was higher than in a car by analyzing the EEG signals of the channels O1, O2, F3, F4.

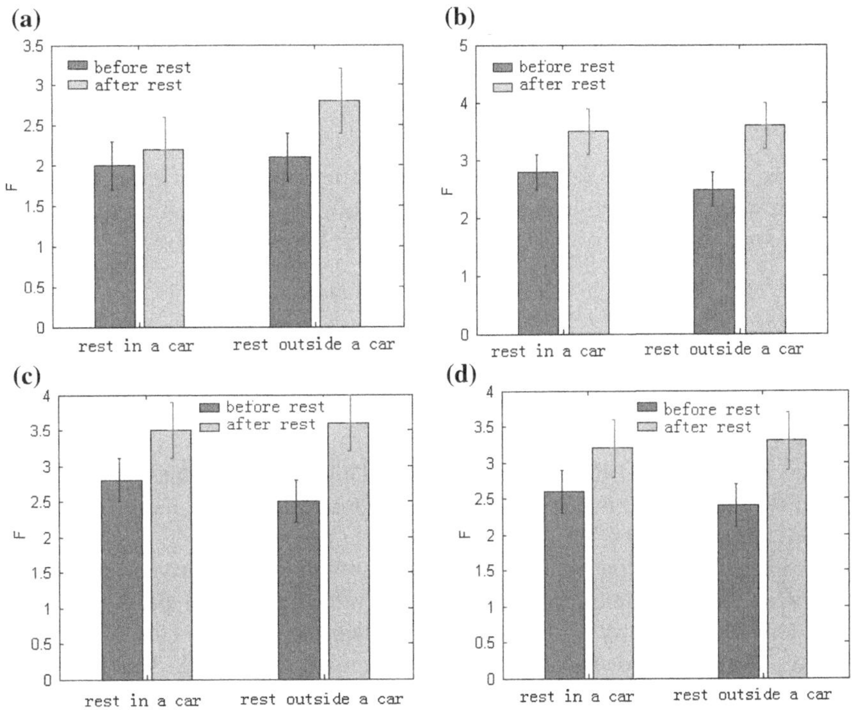

Fig. 59. 4 The change of drivers' EEG fatigue state index F. **a** O1 channel. **b** O2 channel. **c** F3 channel. **d** F4 channel

Table 59.2 The growth rate of drivers' EEG fatigue state index F

	Different channels			
Types of rest	O1	O2	F3	F4
Rest in a car	0.200 ± 0.032	0.045 ± 0.041	0.251 ± 0.035	0.231 ± 0.03
Rest outside a car	0.333 ± 0.057	0.501 ± 0.022	0.440 ± 0.021	0.375 ± 0.028

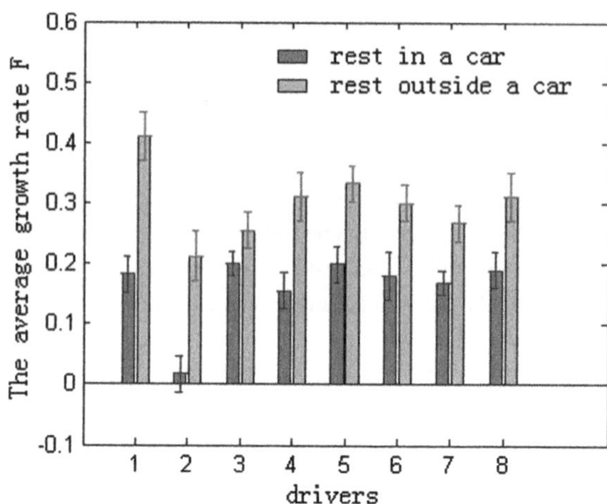

Fig. 59.5 The F average growth rate of eight drivers' two rest mode

59.4 Feature Extraction

In this paper, the growth rate of EEG characteristic fatigue index F was obviously different when taxi drivers chose different types of rest. The growth rate formula of EEG characteristic fatigue index F was shown as Formula (59.6).

$$F_{growth-rate} = \frac{F_{after-rest} - F_{before-rest}}{F_{before-rest}} \tag{59.6}$$

We calculated the growth rate of drivers' EEG fatigue state index F in different channels when drivers chose the two types of rest respectively. The result was shown as follows (Table 59.2).

We could conclude that the growth rate when drivers rest outside a car was higher than in a car. It means that drivers rested outside a car is more effective to relieve fatigue than in a car.

Figure 59.5 shows the comparison that the growth rate that drivers rest outside a car was higher than in a car. It means that it's more helpful to make the body and mind to relax when drivers rest outside a car than rest in a car after a long time driving. The F descended sometimes when drivers rested in a car. That means the drivers brain activity is restrained and the fatigue degree is deepening.

59.5 Conclusion

In this paper, we gathered the EEG signals of eight male taxi drivers using the acquisition equipment called Emotiv in real driving environment. We extracted the EEG features and analyzed the differences of EEG characteristic fatigue index F when taxi drivers chose different types of rest. Through the contrast we can draw a conclusion that drivers rested outside a car is more effective to relieve fatigue than in a car.

Acknowledgments This work was supported by National Science Foundation of China (61071057).

References

1. Lal SKL, Craig A (2001) Electroence phalography activity associated with driver fatigue: implications for a fatigue countermeasure device. J Psychophysiol 15(3):1151–1156
2. Philipa HG, Nathniel SM, Ian J et al (2006) Investigating driver fatigue in truck crashes: trial of a systematic methodology. Transp Res Part F 9(1):65–76
3. Murata A, Uetake A, Takasawa Y (2005) Evaluation of mental fatigue using feature parameter extracted from event-related potential. Int J Ind Ergon 35(4):761–770
4. Anneke H, Rainer G, Aeaeia A (2001) Technologies for the monitoring and prevention of driver fatigue. In: Proceedings of the first international driving symposium on human factor in driver assessment, training and vehicle design. Aspen:81-86
5. Xu BG, Song AG (2009) EEG signal recognition method based on wavelet packet transform and clustering analysis. Chin J Sci Inst 30(1):25–28 (in Chinese)
6. Song GM, Wang HJ, LIU H et al (2010) Analog circuit fault diagnosis using lifting wavelet transform and SVM. J Electr Meas Inst 24(1):17–22 (in Chinese)
7. Zhong GSH, AO LP, Zhao K (2009) Influence of explosion parameters on energy distribution of blasting vibration signal based on wavelet packet energy spectrum. Explosion Shock Waves 29(3):300–305
8. Guo XM, Ding XR, Zhong LS et al (2012) Heart sound feature extraction and classification based on integration of wavelet packet analysis and chaos theory. Chin J Sci Inst 33(9):1938–1944 (in Chinese)
9. Gargoom AM, Ertugrul N, Soong WL (2008) Automatic classification and characterization of power quality events. IEEE Trans Power Deliv 23(4):2417–2425
10. Blankertz B, Tomioka R, Lemm S et al (2008) Optimizing spatial filters for robust EEG single t rial analysis. IEEE Signal Process Mag 25(1):41–56
11. Ye T, Sun YG (2012) A fatigue driving detection method based on wavelet packet sub-band energy ratio of EEG. J Northeast Univ (Nat Sci) 33(8):1088–1092 (in Chinese)

Chapter 60
Landmark Design for Indoor Localization of Mobile Robots

Longhui Wang, Bingwei Gao and Yingmin Jia

Abstract This paper focuses on the landmark design for mobile robot localization in indoor environment. A design method with two types of landmarks, which are color landmarks and black landmarks, is proposed. Particularly, the color landmarks are elaborately encoded by four types of distinctive colors. All the landmarks are installed on the ceiling in a special way so that the mobile robot can uniquely localize itself by observing the ceiling landmarks. In addition, an image processing algorithm is presented to identify the landmarks, and a localization algorithm is proposed to calculate the position and orientation of the mobile robot using the information of the identified landmarks.

Keywords Landmarks · Mobile robots · Localization · Image processing

60.1 Introduction

Autonomous mobile robots are widely applied nowadays in various environments such as rescue in disaster sites, cargo transportation in factories, tourist guidance in museums. To achieve these tasks, the mobile robots should navigate themselves

L. Wang (✉) · B. Gao · Y. Jia
The Seventh Research Division and the Department of System and Control, Beihang University (BUAA), Beijing 100191, China
e-mail: lhui_wang@163.com

B. Gao
e-mail: gaobingwei722@163.com

Y. Jia
e-mail: ymjia@buaa.edu.cn

Y. Jia
Key Laboratory of Mathematics, Informatics and Behavioral Semantics (LMIB), SMSS, Beihang University (BUAA), Beijing 100191, China

Z. Sun and Z. Deng (eds.), *Proceedings of 2013 Chinese Intelligent Automation Conference*, Lecture Notes in Electrical Engineering 255,
DOI: 10.1007/978-3-642-38460-8_60,

autonomously, which requires mobile robots to be located accurately. There are many positioning methods for mobile robots, which can be classified into relative positioning systems and absolute positioning systems. The relative positioning systems use compass and odometer [1] etc. to calculate the position and orientation of the mobile robots. While the absolute positioning systems use sensors, such as sonar [2, 3], laser scanner [4, 5] and vision sensor [6] to localize mobile robot autonomously. Among all the methods, visual positioning systems have attracted more attention due to the rich information, fast speed, low cost and easy operation.

Landmarks play an important role in the visual position systems. They can be detected fast and reliably to provide information for localizing mobile robot and be categorized into natural landmarks and artificial landmarks. Natural landmarks are those features extracted from the environment. However, it is a formidable task to extract them reliably and robustly in a complex environment [7]. Compared with natural landmarks, artificial landmarks are used more frequently [8]. They are usually designed with special patterns and colors so that to be identified easily and to supply enough information for localization, e.g. [9] gives a landmark design utilizing projective invariant.

In this paper, we design two types of artificial landmarks and elaborate how to use of them. The main contributions of this paper are mainly in three aspects. First, a method to design the landmarks for localizing mobile robots is proposed. It can be explanted to other indoor environments. Second, a time-saving and reliable method to search the landmarks is developed. Third, a new image processing method combining segmentation, edge detection and image filling is utilized to extract the landmarks.

60.2 Landmarks Design

There are two types of landmarks used in this paper. One is a black circle as shown in Fig. 60.1a. We use black colored card because it is easy to be segmented from the background. The other one called color landmark is designed as shown in Fig. 60.1b, inspired from the "hat" design in [10]. The color landmark is divided into two parts of inner black circle and outer color ring. The inner circle is smaller than the black landmarks. The outer ring is segmented into four equal sectors denoted as Sector 1 through Sector 4, which can be colored by red, green, blue and yellow respectively. Therefore, each sector is either in white or in the corresponding color. An ID defined by a binary number of four bits is associated with each color landmark. Bit 1 through Bit 4 is associated with Sector 1 through Sector 4, respectively. If one sector is in white, the corresponding bit is 0, otherwise it is 1. For instance, the ID of a color landmark with Sector 1 in red, Sector 3 in blue, and the rest sectors in white is 5. Since at least one sector should be painted in color, the ID ranges from 1 to 15.

As shown in Fig. 60.1b, the reference frame $O_rX_rY_r$ is placed on the color landmark, in which X_r-axis lies on the edge between Sector 1 and Sector 4, and Y_r-

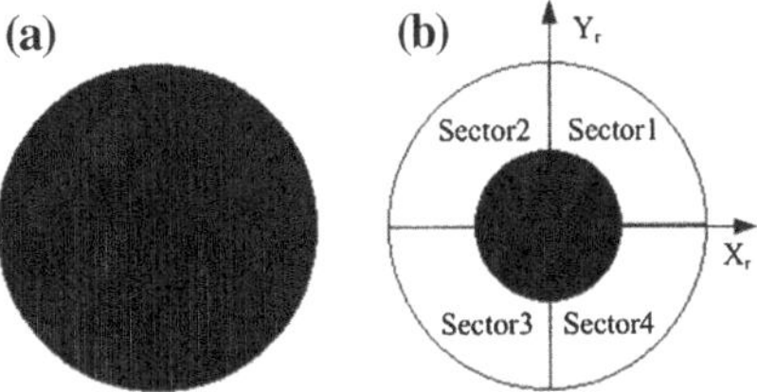

Fig. 60.1 Two types of landmarks

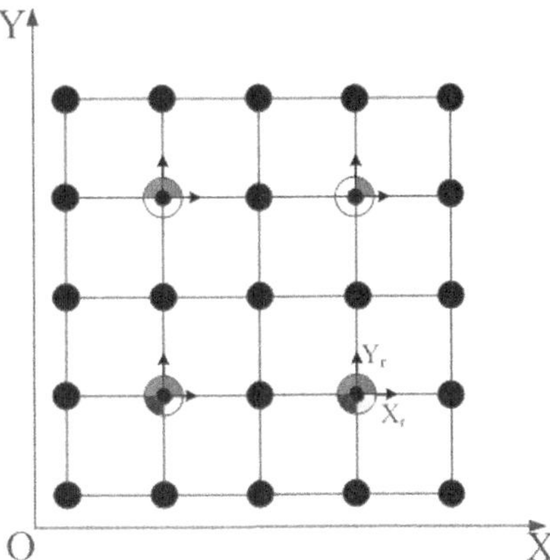

Fig. 60.2 Landmarks attached to the ceiling

axis lies on the edge between Sector 1 and Sector 2. All the landmarks are installed regularly on the ceiling in lattice pattern with constant separation, e.g. a constant of L (see Fig. 60.2). Each color landmark is surrounded by eight black landmarks. Especially, X_r-axis and Y_r-axis of the color landmark are aligned with X-axis and Y-axis respectively, where OXY is the world frame attached on the ceiling. For such a landmark distribution, if the world coordinates of one color landmark are known, the world coordinates of eight black landmarks around can also be figured out. Notice that the separation between any two adjacent landmarks should be suitable. Then no less than four landmarks can be always observed by the camera, which is installed on the mobile robot with its optical axis upward.

60.3 Image Processing

60.3.1 Landmarks Identification

In order to identify the landmarks accurately, they should be segmented out from the captured image. There are numerous image segmentation techniques such as thresholding, edge detection, and region-growing. For instance, thresholding method can be used well to a gray image with a clear contrast between the object

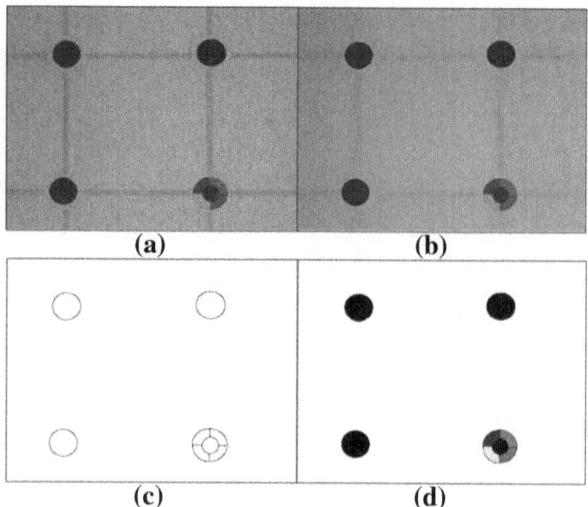

Fig. 60.3 Landmarks identification of a sample image

and the background. But it has an unsatisfactory result here. Considering the fact that most color image segmentation just aims to divide the image into several regions and can't be used to extract the landmarks here, three steps are performed in this paper to segment the landmarks from the background.

An example is provided to illustrate the three steps (see Fig. 60.3). Figure 60.3a is a raw image captured by the camera installed on the mobile robot. Firstly, Mean-Shift Segmentation [11] is used on it. This process aims to segment the image into several parts. Thus those pixels of similar color are classified into the same part (see Fig. 60.3b). Secondly, use edge-detection to find out the edges of image, and use recursive region-growing method to find the connected components. Image morphology such as erosion and dilation operators for binary image is used to make the connected components closed (see Fig. 60.3c). Once the closed components are found, a seed is chosen for each connected component. At last, Flood-Fill [11] is used with those seeds chosen previously. It can be observed from Fig. 60.3d that the landmarks are extracted perfectly.

60.3.2 Pixel Coordinates Calculation of Landmarks

The pixel coordinates of a landmark refer to the position of its center in the image plane. In this paper, sub-sampling method [10] is used to search the centers of landmarks. It examines one pixel every other N-1 pixels in horizontal and vertical directions of the image, so only one pixel is examined in an N $\times$ N pixel square. This method saves scanning time. Furthermore, the search starts from the center of the image and in a clockwise spiral way (see Fig. 60.4a). This is essential as there may be more than four landmarks appear in the view of the mobile robot sometimes. But search in this way guarantees that only four landmarks around the

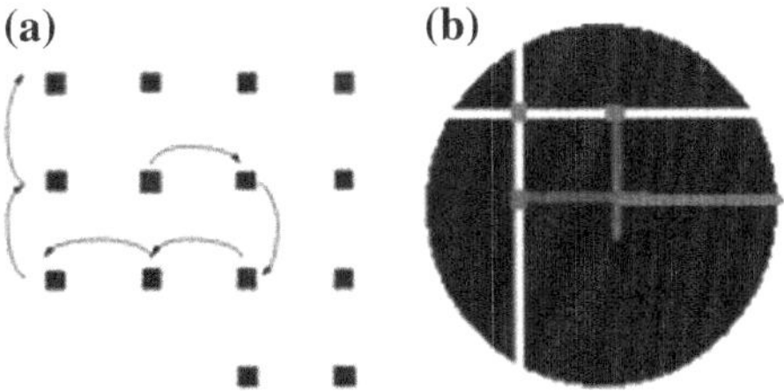

Fig. 60.4 The searching method and center calculation

image center are chosen to localize the robot. Moreover, it doesn't need to scan the whole image so it saves much scanning time. The search continues until a black pixel is found. Once a black pixel is located, we call it a "hit". As shown in Fig. 60.4b, the first "hit" is indicated by a green square. After a "hit", the search expands to the left, right, up and down (see the white pixels in Fig. 60.4b). This search stops when the edges of the black circle are found. This time each pixel is examined thus if one pixel is misclassified, the pixel coordinates calculation does not suffer. It is obvious in Fig. 60.4b that the white lines are perpendicular chords of the black circle, and red squares are the center of each chord which determines the pixel coordinates of the landmark. In addition, the radius of each black circle can be figured out, so that the color landmarks can be distinguished from the black landmarks by the magnitude of the radius. As shown by the red line in Fig. 60.4b, an extra search starting from the detected center of the black circle along a random direction is operated. If it almost reaches the edge of the black circle after a distance of the detected radius, then the detected center and radius of the black circle are confirmed.

60.3.3 World Coordinates Calculation of Landmarks

The world coordinates of the color landmarks are stored in a two-dimensional array in advance. Once the ID of a color landmark is known as N, its world coordinates can be obtained by indexing the Nth element of the array. The next step aims to recognize the ID of each color landmark.

Assume the radius of the inner circle in color landmark is r. We start a search along a circle anticlockwise with radius R (R > r) (see the circle indicated by dotted line in Fig. 60.5), and record the types of colors experienced. According to

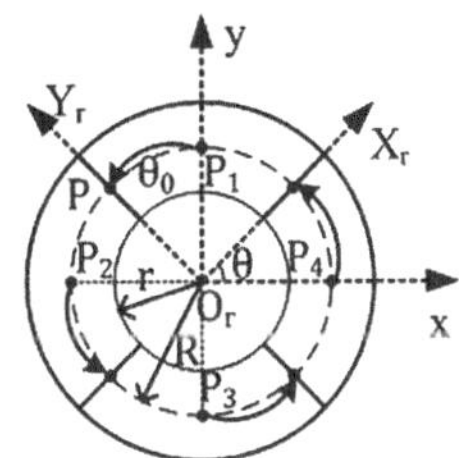

Fig. 60.5 The ID and orientation calculation

the coding rule of ID mentioned in Sect. 60.2, the ID of the color landmark can be uniquely determined by the recorded colors.

The boundary, across which the color changes, is also recorded along the circular search. Then the information of the boundary helps to reconstruct the X_r-axis and Y_r-axis of the reference frame on the color landmark. Now the world coordinates of other three black landmarks around the color landmark can be deduced. Assume the color landmark is at (x_1, y_1) in the world frame. First, pick out one of three black landmarks closest to the Xr-axis. If it is on the positive direction of X_r-axis, its world coordinates are $(\mathrm{X}+\mathrm{L}, \mathrm{Y})$, otherwise $(\mathrm{X}-\mathrm{L}, \mathrm{Y})$. Similarly, calculate out the world coordinates of another black landmark which is nearest to the Y_r-axis. Finally, the world coordinates of the last one can be easily obtained.

60.4 Pose Calculation of the Mobile Robot

60.4.1 Position Calculation

Regard the four identified landmarks as four points A, B, C and D in the image. Their pixel coordinates are (x_1, y_1), (x_2, y_2), (x_3, y_3) and (x_4, y_4) and the corresponding world coordinates are (X_1, Y_1), (X_2, Y_2), (X_3, Y_3) and (X_4, Y_4), respectively. Considering that the camera pointing to the ceiling is mounted on the center of the mobile robot's surface, we can assume that the mobile robot's position is projected onto the center P of the image plane. P is inside the square ABCD obviously. Assume its pixel coordinates are (x, y) and the world coordinates are (X, Y), then the world coordinates are needed to be interpolated. Here, bilinear interpolation [12] is used. Let the interpolation formula be:

$$\begin{cases} X=a_0+a_1x+a_2y+a_3xy \\ Y=b_0+b_1x+b_2y+b_3xy \end{cases} \tag{60.1}$$

The coefficients (a_0, a_1, a_2, a_3) and (b_0, b_1, b_2, b_3) are to be calculated. First substitute the four points' pixel coordinates and their world coordinates into expression (60.1) to obtain linear equations (60.2) and (60.3). Then solve linear equation (60.2) to obtain the coefficients (a_0, a_1, a_2, a_3) and solve the linear equation (60.3) to obtain the coefficients(b_0, b_1, b_2, b_3). Finally substitute $P(x, y)$ and the eight coefficients into expression (60.1) to calculate the X-coordinate and Y-coordinate.

$$\begin{bmatrix} 1 & x_1 & y_1 & x_1y_1 \\ 1 & x_2 & y_2 & x_2y_2 \\ 1 & x_3 & y_3 & x_3y_3 \\ 1 & x_4 & y_4 & x_4y_4 \end{bmatrix} \begin{bmatrix} a_0 \\ a_1 \\ a_2 \\ a_3 \end{bmatrix} = \begin{bmatrix} X_1 \\ X_2 \\ X_3 \\ X_4 \end{bmatrix} \tag{60.2}$$

$$\begin{bmatrix} 1 & x_1 & y_1 & x_1y_1 \\ 1 & x_2 & y_2 & x_2y_2 \\ 1 & x_3 & y_3 & x_3y_3 \\ 1 & x_4 & y_4 & x_4y_4 \end{bmatrix} \begin{bmatrix} b_0 \\ b_1 \\ b_2 \\ b_3 \end{bmatrix} = \begin{bmatrix} Y_1 \\ Y_2 \\ Y_3 \\ Y_4 \end{bmatrix} \tag{60.3}$$

60.4.2 Orientation Calculation

As shown in Fig. 60.5, the $O_rX_rY_r$ is the reference frame mentioned before; the o_rxy is the image reference frame with its x-axis selected as the horizontal direction of the imaging plane from left to right, and y-axis as the vertical direction of the imaging plane from bottom to top. The origins are both at the center of the color landmark. θ is the angle from x-axis to X_r-axis. We define the mobile robot's orientation in the world frame as the angle from Y-axis to mobile robot's heading direction, which is the same as y-axis direction. So the mobile robot's orientation equals $-\theta$. To calculate θ, first, choose a point P_1 along the y-axis and let $OP_1 = R$ $(R > r)$. Then check the color of the pixel at point P_1. If the color is not white, then search at a radius R anticlockwise until the landmark's color changes to others. Without loss of generality, we assume the color is red, and then the search will stop at the Y_r-axis, in Fig. 60.5. This gives a point P which can be used along with the center O_r to calculate a basic orientation θ_0 (see $\angle P_1O_rP$ in Fig. 60.5), which ranges from 0° to 90°. In the case that P_1 is red, θ equals θ_0. But if P_1 is green, θ equals $-90° + \theta_0$, If P_1 is blue, θ equals $-180° + \theta_0$; and if P_1 is yellow then θ equals $90° + \theta_0$. However, if P_1 is in a white color, then choose P_2 as shown in Fig. 60.5 and so on. So there is at least one point whose color isn't NULL can be found to calculate the orientation. Moreover, the orientation calculation is done simultaneously with the ID calculation to save time.

60.5 Conclusion

In this paper, we investigated an indoor localization method for mobile robot. By using the ceiling vision, an integrated landmark design for mobile robot localization was proposed. The main contributions of this paper are tripled. First, a design of landmarks for localizing the mobile robot was proposed. Second, a new technique to search the landmarks was developed. Third, three steps were combined to segment the landmarks out from the background.

Acknowledgments This work was supported by the National Basic Research Program of China (973 Program, 2012CB821200, 2012CB821201) and the NSFC (61134005, 60921001, 90916024, 91116016).

References

1. Bouvet D, Garcia G (2001) Guaranteed 3-D mobile robot localization using an odometer, an automatic theodolite and indistinguishable landmarks. Robot Autom 4:3612–3617
2. Wijk O, Christensen HI (2000) Localization and navigation of a mobile robot using natural point landmarks extracted from sonar data. Robot Auton Syst 31:31–42
3. Lee JS, Nam SY, Chung WK (2011) Robust RBPF-SLAM for indoor mobile robots using sonar sensors in non-static environments. Adv Robot 25:1227–1248
4. Arras KO, Tomatis N (1999) Improving robustness and precision in mobile robot localization by using laser range finding and monocular vision. In: Proceedings of the third European workshop on advanced mobile robots, Zurich, Switzerland, pp 177–185
5. Lingemann K, Nűchter A, Hertzberg J, Surmann H (2005) High-speed laser localization for mobile robots. Robot Auton Syst 51:275–296
6. DeSouza GN, Kak AC (2002) Vision for mobile robot navigation: a survey. IEEE Trans Pattern Anal Mach Intell 24:237–267
7. Jang G, Kim S, Kim J, Kweon I (2005) Metric localization using a single artificial landmark for indoor mobile robots. IEEE Intel Robot Sys, pp 2857–2862
8. Wang ZW, Guo G (2003) Present situation and future development of mobile robot navigation technology. Robot 25(5):470–474 (in Chinese)
9. Yang G, Xin HX (2006) Landmark design using projective invariant for mobile robot localization. IEEE Robot Biomim, pp 852–857
10. Riggs TA, Inanc T, Zhang W (2010) An autonomous mobile robotics testbed: construction, validation, and experiments. IEEE Trans Control Syst Technol 18:757–766
11. Bradski G, Kaebler A (2009) Learning OpenCV. O'Reilly Media, Sebastopol
12. Bilinear interpolation: Available from http://en.wikipedia.org/wiki/Bilinear_interpolation

Chapter 61
A Remote Intelligent Greenhouse Distributed Control System Based on ZIGBEE and GPRS

Ning Su, Taosheng Xu, Liangtu Song and Shu Yan

Abstract In order to improve the production efficiency and automation of the greenhouse, a new remote intelligent greenhouse control system based on ZIGBEE and GPRS technology is put forward in this paper. In this system data and commands are transmitted through ZIGBEE and GPRS which not only realizes higher efficiency for the real-time monitoring of multiple greenhouses, but also achieves the remote cluster control of the shutter and sprinkler irrigation. The system is composed of a central control unit and multiple field monitoring units. The information collecting and field automatic control are implemented by the field monitoring units distributed in each greenhouse, while data display, storage, query, analysis and remote command control are realized by the central control unit. The test is done in Qinghai Province, China. The result indicates that the system bears perfect function and steady performance.

Keywords ZIGBEE · GPRS · Wireless network · Intelligent control · Greenhouse control

61.1 Introduction

The intelligent greenhouse control system has become the tendency of greenhouses management, which can monitor and control the environmental factors such as temperature, humidity by modern sensor and intelligent control

N. Su · T. Xu
Department of Automation, University of Science and Technology of China, Hefei 230026, China

N. Su · T. Xu · L. Song (✉) · S. Yan
Institute of Intelligent Machines, Chinese Academy of Sciences, Hefei 230031, China
e-mail: ltsong@iim.ac.cn

Z. Sun and Z. Deng (eds.), *Proceedings of 2013 Chinese Intelligent Automation Conference*, Lecture Notes in Electrical Engineering 255,
DOI: 10.1007/978-3-642-38460-8_61,

technology, so that the crops can grow in their best status. The intelligent greenhouse system basically has the following kinds: wired data communication based on RS232, RS485 and CAN bus [1], short distance wireless communication based on ZIGBEE, Wi-Fi and Bluetooth [2].

The intelligent greenhouse control system based on RS485 was designed by Yulong and Jiaoqiang in 2011 [3]. By using this system the user can accomplish greenhouse environment control, but the shortage of high costs, break easily and maintenance difficulty exist in the system. Wired data communication system transmits data through cables, but it is hard to place the cables which are easy to be eroded and hard for maintenance [4]. Compared with wired data communication, ZIGBEE with more significant features such as low cost, low power consumption, easy to upgrade and self-organized network is more suitable for using in greenhouse control [5]. In order to overcome the shortcomings of the wired system, the intelligent greenhouse control system is developed based on ZIGBEE. A remote intelligent monitoring system based on ZIGBEE Wireless Sensor Network is presented by Weihong and Shuntian [4]. The system use the wireless mesh network instead of placing the cables which not only avoid various problems brought by wired data communication but also convenient to install and maintain the nodes. However, it still has the shortage of short transmission distance. Those existing greenhouse control systems have the drawbacks, such as single function, hard to expand the structure, high costs, inconvenient operation and difficult to promote.

Therefore, research and development of an intelligent greenhouse remote monitoring system which is well-constructed, low costs, easy to control, suitable for different users and can realize control, intelligent decision and wireless transmission has important practical significance. The system is presented in this paper, which can meet those requirements. ZIGBEE, GPRS technology and distributed control thoughts were adopted in this system. ZIGBEE wireless network with self-organized network function and GPRS with long-distance wireless transmission function are combined together to achieve remote environmental parameters acquisition and long distance control.

61.2 Overall Design of the Remote Intelligent Greenhouse Distributed Control System

The remote intelligent greenhouse distributed control system is consisted of multiple independent field monitoring units distributed in each greenhouse and the central control unit which is composed of computer terminal and mobile phone. The overall structure of the system is shown in Fig. 61.1. The field monitoring units distributed in each greenhouse build up a huge mesh network and through this network the field units which are far away from the coordinator can hop to the coordinator through other field units and they can communicate with each other [6].

Fig. 61.1 Overall structure of remote intelligent greenhouse distributed control system

The function of the field units are similar, which mainly realize data collecting control commands reception, corresponding decision making and field automatic control to ensure the environmental parameters within a appropriate range while transmitting the data to the computer terminal via ZIGBEE and GPRS network. If the environmental parameters are out of the set values, the system will alarm by sound, light and short message. Functionally, the computer terminal as the central control unit is responsible for data display, storage, analysis, historical data inquiry and sending remote control commands.

By using computer terminal, mobile phone sending commands or keyboard selection, users not only can choose system working in three different modes: remote command control mode, field automatic control mode and manual control mode, but also can modify or set system parameters. The central control unit and the field monitoring unit are mutually independent to some extent, that is to say when the system is working in the field automatic control mode, the field unit can achieve automatic intelligent control with its programming control algorithm, doesn't need any control commands from central control unit. The users can send commands by the computer terminal or mobile phone to cluster control of the shutter and sprinkler irrigation when the system is working in remote command control mode. In order to ensure that the greenhouse can be controlled if emergency happens or system failure and improve the system reliability and robustness, the manual control mode is also needed.

61.3 Hardware Design of the Field Monitoring Unit

The hardware design of the field monitoring units make a supportive platform for intelligent control system, the field monitoring unit includes three parts: the data acquisition module, the wireless transceiver module, the execution module. The overall structure of the field monitoring units is shown in Fig. 61.2.

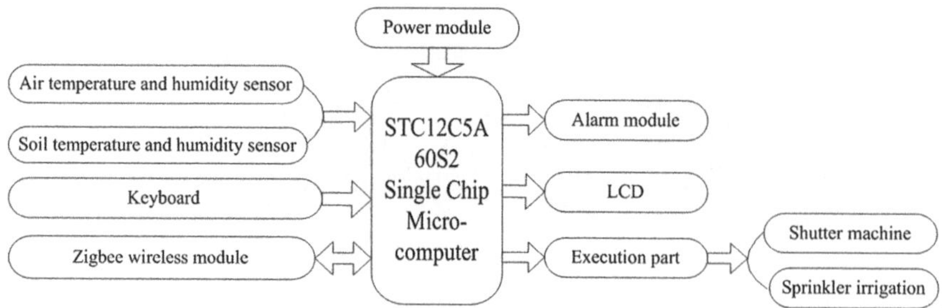

Fig. 61.2 Overall structure of the field monitoring units

61.3.1 Design of the Data Acquisition Module

The main control chip STC12C5A60S2 is a high-speed, low-power, single-chip microcontroller based on a high performance 1T architecture 80C51 CPU, which is produced by STC MCU limited [7]. It not only has 40 I/O ports, internal integrated MAX810 special reset circuit, with EEPROM function, four 16-bit timer/counter, but also has a fully compatible instruction set with industrial-standard 80C51 series microcontroller which can meet the development of this system needs. Parameter signals collected by air sensors and soil sensors are sent to the MCU for processing. We use the integrated temperature and humidity sensors SHT11 and AM2311 respectively as soil sensor and air sensor. The SHT11 is a single chip relative humidity and temperature multi sensor module comprising a calibrated digital output. The device includes a capacitive polymer sensing element for relative humidity and a band-gap temperature sensor. Both are seamlessly coupled to a 14bit analog to digital converter and a serial interface circuit on the same chip [8]. AM2311 is a sensor of low-power, long-term stability and it can output calibrated digital signal. The communication interface uses standard I^2C interfaces which only need two bi-directional signal lines: data line and clock line to complete the serial communications. SHT11 and AM2311 application circuit diagram is shown in Fig. 61.3.

Fig. 61.3 SHT11 and AM2311 application circuit diagram

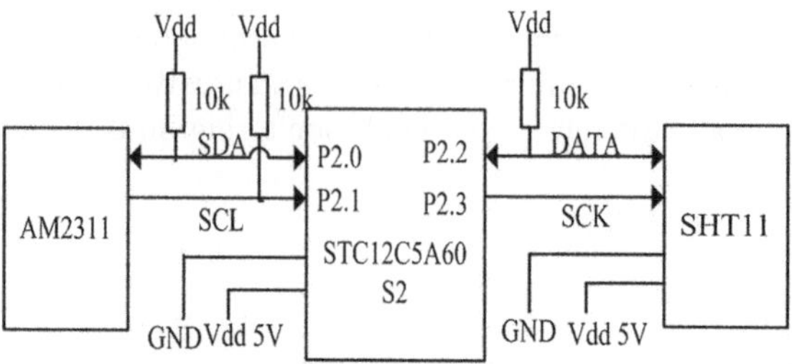

61.3.2 Design of the Wireless Transceiver Module

In order to achieve remote wireless monitoring and control of greenhouse, data and commands are transmitted through ZIGBEE wireless network with the function of self-organized network and GPRS with the function of long-distance wireless transmission. In the field monitoring unit, the ZIGBEE module is connected with STC12C5A60S2 by the serial port. Because the working voltage of STC12C5A60S2 and ZIGBEE module are respectively 5 and 3.3 V, the level conversion circuit is necessary during the processing of serial communication between STC12C5A60S2 and ZIGBEE module. The circuit of ZIGBEE module is shown in Fig. 61.4.

61.3.3 Design of the Execution Module

When the environmental parameters collected by the data acquisition module are out of the set values or receive commands from the central control unit, the field monitoring units will control the electric-relay switch according to its algorithm after analyzing and processing data or commands, adjusting the environmental parameters to the best status and achieving the automatic control and remote control of the greenhouse accordingly. In order to protect the shutter machine and ensure the safety of operation, some software and hardware protection measures were taken such as software timing, thermal overload relay and position limit switch. Since the single chip output current is too small to drive the relay directly, relay driver circuit is necessary. Transistor is used to drive the relay and opto-coupler is used for optical isolation in the circuit design. Relay driver circuit is shown in Fig. 61.5.

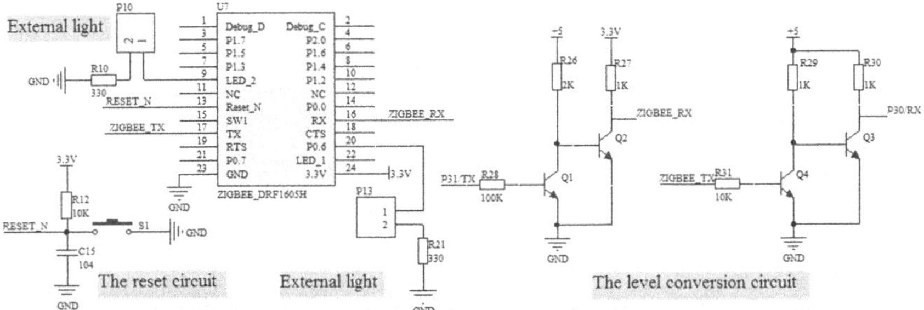

Fig. 61.4 The circuit of ZIGBEE module

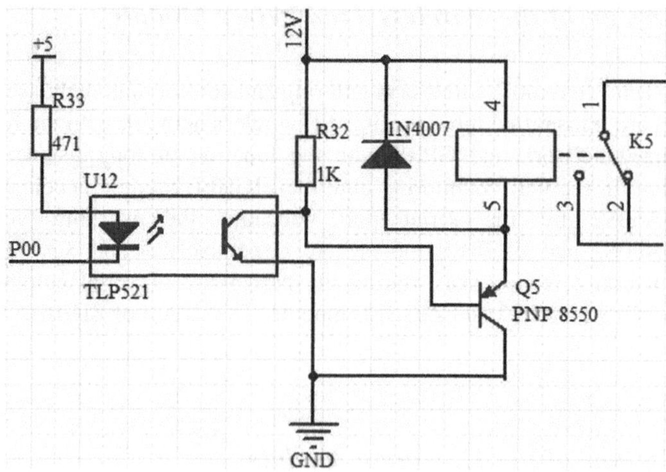

Fig. 61.5 The circuit of relay driver

61.4 Software Design of the Remote Intelligent Greenhouse Distributed Control System

61.4.1 Software Design of Field Monitoring Unit

In order to realize the functions of the field monitoring unit, programming software Keil uVision4, C programming language and modularized program design thoughts were adopted. The main program flow chart of field monitoring unit can be seen in the Fig. 61.6. After the MCU receiving the parameters collected by sensors, the data will be packaged in a certain format during transmission. When the parameters exceed appropriate range, the automatic control response function will be called to adjust the greenhouse environment. As each field monitoring unit has one and only address belonging to its own, when the field monitoring unit receives a packaged command, the command will be accepted until the address extracted by the command is same to its own. Then the corresponding command response function will be called to complete the operation. The data and commands packet are respectively presented in Tables 61.1 and 61.2.

61.4.2 Software Design of Central Control Unit

Central control unit software is the most critical part of the system, which is used to realize environmental parameters display, processing, analysis, inquiry, human–computer interaction, remote command control, database management and communicate with the field monitoring unit. The MySQL database technology is

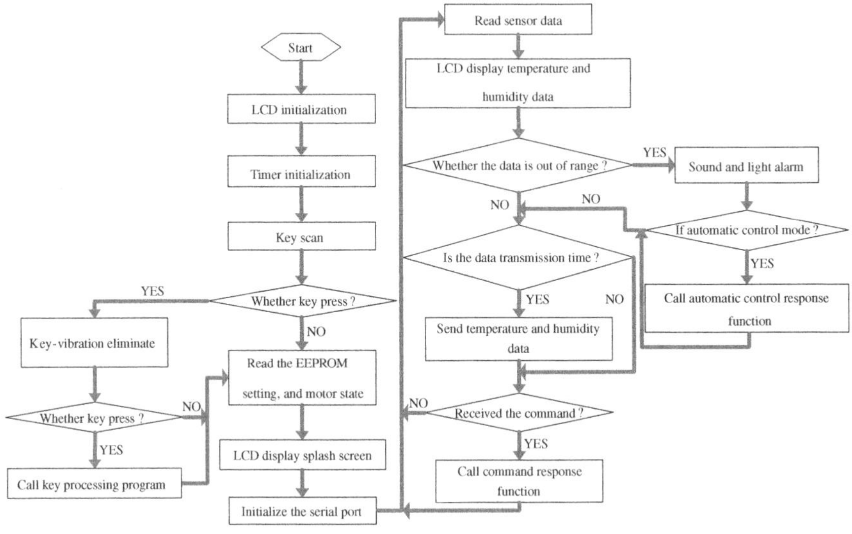

Fig. 61.6 The main program flow chart of field monitoring unit

Table 61.1 Data packet format

Packet head (Byte)	Own address (Byte)	Split symbols (Byte)	Air temperature (Byte)	Split symbols (Byte)	Air humidity (Byte)	Split symbols (Byte)	Soil temperature (Byte)	Split symbols (Byte)	Soil humidity (Byte)	Terminator (Byte)
8	4	1	5	1	4	1	5	1	4	1

Table 61.2 Command packet format

Packet head (Byte)	Destination address (Byte)	Split symbols (Byte)	Command type (Byte)	Split symbols (Byte)	Command (Byte)	Terminator (Byte)
1	4	1	1	1	10	1

also used in this system and the user can realize remote data query, display and commands control through the remote connection of MySQL database by running client program. The remote intelligent greenhouse distributed control interface is composed of real-time data display interface, greenhouse detail display interface, system parameter setting interface, data query interface and system running interface. The real-time data display interface is used for displaying the real-time environmental parameters in each greenhouse. The real-time environmental parameters will be shown as curves and control commands can be selected in the greenhouse detail display interface. The greenhouse detail display interface is shown in Fig. 61.7. The system parameter setting interface is mainly used for realizing the settings of serial port and alarm threshold. The historical data of each greenhouse can be queried in the data query interface. And the system operation

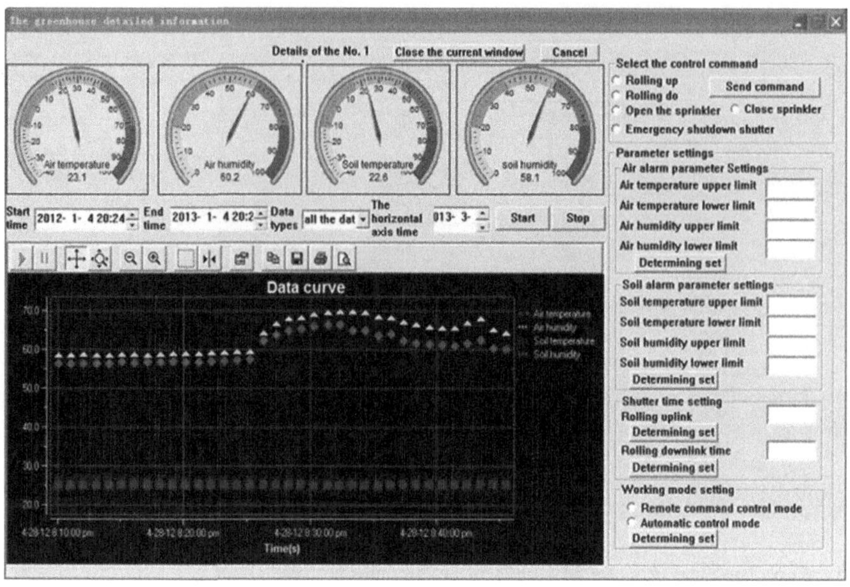

Fig. 61.7 Greenhouse detail display interface

and stop can be controlled in system running interface. As each part of the central control unit software has its own function, so the client interface can be very good for human interaction.

61.5 Conclusion

In this paper, a remote intelligent greenhouse distributed control system based on ZIGBEE and GPRS is developed. Compared with former monitoring and control system, the new system proposed in this paper has more advantages and it is more suitable for promotion. The advantages of the system are shown as: we use the ZIGBEE wireless network instead of placing the cables, which simplifies the system wiring and can reduce the cost of human resource. Because the distributed control ideas were adopted in this system, the other field monitoring units will keep on working normally in case of failure of any field monitoring unit or the central control unit. Therefore, the robustness of the system is improved. Moreover, with the capabilities of self-organizing, lower complexity and lower cost, ZIGBEE wireless communication technology is more suitable for short-distance communication. In this paper ZIGBEE wireless network and GPRS with long-distance wireless transmission function are combined together to achieve remote intelligent monitoring and control which possesses enormous viability for market and wide prospects for application. The system has been put into operation in

Qinghai Province, China and the test results indicated that the system showed high stability and high reliability of data transmission and equipment control.

Acknowledgments This research is supported by the project of Yushu post-disaster reconstruction digital information service key technology integration and demonstration and the National Science and Technology Support Program (No. 2010BAK68B04).

References

1. Guofang L, Lidong C, Yubin Q, Shengtao L, Junyu X (2010) Remote monitoring system of greenhouse environment based on LabVIEW. In: 2010 international conference on computer design and appliations (ICCDA 2010), pp V2-89–V2-92
2. Zhang Q, Yang X, Zhou Y, Wang L, Guo X (2007) A wireless solution for greenhouse monitoring and control system based on Zigbee technology. J Zhejiang Univ Sci A, 8(10):1584–1587
3. Yulong J, Jiaqiang Y (2011) Design an intelligent environment control system for greenhouse baseed on RS485 BUS. In: 2011 second international conference on digital manufacturing & automation, pp 361–364
4. Weihong W, Shuntian C (2009) Application research on remote intelligent monitoring system of greenhouse based on ZIGBEE WSN[J], vol 10, pp 1–5
5. Zhang Q, Yang X, Zhou Y, Wang L, Guo X (2007) A wireless solution for greenhouse monitoring and control system based on ZigBee technology. J Zhejiang Univ Sci A 8(10):1584–1587
6. Yiming Z, Xianglong Y, Xishan G, Mingang Z, Liren W (2007) A design of greenhouse monitoring & control system based on ZigBee wireless sensor network. In: International conference on wireless communications, networking and mobile computing, WiCom 2007, vol 2007. pp 2563–2567
7. STC12C5A60S2 series MCU STC12LE5A60S2 series MCU Data Sheet (2011) STC MCU Limited. Website www.STCMCU.com, vol 7, pp 1–439
8. Sensirion AG (2006) SHT1x/SHT7x humidity & temperature sensor, pp 2–10

Chapter 62
Emergency Response Technology Transaction Forecasting Based on SARIMA Model

Susu Sun, Xinbo Ai and Yanzhu Hu

Abstract Transactions of emergence response technology is considered to be the security of daily life and production. Due to the important role of emergency response technology transaction, a multiplication seasonal autoregressive integrated moving average (SARIMA) model is applied to the monthly emergence response technology transaction forecasting of the Beijing, China. This study demonstrates the usefulness of $\mathrm{SARIMA}(0, 1, 1,) \times (1, 1, 0)_{12}$ in predicting the transaction series with both short- and long-term persistent periodic components. From the analysis of the transaction series, a conclusion has been made that in the next years, the transaction will maintain it growth and fluctuation.

Keywords Emergence response technology transaction · Time series analysis · SARIMA model · R language

62.1 Introduction

Transactions of emergence response technology plays an important role in safeguarding the daily life and production. The truthful prediction of transaction such as its long-term trends or periodic patterns can give effective information for macro-control by government and resource management [1].

During the past years, several studies have developed methods of analysis market series. Perhaps the most widely used model is the ARIMA model [2]. The two general forms of ARIMA models are non-seasonal ARIMA(p, d, q) and multiplicative seasonal ARIMA (p, d, q) × (P, D, Q) in which p and q are

S. Sun · X. Ai (✉) · Y. Hu
School of Automation, Beijing University of Posts and Telecommunications, Room 917, New Keyan Building, NO 10 Xitucheng Road, Haidian District, Beijing, China
e-mail: lovesimple@126.com

Z. Sun and Z. Deng (eds.), *Proceedings of 2013 Chinese Intelligent Automation Conference*, Lecture Notes in Electrical Engineering 255,
DOI: 10.1007/978-3-642-38460-8_62,

non-seasonal autoregressive and moving average, P and Q are seasonal autoregressive and moving average parameters, respectively. The other two parameters, d and D, are required differencing used to make the series stationary.

Beijing is the capital of China. The development of its technology market will provide experiences for other provinces of China. We will look at the volume of emergence response technology transactions 60 consecutive months between 2006 and 2010. The main objective of our study is to develop a valid stochastic model to forecast transactions of Beijing emergence response technology market.

62.2 SARIMA Model Theory

In general, in a serial correlation theory, we will deal with specifications of the form where x_t is a vector of explanatory variables observed at time t, z_{t-1} is a vector of variables known in the previous period, β and γ are vectors of parameters, u_t is a disturbance term and ε_t innovation in disturbance. Vector z_{t-1} may contain lagged values of u and ε or both. If there are no explanatory variables involved in the model, it is possible to replace u_t with y_t in the equations below.

$$y_t = x_t'\beta + u_t \tag{62.1}$$

$$u_t = z_{t-1}'\gamma + \varepsilon_t \tag{62.2}$$

The above mentioned Box-Jenkins methodology, or ARIMA models theory, is developed in work [2]. ARIMA models are generalizations of a simple AR model that uses three tools for modeling serial correlation in disturbance. The first tool is an autoregressive, or AR term. Each AR term corresponds to the use of a lagged value of the residual in a forecasting equation for the unconditional residual. The autoregressive model of order p, AR(p) has the following form:

$$u_t = \varphi_1 u_{t-1} + \varphi_2 u_{t-2} + \ldots + \varphi_p u_{t-p} + \varepsilon_t \tag{62.3}$$

or, alternatively, with the use of a lag operator B:

$$\left(1 - \varphi_1 B - \varphi_2 B^2 - \ldots - \varphi_p B^p\right) u_t = \varphi_p(B) u_t = \varepsilon_t \tag{62.4}$$

where for B holds

$$B^j u_t = u_{t-j} \tag{62.5}$$

The next tool is an integration order term. Each integration order corresponds to the differentiation of the series being forecast. The first-order integrated components means that the forecasting model is designed for first difference of the original series. The second-order component corresponds to the second difference, etc. The third tool is MA, or a moving average term. The moving average forecasting model uses lagged values of a forecast error to improve the current forecast. The first-order moving average term uses the most recent forecast error, the second-order term uses the forecast error from two most recent periods, etc. MA(q) has the form:

$$u_t = \varepsilon_t - \theta_1 \varepsilon_{t-1} - \ldots - \theta_q \varepsilon_{t-q} \quad (62.6)$$

Equivalently, it can be rewritten by the lag operator as follows:

$$u_t = \left(1 - \theta_1 B - \ldots - \theta_q B^q\right)\varepsilon_t = \theta_q(B)\varepsilon_t \quad (62.7)$$

Where modeling time series with systematic seasonal movements—which is the case of monthly average temperatures and monthly precipitation sums—Box and Jenkins in [3] recommended the use of seasonal autoregressive (SAR) and seasonal (SMA) terms. The seasonal autoregressive model of order P can be written as

$$u_t = \Phi_1 u_{t-S} + \Phi_2 u_{t-2S} + \ldots + \Phi_P u_{t-SP} + \varepsilon_t \quad (62.8)$$

The seasonal moving average model of order Q can be as written as

$$u_t = \varepsilon_t - \Theta_1 \varepsilon_{t-S} - \ldots - \Theta_Q \varepsilon_{t-SQ} \quad (62.9)$$

In the two equations above, S denotes the length of seasonality, which is for the time series analysised in this paper equal to number 12.

Finally, we can write the most general SARIMA (p, d, q) × (P, D, Q) with constant model as

$$\varphi_p(B)\Phi_P\left(B^S\right)(1 - B)^d\left(1 - B^S\right)^D u_t = \varphi_0 + \theta_q(B)\Theta_Q\left(B^S\right)\varepsilon_t \quad (62.10)$$

where the constant equals

$$\varphi_0 = \mu[(1 - \varphi_1 - \varphi_2 - \ldots - \varphi_p)(1 - \Phi_1 - \Phi_2 - \ldots - \Phi_P)] \quad (62.11)$$

62.3 Model Construction

R was first written as a research project by Ross Ihaka and Robert Gentleman, and is now under active development by a group of statisticians called 'the R core team' [4]. R is an integrated suite of software facilities for data manipulation, calculation and graphical display. Among other things it has [5]

- an effective data handling and storage facility,
- a suite of operators for calculations on arrays, in particular matrices,
- a large, coherent, integrated collection of intermediate tools for data analysis,
- graphical facilities for data analysis and display either directly at the computer or on hardcopy, and

- a well developed, simple and effective programming language (called 'S') which includes conditionals, loops, user defined recursive functions and input and output facilities,
- it has developed rapidly, and has been extends by large collection of packages.

R has many packages for time series analysis, so we choose the TSA package [6] for the transaction time series analysis.

Monthly transaction time series in the period 2006-2010 really show an increasing linear trend and seasonality with the length equal to 12 which can be seen in Fig. 62.1. We can start modelling by using seasonal differentiation where D = 1 [7]. Further, as the main tool, we use mainly the (residual) correlogram visual analysis method. In the correlogram, the values of the (residual) autocorrelation function (ACF) [8], partial autocorrelation function (PACF) [8] can be seen. An example of R software output is presented in Fig. 62.2.

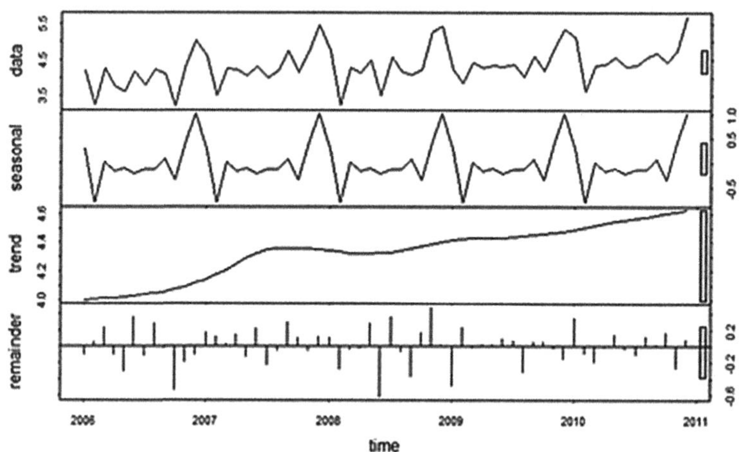

Fig. 62.1 Decomposition of transaction time series

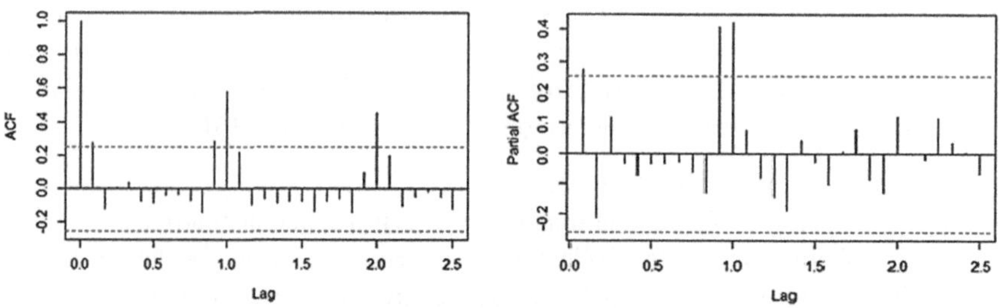

Fig. 62.2 ACF and PACF of seasonally differentiated monthly transaction time series

As can be seen in Fig. 62.1, most statistically significant are the values of ACF and PACF in lag 1 and 12. So the next step is the employment of the SMA(12) term. Or alternatively, Q = 1 and S = 12 in the equation (62.10) term. Now we can speak about the $\text{SARIMA}(0, 1, 1,) \times (1, 1, 0)_{12}$ model. In the next step-according to the residual correlogram-some term for the lag 1 should be added to the model. Two types of criteria were used:

- Primary: statistical significance of the estimated coefficient on the 5 % level considered according to the *t* test.
- Secondary: maximization of the adjusted coefficient of determination and/or minimization of the Akaike information criterion.

Secondary criterions were used when there was no possibility to decide after a primary criterion had been used, or when it was necessary to decide which term was the best one to be added to the model in a certain phase.

62.4 Results and Discussion

The statistical analysis of model parameters is presented in Table 62.1. The bias and standard error of simulations show the validity of the model parameters.

According to the Ljung-Box test [9], the Chi square value with degrees of freedom equals to 11 is 34.65 and the p value is 0.23, the model has caught most of the relations in the transaction time series.

Next we will do goodness-of-fit tests to verify the model. Goodness-of-fit tests verify the validity of the model by some tools. The residuals of the model are usually considered to be time-independent and normally distributed over time.

Figure 62.3 indicates the residuals of the model are time-independent. Residual ACF of the SARIMA do not show significant autocorrelation coefficient. Figure 62.4, histogram and QQ plot of the residual also indicate that the residuals are time-independent. Thus, the $\text{SARIMA}(0, 1, 1,) \times (1, 1, 0)_{12}$ model is accepted for transaction forecasting.

The selected SARIMA model was then used to forecast transaction from January, 2009 to December, 2010. The forecasted and observed series are compared in Fig. 62.5. The $\text{SARIMA}(0, 1, 1,) \times (1, 1, 0)_{12}$ model can show the important feature of the transaction, especially the trend and seasonal periodic patterns.

We used the model to forecast the two years' transaction of Beijing as can be seen in Fig. 62.6. In the case of next two years transactions, the forecast series shows a special period in the end of every year. In general, it could be concluded that the forecasted monthly transaction shows a period of good transaction and the manager of the emergency response technology market should carefully decide on market sources planning and management in the leading time.

Table 62.1 Result of parameter estimation

Parameters	θ	Θ
Values	0.6047	0.9526
Standard error	0.042	0.1106
$\sigma_e^2 = 0.6172$:log likelihood $= -140.58$, AIC $= 294.51$		

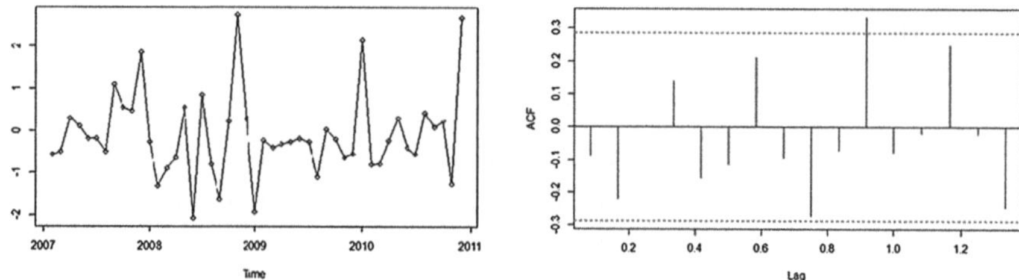

Fig. 62.3 The residual series and ACF of the SARIMA

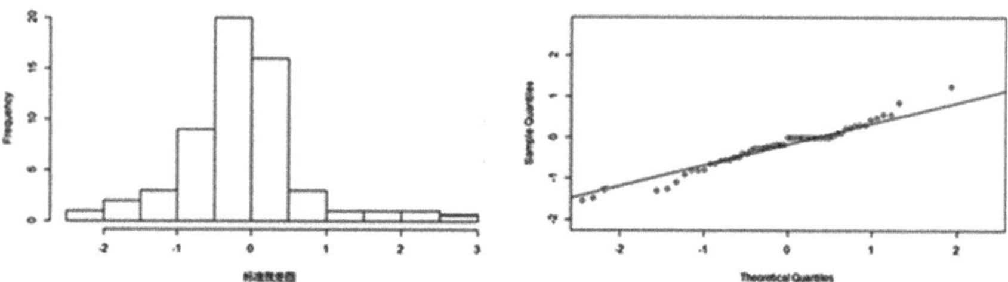

Fig. 62.4 The histogram and QQ plot of the SARIMA residual

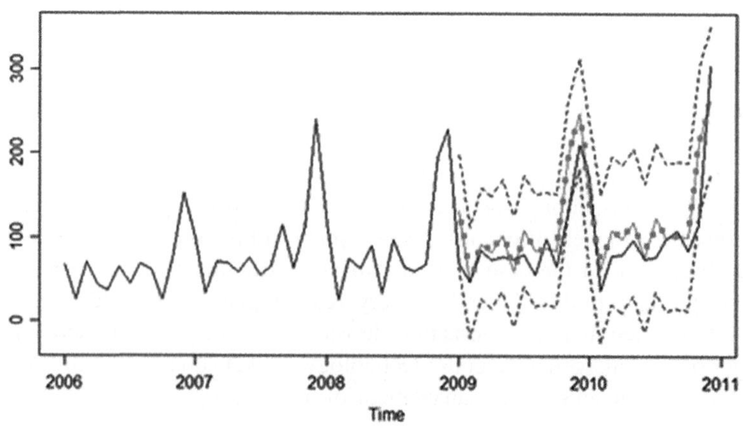

Fig. 62.5 Comparation of forecast series and observed series

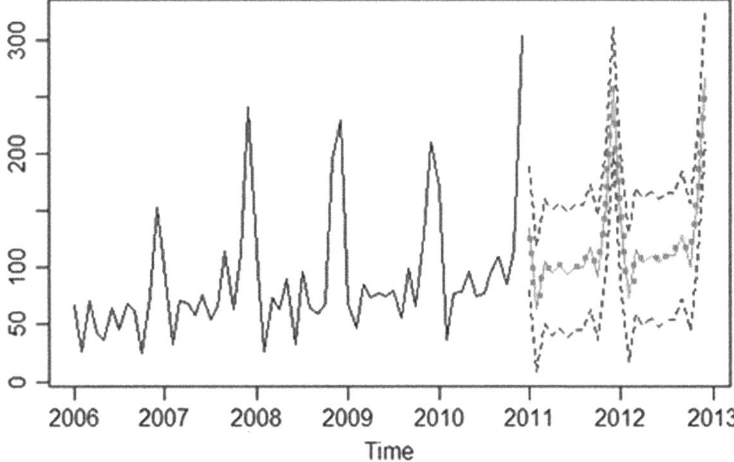

Fig. 62.6 Forecasting series of the transaction

62.5 Conclusion

The study impels the capability of SARIMA model in transaction forecasting. The largest different is observed in the end of every year. The capability of SARIMA model could also be seen from analogy between monthly variation of forecasted and observed values where the highest transaction happens in the end of the year. However, SARIMA model is a short-term forecasting model based on ARIMA added with seasonal change. The past fit error also added as an important factor in the prediction model so that it has higher prediction accuracy. Inevitable, as the time become longer, the prediction error will be gradually increased and the accuracy decreased.

Acknowledgments This work is supported by the National Science and Technology Support Program (NO. 2011BAK01B00).

References

1. Hanssens DM, Parsons LJ, Schultz RL (2002) Market response models: econometric and time series analysis, 2nd edn
2. Box GEP, Jenkins GM (1976) Time series analysis, forecasting and control, revised edn. Holden-Day, San Francisco
3. Bowerman BL, O'Connel RT (1993) Forecasting and time series, an applied approach. Duxbury, Pasific Grove
4. Lopez-de-Lacalle J (2006) The R-computing language: potential for asian economists. J Asian Econ 17.1066–1081
5. Venables WN, Smith DM, The R Development Core Term (2002) An introduction to R: notes on R: a programming environment for data analysis and graphics version 2.6.0

6. Cryer JD, Chan K-S (2008) Time series analysis with applications in R, 2nd edn
7. Saz G (2011) The efficacy of SARIMA models for forecasting inflation rates in developing countries: the case for Turkey. Int Res J Finan Econ 17:1450–2887
8. Brockwell PJ, Davis RA (2002) Introduction to time series and forecasting, 2nd edn
9. Ljung GM, Box GEP (1978) On a measure of a lack of fit in time series models. Biometrika 65:297–303

Chapter 63
Research on Cloud-Based Simulation Resource Management

Qiao Cheng and Jian Huang

Abstract With the development of simulation system, too much simulation resources are accumulated, most of which are reusable. Good management on these resources can fully achieve the reusability of them and facilitate the simulation process, but the distribution and complexity of simulation resource add challenge on the management in aspects like resource storage, computing speed and resource sharing. The cloud computing integrates mature technologies to provide infinite storage and computing resource on users' demand, which is needed in the simulation resource management. To use these merits of cloud, this paper proposed a cloud-based simulation resource management model and produced a detailed design of the prototype system. This model is universal and provides a framework for cloud-based system; the prototype system is flexible for any scale of simulation system, it offer mechanism in data security, storage and usage, and can manage the simulation resources effectively and economically.

Keywords Cloud computing · Simulation resource · Resource management · Resource reusing · Large scale simulation

63.1 Introduction

Nowadays, most simulation systems are large, distributed and complex. Therefore, the quantity and variety of the simulation resources are increasing rapidly. It is critical to organize and manage these simulation resources effectively and efficiently

---

Q. Cheng · J. Huang (✉)
Department of Automatic Control, College of Mechatronic Engineering and Automation, National University of Defense Technology, ChangSha 410073 HuNan, China
e-mail: huang_jian@139.com

Q. Cheng
e-mail: xingdong12@sina.com

Z. Sun and Z. Deng (eds.), *Proceedings of 2013 Chinese Intelligent Automation Conference*, Lecture Notes in Electrical Engineering 255,
DOI: 10.1007/978-3-642-38460-8_63,

in simulation. So we need to seek the best way to achieve this goal. Cloud computing integrates some developed technology like grid computing, cluster computing, distribution computing, etc. [1]. Thus, it indicates a new direction for us.

63.2 The Requirements of Simulation Resources

63.2.1 The Definition and Characteristics of Simulation Resources

Simulation resources refer to all the reusable resources involved in the whole simulation process, such as data, algorithm, model, background information, etc. [2]. These resources are in rich variety of forms, such as text, picture, program, etc. Thus, simulation resources have the following characteristics [2, 3].

- Distribution: due to the largeness of system scale and the complexity of models, most simulation systems are distributed, so are simulation resources.
- Complexity: in a large scale simulation system, there are different sub-systems, system editions, simulation tasks and stages, so the related simulation resources are complex in both types and contents.
- Reusability: because most simulation resources have a lot of similarity and relativity with each other, so they can be reused in similar simulation systems directly or provide helpful reference for other systems.

63.2.2 The Requirements and Problems of Simulation Resource Management

To manage the simulation resource effectively is very meaningful, but there are still a lot of requirements to meet and problems to solve in some aspects.

- Resource storage: there must be enough storage space, good storage mechanism and reliable guarantee in data safety. To add storage hardware will bring heavy economic burden. The distribution and complexity of simulation resource will also add challenge to storage mechanism and data safety.
- Computing speed: it determines the effectiveness in using simulation resource. The supercomputers can solve this problem. But they are too expensive, which will definitely rocket the simulation cost.
- Resource sharing: it helps achieve the reusability of simulation resources. We must think about how to share simulation resources in a distributed system through net, and seek the best resources sharing method to achieve the reusability fully.

Faced with all the requirements and problems in the simulation resource management, the cloud computing provided us a new solution.

63.3 The Cloud-Based Simulation Resource Management (CSRM)

63.3.1 The Definition and Characteristics of Cloud

Cloud computing is developed from Parallel computing, Distributed computing and Grid computing. It assigns the computing tasks on the resource pool, and the resource pool is composed by large amount of computers [1]. As long as connected to the cloud through internet, users can get computing resource, storage space and software service at their own need. They can enjoy these supercomputing and super storage services by any device. The cloud saves resource, reduces cost and provides users a great degree of flexibility [4].

According to the service type, the cloud computing can be classified into three types [1]: Infrastructure-as-a-Service, Platform-as-a-Service and Software-as-a-Service. At present, the cloud computing has many good characteristics [5], such as super scale, virtualization, reliability, universality, scalability, on-demand service and cheap.

63.3.2 The Advantages of the CSRM

Combining the simulation resource management with the cloud [6, 7], we can make full use of the merits of cloud and meet the requirements of simulation resource management very well. Here we list the advantages of the CSRM.

- The CSRM works as if in supercomputers. Because the cloud has enormous server farm to provide infinite storage space and computing capabilities.
- The access to resources will not be constrained by time and area, as long as there are connections to internet. It will also enhance resource sharing.
- The CSRM can response to the change of requirement flexibly and save resource, because users apply the cloud resources on their own demand.
- Data safety problems like data loss, virus attacks can be reduced. Because the cloud server has specific data safety technicians and safety method.
- Reduce the cost in purchasing hardware, system maintenance and upgrading. Because we can get all these from the cloud.
- By using the Platform-as-a-Service from the cloud, developers can concentrate on the development of top immediately.

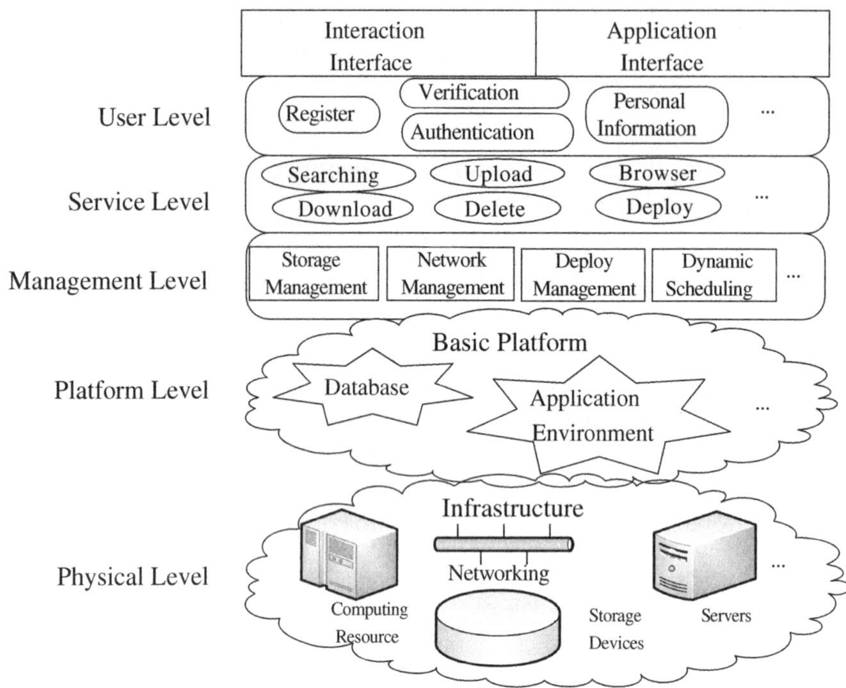

Fig. 63.1 The framework of CSRMM

63.4 Cloud-Based Simulation Resource Management Model (CSRMM)

In order to store and manage the simulation resources collectively, and provide users all kinds of application services on simulation resources, this paper will design the cloud based simulation resource management model (CSRMM) [8, 9]. The framework of this model is shown as Fig. 63.1, which is composed by five levels: physical level, platform level, management level, service level and user level. Therefore, this paper will introduce the model in these five aspects.

63.4.1 The Physical Level

This level provides all kinds of infrastructure for the whole system, like computing device, storage device and network. All the physical resources come from the cloud and located everywhere. Users don't need to care about the exact location and detail configuration information of the devices. The cloud integrates all the hardware devices as a whole, and then provides it to the higher levels. In this way, the upper levels use these physical resources just like using naked device and disk, which can meet all the need of the upper levels. So, this level lays a very good physical foundation for the whole system.

63.4.2 The Platform Level

The platform level utilizes all the physical resource from the physical level, and provides the platform for the upper levels, including database and application environment. Although resources in physical level are very profuse, complex, and have uncertainty in location, the platform level use technologies like virtualization and fault-tolerate to shield them. So, all the application services in the upper levels don't need to care about the related technology and the location of these resources. Thus, users just feel like operating a powerful supercomputer, and will not be constrained by the bottom pattern. On the other hand, it offers a uniform platform foundation for all the applications and services in the upper levels, which can save a lot of repeated work on the system environment configuration and improve the efficiency of the whole system.

63.4.3 The Management Level

The management level is the center level. This level is mainly for management on storage, network, deploy, dynamic scheduling, etc. To realize these functions, it depends on the innate management mechanism in the cloud, as well as the management mechanism in the simulation resource management system.

The storage management includes the management on metadata, resource creation, resource classification, storage space arrangement and so on. The goal of network management is to allocate the network resource properly, so that the network communication can be unblocked and orderly. The deploy management is to control the deployment of simulation resource on users' devices, which includes the deploying configuration and resource downloading. The dynamic scheduling is to help deal with concurrent access. It enhances the reliability, safety and flexibility of resource service, and makes full use of the capacity of computing and storage.

63.4.4 The Service Level

This level provides users all kinds of services on simulation resources, including searching, browsing, downloading, uploading, deleting and deploying. Based on the lower levels, this level accomplishes all the function in the form of service, which is a characteristic of cloud. All the accesses to simulation resource are through the network, which gives great convenience to users.

This level responses the users' service request, and call corresponding algorithm to analyze the request, then apply the virtualized resource from the cloud and provide users the service from the cloud resource pool. It also designed good algorithms about resource browsing and searching, so that the users can enjoy better service on the resource operating.

63.4.5 The User Level

To guarantee the data safety and accessibility, this model designed the user level to provide some corresponding mechanism.

Users need to register their own accounts first. Then, they can store, access and manage their own resources, and can also apply for the permission of accessing other's resources. So, we can guarantee the resource safety, and enhance the resource sharing. The services for users must be verified by the user level. The operations on simulation resources are limited by this level, too. The user level executes uniform management on users' personal information, and customizes service for users based on their own need.

All the levels above guarantee the realization of the CSRM system. This system can organize, store and manage the resource well. It has interaction interface and application interface to accomplish the interaction between users and simulation applications.

63.5 The Design of the Prototype System

According to these requirements, this article will design a simulation resource management prototype system based on the cloud. The prototype system provides a universal platform in managing the resources collectively, which will develop the simulation application and improve the resource reusability. It provides friendly interaction interface for resource searching, browsing and managing, as well as interfaces for other simulation application modules. The structure of the prototype system is shown as Fig. 63.2.

Users submit the access requests, and then the access interface sends the requests to the management control center. After that, the management control center interacts with the authority management center to determine whether the user has the authority on the resource. For those who gained the permission, the management control center will get the computing resource and storage resource from the cloud to provide the required service. The cloud provides the resource accessing services to the users through the access interface.

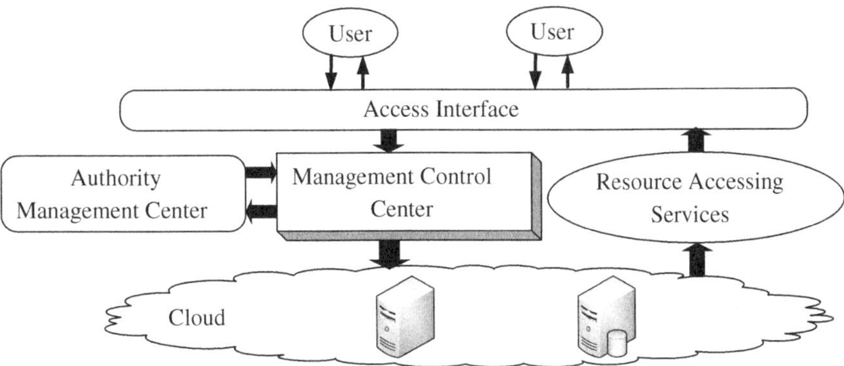

Fig. 63.2 The structure of the CSRM prototype system

The CSRM prototype system mainly provides following service modules: user verification module, resource registering and uploading module, resource retrieving module, resource presentation module, resource downloading and deleting module, resource assistant deploying module.

User verification module: resource safety is the primary requirement in resource management, so we need this module in the system. Users' authorities differ on different resources, so this module has mechanism on user verification and authority management. This module is indispensable especially in public cloud. With it, simulation users can manage simulation resources on their own need, and also helps to transfer resources between different users.

Resource registering and uploading module: this module provides the interface for resource registration and uploading. So it can facilitate timely storing resources, and managing them together. When registering and uploading resources, the interface has wizards to guide users complete the process. Users need to fill the registration information on the interface page. This information is useful in classifying and retrieving the resources.

Resource retrieving module: the function of the system is not only to store resources but also to help users to use these resources. However, the quantity of the resource is very big and we need to find the resource needed in a short time. That's the meaning of retrieving. This module provides us several retrieving methods. Meanwhile, it has several different order methods, which can order the resource at the users' preference after retrieving.

Resource presentation module: this module provides friendly interface to present the basic information of all needed resources and detail information of specific one. The presentation differs by users' needs. For example, the simulation experimenter cares the simulation process, so the resources are presented in the tree form as to the simulation flow. While the simulation analyzer cares more about the resource types, so the resources are presented in tables according to the types.

Resource downloading and deleting module: taking account of the simulation resource safety, users must get the accordingly authority when downloading and deleting the resources. For those who have the proper authority, we provide them the secured downloading and deleting interface.

Resource assistant deploying module: this module helps to deploy the resources in the database onto the users' devices which enhances the resource reusability. According to the users' own settings on the deployment, this module can download corresponding resource to the local computer, or offer the application interface for other simulation application module, even call the computing ability from the cloud to deal with the simulation resource.

Users can login in by browsers, which belongs to the user verification module. After the verification, they can browse the corresponding pages and get all the resource services provided by above modules.

63.6 Applications

By offering good security, storage and usage mechanism, this cloud based simulation resource management system can manage these simulation resources centrally. Users can get lots of operations on simulation resources, such as registering, uploading, retrieving, browsing, downloading and deleting. Due to the flexibility of the cloud, this system can be used in any scale simulation system, no matter the common simulation system or the distributed large scale one. So the system can satisfy the need of simulation users from different application levels.

References

1. Liu P (2010) Cloud computing (in Chinese). Electronic Industry Press, Beijing
2. Zhang YC (2002) Research on the simulation resource management, layering and sharing mechanism (in Chinese). Master Thesis, National University of Defense Technology
3. Skoogh A, Perera T, Johansson B (2012) Input data management in simulation—industrial practices and future trends. Simul Model Pract Theory 29:181–192
4. Hamdaqa M, Tahvildari L (2012) Cloud computing uncovered: a research landscape. Adv Comput 86:41–85
5. Venters W, Whitley EA (2012) A critical review of cloud computing: researching desires and realities. J Inf Technol 27(3):179–197
6. Matsumoto H, Ezaki Y (2011) Dynamic resource management in cloud environment. Fujitsu Sci Tech J, pp 270–276
7. Liu TY, Chai XD, Li BH (2011) Research on key technologies of resource management in cloud simulation platform. In: 23rd European modeling and simulation symposium, EMSS 2011, pp 508–515
8. Rimal BP, Jukan A, Katsaros D et al (2011) Architectural requirements for cloud computing systems: an enterprise cloud approach. J Grid Comput, pp 3–26
9. Chen Y, Wo TY, Li JX (2009) An efficient resource management system for on-line virtual cluster provision. In: Cloud: 2009 IEEE international conference on cloud computing, pp 72–79

Chapter 64
Research on the Design of Railway Passenger Traffic Decision Support System

Zhuomin Wei and Hongchao Song

Abstract Considered the insufficient usage of railway passenger traffic information system, we use data mining, online analytical processing, knowledge management and other technologies to develop a railway passenger traffic decision support system. This paper stated the model components and functional module of the decision support system.

Keywords Decision support · Railway traffic · Data mining · Knowledge management

64.1 Introduction

With the implement of railway ticket sale system and real-name ticketing system, multiple ways of getting tickets such as sale windows, phone booking and Internet sales so on, make the railway ticket's getting more and more convenient. Also, the diversification of ticket sale ways makes the related working changes a lot. In this condition, many useful customer information data stored in the databases of railway stations and centers, but have not be analyzed for the improvement of railway passenger service. The railway information systems at present try to solve the problem of railway passenger transport capacity's shortage, and improve the services of railway passenger system. These systems considered only the optimization of the train operation plan but not marketing and decision. The current railway passenger service system has several problems:

1. The system can't divide different influences to railway traffic by different customer groups such as students, migrant workers, travelers and so on.

Z. Wei (✉) · H. Song
College of Information Engineering, Capital Normal University, Beijing, China
e-mail: pinkie_618@163.com

Z. Sun and Z. Deng (eds.), *Proceedings of 2013 Chinese Intelligent Automation Conference*, Lecture Notes in Electrical Engineering 255,
DOI: 10.1007/978-3-642-38460-8_64,

Without this consideration, the prediction of passenger flow volume will be inaccurate.

2. Ticket data contains much new information, for example, customers' name, passengers' ID number, way of purchase, tickets-getting location, etc. This information can be used in customer clustering, and these analyses will be helpful to solve the difficulty of getting tickets.
3. Railway traffic changes with time, the information system cannot make real-tine analyze. And that brings a lot of problems of station resources' allocation and passenger flow's management.
4. The railway passenger information systems have been built, and have many customers' personal information. But have not been used to supply personal service [1].

To solve the problems we mentioned above, we designed the Railway Passenger Traffic Decision Support System (RPTDSS) with the technology of data mining, data warehouse, knowledge base, information management and online analytical processing (OLAP) so on. The system can analyze the operation data in different time with suitable method to meet with different needs, the analyze result will turn into the corresponding management decision support information [2]. In addition, the system can record customer's behaviors and analyze customer demand and consumption rules, which will help the management section to provide personalized service [3, 4]. RPTDSS makes full use of tickets data, gives more scientific and comprehensive decision support to the railway transport organization sector.

64.2 Architectures of RPTDSS

The main section of RPTDSS is the decision support information hierarchy system, the processes of the hierarchy system includes data collection and preprocessing, data storage and management, data processing, information representation and so on. For the ticket information contains much information, the data storage sector is divided in two branches according to different demands: database processing hierarchy system and warehouse processing hierarchy system. In the processing of database, the model base and model base management system can select proper model based on different command, after the treatments and analyses of the given data, we can get the support decision information we needed. In the processing of warehouse, the data comes from various data resources are divided into different subjects, these data saved in warehouse after data preprocessing, OLAP use the information in warehouse to establish a decision support assumption in multi-angles, and display the result in the form of tables and graphs. In addition, RPTDSS contains the knowledge base hierarchy system, this system will storage the discovered knowledge from database and warehouse, and the system will help us in decision optimization and knowledge sharing. The architectures of RPTDSS are showed in Fig. 64.1.

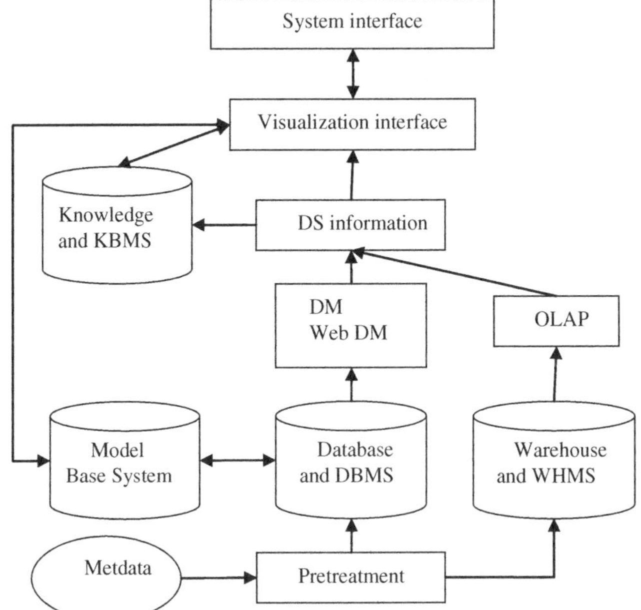

Fig. 64.1 The architectures of RPTDSS

64.2.1 Database and Database Management System

The storage of data is a basic problem of decision support system. Database and database management system are used to keep and operate the date in RPTDSS. The data contains train data, ticket data, passenger data and so on. Every kind of data includes many items and these data all need a data form to store the original data and processed data. The database management system can operate the data in database by the usage of data definition language (DDL) and data manipulation language (DML) [4].

64.2.2 Warehouse and OLAP

Warehouse is built for the fore-end query, and is the basic of online analyze. For this reason, the shortage of warehouse must be efficient; the processing of data needs to be precise. The central problem of warehouse is the storage and management of data; the first step of designing is the definition of system subjects. The warehouse of RPTDSS has five subjects, they are ticket subject, passenger subject, passenger flow subject, transport capacity subject and train schedule subject. Every subject contains fact tables and dimension tables, fact tables are used to store the measure of fact and the code value of dimension, and dimension tables are used to save the metadata of dimension [5]. The dimension tables can contact with the fact tables by the linkage of key words in dimension tables and foreign keys in fact tables. The original data is processed according to different subject for further analysis.

OLAP can imitate the users' multi-thinking model to make multidimensional analysis. The data in RPTDSS is very complex and contains various kinds of information, so the decision analysis needs multi-thinking of various sides. OLAP can switch between different angles to make comprehensive analysis [6], and turns the results into the information that is needed by railway passenger transport decision makers.

64.2.3 Knowledge Base and Inference Engine

The knowledge base of RPTDSS is used to keep and manage the decision-related knowledge. The system knowledge base has the knowledge from experts of railway passenger transport, and also has the knowledge got from the processing of data mining and OLAP. The store of knowledge base is lamellate: the low-level is fact knowledge, middle- level is rule, and the high-level is policy. The knowledge in these three levels is interdependent. Inference engine is the realization of knowledge reasoning in computer field. It contains two parts: reason and control. Inference engine selects the appropriate control strategies according to the semantics of those knowledge, got new rules and knowledge after reasoning and consistency test [7]. Inference engine gives more scientific decision support to railway decision maker.

64.2.4 Model Base System

Model base system is used to manage and maintain models; it contains model base and model base management system. Model base system is an important part of the decision support systems [8]. The model base in RPTDSS has various kinds of data-mining models such as mathematical models, data processing models, image models, report models, spatial analysis models, etc. These models are stored by a certain structure; we need a model dictionary to index and descript the corresponding model files. Through the model dictionary, we can pick up the models we needed by model name, modeling method, model function and other model information, then we can make further operation such as visit, update, compose and soon. In addition, model base management system can compose several basic models in order structure or choice structure or looping structure, in this way, data base management makes data processing more flexible.

64.3 Function Modules of RPTDSS

Railway Passenger Transport Decision Support System adopts several technologies such as: information management, data management, data mining, behavior analysis, knowledge reasoning, visualization, etc. The system fully dig out the knowledge from information of tickets and customers about passenger flow, customer behavior and other railway field, and reflect to the user in the form of report or graphic. RPTDSS can provide a scientific decision support to the manager of railway transport. To realize the functions of RPTDSS, the system is divided in seven parts, each function modules have several child functions, and all these functions will be helpful to the work in railway system and the improvement of railway services. The function modules of RPTDSS are showed in Fig. 64.2.

The descriptions of function modules of RPTDSS are showed as below.

64.3.1 The System Management Module

This module is used to manage the information and permission of system users; managerial staff can view the daily operation conditions of RPTDSS, and maintain the daily operation of the system.

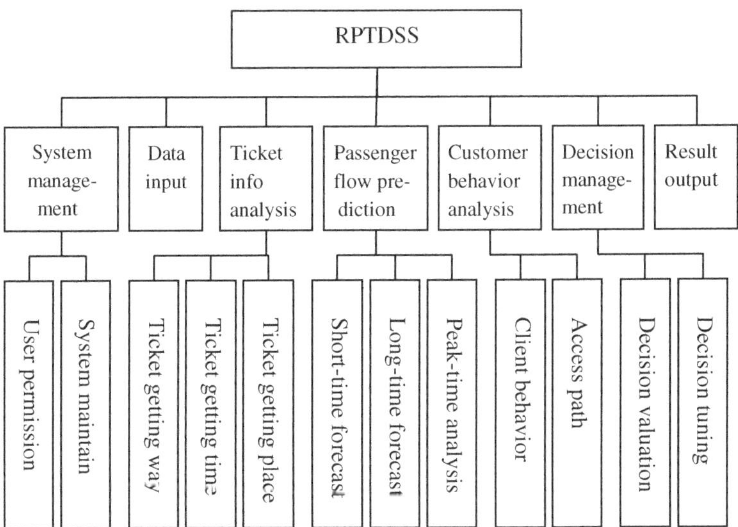

Fig. 64.2 The function modules of RPTDSS

64.3.2 The Data Input Module

The main purposes of this module are collecting the input ticket information and building a suitable database to store the entered information for the further processing. This module can also collect the analysis result of other system to assistant the decisions' making.

64.3.3 The Ticket Information Analyze Module

The function of this module is mining the sales information from ticket information, these information supply decision support for ticket the resource allocation of ticket sale system. This module contains ticket-getting way analysis, ticket-getting time analysis, and ticket-sale location analysis so on. Ticket-getting way analysis is used to count the amount of tickets getting in each way, and according to the result and change trend to adjust the distribution of ticket system resource. Ticket-getting time analysis is used to calculate the proportion of the daily ticket-sale volume in ticket presale period. The result will help the railway manager to adjust the presale period. Ticket-sale location analysis is used to analyze the ticket sales distribution of a train in different station and different time, and it will be helpful to adjust the distribution of tickets' amount in station.

64.3.4 Passenger Flow Analysis Module

The main task of this module is the analysis and prediction of railway passenger flow, the prediction can divide into two parts: long-time prediction and short-time prediction. Otherwise, this module also provide peak time prediction, it can calculate the length of peak time and daily passenger flow volume. These analysis and prediction will be helpful to railway work organizers in transport plan's making.

64.3.5 Customer Behavior Analysis Module

This module is based on the user base of online sale system; it contains customer purchasing power analysis and access path analysis. Customer purchasing power analysis can get customers' preferences in train and seat type through the analysis of customers' trade records. It provides the foundation of personal service. Access path analysis is used to optimize the structure of website, and improve the user experience of the website. Customer behavior analysis give advice to the ticket sector and website operation sector, and it is the basic of personal services.

64.3.6 Decision Management Module

This module is used to record and manage the decision method and plan, it makes evaluation of the performed decision method and gives tuning recommendation. This function gives the system user a visual feedback of the decision.

64.3.7 Result Output Module

This module can output the analytical result of the function modules we mentioned above, this function provides two form of analytical result: report and figure, the railway manager will get paper-based materials. It makes the communication between different sectors more convenient.

64.4 Conclusion

Nowadays the reference of railway passenger transport decision is only the decision makers' experience [8]. The decision support is not very accurate and reliable. To change this situation, this paper adopts several technologies such as data mining, database, and knowledge base management so on to develop a decision support system called Railway Passenger Traffic Decision Support System (RPTDSS). The system concludes ticket information analysis, passenger flow prediction, customer behavior analysis and other auxiliary functions. This system can supply scientific support decision information to the railway transport manager; it is meaningful to the improvement of railway passenger service. Now the design work is complete, but design's turning into practical application still need further research.

References

1. Hu H, Zhou D (2009) A study on the problems and solutions in China railway passenger service system. Autom Panorama 11:74–77
2. Ji P, Shan X, Wang W (2001) The implement design of railway passenger Dss. Railway Comput Appl 2001.04, 20–22 (in Chinese)
3. Zhang M (2005) Discussion on client relationship management in railway passenger transport system. Railway Transp Econ (2005.11) 59–60
4. Mckenney JL, Scott MM Management decision systems: computer-based support for decision making. Harvard Business School Press, Boston, MA, USA, p 35
5. Tan X, Liu B (2003) Research on the application of DM in CRM. Accounting Oper Immaterial Assets 02:145–147
6. Liu Y, Wang Y (2003) Build of CRM's data warehouse. Inf Technol 01:20–22 (in Chinese)

7. Rizzi S, Abello A, Lechtenbörger J, Trujillo JJ Research in data warehouse modeling and design: dead or alive? In: DOLAP '06 proceedings of the 9th ACM international workshop on data warehousing and OLAP pp 3–10
8. Wang R, Du J, He Y, Yang X (2007) Study of fuzzy knowledge-based system and fuzzy inference machine. J Comput Technol Dev 03 (in Chinese)

Chapter 65
State Identification of Automatic Gauge Control Hydraulic Cylinder Using Acoustic Emission

Hongzhi Chen, Chao Wu, Yanguang Sun and Hua Zhao

Abstract The growing requirements of steel products qualities bring even tricky restrictions to the hydraulic cylinder which acts as a screw down to maintain the rolling spaces precisely. However, faults and disfunctions are unavoidable and bring underlying safety issues. Working with un-expected loading is one of major impact that may cause catastrophic consequence happen to the cylinder, as the classical diagnostic process to-some-extent lack of cross-validating and time-consuming, the paper proposes the potentials of using acoustic emission to fill the dilemmas. The works include the data acquisition process to record the ultrasound acoustic signals from the hydraulic cylinder, application of modified image based acoustic emission approach to generate visual effects, as well as the application of principal component analysis to project the profiles onto the 3D plane for further analyses. Through the analysis of image based profiles, the subtle differences of various cylinder conditions can be examined using the pixel and intensity values. By assessing the projection of profiles in the principal component space, a clear trajectory can be observed with normal and overload conditions allocated upon the positive and negative sides of the axis. The result provided not only the potential of using acoustic emission for dynamic state identification of the subtle changes, but also opens up the possibility of preventive measures to the cylinder at risks.

Keywords State identification · Acoustic emission · Automatic gauge control

H. Chen (✉) · Y. Sun · H. Zhao
State Key Laboratory of Hybrid Process Industry Automation System and Equipment Technology, Automation Research and Design Institute of Metallurgical Industry, 72, Xi Si Huan Nan Lu, Beijing 100071, China
e-mail: Shadow_c3186@msn.com

C. Wu
Drive Department, Beijing Aritime Intelligent Control Co., Ltd, 6, FuFeng Rd, Feng Tai Ke Xue Cheng, Beijing 100070, China

Z. Sun and Z. Deng (eds.), *Proceedings of 2013 Chinese Intelligent Automation Conference*, Lecture Notes in Electrical Engineering 255,
DOI: 10.1007/978-3-642-38460-8_65,

65.1 Introduction

The growing industry requires extreme quality of rolling steel products, which brings even tricky accuracy to the rolling process to maintain the manufacturing tolerances under the acceptable condition [1]. According to this situation, the precision of the space between the rolls is considered to be one of key issue to minimize the product manufacturing tolerances. With the fast response and high level of control accuracy performance under the complicate force conditions, the hydraulic automatic gauge control (HAGC) systems have been widely applied in the steel rolling processes [2].

With the non-stop working of rolling mills, the abnormal and faults of HAGC system are unavoidable. Previous studies showed a set of researches and contributions regarding to HAGC state identifications [3–7], the works include the object-oriented approach for obtaining fault knowledge on HAGC system followed by the development of enhanced neural network based adaptive classifier for dealing with the pressure change forecasting problem on the strip mill HAGC system [3, 4]; the design of a decoupling subsystem based on differential geometric approach for robust diagnostic of load uncertainties [5]; as well as the innovative design of optimized BP neural work for dealing with the multi-sensors faults [6, 7].

According to the official statistics revealed from one of Chinese major steel-supplier, the hydraulic cylinder dominants over 33 % of rolling mill hydraulic components faults [8], however, only a little works focused on the state identification of the cylinder [9, 10], as the executing device of HAGC system that directly justify the working roll space, the abnormal of the cylinder could directly influence the product quality and bring underlying safety issues, a nonlinear model based adaptive robust observer was designed for cylinders' typical fault detection with verification supported by the simulation results [9]. As the classical diagnostic technique is sensitive to fluctuation, a wavelet energy [11] based approach was derived to deal with this issue and improve the inputs of neural network classifier [12] with unexpected loading caused by the inner leakage being correctly identified [10], however, with wavelet energy and neural network approach employed to deal with the noise issue, large computational burden could be charged to the diagnostic system.

Acoustic emission (AE) has been applied in several areas for pattern recognition and health monitoring [13] regarding to its subtle changes sensitivity and noise robustness in the microscopic level [14]. With identifying the loading developments in a hydraulic cylinder until failure can be considered as to determine the subtle tolerances of the cylinder during its still under health condition, AE could be a useful option to meet the criterion. The paper presents an image based AE signal processing approach for state identification over the hydraulic cylinder, thereby demonstrating the potential of using AE for quick state examination as well as providing a potential cross-validation option for dynamic fault detection over the cylinder. The experimental setup is introduced in Sect. 65.2. Section 65.3 devotes to the image based AE signal processing approach. By further analyzing the image based profiles using the principal component analysis (PCA), the performance of the AE for cylinder state identification is discussed in Sect. 65.4. Concluding remarks is given in Sect. 65.5.

65.2 Experimental Setup

The object under examination is a HAGC hydraulic cylinder with dimensionality R = 550 mm × I = 600 mm × O = 850 mm (R, I, O are referred to as diameter of piston rod, as well as the inner and outer diameter of the cylinder, respectively) and with maximum 25 MPa working load designed. It is used for 660 mm hot rolling mill screw down, the cylinder is made by the carbon steel except the piston seal which is made by the polytetrafluoroethylene.

The experimental set-up is shown in Fig. 65.1. The AE signals emitted from the cylinder were captured and amplified by five piezoelectric transducers (PAC-R15i, frequency response 100–400 kHz, 40 dB gain pre-amplifier integrated) were fixed on the cylinder by using the magnetic clips, with four sensors were mounted on the surface of cylinder with 25 mm apart from the ground, where is the closest place towards the cylinder piston. Adjacent sensors having around 90° with reference to the cylinder center if considering the cylinder surface where the sensors attached as a circle. The remaining one sensor that acts as a filtering sensor was attached on the place just nearby the fuel inlet. Signals were then recorded and pre-processed by using the SH-II AE condition monitoring system supported by the Physical Acoustic Corporation (PAC), and finally stored in a Lenovo-Thinkpad-T420 laptop computer.

Each individual AE hit was recorded with 1 MHz sampling frequency within 2 ms, a 40 dB detection threshold was employed and the peak definition time

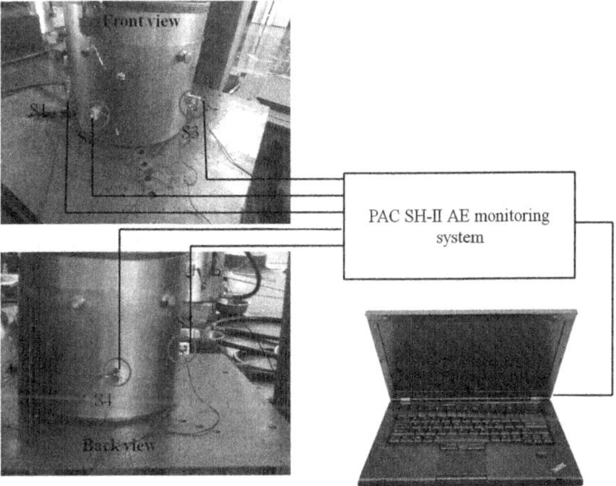

Fig. 65.1 Data acquisition system

(PDT), hit definition time (HDT) and hit lockout time (HLT) are fixed to 20, 800 and 1,000 μs, respectively, to enable convenient data acquisition. As it is an exploratory study of AE emitted from the hydraulic cylinder, the data acquisition protocol are determined according to the ASTM E650/650 M-12 and ASTM E1930/1930 M-12 standards [15, 16]. Prior to record the AE, the cylinder is placed on the standard 45 MPa hydraulic testing bed which consists of a rolling mill housing on top of the testing kit and a bearing cylinder on the bottom as well as the corresponding supplies to simulate the actual working environment of steel roller screw down. The AE events were created by pumping the hydraulic oils into the cylinder to make it reaching 5, 10, 15, 20, 25, and 35 Mpa, respectively, and holding each condition for 100 s individually. To ensure the cylinder is under precise loading, the actual pressure were monitored by the pressure scale and cross validated by the actual volumes of hydraulic oils pumped into the cylinder.

65.3 Introduction of Image Based Acoustic Emission Profile

As there are vast majorities (maximum number of AE events > 14,000 hits) of AE bursts created by the cylinder under various loading conditions, direct separation of different loading conditions using the whole waveforms is difficult. In order to reduce the dimensionalities of the dataset, and convert the cylinder AE signals from each loading condition to a uniform format thereby creating the visual effect for further inspections, the image based AE signal processing method, namely hits density approach is proposed. The hits density approach and its applications were previously investigated by the authors since 2009 [17–21], it is a multivariate statistical based methodology that re-classifying the number of AE events by using the specified AE feature combinations. The mathematical representation of this approach is given by (65.1)

$$\mathbf{N}(F, \psi) = F_{\psi}(V_1, V_2, \ldots, V_i) \tag{65.1}$$

where, **N** denotes the number of AE events constrained by the specific AE feature combination, V_i are referred to as the selected features; $F\psi_j$ denotes the number of AE events which satisfies the particular AE feature combinations [17].

The visual prototype of hits density is shown in Fig. 65.2, whereby the image represents the cylinder under a particular working status can be segmented into four quadrants by using two axis radiated via the horizontal and vertical directions from the centre of image, with each quadrant represents one of particular AE feature combination. The 2D image based representation allows a maximum of 4 AE feature combinations determined by ψ_j to be displayed, each pixel within ψ_j represents the number of AE events retained by a particular feature class determined by not only the feature types, but also the value intervals. The directions of feature classes with increasing values are oriented outwards from the image centre

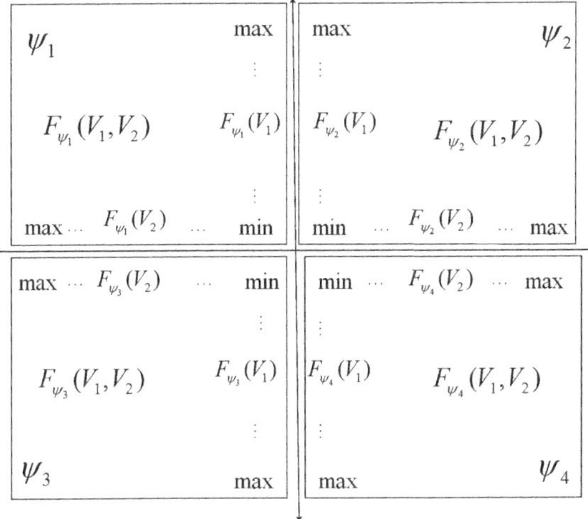

Fig. 65.2 Image based representation of AE signals

to enable visualisation of pattern symmetry. The AE hits density in each quadrant can be represented by using (65.2)

$$F_{\psi j}(V_1, V_2) = F_{\psi j}(V_1) \wedge F_{\psi j}(V_2) \tag{65.2}$$

where, $F_{\psi j}(V_1)$ and $F_{\psi j}(V_2)$ are referred to as the selected feature vector in quadrant ψ_j, respectively.

65.4 Experimental Result

This section discusses the signal processing result of AE emitted from the cylinder using the methodology discussed in Sect. 65.3. Prior to create the image based profiles, the AE feature vector consists of traditional AE descriptors [13] and statistical moments [22] of each individual AE waveform were calculated. Through the statistical distribution analysis which aids the feature selection for AE hits density creating, eight waveform features showing statistical significancy are selected, namely, amplitude, average signal level above the detection threshold (ASLOT), counts, duration, energy, average signal level over the whole AE wave (ASLOW), rise-time, and counts-to-peak (CTP), respectively. The amplitude is the peak magnitude of each AE waveform; duration and rise-time are the time interval of the AE pulses exceeding the detection threshold and between the 1st threshold crossing and the AE peak, respectively; counts and CTP are the number of AE pulses exceeding the detection threshold and between the 1st threshold crossing and the AE peak, respectively; The calculation of two ASL terms (i.e., ASLOT and ASLOW) are focused on the instantaneous varying, with ASLOT represents the average amplitude over the duration and ASLOW represents the average amplitude over the whole waveform while an AE event is being identified. The suggestions of feature selection can be found in authors' previous publications [20, 21], in order to save spaces in this paper, feature selection advices are omitted here.

65.4.1 Granularities Setting for the Selected Feature Combinations

By dividing the selected AE features into four pairs, the AE signals generated by the cylinder under each loading condition mentioned in Sect. 65.2 can be transformed to a 2D image, with the left top quadrant of the image represents number of AE events constrained using amplitude and ASLOT, the feature combinations are changed to counts and duration, energy and ASLOW, as well as rise-time and CTP, respectively when the image sub-field moving from the 1st quadrant to the 4th quadrant, respectively. The feature granularities can be determined as $k(n) \leq F(V) \leq k(n+1)$ when $n \in [1, 6]$, and $F(V) > k(n)$ when $n > 6$, where, $k(n)$ is a vector consists of the value intervals of selected features. In ψ_1, the feature interval of amplitude is set to $k_{amp}(n) = [40, 50, 60, 70, 75, 80, 85]^T$, as the amplitude values showing more separation among various loading at the tail of distribution, the value interval between the 4 and 7th terms are set to 5 dB; the feature interval of ASLOT is set to $k_{ASLOT}(n) = [0, 10, 20, 30, 40, 50, 60]^T$. In ψ_2, the feature interval for AE counts and duration are set to $k_{counts}(n) = [1, 250, 500, 750, 900, 1250, 1500]^T$, and $k_{duration}(n) = [0, 300, 600, 900, 1200, 1500, 1800]^T$, respectively. In ψ_3, the feature interval for energy is fixed to $k_{energy} = [0, 10^5, 2 \times 10^5, 3 \times 10^5, 4 \times 10^5, 5 \times 10^5, 6 \times 10^5]^T$ and $k_{ASLOW} = [0, 10, 20, 30, 35, 40, 45]^T$ with the intervals within the last three terms set to 5 dB according to the statistical significant observed from the univariate distribution [20]. In ψ_4, the feature interval for rise-time and CTP are set to $k_{rise\text{-}time} = [0, 300\ 600, 900, 1200, 1500, 1800]^T$, and $k_{CTP} = [1, 250, 500, 750, 1000, 1250, 1500]^T$, respectively. Also with the symmetric visualisation purpose, the intervals are set to the same value in each quadrant in this study, the robustness of feature granularities were discussed by [17], which proofs the value interval affects little in the hits density model.

65.4.2 Construction and Evaluation of the Cylinder AE Profiles

By constrain the AE events using (65.1) and (65.2) based on the value intervals mentioned, the AE profiles of the cylinder under six various loading conditions are shown in Fig. 65.3. With the paper reports the possibility of using AE for hydraulic cylinder state identification, the comparative study of the signal emitted from various sensors will be discussed in the future publications, AE signals from sensor S3 with the maximum amount of AE events acquired are picked up for demonstration, as the hits density created by using the signals collected by the

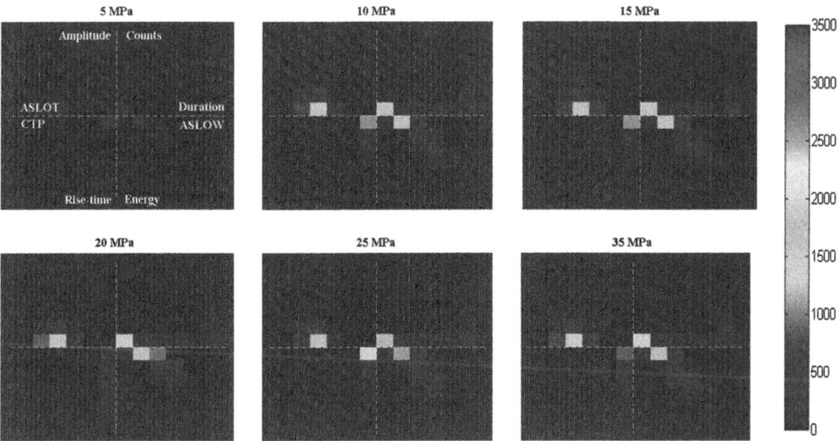

Fig. 65.3 Image based profiles for the cylinder under various loadings

other sensors apart from S1 showed similar results. In order to simplify the analysis, AE signals generated by the transition of two adjacent pressure statuses (e.g., transition of 5–10 MPa) are removed.

As shown in Fig. 65.3, it is seen that 5 MPa performed the minimum level of AE events in terms of types and amount. Signal levels and amounts are increased since the cylinder has been loaded to 10 MPa, and the main types of waveform are seen to remain unchanged when the loadings keep increasing. However, the subtle differences can still be observed through the visual inspection over a portion of image intensity. Although all conditions provide similar type of AE signals, the number of AE events with minimum duration and counts are increased from 10 to 20 MPa and decreased from 25 to 35 Mpa, with the maximum amount of AE events reach to 2,500 during the cylinder is loaded at 20 MPa. Another distinctive difference can be observed in ψ_4 of the image, the number of AE events with the minimum CTP and the shortest rise-time are seen to increase slightly from 10 to 20 MPa, and decreased when the cylinder is pumped to 25 MPa, and re-increased when the cylinder is loading at 35 MPa, with the maximum appeared in 20 MPa. This scenario can be linked to the actual specification of the cylinder, as a maximum 25 MPa working load designed, the aforementioned scenario could be understood as some underlying mechanical disfunctions of the cylinder that causes by the overloading, which may further brings failure to the cylinder during the rolling process.

65.4.3 Cylinder States Identification by Principal Component Analysis

In order to further assess the sensitivity of AE as a potential signature for state identification of the HAGC hydraulic cylinder, the further multivariate analysis is needed. According to the previous studies produced by the authors, PCA [22] could be a good option for highlighting the dissimilarities between the hits density profiles of the testing object under various conditions [17–21].

With the image constructed by amplitude and ASLOT showing more clear trajectory in the PCA space than the other feature combinations through the comparative study, Fig. 65.4 shows the 3D PCA projection of the AE image profiles constrained using amplitude and ASLOT, with 99.88 % of variances remained. The projection shows a clear trajectory of pressure changes from 5 to 35 MPa, it starts from 5 MPa at the left of the PCA space, pattern is moving towards the positive direction through the 1st PC and the negative direction through the 2nd and 3rd PC with the longest displacement when the cylinder pressure is increasing to 10 MPa, and it is moving through the positive direction over the 1st three PCA spaces when the cylinder status is increasing to 15 MPa. The pattern is keep ascending mainly through the 2nd and 3rd PC domain while the loads are increasing from 15 to 20 MPa with the 1st PC scores remained nearly unchanged. The pattern is seen to increase in the 1st PC domain but decrease through the 2nd and 3rd PC when the pressure is changing from 20 to 35 MPa. With the pressure changes, the 1st PC score only crossover the positive side once (35 MPa), with the projection formed by the 1st PC weighted around 98 % of total variance, this should show some statistical significancy. It could be related to the properties of PCA and the previous findings provided by the authors [17, 20, 21], as well as the actual mechanisms of the cylinder, with the overload condition distributed on another extreme of the PC plane.

The result demonstrates the potential of using AE for dynamic state identification and assessment of HAGC hydraulic cylinder. By increasing the majority of AE signals related to more cylinder conditions, it will have possibility to establish a reference trajectory of the cylinder statuses developing from normal working condition to various abnormal or fault conditions through the further modification of image based AE profile, a 3D cube based approach will be applied to bearing more AE feature combinations for describing the detail conditions as well as even subtle differences. The dynamic working condition of a cylinder and the possible

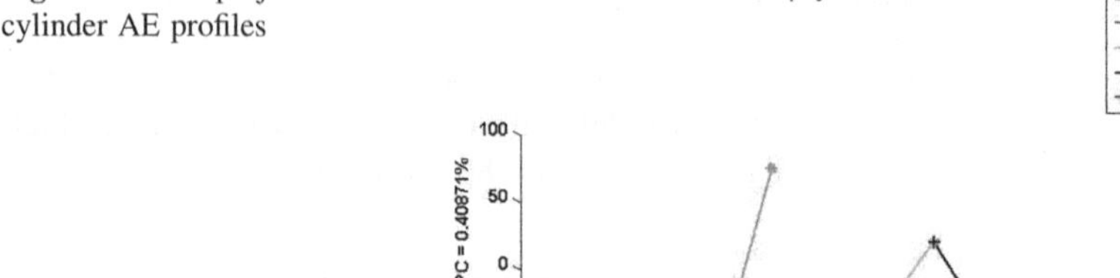

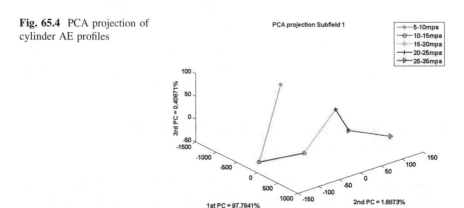

Fig. 65.4 PCA projection of cylinder AE profiles

developments could be examined based on the position of its projection on the PC domain along the trajectory and the distance of it with respect to each reference condition. Use of AE to assess the hydraulic cylinder statuses opens up a possibility to introduce not only the potential tool for dynamic monitoring of subtle state variation in the executing component of HAGC system, but also the preventive measures to the cylinder at risks.

65.5 Conclusion

The works presented the potential of using AE as a signature to identify the dynamic statuses of the HAGC cylinder. By continuously loading six pressure statuses to the cylinder to generate AE and applying the hits density based approach to the signals acquired, the image based cylinder AE profiles could be created for state identification, with the cylinder under 5 MPa generated minimum level of AE events; existing signal classes were increased while the pressure was increased from 5 to 10 MPa, the waveform types which determined by selected feature combinations remain unchanged since the cylinder is working with the pressure higher than 10 MPa, whereas the majority of AE events in two particular waveform classes determined by counts and duration, as well as CTP and rise-time were seen to increase corresponding to the increasing pressures from 10 to 20 MPa and decrease from 20 to 35 MPa, which could be related to the underlying mechanisms of the cylinder under examination. By projecting the AE profiles onto the PCA space, a clear trajectory showed consistent result could be observed with the main trend of the patterns moving from the negative axis towards the positive axis with overload condition distributed in the positive side of the space formed by the 1st PC.

The above mentioned results showed the pioneering work of using AE for state identification of HAGC hydraulic cylinder. In the future studies, additional tests will be applied to more types and conditions of cylinders, in order to acquire more AE signals for further studies; sensor attachment optimisation protocol for the HAGC hydraulic cylinder will be investigated, thereby discovering the most suited sensor attachment solution for the cylinder; unsupervised feature selection criterion will be established based on the application of intelligent searching algorithms on the AE features derived, thereby selecting the most appropriate feature combinations to build the hits density profiles relating to even more conditions of the cylinder, and clustering analysis will be applied to dig out the dominant and distinctive AE waveforms for further supports of the cylinder states identified by AE, thereby building up a complete model and system for cylinder subtle changes monitoring during the rolling process.

Acknowledgments The work is supported by state key development program for basic research of China (Grant ref: 2012CB724304), and we would also like to extend our gratitude to Prof. L-K Shark from University of Central Lancashire, UK, Dr. Benoit Mascaro from France, and Physical Acoustics Corporation Beijing office for their technique inputs.

References

1. Stepanov A, Zinchenko S, Efimov S et al (2005) Main trends in the growth of converter steelmaking at the company severstal. Metallurgist 49:380–382
2. Liu H (2012) Application of predicted extrapolation control strategy in hot strip rolling mills gauge system. Proc World Automat Cong (WAC) 1–4
3. Wang H, Rong Y, Cui J et al (2010) Study on knowledge processing of fault diagnosis for hydraulic AGC system. In: 2nd IEEE international conference on information management and engineering (ICIME), pp 1–3
4. Wang H, Rong Y, Liu S et al (2010) Identification for hydraulic AGC system of strip mill based on neural networks. Intl Conf Comput Desgn Appl (ICCDA) 22:377–380
5. Dong M, Liu C, Li G (2010) Robust fault diagnosis based on nonlinear model of hydraulic gauge control system on rolling mill. IEEE Trans Contr Syst Tech 18:510–515
6. Wang X, Liu C, Li M (2010) Sensors fault diagnosis of hydraulic automatic gauge control system based on wavelet neural network. In: IEEE international conference on electrical and control engineering (ICECE), pp 3009–3012
7. Wang X, Zhang K (2012) Sensors fault diagnosis of hydraulic automatic gauge control system based on neural network optimized by genetic algorithm. In: IEEE international conference on oxide materials for electronic engineering (OMEE), pp 114–117
8. Ye W, Xu Z, Deng K (2008) The fault and diagnosis for hydraulic system of metallurgical machinery. China High Tech Enterp 12:124–125 (in Chinese)
9. Garimella P, Yao B (2005) Model based fault detection of an electro-hydraulic cylinder. Proc Amer Contr Conf 1:484–489
10. Zhang L, Zhang C, Shi T (2010) Inner leakage fault diagnosis of hydraulic cylinder using wavelet energy. Adv Mater Res 139–141:2517–2521
11. Qian S, Chen D (1996) Joint time frequency analysis: methods and applications. Prentice-Hall, London
12. Bishop CM (1995) Neural networks for pattern recognition. Clarendon Press, Oxford
13. Chen H (2011) Discovery of acoustic emission based biomarker for quantitative assessment of knee joint ageing and degeneration. Ph.D Thesis, University of Central Lancashire, Preston
14. Tan CK, Irving P, Mba D (2007) A comparative experimental study on the diagnostic and prognostic capabilities of acoustics emission, vibration and spectrometric oil analysis for spur gears. Mech Syst Signal Proc 21:208–233
15. ASTM Standard E650/650 M (2012) Standard guide for mounting piezoelectric acoustic emission sensors. ASTM International, West Conshohocken
16. ASTM Standard E1930/E1930 M (2012) Standard practice for examination of liquid-filled atmospheric and low-pressure metal storage tanks using acoustic emission. ASTM International, West Conshohocken
17. Mascaro B, Prior J, Shark LK et al (2009) Exploratory study of a non-invasive method based on acoustic emission for assessing the dynamic integrity of knee joints. Med Eng Phy 31:1013–1122
18. Shark LK, Chen H, Goodacre J (2010) Discovering differences in acoustic emission between healthy and osteoarthritic knees using a four-phase model of sit-stand-sit movements. J Open Med Infor 4:116–125
19. Shark LK, Chen H, Goodacre J (2010) Knee acoustic emission: a clue to joint ageing and failure. Rheumatology 49:I79–I82
20. Shark LK, Chen H, Goodacre J (2011) Knee acoustic emission: a potential biomarker for quantitative assessment of joint ageing and degeneration. Med Eng Phy 33:534–545
21. Chen H, Lu Y, Wang L (2013) Analysis of dynamic acoustic emission signals using multivariate statistical technique for smaller dataset J Vibr Meas Diag 2 (in press) (in Chinese)
22. Jolliffe IT (1986) Principal component analysis. Springer, New York

Chapter 66
Small-Signal Model and Control of PV Grid-Connected Micro Inverter Based on Interleaved Parallel Flyback Converter

Qiqi Zhao, Yu Fang, Zhibin Wang and Yong Xie

Abstract In this paper, interleaved parallel flyback grid-connected micro inverter was focused on, and its grid-connected operation principle and control strategy were presented; what is more, the key techniques were also discussed in the paper. In photovoltaic (PV) grid-connected micro-inverter system, the tracking control is the core and key technology of the system, and directly affects the output power quality and system efficiency. The direct current control has been chosen to synchronize the current frequency and phase with the grid. The current loop control parameters was designed and presented in the paper, and finally the simulation and experimental results verify the feasibility of the control strategy.

Keywords Interleaved parallel flyback · Control strategy · Current source · PV grid-connected power generation

66.1 Introduction

With the development of economy, the energy crisis and environment pollution become more serious, the development and utilization of new energy are taken into consideration. Solar energy has become the most promising energy [1]. Solar

---

Q. Zhao (✉) · Y. Fang · Z. Wang · Y. Xie
Yangzhou University, Yangzhou 225127, China
e-mail: zhao_qiqi@126.com

Y. Fang
e-mail: yzfangyu@126.com

Z. Wang
e-mail: wzbin_cool@126.com

Y. Xie
e-mail: yzxiey@yahoo.com.cn

Z. Sun and Z. Deng (eds.), *Proceedings of 2013 Chinese Intelligent Automation Conference*, Lecture Notes in Electrical Engineering 255,
DOI: 10.1007/978-3-642-38460-8_66,

photovoltaic generation is developing rapidly for its no pollution and high reliability. As a technical direction of solar photovoltaic generation, PV grid-connected micro inverters are gradually attracting the attention of people for their small size, easy installation, low cost, and high efficiency of power generation [2].

The performance of PV grid-connected inverter system largely depends on the choosing of control strategy [3–7]. In this paper, direct current control mode is adopted. Direct current control mode generally includes a regulator whose output is the reference current of ac side in current control loop. Direct current control mode can control current error and current waveform via feedback control, so the speed of dynamic response is much faster, and direct current control can suppress current harmonic effectively which is caused by grid voltage distortions [2].

The Mathematical model of main power circuit is made in this paper, and the small-signal control model also deduced. As the result that the control parameters are designed to make the system presented in this paper stable, and achieve fast dynamic response. The specific control principles and the design method of current control loop will be described and instructed in the following. The simulation and experimental results show that the system has a good static performance as well as the tracing performance.

66.2 The Structure and Principle of Photovoltaic Grid-Connected Micro-Inverter System

The structure and control principle of the studied interleaved parallel flyback PV grid-connected micro inverter in the paper are shown in Fig. 66.1. It has advantages of less components and isolation between PV modules and the bus. The main

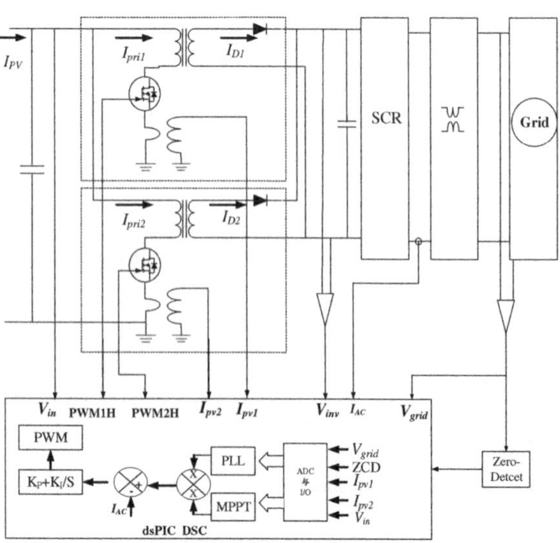

Fig. 66.1 The structure of the system

circuit is mainly composed of PV modules, a high-frequency converter, full-bridge circuit, filter circuit and a digital controller. The digital Phase Locked Loop (PLL), the Maximum Power Point Tracking (MPPT) control loop, and the current control loop are all included in the controller of the system.

The two flyback converters are operated 180° out of phase to accomplish interleaved parallel operation. The total current ripple of the PV side reduces for mutually offsetting efforts. Then, the current ripple through electrolytic capacitor is relatively small, which increases the life of the capacitor. Besides, the THD of output current ripple also reduces [3, 8].

66.3 The Control Strategy of Flyback Grid-Connected Micro Inverter

66.3.1 The Modeling of Flyback Grid-Connected Micro Inverter

The principle diagram of main circuit is as shown in Fig. 66.2

The input voltage (Vin), output voltage (Vo) and switches duty cycle (D) of flyback converter have the following relation:

$$V_O \cdot (1 - D) = N \cdot V_{in} \cdot D \tag{66.1}$$

The output current *(Io)* and the current of magnetizing inductance *(I_P)* have the following relation:

$$I_O = I_P(1 - D)/N \tag{66.2}$$

In flyback converter, the magnetizing inductance of transformer is involved in the processes of energy storage and transfer, the rectified output voltage of flyback transformer can be reflected to primary side, then average equivalent circuit models can be obtained as shown in Fig. 66.3.

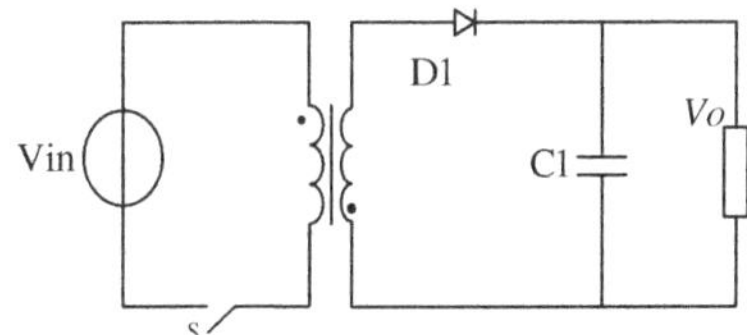

Fig. 66.2 Flyback converter circuit

Fig. 66.3 The average equivalent circuit of grid-connected flyback inverter

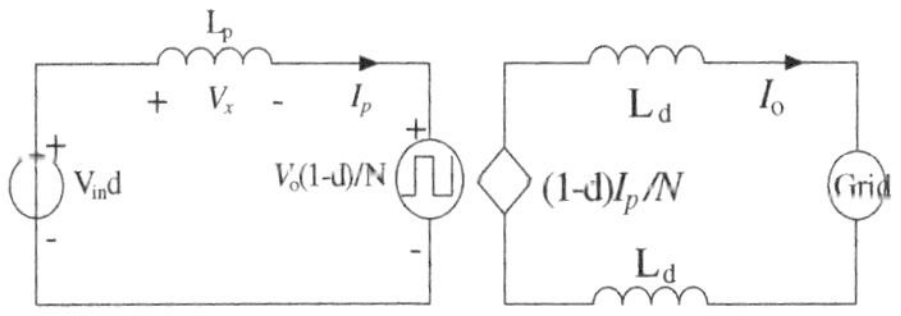

The classical PI controller has a simple structure and is a kind of linear controller, while the flyback converter at high-frequency is nonlinear [2], it is not easy to generate sine current waveform. Then small signal average model of flyback grid-connected micro inverter will be established around steady state operation point in the paper, on the basis of it the linear transfer function from the input to the output are calculated. The transfer function of the PI regulator is shown as Eq. (66.3).

$$G_c(s) = \frac{V_x(s)}{E_{rr}(s)} = \left(K_p + \frac{K_i}{s}\right) \tag{66.3}$$

where Err(s) is the amplified difference between given value and feedback value of grid-connected current. In order to improve the dynamic response speed further, the decoupling control will be realized. Namely, the PI regulator output corresponds to the voltage of the inductance, which means that a voltage Vx linearly with the current error is applied on the inductance, as is shown in Eq. (66.4):

$$V_x = G_c \cdot \left(I_P^* - I_P\right) \tag{66.4}$$

According to Fig. 66.3, Eq. (66.5) can be obtained:

$$V_x = V_{in} \cdot d - (1 - d) \cdot V_O/N \tag{66.5}$$

where d is the needed duty cycle. Then d can be calculated as is shown in Eq. (66.6).

$$d = G_c \cdot \frac{\left(I_O^* - I_O\right) \cdot N}{V_{in}} + \frac{V_O}{N \cdot V_{in} + V_O} \tag{66.6}$$

The former is the contribution of PI regulator output V_X (component of duty cycle $\hat{d}$), the second is the contribution of decoupling (component of duty cycle D). D is the duty cycle of steady state operation point, which can be understood as the needed duty cycle to meet the input and output voltage relationship when operating without load. D is also called the feed forward component or preset duty cycle, which is helpful to reduce the steady error of the grid-connected current [9].

66.3.2 The Small Signal Modeling of Flyback Grid-Connected Micro Inverter

In this paper, perturbation method is used to establish small signal AC model which reflects a dynamic behavior when the circuit operating around a steady state working point and is the basis of feedback control. According to the equivalent circuit in Fig. 66.3, Eq. (66.7) can be obtained.

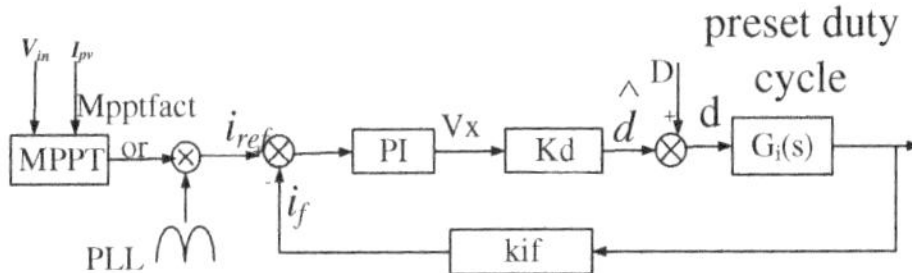

Fig. 66.4 The design of control loop

$$\begin{cases} L_p \frac{d(I_P+\hat{i}_p)}{dt} = V_{in} \cdot d - \frac{V_O(1-d)}{N} = V_{in} \cdot (D+\hat{d}) - \frac{V_O(1-D-\hat{d})}{N} \\ I_o + \hat{i}_o = (I_p + \hat{i}_p)\left(1 - D - \hat{d}\right)/N \end{cases} \tag{66.7}$$

Finally the transfer function $G_i(s)$ between the duty cycle and output current of the inverter can be obtained as is shown in Eq. (66.8). Where N = 12, Vin = 25, Vo = 311, L_P = 28 uH

$$G_i(s) = \frac{\hat{i}_o(s)}{\hat{d}(s)} = \frac{\hat{i}_o(s)}{\hat{i}_P(s)} \cdot \frac{\hat{i}_p(s)}{\hat{d}(s)} = \frac{(V_{in} + V_O/N) \cdot (1-D)}{sNL_P} \tag{66.8}$$

The structure of the control system is shown in Fig. 66.4. A digital chip is used to realize control algorithm in the system described in the paper. In the program, the input and output of the regulator has been normalized, so $K_d = 1$. In control algorithm, the perturbation and observation method of MPPT is performed. The output of the MPPT controller is the given output current amplitude of inverter. Phase lock control is realized by the capture of digital chip, namely the digital phase locked loop. The given output current of inverter can be obtained by the output product of MPPT control loop and PLL control loop.

66.3.3 The Design of the Control Parameters

Because the preset duty cycle is relative stable, in the adjustment process of current regulator, the dynamic effect of preset duty cycle can be neglected. Considering sampling time delay in the digital control system, the block diagram of current closed-loop control is as shown in Fig. 66.5.

Where, D is taken the maximum 0.62, Ts = 1/172 k, $K_{if} = 0.5$. The transfer function of current open-loop system without regulator is shown in Eq. (66.9):

$$G_{i\_open}(s) = \frac{268.727}{3.38 \times 10^{-11}s^3 + 1.454 \times 10^{-5}s^2 + s} \tag{66.9}$$

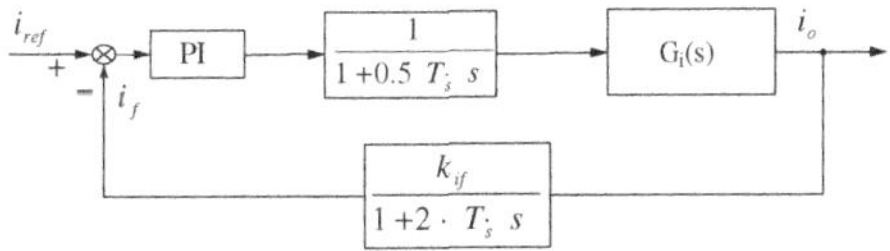

Fig. 66.5 The block diagram of current closed-loop control

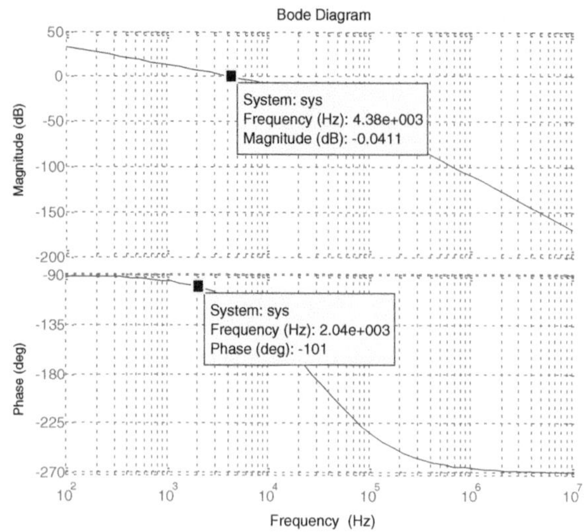

Fig. 66.6 Bode plot of current open-loop without regulator

The bode plot is shown in Fig. 66.6.

The cutoff frequency is 4.38 kHz. The transfer function of current open-loop system with PI regulator is shown in Eq. (66.10).

$$G_{i\_open\_reg}(s) = (k_{ip} + \frac{k_{ii}}{s})(\frac{1}{1 + 0.5 \cdot T_s \cdot s})(\frac{k_{if}}{1 + 2 \cdot T_s \cdot s})G_i(s) \tag{66.10}$$

In this paper, the phase margin of corrected open -loop transfer function is set as 45° at the point of 2 kHz (the maximum phase margin at this point can be reached 79°). Then, Eq. (66.11) can be obtained.

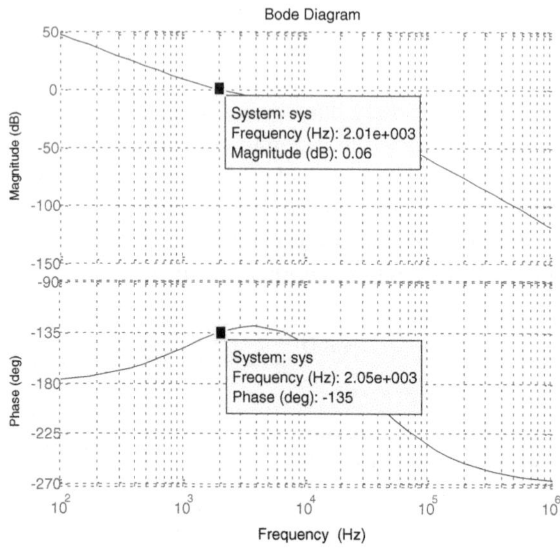

Fig. 66.7 Bode plot of current open-loop with regulator

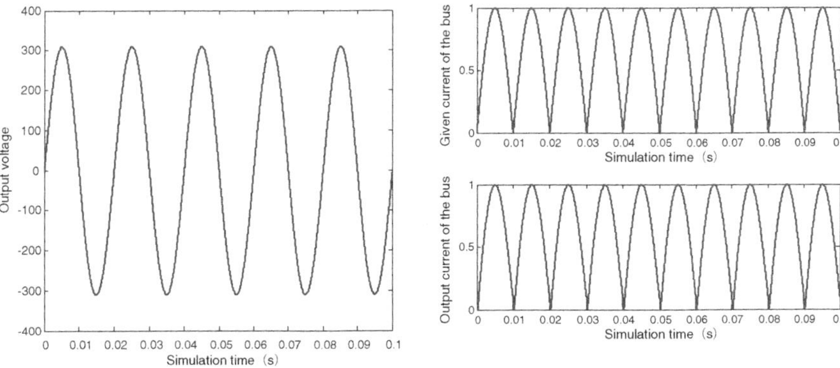

Fig. 66.8 Simulation waveforms, **a** Output voltage of inverter. **b** Given current, output current

$$\begin{cases} |G_{i\_open\_reg}(s)| = 1 \\ \gamma = 180^0 + \angle G_{i\_open\_reg}(s) = 45^0 \end{cases} \Rightarrow \begin{cases} k_{ip} = 0.363 \\ k_{ii} = 3149 \end{cases} \tag{66.11}$$

The bode plot are shown in Fig. 66.7.

66.4 Simulation and Experimental Validation

In order to verify the feasibility of the control strategy, a prototype is designed and manufactured, simulation and experiment are done. Simulation waveforms are shown in Fig. 66.8 and experimental waveforms are shown in Fig. 66.9.

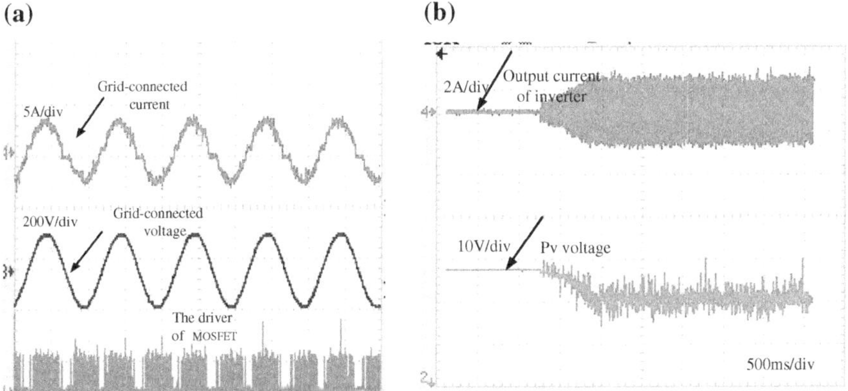

Fig. 66.9 Experimental waveforms, **a** Grid-connected waveforms. **b** Slow start waveforms of MPPT

Simulation is realized with the help of Matlab. The simulation and experimental waveforms show that grid-current can track grid voltage well, and the system has good static and dynamic performance, which verifies the feasibility of the control strategy.

66.5 Conclusion

Grid technology is the key of photovoltaic grid-connected inverter system. In distributed power, the method of current-control mode is usually adopted to connect to grid in grid-connected inverter. In this paper, the transfer function of the flyback micro inverter is concluded and then the design method of PI regulator coefficients in interleaved parallel flyback grid-connected micro-inverter is given. In the system, the digital control mode is used, which makes the control more agile, and simple to debug.

References

1. Thang TV, Thao NM, Kim DH, Joung Hu, Park JH (2012) Analysis and design of a single—phase flyback micro inverter on CCM operation. Power Electron Motion Control Conf 2:1229–1234
2. Kamil M (2010) Grid-connected solar micro-inverter reference design using a dsPIC ® Digital Signal Controller. http://ww1.microchip.com/downloads/en/AppNotes/01338D.pdf
3. Ping X, Yongqiang L, Shichao MA, Junyan H (2010) Control strategy of grid connected photovoltaic inverter. Low Voltage Apparatus 14:24–27 (in Chinese)
4. Zhou X, Song D, Ma Y, Guo R, Cheng D (2010) Control strategy of photovoltaic grid connected inverter. East China Electr Power 38(1):80–83 (in Chinese)
5. Chen Z, Liu X, Pan J (2011) Current control strategy for grid connected inverter based on IDA-PBC. Trans China Electrotechnical Soc 26(8):99–105 (in Chinese)
6. Deng X, Chao Q, Yuan T, Tuerxun Y, Zhang X (2011) Current control strategies for three-phase grid-connected inverters. Power Syst Clean Energy 27(5):31–35 (in Chinese)
7. Yunya W,Jiarong K, Shaojun X (2012) Control strategy design for inverters in low voltage microgrids. Autom Electr Power Syst 36(6):39–44 (in Chinese)
8. Mao C, Chen W, Lu Z (2011) Small signal analysis of interleaved flyback converter with magnetic integration. Adv Technol Electr Eng Energy 30(4):26–29 (in Chinese)
9. Zhang Y, Yishu Z, Yu M, Xiong J (2009) Research on current control of photovoltaic grid-connected inverter. Power Electron 43(5):29–31 (in Chinese)

Chapter 67
An Improved Total Variation Regularization Method for Electrical Resistance Tomography

Xizi Song, Yanbin Xu and Feng Dong

Abstract Electrical resistance tomography (ERT) attempts to image the conductivity distribution of an object by measuring the voltage on its boundary. The reconstruction of ERT is an ill-posed inverse problem, and little noise in the measured data can cause large errors in the estimated conductivity. In this paper, an improved TV regularization method for ERT is introduced with iterations updated by the projected Gauss-Newton steps. It replaces the conventional TV regularization penalty term $\int_{\Omega} |\nabla g| d\Omega$ by $\int_{\Omega} |\nabla g|^p d\Omega$, in which p is selected as 1.5. The choice of such a penalty compromises both the conventional smoothness and discontinuities of the imaged conductivity. The improved approach can reconstruct images with sharp edges as well as reducing the suffering of the staircase effect. Simulation and experimental results of the improved method, TV regularization and Tikhonov regularization are compared, which show that this improved TV regularization can endure a relatively high level of noise in the measured voltages.

Keywords Electrical resistance tomography (ERT) · Image reconstruction · Tikhonov regularization · Total variation regularization

67.1 Introduction

Electrical resistance tomography (ERT) is a visualization and measurement technique, investigated extensively during the past decades, which has been widely used in the industrial process imaging [1, 2]. It has the advantages of portability,

X. Song · Y. Xu (✉) · F. Dong
Tianjin Key Laboratory of Process Measurement and Control, School of Electrical Engineering and Automation, Tianjin University, 300072 Tianjin, China
e-mail: xuyanbin@tju.edu.cn

Z. Sun and Z. Deng (eds.), *Proceedings of 2013 Chinese Intelligent Automation Conference*, Lecture Notes in Electrical Engineering 255,
DOI: 10.1007/978-3-642-38460-8_67,

safety, low-cost, non-invasiveness and rapid response, which makes it a promising imaging technique.

However, the ill-posedness of the inverse problem is the principal difficulty in image reconstruction for ERT [3]. And regularization methods are good ways to address this problem, among which the Tikhonov regularization and the TV regularization have been generally accepted as important algorithms. As Tikhonov regularization is a kind of l_2 regularization, it makes the image edge over-smoothed. Although the TV regularization penalizes the transition amplitude, which makes it suitable for sharp transitions and better than Tikhonov regularization, it often suffers the staircase effect and the loss of fine details. Some improvements of TV regularization have been made, including using Primal Dual Method for solving the nondifferentiability of regularization function [4, 5], adaptive mesh refinement based on TV in the process of inverse problem [6] and the adaptive TV regularization that adaptively modifies the regularization factor according to the uncontinuity of the image boundaries [7]. However they do not consider the function of TV itself, which characterize the measurement of smoothness and leads to staircase effect.

In this paper, an novel TV function replaces the conventional TV regularization penalty term $\int_{\Omega} |\nabla g| d\Omega$ by $\int_{\Omega} |\nabla g|^p d\Omega$, in which p is selected as 1.5. By means of relaxing the smoothness constraints, this novel algorithm can preserve the edges of reconstruction images, which are more informative about the sizes and shapes of the inclusions than the smoothness approach (such as Tikhonov). And it also can overcome the shortage of TV regularization suffering from the staircase effect. Besides its reconstruction images are less sensitive to noise.

67.2 Theoretical Background

In the problem of ERT, let Ω be the support of the object, and $\partial\Omega$ its surface (boundary). We assume Ω contains material with electrical conductivity $\sigma(x)$ satisfying $\sigma(x) > 0$. The electrical potential $u(x)$ inside Ω satisfies

$$-\nabla \cdot (\sigma \nabla u) = 0 \text{ in } \Omega \tag{67.1}$$

$$\sigma \frac{\partial u}{\partial n} = j \text{ on } \partial\Omega \tag{67.2}$$

where j is the applied current density $\partial\Omega$ on such that the following conservation of charge relation holds:

$$\int_{\partial\Omega} j ds = 0 \tag{67.3}$$

The inverse problem is to estimate the conductivity σ from experiments.

In the problem of ERT, the relationship between the measured edge voltage (u) and the conductivity distribution (σ) of measured object is non-linear:

$$u = F(\sigma) \tag{67.4}$$

Based on the perturbation theory, functional (67.4) is

$$\Delta u = \frac{dF}{d\sigma}(\Delta\sigma) + o\left((\Delta\sigma)^2\right) \tag{67.5}$$

If $\Delta\sigma$ is small enough, then $o\left((\Delta\sigma)^2\right)$ can be ignored. With the aid of physical modeling and FEM discretization, functional (67.4) becomes

$$\Delta u = J\Delta\sigma \tag{67.6}$$

where, J is the Jacobian matrix.

To simplify the reconstruction equation, Eq. (67.4) is normalized as

$$b = Sg \tag{67.7}$$

where, S is the Jacobian matrix, b the measured voltage, g the conductivity of the object measured [3].

67.2.1 Tikhonov Regularization

Regularization methods have been developed for the solution of ill-posed inverse problems. Tikhonov regularization is one of the most popular regularization tools for solving ill-posed problems and has been applied to electrical tomography by Dobson and Santosa [8]. The general form of the Tikhonov regularization for conductivity distribution is given by:

$$\min\left\{ \|Sg - b\|_2^2 + \lambda^2 \|g\|_2^2 \right\} \tag{67.8}$$

Here, the regularization factor λ is a positive that controls the weighting between the corresponding residual norm and the solution norm of the criterion function. Tikhonov regularization obtains stability in the reconstruction process, but it always imposes excessive smoothness to the edge of the reconstruction images.

67.2.2 Total Variation Regularization

The TV function was firstly employed by Rudin, Osher and Fatami for regularizing the restoration of noisy images [9]. TV regularization allows a much broader class of functions to be the solution of the inverse problem, including functions with discontinuities [10]. The total variation function, formulated as an unconstrained problem, is expressed as:

$$\min\left\{\|Sg-b\|_2^2+uTV(g)\right\}=\min\|Sg-b\|_2^2+u\int_{\Omega}|\nabla g|d\Omega \tag{67.9}$$

Here, Ω is the interested region. u is chosen to be a small positive constant as the regularization factor. The regularization factor u determines the balance between the solution norm and the corresponding residual norm. Suppose that the conductivity is described by piecewise constant elements, the discretized version of $TV(g)$ can be expressed as:

$$TV(g)=\sum_i\|L_ig\| \tag{67.10}$$

where, L is a sparse matrix [6].

Usually, a smooth approximation is used to solve the nondifferentiability of $|\nabla g|$ in the TV regularization function. Then function (67.5) becomes

$$J(g)=\frac{1}{2}\|Sg-b\|^2+uTV_\beta(g)=\frac{1}{2}\|Sg-b\|^2+u\sum_i\sqrt{\|L_ig\|^2+\beta} \tag{67.11}$$

where $\beta>0$ is a predetermined small constant.

Although TV regularization can overcome the over-smoothed shortages from Tikhonov regularization and preserve discontinuities in reconstruction images, it often suffers the staircase effect and the loss of fine details.

67.3 The Improved Total Variation Regularization

An improved TV regularization is introduced to overcome the foregoing shortages. This improved TV function replaces the conventional TV regularization penalty term $\int_\Omega|\nabla g|d\Omega$ by $\int_\Omega|\nabla g|^p d\Omega$, in which p is selected as 1.5. By means of relaxing the smoothness constraints, the algorithm describes rapid variations in the object, as well as reducing the suffering of the staircase effect.

This improved TV function, formulated as an unconstrained problem, is expressed as:

$$\min\left\{\|Sg-b\|_2^2+u\int_{\Omega}|\nabla g|^{\left(\frac{3}{2}\right)}d\Omega\right\} \tag{67.12}$$

Similarly to TV regularization, the object function is

$$J(g) = \frac{1}{2}\|Sg - b\|^2 + u\int_{\Omega} |\nabla g|^{\left(\frac{3}{2}\right)} d\Omega = \frac{1}{2}\|Sg - b\|^2 + u\sum_{i} \sqrt{\|L_i g\|^2 + \beta}^{\left(\frac{3}{2}\right)} \tag{67.13}$$

The gradient and Hessian of J can be calculated as follows:

$$J'(g) = S^T(Sg - b) + uL_\beta(g)g \tag{67.14}$$

$$H(g) = S^T S + uL_\beta(g) \tag{67.15}$$

where, S^T is the transpose of S and

$$L_\beta(g) = \frac{3}{2}L^T\left(diag\left(\sqrt[\frac{1}{2}]{\|L_i g\|^2 + \beta}\right)\right)^{-1} L \tag{67.16}$$

In this paper, the Newton-Raphson iteration method is also used to solve the object function, as same as TV regularization.

67.4 Results and Discussion

67.4.1 Simulation Results

Simulations were carried out to evaluate the performance of this improved TV regularization, TV regularization and Tikhonov regularization. The forward problem is solved using a finite element method. The measured voltages were simulated using the complete electrode model and adjacent current patterns. A mesh of adaptive first-order triangular elements is used for the forward calculations. The reconstruction images present conductivity distribution for inverse problem using another mesh with 812 square elements. In the reconstruction images, the conductivity of background and the body in the object are set as 0.01 and 0.03 s/m. The regularization factor of the improved TV and TV are set as 1e-3, and the L-Curve is used to decide the regularization factor of Tikhonov regularization.

The simulation results of those three algorithms are shown in Fig. 67.1, in the first column of which are the five test distributions. And the noise level was 0.5 %, which is corresponding to the typical noise level in practical systems. It can be seen from Fig. 67.1 that the improved TV regularization method can give better quality of reconstruction than the TV regularization method and Tikhonov regularization.

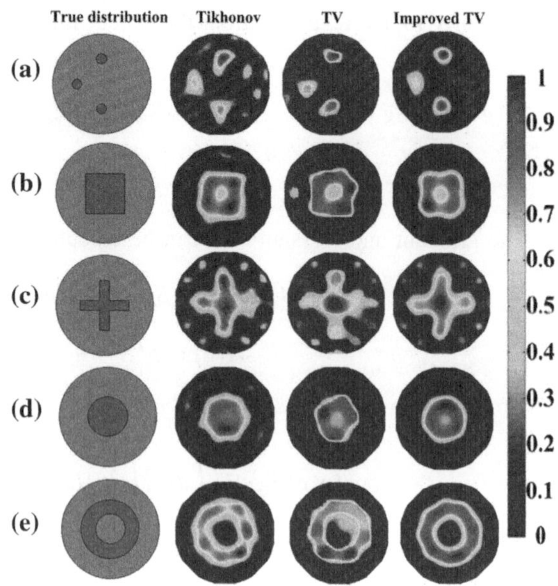

Fig. 67.1 Results from the reconstruction with three methods with noise level of 0.5 %

The relative image errors (ε_{image}) and the correlation coefficient (ε_{corr}) of the reconstruction images compared to the true conductivity distribution are compared, which are shown in Table 67.1.

$$\varepsilon_{image} = \frac{\|\sigma - \sigma^*\|_2}{\|\sigma^*\|_2} \tag{67.17}$$

$$\varepsilon_{corr} = \frac{\sum_{i=1}^{N} (\sigma_i - \bar{\sigma})(\sigma_i^* - \bar{\sigma}^*)}{\sqrt{\sum_{i=1}^{N} (\sigma_i - \bar{\sigma})^2 \sum_{i=1}^{N} (\sigma_i^* - \bar{\sigma}^*)^2}} \tag{67.18}$$

Table 67.1 Comparison of relative image errors and correlation coefficient for the reconstruction displayed in Fig. 67.1

	Relative error			Correlation coefficient		
Model	Tikhonov	TV	Improved TV	Tikhonov	TV	Improved TV
(a)	0.1889	0.1760	*0.1510*	0.5356	0.6360	*0.7444*
(b)	0.2279	0.2363	*0.1860*	0.8633	0.8355	*0.9059*
(c)	0.4356	0.4388	*0.4258*	0.6231	0.6259	*0.6501*
(d)	0.1850	0.1653	*0.1458*	0.9168	0.9082	*0.9322*
(e)	0.2609	0.2467	*0.2156*	0.8846	0.8885	*0.9129*

The italics presents the relative image errors and correlation coefficient of the Improved TV, which is the new algorithm proposed in this paper. It can be seen that the results from the new algorithm is better than the other algorithms.

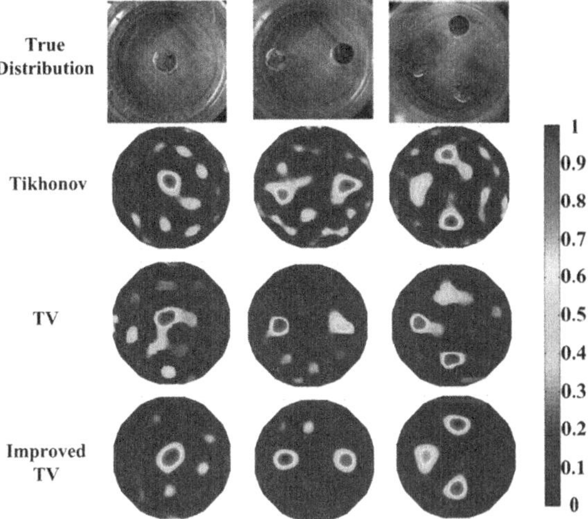

Fig. 67.2 Reconstruction images of plastic rods by using three methods

Where σ is the calculated conductivity, σ^* is the actual one, $\bar{\sigma}$ is the average of the calculated one, $\bar{\sigma}^*$ is the average of the actual one. The simulation results also demonstrate that the improved TV regulation method performs better.

67.4.2 Experimental Results

In practical measurements, the measured voltages would be different from the numerical ones in simulation due to the existence of the systematic and random errors together. An ERT system with 16 electrodes was used. Experiments have been conducted for plastic rods positioned in saline water. The true conductivity distribution and reconstructed images are shown in Fig. 67.2.

67.5 Conclusion

In this paper, an improved TV regularization algorithm is presented for ERT image reconstruction. Both simulation and experimental results have shown that the proposed method not only can overcome the corresponding shortages of Tikhonov regularization and TV regularization, but also its reconstruction images are less sensitive to noise than them. Based on this work, this improved TV regularization can perform better in the image reconstruction, it's feasible to make a adaptive TV regularization, which can adjust the regularization function according to the smoothness of the edge automatically.

Acknowledgments The authors would like to thank the National Natural Science Foundation of China for supporting this work (61227006) and Natural Science Foundation of Tianjin Municipal Science and Technology Commission (11JCYBJC06900).

References

1. Tapp HS, Peyton AJ, Kemsley EK, Wilson RH (2003) Chemical engineering applications of electrical process tomography. Sens Actuators B Chem 92(1–2):17–24
2. York T (2001) Status of electrical tomography in industrial applications. J Electron Imaging 10(3):608–619
3. Yang WQ, Peng LH (2003) Image reconstruction algorithms for electrical capacitance tomography. Meas Sci Technol 14:R1–R13
4. Borsic A (2002) Regularisation method for imaging from electrical measurement. PhD Thesis Oxford Brookes University
5. Borsic A, Graham BM, Adler A, Lionheart WR (2009) In vivo impedance imaging with total variation regularization. IEEE Trans Med Imaging 29(1):44–54
6. Wang HX, Tang L, Cao Z (2007) An image reconstruction algorithm based on total variation with adaptive mesh refinement for ECT. Flow Meas Instrum 5–6(18):262–267
7. Fan ZY, Gao RX (2012) An adaptive total variation regularization method for electrical capacitance tomography. IEEE international instrumentation and measurement technology conference (12MTC), Graz, 13–16 May 2010, pp 2230–2235
8. Dobson DC, Santosa F (1994) An image-enhancement techniques for electrical impedance tomography. Inverse Prob 10:317–334
9. Rudin L, Osher S, Fatemi E (1992) Nonlinear total variation based noise removal algorithms. Physica D 60:259–268
10. Strong D, Chan T (2003) Edge-preserving and scale-dependent properties of total variation regularization. Inverse Prob 19:165–187

Chapter 68
Modeling and Prediction of Pressure Loss in Dilute Pneumatic Conveying System with 90° Bend

Chao Wang, Yakun Zhao and Hongbing Ding

Abstract Dilute gas–solid flow exits widely in industrial applications and nature process. It is of great value to estimate pressure loss during conveying. Because of the complex mechanism in gas–solid flow, accurate estimation of pressure loss can hardly be made. Based on the experimental condition in Tianjin University, a Computational Fluid Dynamics (CFD) model has been built, considering different states of motion during collisions between wall and particles with different properties. CFD model has been validated with experiment data and an automatic modeling system has been built. This paper has engaged in studying pressure loss in 90° bend with different curative radius.

Keywords Dilute gas–solid flow · Pressure loss · Computational fluid dynamics · 90° bend

68.1 Introduction

Dilute gas–solid flow exits not only in industrial applications, but also occurs in nature process. Accurate and quick measurement of the parameters in this process is of great importance [1–3]. Pipelines in industrial field contain straight lines and bends. Bends provide more flexibility on pipeline arrangement, however, which leads to issues like greater pressure loss, erosion of the supplies and so on [4].

Accurate estimation of the pressure loss in a pipe is important for the design of the conveying system and the related security, reliability and economic efficiency.

C. Wang (✉) · Y. Zhao · H. Ding
School of Electrical Engineering and Automation, Tianjin University,
No. 92 Weijin Road, Nankai District, Tianjin 300072, China
e-mail: wangchao@tju.edu.cn

Z. Sun and Z. Deng (eds.), *Proceedings of 2013 Chinese Intelligent Automation Conference*, Lecture Notes in Electrical Engineering 255,
DOI: 10.1007/978-3-642-38460-8_68,

It is of great significance to evaluate the pressure loss in a 90° elbow, which is the most common component in pipeline system. Many scholars have been engaged in the studying of the pressure loss in a pipeline; however, most of them aim at single phase flow [5, 6]. As flow characteristics in the process of two-phase flow are rather complex, it is rather difficult to consider flows under different conditions (laminar flow, turbulent flow or transitional flow) with various solids properties by theoretical calculation. On the other hand, experimental research contains vast manual efforts, material requirements, tedious work and long study period. What's more, it is rather inconvenient to alter the experimental equipment. As a result, there is a great limit in the launch of such experimental researches. With high-speed computers, better solutions can be achieved in an efficient, safe and clean way by using CFD method.

Based on the experimental rig in Tianjin University, suitable CFD model for pneumatic conveying of solids are constructed. This model could be used to predict pressure loss of pipeline with different bend and offer valuable information to practical process.

68.2 Experimental Conditions

Figure 68.1 shows a simplified block diagram of the gas–solid two-phase flow rig in Tianjin University. Solids discharged from the hopper are pneumatically conveyed through the experimental section and finally delivered to the receiving tank. The mass flow rate of solids is controlled by screw feeder. Time averaged mass flow rate is measured by the load cell constructed under the hopper. Quantity of air flowing into the system is controlled by computer based regulating valves and could be measured by a group of standard flow meters (vortex flow meter and turbine flow meter) in order to obtain high accuracy and wide measurement range. Besides, there are pressure and temperature sensors at the experimental section, allowing the volumetric flow rate of air through this section to be evaluated.

The air flow rate of working condition in the experimental section varies from 160 to 240 m^3/h and the solids to gas mass ratio changes from 0.4 to 1.4, the flow condition of which can be categorized to dilute phase flow. The solid particles used in this experiment are quartz powders, the density of which is 2.65 g/cm^3. The particle size is around 100 μm. By using this two-phase flow rig, the electrostatic signals of different air flow rates and different solids to gas mass ratios are obtained from the experiments.

Experimental section is located on the horizontal pipeline, which contains two 90° bends. Two pressure transmitters (EJA530A) are installed. The measurement range of EJA530A is 10–200 kPa with precision of ±0.2 %. Figure 68.2 shows the research part of the pipeline. The cross section of the pipeline is circular, with inner diameter d1 = 0.05 m, outer diameter d2 = 0.06 m. Research line is

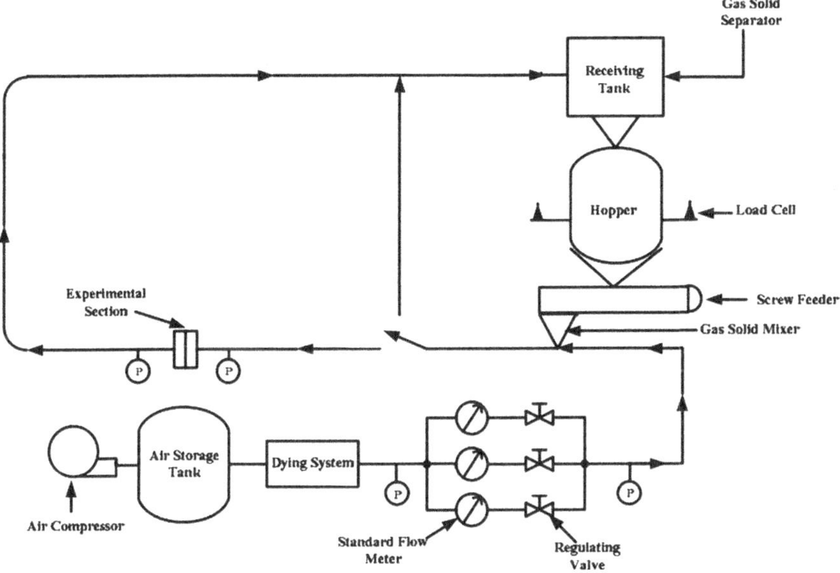

Fig. 68.1 Gas-solid two-phase flow rig

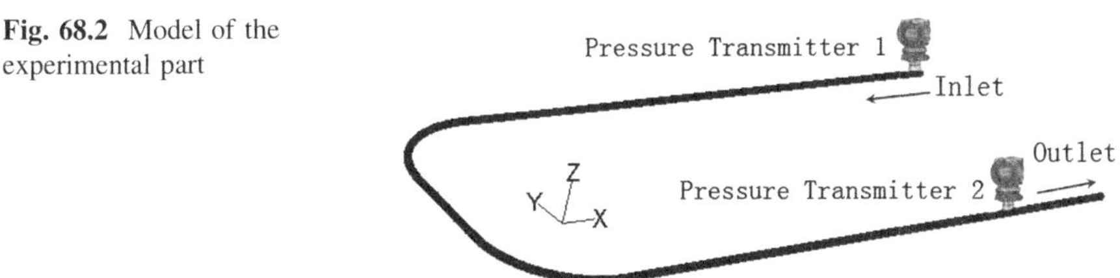

Fig. 68.2 Model of the experimental part

composed of two bends and three straight lines. Lengths of the straight line are 4, 0.75 and 3.5 m respectively and radius of the bend R is 0.7 m.

68.3 Modeling of CFD

CFD model is built by software FLUENT. The solid–gas ratio is less than 2 in this paper, belonging to dilute gas–solid flow. Velocity of air through this section is 20–36 m/s. Flow can be defined as turbulent flow by calculating the Reynolds number. RNG k-ε turbulent model is chosen as the air flow model [7].

Solid phase in the solid–gas two phases flow research is quartz powder with average diameter d = 100 μm and density $\rho = 2650$ kg/m^3. In discrete solver, continuous phase is the gas phase, which is solved in Eula framework; discrete phase is the solid phase, which is solved in Lagrange framework [7].

Collision process between solid particles and wall can hardly be ignored for the average diameter of the particles is 100 μm. Particle-wall collision model has been offered by FLUENT, considering momentum loss in this process. However, the collision behavior is calculated with a constant number of momentum losses, ignoring the complex mechanism in the process of collision, which could lead to errors in the simulation.

In this essay, a practical wall-particle collision model has been introduced into the numeric simulation, leading to better results. Specific collision process will be described below.

According to Hertzian theory [8], different motion states of particles are taken into account in the particle-wall collision model in this paper. Particle motion can be described with its normal momentum (F_n) and tangential momentum (F_t), relationship of which can be defined by Coulomb friction coefficient (μ) $F_t = \mu F_n$.

Under condition of elastic collision, equations of motion for a particle with radius of R can be described in three directions as

Normal direction

$$m \cdot \ddot{n} + K_n \cdot n = 0 \tag{68.1}$$

Tangential direction

$$m \cdot \ddot{t} + K_t \cdot (t + R\theta) = 0 \tag{68.2}$$

Peripheral direction

$$I \cdot \ddot{\theta} + RK_t \cdot (t + R\theta) = 0 \tag{68.3}$$

Displacement of the junction point is defined as: $s = t + R\theta$. According to Coulomb law of friction, when tangential force exceeds limit of friction ($K_t|s| > \mu K_n|n|$), elastic force would be replaced by frictional force. Motion of the particle would change from elastic collision to sliding friction. Equation of junction point converts to

$$m \cdot \ddot{s} + r\mu K_n \cdot s = 0 \tag{68.4}$$

where r is the zero dimension of radius for particle, which is 7/2 for sphere.

To express the switching time, some parameters should be introduced. κ is the ratio of tangential rigidity and contact rigidity for material with Poisson's ratio of v.

$$\kappa = \frac{2(1-v)}{2-v} \tag{68.5}$$

Table 68.1 Description of three regions of collision process

Range	Description
$\Psi_i \leq 1$	To start with, motion is controlled by elastic force until tangential force exceeds Coulomb friction limit. Then motion is controlled by frictional force
$1 < \Psi_i < 4\chi - 1$	To start with, governing equation changes from frictional force to elastic force and at the end of collision, it goes back to frictional force
$\Psi_i \geq 4\chi - 1$	The whole collision is controlled by frictional force

Two scalars: radius of gyration (χ) and incidence angle (Ψ_i) are defined as

$$\chi = 7\kappa/4 \tag{68.6}$$

$$\psi_i = \frac{\kappa}{\mu}\tan(\theta_i) \tag{68.7}$$

According to the relationship between Ψ_i and χ, the collision process could be divided into three regions, described in Table 68.1.

Using C language, the process above has been introduced into the CFD model by User Defined Function (UDF).

68.4 Calculation and Validation of CFD Model

Inlet air velocity can be calculated according to data (pressure at inlet, temperature and standard air flow) obtained in experiment. Set the inlet boundary condition as velocity-inlet and initial it by air velocity calculated before. Set the outlet boundary condition as pressure-outlet and initial it by 0. Since actual pressure distribution of a section is not uniform, outlet is set 0.5 m downstream of the pressure transmitter 2. Simulation work is done with gravity directed at − Z. It has been validated that problems exist when setting residue as convergence factor [9], so velocity at pressure transmitter 2 is chosen as the criterion of simulation.

CFD simulation is done under conditions of air flow Q = 160, 180 and 200 m^3/h, solid–gas ratio 0.4, 0.6, 0.8, 1.0, 1.2, 1.4. Pressure loss of experiment data and simulation result is compared in Fig. 68.3.

Results of experiment and simulation fit well in Fig. 68.3, which validates the CFD model preliminarily. This model would be used to predict pressure loss of different pipeline and offer valuable information for actual use.

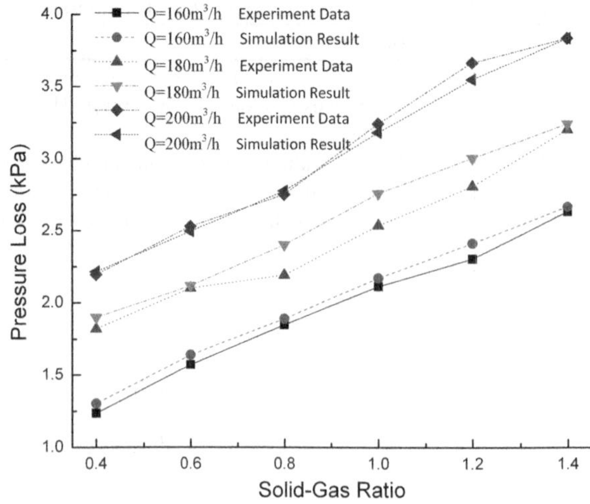

Fig. 68.3 Comparison between experiment and simulation

68.5 Prediction of Pressure Loss in Pipeline with Bend of Different Ratio of Curative

In order to predict pressure loss of different pipelines, numerous pipe models should be built. These pipes have similar structure, except for several parameters. As the constructive process is rather intricate, a program is developed with Visual C++, based on the CFD model above. This program achieves flexible modeling process by linking journal file created by Gambit and case file created by FLUENT. Figure 68.4 is the interface of this program.

Simulation work is done under the condition of Q = 200 m^3/h, solid–gas ratio of 0.4, 0.6, 0.8, 1.0, 1.2, 1.4 by using of such program. Figure 68.5 shows the results of pressure loss in pipeline with bend radius of curvature 1, 2, 5.

Traditional theory states that particles move along the outer wall of the bend. As a result, the bigger the curvature radius is, the more similar the bend is to and the less pressure loss it would be. However, experiment conducted by Mason confirmed that it fits not well with actual circumstance [10]. Simulation work accords well with Mason's theory.

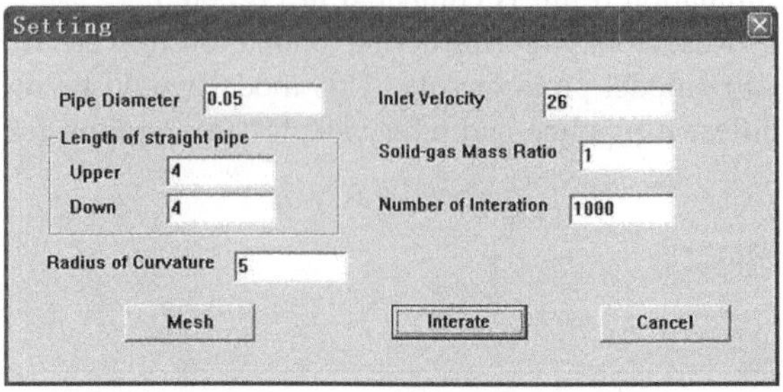

Setting

Pipe Diameter 0.05
Length of straight pipe
Upper 4
Down 4
Radius of Curvature 5
Inlet Velocity 26
Solid-gas Mass Ratio 1
Number of Interation 1000
Mesh Interate Cancel

Fig. 68.4 Interface of modeling program

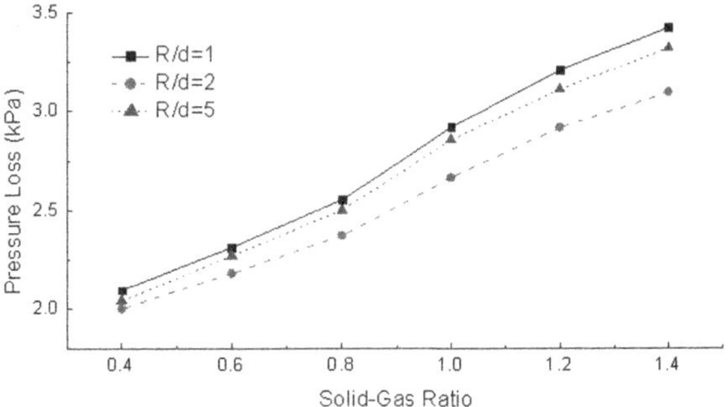

Fig. 68.5 Pressure loss of bend with different radius of curvature

68.6 Conclusion

Dilute gas–solid CFD model is built. Validation is made by comparing the simulation results with experiment data obtained at Tianjin University. An interface program is designed to simplify the modeling procedure, based on which investigations on pressure loss in different kind of bends have been done. The CFD model can be used to analyze flow condition in industrial field, guide the method of measurement and optimize the pipeline design.

Acknowledgments This work is supported by National Natural Science Foundation of China under Grant 61072101 and Program for New Century Excellent Talents in University under Grant NCET-10-0621.

References

1. Yan Y et al (1999) Velocity measurement of pneumatically conveyed solids using electrodynamic sensors. Meas Sci Technol 6(5):515
2. Benavides A, van Wachem B (2008) Numerical simulation and validation of dilute turbulent gas–particle flow with inelastic collisions and turbulence modulation. Powder Technol 182(2):294–306
3. Chu K, Yu A (2008) Numerical simulation of the gas–solid flow in three-dimensional pneumatic conveying bends. Ind Eng Chem Res 47(18):7058–7071
4. Cun-gang J, Yong L, Jing-peng Y (2005) Study on pressure drop in pneumatic transfer elbow. SP & BMH Relat Eng 51(2):15–19 (in Chinese)
5. Rongxian Y, Wenqi W (1988) Pressure loss in the pneumatic transport of powder and particle through the pipe bend. J Southeast Univ 19(6):49–54 (in Chinese)
6. Henthorn KH, Park K, Curtis JS (2005) Measurement and prediction of pressure drop in pneumatic conveying: effect of particle characteristics, mass loading, and Reynolds number. Ind Eng Chem Res 44(14):5090–5098
7. Manual FU (2005) Fluent user's guide. Fluent Inc, Lebanon

8. Maw N, Barber J, Fawcett J (1976) The oblique impact of elastic spheres. Wear 38(1):101–114
9. Huber N, Sommerfeld M (1998) Modelling and numerical calculation of dilute-phase pneumatic conveying in pipe systems. Powder Technol 99(1):90–101
10. Mills D, Mason J (1982) Learning to live with erosion of bends. In: First international conference on the inter and external protection of pipes. BHRA Fluid Engineering, Durham

Chapter 69
The Coverage Problem in Heterogeneous Wireless Sensor Network: An Improved Algorithm of Virtual Forces

Jie Chen and Xianjin Wang

Abstract At the present time, many algorithms use "virtual forces" method to solve the sensor nodes coverage problem. However, all of them studied the homogeneous sensor networks. This paper proposed an extended virtual force algorithm for the coverage of heterogeneous sensor networks. It adopted the probability sensor model, combined the static coverage strategy with the dynamic coverage strategy according to the difference of sensor radius, analyzed the threshold distance based on the difference of sensor radius. Additionally, it proposed a novel sensor "relay move" technique during the movement of sensors. Finally, we take a simulation. Simulation results show that using the algorithm is able to deploy the heterogeneous sensors in the aimed region suitably to solve the coverage problem of sensor nodes and it can balance the dissipative energy efficiently which benefits for prolonging the lifetime of network.

Keywords Wireless sensor networks · Virtual force · Probabilistic sensor detection model · Heterogeneous · Coverage

69.1 Introduction

Wireless sensor networks' coverage is an important issue in wireless sensor network research. In order to achieve the target task of monitoring and gathering information, we must ensure that the deployment of sensor nodes can cover the aimed area effectively [1]. However, due to wireless sensor networks often work in the unknown, no people watch or even the battlefield environment, it's difficult to deploy the configuration manually. And frequently used means of random throw is difficult to

J. Chen (✉) · X. Wang
Science and Technology on Information Systems Engineering Laboratory of CETC 28th, No 69th sub-post mail, No 1406th post mail,Bai Xia Zone, Nanjing, JiangSu, China
e-mail: chenjie600006@126.com

Z. Sun and Z. Deng (eds.), *Proceedings of 2013 Chinese Intelligent Automation Conference*, Lecture Notes in Electrical Engineering 255,
DOI: 10.1007/978-3-642-38460-8_69,

deploy in the right place, so the deployment of sensor nodes has become one of the key issues in the field of wireless sensor network research [2–4]. This paper studies heterogeneous sensor network nodes' deployment method based on probabilistic sensor detection model, proposes an extended virtual force algorithm (EX-VFA) which can achieve heterogeneous sensor network nodes' effective deployment.

69.2 Related Work

Researchers who first proposed the virtual force algorithm (VFA) [5] are Zou and Chakrabarty. The algorithm's theoretical ideas come from the disc packing theory [6] and the virtual potential field in mobile robot field [7]. In the sensor field, each sensor behaves as a "source of force" for all other sensors. This force can be either positive or negative. If two sensors are placed too close to each other, the "closeness" being measured by a pre-determined threshold, they exert negative forces on each other. This ensures that the sensors are not overly clustered, leading to poor coverage in the other parts of the sensor field. On the other hand, if a pair of sensors is too far apart from each, they exert positive forces on each other. The algorithm uses the vector sum of attractive force and repulsive force exerting on the node to determine the moving distance and direction of mobile nodes, hence it ensures that a globally uniform sensor placement is achieved [5]. The Algorithm defines a threshold distance to control node density, avoiding the deployment of sensors too dense or sparse. Reference [8] derived the threshold distance between nodes. It considered the factors of boundary effect of covering, the inefficacious movement of node and the threshold mobile node distance. Reference [9] focused on the energy consumption problems of wireless sensor network during the nodes move after the implementation of the VFA algorithm, valuing the effect of the energy consumption on residual energy balance during the sensor nodes move. Reference [10] proposed a new sensor network deployment optimization strategy based on virtual force algorithm taking the background of target tracking.

However, these algorithms are based on homogeneous sensor network's research [5, 8–10]. Reference [8] theoretically analyzed the homogeneous state of sensor networks in the ideal maximum efficiency of the node distribution coverage, and deduced that the threshold distance between nodes is $\sqrt{3}$ times of the node's radius. However, heterogeneous sensor network's derivation is more complicated and sensor networks must consider the heterogeneity of sensor nodes in the actual sensor deployment. In addition, these algorithms are mainly based on Binary sensor model [5, 8, 9], that is when monitoring objects within the sensor's sensing radius, the probability be detected is 1, and 0 otherwise. Reference [10] has pointed out the probability sensor model but it's not really applied to the virtual force algorithm. There is no direct study on threshold distance which impact on the performance of algorithm, and no in-depth research the relationship between probabilistic sensor detection model and the virtual force algorithm.

69.3 Probabilistic Sensor Detection Model

Consider a square sensor field and assume that there are k sensors deployed in the random deploy stage. Each sensor has a detection range r. Assume sensor s_i is deployed at point (x_i, y_i). For any point p at (x, y), we denote the Euclidean distance between s_i and P as d_{ip}, the coverage probability of a point p detected by s_i as $C(s_i, p)$. In the binary sensor model, if $d_{iP} < r$, the $C(s_i, p)$ is 1, otherwise 0.

The binary sensor model assumes that sensor reading have no associated uncertainty. In reality, sensor detections are imprecise, hence the $C(s_i, p)$ needs to be expressed in probabilistic terms.

Probabilistic sensor detection model describes the perception of the node and the probability of target node is perceived more accurately. However, the traditional probability sensor model only considers the impact of distance on detection probability. In this work, we provided an improved probabilistic sensor detection model. We add normally distribution disturbance in the traditional probabilistic sensor detection model. We also take into account the reality factors of electromagnetic interference, the background noise and so on. In addition, with the progress of detection work, the residual energy of each sensor node must be some attenuation over time, so sensor's detected probability of target point p will decrease. Therefore, we considered the effect of the node residual energy on the perceived probability. The improved probabilistic sensor model makes the detection closer to reality. Improved probabilistic sensor detection model is [see Eq. (69.1)]:

$$C(S_i, P) = \begin{cases} 0, & r + r_e \leq d_{iP} \\ \frac{E_{ir}}{E_i} e^{(-\lambda\alpha^\beta + \alpha\sigma)}, & r - r_e \leq d_{iP} \leq r + r_e \\ 1, & r - r_e \geq d_{iP} \end{cases} \tag{69.1}$$

where r_e ($r_e < r$) is a measure of the uncertain in sensor detection, E_i is the initial energy of node s_i, E_{ir} is the residual energy, $\alpha = d_{iP} - (r - r_e)$, and λ and β are parameters that measure detection probability when a target is at distance greater than r_e but within a distance from the sensor. σ is a normally distributed random numbers, expresses the reality effect of disturbance on the detection probability. Based on improved probabilistic detection sensor mode, we give the following definition:

Definition 69.1 Detection Probability. Let p be a point in the target area, and let s_{ov} is a set of sensors in the network, and $k_{ov} = | s_{ov} |$, that is k_{ov} describes the sensor number of s_{ov}. If point p is in the k_{ov} sensors' overlap detection region, then we define the detection probability $C(s_{ov}, p)$ of point p by s_{ov} sensors is [see Eq. (69.2)]:

$$C(S_{ov}, P) = 1 - \prod_{S_i \in S_{ov}} (1 - C(S_i, P)) \tag{69.2}$$

Definition 69.2 Network Coverage Probability. Let C_{th} be the threshold of region's required detection probability. We define the coverage probability of entire sensor network deployment area is the ratio of the region area which target detection probability is greater than C_{th} and the entire deployment area [see Eq. (69.3)].

As a flat area is continuous geometric plane, it's impossible to calculate the coverage probability of all points in the plane, we use an approximate calculation method for solving the entire network coverage probability. We make the sensor network area is divided into equally spaced grid. If the division of the grid spacing is small enough, the network coverage probability can be approximately defined as: the ratio of the number of grid points which detection probability is greater than C_{th} and the entire area's grid point number.

$$C(S,A) = \frac{\sum_{i=1}^{k} \{P_i | C(S,P_i)_{\text{th}}\}}{K} \tag{69.3}$$

And, $C(S, A)$ is the network coverage probability, S represents the set of all sensor nodes in the network, A represents the deployment stage of entire sensor network, k represents the number of grid points in the divided target sensor network area, $C(S, P_i)$ represents the detection probability of mesh point P_i by S.

69.4 Extended Virtual Forces Algorithm for Heterogeneous Wireless Sensor Network Nodes Deployment

There may be various types of sensor nodes in the practical application of sensor networks. We need to consider many heterogeneous features in locating these sensor nodes, such as sensor energy, sensor detection model, sensor detection radius, communication ability, computing ability, storage ability and so on. This work will focus on the difference of heterogeneous sensor nodes' sensor detection radius.

69.4.1 Description of Algorithm Problem

This paper considers deploying the heterogeneous sensors in a square field which exist hot spots and obstacles. And hot spots are the areas in the network which are more important than other areas and need to be covered by more sensors, obstacles are the terrain circumstance which need keeping away from when deploying sensors, like mountain and swamp. We premise each sensor node can obtain its and other nodes' location information (nodes can use GPS or some existent positioning algorithm to obtain the information), and compute the distance between other nodes, hot spots and obstacles. We assume that the sensor's sensing radius is different; the sensing region of sensor node is a circle which takes the node-self as the center and takes sensor sensing radius as the circle's radius. Supposing node communication radius is large enough to ensure sensor network connectivity between nodes after deployment. Each node has the ability to move. All nodes have equal status, no differentiation of cluster head node and ordinary node. The sensor node uses the probabilistic sensor detection model which Eq. (69.1) shows. Initially, the sensor nodes which with different sensing radius be tossed in the plane square area by a certain percentage randomly.

As sensors sensing radius is different in heterogeneous sensor network, the same node distance *d* may appear in several cases of nodes' relation in heterogeneous sensor networks, as shown in Fig. 69.1.

It can be seen from Fig. 69.1 that nodes in the same distance in heterogeneous sensor network suffered vary widely virtual force which may be attraction, repulsion or no virtual force. It can be concluded that: (1) the virtual force between the nodes in heterogeneous sensor networks is not only determined by the distance between nodes but also the sensing radius; (2) the threshold distance between nodes in heterogeneous network is determined by "diversity degree" of the node sensing radius. The diversity degree of node sensing detection radius is defined as follows:

Definition 69.3 Diversity Degree. Let r_i and r_j be sensing radius of nodes s_i and s_j respectively, the Diversity degree of diff(i, j) as follows:

$$\mathrm{diff}(i,j) = |r_i - r_j| / \sqrt{r_i^2 + r_j^2} \tag{69.4}$$

Secondly, if r_i and r_j are two different nodes with different sensing radius respectively, the relationship of location between them is more complex than homogeneous nodes shown in Fig. 69.2. It increases the difficulty to determine the threshold distance.

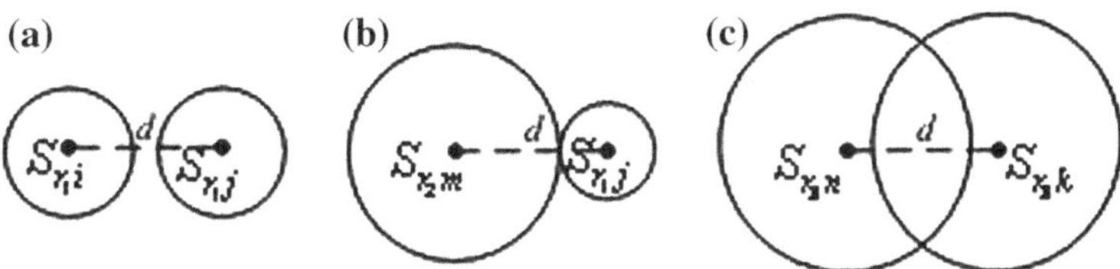

Fig. 69.1 Relation between the nodes with different radius in same distance **a** off **b** tangency **c** intersect

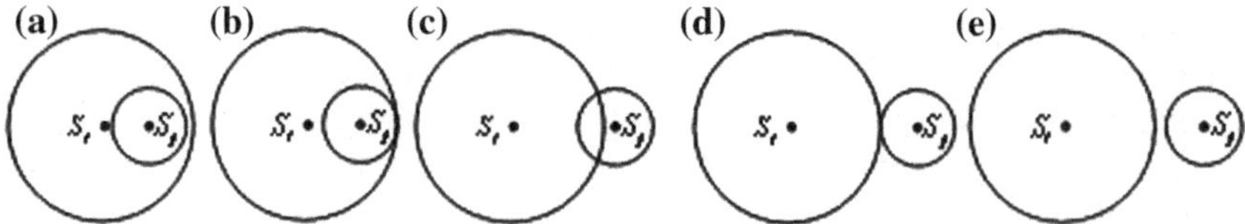

Fig. 69.2 Position relation between nodes with different sensing radius **a** include **b** inner tangency **c** intersect **d** outside tangency **e** off

69.4.2 Calculation of Virtual Force

We take into account the sensing capabilities and the hardware cost diversity of the high/low energy nodes in heterogeneous sensor network. We adopt static deployment combines with the random dynamic deployment. The hot spot area takes static deployment using high-energy nodes to increase the coverage probability. There is no virtual force exert on high-energy static nodes. The other ordinary area takes random dynamic deployment using low-energy nodes. The calculating of virtual force between low-energy nodes according to the equation below: Eqs. (69.5–69.7), which can calculate the virtual force and virtual force's direction. The method of calculating the virtual force of nodes in heterogeneous sensor networks is as follows.

1. Virtual force between sensors.

Based on the analysis of Sect. 69.4.1. We now describe the virtual force and virtual force calculation in the EX-VFA algorithm. Let total force on sensor s_i be denoted by F_i. Note that F_i is a vector whose orientation is determined by the vector sum of all virtual forces on s_i. Let the force exerted on s_i by another sensor s_j be denoted by F_{ij}. We focused on the different sensing radius between nodes and gave a reasonable force according to the Euclidean distance between nodes [see Eq. (69.5)].

$$F_{ij} = \begin{cases} (\omega_{att}(d_{ij} - \sum_{i=1}^{k} \omega_i d_{th,i}), \alpha_{ij}) & d_{ij} > (d_{th,i} + d_{th,j})/2 \\ 0 & d_{ij} = (d_{th,i} + d_{th,j})/2 \\ (\omega_{neg} d_{ij}^{-\beta}, \alpha_{ij} + \pi) & d_{ij} < (d_{th,i} + d_{th,j})/2 \end{cases} \tag{69.5}$$

where d_{th} is the threshold distance between s_i and s_j. Suppose that there are k different sensing radius in heterogeneous sensor network, then the threshold distance are d_{th1}, d_{th2}, …, d_{thk} respectively. When the sensing radius of two sensors are the same, let it denoted by r_i, and their threshold distance is d_{thi}. If the sensor s_i and another sensor s_j have the same sensor radius r_m, means the "Diversity Degree" (see definition 69.3) is 0. Let the value of virtual force factor ω_m is 1 of threshold distance d_{thm}, the ω_i ($i \in [1, k]$ and $i \neq m$) of other threshold distance is 0. If the sensing radius of a pair of sensors is different, denoted as r_i and r_j

respectively. The value of virtual force factor ω_i of d_{thi} ($i \in [1, k]$) depends on the "Diversity Degree" of sensing radius r_i and r_j and satisfied the equation $r_i - r_j < \sum_{i=1}^{k} \omega_i d_{th,j} < r_i + r_j$. The d_{ij} is the Euclidean distance between sensor s_i and s_j. The $\omega_{att}(\omega_{neg})$ is a measure of the attractive(repulsive) force which are constant. The α_{ij} is the vector angle of a line segment from s_i to s_j. The β is parameter that measure detection probability when a target is at distance greater than the average of d_{th1} and d_{th2} and it can modulate depends on different sensor type, the value extent is 1 to 4.

Let the attractive force of the hot spots (the preferential region) exerting on sensors be denoted as F_{iO} and the repulsive force of the obstacles exerting on the sensors be denoted as F_{iR}. We assume that the hot region is the anomalous polygon. For the obstacle is tiny related to the hot spots, we regard the obstacle as a dot. Let point O is the hot region geometry central point. The F_{iO} and F_{iR} can be expressed as Eqs. (69.6), (69.7) shows, where β shows the relation between the virtual force and sensors' distance, generally the value is 1. Other parameter's meaning see the preceding text.

$$F_{iO} = \begin{cases} 0, & r - r_e > d_{iO} \\ (\omega_{attO}(d_{iO} - r), \alpha_{iO}), & r - r_e \le d_{iO} \le r + r_e \\ (\omega_{attO} d_{iO}, \alpha_{iO}) & r + r_e < d_{iO} \end{cases} \tag{69.6}$$

$$F_{iR} = \begin{cases} 0, & r + r_e < d_{iR} \\ (\omega_{negR}(d_{iR} - r)^{-\beta}, \alpha_{iR} + \pi) & 0 \le d_{iR} \le r + r_e \end{cases} \tag{69.7}$$

2. Sum virtual force on sensor.

Finally, the total force on sensor s_i is the sum of virtual force between the sensors, the virtual force of hot spots and the obstacles on the sensor s_i. The mathematical expression is:

$$F_i = \sum_{j=1, j \neq i}^{k} F_{ij} + F_{iR} + F_{iO} \tag{69.8}$$

69.4.3 Moving Distance and the Balance of Move Energy

After the calculation of virtual force on sensor s_i is finished, the next work is moving the node. The node's moving distance is direct ratio with the sum virtual force on s_i. We denote the direction of sum virtual force as α_i and denote the moving distance in the direction of α_i as $d_{move,\ i}$. Equation (69.9) shows the calculation of $d_{move,\ i}$.

$$d_{move,i} = \begin{cases} \omega_f \times |F_i|, & if \quad |F_i| > F_{\text{th}} \\ 0, & if \quad else \end{cases} \tag{69.9}$$

where ω_f is the factor of moving distance that according to the target area's size, F_{th} is the threshold of resultant force that exerted on sensor node.

After all sensors' moving distance and direction have been computed finished, then they can be moved to the new position. For example, sensor s_i moves a distance of $d_{move,\ i}$ in the resultant force's direction α_i. In this work, we bring forward a new node move method called "Relay move". For an example, the relay move process of nodes is showed in Fig. 69.3. Node 1 aimed position is "finally" in Fig. 69.3. It doesn't move to "finally" position directly in our "Relay move" method. The moving process of node 1 is: node 1 moves to the position of node 2 (which position is near to node 1 and sensor radius is same as node 1), node 2 moves to the position of node 3,…, node 6 moves to the position of node 7, node 7 moves to "finally" position. In this way, the moving energy is divided to other sensors, so not only preserve the single sensor's energy but also balance the energy consumption of entire wireless sensor network. At the same time, the exertion of EX-VFA is achieved.

EX-VFA algorithm is distributed. It's complicacy is $O(nmk)$, n is the number of low-energy sensors, m and k express the number of rows and columns of aimed region grid respectively.

69.5 EX-VFA Algorithm's Simulation and Performance Evaluation

This section will verify and analyze the EX-VFA algorithm through simulation experiment. The simulation is on a 100 × 100 m square area which contains an irregular hot spots and an obstacle (a hexagram in Fig 69.4). Because sensor radius of heterogeneous nodes may arise complex situation, we limit three different sensor nodes radius (7, 10, 15 m), number of nodes is 50, 8 and 3 respectively, $r_e = 1$, $\lambda = 0.5$, $\beta = 0.5$. Because when threshold distance between two sensors (their sensor radius are the same) is equal to $\sqrt{3}$ times radius values, the coverage

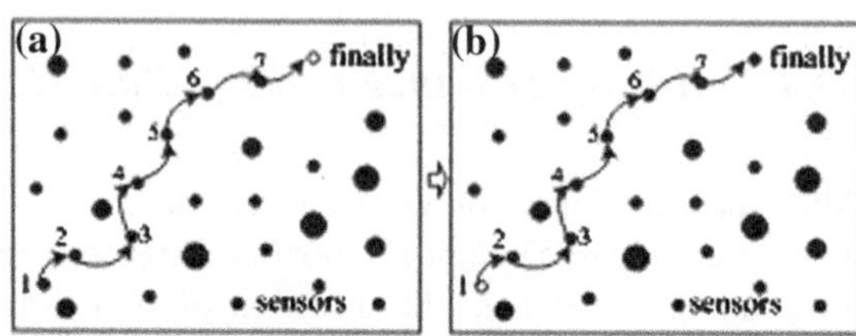

Fig. 69.3 Relay move of nodes

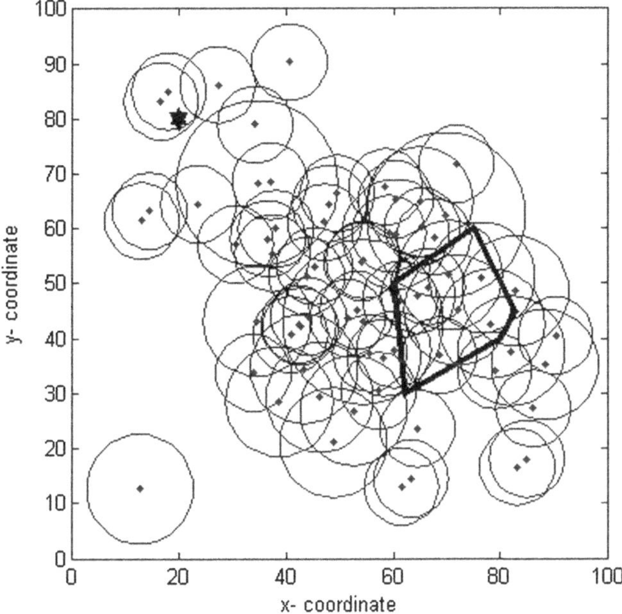

Fig. 69.4 Initial nodes deployment

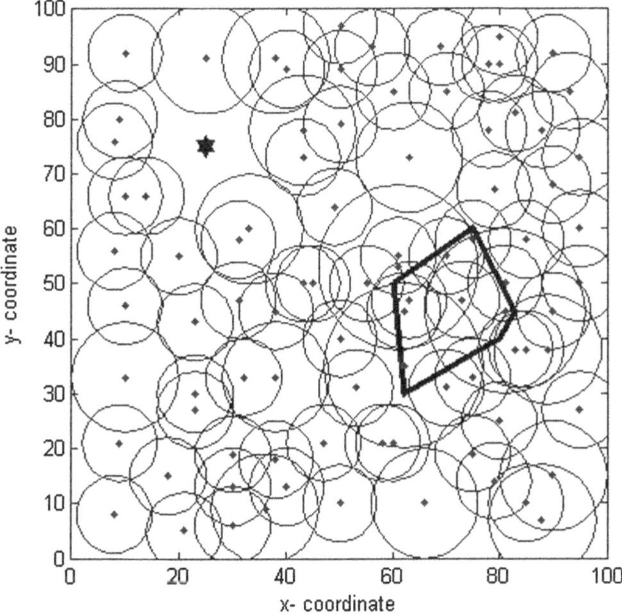

Fig. 69.5 Nodes deployment after executed the EX-VFA

efficiency is maximum [5], take $d_{thi} = \sqrt{3}\, r_i$. Before the experiment, the sensors were distributed in the target area randomly. In this experiment, C_{th} takes 0.6, the coverage probability of application takes 0.5. Figures 69.4 and 69.5 show the sensors distributed situation before and after executing EX-VFA algorithm.

Comparing Figs. 69.4 and 69.5 shows, EX-VFA algorithm not only can deploy heterogeneous sensors in the target area, but also can avoid the obstacle (the figure the black points of the top left corner), and focus on the hot spot area's coverage (the non-rule pentagon of the lower right part).

Figure 69.6 shows that how the value that threshold distance takes in heterogeneous network affects on network coverage probability. We can see that taking into account the differences in the Diversity degree of radius between the heterogeneous sensors ($\omega_1 = 0.2$, $\omega_2 = 0.5$, $\omega_3 = 0.3$) compared to base on single sensor radius (7, 10, 15 m) has an increase by 20.6, 16.6, 30.4 % on average coverage probability in 10 cycles. So considers the Diversity degree of sensor radius is valid when we value the threshold distance in heterogeneous.

Assuming the sensor's initial energy is E and energy consumption of sensor's mobility is $0.01E$ per meter, Fig. 69.7 shows how the "Relay move method" balances the energy consumption of 30 sensor nodes. Node energy consumption of the move is a ratio relative to the initial energy. It can be seen from Fig. 69.8, using the maximum distance method limits the sensor nodes' moving distance floating range between the 0.08 and $0.81E$, using the "Relay move method" limits

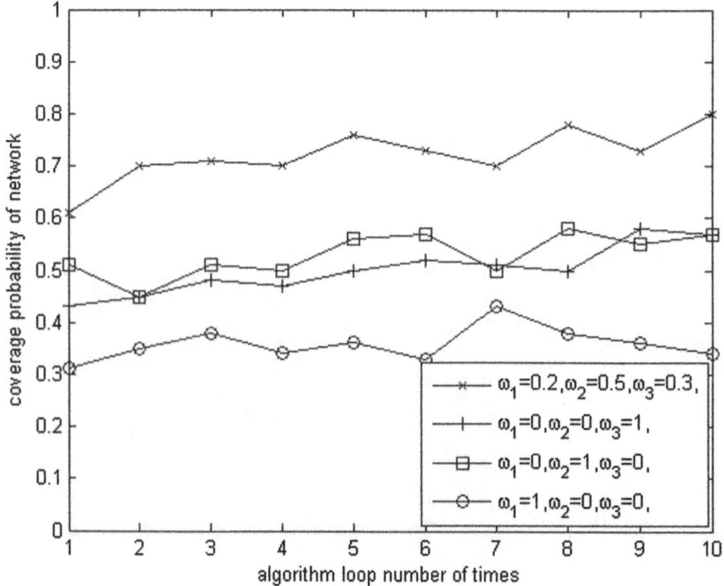

Fig. 69.6 Relation between d_{th} and coverage probability

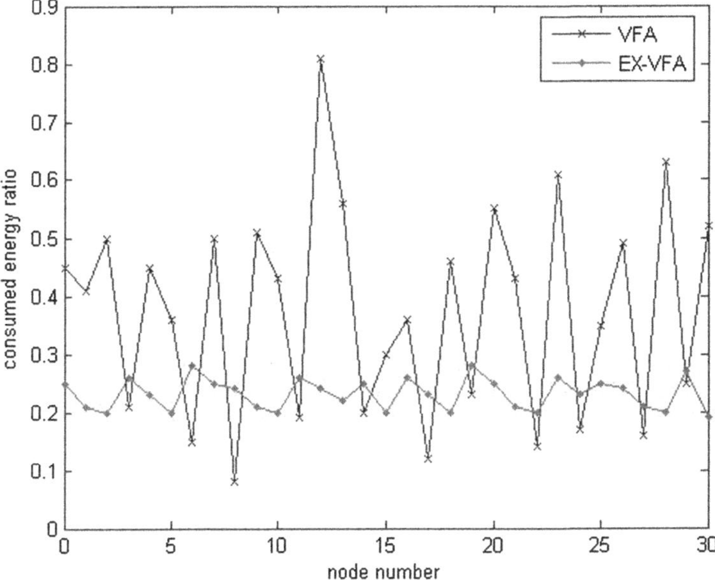

Fig. 69.7 Nodes energy cost when moving

the energy consumption of sensor nodes fluctuate around $0.25E$ that is relatively balanced which can reduce the early end up because of energy exhaustion, extending the network lifetime.

69.6 Conclusion

This paper studied deployment of heterogeneous sensor nodes based on probability sensor model, Firstly focused on analyzing the threshold distance values which affect the performance, then adopted a combination methods of static and dynamic deployment, at last improve the node moving method and then proposed the EX-VFA algorithm. The EX-VFA algorithm also can apply to homogeneous sensor networks not only to meet the needs of the application's deployment, but also to fully balance the energy of each node in the network and extend the network lifetime. Of course, the sensor radius is just one factor in heterogeneous sensor networks. There are many different factors needing to be considered in conjunction with different radius that will be in-depth study later.

References

1. Wang B (2006) A survey on coverage problems in wireless sensor networks. ECE Technical Report, ECE Dept, National University of Singapore
2. Nojeong V, Heo PK et al (2003) An intelligent deployment and clustering algorithm for a distributed mobile sensor network. In: Proceedings of the 2003 IEEE international conference on systems, Man and Cybernetics, pp 4576–4581
3. Meguerdichian SK, Potkonjak M (2001) Coverage problems in wireless ad-hoc sensor networks. In: Proceedings of the 20th IEEE INFOCOM, pp 1380–1387
4. Ke-zhong L, Liu-sheng H, Ying-yu W et al (2010) Deploying wireless sensor networks. Min Micro Syst 27(11):2003–2006 (in Chinese)
5. Zou Y, Chakrabarty K (2003) Sensor deployment and target localization based on virtual forces. In: Proceedings of IEEE INFOCOM conference, pp 1293–1303
6. Locateli M, Raber U (2002) Packing equal circles in a square: a deterministic global optimization approach. Discrete Appl Math 12(2):139–166
7. Howard A, Matari′c MJ, Sukhatme GS (2002) Mobile sensor network deployment using potential field: a distributed scalable solution to the area coverage problem. In: Proceedings of the 6th international conference on distributed autonomous robotic system, pp 299–308
8. Li-ping L (2009) Energy-efficient in wireless sensor network. Zhe jiang University (in Chinese)
9. Shi BB, Qin NN, Xu BG (2008) Deployment of energy and distance efficient sensor for wireless sensor networks. Comput Simul 5(26):146–149 (in Chinese)
10. Li SJ, Xu CF, Wu CH (2010) Optimal deployment and protection strategy in sensor network for target tracking. Acta Electronica Sinica 34(1):71–76(in Chinese)

Chapter 70
Object Localization with Wireless Binary Pyroelectric Infrared Sensors

Baihua Shen and Guoli Wang

Abstract A top-view sensing approach is proposed in this paper for infrared radiation object motion detecting and localization with binary pyroelectric infrared detectors. We develop a modulation strategy based on the radial distance and Gray code to obtain the horizontal distance between the sensor node and object. Firstly, the field of view of Fresnel lens array in a sensor node is modulated to achieve single degree of freedom spatial partition, and then the localization algorithm is proposed to coordinate multiple sensors nodes to achieve two degrees of freedom spatial partitions. The tracking experiment shows that the method proposed here has the advantages in computational efficiency and high accuracy.

Keywords Pyroelectric infrared detector · Wireless sensor networks · Motion detecting · Localization and tracking

70.1 Introduction

Motion detecting and object localization are the common key technology and the controversial research hotspot in intelligent monitoring, advanced human–machine interaction [1], motion analysis [2], activity understanding [3]. They are widely used in calamities aiding, security monitoring, medical monitoring, etc. Binary Pyroelectric Infrared (BPIR) detector, in a form of non-contact, detects infrared radiation variation in the specific environment and it is of very high sensitivity to

B. Shen · G. Wang (✉)
School of Information Science and Technology, Sun Yat-sen University,
Guangzhou, China
e-mail: isswgl@mail.sysu.edu.cn

B. Shen · G. Wang
School of Information Engineering, Guangdong University of Technology,
Guangzhou, China

Z. Sun and Z. Deng (eds.), *Proceedings of 2013 Chinese Intelligent Automation Conference*, Lecture Notes in Electrical Engineering 255,
DOI: 10.1007/978-3-642-38460-8_70,

human motion and advantages of low cost, non-invasion, strong concealment and little interference by ambient light. The localization and tracking by PIR detector have been receiving increased attention.

There are two major types of PIR sensing method adopted in pre-existing research, namely, side view [4–7] and top view [8, 9] sensing.

In side view sensing research in paper [5], to obtain the object's position, the PIWSNTT system synthesizes many object azimuth data sensed by PIR sensor nodes. But this method, relying on restricted condition of object keeping uniform linear movement, cannot manage the situation of constant turn-back movement. In Paper [6, 7], the horizontal angle modulation method is adopted to obtain object's local fine granularity localization, tracking and identification, but these side-view sensing methods are not applicable when the object is shielded by obstacles, therefore, their applicability are quite limited.

Compared with side view sensing, top view sensing can effectively overcome the problem of obstacles shielding. In paper [9] the sensed granularity is depended on the scale of field of view (FOV) crossed area, in order to obtain fine granularity sensibility, the numbers of sensors should be increased by leaps and bounds. For example, if the location accuracy reached 0.5 m, sensor nodes should be placed every 1 m. That is far from the efficiency of side view sensing, therefore, sensibility improving is restricted by coarse granularity sensing.

In order to solve the above-mentioned problems, FOV modulation strategy is introduced into top view sensing mode to make it possess fine granularity sensibility. For different sensing view angle, the horizontal angle modulation method in side view sensing mode is not suitable for top view sensing mode. Therefore, this paper proposes radial distance modulation method. Particularly, this paper applies hierarchical structure and multiplexing system to spatially modulate the FOV of Fresnel lens array, to establish FOV modulation mode and localization model in spatial juxtaposition and to attribute movement spatial localization to multivariant FOV subdivision, which is formed by many collaboration sensors of single degree of freedom (SDOF) FOV subdivision for object localization. The method, extracting information directly from the radiation source (object) movement characteristics and spatial position, apply to large-scale wireless sensor networks (WSN) space deployment, because of low cost data transport and processing.

70.2 Pyroelectric Sensor Model

Define $s(r,t)$ as the infrared radiation field function of radiation source, where r represents position, and t represents time, state space (object space) as the two dimensions plane space where infrared radiation source moving, and measurement space as the point set of the location of each sensor node. Reference structure modulates the visibility from source space to measurement space.

Define $v(r_i, r)$ or $v_i(r)$ as the visibility function between point r in source space to point r_i in measurement space. $v(r_i, r) \in [0, 1]$, the bigger the value of $v(r_i, r)$, the better the visibility, in which, “1” stands for fully visible, while “0” completely invisible. The response signal of a PIR detector at point r_i is given by

$$m(r_i, t) = \int v(r_i, r)s(r, t)dr \tag{70.1}$$

Formula (70.1) can be described as the discrete non-isomorphic model

$$\mathbf{m} = \mathbf{V}\mathbf{s} \tag{70.2}$$

where $\mathbf{m} = [m_i](i = 1, 2, \ldots, M)$. It is a M dimensional vector of measurement space; **V** represents visibility matrix, decided by the modulated strategy of radiation field, $\mathbf{V} = [v_{ij}]$ $(i = 1, 2, \ldots, M;\ j = 1, 2, \ldots, L)$; and **s** is the status vector, its dimension is decided by the cells number of state space, $\mathbf{s} = [s_j]$ $(j = 1, 2, \ldots, L)$..

In Boolean sensing model, m_i, v_{ij}, and sj are binary value. v_{ij}=1, if and only if Cj is visible to ith (number i) BPIR detector, otherwise, v_{ij}=0; sj=1, if and only if the position of the object (radiation source) is in Cj, otherwise, s_j=0. The output of ith detector is $m_i = \vee v_{ij}s_j$, where “$\vee$” represents Boolean sum. This is called Boolean sensing model in this paper, whose output is a binary value.

Boolean sensing model improves the reliability of infrared sensing, and decreases the data size of data gathering and transmission process. The lightweight non-isomorphic sensing model turns complicated sampling of object space into simple binary state sampling process, in which, what need to be obtained is only to know in which sampling cell the object appears. This direct measurement of motion states can meet the needs of most motion detecting and localization task.

70.3 Modulation Strategy Based on Radial Distance

A single BPIR detector can only output two different states and cannot be used to measure distances between the object and detector. In order to estimate the position of an infrared radiation object, the problem of the location of object state space is down to multi-degree of freedom FOV division in this paper. At first a radial distance modulated method is proposed, therefore the distance between object and detector can be measured by several BPIR detectors in a single sensor node, and then with several sensor nodes data fusion, the position of the object can be estimated.

The Fresnel lens array equipped on BPIR detector cannot only enhance perception sensitivity, but also modulate its infrared visibility by infrared absorption or mask material. By designing the visibility of Fresnel lens array properly, we not only acquire motion characteristic information directly and effectively, but also simplify the data processing greatly.

The FOV of a BPIR detector with hemispherical Fresnel lens array is a cone-shape space, and its projection on the ground plane is an area with several

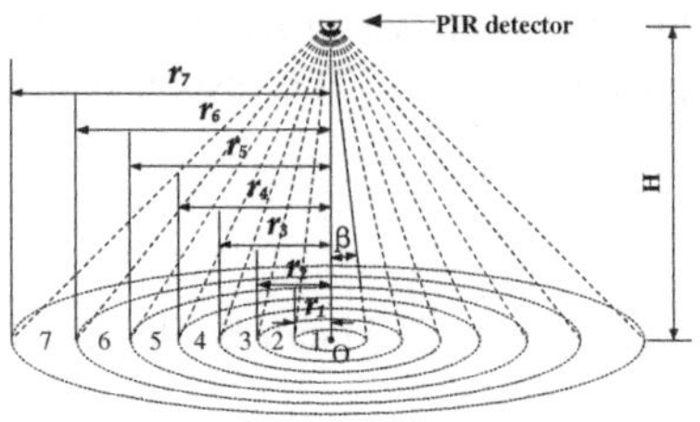

Fig. 70.1 FOV of a PIR detector with hemispherical Fresnel lens array

concentric rings. The FOV of a BPIR detector with 7-ring hemispherical Fresnel lens array is showed in Fig. 70.1.

Due to the restrictions of physical characteristics of Fresnel lens array, a n-ring hemispherical lens array divides its FOV into n concentric rings, therefore, we can achieve this functionality by using M ($M \geq [\log_2(n+1)]$) PIR detectors.

As an example, if $n = 7$, $M \geq \log_2(7+1) = 3$.. The sensing model reaches its maximum efficiency when $M = 3$.

Gray code was applied in our model to encode the output of M detectors, which is a minimize-error code where two successive values differ in only one bit, thus Gray code reduces the impact of error code on measuring result. An example of 3-bits wide Gray code is 001,011,010,110,111,101,100,000.

According to the order of Gray codes, we use masks to restrict the FOV of each Fresnel lens array. The restriction region of each Fresnel lens array is the shading pattern as showed in Fig. 70.2. The object can't be detected by BPIR detector if it is in the shading areas. The outputs of 3 detectors are listed in Table 70.1 when object is located in each different ring area, thus on the basis of the modulation strategy, the horizontal distance between the object and the detectors can be estimated.

70.4 Cooperative Localization Algorithms

According to the theory of trilateration, when the distance between the object and each sensor node has been estimated by at least three nodes, the position of the object can be estimated too. Due to the limitation of the nodes energy, memory resources and computation ability in WSN, least square algorithm is not applicable

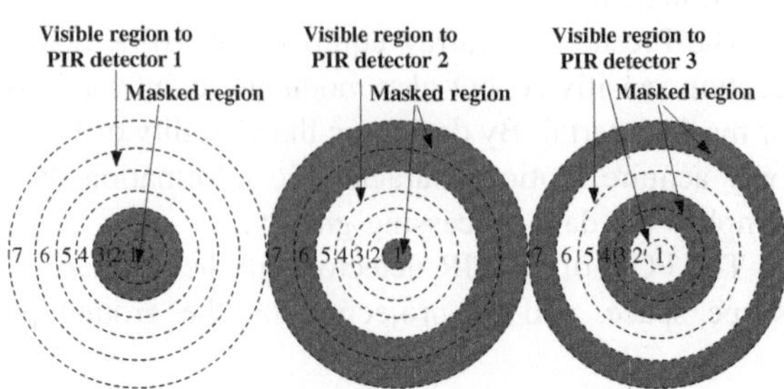

Fig. 70.2 Modulation strategy in a WSN node based on radial distance

Table 70.1 Output parameters of each detector

Rings no.	Detectors outputs			Horizontal distance to detector center	Estimated horizontal distance	Which detectors can sense object
	s1	s2	s3			
1	0	0	1	$0 \sim r_1$	$r_1/2$	s3
2	0	1	1	$r_1 \sim r_2$	$(r_1 + r_2)/2$	s2 s3
3	0	1	0	$r_2 \sim r_3$	$(r_2 + r_3)/2$	s2
4	1	1	0	$R_3 \sim r_4$	$(r_3 + r_4)/2$	s1 s2
5	1	1	1	$r_4 \sim r_5$	$(r_4 + r_5)/2$	s1 s2 s3
6	1	0	1	$r_5 \sim r_6$	$(r_5 + r_6)/2$	s1 s3
7	1	0	0	$r_6 \sim r_7$	$(r_6 + r_7)/2$	s1

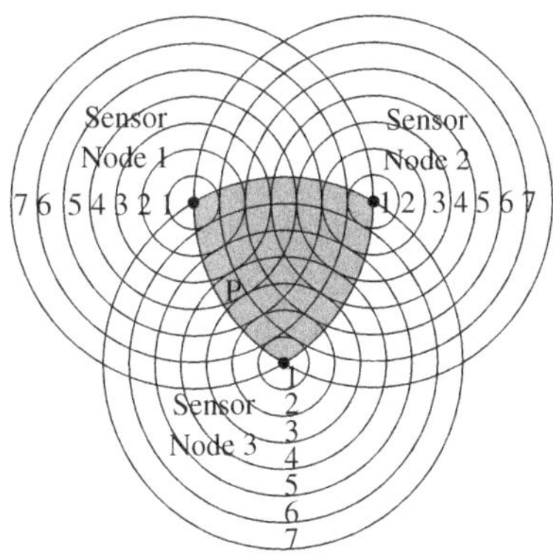

Fig. 70.3 Division of common FOV of three sensors nodes

to WSN nodes which are supplied by batteries. As mentioned before, each WSN node divides its FOV into 7 ring areas, and three WSN nodes divide their common FOV into at most 343 (7 cubed) small cells based on permutation and combination theory. Actually, most of these cells are so small or even don't exist, and there are only about 40 cells (Fig. 70.3) which have the practical value for position estimation. Thus, this greatly simplifies the calculation of the sensor node.

The size, shape and position of each cell are confirmed uniquely once the positions of three sensors nodes are confirmed. Define geometric center of each cell as its position, we can build a three-dimensional array of size [7][7][7], whose elements indicate the position of each cell, and array subscripts indicate measurement result of each WSN node. Measurement result in here means the ring number the object located. Thus, we can estimate the position of the object by looking up tables after combining measurement data of three sensors nodes.

Fig. 70.4 Hardware prototype of a WSN node

70.5 Experiments

70.5.1 Experimental Environment and Parameters

A WSN node consists of three modules: CPU and radio communication module, PIR detectors module and power module. Chip CC2530 is used in CPU and radio communication module, which is the second generation system-on-chip solution for 2.4 GHz IEEE 802.15.4. PIR detectors module include three BPIR detectors and three Fresnel lens arrays. Photos of a sensor node are listed in Fig. 70.4.

The network including 7 WSN nodes is deployed under the ceiling, 300 cm height from ground, in an indoor environment with the size of 600*520 cm. Figure 70.5 is the layout of WSN nodes. The black circle represents the locations of sensor nodes, the number in the black circle represent the serial number. Related parameters in Figs. 70.1 and 70.5 are listed in Table 70.2, and the maximum theoretical error of radial distance is $(r_7 - r_6)/2 = 31.5$ cm.

70.5.2 Tracking Experiment of a Smart Toy Car

An autonomous tracing smart toy car, at uniform speed v and act as the tracked object by WSN, is used in this experiment, and a bottle of hot water, mounted on the smart car, is chosen as the infrared radiation source.

In Fig. 70.6 red solid line represents the actual motion path of the smart car (at the speed of 100 cm/s), red solid line with circle represents the original localization result of WSN nodes with least-square algorithm, and black solid line with asterisk represents the tracking result of block Extended Kalman Filter (EKF) algorithm, and blue solid line with box represents the localization result of mean filter algorithm. Table 70.3 shows the maximum error, average error and standard

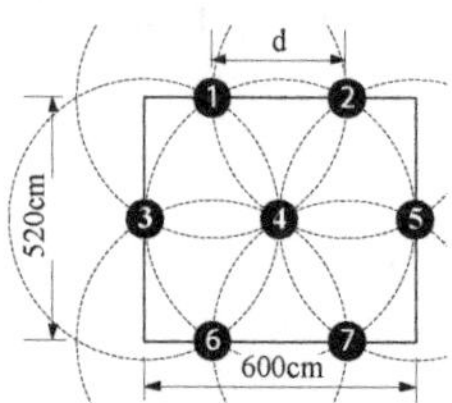

Fig. 70.5 Layout of WSN nodes

Table 70.2 Experimental parameters

Parameter	Value	Unit	Parameter	Value	unit
H	300	cm	r_3	106	cm
d	300	cm	r_4	145	cm
β	6.5	degree	r_5	191	cm
r_1	34	cm	r_6	242	cm
r_2	69	cm	r_7	305	cm

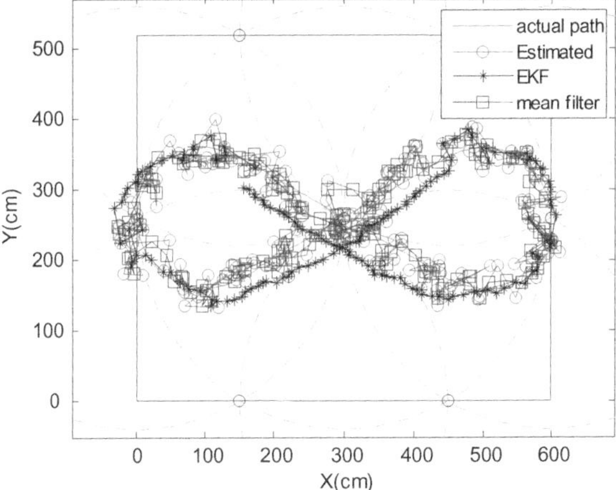

Fig. 70.6 "8"-shape route tracking

Table 70.3 Location errors under three different algorithms (*unit* cm)

	Estimated	EKF	Mean filter
Maximum	60	48	43
Mean	26	26	18
Standard deviation	10	12	9

deviation with three different algorithms. It is obvious that the effect of mean filter algorithm is the best among these three algorithms in this curve tracking experiment.

70.6 Conclusions and Prospection

Due to the disadvantages of traditional localization and tracking in these aspects of users' privacy protection, system configuration and maintenance, this paper presents a new method, based on radial distance modulation, to detect and locate moving object from top view angle. Our method has advantages of extracting information directly from the moving object's characteristics of movement and spatial position, small computation, good robustness, convenient configuration, non-contact etc. The experiments demonstrate that although the output of PIR detectors only has two forms, "0" and "1", we can locate the moving object with simple information after modulating and encoding the perception area of sensors.

The validity and feasibility of the method have been proved by the localization experiments. Since the study of the technology is still in a fledging period, these aspects of location accuracy, real-time capability and multi-object tracking, etc. need to be studied further. On the basis of preliminarily accomplished localization and tracking of moving object, what can be carried on for further studies include: human height sensing, attitude perception, typical behavior (e.g. tumbling) detection and analysis, incident recognition and so on.

Acknowledgments This paper was supported by the National Natural Science Foundation of China (Grant no. 60775055).

References

1. Turaga P, Chellappa R, Subrahmanian VS, Udrea O (2008) Machine recognition of human activities: a survey. IEEE Trans Circuits Syst Video Technol 18(11):1473–1488
2. Ji X, Liu H (2010) Advances in view-invariant human motion analysis: A review. IEEE Trans Syst Man Cybern Part C Appl Rev 40(1):13–24
3. Poppe R (2010) A survey on vision-based human action recognition. Image Vis Comput 28(6):976–990
4. Xue-Bin G, Zhi-Qiang Z, Shi-Wei Y (2007) Research of passive infrared sensor model for wireless sensor networks. J Comput Appl 27(5):1086–1089
5. Sen W, Ying-Wen C, Ming X (2008) Research of passive infrared wireless sensor network target tracking. J Transduction Technol 21(11):1929–1934
6. Hao Q, Hu F, Lu J (2010) Distributed multiple human tracking with wireless binary pyroelectric infrared (PIR) sensor networks. IEEE
7. Hao Q, Brady DJ, Guenther BD, Burchett JB, Shankar M, Feller S (2006) Human tracking with wireless distributed pyroelectric sensors. IEEE Sens J 6(6):1683–1695
8. Luo RC, Chen O, Lin PH (2012) Indoor robot/human localization using dynamic triangulation and wireless Pyroelectric Infrared sensory fusion approaches. IEEE
9. Suk L, Kyoung NH, Kyung CL (2006) A pyroelectric infrared sensor-based indoor location-aware system for the smart home. IEEE Trans Consum Electron 52(4):1311–1317

Chapter 71
Petri Net Based Research of Home Automation Communication Protocol

Guangxuan Chen, Yanhui Du, Panke Qin, Jin Du and Na Li

Abstract The popularity of home automation has been increasing greatly in recent years. However, during the designation of communication protocol (especially for asymmetric, parallel, distributed, uncertain or randomized protocol model) of home automation, many questions concerned. For instance, is the protocol correct? Is every message eventually delivered and received in the right order? Can two messages with the same alternating bit value be confused? In this paper we adopt Petri net to describe and analyze the protocol of home automation. Through the analysis of dynamic properties (such as reachability, boundedness, coverability and liveness) and invariants on the base of *P/T_* system modeling and its reachable marking graph, we verified the correctness of the communication protocol and finally optimized the protocol in practice.

Keywords Communication protocol · Formal method · Petri net · Dynamic properties

71.1 Introduction

The popularity of home automation has been increasing greatly in recent years due to its features of ease, amenity, security and energy efficiency. Home automation (or called smart home) usually refers to the adoption of computer and information technology to control home appliances and features [1, 2]. However, the data transmitted by the protocol may be the main concern of the end users. However, many data communication systems are based on unreliable connections that may distort, lose or duplicate message. When a new protocol is designed, some questions

G. Chen (✉) · Y. Du · P. Qin · J. Du · N. Li
People's Public Security Univesity of China, No. 1 Muxidi South Lane, Xicheng District, Beijing, China
e-mail: ericcgx@163.com

Z. Sun and Z. Deng (eds.), *Proceedings of 2013 Chinese Intelligent Automation Conference*, Lecture Notes in Electrical Engineering 255,
DOI: 10.1007/978-3-642-38460-8_71,

concerned. Is the protocol correct? Is every message eventually delivered, assuming that the channels are not permanently faulty? Are message received in the right order? Can two messages with the same alternating bit value be confused?

In order to guarantee the reliability, validity and ease of the communication between the control unit and home appliances, we adopt formal methods to describe and analyze the RFID communication protocol and infrared communication protocol. Formal methods are a particular kind of mathematically based techniques for the specification, development and verification of hardware and software systems in computer science, specifically hardware engineering and software engineering [3].

In this paper, we use Petri net, one of the important formal methods, which is suitable for the description and analysis of asymmetric, parallel, distributed, uncertain or randomized protocol model, to modeling, verify and optimize the communication protocol of the home automation.

The remainder of this paper is organized as follows: Sect. 71.2 provides brief introduction about the communication protocol model of the home automation we designed and its Petri net model. Sections 71.3 and 71.4 provide the analysis of the dynamic properties and invariant of the Petri net model respectively. Finally, Sect. 71.5 concludes with a summary of the modeling and analysis of the home automation communication protocol through Petri net and suggested future work.

71.2 Construction of the CCU and its Protocol

A series of technologies are involved in home automation, range mainly from computer and information technology, RFID technology, wireless infrared technology and internet of things. A home automation system integrates electrical devices in a house with each other [4]. As shown in Fig. 71.1, the home automation system we designed includes controller (or called central control unit, CCU) which is a general-purpose PC, repeaters (used for transiting the instructions), sensors (such as motion detection, daylight, and temperature), actuators (such as light switches), and TEs (abbreviation for terminal equipments, all the appliances that used for vary purpose, such as HVAC, lighting, audio-visual, shading, security). Meanwhile, several all-purpose remoter controllers are required for the residents of the home that can interact with the system for monitoring and control.

The running mechanism among CCU, TEs and repeaters can be described as follows:

CCU controls the TEs and monitors their running state; TEs receive the instructions from CCU which is operated by the users and thus provide necessary services they need; while repeaters transit the instructions to extend the communication distance.

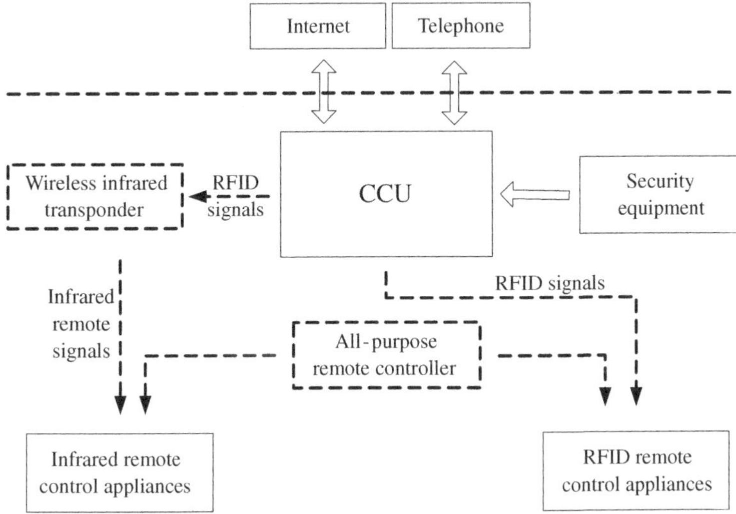

Fig. 71.1 Home automation system

In order to analyze and optimize the running mechanism of the home automation, we first depict the simplified communication protocol of CCU, as shown in Fig. 71.2.

The running state of the communication protocol can be divided into three cases:

1. CCU sends instructions to the TEs and receives a reply within $T_{\max}$ the communication between CCU and TEs succeed and relative services will be provided to the users by the TEs according to the instructions, shown in Fig. 71.2a. ($T_{\max} = M * T$, means the maximum transmitting time between

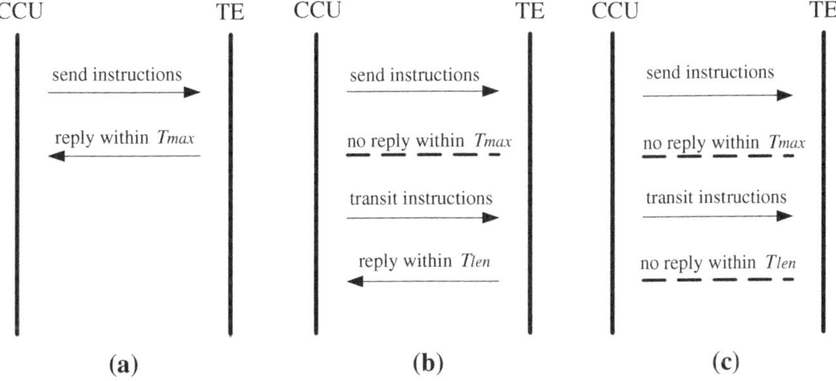

Fig. 71.2 Simplified communication protocol of CCU. **a** Succed. **b** Succed. **c** Failed.

CCU and TEs; M means the overtime index of sending the initial instruction packet and T means the time required to transmit a single instruction packet).

2. CCU sends instructions to the TEs and receives no reply from TEs within $T_{\max}$, then the instructions will be transited by the repeaters to the TEs, if CCU receives reply within T_{len} ($T_{len} = N * T$, means the time needed to transit the instructions along the longest route through the repeaters; N means the overtime index of transiting instruction packet), the communication is succeed and TEs then will provide relative services to the user as described in Fig. 71.2a, shown in Fig. 71.2b.
3. CCU sends instructions to the TEs and receives no reply from TEs within $T_{\max}$, then, the instructions will be transited by the repeaters to the TEs, if CCU still receives no reply within T_{len}, the communication is failed and the task will be aborted, shown in Fig. 71.2c.

Petri net is a basic model of parallel and distributed system, designed by Carl Adam Petri in 1962 in his PhD Thesis. The basic idea of Petri net is to describe state changes in a system with transitions [5]. Place/transition system (or called P/T_system) is one the Petri nets that suitable for describing the flow of resources. Here, P represents places and T transitions.

Communication protocols are another area where Petri nets can be used to represent and specify essential features of a system. The liveness and safeness properties of a Petri net are often used as correctness criteria in communication

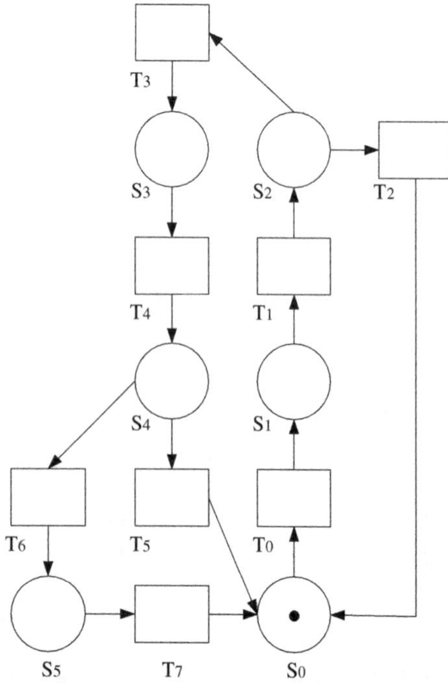

Fig. 71.3 P/T_system model of the communication protocol of CCU

Table 71.1 Description of places

Places	Descriptions
S_0	CCU is waiting for user's instructions
S_1	CCU receive user's instructions and prepare to send instructions packets to TEs
S_2	CCU is waiting for the reply within T_{max}
S_3	CCU prepare to transit the instructions through repeaters
S_4	CCU is waiting for the reply within T_{len}
S_5	CCU prepare to abort the task

Table 71.2 Description of transitions

Transitions	Descriptions
T_0	User enter instructions to CCU
T_1	CCU sends user's instructions
T_2	TEs reply succeed
T_3	TEs reply failed
T_4	CCU re-send instructions
T_5	TEs reply succeed
T_6	TEs reply failed again
T_7	Abort the task

protocols [6]. According to the communication protocol, we designed the P/T_system model of the CCU, shown in Fig. 71.3.

In the diagram of this Petri net model, transitions are conventionally depicted with rectangles, places with circles and arcs as one-way arrows that show connections of transitions to places or places to transitions [7]. The descriptions of places and transitions are shown in Tables 71.1 and 71.2 respectively.

Graphically, places in a P/T_system may contain a discrete number of marks called tokens. For example, there is one token in the P/T_system we designed above. The tokens distributed over the places represent a configuration of the net called marking. A transition of a P/T_system may fire whenever there are sufficient tokens at the start of all input arcs; when it fires, it consumes these tokens, and places tokens at the end of all output arcs. The process of firing is atomic, i.e., a single non-interruptible step [7].

71.3 Analysis of Dynamic Properties

In the Petri net model, the data has been abstracted away. The dynamic properties of a Petri net, such as liveness, reachability, coverability, boundedness, safeness and so on, are often used as correctness criteria in communication protocols. The main analytic methods of Petri net model include analysis of reachability graph, incidence matrix, state equation and invariants. Given a model of a dynamic system, the behavior of the model should be in a close relationship to the system's behavior [8].

In order to research the construction and dynamic behavior of the Petri net model of the protocol, we focus on the states that may change and their relationship and adopt the analytical methods mentioned above to reveal the conditions that lead to the occurrence of the changes and the state of the system after the changes.

71.3.1 Analysis of Reachability

Reachability is a fundamental basis for studying the dynamic properties of Petri net system, which refers to all the reachable states of the Petri net. According to transition rule, the firing of an enabled transition will change the token distribution (marking) in a net. A sequence of firings will result in a sequence of markings [6]. The variations of the system's state and sequence of the transition can be analyzed through reachable marking graph or reachable marking tree. The basic method of reachability analysis is exhaustion, i.e., build reachable marking tree or reachable marking graph to traverse all the progresses and states.

Firstly, we give now a formal definition of a reachable marking graph.

Definition 1 Let $\Sigma = (S, T; F, M_0)$ be a bounded Petri net, 3-tuple $RG(\Sigma)$ is a reachable marking graph,

$$RG(\Sigma) = (R(M_0), E, P)$$

where

$$E = \{(M_i, M_j)|M_i, M_j \in R(M_0)$$

exists $t_k \in T : M_i[t_k > M_j\}$, $P : E \to T, P(M_i, M_j) = t_k$ iff $M_i[t_k > M_j$.

1. $R(M_0)$ is called the vertex-set of $RG(\Sigma)$.
2. E is called the arc-set of $RG(\Sigma)$.
3. If $P(M_i, M_j) = t_k$, t_k is called the side marking of arc (M_i, M_j).

According to the Petri net model, we structured the reachable marking set $[M_0 >$ and reachable marking graph of the system (Fig. 71.3), where

1. M is a marking of the Petri net, described by a 6-dimensional vector.
2. $M_0 = [1, 0, 0, 0, 0, 0]^T$ is called the initial marking of the Petri net.

According to Fig. 71.4, we are ascertainable that the reachable marking graph is a strongly connected graph, i.e., any two vertexes $M_i, M_j \in R(M_0)$ are

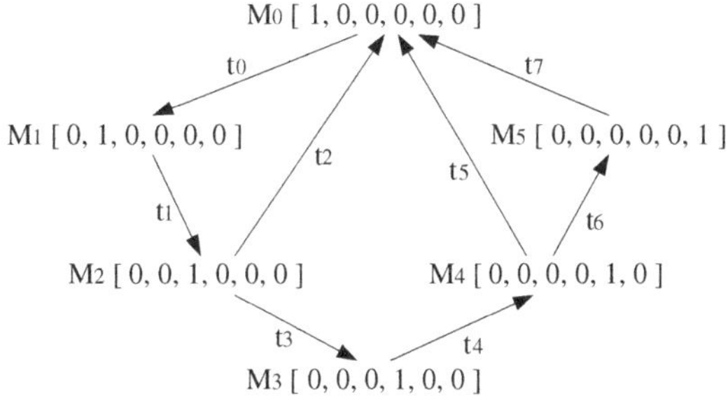

Fig. 71.4 Reachable marking graph of the Petri net model of the CCU

interconnected, which proved that every state of the CCU is reachable and no redundant and isolated state exists in the protocol.

71.3.2 Analysis of Coverablility

Problems like whether the P/T_system is bounded or whether has dead transitions can be learned from the analysis of coverability, i.e., given a marking M or a group of markings $\{M_i\}$ to verify whether there any reachable marking can cover M or $\{M_i\}$.

Definition 2 Let M' and M' be the two markings of the base net $N = (S, T; F)$ of P/T_system $\sum$,

1. if $\forall s \in S : M(s) \leq M'(s)$, it is said that M is covered by M', written as $M \leq M'$
2. if $M \leq M'$ and $M \neq M'$, it is said M is less than M', written as $M \leq M'$
3. if $M \leq M'$ and $M(s) < M'(s)$, it is said $M \leq M'$ is tenable in place $s \in S$.

According to the model, given a marking M, it can be known from the analysis of reachability of the model above, we have $M' \geq M$, i.e., the coverability of the Petri net model of the CCU protocol is in evidence.

71.3.3 Analysis of Boundedness

In a Petri net, places are often used to stand for registers and buffers for storing intermediate data. Problems like whether the system is bounded, whether the system has dead transitions, can be reduced to coverability problems. It is guaranteed that there will be no overflows in the registers or buffers, no matter what firing sequence is taken, if the Petri net is verified to be bounded or safe [6].

For the communication protocol, CCU sends an instruction and then waits for the reply from the TEs. Both capacities of receiving buffer and sending buffer are 1.

For the Petri net model, $\forall M \in R(M_0)$, the marking numbers of every place is not exceeding than 1, i.e., $s \in S, M(s) \leq 1$. The bound of the P/T_system is 1, i.e. $B(\sum) = 1$. So, this is a security system, which verified that the capacities of all buffers of the CCU are 1.

71.3.4 Analysis of Liveness

The concept of liveness is closely related to the complete absence of deadlocks in operating systems [6]. For the communication protocol of the CCU, we are not only focus on certain "result", but also the intermediate states of the system and the occurrence of the events that lead to the variation of the intermediate states. The liveness of the Petri net system is just focus on all the events and their probability of occurrence, but not certain result or target.

Definition 3 The transition $t \in T$ is live satisfying the following requirement:

$\forall M \in [M_0 >$, exists $M' \in [M >$ that enable $M'[t >$. M is a reachable marking.

If $\forall t \in T$ is live, the system $\sum$ is live.

If there are no deadlocks in the Petri net, the Petri net is live. When a Petri net goes into deadlock, the markings is dead. So, if there are end nodes in the reachable graph, the Petri net may go into deadlock. According to Fig. 71.4, For $\forall t \in T$ and $\forall M \in R(M_0)$, exist $M' \in R(M)$ that enable $M'[t >$, i.e., no end nodes exist in the reachable graph, which verify the Petri net model of the protocol is live and the CCU will not go into deadlocks.

71.4 Invariants Analysis

Invariants are important construction characteristics of Petri net system. The properties of the base net and weight function are irrelevant with initial markings. P/T_ invariants of the Petri net are significance to the analysis of the performance of the P/T_system, such as boundedness, liveness and cyclicity. The incidence matrix of the Petri net is a common method to describe the construction features of system.

Analysis of invariants means the research of the properties of the invariants in a special execution mode. S_invariant and T_invariant are the two most comprehensive invariants in the research of Petri net. S_invariant refers to the totality of resources remain unchanged under any reachable marking, that reflect the conservation of the protocol, i.e., S_invariant stands for the flow range of a certain amount resources in $\sum$. T_invariant is an antithesis concept to S_invariant, which stands for the transition sequence that keep marking unchanged in the Petri net. T_invariant reflects the repeatability and cyclicity of the protocol.

According to the reachable marking graph shown in Fig. 71.4, we build the incidence matrix C of the protocol, where

$$C = \begin{array}{c} \\ S0 \\ S1 \\ S2 \\ S3 \\ S4 \\ S5 \end{array} \begin{array}{c} \begin{array}{cccccccc} t_0 & t_1 & t_2 & t_3 & t_4 & t_5 & t_6 & t_7 \end{array} \\ \begin{pmatrix} -1 & 0 & 1 & 0 & 0 & 1 & 0 & 1 \\ 1 & -1 & 0 & 0 & 0 & 0 & 0 & 0 \\ 0 & 1 & -1 & -1 & 0 & 0 & 0 & 0 \\ 0 & 0 & 1 & 1 & -1 & 0 & 0 & 0 \\ 0 & 0 & 0 & 0 & 1 & -1 & -1 & 0 \\ 0 & 0 & 0 & 0 & 0 & 0 & 1 & L-1 \end{pmatrix} \end{array}.$$

As for initial making M_0 and certain marking M, it is verified by matrix multiplication that $M_0 + C * U = M$ is valid.

71.4.1 S_invariant Analysis

Here, we need to calculate S_invariant. According to Eq. (71.1)

$$C^T \bullet X = \Theta_T \tag{71.1}$$

where

Θ_T is a T_vector whose components are 0, i.e., $\Theta T = [0, 0, 0, 0, 0, 0]^T$.

X is a S_vector composed by variables and

$$X = [x_1, x_2, x_3, x_4, x_5, x_6]^T \tag{71.2}$$

C and Θ_T are substituted in Eq. (71.1),we have

$$S\_\text{invariant } S = [1, 1, 1, 1, 1, 1]^T,$$

i.e., $M(s_0) + M(s_1) + M(s_2) + M(s_3) + M(s_4) + M(s_5) = 1$

This is a place invariant which indicates the number of token of the CCU model is 1. The model of the protocol neither loses markings nor gets markings, but only move them in the net. During the process, the protocol will accept no new instructions before the current instruction is finished no matter what state the protocol is proceeding to. This guarantees the reliability of the protocol of the CCU and reflects its conservation.

71.4.2 T_invariant Analysis

According to T_invariant solving equation

$$C \bullet J = \Theta s \tag{71.3}$$

where

$$\Theta s = [0,0,0,0,0,0,0,0]^T$$

Θs and C are substituted into Eq. (71.3)
We have T_invariant J,where

$$J = [1,1,0,1,1,0,1,1]^T$$

For any firing sequence Q_q that starts from initial state M_0, where $M_0 = [1,0,0,0,0,0]^T$, it can be acknowledged from the reachable marking equation

$$M_q = M_0 + W \bullet Q_q \tag{71.4}$$

that CCU will return to initial state under the effect of sequence Q, where

$$Q = (t_0 t_1 t_3 t_4 t_6 t_7) \tag{71.5}$$

i.e., $M_0[t_0 t_1 t_3 t_4 t_6 t_7] > M_0$, which indicates that firing sequence Q is a repeatable sequence in the protocol of CCU, thereby, reflects the cyclicity of the protocol.

71.5 Conclusions

In this paper, we disassembled the communication protocol of home automation and adopt Petri net to describe and analyze the model. Through the analysis of dynamic properties and invariants of the base of P/T_system modeling and its reachable marking graph, we resolved many problems commonly exist in the designation of the protocol. Finally, we verified the correctness of the protocol and optimized the protocol in practice.

Future work will focus on improving the performance of the home automation model, and related communication protocols will be proposed in order to extend and improve the complete set of communication protocols of home automation.

References

1. Gerhart J (1999) Home automation and wiring. McGraw-Hill, New York (Professional)
2. Harper R (2003) Inside the smart home. Springer
3. Butler RW (2006) What is formal methods? http://shemesh.larc.nasa.gov/fm/fm-what.html
4. Wikipedia (2012) Smart home. http://en.wikipedia.org/wiki/Smart_home
5. Yuan C (2005) Theory and application of petri net. Publishing House of Electronics Industry (in Chinese)
6. Murata T (1989) Petri nets: properties, analysis and applications. Proc IEEE 77(4):541–580
7. Wikipedia (2012) Petri net. http://en.wikipedia.org/wiki/Petri_net

8. Desel J, Juhás G (2002) "What is a petri net?" Informal answer for the informed reader. Unifying Petri Nets, LUNCS 2128:1–25
9. Liu Y, Zheng H (2011) Study on digital home wireless RF protocol based on petri net. Comput Appl Softw 28(10):89–92 (in Chinese)
10. Kleijn J, Koutny M (2009) A petri net model for membrane systems with dynamic structure. Nat Comput 8(4):781–796
11. Masria A, Bourdeaud' huya T, Toguyenia A (2009) Performance analysis of IEEE 802.11b wireless networks with object oriented petri nets. Electron Notes Theor Comput Sci 242(2):73–85

Chapter 72
Research and Implementation of Data Link Layer in KNX Communication Protocol Stack

Xiajing Wang and Yan Wang

Abstract KNX has become worldwide open standard in home and building automation systems. In this paper, the function of data link layer in KNX Protocol Stack is presented in details. And the finite state stack (FSM) is applied to realize the data link layer which provides the medium access control and the logical link control. There are three problems in the implementation process of the data link layer: the interaction of data link layer, the upper and lower layer; the implementation of data link layer protocol; and the definition of the data structure. The first two are solved using the FSM theory and the rest is designed. Finally, we design the software of data link layer protocol by the above method. And the software testing indicates that it meets the data link layer functionality.

Keywords KNX · Data link layer · FSM

72.1 Introduction

KNX is a distributed bus control technology, which originated from Europe and is mainly used in building and home automation [1]. KNX results from the formal merger of the 3 leading systems (EIB, EHS, BatiBUS) into the specification of the new KNX Association [2]. On the KNX Device Network, all the devices come to life to form distributed applications in the true sense of world [3]. The Communication System (KNX Common Kernel) must tend to the needs of the Application Models, Configuration and Network Management [4]. On top of the Physical Layer and their particular Data Link Layer, a Common Kernel model is shared by all the devices of the KNX Network [5]; in order to answer all the requirements, it

X. Wang (✉) · Y. Wang
School of Automation Science and Electrical Engineering, Beihang University,
XueYuan Road No.37, HaiDian District, Beijing, China
e-mail: wxjqhd@126.com

Z. Sun and Z. Deng (eds.), *Proceedings of 2013 Chinese Intelligent Automation Conference*, Lecture Notes in Electrical Engineering 255,
DOI: 10.1007/978-3-642-38460-8_72,

includes a 7 Layers OSI model compliant communication system [6]. The layers include Physical Layer, Data Link Layer General, Network Layer, Transport Layer, Application Layer while the Session and presentation Layer are empty [7].

A finite state machine (FSM) is a mathematical model of computation used to design both computer programs and sequential logic circuits. It is conceived as an abstract machine that can be in one of a finite number of states. The machine is in only one state at a time; the state it is in at any given time is called the current state. It can change from one state to another when initiated by a triggering event or condition, this is called a transition. A particular FSM is defined by a list of its states, and the triggering condition for each transition [8]. This paper adopts the FSM method to implement data link layer in KNX protocol stack.

Throughout the paper, data link layer protocol software is taken as the example to analyse and implement KNX protocol stack. Then data link layer protocol specification is analysed and the FSM theory is used to realize its software. In the implementation process, there are three main problems: the definition of the data structure, the interaction of data link layer, the upper and lower layer, the implementation of data link layer protocol.

72.2 Specifications of the Data Link Layer

The data link layer is the layer between the data link layer user and the physical layer. The KNX data link layer conforms to the definitions of the ISO/OSI model (ISO 7498) data link layer [9]. It provides the medium access control and the logical link control [7]. The data link layer shall be concerned with reliable transport of single frames between two or more devices on the same subnetwork [1].

When transmitting, it shall be responsible for:

- Building up a complete Frame from the information passed to it by Network Layer,
- Gaining access to the medium according to the particular medium access protocol in use and
- Transmitting the Frame to the Data Link Layer in the peer entity or entities, using the services of Physical Layer.

When receiving, data link layer shall be responsible for:

- Determining whether the Frame is intact or corrupted,
- Deciding after destination address check to pass the frame to upper layers,
- Issuing positive or negative acknowledgements back to the transmitting data link layer entity.

The data link layer shall use the services of the physical layer and shall provide services to the data link layer users (network layer and others).

72.3 Implementation of the Data Link Layer

The goal of software is to achieve the functions of data link layer based on its specification. The operation of the protocol stack is actually a process of state transition and data transition which can be achieved by FSM [10]. State transition in data link layer concludes two aspects: the interaction between it and the upper and lower layer, and its internal state transition. Data transmit from upper Layer to the lower one and during which there are many operations on them, so the main data structure is defined in the paper.

72.3.1 Interaction Between Data Link Layer and Upper and Lower

The data link layer serves as the second layer of the protocol stack, whose main function is to plus or minus the control information of the present layer and then transfer the data.

This paper introduces the FSM to achieve the interaction between layers which is achieved through the transition of the states. For example, to transmit the data from the upper layer to the lower one, the network layer (the upper layer than the data link layer) transfers the information to the data link layer, and the data link layer adds the control information to form the data frame and pass it to the physical layer (the lower layer than the data link layer), then waits for the physical layer to send back a response massage to confirm whether the data frame has been transmitted correctly. Such a process can be achieved by the three finite state machines: nwkFSM, dlkFSM and phyFSM which is the FSM of the network layer, the data link layer and the physical layer respectively. And the current state of the nwkFSM, dlkFSM and phyFSM is shorted by nwkState, dlkState and phyState respectively.

Table 72.1 shows the interactive process between each of the three finite state machines in Fig. 72.1

72.3.2 Internal State Transition

The internal functions of data link layer include adding control information, transferring data and sending ACK response. Figure 72.1 shows the process of state machine in Data Link Layer. According to the functions of the data link layer, the status of it can be abstracted to five states: the idle state, IDLE; the transmission data request status, L_DATA_REQ; the status waiting for addressed node or nodes to send acknowledge Frame and feedback the information whether the data has been transferred successfully or not to the network layer, L_DATA_CON;

Table 72.1 The Interactive process between each of the three finite state machines

The interactive process from upper layer to lower layer

#network layer: nwkFSM
    Send NSDU[1] and arguments to data link layer
    Set dlkState = L_DATA_REQ
    Run dlkFSM
        #data link layer: dlkFSM
            Receive NSDU and arguments from network layer
            Form LPDU[2] with NSDU and arguments
            Send LPDU to physical layer
            Set phyState = PHY_DATA_REQ
            Run phyFSM
                #physical layer: phyFSM
                    Receive LPDU from data link layer
                    Send LPDU to KNX bus
                End physical layer
        End data link layer
End network layer

The interactive process from lower layer to upper layer

#physical layer: phyFSM
    Receive acknowledgment sent by addressed nodes from KNX bus
    Send acknowledgment to data link layer
    Set dlkState = L_DATA_CON
    Run dlkFSM
        #data link layer: dlkFSM
            Receive acknowledgment from physical layer
            Confirm whether LPDU has been sent successfully
            Send status[3] to network layer
            Run nwkFSM
                # network layer :nwkFSM
                    Receive status from data link layer
                    Send status to upper layer
                End network layer
        End data link layer
End physical layer

[1] NSDU, network service data unit provided by network layer
[2] LPDU, data link layer protocol data unit provided by data link layer
[3] status = OK, LPDU sent successfully
status = not_ok, transmission of LPDU didn't succeed

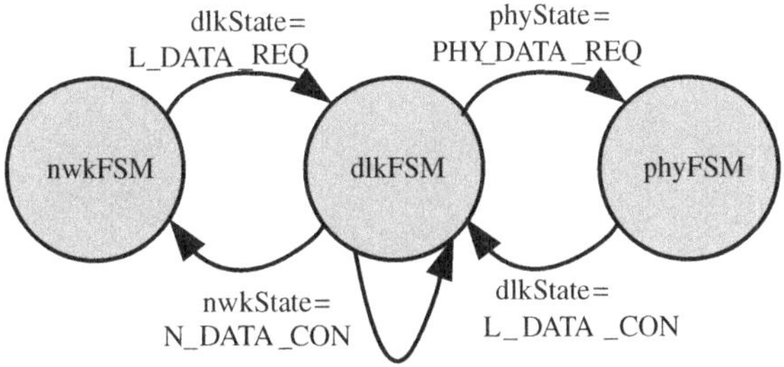

Fig. 72.1 Interactive process between each of the three finite state machines

the status of that data has been transferred from the physical layer and needing to transfer acknowledge Frame to data sender, L_DATA_SEND_ACK, and the status of having received the data that needs to be passed to the upper layer, L_DATA_IND (Fig. 72.2).

72.3.3 Design of Data Structure

In this paper, we define the data unit structure of the received and sent data in the data link layer. Take the sent data unit structure as an example, its code is as Table 72.2:

72.4 Testing and Analysis

Under VS2005 environment, data link layer software is implemented by the method above. The input and output test data are shown in Table 72.3:

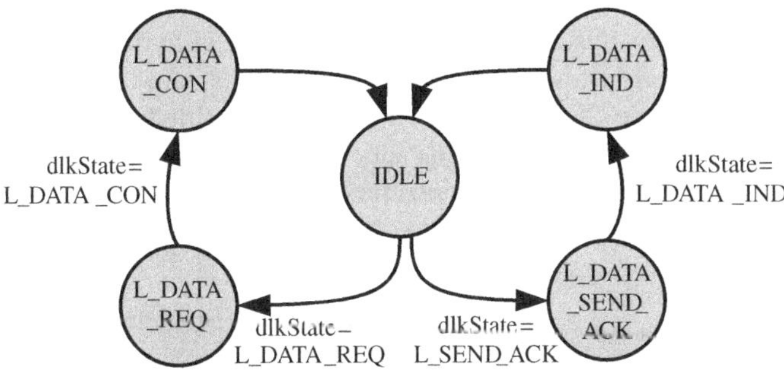

Fig. 72.2 Process of state machine in data link layer

Table 72.2 Code of sent data unit structure

Typedef struct_DLK_TX_DATA
{
BYTE FrameType;
BYTE RepeatFlag;
BYTE pritority;
BYTE SourAddrHigh;
BYTE SourAddrLow;
BYTE DestAddrHigh;
BYTE DestAddrLow;
BYTE AddrType;
BYTE length;
BYTE *Plpdu;
BYTE lpdu [dlkFrameMaxLen];
BYTE ChkSum;
}DLK_TX_DATA;

Table 72.3 Input and output

Input	Output
a_dlk_tx_data.FrameType = 0x80;	Lpdu = B0 08 F0 08 01 6B 6A 88 40 03 10 00 10 01 10 02 10 03 10
a_dlk_tx_data.RepeatFlag = 0x20;	
a_dlk_tx_data.pritority = 0x00;	
a_dlk_tx_data.SourAddrHigh = 0x08;	
a_dlk_tx_data.SourAddrLow = 0xF0;	
a_dlk_tx_data.DestAddrHigh = 0x08;	
a_dlk_tx_data.DestAddrLow = 0x01;	
a_dlk_tx_data.AddrType = 0x00;	
a_dlk_tx_data.length = 0x0B;	
lpdu = {0x60,0x6A,0x88,0x40,0x03,0x10,0x00,0x10, 0x01,0x10,0x02,0x10,0x03}	
plpdu = lpdu;	

Analysis of the output data is shown in Fig. 72.3 where all data are in hexadecimal notation. From Fig. 72.3, we can see that the Data Frame is not a repeated one, and the priority is System Priority, while the value of Source Address is 08F0, and the Destination Address is physical address with its value 0801, the Length is 0B and the value of the Check Octet is 10. Comparing with the input, the data link layer software can achieve the sending function.

It can be tested by the method mentioned above that the software can also achieve the receiving function correctly.

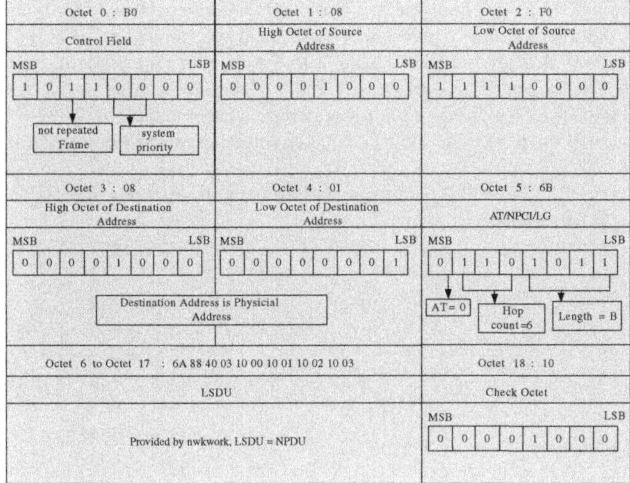

Fig. 72.3 Analysis of output data

72.5 Conclusion

This paper analyzed the implementation process of the data link layer in the KNX protocol stack, achieved the interaction and internal agreement between the data link layer and the upper or the lower layer with the finite state machine. In additional, the data structure was designed. Though only the implementation process of the data link layer was given, and the same design method can be applied to the other layers of the protocol stack. In this way the entire protocol stack can be implemented by the same implementation method of the data link layer.

References

1. Konnex Association (2009) KNX Specification Version 3.0
2. Kastner W, Neugschwandtner G, Soucek S, Newmann HM (2005) Communication systems for building automation and control. Proc IEEE 93(6):1178–1203
3. Bujdei C, Moraru SA (2011) Ensuring comfort in office buildings: designing a KNX monitoring and control system. In: 7th International conference on intelligent environments, pp 222–229
4. Ning H, Ya-Hu W, Yi T (2010) Research of KNX device node and development based on the bus interface module. In: 29th Chinese control conference (CCC), pp 4346–4350
5. Lázaro J, Abejón S, Astarloa A, Chamorro F, Bidarte U (2008) SoPC implementation of the TP-KNX protocol for domotic applications. In: International conference on advances in electronics and micro-electronics, pp 115–120
6. Yao S (2007) Development of KNX/EIB bus device and research on BCU communication core. Zhejiang University, Hangzhou, China, pp 53–70 (in Chinese)

7. Lee WS, Hong SH (2009) Implementation of a KNX-ZigBee gateway for home automation. In: IEEE 13th international symposium on consumer electronics, pp 545–549
8. Reese RB (2009) A Zigbee-subset/IEEE 802.15.4 multi-platform protocol stack. 2009-11-11. http://www.ece.msstate.edu/reese/msstatePAN
9. Neugschwandtner G, Kastner W (2009) Congestion control in building automation networks: Considerations for KNX. In: 35th Annual conference of IEEE industrial electronics, pp 4149–4154
10. Hao J, Huang J, Jia F (2004) Implementation methods of protocol stacks. Comput Eng 30(14):93–94 (in Chinese)

Chapter 73
Experimental Research of Vortex Street Oil–Gas–Water Three-Phase Flow Base on Wavelet

Hongjun Sun, Jian Zhang and Xiao Li

Abstract Vortex flowmeter was chosen to measuring the oil-gas-water three-phases flow due to its various of characteristics, experiment has been reported in the conditions as follows: vertical upward pipeline with 50 mm diameter, the volumetric flow of water is 5–8 $m^3 h^{-1}$, oil account for 5–30 % of the total mixed flow and the gas proportion of the gas void fraction is from 1 to 5 %. The signals collected by vortex sensor are compressed and de-noised by ddencmp function. As PSD (power spectral density) peak values and variations in water flow have a very clear relationship, based on the feature of wavelet transformation, criteria may be found to distinguish different flow. The result shows that low-frequency energy is greatly influenced by the oil content and low-frequency energy is declined with the oil content is increased while low-frequency energy is increased with the gas void fraction is increased. There is a flow threshold, when the three-phase flow does not reach the threshold, the influence on Low-frequency energy by the gas void fraction can be ignored. The volume measured by the vortex flow-meter was less influenced by media with low gas void fraction.

Keywords Vortex flowmeter · Oil-gas-water · Three-phase flow · Wavelet transformation

73.1 Introduction

Oil-gas-water three phases flow appeared in the petroleum, chemical and other related industry fields widely [1]. The problem of how to meter oil-gas-water mixtures has been of interest to the petroleum industry since the early 1980s,

H. Sun (✉) · J. Zhang · X. Li
College of Electrical Engineering and Automation, Tianjin University,
Tianjin 300072, China
e-mail: sunhongjun@tju.edu.cn

Z. Sun and Z. Deng (eds.), *Proceedings of 2013 Chinese Intelligent Automation Conference*, Lecture Notes in Electrical Engineering 255,
DOI: 10.1007/978-3-642-38460-8_73,

Thron and Johansen described some of principal technologies and strategies which have been developed over last ten years for three phases flow measurement, and considered future developments in this area [2].

Vortex flow-meter is suited to measurement multi-phases flow, because it is a kind of novel velocity type flowmeter with the characteristics of stabilization, low pressure loss and wide measurement range and has no relation with fluids' physical properties like: temperature, pressure, density, viscosity [3]. It is widely used in chemical industry, petroleum, metallurgy, light industry and other flow industries [4]. So vortex flow-meter was chosen to measure the rules of oil-gas-water three-phase flow in a vertical pipeline.

At present, the study and application of theory of wavelet transform is a most active domain in signal processing because of its characteristics such as time shift and multi-scale resolution. The wavelet de-noising method has the adaptive and multi-resolution feature fit for the filtering of non-stationary signal [5]. Vortex signal is a kind of single frequency sine wave signal which is often noised. Comparing with the algorithms of time–frequency analysis, Wavelet analysis method is perfected in process of filtering and frequency detecting. This paper use wavelet analysis method to de-noise and detect vortex signal, the Biorthogonal wavelet is selected to be the original wavelet [6].

73.2 Experimental Device and Methods

73.2.1 Experimental Device

This experiment was carried out at the national key laboratory of process measurement and control in Tianjin University. As Fig. 73.1a shown, the experimental device consists of five parts: the water piping system, air piping system, oil piping system, test section and oil-gas recovery system. Contain air compressors, gas tank, water storage tanks, oil storage tanks, separation tank, flow meter, pressure transmitter, temperature transmitters, test section and a variety of valve.

The IPC realizes flow regulating by controlling all electrically operated valve, and it can also real-time display the data from all the sensors. The water storage tank provides the required liquid and the water flow is measured by the standard flowmeter (1) The oil is provided by the oil storage tank and the oil flow is measured by the standard flowmeter (2) Air compressor pumps the atmosphere into gas storage and the flow velocity of the gas is measured by the standard flowmeter (3) The air, water and oil as the experimental medium are pumped into the entrance ejector device separately and well mixed at the beginning of the pipeline. Then the mixed three-phase liquid flowed into the testing section which contains plexiglass tube, piezoelectric vortex flow sensor with signal processing unit and data acquisition unit, as Fig. 73.1b shown. After testing, the three-phase flow is separated by separation tank. Atmosphere is emitted by emission switch, at

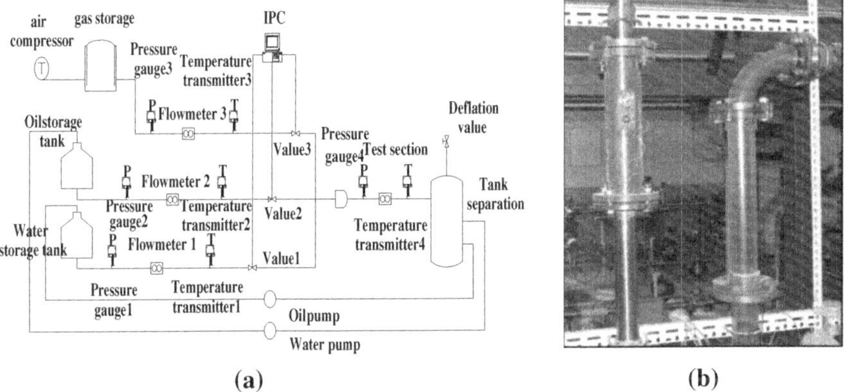

Fig. 73.1 Mulitphase measuring devices. **a** The schematic diagram of the system. **b** Real photo of test section

last, oil and water separate by themselves. The water is pumped into water storage tank and oil to oil storage tank for recycle [7, 8].

73.2.2 Experimental Methods

The volumetric flow of water is 5–8 $m^3 h^{-1}$, oil account for 5–30 % of the total mixed flow and the gas proportion of the void fraction is from 1 to 5 %. Steps of experiment as follows: Control the water flow through IPC to maintain a certain value (such as 5, 6, 7, 8 $m^3 h^{-1}$), and the oil flow rate begins at 5 %, the actual volumetric flow can be calculated by the proportion of water and oil, increase the oil flow by 5 % each time until to 30 %. For each fixed oil level, the void fraction starting at 1 %, increase the gas void fraction level by 1 % each time until to 5 %. Then adjust a new value of water flow and repeat the step again.

The experimental flowmeter is a Super Dew with piezoelectric vortex flow sensor installed in the middle of the vertical pipe section after the bluff body, which includes signal processing unit and data acquisition unit. The diameter of the vertical pipe is 50 mm. Periodic changes of the pressure field in the pipe can produce periodic charges in the piezoelectric crystal of the vortex sensor probe, then the electric charge signal is enlarged by the charge amplifier, and the low-pass filter eliminates the high frequency interference. The amplification and rectification circuit converts the sinusoidal signal of the vortex to impulse signal for output. In the experiment, the sampling frequency is 5,000 Hz. Each experimental point relates to four groups of data and every group of data has 50,000 sampling points. The frequency of each experimental point can be obtained by Fourier transformation.

73.3 PSD Peak Values and Energy of Wavelet Decomposition

In this experiment, the fluid velocity rate is relatively low. Useful stationary signals locate at low frequency while the noise signals are showed as high frequency mutations in the signals. Based on the feature of wavelet transformation, the signals are compressed and de-noised by ddencmp function. The function can returns default values for de-noising or compression, using wavelets or wavelet packet, of an input vector or matrix X which can be a 1-D or 2-D signal. After that, the PSD (power spectral density) peak values and the energy of wavelet decomposition are calculated.

73.3.1 Original Signal and Compressed De-noised Signal

It can be seen that the de-noised signal is more regular and smoother than the original signal. The compressed and de-noised signal is closed to sine wave (Fig. 73.2).

73.3.2 PSD Peak Values

The results of the experiment and power spectrum analysis (PSD) shown in Fig. 73.3, β_o is oil content. The input gas which from standard conditions, convert by the equation of state, obtained the actual volume in the flow pipeline. Horizontal axis was gas void fraction (1, 2, 3, 4, 5 % five points) and vertical axis stands for the PSD peak value. We can see from Fig. 73.3a, b, c, and d that:

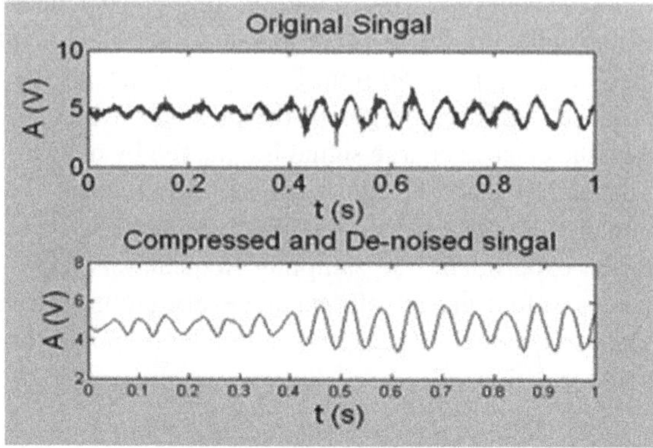

Fig. 73.2 The compare of original signal and de-noised signal

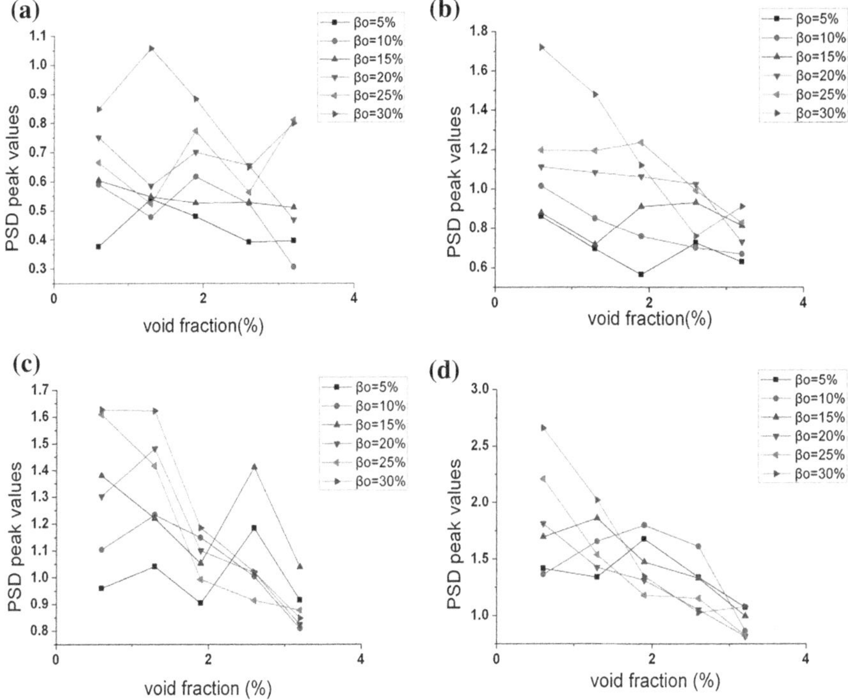

Fig. 73.3 Variation of the PSD peak values. **a** Water flow is 5 $m^3 h^{-1}$. **b** Water flow is 6 $m^3 h^{-1}$. **c** Water flow is 7 $m^3 h^{-1}$. **d** Water flow is 8 $m^3 h^{-1}$

1. In a fixed water flow, PSD peak values show an increasing trend with the increase in oil content. On the other hand, PSD peak values show a decline trend with the increase in gas void fraction.
2. With the increase of water flow, the PSD peak values increase. When the water flow is 5 $m^3 h^{-1}$, PSD peak values range between 0.24 and 1.26. When the water flow is 6 $m^3 h^{-1}$, PSD peak values range between 0.52 and 1.8. When the water flow is 7 $m^3 h^{-1}$, PSD peak values range between 0.69 and 1.93. When the water flow is 8 $m^3 h^{-1}$, PSD peak values range between 0.72 and 2.59. PSD peak values are affected more obviously by the gas void fraction, and the values decline faster with the water flow increasing.

73.3.3 The Percentage of Energy Corresponding to Approximation Ea

In this paper, the wavelet function bior6.8 is selected to decompose the compressed and de-noised signals. For a one dimensional wavelet decomposition, energy function returns Ea, which is the percentage of energy corresponding to the

Table 73.1 The percentage of energy corresponding to the approximation (Ea)

	$\beta_o = 5\ \%$	$\beta_o = 10\ \%$	$\beta_o = 15\ \%$	$\beta_o = 20\ \%$	$\beta_o = 25\ \%$	$\beta_o = 30\ \%$
$Q_w = 5\ m^3\ h^{-1}$ $\beta g = 1\ \%$	99.586	99.267	98.867	98.202	96.201	93.502
$Q_w = 5\ m^3\ h^{-1}$ $\beta g = 2\ \%$	99.599	99.308	98.885	97.597	96.533	94.139
$Q_w = 5\ m^3\ h^{-1}$ $\beta g = 3\ \%$	99.563	99.282	98.741	98.015	96.467	95.264
$Q_w = 5\ m^3\ h^{-1}$ $\beta g = 4\ \%$	99.588	99.305	98.735	97.927	96.872	95.608
$Q_w = 5\ m^3\ h^{-1}$ $\beta g = 5\ \%$	99.595	99.401	98.796	97.733	97.082	95.481
$Q_w = 6\ m^3\ h^{-1}$ $\beta g = 1\ \%$	99.361	99.348	99.246	98.535	96.304	93.844
$Q_w = 6\ m^3\ h^{-1}$ $\beta g = 2\ \%$	99.277	99.235	98.7981	98.5916	97.1407	94.3904
$Q_w = 6\ m^3\ h^{-1}$ $\beta g = 3\ \%$	99.174	99.215	98.826	98.571	97.201	94.625
$Q_w = 6\ m^3\ h^{-1}$ $\beta g = 4\ \%$	99.115	99.224	98.807	98.698	97.4623	94.964
$Q_w = 6\ m^3\ h^{-1}$ $\beta g = 5\ \%$	99.126	99.186	98.821	98.6715	95.339	95.032
$Q_w = 7\ m^3\ h^{-1}$ $\beta g = 1\ \%$	98.019	96.897	95.195	92.176	89.922	87.137
$Q_w = 7\ m^3\ h^{-1}$ $\beta g = 2\ \%$	98.039	96.901	95.243	93.032	90.604	87.362
$Q_w = 7\ m^3\ h^{-1}$ $\beta g = 3\ \%$	97.461	96.779	95.029	93.729	91.836	88.439
$Q_w = 7\ m^3\ h^{-1}$ $\beta g = 4\ \%$	98.168	97.089	95.056	93.903	92.868	89.061
$Q_w = 7m^3\ h^{-1}$ $\beta g = 5\ \%$	97.676	97.123	95.485	94.703	93.247	90.149
$Q_w = 8\ m^3\ h^{-1}$ $\beta g = 1\ \%$	94.616	91.511	88.429	85.107	80.365	78.124
$Q_w = 8\ m^3\ h^{-1}$ $\beta g = 2\ \%$	94.332	91.901	88.243	85.032	81.604	79.543
$Q_w = 8\ m^3\ h^{-1}$ $\beta g = 3\ \%$	93.534	91.773	88.029	85.729	82.839	79.939
$Q_w = 8\ m^3\ h^{-1}$ $\beta g = 4\ \%$	94.663	92.089	88.056	85.904	83.868	81.685
$Q_w = 8\ m^3\ h^{-1}$ $\beta g = 5\ \%$	95.103	92.123	89.485	86.703	84.248	82.733

approximation and Ed, which is a vector containing the percentages of energy corresponding to the details. The calculated Ea values are showed in the Table 73.1. Throughout the experiment, there are three variables, oil content, gas void fraction and water flow. We can analyse the Table 73.1 from three aspects by

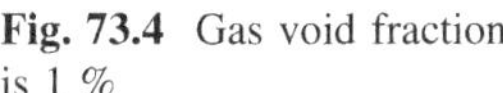

Fig. 73.4 Gas void fraction is 1 %

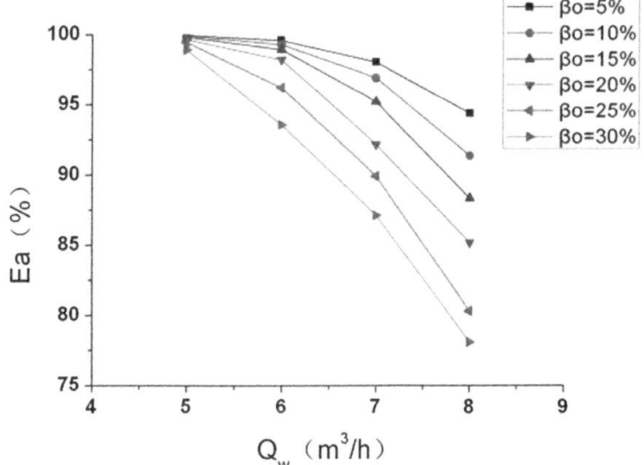

fixing one of these variables. Three representative cases of each aspect are shown in Figs. 73.4, 73.5 and 73.6.

When the gas void fraction was fixed in 1 %, the obvious ray-like graphics can be seen in the Fig. 73.4. Looking vertically, Ea values decline with the increase in oil content at the same water flow rate. Looking horizontally, Ea values decline with the increase in water flow at the same oil content and the rate of decrease is growing.

In the case of the oil content is 5 %, longitudinal data shown that Ea values decline with the water flow increasing, the rate of decline increases with the increase of water flow. When the water flow is 5 and 6 $m^3\ h^{-1}$, Ea values are almost constant and do not change with gas void fraction. When the water flow is 7 and 8 $m^3\ h^{-1}$, Ea values fluctuates with the gas void fraction increased.

From the Fig. 73.6, when the water flow is fixed in 5 $m^3\ h^{-1}$, at the same gas void fraction, Ea value increased with the oil content increasing. In the cases of oil content is 5 $\sim$ 15 %, Ea values are almost constant, in other cases, Ea values are increasing with the void fraction increase.

All the data of the Table 73.1 indicates that there is a flow threshold. When the three-phase flow does not reach the threshold, the influence on Low-frequency energy by the gas void fraction can be ignored. The influence of gas void fraction increasing once the three-phase flow beyond the threshold.

In order to research the disciplines of the vortex street oil-gas-water three-phase flow, experiments were done in the conditions of different rate of oil and gas, wavelet transformation method was used to process the data, finally obtaining the following conclusions:

1. Through analyzing the PSD peak values, oil and gas ratio can confirm that the oil phase components have little influence on the vortex flowmeter, but the gas phase makes a great impact of vortex flowmeter. PSD peak values and variations in water flow have a very clear relationship. A criteria may be found to distinguish different flow pattern.

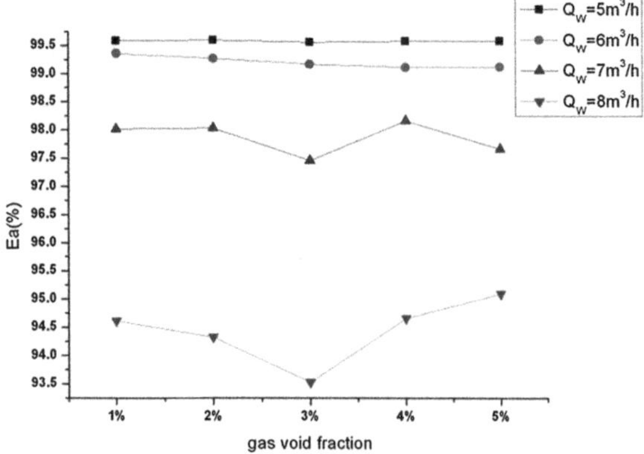

Fig. 73.5 Oil content is 5 %

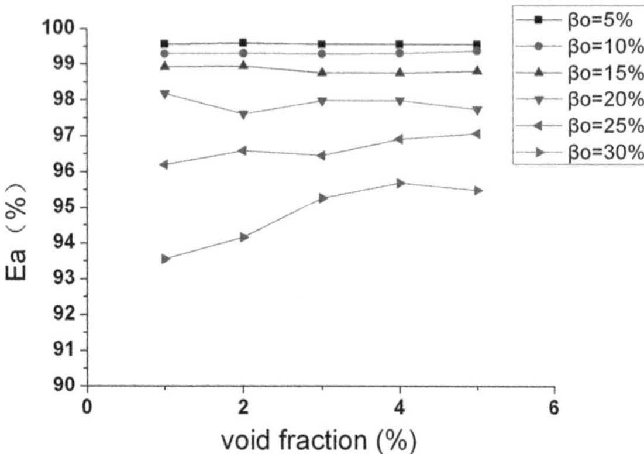

Fig. 73.6 The water flow is $5\ m^3\ h^{-1}$

2. Ea as the percentage of energy corresponding to the approximation. Ea reflects the low frequency signal energy ratio. In contrast, void fraction plays a significant role in the impact of measuring vortex signals. In the three-phase flow, when the mixed flow is low, oil content and void fraction do not significantly affect the signals. Low-frequency energy is greatly influenced by the oil content.

Acknowledgments The research work is supported by National Youth Science Foundation (No. 50906061).

References

1. Fang LD, Jiang QY, Zhang T, Xu Y (2008) Based on the simple separation of the oil-gas-water three-phase meter. Meas J 29(5):445–448
2. Thorn R, Johansen GA, Hammer EA (1997) Recent developments in three-phase flow measurement. Meas Sci Technol 8(1997):691–701

3. Amadi-Echendu JE, Zhu HJ (1992) Signal analysis applied to vortex flow meters. Trans Instrum Meas 41(6):1001–1004
4. Szalinski L, Abdulkareem LA, Da Silva MJ, Thiele S, Beyer M, Lucas D, Hernandez Perez V, Hampel U, Azzopardi BJ (2010) Comparative study of gas-oil and gas-water two-phase flow in a vertical pipe. Chem Eng Sci 65(12):3836–3848
5. Sun HJ, Qu WJ (2009) Wavelet analysis of vortex signal. In: Proceedings of the 8th international conference on machine learning and cybernetics, Baoding, pp 12–15
6. Sun HJ, Zhang T (2003) Wavelet denoising method used in the vortex flowmeter. In: Proceedings of IEEE international conference on machine and cybernetics
7. Sun HJ, Jiang XY, Sun M, Wang HX, Zhang C, Zhang T (2010) The experimental research of vortex flowmeter in vertical upward oil-gas-water three-phase flow. In: Proceedings of I^2MTC 2010—international instrumentation and measurement technology conference
8. Li YM, Xu Y, Zhang LW, Zhang T, Li Q (2009) Simulation and experiment investigation on effect of upstream pipe single elbow on the performance characteristics of V-Cone flowmeters. Yi Qi Yi Biao Xue Bao J Sci Instrum 30(2009):1195–1201 (in Chinese)

Chapter 74
Multi-Phase Kernel Based Adaptive Soft Sensor Approach for Fed-Batch Processes

Kun Chen and Yi Liu

Abstract An adaptive soft sensor, multi-phase kernel based regression (MPKR) method, is proposed for online modelling of fed-batch processes. By introducing the multi-phase kernel and online phase estimation, the proposed MPKR can handle the multi-stage nature of fed-batch processes to achieve an accurate model. Simulation on biomass concentration online prediction in penicillin fermentation indicates that MPKR can obtain a promising accuracy and robust results compared to traditional kernel regression method.

Keywords Multi-kernel learning · Fed-batch processes · Multi-phase processes

74.1 Introduction

Fed-batch processes have been widely employed to product high value but low volume products in chemical and biological industries [1, 2]. However, owing to the non-steady-state operation and instinct time-variation characteristics, it is quite difficult to ensure the quality variables within acceptable regions during the whole production process [3]. Hence, it is crucial to obtain the product quality measurement online. In the past two decades, data-driven approaches [3–8] have been successful applied to model those continuous processes. Among these methods, kernel regression (KR) methods [7, 8] are believed to attain a promising performance under

K. Chen (✉)
Shaoxing University, Shaoxing, China
e-mail: kchen@usx.edu.cn

Y. Liu
Institute of Process Equipment and Control Enginnering,
Zhejiang University of Technology, Hangzhou, China

Z. Sun and Z. Deng (eds.), *Proceedings of 2013 Chinese Intelligent Automation Conference*, Lecture Notes in Electrical Engineering 255,
DOI: 10.1007/978-3-642-38460-8_74,

small-sample condition due to its structural risk minimization criterion, but still fail to handle the multi-phase nature of fed-batch processes.

Recently, multi-kernel learning (MKL) methods [9–12] have shown advantages in the area of pattern recognition. Instead of using a single kernel, MKL adopts an alternative way of fusing secondary variables and kernels [9–12]. However, most MKL methods, e.g., the canonical MKL [9, 10], adopted a uniform similarity over the whole input domain, which may fail to cope with the complexity of data distribution in multi-phase processes. On the other hand, the sample-specific method [11], which reflected the relative importance of different kernel for each sample, achieved certain performance improvements. Nevertheless, too many parameters were needed, as well as the expensive computation load and the high risk of over-fitting. Yang et al. presented a group-sensitive MKL [12] to learn a single classifier using group-sensitive kernel combinations, but the iterative alternant optimization nature made it uneasy to extend to online modelling.

In this paper, we explore the potential to improve MKL, by introducing a multi-phase kernel with phase and input information simultaneously, and a k-nearest-neighbourhood based kernel density estimation method to estimate the phase probability of each new-coming sample. The proposed algorithm, multi-phase kernel based regression (MPKR) is applied to model the biomass concentration in a penicillin process [13] to evaluate the performance. The simulation results demonstrate that the proposed MPKR method can obtain a more accurate prediction under batch-to-batch variations.

74.2 Kernel Regression

Given the training data $L = \{(x_1, y_1), \ldots (x_l, y_l)\}$, the KR framework can be described as the following optimization problem [8]:

$$\begin{gathered} \min_{\omega,b,e} J(\omega, b) = \frac{1}{2}\omega^T\omega + \frac{r}{2}\sum_{i=1}^{l} e_i^2 \\ s.t.\ y_i = \omega\varphi(x_i) + b + e_i, \quad i = 1, 2, \ldots, l \end{gathered} \tag{74.1}$$

where ω, b and e_i are weight vector, bias, residual for the ith sample respectively, with φ a nonlinear operator to map the input x into a high-dimensional space, and r the trade-off parameter to balance the model complexity and approximation error.

To solve such problem, the Lagrangian is constructed as:

$$L(\omega, b, \mathbf{e}, \boldsymbol{\alpha}) = J(\omega, b) - \sum_{i=1}^{l} \alpha_i \{\omega^T \varphi(x_i) + b + e_i - y_i\} \tag{74.2}$$

Then, the Karush-Kuhn-Tucker system can be obtained as:

$$\begin{bmatrix} 0 & \mathbf{1}^{\mathrm{T}} \\ \mathbf{1} & \mathbf{\Omega}+\mathbf{V} \end{bmatrix} \begin{bmatrix} b \\ \boldsymbol{\alpha} \end{bmatrix} = \begin{bmatrix} 0 \\ \mathbf{y} \end{bmatrix} \tag{74.3}$$

where $\mathbf{\Omega}(i, j) = \varphi(x_i)^{\mathrm{T}}\varphi(x_j) = \mathrm{K}(x_i, x_i)$ with K(*,*) the positive definite kernel function satisfying the mercer condition [7], $\boldsymbol{\alpha} = [\alpha_1, \alpha_2, \ldots, \alpha_l]^{\mathrm{T}}$, $\mathbf{y} = [\mathrm{y}_1, \mathrm{y}_2, \ldots, \mathrm{y}_l]^{\mathrm{T}}$ and $\mathbf{V}$ is a diagonal matrix given by $\mathbf{V}(i, i) = 1/r$. By solving (74.3), b and $\boldsymbol{\alpha}$ can be obtained and the soft sensing model can be written as:

$$f(x^*) = \sum_{i=1}^{l} \alpha_i \mathrm{K}(x_i, x^*) + b = \boldsymbol{\alpha}^{\mathrm{T}}\mathbf{k} + b \tag{74.4}$$

with $\mathbf{k}$ a kernelized vector between the query data x^* and the training samples.

It is worth to note that, how to select the kernel form is still an open question in KR. Traditional way is to adopt some basis functions such as Gaussian kernel, Polynomial kernel or a combination of these functions as multi-kernel [9–12]. But for fed-batch processes, such as fermentation, which contains several production stages with totally different nonlinear relations, these kernels may fail to establish an accurate global model.

74.3 The Proposed MPKR Method

In the proposed MPKR method, a multi-phase kernel utilized both phase and input information is proposed first, which would help model the multi-phase fed-batch processes in a universal form. Then, a k-nearest-neighborhood based kernel density estimation (kNN-KDE) with the same kernels in multi-phase kernel is employed to estimate the label probability before prediction, which reduces the computation load and helps remove the effect of uneven distribution of different classes. Furthermore, the sample phase label will be revised.

74.3.1 Multi-Kernel Construction

Instead of using the same kernel form for all training data, a multi-phase kernel is proposed to handle multi-phase ($\mathrm{G} \geq 2$) characteristics. As the scheme of $\mathrm{G} = 2$ is the same as that of $\mathrm{G} > 2$, our work will focus on the two-phase case ($\mathrm{G} = 2$) only. Let $L_0 = \{(x_1, y_1, 0), \ldots, (x_m, y_m, 0), (x_{m+1}, y_{m+1}, 1), \ldots, (x_{m+n}, y_{m+n}, 1))$ denotes the initial labeled training samples where the first m samples belong to phase 0 and the later n samples are labeled as phase 1. A new multi-phase kernel to incorporate different phase models into a unified frame for fed-batch processes is proposed as:

$$K_{MP} = K_{p0} \cdot K_0 + K_{p1} \cdot K_1 \tag{74.5}$$

where K_{MP} is the proposed kernel, K_{P0} and K_{P1} are the similarity measurement between the query data phase and phase 0, phase 1 respectively, K_0 and K_1 are the kernel form chosen for data in class 0 and class 1. As proved in [9], if K_{P0}, K_{P1}, K_0 and K_1 satisfy the Mercer condition, K_{MP} will also be kept as a kernel function.

By replacing the kernel K with K_{MP} in (74.3) and (74.4), the multi-phase kernel regression model can be obtained. In this work, the linear kernel is employed for K_{p0} and K_{p1} and the Gaussian kernel is utilized for K_0 and K_1 as:

$$K_{p0}(i,j) = K(1 - l(x_i), 1 - l(x_j)) = (1 - l(x_i))(1 - l(x_j)) \tag{74.6}$$

$$K_{p1}(i,j) = K(l(x_i), l(x_j)) = l(x_i)l(x_j) \tag{74.7}$$

$$K_0(i,j) = \exp\left(-\frac{\|x_i - x_j\|}{2\sigma_0^2}\right), \quad K_1(i,j) = \exp\left(-\frac{\|x_i - x_j\|}{2\sigma_1^2}\right) \tag{74.8}$$

where σ_0 and σ_1 are the corresponding kernel width obtained from data in phase 0 and 1 respectively, and $l(x_i)$ is the phase label of sample x_i.

74.3.2 Prior Probability Estimation

To predict the output for a query sample, the phase label needs to be estimated first. It is common known that the probability estimation methods can be mainly divided into two categories: parametric and non-parametric methods. Parametric methods first assume a certain probability distribution for each class, and estimate the parameters using the labeled data. Unfortunately, it is quite difficult to determine the distribution function form and the estimated probability may deteriorate a lot if a wrong distribution function is used. For non-parametric methods, no particular distribution form is assumed there, e.g., *k*-nearest-neighborhood (*k*NN) method and the kernel density estimation (KDE) method [14]. However, the traditional *k*NN treats all nearest samples with equal weights, and the kernel density estimation methods are time-consuming as the shape of the density

$$p_h(x) = \frac{1}{l}\sum_{i=1}^{l} K_h(x - x_i) \tag{74.9}$$

need to be calculated for every query sample, where K_h is a symmetric function that integrates to one and l is number of training samples.

Inspired by *k*NN and KDE, and to release the computation load, a *k*NN-KDE method is adopted in the proposed algorithm. First, the similarity $\mathbf{k}_0$ (or $\mathbf{k}_1$) using K_0 (or K_1) for data with phase 0 (or phase 1) is calculated. Then, the largest k values in $\mathbf{k}_0$ and $\mathbf{k}_1$ are selected to construct the density belongs to class 0 and class 1defined as Eq. (74.9). Furthermore, the class label is obtained by p = p1 / (p0 + p1). Finally, a small adjustment is introduced as:

$$p = \begin{cases} 0 & \text{if } p < T_1 \\ p & \text{if } T_1 \leq p \leq 1 - T_1 \\ 1 & \text{if } p > 1 - T_1 \end{cases} \tag{74.10}$$

where T_1 is a predefined threshold. Thus, two improvements will be achieved:

- The same form of the used kernel in the multi-phase kernel is adopted, which reduces the computation load.
- Only the k biggest value of $\mathbf{k}_0$ and $\mathbf{k}_1$ is used to calculate the prior probability, regardless the number of samples belong to different classes, which helps remove the effect of the sample number.

74.3.3 Posterior Probability Evaluation

Once the output of the new-coming data is obtained, it is essential to re-evaluate the phase probability. Define the approximation error between the actual process output y and the one-step-ahead prediction $\hat{y}$ at instance i as below:

$$e_i = y_i - \hat{y}_i \tag{74.11}$$

If the approximation error is significant, that is $\|e_i\| > \delta$, the estimated prior class label is treated as wrong and the posterior probability will be re-estimate. Let $e_{i,0}$ and $e_{i,1}$ denote the prediction error using phase label 0 and 1 for the query sample respectively, the posterior probability is obtained as:

$$p_{posterior} = \frac{\|e_{i,1}\|}{\|e_{i,1}\| + \|e_{i,0}\|} \tag{74.12}$$

Similar to the prior probability, the posterior label $p_{posterior}$ is adjusted according to (74.10). Then the new kernelized vector $\mathbf{k}_{posterior}$ and approximation error $e_{posterior}$ is computed with the class label $p_{posterior}$.

In summary, the main steps of the proposed algorithm can be described as:

Input: initial labeled training dataset L, and sequential input L' *without label*

1. Offline training with the labeled L according to Eqs. (74.1)–(74.4)
2. For sequential testing data L'
   a. Phase probability estimation using the input information in Eqs. (74.9)–(74.10);
   b. Output prediction using Eq. (74.4);
   c. Approximation error calculation after obtaining the output. If $\|e_i\| > \delta$, re-estimate the posterior probability of class label with Eqs. (74.11)–(74.12), and compute the approximation error with the posterior label.

74.4 Experimental Results

In this section, MPKR is applied to online predict the biomass concentration of a penicillin production, a typical fed-batch process with multistage nature. Generally, the penicillin production can be divided into two operational phases: the rapid-growth batch operation phase and the slow-growth fed-batch operation phase. A well-known benchmark platform, PenSim [15], which is available at http://216.47.139.198/pensim, is adopted to evaluate the proposed method. According to [15], the following features are considered as secondary variables at time t to estimate the biomass concentration (C_B): the culture volume (V), the dissolved oxygen concentration (C_{DO}), the carbon dioxide concentration (C_{CO2}), both at time t and $t-1$, and the biomass concentration C_B at time $t-1$. The detail of the initial condition and the set point of PenSim can be found in [15].

To make the simulation more practical, the simulation time is set at 300 h, and data are sampled per 5 h. Meanwhile, only 10 batches are generated under different initial condition and set points to evaluate the modeling ability on a small size dataset. The first batch is labeled manually, and used as the initial training samples to select the kernel and trade-off parameters. In order to investigate the model performance, 2 common indices, the Root-Mean-Square Error (*RMSE*) and Mean Absolute Error (*MAE*) are adopted, which are defined as follows:

$$RMSE = \sqrt{\frac{\sum_{i=1}^{N} (y_i - \hat{y}_i)^2}{N}} \tag{74.13}$$

$$MAE = \frac{\sum_{i=1}^{N} |y_i - \hat{y}_i|}{N} \tag{74.14}$$

where $\hat{y}_i$ is the estimated value of y_i.

To show the effectiveness of MPKR, two cases are considered in the experiments: noise-free and noisy case. In the noise-free case, 3 methods are applied: (1) KR, with one Gaussian kernel; (2) KR2, where 2 kernel regression models are employed on 2 operational phases, respectively; (3) MPKR. Figure. 74.1 shows the result of predictive error on the first 4 batches and Table 74.1 lists the *RMSE* and *MAE* values. It is noticeable that the result of MPKR is related to the value of T_1 and k. Thus, experiments on the effect of different T_1 (0, 0.25, 0.5) and k (5, 10, 15, 20, 25 and 30) are illustrated in Table 74.2.

From Table 74.1 and Fig. 74.1, it is clear that MPKR can achieve a promising performance ($RMSE = 0.115$, $MAE = 0.052$), compared to traditional KR method ($RMSE = 0.240$, $MAE = 0.109$), but less accuracy than KR2 due to the fact that errors may happen during the phase estimation procedure. From Table 74.2, a conclusion can be drawn that a soft classification threshold $T_1 = 0.25$ can improve the predictive accuracy compared to the hard classification manner $T_1 = 0.5$, and the fuzzy class label $T_1 = 0$. Furthermore, the proposed back-learning refinement can improve the result as well.

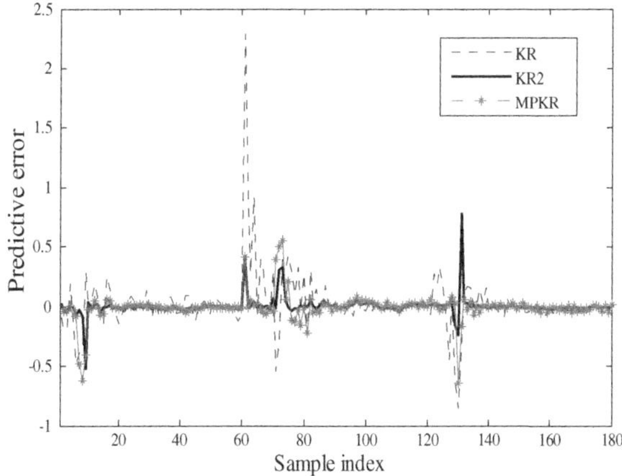

Fig. 74.1 Predictive error of the first 4 batches using different methods in noise-free case: (1) KR; (2) KR2, i.e., 2 kernel regression models to estimate different phase model respectively; and (3) MPKR. It seems that the proposed MPKR achieves a similar result compared to KR2

Table 74.1 RMSE and MAE of different modeling methods on biomass concentration prediction

Methods	KR	KR2	MPKR
RMSE	0.240	0.078	0.115
MAE	0.109	0.036	0.052

Table 74.2 RMSE values via different parameters T_1 and k

Criterion	RMSE						
k	5	10	15	20	25	30	Mean
$T_1 = 0$	0.158	0.162	0.162	0.167	0.172	0.177	0.167
$T_1 = 0.25$	0.104	0.120	0.115	0.120	0.120	0.113	0.115
$T_1 = 0.5$	0.147	0.147	0.150	0.150	0.150	0.150	0.149

In the noisy case, 3 % Gaussian noise is added into both the primary and secondary variables. Traditional KR and MPKR method are implemented to estimate the soft sensor model. Notice that MPKR here is implemented with parameters $T_1 = 0.25$. As shown in Table 74.3, the same conclusion can be drawn that the proposed MPKR method performs better than traditional single kernel KR due to effect of the multi-phase kernel, and back-learning refinement.

Table 74.3 RMSE and MAE values of LSSVR, MK-LSSVR and MK-LSSVR2 of online biomass concentration prediction in noisy case

Methods	LSSVR	MK-LSSVR	MK-LSSVR2
RMSE	0.360	0.296	0.311
MAE	0.272	0.224	0.229

74.5 Conclusion

In this work, a new multi-phase kernel based regression method (MPKR) has been developed to estimate the soft sensor for fed-batch processes by introducing techniques such as multi-phase kernel to combine the class label and the input information, the recursive posterior probability estimation to ensure the accuracy of class labels, and online model updating. Distinct from traditional single kernel modeling, the proposed MPKR can trace the multi-stage fed-batch processes more accurately. Moreover, MPKR achieves a universal model with similar accuracy compared to the piecewise kernel regression models. The experimental results on online prediction of the biomass concentration demonstrate that the proposed MPKR modeling method can handle the fed-batch processes more accurately and robust to noise. It is worthy mentioning that the accuracy of the proposed method mostly depends on the estimated class label. Our future works, including automatic clustering of initial training samples and online label re-signing still need be further studied.

Acknowledgments This study is supported by the National Natural Science Foundation of China under Grant 61004136, which is gratefully acknowledged.

References

1. Alford J (2006) Bioprocess control: advances and challenges. Comput Chem Eng 30:10–12
2. Dochain D (2003) State and parameter estimation in chemical and biochemical processes: a tutorial. J Process Control 13(8):801–818
3. Yu J, Qin SJ (2009) Multiway Gaussian mixture model based multiphase batch process monitoring. Ind Eng Chem Res 48(18):8585–8594
4. Kadlec P, Gabrys B, Strandt S (2009) Data-driven soft sensors in the process industry. Comput Chem Eng 33(4):795–814
5. Yao Y, Gao FR (2009) A survey on multistage/multiphase statistical modeling methods for batch processes. Annu Rev Control 33(2):172–183
6. Gonzaga JCB, Meleiro LAC, Kiang C, Maciel Filho R (2009) ANN-based soft-sensor for real-time process monitoring and control of an industrial polymerization process. Comput Chem Eng 33(1):43–49
7. Schölkopf B, Smola AJ (2002) Learning with kernels. The MIT Press, Cambridge
8. Suykens JAK, Van Gestel T, De Brabanter J (2002) Least squares support vector machines. World Scientific, Singapore
9. Bach FR, Lanckriet GRG, Jordan MI (2004) Multiple kernel learning, conic duality and the SMP algorithms. In: Proceedings of ICML 2004
10. Lanckriet GRG, Bie TD, Cristianini N, Jordan MI, Nobel WS (2004) A statistical framework for genomic data fusion. Bioinformatics 20:2626–2635
11. Gonen M, Alpaydin E (2008) Localized multiple kernel learning. In: Proceedings of ICML 2008
12. Yang JJ, Li YN, Tian YH, Duan LY, Gao W (2009) Group-sensitive multiple kernel learning for object categorization. In: Proceedings of ICCV 2009

13. Birol G, Ündey C, Cinar A (2002) A modular simulation package for fed-batch fermentation: penicillin production. Comput Chem Eng 26(11):1553–1565
14. Postaire JG, Vasseur C (1982) A fast algorithm for nonparametric probability density estimation. IEEE Trans Pattern Analy Mach Intell 4(6):663–666

Chapter 75
Application of Ultrasonic Phased Array Testing Technology on Ladle Trunnion

Dong Hu, Qiang Wang and Changming Yuan

Abstract The welding quality of the trunnion's root welds is critical to the safety operation of ladles. However, due to the complex structure of the detecting surface, the non-destructive testing of these welds has been a significant problem. Ultrasonic phased array technology can achieve flexible and controllable ultrasound beam by changing the time delay of each element by which ultrasonic phased array technology has been widely used for the inspection of complex structure weld. In this paper, Omniscan MX with a 64-element phased array probe was employed in inspecting a big ladle trunnion. The results illustrate that: Ultrasonic phased array obtained high signal-to-noise ratio, accurate positioning and high detection efficiency, it is proved feasible in ladle trunnion's inspection.

Keywords Ultrasonic phased array · Ladle trunnion inspection · Deep weld · Complex structure

75.1 Introduction

Ladle is a device widely used in the steelworks, its role is to carry and transfer molten steel for casting job. Trunnion is the key component for lifting the ladle [1]. As ladles are lifted frequently and always under complex loads, it is easy to generate defects, especially the welds in the root of the trunnion. In order to ensure the ladle operating safety, timely inspection should be carried out.

Ultrasonic phased array (UPA) is a special ultrasonic testing technology which has excellent abilities of electronic steering, beam deflection and beam focusing [2]. However, due to the complexity of the system and high costs of the equipment,

D. Hu · Q. Wang (✉) · C. Yuan
College of Quality and Safety Engineering, China Jiliang University,
No 258 XueYuan Road, Hangzhou, Xiasha, China
e-mail: Qiangwang@cjlu.edu.cn

Z. Sun and Z. Deng (eds.), *Proceedings of 2013 Chinese Intelligent Automation Conference*, Lecture Notes in Electrical Engineering 255,
DOI: 10.1007/978-3-642-38460-8_75,

it was mainly used in the medical field at the early stage. As the rapid development of electronic and computer technology, it has been widely employed in industry non-destructive testing (NDT) [3], especially the complex components or deep weld that conventional detection technologies cannot solve, and amount of application cases such as the application on the train wheel rim [4], the application on the fillet butt weld of plant boiler [5] and the application on thick wall pressure vessel [6] have proved it.

In this paper, structural features and inspection difficulties of ladle's trunnion are analyzed. Combined with the theory of UPA, an experiment was taken out on a big ladle trunnion.

75.2 Structural Features and Inspection Difficulties of Ladle Trunnion

Seen from Fig. 75.1, the trunnion is nested in the trunnion box and they are welded with the ladle's ektexine, Table 75.1 shows the materials of them. The weld is located at the root of the trunnion which is apart from the surface above 200 mm, so it is difficult to penetrate for most conventional detection technologies. Although ultrasound have strong penetration, its angle is fixed, resulting that the space for placing the probe becomes narrow, so ultrasonic testing can only cover a small area of the trunnion.

75.3 Ultrasonic Phased Array Technology

Ultrasonic phased array probe UPA testing instrument can electronically steer and focus the transmission ultrasonic beams through the delay rules, Take linear UPA for example, if the deflection angle of the ultrasound beam is γ, the delay time of random adjacent elements $\Delta\tau_0$ satisfies [11]:

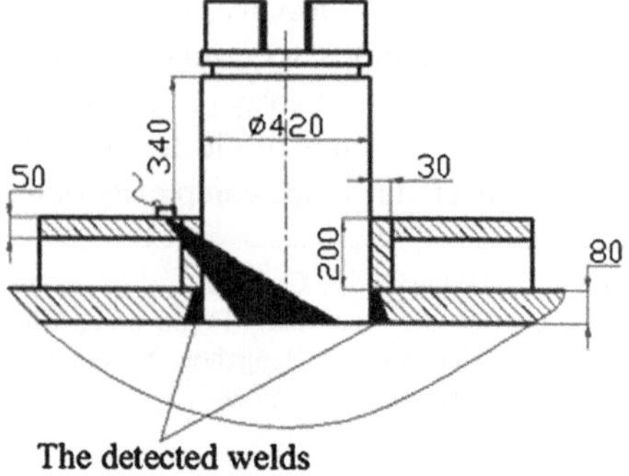

Fig. 75.1 Structure of ladle's trunnion

Table 75.1 Material of the ladle's trunnion

Name	Material
Trunnion	18MnMoNb [7]
Trunnion box	Q235 [8]
Ladle's ektexine	35# [9]
Weld	E5016 [10]

$$\Delta\tau_0 = d\sin\gamma/c \tag{75.1}$$

where c is the ultrasound velocity in the medium, d is the center distance between two adjacent elements.

Supposing the ultrasonic beam focuses along the line that has an angle of γ with the normal line. The line is called the beam direction. There are N elements are excited. Relative to the middle element, the delay of element is $n(1 \le n \le N)$:

$$t_n = t_0 + \frac{f}{c}\left\{1 - \sqrt{1 + \left[\frac{d}{f}(n - \frac{N-1}{2})\right]^2 - 2\sin\theta\frac{d}{f}(n - \frac{N-1}{2})}\right\} \tag{75.2}$$

Figure 75.2 shows the beam focusing and beam deflection of liner UPA, with these features, UPA technology can provide an efficient solution for the ladle's trunnion's detection problem [12].

75.4 Experimental Section

75.4.1 Detection Parameters

Omniscan MX portable phased array ultrasonic detector is selected as the detection equipment. The probe is linear probe with wedge, its model is 5L64-A2, its center frequency 5 MHz and it have 64 elements, but only 16 of them can be stimulated at one time. Specific detection parameters are shown in Table 75.2.

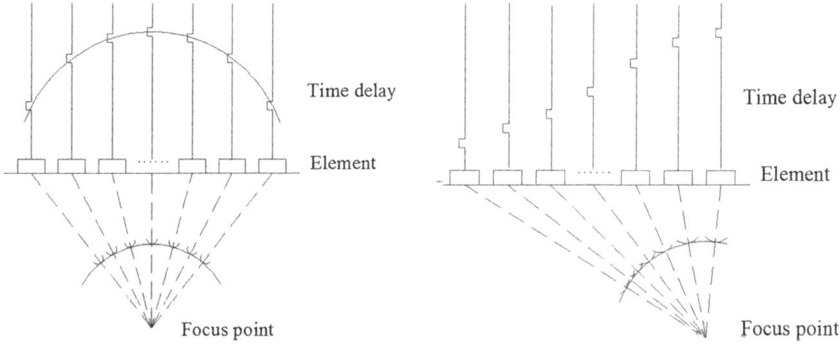

Fig. 75.2 Beam focusing and beam deflection of liner UPA

Table 75.2 Detection parameters

Name	Material
Couplant	Engine oil
Detection type	Pulse-echo
Ultrasound type	Transverse wave
Imaging type	A scan and S scan
Scanning angle	40°–70°
Scale	True depth

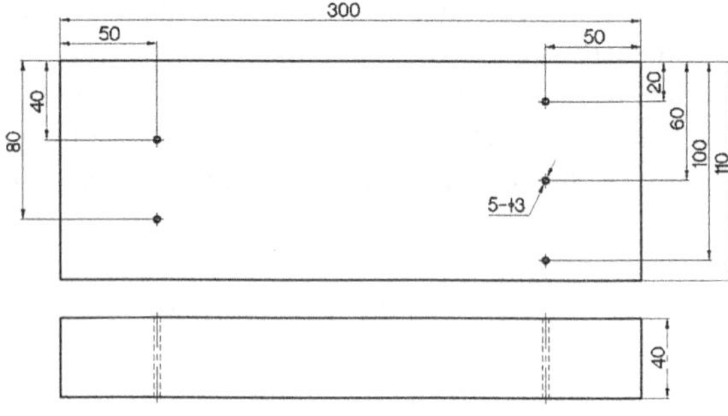

Fig. 75.3 RB-2/20

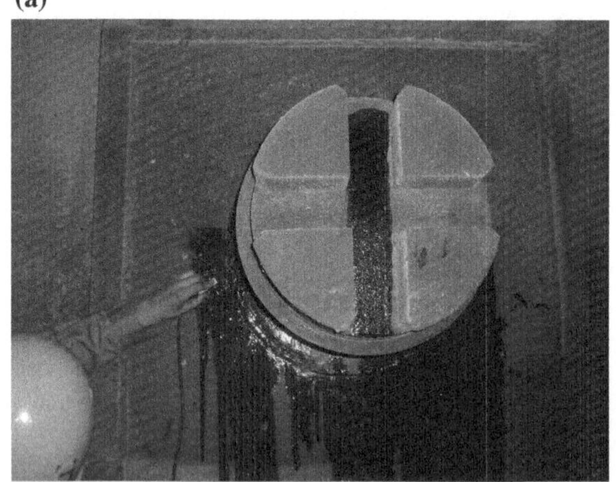

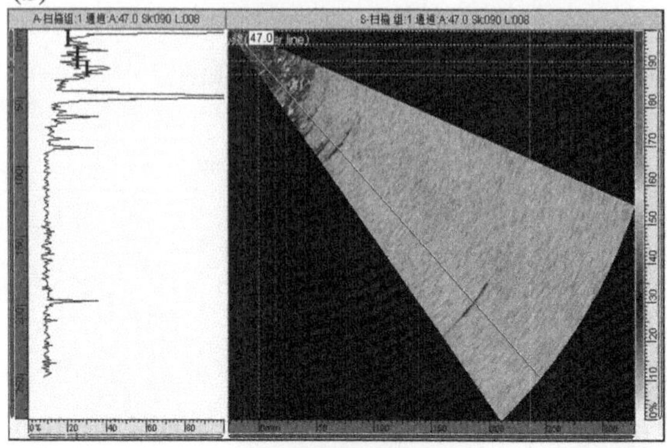

Fig. 75.4 **a** Detection position 1. **b** Test result 1

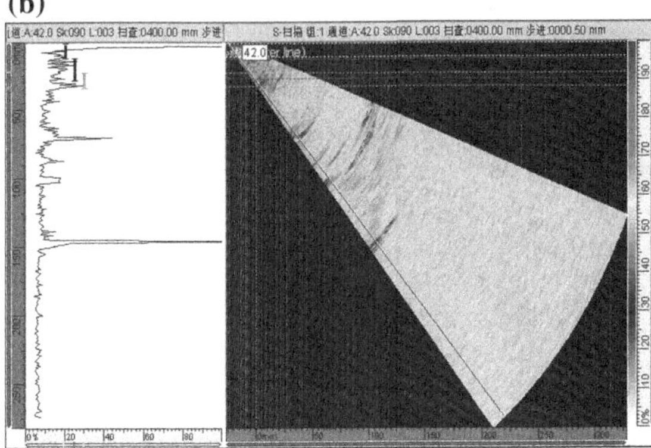

Fig. 75.5 **a** Detection position 2. **b** Test result 2

Before the testing experiment, calibration of the detector is needed. As the ultrasonic property is similar to the ladle's trunnion, it was used for sound velocity calibration. Figure 75.3 shows the specific information of RB-2 test block. The calibrated sound velocity is 3,223 m/s.

75.4.2 Results and Discussion

Figure 75.4a shows the first detection position, the shortest distance between the probe and the side of the trunnion is 50 mm. Seen from Fig. 75.4b, a large signal occurs in 50 mm depth, 47°, it must be the echo of the trunnion box's outer steel plate. The signal in 200 mm depth is the echo of the ladle's ektexine, between the two signals, there are two small signals, it illustrates that there are two small defects in the trunnion sleevc.

Figure 75.5a shows the second detection position, the shortest distance betwcen the probe and the side of the trunnion is 30 mm. Seen from Fig. 75.5b, a small signal occurs in 30 mm depth, 43°, it must be the echo of the trunnion's sidewall. The signal in 70 and 150 mm depth illustrates that there are two defects inside the trunnion.

75.5 Conclusion

UPA technology has the advantages of convertible beam direction and focus, the results is easy imaging and the detection has better accessibility and applicability. Therefore, it can solve the inspection difficulties of ladle trunnion; similarly, it can be generalized to other complex components or deep weld.

Acknowledgments This work is supported by National Natural Science Foundation of China under Grant 51275498 and Key Program of Science and Technology Planning Project of Zhejiang Province, China under Grant 2011C11079.

References

1. Lu X (2010) Trunnion design of large steel and iron ladle. J Mech Eng 9:168–169 (in Chinese)
2. Ye J, Kim HJ, Song SJ et al (2011) Model-based simulation of focused beam fields produced by a phased array ultrasonic transducer in dissimilar metal welds. J NDT&E Int 44:290–296
3. Steve M, Jean LJ, Olivier R et al (2004) Application of phased array techniques to coarse grain components inspection. J Ultrason 42:791–796
4. Wang CX, Zhang H, Gao XR et al (2009) The application of ultrasonic phased array in wheel rim flaw detection. J Chin Railways 5:69–71 (in Chinese)
5. Mou YC, Jin NH, Ge X (2011) Ultrasonic phased array inspection of fillet pipe butt weld for plant boiler. J NDT 33:75–78 (in Chinese)
6. Tian AD (2008) The principle and its application of phased array ultrasonic inspeciton technology. J NDT 30:58–62 (in Chinese)
7. GB 713-2008 (2008) Steel plates for boilers and pressure vessels. Trade standards of the People's Republic of China
8. GB/T 700-2006 (2006) Carbon structural steels. Trade standards of the People's Republic of China
9. GB/T 699-1988 (1999) Quality carbon structural steels. Trade standards of the People's Republic of China
10. GB/T 5118-1995 (1995) Low alloy steel covered electrodes. Trade standards of the People's Republic of China
11. Azar L, Shi Y, Wooh SC (2000) Beam focusing behavior of linear phased array. J NDT & E Int 33:189–198
12. Hu D, Wang Q, Xiao K, Ma YH (2012) Ultrasonic phased array for the circumferential welds safety inspection of urea reactor. J Procedia Eng 43:459–463

Chapter 76
Optimal Tracking Design for Dry Clutch Engagement

Taotao Jin, Pingkang Li and Zhizhou Jia

Abstract To improve the shift quality of a vehicle equipped with an automated manual transmission, the dry clutch engagement process needs to be controlled precisely. This paper proposes to generate the optimal clutch engagement reference speed based on the optimal control theory. The gear-shift duration weighting parameters are used to balance the performance of gear-shift duration and shift smoothness. A simplified vehicle system model is developed for the vehicle level simulation validation; the vehicle launch process is studied by using proposed control strategy. Numerical simulation results indicated that the balanced shift performance can be achieved.

Keywords Automatic manual transmission · Clutch engagement · Optimal tracking curve design · Shift quality optimization

76.1 Introduction

Automatic manual transmission (AMT) is an inexpensive up-grade solution to the classical manual transmission (MT) that can be easily integrated with a conventional multi-speed MT using electric, pneumatic or hydraulic actuator to operate the gear-shift lever and clutch [1]. Due to the engine and clutch speed difference, shifting from one gear to another causes shock easily, which deteriorates the vehicle driving comfortableness [2, 3]. Therefore, controlling clutch engagement process during gear-shift with good driving comfort is a key problem for AMTs [4]. Fuzzy logic control was used to find the optimal gear-shift schedule in [5];

T. Jin (✉) · P. Li · Z. Jia
School of Mechanical, Electronic and Control Engineering,
Beijing Jiaotong University, Beijing 100044, China
e-mail: jintt@msu.edu

Z. Sun and Z. Deng (eds.), *Proceedings of 2013 Chinese Intelligent Automation Conference*, Lecture Notes in Electrical Engineering 255,
DOI: 10.1007/978-3-642-38460-8_76,

Genetic algorithm [6] is realized to optimize the shift quality of a dual clutch transmission; and optimal control technique [7] was used for tracking the reference signal, however, in most of the research, the clutch engagement reference speeds are determined based on the experience [7, 8].

This paper discussed the clutch engagement reference speed generation based on the optimal control theory, a dynamic AMT vehicle model is derived and the control problem of dry clutch engagement is studied; then, a tracking reference signal calculation algorithm is proposed and applied to generate the clutch reference speed during the engagement process. The gear-shift duration weighting parameter is used to balance the requirements of gear-shift duration and smoothness, the vehicle launch process is simulated as an example.

76.2 System Modeling

Using the theories of multi-body dynamics, an AMT vehicle can be simplified into a dynamic system as shown in Fig. 76.1.

Torque produced by the engine is transmitted through the dry clutch, gearbox, final driveshaft to the driving wheels. The engine dynamics can be modeled as

$$J_e\dot{\omega}_e = T_e - \beta_e\omega_e - T_c \tag{76.1}$$

where ω_e is the engine rotation speed; T_c is the transmission clutch load torque. The clutch operation modes can be divided into two: slipping and locked.

- Slipping clutch

  Dynamic model of the clutch under the slipping operation can be expressed as:

$$J_c\dot{\omega}_c = T_c - k_{cm}\theta_{cm} - \beta_{cm}(\omega_c - \omega_m);\ \ \dot{\theta}_{cm} = \omega_c - \omega_m \tag{76.2}$$

where J_c is the clutch disc inertia; ω_c and ω_m are the corresponding rotation speeds of the clutch disc and main shaft, respectively.

- Locked clutch

When the clutch is locked, $\omega_c = \omega_e$ and two differential Eqs. (76.1) and (76.2) can be reduced down to a single differential equation:

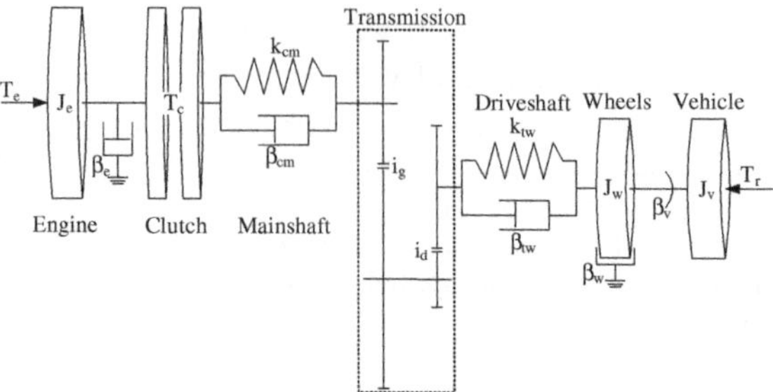

Fig. 76.1 Schematic diagram of the AMT system

$$(J_e + J_c)\dot{\omega}_c = T_e - \beta_e \omega_e - k_{cm}\theta_{cm} - \beta_{cm}(\omega_c - \omega_m) \tag{76.3}$$

The clutch remains locked as long as the torque transmitted through clutch is below the maximum transmittable torque.

The selected gear ratio of the transmission is denoted by i_g, the final differential gear ratio is i_d. The main shaft dynamics can be described as follows:

$$\begin{aligned} J_{eq}\dot{\omega}_m &= k_{cm}\theta_{cm} + \beta_{cm}(\omega_c - \omega_m) - \frac{1}{i}\left[k_{tw}\theta_{fw} + \beta_{tw}(\omega_f - \omega_w)\right] \\ \dot{\theta}_{fw} &= \omega_f - \omega_w; \quad \omega_f = \frac{1}{i}\omega_m \end{aligned} \tag{76.4}$$

where $J_{eq}(i_g, i_d) = J_m + J_t/i^2$; $i = i_g i_d$; J_t and J_m are the transmission inertia and main shaft inertia. The vehicle driving wheels and tires are modeled as an inertia connected to the vehicle through a linear damper with coefficient β_v and the tires also experience a rolling resistance torque $\beta_w \omega_w$. The differential equation governing the wheel and tire dynamics can be expressed as:

$$\begin{aligned} J_w\dot{\omega}_w &= k_{tw}\theta_{fw} + \beta_{tw}(\omega_f - \omega_w) - \beta_w\omega_w - \beta_v s \\ s &= \omega_w - v_v/r_w = \omega_w - \omega_v \end{aligned} \tag{76.5}$$

where J_w is the inertia of the driving wheels; ω_w and v_v are the wheel rotational speed and vehicle speed separately; r_w is the wheel radius. So,

$$\begin{aligned} J_v\frac{\dot{v}_v}{r_w} &= \beta_v s - T_r \\ T_r &= (F_a + F_g)r_w = \left(\frac{1}{2}A\rho C_D v_v^2 + Mg\sin(\theta)\right)r_w \end{aligned} \tag{76.6}$$

where $J_v = Mr_w^2$ is the equivalent vehicle mass inertia; F_a is aerodynamic force; A is vehicle front area; ρ is the ambient air density; C_D is aerodynamic drag coefficient; F_g is gradient resistance force; M is vehicle mass; g is the gravity acceleration; θ is gradient due to road elevation.

76.3 Optimal Tracking Reference Design

To simplify the control design process, a control oriented reduced order model is used. Assume that the driveline is rigid and that the wheel inertia and equivalent vehicle inertia can be lumped together into a single inertia as follows

$$\omega_c = \omega_m = \omega_t = i\omega_w = iv_v/r_w \tag{76.7}$$

Substituting (76.7) into the Eqs. sets from (76.2) to (76.6), the entire driveline dynamics from the clutch disk to the wheels can be approximated by

$$J_a\dot{\omega}_c = T_c - \beta_a\omega_c - T_L \tag{76.8}$$

where $J_a = J_c + J_{eq} + J_w/i^2$, $\beta_a = \beta_w/i^2$ and $T_L = T_r/i^2$.

76.3.1 Design Constrains

It is obvious that the requirements listed above are in conflict. Compromise between them, the desired engagement duration should be made, the following cost function is proposed for the problem solving

$$J = \int_{t_0}^{t_f} (W + \dot{a}^2/2)dt \tag{76.9}$$

where W represents the slipping duration weighting coefficient and $\dot{a}$ is the rate of vehicle longitudinal acceleration that can reflect the gear-shift shock.

76.3.2 Optimal Tracking Reference Design

The state space model shown in (76.8) can be simplified as follows

$$\dot{x}_1 = \frac{1}{J_a}x_2 - \frac{T_L'}{J_a}; \quad \dot{x}_2 = u_1 \tag{76.10}$$

where $x_1 = \omega_c$, $x_2 = T_c$, $T_L' = T_L + \beta_a\omega_c$ and $u_1 = \dot{T}_c$. In order to avoid discontinuity on the clutch torque, the time derivative of the clutch torque is also considered as an input. The vehicle acceleration a can be rewritten as

$$\dot{a} = \ddot{\omega}_c r_w/i \tag{76.11}$$

The engine speed also needs to be controlled precisely to minimize the clutch slipping duration. Supposing the transmission rate before and after the gear-shift are $i_{g(k)}$ and $i_{g(k+1)}$ separately, the output shaft speed before the gear-shift is

$$\omega_{o(k)} = \omega_{e(k)}/\left(i_{g(k)}i_d\right) \tag{76.12}$$

and the speed after the gear-shift is

$$\omega_{o(k+1)} = \omega_{e(k+1)}/\left(i_{g(k+1)}i_d\right) \tag{76.13}$$

The vehicle speed during the gear-shift is assumed to be constant ($\omega_{o(k)} = \omega_{o(k+1)}$), combined Eqs. (76.12) and (76.13), the target engine speed is

$$\omega_{e(k+1)} = \omega_{o(k)}i_{g(k+1)}i_d \tag{76.14}$$

In the shifting process, engine speed should be controlled to minimize the speed difference between the clutch disc plates, which can not only speed up the clutch engagement, but also reduce the impact.

Substituting (76.11) into (76.9) and introducing the terminal conditions of the clutch engagement ($\omega_e(t_f) = \omega_c(t_f) = r_2$), the object function can be rewritten as

$$J^* = \phi + \int_{t_0}^{t_f} \left(\psi + \frac{1}{2}\ddot{x}_1^2\right) dt = v_1 (r_2 - x_1(t_f)) + \int_{t_0}^{t_f} \left(\psi + \frac{1}{2}\ddot{x}_1^2\right) dt \qquad (76.15)$$

where $\psi = i^2 W / r_w^2$. Therefore, the optimal control problem is to minimize the cost function described in (76.15) and the Hamiltonian equation for this problem is

$$H = \psi + \frac{1}{2}\ddot{x}_1^2 + \lambda_1 \left(\frac{1}{J_a} x_2 - \frac{T_L'}{J_a}\right) + \lambda_2 u_1 \qquad (76.16)$$

where λ_m $(m = 1, 2)$ is the Lagrangian multiplier. Since the clutch engagement duration is very small, it is assumed that driveline torque load T_L' during the clutch engagement process is constant. Substitute system Eqs. (76.10) into (76.16)

$$H = \psi + \frac{1}{2J_a^2} u_1^2 + \lambda_1 \left(\frac{1}{J_a} x_2 - \frac{T_L'}{J_a}\right) + \lambda_2 u_1 \qquad (76.17)$$

Using the relevant optimal control theory [9], the optimal control satisfying

$$\frac{\partial H}{\partial u_1} = 0, \quad \frac{1}{J_a^2} u_1 + \lambda_2 = 0, \quad \dot{\lambda}_1 = -\frac{\partial H}{\partial x_1} = 0, \quad \dot{\lambda}_2 = -\frac{\partial H}{\partial x_2} = -\frac{1}{J_a}\lambda_1 \qquad (76.18)$$

with the following boundary conditions

$$\begin{cases} \lambda_1(t_f) = -v_1 \\ \lambda_2(t_f) = 0 \end{cases}; \quad \begin{cases} x_1(0) = r_1 \\ x_2(0) = 0 \end{cases}; \quad r_2 - x_1(t_f) = 0$$
$$H_{t_f} = \psi + \tfrac{1}{2J_a^2} u_1^2(t_f) + \lambda_1(t_f)\left(\tfrac{1}{J_a} x_2(t_f) - \tfrac{T_L'}{J_a}\right) + \lambda_2(t_f) u_1(t_f) = 0 \qquad (76.19)$$

Solving the above Two Point Boundary Problem (TPBP), the following optimized clutch speed and torque can be derived

$$\begin{cases} t_f = -\frac{J_a c_2}{v_1}; \quad v_1 = \frac{J_a}{T_L'}\left(-\psi + \frac{J_a^2 c_2^2}{2}\right) \\ \frac{J_a^5 c_2^4 (r_2 - r_1)}{4T_L'^2} - \frac{J_a^2 c_2^3}{6} - \frac{J_a^3 \psi (r_2 - r_1)}{T_L'^2} c_2^2 + \psi c_2 + \frac{J_a (r_2 - r_1)}{T_L'^2} \psi^2 = 0 \end{cases} \qquad (76.20)$$

$$\begin{cases} x_1 = -v_1 \frac{t^3}{6} - J_a c_2 \frac{t^2}{2} - \frac{T_L'}{J_a} t + r_1 \\ x_2 = -J_a v_1 \frac{t^2}{2} - J_a^2 c_2 t \end{cases} \qquad (76.21)$$

where C_2 and v_1 are constant and can be calculated using the boundary conditions shown above. From (76.11) and (76.21), it can be find that the optimal acceleration is proportional to the optimal friction torque with the following form

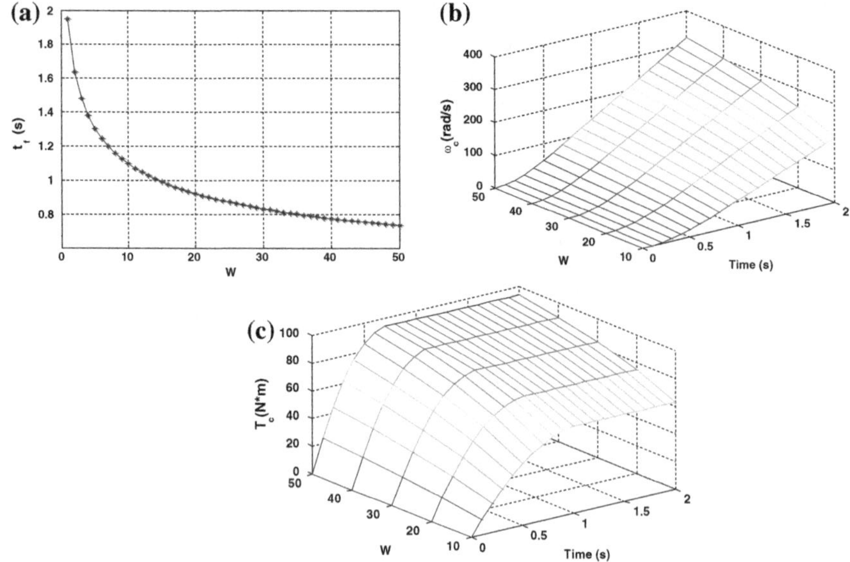

Fig. 76.2 Simulation results for different W

$$\dot{a}_{\max} = -J_a r_w c_2 / i \tag{76.22}$$

The rate of the vehicle longitudinal acceleration, treated as the shock intensity, needs to be limited to $|\dot{a}| \leq 10\,\text{m/s}^3$. This is achieved by selecting a proper W in (76.9). Normally, its upper bound is determined by the driving comfortableness and the lower bound by the requirement of minimal clutch plate wear. From above analysis, the weighting coefficient W can be chosen between 0 and 50, and simulation results for different W are as shown in Fig. 76.2.

Figure 76.2a and b show the clutch engagement duration (t_f) and speed (ω_c) as a function of W; Fig. 76.2c presents the clutch engagement torques (T_c) under different W. It can be seen that the larger the slipping duration weighting coefficient W, the shorter the engage time. Large slipping duration coefficient W also leads to large rate of the vehicle acceleration speed. Based on the analysis results, coefficient W can be used as a tuning parameter associated with the driving comfortableness. If the sport vehicle performance is required, large W should be selected subject to the clutch torque limitation constraints; while if smooth driving performance is required, smaller W should be selected.

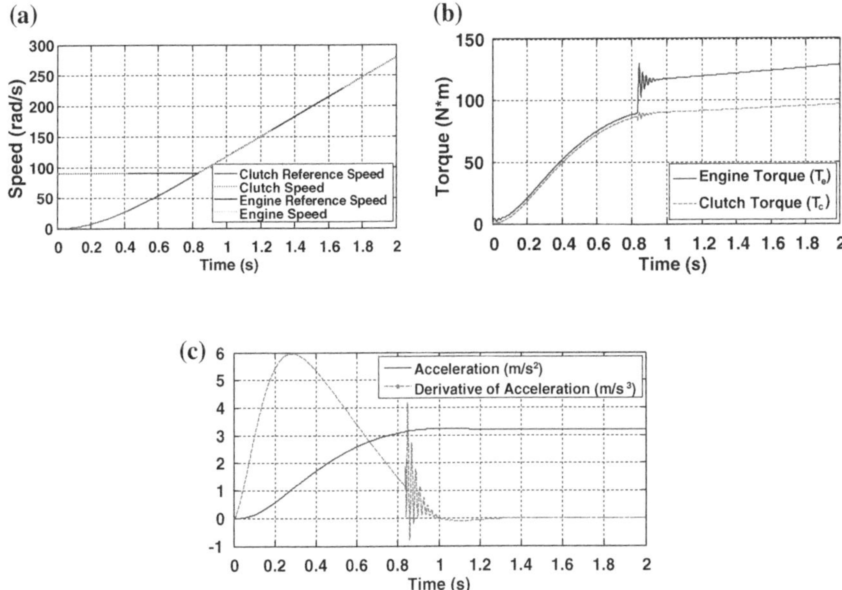

Fig. 76.3 Tracking of the optimal reference. **a** Reference signal tracking. **b** Engine and clutch torque. **c** Vehicle acceleration and its derivation

76.4 Tracking of the Optimal Reference Simulation

In order to track the optimized trajectories with minimal driveline vibrations, a PID controllers are implemented to independently tracks the optimized engine speed ω_e and clutch speed slip speed ω_c.

In order to show the effectiveness of the derived control strategy and to highlight the undesired oscillatory behavior of the driveline, the full order dynamic model is used in the simulation. The clutch engagement process during the vehicle launch is considered. Clutch engagement speed is $r_1 = 0$ rad/s and the engine references speed is chosen at $r_2 = 90$ rad/s for the vehicle launch in the first gear. After the clutch engaged, the vehicle start accelerating based on the opening of the acceleration pedal. As shown in Fig. 76.3a and b, all state variables, such as ω_e and ω_c, follow their respective references very well, and clutch torque increase smoothly, indicating good vehicle launch performance achieved. Figure 76.3c shows the vehicle acceleration and its derivation, the curves changed smoothly and indicating good vehicle launch performance achieved. The acceleration change at the end of the engagement also in the acceptable range.

For the operational conditions other than the vehicle launch, such as gear-shift from 2nd to 3rd gear, properly select the initial and final conditions of the engine and clutch; and the optimized reference signal can be obtained following the similar procedure.

76.5 Conclusion and Future Works

This paper considers the clutch engagement control for vehicle equipped with the AMTs. The speed tracking reference signals are derived based on the optimal control theory. The engagement duration weighting parameter is used to balance the gear-shift duration and driving comfortableness; the simulation results indicated that the proposed algorithm can provides fast clutch engagement with reduced slipping losses. Future work will consider the real vehicle application.

Acknowledgments This Research was supported by National Science Fund of China (61074104).

References

1. Zongxuan S, Hebbale K (2005) Challenges and opportunities in automotive transmission control. In: Proceedings of the 2005 American control conference, USA, pp 3284–3289
2. Hyeoun-Dong L, Seung-Ki S, Han-Sang C (2000) Advanced gear-shifting and clutching strategy for a parallel-hybrid vehicle. Ind Appl Mag 6(6):26–32
3. Lucente G, Montanari M, Rossi C (2007) Modelling of an automated manual transmission system. Mechatronics 17(2–3):73–91
4. Van Der Heijden A, Serrarens A, Camlibel M (2007) Hybrid optimal control of dry clutch engagement. Int J Control 80(11):1717–1728
5. Guihe Q, Anlin G, Ju-Jang L (2004) Knowledge-based gear-position decision. IEEE Trans Intell Transp Syst 5(2):121–125
6. Hongwei H, Xianghong W, Yimin S (2010) Optimization of the shift quality of dual clutch transmission using genetic algorithm. In: Proceedings of 6th ICNC, pp 4152–4156
7. Garofalo F, Glielmo L, Iannelli L (2002) Optimal tracking for automotive dry clutch engagement. In: Proceedings of 15th IFAC world congress, Barcelona
8. Glielmo L, Iannelli L, Vacca V (2006) Gearshift control for automated manual transmissions. IEEE/ASME Trans Mechatron 11(1):17–26
9. Athans M, Falb PL (2007) Optimal control: an introduction to the theory and its applications. Dover Publications, New York

Chapter 77
Automatic Generation of User Interface Method Based on Automated Planning

Jie Gao, Fu Wang and Lei Li

Abstract We introduce the thought of intelligent planning into interface design and development in software engineering, and propose a planning-based approach for automatic interface generation. The advantages of this approach for interface design are that interface design and code realization can be separated effectively, coupling of programming can be reduced, so that developers can invoke interfaces in logical sequence, not necessarily by manual programming. It can be proved that this approach can greatly reduce manpower of developers, shorten development cycle and save costs.

Keywords Intelligent planning · Software engineering · Automatic interface generation

77.1 Introduction

In software engineering, as the media between system and users, interface design is getting highly attention from software engineer and users. If an interface of a software product is not friendly and beautiful enough, it will not be accepted by users, even if the system has powerful function. As for the variability of user's demand, designing and realizing interfaces is random and variable, which makes

J. Gao (✉)
Zhuhai College, Jilin University, Zhuhai 519000, China
e-mail: jiegao26@163.com

J. Gao · F. Wang · L. Li
School of Information Science and Technology,
Sun Yat-Sen University, Guangzhou 510275, China
e-mail: 617374250@qq.com

L. Li
e-mail: lnslilei@mail.sysu.edu.cn

Z. Sun and Z. Deng (eds.), *Proceedings of 2013 Chinese Intelligent Automation Conference*, Lecture Notes in Electrical Engineering 255,
DOI: 10.1007/978-3-642-38460-8_77,

interface designers under great pressure. Automatic interface generation technique attaches more and more attention for reducing pressure from designers.

Automatic interface generation technique has been used to separate user interface from the application logic. Although this approach is not successful in general application, we believe it can be a valuable tool in certain situations. Automatic interface generation also allows specific information about users and current situation to be incorporated into designing of user interface. There has been a rich history of research in the area of automatic generation, mainly in the model-based user interface community. Model-based systems attempt to formally describe the tasks, data, and users and then use these formal models to guide the generation of the user interface. Some system automatically designs user interface, such as UIDE [1] and Jade [2], and others provide design assistance to a human, such as Trident [3] and Mobi-D [4]. Despite a lot of research, model-based user interface tools have not become common, partially because building models is an abstract process and better results are often achievable by a human designer in less time [5]. There have been success in limited domains such as dialog box design [1, 2] and remote controls [6], and we believe that model-based techniques show promise provided that the scope of the generated interfaces is kept manageable.

The paper proposes a novel approach of automatic interface generation based on intelligent planning, called Planning method. The main idea of this method is that we define object classification, states set and actions set for modeling the system, and describe the model by PDDL, generate PDDL description of user's operation, then solve for a planning solution by planner. Finally, for each action in the sequence, an interface is generated automatically.

The main contribution of this method can be summarized as follows. Firstly, domain definition, planning and programming for interface generation are separated. Once domain information needs to be changed, only these files need to be modified. It solves the mentioned problem that system design and code realization has high degree of coupling. Secondly, since each interface corresponds to each action in sequence, the sequence of invoking interfaces is decided by that of corresponding actions. Therefore, invoking interfaces needs not to be programmed by interface designer manually. It proves that this method can significantly reduce work intensity of software engineer, shorten development cycle, and reduce development cost, which is another point of contribution.

77.2 Related Work

77.2.1 Intelligent Planning

Intelligent planning systems achieve goals by producing sequences of actions from given action models that are provided as input. In 1971, Fikes and Nils designed STRIPS system [7] to introduce definitions of STRIPS operators, which made significant difference in the research of automated planning. In 1991, Soderland and Weld [8] designed the first nonlinear planning system SNLP of the world. In 1996, Kautz [9] converted planning into SAT problem, which effectively solved partial planning problem and showed new direction of automated planning. In 1995, Avrim and Merrick [10] designed the first graph planner system Graph plan to solve planning problem, and proposed concept of graph plan. In 1998, Malik proposed Plan Domain Definition Language (PDDL) [11], then PDDL gradually became a general standard of representing domain models and was applied broadly in international planning competitions.

77.2.2 User Interface Generation Technique

Current user interface generation technique mainly include three kinds of methods, respectively based on visual sense, model, and POMDP.

In 1990, Zanden and Myers from Carnegie Mellon University, proposed one of the early methods of interface generation based on visual sense, called Jade [2], which means Judgement-Based Automatic Dialog Editor. Jade is a new interactive tool that automatically creates graphical input dialog such as dialog boxes and menus. This specification contains absolutely no graphical information and thus is look-and-feel independent. Jade combines the application programmer's specification with the look-and-feel database to automatically generate a graphical dialog.

Model-based systems attempt to formally describe the tasks, data, and users that an application will have, and then use these formal models to guide the generation of the user interface. Some systems automatically design the user interface, such as UIDE [1] and Jade [2], and others provide design assistance to a human, such as Trident [3] and Mobi-D [4]. Despite a lot of research, model-based user interface tools have not become common, in part because building models is an abstract process and better results are often achievable by a human designer in less time [5]. There have been successes in limited do-mains such as dialog box design [1, 2] and remote controls [6], and we believe that model-based techniques show promise provided that the scope of the generated interfaces is kept manageable.

POMDP-"Partially Observable Markov Decision Processes" [12] is a mathematical model to solve decision problems with partial observation. Jaeyoung Park from South Korea research college applied POMDP to decide the most favorable interface users need to obtain. By applying POMDP model, we can calculate the probability of each interface obtained by users, and then sort all the interface elements from high to low according the probabilities. POMDP model of approximation version is actually applied in research, like Witness algorithm. But, approximation algorithm will lower correctness of algorithm. Therefore, how to choose between complexity and correctness is one of research direction of POMDP model.

77.3 Automatic Interface Generation Method Based on Intelligent Planning

This paper focuses on how to apply intelligent planning to solve automatic interface generation problem, called planning method. In this section, we introduce main steps of Planning method, including setting up xml file to store domain information; producing domain description file according to domain information; solving for a planing solution by forward search algorithm in space states space; renewing init file according to action sequence of the planning solution. We will introduce each step of planning method in detail, and analyze complexity of planning method. Framework of planning method is shown in Table 77.1.

Firstly, we will define four xml files, named as types.xml, predicates.xml, actions.xml and init.xml, which are used to save information of objects set, states set, action set, and initial state set respectively. The operation procedure is shown as follows:

1. Write information of *types.xml*, *predicates.xml*, *actions.xml* into *domain.pddl*, and generate description file of domain.
2. Users execute some operation (e.g. Click on menu operation).
3. Write information in *init.xml* and goal information of user's operation into *problem.pddl*, and produce description file of planning problem.
4. Input *domain.pddl* and *problem.pddl* into planner, and solve for a planning solution after calculation.
5. Put actions in the planning solution into *queue* in sequence.
6. Go through the queue, and generate interfaces according to each action and parameter.
7. Write change of initial states into *init.xml*.

Table 77.1 Framework of planning method

Input Types.xml, predicates.xml, actions.xml
Step 1 Write information in types.xml, predicates.xml, actions.xml into domain.pddl, and produce description file of domain
After step 1, users execute some operation (e.g. Click on menu operation)
Step 2 Write information in init.xml and goal information of user's operation into problem.pddl, and produce description file of planning problem
Step 3 Input domain.pddl and problem.pddl into planner, and solve for a planning solution after calculation
Step 4 Put actions in the planning solution into queue in sequence
Step 5 Go through the queue, and produce interface according to each action and parameter
Step 6 Write change of initial states into init.xml
Output Interfaces generated corresponding to actions in the planning solution in sequence

77.3.1 Genertating pddl File

77.3.1.1 Setting up xml File

Information of objects classification is saved as *types.xml*, with the following storage structure,

$$
\begin{gathered}
<type> \\
<name>\ selfname </name> \\
<fathertype>\ typename </fathertype> \\
</type>
\end{gathered}
$$

where *name* term denotes the name of type, *fathertype* term denotes the name of father type, and *fathertype* must be a type defined in type.xml.

All the atom propositions are saved in predicate.xml, namely information of states set. Atom proposition is saved as follows.

$$
\begin{gathered}
<predicate> \\
<name>\ predicatename </name> \\
<parameter\ type = typename>\ variablename </parameter> \\
\ldots \\
<parameter\ type = typename>\ variablename </parameter> \\
</predicate>
\end{gathered}
$$

where *name* term denotes the name of predicate, *parameter* term denotes the parameter of predicate, and type of parameter must be defined in *types.xml*.

All the actions are saved in *action.xml* with the following storage structure.

$$
\begin{gathered}
<action>\ <name>\ actionname </name> \\
<parameter\ type = typename>\ parametername </parameter> \\
\ldots \\
<parameter\ type = typename>\ parametername </parameter> \\
<precond\ type = predicate\ valid = true\ or\ false>\ predicatename </precond> \\
\ldots \\
<precond\ type = predicate\ valid = true\ or\ false>\ predicatename </precond> \\
<effect\ type = predicate\ valid = true\ or\ false>\ predicatename </effect> \\
\ldots \\
<effect\ type = predicate\ valid = true\ or\ false>\ predicatename </effect> \\
</action>
\end{gathered}
$$

where *name* denotes name of actions, *parameter* denotes parameters list of action, type of parameters must be defined in *types.xml*, and *precond* and *effect* denote preconditions and effects of action respectively. Their types are predicates, and names must be defined in *predicates.xml*, *valid* is used to denote true or false of predicates.

Init.xml is a subset of *predicates.xml*, having the same definition with *predicates.xml*.

77.3.1.2 Generating Domain File

Domain file includes domain.pddl and problem.pddl. For domain.pddl, we transform types.xml, predicates.xml, actions.xml into pddl file. Problem file is produced by user's operation. Firstly, we build up a problem.pddl file manually and define problem name, related fields and objects. Then we make a user operation and target map. Each time when user performs an operation, the system will write corresponding target into problem.pddl according to target map, and write content of init.xml into problem.pddl. The genearted files including domain.pddl and problem.pddl will be submitted into planner as input.

77.3.2 Obtaining a Plan Solution

For STRIPS planning, the set of propositions is a limited set, which means the size of states set in STRIPS is limited. Exhaustive search method can be applied to determine the existence of solution, therefore forward states space search algorithm can be adopted. The input of algorithm is a planning problem P = (A, s0, g), if P is solvable, then returns a planning solution, otherwise returns failure. The main idea of algorithm is to search for subsequent actions to be suitable for current status of the system. Pseudo code of the algorithm is shown in Table 77.2. For a planner, action set, initial states and goal states are denoted in domain.pddl and problem.pddl respectively, therefore the two files are used as input of planner, then acquire a planning solution by a planner, which is a set of actions.

Table 77.2 Forward states space search algorithm

```
ForwardSearch(A,s0,g)
s ⇐ s0
π⇐the empty plan
if s satisfies g then return π then
   applicable⇐{a—a is a ground instance of an operator in A,and precond(a) is true
   in s}
end if
if applicable=∅ then
   return failure
   choose an action a∈ applicable
   s⇐γ(s,a)
   π⇐π.a
end if
```

77.3.3 *Update init File According to Plan Solution Interface to be Generated*

77.3.3.1 Generating Interface According to Plan Solution

The basic idea of generating interface is explained as follows. Interface elements correspond with action parameters, and type of action parameter corresponds with type of object. Therefore, we use a data table to store types of all objects. For each type, range and parameter constraints should be defined. The data table is named as a parameter table, with storage structure shown as follows.

Parameter ID	Parameter type	Range	Parameter constraints

Parameter ID represent types of different parameters. Parameter type corresponds to name of each type. Range is expressed in the form of set. If range of some parameter is limited, then it can be described by brute-force method; if range of some parameter is wide, then it can be defined by description method.

The file type corresponds to a button of opening the file. The correspondence between object type and control can be described as a table. For types with constraints, we should attach constrained regular expressions to the generated control. The relationship between object type and control is described in Table 77.3. In addition, storage mechanism can be built up for interfaces, then it will not be necessary to create a new interface when the same action appears in the action sequences, but only to search database table and reshape interfaces according to obtained information.

It is necessary to build up two database tables for storage of interfaces, where the first database table stores the correspondence relationship between interfaces and actions, named as mapping table of interface and action, shown in Table 77.4. Specific elements in the interface are stored in the table of interface elements, shown in Table 77.5, where control name is static text labels corresponding to the control in the interface; control ID is used to label different controls. Control type is the type of control, such as a single box, check box, etc. Control constraints include constraints of range and expression for controls, and regular expressions are required in the constraints of expression.

Table 77.3 The relationship between object type and control

Object type	Control
1. The number of selectable values is less than 6	1. Single box or multiple box
2. The number of selectable values is between 7 and 30	2. Check box
3. The number of selectable values is greater than 30	3. Define text input box with constraint
4. File type	4. Icon of opening a file

Table 77.4 Storage construction of action interfaces

Action ID	Action name	Interface ID

Table 77.5 Storage structure of interface elements

Interface ID	Control ID	Control name	Control type	Control constraint

The basic idea of interface generating algorithm is summarized as follows. The algorithm traverses the queue consisting of actions in the action sequence, so that actions get out of the stack sequentially. If the action has already been located in the mapping table of actions and interfaces, then control elements corresponding to this action are acquired and interfaces are generated; otherwise, the algorithm traverses parameter list of the action and acquires range of each parameter from parameter table, then corresponding interface controls and interface are generated.

77.3.4 Updating init File

After generating interfaces, init file is updated according to action sequence. The basic idea is to find the last action in the action sequence, and traverse the subsequent predicates affected by the action. If the predicates are true and not saved in init file, then they are added into init file. Otherwise, if the predicates are false and already stored in init file, then they will be deleted from init file.

77.4 Comparison with Other Methods and Analysis

Currently, automatic interface generation method mainly include the method based on visual sense, based on models. These methods have similar basic idea, namely, describing system interfaces by files or building up models, and transform the description into interface prototype. Compared with these methods, planning methods have the following two aspects of advantages:

1. Separating code realization and design completely, and reducing the coupling degree of the program.

Here we consider a classic method of automatic interface generating algorithm based on models-Model-Driven User Interface generation Method. One instance of this approach is to automatically generate interfaces of library management system. Firstly, we create a class diagram for library management system, each class corresponding to an interface, and then we transform class diagram into interface prototype by tools. The attribute element in the class corresponds to the control element in the interface.

From diagram, it can be seen that if any change is happened in the system, inaddition to modifying original design, we should also modify codes, therefore there exists high level of redundancy between design and code. In this paper, interface element corresponds to parameter in action sequence, and each action is defined independently in xml file. If any change is happened in some action or parameters of some action, the only thing to be modified is xml file and data table. Code part do not necessarily need to be modified. Therefore, coupling degree of design and code can be reduced, and furthermore, independence of system is relatively enhanced.

2. Automatically establish logic relationship between interfaces

Take interface generation method based on visual sense as example. This method is based on the interface description as input. The requirement of users is understood by system, through reading files and generating corresponding interfaces. This method has similarities with planning method we proposed in this paper. The difference is that only one interface is generated each time for the method based on visual sense, and interfaces are independent respectively. The logic relationship between interfaces needs interface designer to make programming manually. In planning method, interface corresponds with action in action sequence, so that the logic relationship between interfaces are determined by those between actions in action sequence. Therefore, it is not necessary for designers to determine logic relationship between interfaces by programming manually. Obviously, this property will make system more dynamic and self-adaptive.

77.5 Conclusion

In this paper, we combine intelligent planning with traditional software design method, and propose a novel approach of automatic interface generating. As shown in this paper, the advantages of this approach are that interface design and programming can be separated efficiently, and degree of coupling can be reduced, so that developers do not need to call interfaces logically by manual programming. It can be proved practically that the strength of the interface development work can be reduced greatly and software development cycle can be reduced, by applying this approach.

References

1. Kim WC, Foley JD (1993) Providing high-level control and expert assistance in the user interface presentation design. In: Proceedings of INTERCHI'93. Amsterdam, The Netherlands, pp 430–437
2. Zanden BV, Myers BA (1990) Automatic, look-and-feel independent dialog creation for graphical user interfaces. In: Proceedings of SIGCHI'90. Seattle, pp 27–34

3. Puerta AR (1997) A model-based interface development environment. IEEE Softw 14(4):41–47
4. Vanderdonckt J (1995) Knowledge-based systems for automated user interface generation: the TRIDENT experience. In: Technical report RP-95-010. Namur: Facultes Universitaires Notre-Dame de la Paix, Institut d' Informatique
5. Myers BA, Hudson SE, Pausch R (2000) Past, Present and Future of User Interface Software Tools. ACM Transactions on Computer Human Interaction 7(1):3–28
6. Nichols J, Myers BA, Higgins M, Hughes J, Harris TK, Rosenfeld R, Pignol M (2002) Generating remote control interfaces for complex appliances. In UIST 2002. Paris, France, pp 161–170
7. Fikes R, Nils NJ (1971) Strips: a new approach to the application of theorem proving to problem solving. Artif Intell 2(3):189–203
8. Soderland S, Weld D (1991) Evaluating nonlinear planning. Technical Report TR 91-02-03. University of Washington CSE
9. Kautz H, McAllester D, Selman B (1996) Encoding plans in propositional logic. In: Proceedings of the 5th international conference of principles of knowledge representation and reasoning, pp 1084–1090
10. Avrim LB, Merrick LF (1995) Fast planning through planning graph analysis. In: Proceeding of the 14th International joint conferences on artificial intelligence, pp 1636–1642
11. Malik G, Adele H, Craig K, Drew M, Ashiwin R, Manuela V, Daniel W, David W (1998) PDDL-the planning domain definition language, http://www.informatik.uni-ulm.de/ki/Edu/Vorlesungen/GdKI/WS0203/pddl.pdf
12. Jaeyoung P, Kee EK, Sungho J (2010) A POMDP approach to P300-based brain-computer interfaces. In: Proceedings of the 15th international conference on intelligent user interfaces, Hong Kong, China, pp 1–10

Chapter 78
Coordinated Passivation Techniques for the Control of Permanent Magnet Wind Generator

Bing Wang, Yanping Qian and Honghua Wang

Abstract With coordinated passivation techniques, the nonlinear controller of permanent magnet wind generator is designed to achieve the stability of the closed-loop system. The design method combines the control Lyapunov function with the passivity-based approach, which can ensure that the states converge to the equilibrium point quickly. Simulations have shown the effectiveness of the obtained controller and the ideal dynamic property of the controlled system.

Keywords Permanent magnet wind generator (PMWG) · Coordinated passivation technique · Control Lyapunov function (CLF) · Passivity-based control

78.1 Introduction

Recently, a lot of on-shore or off-shore wind farms are being built on behalf of the policy of good tariff for electricity from natural renewable energy source. At the same time, the control problems in wind power systems have been the focus in the field of research [1, 2]. Permanent magnet generator is a class of important wind generator. Permanent magnet wind generator (PMWG) has attracted a lot of attentions and many advanced methods are used to design the controller for it [3, 4].

The feedback passivation technique [5, 6] has become a popular approach to nonlinear control design. Passivity provides us with a physical insight for the analysis of nonlinear systems. And coordinated passivation [7, 8] is an improvement for the passivity method. In this paper, the coordinated passivation techniques are employed for the third-order model of PMWG, which have two control inputs. By combining control Lyapunov function (CLF) with coordinated

B. Wang (✉) · Y. Qian · H. Wang
College of Energy and Electrical Engineering, Hohai University, 211100 Nanjing, China
e-mail: icekingking@hhu.edu.cn

Z. Sun and Z. Deng (eds.), *Proceedings of 2013 Chinese Intelligent Automation Conference*, Lecture Notes in Electrical Engineering 255,
DOI: 10.1007/978-3-642-38460-8_78,

passivation approach, the nonlinear controller is designed in two steps. Finally, simulations show the effectiveness of the proposed methods.

78.2 Passivity and Coordinated Passivation

78.2.1 Passivity

Consider a dynamical system represented by the model

$$\begin{cases} \dot{x} = f(x,u) \\ y = h(x,u) \end{cases} \tag{78.1}$$

where $f : R^n \times R^p \to R^n$ is locally Lipschitz, $h : R^n \times R^p \to R^q$ is continuous, $f(0,0) = 0$, and $h(0,0) = 0$.

Definition 1 [5]: The system (78.1) is said to be passive if there exists a continuously differentiable positive semi-definite function $V(x)$ (called the storage function) such that

$$u^T y \geq \dot{V} = \frac{\partial V}{\partial x} f(x,u), \quad \forall (x,u) \in R^n \times R^p \tag{78.2}$$

Moreover, it is said to be

- lossless if $u^T y = \dot{V}$;
- input strictly passive if $u^T y \geq \dot{V} + u^T \varphi(u)$ and $u^T \varphi(u) > 0, \ \forall u \neq 0$;
- output strictly passive if $u^T y \geq \dot{V} + y^T \rho(y)$ and $y^T \rho(y) > 0, \ \forall y \neq 0$;
- strictly passive if $u^T y \geq \dot{V} + \psi(x)$ for some positive definite function ψ.

From Definition 1, we know that if the system (78.1) is passive, it is easy to find a controller u to achieve $\dot{V} < 0$, which ensure stability. To get asymptotic stability the following give the definition of zero-state observability and a lemma.

Definition 2 [5]: The system (78.1) is said to be zero-state observable if no of solution $\dot{x} = f(x,0)$ can stay identically in $S = \{x \in R | h(x,0) = 0\}$, other than the trivial solution $x(t) \equiv 0$.

Lemma 1 [5]: *Consider the system* (78.1). *The origin of* $\dot{x} = f(x,0)$ *is asymptotically stable if the system is*

- strictly passive or
- output strictly passive and zero-state observable.

Furthermore, if the storage function is radially unbounded, the origin will be globally asymptotically stable.

78.2.2 Coordinated Passivation

Based on the definition of passivity, a few methods are proposed to solve the problems of nonlinear controller design. The coordinated passivation approach [4] gives a way to design the control of dual-input systems, which is presented as follows:

Lemma 2 [8]: *By choosing the input–output pair* (u_1, y)*, which the vector relative degree is one, the dual-input system can be rewritten into*

$$\dot{z} = q(z, y) + p(z, y)u_2 \tag{78.3}$$

$$y = \alpha(z, y) + \beta_1(z, y)u_1 + \beta_2(z, y)u_2 \tag{78.4}$$

where the state $z \in R^n$, $y \in R^m$ is treated as output, input $u_1 \in R^p$, $u_2 \in R^q$. Design the control law as follows:

$$u_1 = \beta_1^{-1}(z, y)\left(-\beta_2(z, y)u_2(z) - \alpha(z, y) - \frac{\partial W}{\partial z}\tilde{p}(z, y) + v\right) \tag{78.5}$$

$$u_2 = \gamma(z) \tag{78.6}$$

where $\tilde{p}(z, y)y = q(z, y) - q(z, 0) + p(z, y)\gamma(z) - p(z, 0)\gamma(z)$. The control law makes the CLF $W(z)$, which is positive definite, satisfying

$$\dot{W} = \frac{\partial W(z)}{\partial z}(q(z, 0) + p(z, 0)\gamma(z)) < -\alpha(\|z\|) \tag{78.7}$$

for some class κ function α. Then, the closed-loop system of (78.3–78.4) is passive, and $y(t) \to 0$ when $t \to \infty$. Moreover, if we choose $v = -\phi(y)$ satisfying the sector-nonlinearity

$$y^T\phi(y) > 0 \qquad \text{for } y \neq 0 \text{ and } \phi(0) = 0 \tag{78.8}$$

Then, the system is asymptotically stable.

78.3 Nonlinear Controller Design of PMWG

As shown in Fig. 78.1, the dynamic model of PMWG consists of the models of gearbox and permanent magnet synchronous generators, which can be described by the following third-order equations [4].

$$\begin{cases} J_t\frac{d\omega_r}{dt} = T_m - K_t\omega_r - n_gT_e \\ L_q\frac{di_q}{dt} = u_q - R_Si_q - L_dp\omega_gi_d - p\omega_g\psi \\ L_d\frac{di_d}{dt} = u_d - R_Si_d + L_qp\omega_gi_q \end{cases} \tag{78.9}$$

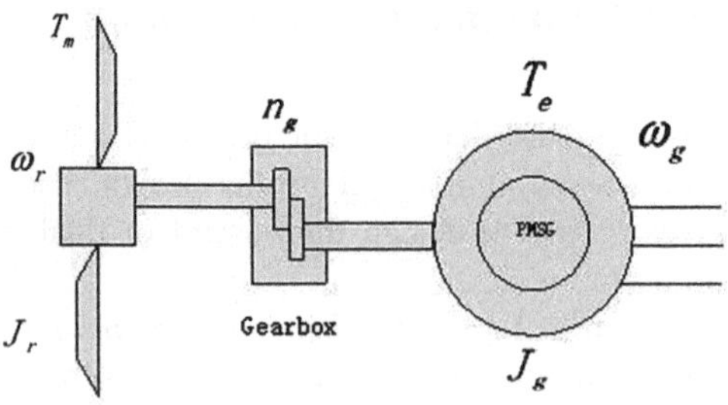

Fig. 78. 1 Mechanical model of permanent magnet wind generator

where T_e is the electromagnetic torque applied on the generator shaft, which satisfies

$$T_e = p(L_d - L_q)i_d i_q + p\psi i_q \tag{78.10}$$

$J_t = J_r + n_g^2 J_g$, $K_t = K_r + n_g^2 K_g$, $n_g = \frac{\omega_g}{\omega_r}$, ω_r is the turbine rotor speed, ω_g is the generator rotor speed, n_g is the gearbox ratio, T_m is the effective input mechanical torque to the wind turbine, J_r is the turbine rotor inertia, J_g is the generator rotor inertia, K_r is the turbine friction coefficient, K_g is the generator friction coefficient, ψ is magnet flux, p is the number of pole pairs, R_S is the resistance of stator, L_d and L_q are d-axis and q-axis generator inductances, u_d and u_q are d-axis and q-axis voltages, i_d and i_q are d-axis and q-axis currents, respectively. It is known that the operating point of system is $(\omega_{r0}, i_{q0}, i_{d0})$. The model (78.9) and (78.10) are transformed into the incremental form as follows:

$$\begin{cases} \Delta\dot{\omega}_r = -\dfrac{K_t}{J_t}\Delta\omega_r - \dfrac{n_g p(L_d - L_q)}{J_t}\left(i_{d0}\Delta i_q + i_{q0}\Delta i_d + \Delta i_d \Delta i_q\right) - \dfrac{n_g p\psi}{J_t}\Delta i_q \\ \Delta\dot{i}_q = -\dfrac{R_S}{L_q}\Delta i_q - \dfrac{L_d p\omega_g}{L_q}\Delta i_d + \dfrac{1}{L_q}u_q \\ \Delta\dot{i}_d = -\dfrac{R_S}{L_d}\Delta i_d + \dfrac{L_q p\omega_g}{L_d}\Delta i_q + \dfrac{1}{L_d}u_d \end{cases} \tag{78.11}$$

Define the states $x_1 = \Delta\omega_r$, $x_2 = \Delta i_q$, $x_3 = \Delta i_d$, the inputs $u_1 = u_d$,$u_2 = u_q$, the output $y = x_3 = \Delta i_d$. It is easy to know that the relative degree of input–output pair (u_1, y) is one. The Equ. (78.11) is represented by

$$\begin{bmatrix} \dot{x}_1 \\ \dot{x}_2 \end{bmatrix} = \begin{bmatrix} -\dfrac{K_t}{J_t}x_1 - (a_1 i_{d0} + a_2)x_2 \\ -\dfrac{R_s}{L_q}x_2 \end{bmatrix} + \begin{bmatrix} -(a_1 i_{q0} + a_1 x_2) \\ -\dfrac{L_d}{L_q}a_3 \end{bmatrix} y + \begin{bmatrix} 0 \\ 1 \\ L_q \end{bmatrix} u_2 \triangleq Q_1(x) + Q_2(x)y + Pu_2 \tag{78.12}$$

$$\dot{y} = -\frac{R_s}{L_d}y + \frac{L_q}{L_d}a_3x_2 + \frac{1}{L_d}u_1 \tag{78.13}$$

where $a_1 = n_g p(L_d - L_q)/J_t$, $a_2 = n_g p\psi/J_t$, $a_3 = p\omega_g$.

Next, we will design the nonlinear controller of PMWG by using the coordinated passivation techniques.

78.3.1 CLF Design of x-Subsystem

The zero dynamics of the x-subsystem can be written as follows:

$$\begin{cases} \dot{x}_1 = -\dfrac{K_t}{J_t}x_1 - b_1x_2 \\ \dot{x}_2 = -\dfrac{R_s}{L_q}x_2 + \dfrac{1}{L_q}u_2 \end{cases} \tag{78.14}$$

where $b_1 = a_1 i_{d0} + a_2$.

Based on CLF method, let the Lyapunov function $W(x) = (x_1^2 + x_2^2)/2$. The derivative of $W(x)$ is

$$\begin{aligned} \dot{W}(x) &= \frac{\partial W}{\partial x}\dot{x}\bigg|_{y=0} = x_1\left(-\frac{K_t}{J_t}x_1 - b_1x_2\right) + x_2\left(-\frac{R_s}{L_q}x_2 + \frac{1}{L_q}u_2\right) \\ &= -\frac{K_t}{J_t}x_1^2 - b_1x_1x_2 - \frac{R_s}{L_q}x_2^2 + \frac{1}{L_q}x_2u_2 \\ &= -\frac{K_t}{J_t}x_1^2 - \frac{R_s}{L_q}x_2^2 + \frac{x_2}{L_q}\left(u_2 - b_1L_qx_1\right) \end{aligned} \tag{78.15}$$

Design the controller as

$$u_2 = u_q = b_1L_qx_1 = (a_1i_{d0} + a_2)L_qx_1 \tag{78.16}$$

Taking the controller (78.16) into (78.15) yields

$$\dot{W}(x) = \frac{\partial W}{\partial x}\dot{x}\bigg|_{y=0} = -\frac{K_t}{J_t}x_1^2 - \frac{R_s}{L_q}x_2^2 < -\alpha(\|x\|) \tag{78.17}$$

Therefore, the zero dynamics of x-subsystem are stable under the controller (78.16).

78.3.2 Coordinated Passivation Design

In this part, we proceed to the feedback passivation step of u_1, that is to design u_1 to stabilize the whole system. Select the storage function $V = W + y^2/2$ and design the control law

$$u_1 = u_d = L_d\left(-\frac{L_q}{L_d}a_3x_2 - \frac{\partial W}{\partial x}Q_2(x) + v\right) \tag{78.18}$$

Then, the derivative of storage function is

$$\dot{V} = \dot{W} + y\dot{y} = \frac{\partial W}{\partial x}\dot{x}\bigg|_{y=0} + \frac{\partial W}{\partial x}\begin{bmatrix} -(a_1 i_{q0} + a_1 x_2) \\ -\frac{L_d}{L_q}a_3 \end{bmatrix} y + y\left(-\frac{R_s}{L_d}y + \frac{L_q}{L_d}a_3x_2 + \frac{1}{L_d}u_1\right)$$
$$= \frac{\partial W}{\partial x}\dot{x}\bigg|_{y=0} - \frac{R_s}{L_d}y^2 + vy \leq -\frac{R_s}{L_d}y^2 + vy \tag{78.19}$$

From Definition 1, it is known the closed-loop system is output strictly passive. Moreover, it is easy to show the model (78.11) is zero-state observable. So the whole system is asymptotically stable according to Lemma 1.

Now, we summarize the design result into the following theorem.

Theorem 1: *Consider the PMWG system* (78.9–78.10). *Under the controller* u_q and u_d *as in* (78.16) *and* (78.18), *the close-loop system is output strictly passive. Moreover, the whole system is asymptotically stable for the system is zero-state observable.*

78.4 Simulations

In order to show the effectiveness of the proposed controller, the simulations are operated in MATLAB. In the simulations, the parameters of the PMWG are give as follows: $J_r = 59.5 \times 10^5$, $J_g = 350$, $n_g = 12$, $K_r = 0.14$, $K_g = 0.005$, $p = 10$, $\psi = 1.67$, $L_d = 0.2446$, $L_q = 0.4778$, $R_s = 0.0012$.

From Figs. 78.2 and 78.3, we can know that the response curves of all states of the closed-loop system converge to the operating point quickly under the nonlinear controller. So the coordinated passivation controller is effective for the PMWG system.

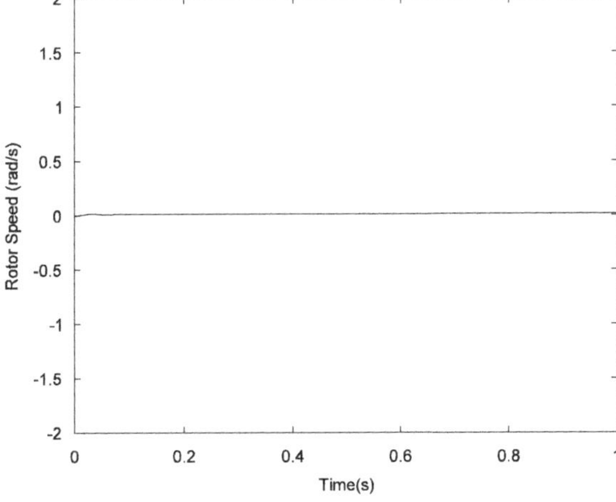

Fig. 78.2 The response curse of the turbine rotor speed

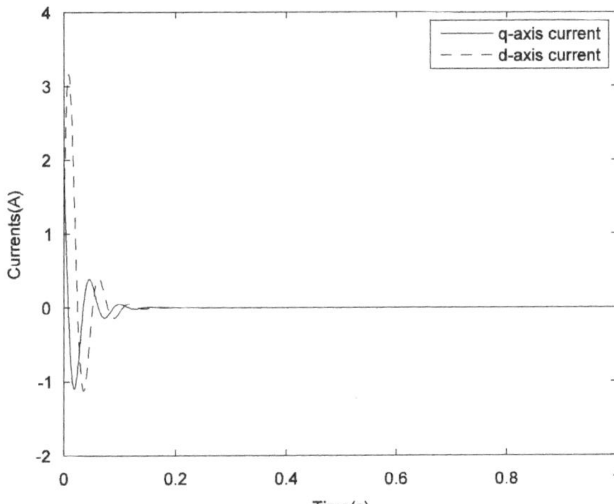

Fig. 78.3 The response curses of *q-axis* and *d-axis* currents

78.5 Conclusion

A nonlinear controller is proposed for PMWG by using coordinated passivation approach, which releases some constraints for multi-input systems. At first, the control input u_q is obtained by CLF method to stabilize the zero dynamics of the x-subsystem. Next, the control law of u_d is gotten by using the coordinated passivation techniques to make the whole system asymptotically stable. At last, the simulation results show the nonlinear controller is effective for PMWG.

Acknowledgments This work was financially supported by the National Natural Science Foundation of China (51007019).

References

1. Bianchi FD, Battista HD, Mantz RJ (2007) Wind turbine control systems: principles, modelling and gain scheduling design. Springer, London
2. Munteanu L, Bratcu AL, Cutululis N, Ceanga E (2008) Optimal control of wind energy systems. Springer, London
3. Chinchilla M, Arnaltes S, Burgos JC (2006) Control of permanent-magnet generators applied to variable-speed wind-energy systems connected to the grid. IEEE Trans. on Energ Conversion 21(1):130–135
4. Zhang JZ, Chen M, Chen Z (2007) Nonlinear control for variable-speed wind turbines with permanent magnet generators. In: Proceeding of international conference on electrical machines and system. Seoul, Korea, pp 324–329
5. Khalil HK (2002) Nonlinear systems, 3rd edn. Prentice-Hall, New Jersey
6. Kokotovic PV, Arcak M (2001) Constructive nonlinear control: a historical perspective. Automatica 37(5):637–666
7. Larsen M, Jankovic M, Kokotovic PV (2003) Coordinated passivation designs. Automatica 39(2):335–341
8. Wang B, Ji HB, Chen H, Xi HS (2004) The coordinated passivity techniques for the excitation and steam-valving control of generator. Proceeding of the CSEE (in Chinese) 24(5):104–109

Chapter 79
The Electromagnetic Field of a Horizontal and Time-Harmonic Dipole in a Two-Layer Medium

Lu Xiong and Shenguang Gong

Abstract To investigate the modeling of Shaft-Rate Electric Field, the analytic expressions of the electromagnetic fields of time-harmonic, horizontal electric dipole (HED) embedded in the sea have been derived by using image method according to uniqueness theorem of the electromagnetic field with two layer model. The distribution of electromagnetic field produced by a horizontal electric dipole embedded in seawater has been numerically calculated and analyzed by applying fast Hanker transform and fast Fourier transform. Then the case of shallow sea and harmonic electric dipole has been simulated in laboratory and the experimental results are accorded with the simulation results, thus the correctness of the solution and the derivation are validated. The electromagnetic field of a horizontal and time-harmonic dipole is put in a two-layer medium.

Keywords Shaft-rate electric field · Harmonic electric dipole · Image method · Hankel transform · Fast Fourier transform

79.1 Introduction

Time-harmonic electric dipole is a fundamental unit that is commonly used to analog the characteristics of electric field from a ship in the sea, which has attracted many researchers for a long time. Thus, many researchers have been working so hard to obtain the analytic expressions of electromagnetic fields from time-harmonic, horizontal electric dipole over the years [1].

Image method is a classical method to absolve electromagnetic conductor boundary value problem on the basis of uniqueness theorem [2]. For parallel

L. Xiong (✉) · S. Gong
Department of Weaponry Engineering, Naval University of Engineering,
No717 JieFang Road, Wuhan, China
e-mail: litubaier@163.com

Z. Sun and Z. Deng (eds.), *Proceedings of 2013 Chinese Intelligent Automation Conference*, Lecture Notes in Electrical Engineering 255,
DOI: 10.1007/978-3-642-38460-8_79,

medium of two layers, the distribution of magnetic vector potential from time-harmonic horizontal electric dipole under deep sea environment which is composed of sea and air is derived in seawater and air, the analytical expression of electromagnetic field distribution produced by a horizontal time-harmonic electric dipole embedded in seawater can be achieved finally. Distribution of electric field underwater is simulated and analyzed. Then the case of deep sea and a time-harmonic electric dipole source have been simulated in laboratory and the electric potential in seawater has been measured. The research results provide a theoretical base for the mathematical modeling and the extrapolation of Shaft-Rate Electric Field of a ship or other underwater weaponry.

79.2 Two Layer Medium Model

Under the deep-sea environment, the whole space is filled with air and sea. A Cartesian coordinate system of deep sea model is established with parameters in Fig. 79.1. The origin O is located in the water, the x-axis and y-axis is parallel to the surface of the water, while the z-axis is vertical to the surface and the downward direction is positive, $z < 0$ represents the region for air, $z \geq 0$ represents the area for seawater. The horizontal electric dipole in seawater is supposed to locate at (x_0, y_0, z_0), which points to the x-axis in the positive direction, electrical dipole moment is taken as $I\vec{dl} = I_x\vec{dl}$. The coordinate of field point is (x, y, z) in the sea.

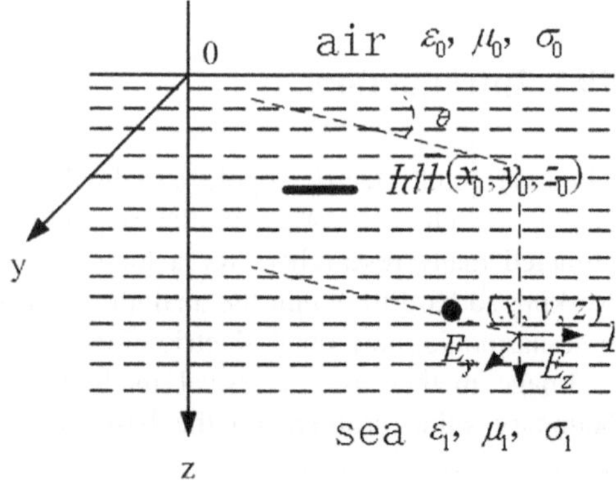

Fig. 79.1 The modeling of *two*-layer medium

79.3 The Application of Image Method with Two Layer Model

79.3.1 *Magnetic Vector Potential of Horizontal Time Harmonic Electric Dipole*

In linear medium, the expressions of magnetic vector potential A and electric field strength E are described as follows [3]:

$$\nabla^2 A + k^2 A = -\mu J_s \tag{79.1}$$

$$E = -j\omega\left[A + \frac{1}{k^2}\nabla(\nabla \cdot A)\right] \tag{79.2}$$

Where, k represents propagation constant, and $k^2 = -j\omega\mu\sigma + \omega^2\mu\varepsilon$.

From Eqs. (79.1) and (79.2), it can be seen, as long as magnetic vector potential A is determined uniquely, the electric field component of electromagnetic field E can be fixed uniquely.

79.3.2 *Image Method for Solving the Magnetic Vector Potential*

When spot point locates in seawater, the horizontal component of the harmonic electric dipole in seawater can be calculated by image method. Then, suppose that the image located at $(x_0, y_0, -z_0)$, when the whole space is seen to be filled with sea water. According to the principle of superposition, the horizontal component of magnetic vector potential which is composed of the harmonic electric dipole and image are as follows:

$$A_{1x} = \frac{\mu_1 Il}{4\pi}\int \frac{\xi}{\upsilon_1}J_0(\rho\xi)\left(e^{-\upsilon_1|z-z_0|} + D(\xi)e^{-\upsilon_1|z+z_0|}\right)d\xi. \tag{79.3}$$

Where $\rho = \sqrt{(x-x_0)^2 + y^2}$, $\upsilon_1 = \sqrt{\xi^2 - k_1^2}$.

$J_0(\rho\xi)$ represents the first kind of zero order Bessel function, $D(\xi)$ is undetermined coefficient $z > 0$.

When spot point locates in air, $z > 0$, the horizontal component A_{0x} of the harmonic electric dipole in air can be calculated by image method. Then, suppose that image locate at (x_0, y_0, z_0) when the whole space is seen to fill with air, resource and image are located in the same position. According to the principle of superposition, the horizontal component of magnetic vector potential which is composed of the harmonic electric dipole and image are as follows:

$$A_{0x} = \frac{\mu_0 Il}{4\pi} \int \frac{\xi}{\upsilon_0} C(\xi) J_0(\rho\xi) e^{-\upsilon_0 |z - z_0|} d\xi \tag{79.4}$$

The image method requires the magnetic vector potential to satisfy boundary conditions, so A_{0x} and A_{1x} should meet the following four conditions [4, 5]:

Substitute (79.3, 79.4) into the boundary conditions and we can get:

$$C(\xi) = 2\frac{\mu_1 \upsilon_0}{\mu_0 \upsilon_1 + \mu_1 \upsilon_0} e^{(\upsilon_1 - \upsilon_0) z_0} \tag{79.5}$$

$$D(\xi) = \frac{\mu_0 \upsilon_1 - \mu_1 \upsilon_0}{\mu_0 \upsilon_1 + \mu_1 \upsilon_0} \tag{79.6}$$

Similarly, vertical components and of the magnetic vector potential in both air and water also can be solved as follows.

$$A_{0z} = \sum_{n=0}^{\infty} \cos n\theta \int_0^{\infty} A_n(\xi) J_n(\xi\rho) e^{-\upsilon_0 z} d\xi \tag{79.7}$$

$$A_{1z} = \sum_{n=0}^{\infty} \cos n\theta \int_0^{\infty} B_n(\xi) J_n(\xi\rho) e^{-\upsilon_1 z} d\xi \tag{79.8}$$

where, $J_n(\xi\rho)$ is the first class of n order Bessel function, $A_n(\xi)$ and $B_n(\xi)$ are undetermined coefficients, which is the first class of n-order Bessel Function $\theta = \arctan\frac{y - y_0}{x - x_0}$.

Similarly, the vertical component of the magnetic vector potential generated by time-harmonic horizontal electric dipole in deep sea can be obtained as:

$$A_{1z} = \frac{\mu_1 Il}{2\pi} \cos\theta \int_0^{\infty} M J_1(\xi\rho) e^{\upsilon_1 (z + z_0)} d\xi \tag{79.9}$$

From

$A = A_x i + A_z k$

The relationship between three components of both the vector A and the electric field E can be solved as:

$$E_x = -j\omega A_{1x} - \frac{j\omega}{k_1^2}\left(\frac{\partial^2 A_{0x}}{\partial x^2} + \frac{\partial^2 A_{1z}}{\partial x \partial z}\right) \tag{79.10}$$

$$E_y = -\frac{j\omega}{k_1^2}\left(\frac{\partial^2 A_{1x}}{\partial x \partial y} + \frac{\partial^2 A_{1z}}{\partial y \partial z}\right) \tag{79.11}$$

$$E_z = -j\omega A_{1z} - \frac{j\omega}{k_1^2}\left(\frac{\partial^2 A_{1x}}{\partial x \partial z} + \frac{\partial^2 A_{1z}}{\partial^2 z}\right) \tag{79.12}$$

The electric field distribution of horizontal electric dipole in seawater can be directly calculated as follows, where $G_1 = \frac{\exp(-jk_1R)}{R}, G_2 = \frac{\exp(-jk_1\bar{R})}{\bar{R}}$, $W = \int_0^{\infty} \frac{2\xi^2(k_1^2-k_0^2)e^{-\upsilon_1(z+h)}}{(\upsilon_0+\upsilon_1)(\upsilon_0k_1^2+\upsilon_1k_0^2)} J_1(r\xi)\cos\theta d\xi$, $U = \int_0^{\infty} \frac{2\xi}{\upsilon_1+\upsilon_0}\exp[-\upsilon_1(z+h)]J_0(r\xi)d\xi$

$$E_{1x} = -\frac{j\omega\mu_0 Il}{4\pi}\left[G_1 - G_2 - \frac{1}{k_1^2-k_0^2}\left(\frac{\partial^2 V}{\partial r^2} + \frac{1}{r}\frac{\partial V}{\partial r} + 2\frac{\partial^2 G_2}{\partial z^2}\right)\right]$$

$$-\frac{j\omega\mu_0 Il}{4\pi k_1^2}\left(\cos^2\theta\frac{\partial^2}{\partial r^2} + \frac{\sin^2\theta}{r}\frac{\partial}{\partial r}\right)(G_1 - G_2 + V) \quad (79.13)$$

$$E_{1y} = -\frac{j\omega\mu_0 Il}{4\pi k_1^2}\sin\theta\cos\theta\left(\frac{\partial^2}{\partial r^2} - \frac{1}{r}\frac{\partial}{\partial r}\right)(G_1 - G_2 + V) \quad (79.14)$$

$$E_{1z} = -\frac{j\omega\mu_0 Il}{4\pi k_1^2}\frac{\partial^2}{\partial r\partial z}\left(G_1 + G_2 - \frac{k_0^2}{k_1^2}V\right)\cos\theta \quad (79.15)$$

79.4 Electric Field Distribution of Horizontal Electric Dipole in Seawater

According to Eqs. (79.13), (79.14), (79.15), the expression of field strength of the harmonic horizontal electric dipole can be obtained in seawater. Because the expression contains infinite summation, it is difficult to reflect the impact of various factors on the electric field distribution from the equation directly, therefore, only numerical simulation can be used to analysis the distribution, on the one hand, it can express the variation of the electric field with various factors intuitively, on the other hand, it is convenient to make comparative analysis of results [6].

Solving field strong expression can be adopted by a variety of numerical calculation methods; fast Hanker transform and fast Fourier transform numerical calculation are used here.

The distribution of field is simulated, with coordinate axis of the electric dipole shown in Fig. 79.1. Assuming that the depth of water is 100 m, the coordinate of electric dipole is (0, 0, 5), the electric dipole moment is 10 Am. The seawater conductivity is set to 4 S/m, the relative permittivity is 80, and the sea-bed conductivity is 0.04 S/m. The simulation results of three directions are shown in Figs. 79.2, 79.3.

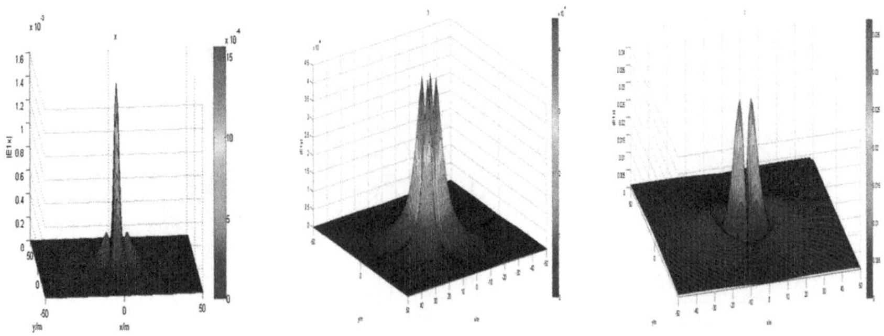

Fig. 79.2 *Three* components of electric field strength generated by horizontal electric dipole

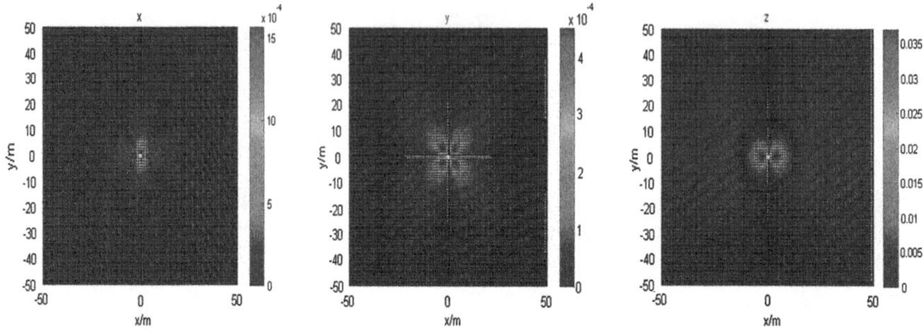

Fig. 79.3 *Three* components of electric field strength generated by horizontal electric dipole

79.5 The Test of Electric Field Distribution of Electric Dipole Underwater

The electromagnetic field in deep sea generated by horizontal time-harmonic electric dipole is investigated, aiming to apply it to shaft-rate electric field modeling. In order to verify the practicality of this article, the calculated results are compared with field distribution of horizontal electric dipole in deep sea under laboratory conditions.

In a nonmagnetic experimental pool, whose length width and depth are 8, 5, 1.5 m respectively. Electric dipoles are simulated by platinum-niobium wire with the low frequency alternating current passing into.

The diameter of platinum-niobium wire is 0.2 mm; the length of bare segment is 1 mm and a pitch of 1 mm. The remaining is sealed by insulating material to ensure insulate it with seawater.

Series resistance of seawater and system is 26 ohm. Output voltage of signal generator is 0.4 v, frequency is 2 Hz. The Pool is filled with artificial seawater modulated by industrial salt, the conductivity was 0.027 S/m, and the depth of water is 1 m. The bottom of pool is covered with tempered glass. Triaxial measurement system is used in the experiment.

In order to get the electric field distribution of the entire plane, as well as to save the number of electrodes, the measuring electrode is fixed, when the electric dipole source is dragged by stepping motor of uniform speed, that is to say, electric dipole move from one side to the other side of the measuring electrode, the output signal of measuring electrode is continuously collected by the data acquisition system and recorded, as to obtain the distribution of electric field in the entire plane.

Electric dipoles are driven by stepper motor forward with a speed of 4 cm/s. In order to reduce electromagnetic noise interference of the external environment, we make use of coaxial cable to connect the Ag/AgCl electric field sensor and acquisition system, the system sampling frequency is 40 Hz. Figure 79.4 shows the experiment schematic diagram. Cross range represents the distance between the platinum niobium wire and the plane consisted by longitudinal and vertical measuring electrode.

The results are shown in Fig. 79.5, compared with simulation results, the results are very consistent, indicating that the derivation methods and results are credible. At the same time the conclusion can be got:

Symmetry center of the envelopes of passing characteristic curve of three components of the electric dipole generated by the horizontal time-harmonic is not overlapped. The center of lateral component and vertical component is coincident, the center of longitudinal component lag behind than the horizontal component and the vertical component. It is considered that, when electric dipole goes through the three-axis measurement system, the distance between the pair of longitudinal electrodes is 12 cm, while the lateral and vertical components of the electrode overlap on a plane.

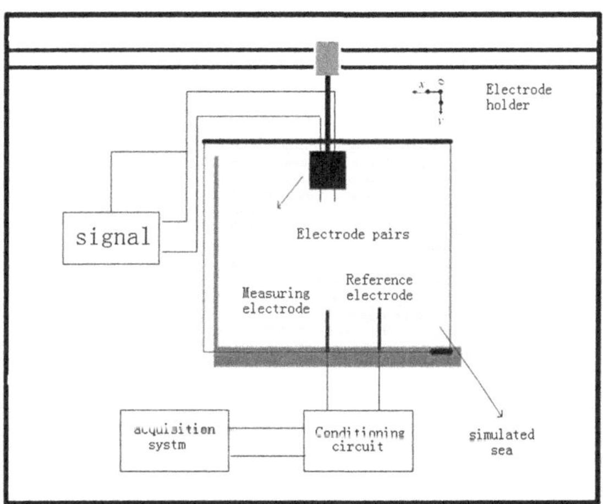

Fig. 79.4 The sketch map of experiment

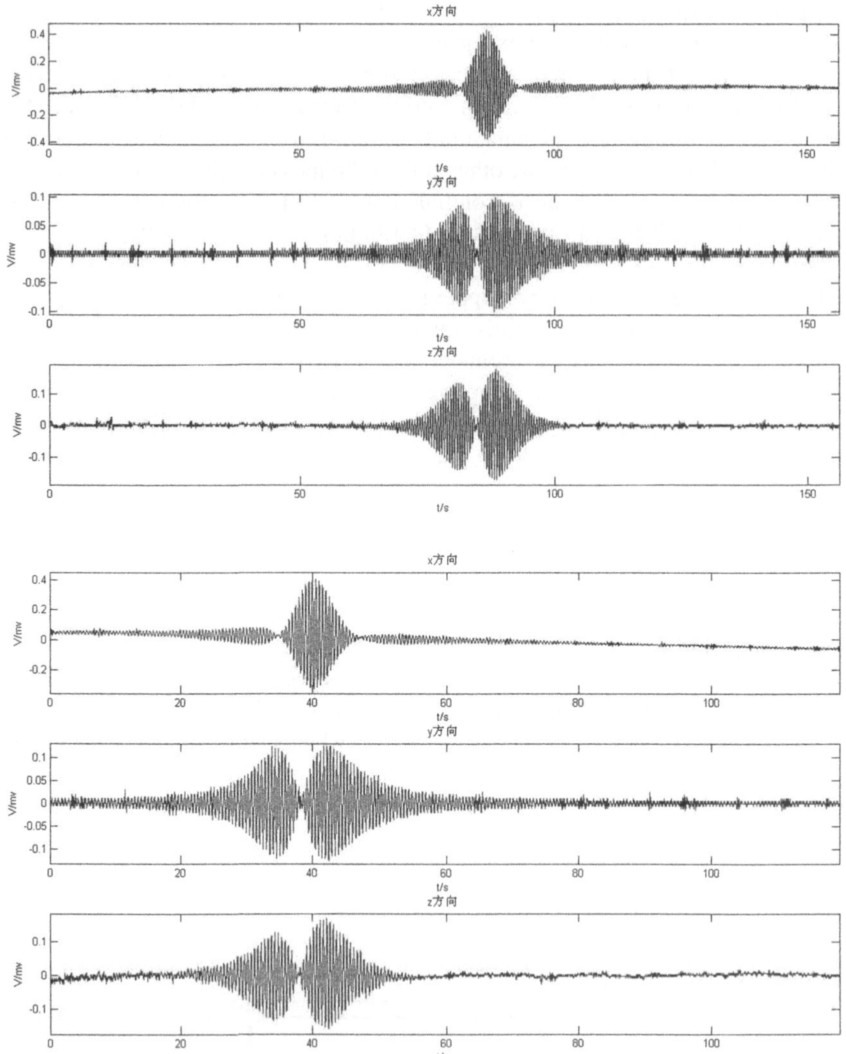

Fig. 79.5 The distribution of measuring Electric Field (cross range of 3 and 10 cm)

79.6 Conclusion

In this paper, the image method, which is based on the uniqueness principle of electromagnetic field, is used in the air–sea model to derive analytical expression of electric field generated by time-harmonic electric dipole in seawater. Simulation and experimental verification are performed as well, comparing experimental results with theoretical results, the results show that the mirror method is used to derive the electric field distribution of time-harmonic electric dipole in deep sea, the process is clear, and easy to calculate, the theoretical results are coincided with the experimental results at the same time. The analysis of the results will be very valuable for the modeling of the electric field distribution in the sea. Therefore, how to get more convenient, reasonable, scientific and more accurate models will be the next working points.

References

1. Fco JR, Antonio S (1999) Using electric signature for extracting target navigation parameters. In: Martin C, Amsterdam, Undersea defence technology, pp 12–18
2. Lu X, Gong S (2004) Measurement of extremely low frequency field caused by shaft-rate modulated corrosion current. ACTA Armamentarh, 9:544–546 (in Chinese)
3. Certenais J, Perious J (1997) J Electromagnetic measurements at sea. In: larke C, Hamburg T, Undersea defence technology, p 433
4. Hubbard JC, Brooks SH, Torrance BC (1996) Practical measures for reduction and management of the electro-magnetic signatures of in-service surface ships and submarines. In: Bantel T, Malmo undersea defence technology, pp 64–65
5. Davidson SJ, Rawlins PG, Jones H (2006) The choice of sensor type for electric field measurement applications. Ultra Electron, pp 105–121
6. Broomhead DS, Lowe D (1988) Multivariate functional interpolation and adaptive networks. Complex Syst, 16(2):321–355

Chapter 80
Analysis and Application of Sensor in Index

Chenghui Yang

Abstract The paper mainly analyzes the research and application of the sensor index in the networking, and makes a deep understanding of application area and application mode of networking. It mainly focuses on the research and application of the sensor and the mastery of the qualifications of static and dynamic characteristics of the sensor by analyzing the static and dynamic characteristics, and then describes the application of sensor in the detecting techniques and analyzes the characteristic of the sensor, making the sensor more accurate in the detecting techniques. The design studies the research and application of eddy-current sensor, and then gives rise to the application of index research in the nonlinear compensator of eddy-current sensor, by the analysis of the eddy-current sensor characteristic. In the design, the author utilizes eddy current displacement sensor, and then uses experiments to illustrate, by the design analysis of index compensation circuit, that it can effectively improve the original transmission characteristic of sensor and expand linear measurement range to take index operation circuit as nonlinear compensation link of eddy current displacement sensor.

Keywords Internet of things · Sensor index · Eddy current · Nonlinear compensation · Compensation circuit

C. Yang (✉)
College of Electrical Engineering, Northwest University for Nationalities, Lanzhou, China
School of Automation and Electrical Engineering, Lanzhou Jiaotong University, Lanzhou, China
e-mail: yangchenghui36@163.com

Z. Sun and Z. Deng (eds.), *Proceedings of 2013 Chinese Intelligent Automation Conference*, Lecture Notes in Electrical Engineering 255,
DOI: 10.1007/978-3-642-38460-8_80,

80.1 Introduction

The contents of this study is the analysis of the research and application of the sensor index in the Internet of Things Internet of Things applications, as well as mode-depth understanding of key research on the research and application of the sensor through the sensor static and dynamic characteristics of analysis, the sensor static and dynamic characteristics of the technical indicators, and finally the combination of sensor and detection technology, the sensor applied to the detection technology, more accurate analysis of sensor characteristics, sensor detection technology. Selected research and application of eddy current displacement sensor, through the analysis of the characteristics of the eddy current displacement sensor, which leads to the Index in the nonlinear compensation of the eddy current displacement sensor. In the design, we refer to the eddy current displacement sensor, index compensating circuit analysis, nonlinear compensation aspect of the exponentiation circuit as an eddy current displacement sensor can effectively improve the sensor's original transmission characteristics and the expansion of the sensor linear measurement range.

80.2 The Basic Structure of the Internet of Thing

Technology architecture point of view, the Internet of Things can be divided into three layers: the perception layer, network layer and application layer [1].

The perception layer consists of a variety of sensors and sensor gateways, including the concentration of carbon dioxide sensor, temperature sensor, humidity sensor, two-dimensional code label, RFID tag and reader, camera, GPS-aware terminal. The perception layer acts as the human eye, ENT and skin nerve endings, which are the Internet of things by the object recognition, collection of the source of the information, its main function is to identify objects, collecting information.

Network layer composed of a variety of private networks, Internet, wired and wireless communication networks, network management systems and cloud computing platforms, equivalent to the nerve center and brain of the person responsible for the transmission and processing of information obtained by the perception layer.

The application layer is the Internet of Things and the user interface (including people, organizations and other systems), combined with industry needs, the intelligent application of the Internet of Things.

80.3 Analysis of the Characteristics of the Sensor

80.3.1 Static Characteristics

The static characteristics of the sensor is the input–output relationship when the values measured in the steady state, consider only the static characteristics of the sensor, the relationship between the output and input does not contain any variable to measure the sensor static important indicator of the characteristics of linearity, sensitivity, hysteresis and repeatability parameters [2].

1. Linearity
   Linearity is expressed as a percentage of the detection system using the measured input—the maximum deviation between the output characteristic curves and fitting a straight line with full-scale output.

$$r_1 = \pm \frac{\Delta S_{\max}}{Y_{FS}} \times 100\,\% \tag{80.1}$$

2. Sensitivity
   The sensitivity S is the ratio of the sensor output increment caused by the incremental input increment, That is,

$$S = \frac{\Delta y}{\Delta x} \tag{80.2}$$

3. Sluggish
   Hysteresis size is generally decided by the experiment. Hysteresis error by the (80.3) calculation

$$r_L = \pm \frac{1}{2}\frac{\Delta L_{\max}}{Y_{FS}} \times 100\,\% \tag{80.3}$$

4. Repeatability
   Repeatability is the degree of inconsistency of the income curve repeatedly changes the sensor in the same direction in the input full scale.

80.3.2 The Dynamic Characteristics

The dynamic characteristics of the sensor is the output of the sensor response characteristics, the amount of time-varying input to reflect the value of the output is a true reproduction of the ability of changing the amount of input [3]. Good dynamic characteristics of the sensor, its output will be as much as possible to reproduce the variation of the input, i.e. have the same function of time. In fact, in addition to the ideal ratio features link, the sensor inherent factors, the output signal will not be with the input signal has the same function of time, the difference of the output and input is the so-called dynamic error. Study the dynamic characteristics of the sensor is a dynamic error analysis from the perspective of measurement error causes and improvement measures. Because most of the sensor can be simplified as a first-order or second order system, so the first and second sensor is essential. Study the dynamic characteristics of the sensor can be two aspects from the time domain and frequency domain, transient response method and frequency response analysis.

1. Transient response characteristics an order of the sensor unit step response in the works will generally be under the general formula (80.4) as an order of the sensor unit step response.

$$\tau \frac{d_y(t)}{dt} + y(t) = x(t) \tag{80.4}$$

where x (t), y (t) respectively for the input and output of the sensor are a function of time, the characterization of the sensor time constant has the dimension of time "seconds".
Second-order sensor unit step response
Second-order sensor unit step response of the general formula as follows:

$$\frac{d^2y(t)}{dt^2} + 2\xi\omega_n \frac{dy(t)}{dt} + \omega_n^2 y(t) = \omega_n^2 x(t) \tag{80.5}$$

2. Frequency response characteristics
   Zero-order frequency characteristics of the sensor Zero-order transfer function of the sensor are:

$$H(s) = \frac{Y(s)}{X(s)} = k \tag{80.6}$$

Frequency characteristics:

$$H(j\omega) = k \tag{80.7}$$

Therefore, the zero-order input and output of the sensor is inversely proportional to, and has nothing to do with the signal frequency. Therefore, the amplitude and phase distortion with the dynamic characteristics of the ideal. Potentiometer sensor is an example of the zero-order system. In practical applications, much high-end system when the change is slow, the frequency is not high, can be approximated as a zero-order system for processing.
One order of frequency characteristics of the sensor S use $j\omega$ in the first-order transfer function of the sensor instead of the expression can be obtained by the frequency characteristics of

$$H(j\omega) = \frac{1}{\tau(j\omega) + 1} \tag{80.8}$$

Amplitude-frequency characteristics

$$A(\omega) = \frac{1}{\sqrt{1 + (\omega\tau)^2}} \tag{80.9}$$

Phase frequency characteristics

$$\Phi(\omega) = -\arctan(\tau\omega) \tag{80.10}$$

The frequency characteristics of the second-order sensor.
Expressions of the second-order frequency characteristics of the sensor, the amplitude-frequency characteristics of phase-frequency characteristics were

$$H(j\omega) = \left[1 - \left(\frac{\omega^2}{\omega_n}\right) + 2j\xi\frac{\omega}{\omega_n}\right]^{-1} \tag{80.11}$$

$$A(\omega) = \left\{\left[1 - \left(\frac{\omega^2}{\omega_n}\right)\right]^2 + \left(2\xi\frac{\omega}{\omega_n}\right)^2\right\}^{-\frac{1}{2}} \tag{80.12}$$

$$\phi(\omega) = -\arctan\left[\frac{2\xi\frac{\omega}{\omega_n}}{1 - \left(\frac{\omega}{\omega_n}\right)^2}\right] \tag{80.13}$$

80.4 Internet of Things in the Sensor of Study

80.4.1 The Composition of the Sensor

Sensors are generally sensitive components, conversion devices.

- Sensitive components—it is a direct experience is measured, and the output is measured to determine the relationship between the components of a physical quantity.
- Conversion components—sensitive components of the output is converted to the component input, it is converted into the input circuit parameters.
- Conversion circuit—the conversion element to convert the access conversion circuit of the circuit parameters can be converted into power output.

80.4.2 The Composition of Automatic Detection Systems

Automatic detection system, the various components often in the process of division of the flow of information, and generally can be divided into: information extraction, conversion, processing and output sections [4].

The monitoring system must first obtain detection information (input), through the sensor will be available to non-electricity conversion into electricity and then converted into electricity through the detection circuit information amplification, shaping the conversion processing, the output unit to display the information they need, or the processed information is sent to the output unit to other units of the control system use.

80.5 Exponentiation in the Eddy Current Sensor

80.5.1 Exponentiation in the Measuring Circuit

The measurement circuit of the eddy current sensor can be summarized as two kinds of amplitude modulation and frequency modulation type. The amplitude modulation circuit can be divided into two kinds of amplitude modulation of the constant frequency amplitude modulation and frequency change, constant frequency amplitude modulation circuit is characterized by the output can be conditioning for DC voltage, the advantage of the DC voltage, the speed of data acquisition time is short can reduce power consumption [5].

Eddy current displacement sensor nonlinear basic aspects of the compensation principle [6].

If the set of sensor input x, output m, then m = f (x) for the non-linear relationship. If a compensation part series in the sensor input, so that y = g (m) = kx, the nonlinear compensation of the sensor, when k = 1, y = x = g (m) is called the reverse mode of the sensor.

For eddy current displacement sensor, the displacement of 50 mm outside changes, the output voltage will change, but change is slow [7]. To increase the measurement range, to compensate for the nonlinear segment in series to a compensation part, compensation aspects of the transmission characteristics shown in Fig. 80.4 in the third quadrant as shown in Curve 2, it first quadrant sensor output characteristics curve (curve 1) together to achieve a final fourth quadrant curve linear [8].

In the third quadrant compared with the distance, when the input (horizontal direction) gradually increases, the output (vertical direction) the rate of change is increasing, and this curve is similar to exponentiation. Therefore, using two bipolar transistors with precision, low-noise op amp AD704 constitute exponentiation circuit can meet these requirements.

Curve 1: compensation link transmission characteristics
Curve 2: the output characteristics of the sensor
Curve 3: after compensation curve

80.5.2 Nonlinear Compensation Starting Point Voltage for the Circuit Design

Before the exponential back off in Fig. 80.4, the corresponding change in voltage by subtracting the displacement x; and the role is to select the starting point of the index of compensation voltage, that is, adjust, select the corresponding displacement of the x = 40 mm the threshold voltage, the voltage compensation circuit computing the relationship:

$$V_2 = V_{ref} - \left(V_{in} - 2V_{ref2}\right) \tag{80.14}$$

Circuit 5 V voltage reference A, C, and the back of B by the low noise, low drift, Precision Voltage Reference MAX6250; switching diode to ensure that the output voltage of a single direction (is, D > 0) protective effect from the index compensation circuit.

80.5.3 Index Compensation Circuit Design

Index compensation circuit shown in Fig. 80.1, the characteristic curve shown in Fig. 80.2. Where a, b, c and d, the composition of the index arithmetic circuit, its input and output relations are analyzed as follows:

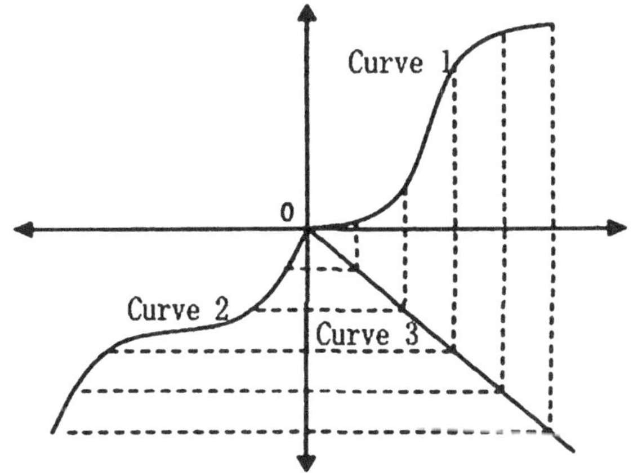

Fig. 80.1 Nonlinear compensation of transmission characteristics

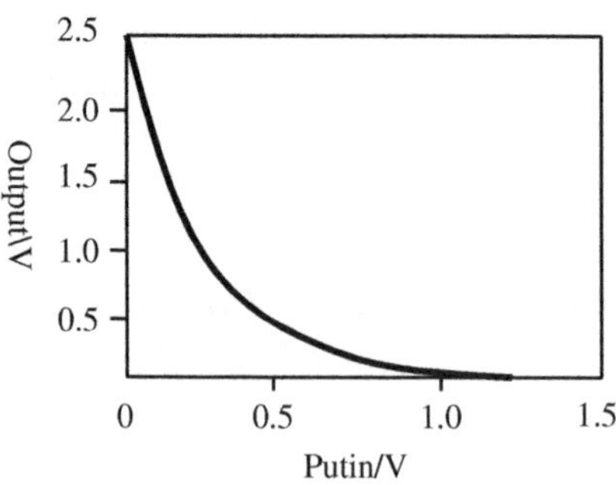

Fig. 80.2 Index compensation circuit characteristic curves

Q tube base current is ignored, the potential of point A

$$V_{p1} \approx \frac{R_{11}}{R_{10} + R_{11}} \cdot V_2 \tag{80.15}$$

Q tube collector current:

$$I_{cl} = I_{ref} \approx I_s e^{\frac{V_{BE1}}{V_T}} \tag{80.16}$$

E-point potential:

$$V_E = V_{p1} - V_{BE1} = -V_{BE2} \tag{80.17}$$

$$V_{BE2} = -V_{p1} + V_{BE1} \tag{80.18}$$

Output voltage:

$$\begin{aligned} V_3 &= i_{c2} R_{13} = I_s e^{\frac{V_{BE2}}{V_T}} R_{13} = I_s e^{\frac{VBE1}{V_T}} e^{-\frac{R_{11}}{R_{10}+R_{11}} \cdot \frac{V_2}{V_T}} R_{13} \\ &= I_{ref} e^{\frac{R_{11}}{R_{10}+R_{11}} \cdot \frac{V_2}{V_T}} R_{13} \end{aligned} \tag{80.19}$$

Circuit to the introduction of deep negative feedback, can be considered

$$V_{N1} \approx V_{p2} \tag{80.20}$$

$$I_{N2} \approx I_{p2} \approx 0 \tag{80.21}$$

Therefore, the current equation for the nodes of a:

$$\frac{V_4 - V_{N1}}{R_{18}} = \frac{V_{N1} - V_5}{R_{20}} \tag{80.22}$$

Thus, the point potential

$$V_{N1} \approx \left(\frac{V_4}{R_{18}} + \frac{V_5}{R_{20}} \cdot R_N \right) \tag{80.23}$$

of which:

$$R_N = R_{18}//R_{20} \tag{80.24}$$

Junction point P2 current equation:

$$\frac{V_{p2}}{R_{17}} + I_{ref} = \frac{V_5 - V_{p2}}{R_{19}} \tag{80.25}$$

Thus, the P2 potential:

$$V_{p2} = \left(\frac{V_5}{R_{19}} - I_{ref}\right)(R_{17}//R_{19}) \tag{80.26}$$

Since, $V_{N2} \approx V_{P2}$ so the use of type 4-4-10 and style 4-4-12, expand finishing can be

$$\frac{R_{20}}{R_{18} + R_{20}} V_4 + \frac{R_{18}}{R_{18} + R_{20}} V_5 \approx \frac{R_{17}}{R_{17} + R_{19}} V_5 - I_{ref} \frac{R_{17} R_{19}}{R_{17} + R_{19}} \tag{80.27}$$

If $\frac{R_{20}}{R_{18}} = \frac{R_{19}}{R_{17}}$ can be trimmed away the common factor to get

$$I_{ref} \approx -\frac{V_4}{R_{17}} \tag{80.28}$$

and because

$$V_1 = -\frac{R_{16}}{R_{15}} V_{ref3} \tag{80.29}$$

Then, $R_{15} = R_{16}$ the output current

$$I_{ref} \approx \frac{V_{ref3}}{R_{17}} \tag{80.30}$$

80.5.4 Nonlinear Index Compensation Circuit Design

Although the closer distance, but the negative effects of attenuation the sum of x corresponding to changes in voltage V1.

80.6 The Experimental Method and Summary

80.6.1 Test Method

Experiment with the design of large-displacement eddy current measurement circuit, the eddy current sensor is fixed on one side, the other to place a thickness 2 cm area of 200 × 200 mm steel plate as the conductor under test, the steel plate can be moved on the device, without the compensation circuit, steel displacement read with a venire caliper, measuring circuit output voltage with a digital oscilloscope directly read out.

80.6.2 The Experimental Results

We can obtain the data of the relationship between displacement and voltage thought the experiments; it is showed in the Table 80.1.

Compensation before displacement.

Compensation prior to the voltage data in the Multiuse software to design the compensation circuit to compensate for voltage measurement to be compensated after the compensation circuit to compensate for the displacement voltage data shown in Table 80.2.

Voltage measurement in the multisim shown in Figs. 80.3 and 80.4

1. change the input voltage of the circuit.
2. Corresponding to the output voltage as follows:
   The experimental results shows that the displacement x = 50–90 mm range of linearity has been well compensated by the measurement range of the sensor is able to expand.

The experimental results show that: the exponentiation circuit as eddy current displacement sensor nonlinear compensation link later, after compensation, can

Table 80.1 Compensation displacement—voltage data table

Displacement/mm	0	10	20	30	40	50
Voltage/V	0.33	0.61	1.09	2.18	2.98	3.81
Displacement/mm	60	70	80	90	100	110
Voltage/V	4.32	5.33	6.03	6.58	6.92	6.93

Table 80.2 Compensation after displacement—voltage data table

Displacement/mm	0	10	20	30	40	50
Voltage/V	10.31	10.51	11.10	12.21	12.92	13.52
Displacement/mm	60	70	80	90	100	110
Voltage/V	13.91	14.06	14.23	14.24	14.24	14.24

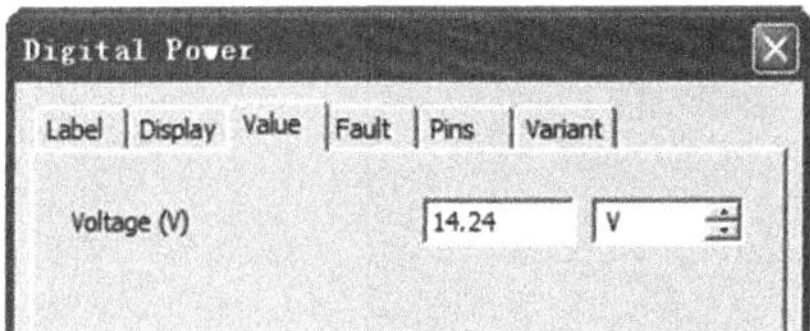

Fig. 80.3 Change the input voltage

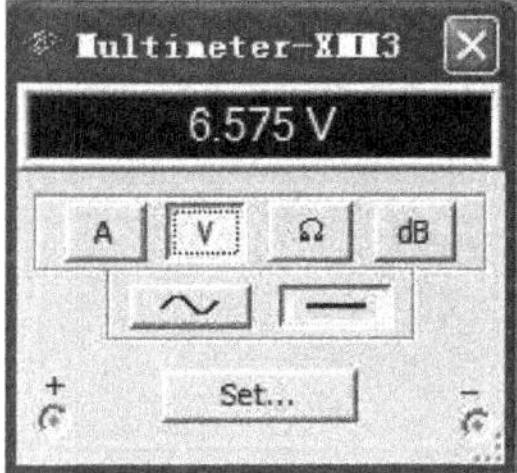

Fig. 80.4 Output voltage

effectively improve the sensor's original transmission characteristics can be seen from the experimental data, the linear range of compensation between 30–50 mm, less than the diameter of the linear measurement range of 1/2; compensated after the linear range of between 30–90 mm, expanded to about 1 times the diameter can be seen, exponentiation circuit nonlinear compensation part of the eddy current displacement sensor can satisfy the measurement needs of the large displacement.

References

1. Internet of things architecture and technology roadmap 2009, 12
2. Wang C, Zhang XQ, Dong J (2008) sensor applications, and circuit design. Chemical industry press, Beijing, 5:2–4
3. Of Scientific Matthew water, Xiaolin sensor and detection technology (2005). Publishing House of Electronics Industry, Beijing, 7:19–23
4. Of Scientific Matthew water, Xiaolin sensor and detection technology (2005) Publishing house of electronics industry, Beijing, 7:2–3
5. Xia S, Hui Z, Liu W (2005) Frequency amplitude modulation of eddy current sensor circuit. Comput Measure Control 1:508–510
6. JianYang T, Huizhong Z (1998) process eddy current displacement sensor linearization circuit. Instrument Technique and Sensor, 6:27–30
7. Bernyk ZA, Uchanin VM, Bilokur IP (1994) Eddy current testing of mechanical characteristics of pressure vessel metals at high frequencies. Mater Sci 30(2):248–251
8. Hongyi W, Xinquan L, Yushan L et al (2004) A piecewise-linear compensated band gap reference. Chin J Semiconductors 25(7):771–777

Chapter 81
Design of Digital Switching Power Amplifier for Magnetic Suspended Bearing

Jingwen Gong, Geng Zhang, Jinguang Zhang, Huachun Wu and Xin Cheng

Abstract Digital power amplifier used to drive magnetic bearings has gradually replaced analog power amplifier due to its advantages such as high hardware integration, excellent control performance, anti-jamming capability, fast response speed, convenient debugging and so on. In this paper, TMS32028335 is selected to be the main controller as well as using two-level current control modulation to design digital amplifier system. Experiments results indicate that the digital amplifier can effectively track coil current.

Keywords Digital power amplifier · TMS320F28335 · Two-level current control

81.1 Introduction

Magnetic bearings can generate electromagnetism to suspend the shaft without mechanical contact, which is a kind of new-type bearing with the advantages of frictionless, no abrasion, without lubricating, simple maintenance and so on. The magnetic bearing system is consist of rotor, electromagnetic coil, sensor, controller, power amplifier, etc. Power amplifier is one of the main components of active magnetic bearing system, it has a function of amplifying the control signal or converting it into a power signal with enough energy to drive the electromagnetic coil [1]. The performance of power amplifier plays a very important role on the overall technical characteristics of the maglev system [2]. Furthermore, the

J. Gong · G. Zhang · J. Zhang (✉) · H. Wu · X. Cheng
School of Mechanical and Electrical Engineering, Wuhan University of Technology, Wuhan, China
e-mail: jingwen_gong@163.com

J. Gong
Department of Control Science and Engineering, Huazhong University of Science and Technology, Wuhan, China

Z. Sun and Z. Deng (eds.), *Proceedings of 2013 Chinese Intelligent Automation Conference*, Lecture Notes in Electrical Engineering 255,
DOI: 10.1007/978-3-642-38460-8_81, © Springer-Verlag Berlin Heidelberg 2013

power-amplifier's voltage and current has a decisive influence on the control accuracy (especially high spindle speed situation).

Traditional analog power amplifier circuit is relatively huge with the disadvantage of high heating value, high power consumption and poor reliability [3]. Temperature drift and zero drift of the analog devices will also reduce the accuracy of the power amplifier. In addition, if it's necessary to modify or improve the control algorithm of the system, it needs to redone the hardware circuit which is not conductive to the debugging of our research. Compared to analog amplifier, digital amplifier has the advantage of higher hardware integration, better control performance, anti-interference ability, fast response and convenience in debugging.

In this paper digital power amplifier based on TMS32028335 [4] is proposed and built to realize current control of magnetic bearings. Experimental results demonstrate that the digital amplifier is feasible in enhancing the overall performance of the magnetic bearing.

81.2 Analysis of Implementation Method of Digital Power Amplifier

81.2.1 Analysis of Systematic Principle

Figure 81.1 illustrates block diagram of the power amplifier, it mainly includes the following submodules:

1. Power supply
2. DSP
3. Gate driver
4. Power bridge circuit
5. Current sensor.

An analog signal is given to a power amplifier as given signal (so-called instruction). For DSP submodule, the controller calculates the error signal Δ by formula (81.1):

$$D_{in} - D_{feedback} = \Delta \tag{81.1}$$

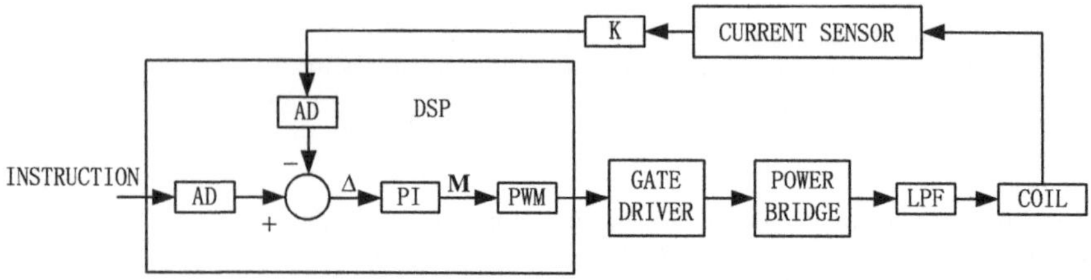

Fig. 81.1 Block diagram of power amplifier

where D_{in} represents the input digital signal, $D_{feedback}$ represents the feedback digital signal. Input signal as Δ is, DSP calculates the output control signal M by using PI algorithm. M is send to the DSP PWM module as an input to achieve a full 0–100 % duty cycle control. The PWM wave must get through the optocoupler to get into corresponding drive pin of the digital driver for the isolation of DSP and power circuit. The digital driver's capability should reach a high level to ensure the absolute turn-on and turn-off actions of the MOSFET. Thus the load current can be well controlled.

81.2.2 Topological Structure of Power Amplifier Circuit

Since the unidirectional current can produce enough electromagnetic suction, H semi-bridge structure is adopted and shown in Fig. 81.2. Predetermined current direction from a to b is defined to be the positive direction, the related two-level current mode is shown below:

- Charge state
  The power supply charges the load just after the simultaneous turn-on of both S1 and S4, then the coil current increases.
- Discharge state
  The load current charges the power supply through flywheel diode D2 and D3 just after the simultaneous turn-off of both S1 and S4, then the coil current decreases.

81.2.3 Impact of Power Amplifier's Current Response Rate and Current Ripple on System's Performance

As shown in Fig. 81.2, The voltage-current relationship of the solenoid coil of the magnetic bearing can be expressed in formula (81.2):

$$U_{in} - \Delta U = L\frac{di}{dt} + iR \tag{81.2}$$

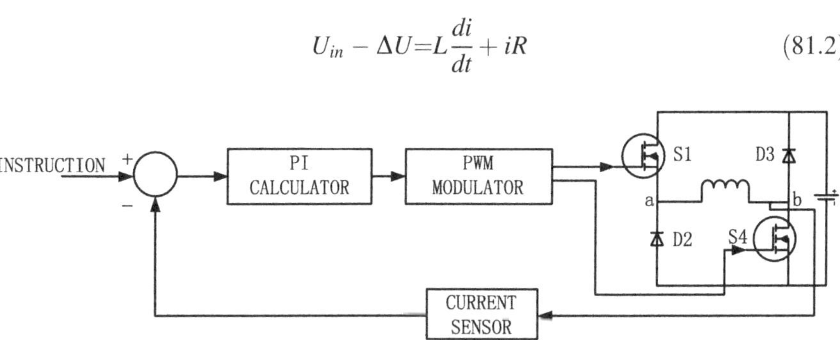

Fig. 81.2 Topological structure of power amplifier circuit

where U_{in} is the supply voltage, ΔU is the voltage drop of the turned-on MOSFET, R and L are resistance and inductance of the coil respectively.

Since R is less than 1 Ω and ΔU is only about 0.5 V, the two parameters are ignored to simplify formula (81.2):

$$\frac{di}{dt} = \frac{U_{in}}{L} \tag{81.3}$$

As current given is $I = I_m \sin wt$, formula (81.3) will get the maximum current rate by substituting I for i:

$$\left[\frac{di}{dt}\right]_{\max} = I_m w \leq \frac{U_{in}}{L} \tag{81.4}$$

Only when the $I_m w$ which stands for the product of current amplitude and angular frequency [5] meets the condition of formula (81.4), can the output current undistortedly tracks a given signal. If not, the output of the power amplifier with respect to the given current signal will have amplitude change and phase lag [6]. Generally L is considered as a constant, so rising power supply is taken to obtain an excellent output current wave.

Traditional two-level switching amplifier has only two voltage status: either +V or −V. The rapid rise and fall of the current will lead to a relatively large current ripple. Reference [7] provides the formula of the approximate current ripple that is shown below:

$$I_V \approx \frac{V_{in}}{2f_s L_m} \tag{81.5}$$

where I_V is the current ripple, V_{in} is the power supply voltage, f_s represents the switching frequency of MOSFET, L_m represents the load inductance.

Formula (81.5) shows that I_V is proportional to V_{in}, on the other hand, I_V is inversely proportional to both f_s and L_m.

As a result of the relatively large current ripple, the control current of magnetic bearing system will enter into nonlinear region, thus will cause the mechanical concussion of the rotor [8]. Therefore it is necessary to take appropriate measures to reduce the current ripple of the magnetic bearing system. Generally the equivalent inductance of the magnetic bearings stays unchanged, therefore, in order to reduce the current ripple, the power voltage should be reduced or the switching frequency of MOSFET must be increased according to formula (81.5).

In summary, for obtaining a higher current response speed, it is necessary to increase the supply voltage when two-level switching power amplifier is chosen. As heightening the supply voltage increases the current ripple correspondingly, the stability of the system declines. In order to ensure that the amplifier has both a certain current response speed and a smaller current ripple, the method of giving a high supply voltage and increasing the switching frequency simultaneously is adopted.

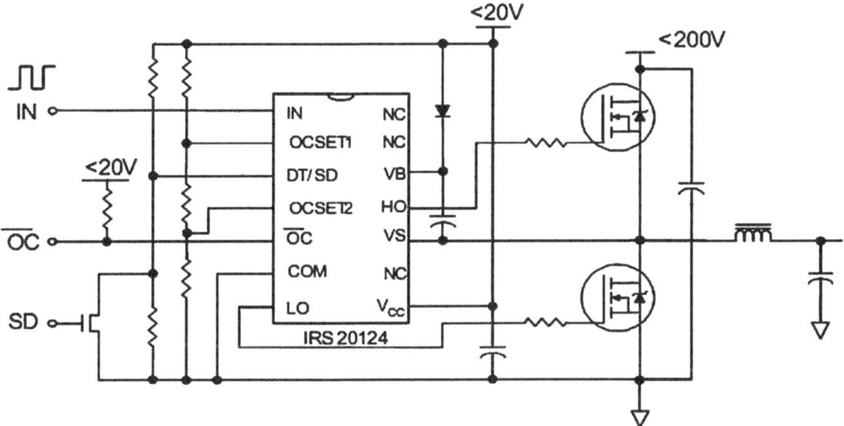

Fig. 81.3 Application circuit of IRS20124

81.3 Type Selection of Power Device

81.3.1 Type Selection of MOSFET Driver

After comprehensive comparison of the performance of a variety of driver chips, IRS20124 meets all the requirements. The IRS20124 is a high voltage, high speed power MOSFET driver with internal dead-time and shutdown functions which is specially designed for Class D audio amplifier applications. Its supply voltage can achieve 200 V, and its output drive current ranges from 1 to 1.2 A and output drive voltage ranges from 10 to 20 V. In addition, the switching time is only dozens of nanoseconds. All these features mentioned previously will ensure the effective drive of the MOSFET. The nonlinear area of the output waveform can be adjusted by setting the dead-time so that the waveform distortion can be reduced as well as preventing simultaneous conduction of the upper and lower bridge arm. Besides, the IRS20124 is able to sense negative as well as positive current flow, enabling bidirectional load current sensing without the need for any additional external passive components (Fig. 81.3).

81.3.2 Type Selection of MOSFET

After comprehensive comparison of various MOSFETs, IRFS5620 is ultimately chosen. Its main parameters are listed: 14.6 ns turn-on time, 8.6 ns turn-on delay time, 9.9 ns turn-off time and 17.1 ns turn-off delay time. All the parameters are fully capable of matching with IRS20124's drive ability. Moreover, the drain-source voltage is up to 200 V and the turn-on voltage is 5 V, these parameters are within the scope of IRS20124's drive capability too. To sum up, it's very much convenient for the circuit design when the MOSFET and driver chip can match each other.

81.4 Experimental Results Analysis

The digital-power-amplifier's performance is shown by figures. The switching frequency of the drive waveform is set to 200 kHz by program, supply voltage is 100 V and the maximum output current is 10 A. Figure 81.4 shows a PWM signal which is used to drive the MOSFET. Figure 81.5 shows the amplifier's current output when a sine wave is given. Figure 81.6 shows the amplifier's current output when a square wave is given.

In Fig. 81.4 the PWM driving signal's frequency is equal to the given signal's frequency. Seen from Figs. 81.5 and 81.6, the output of the amplifier can efficiently track the given signal. The current ripple is very small while there is a certain frequency switching noise spreading all over the system that is remained to be improved afterwards.

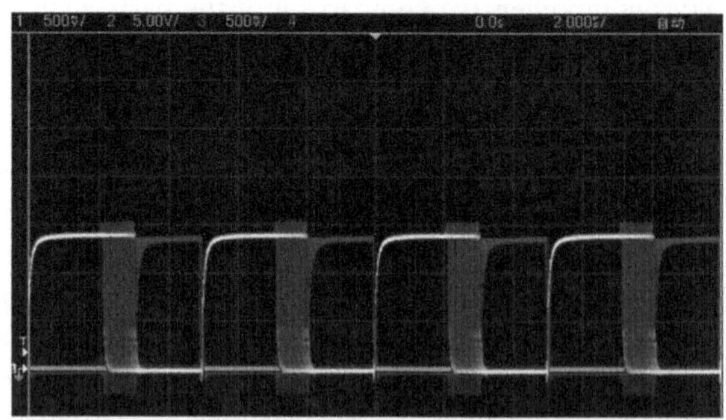

Fig. 81.4 PWM driving signal of the MOSFET

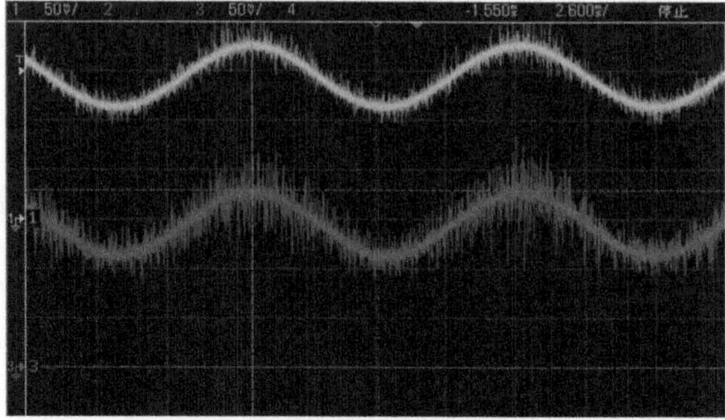

Fig. 81.5 Output current signal when given a *sine wave*

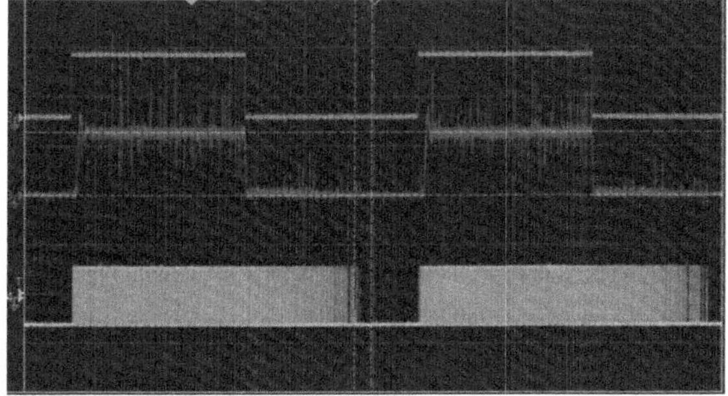

Fig. 81.6 Output current signal when given a *square wave*

81.5 Conclusion

Switching-Power-Amplifier's driving ability has an important impact on the overall performance of magnetic bearing systems as the actuator. In this paper, the design is based on traditional two-level current modulation. By way of the experiment, both a high current response speed and small output current ripple are realized by using the method of heightening the supply voltage as well as raising the switching frequency.

Acknowledgments This research is supported by the Natural Science Foundation of China (No. 51275368, No. 51275371, No. 51205296, No. 51205300).

References

1. Hu Y, Zhou Z, Jiang Z (2006) The basic theory and application of magnetic bearings. Machinery Industry Press, pp 84–85 (in Chinese)
2. Jun W, Longxiang X (2010) The system modeling and control of the switching power amplifier of magnetic bearings. Chin J Mech Eng 21(4):477–481 (in Chinese)
3. Yu Y, Hu Y (2005) The research of the design of digital power amplifier used for magnetic bearings. J WUT (Inf Manage Eng) 27(5):230–233 (in Chinese)
4. Chang X, Xu L, Dong J (2010) The digital power amplifier of magnetic bearings. J Mech Eng 46(20):9–14 (in Chinese)
5. Zhang D, Zhao L, Zhao H (2001) Effect of current response rate and force response rate on performance of magnetic bearing systems. J Tsinghua Univ (Nat Sci) 41(6):23–26 (in Chinese)
6. Gu H, Zhao H, Zhao L (2006) The parameter design of power amplifier for electromagnetic bearing. Chin J Mech Eng 42(2):208–211 (in Chinese)
7. Zhang J (1995) Power amplifier for magnetic bearing. Swiss Federal Institute of Technology, Switzerland
8. Zang X, Wang X, Qiu Z, Deng Z (2004) Research on current control mode tri-state modulation technology in switching power amplifier for magnetic bearings. Proc CSEE 24(9):167–172 (in Chinese)

Chapter 82
Weld Pool Surface Model Establishment for GTAW Based on 3D Reconstruction Technology

XueWu Wang

Abstract In Gas Tungsten Arc Welding (GTAW) process, weld pool contains information that can be used to establish process model, realize desired penetration control and study the weld pool characteristic. Hence, various sensing methods are used to sense weld pool status. At the same time, many researches are conducted to establish the weld process model, and obtain desired weld process control. In this paper, a three dimensional weld pool sensing system was introduced, and the image processing, weld pool reconstruction results were presented. Then, based on the reconstructed 3D weld pool geometry, an artificial neural network (ANN) model was established to describe the weld pool geometry. Simulation results show that the ANN model can reflect weld pool geometry accurately, which will benefit the GTAW process analysis and control.

Keywords GTAW · 3D · Weld pool · Description model · ANN

82.1 Introduction

In manual welding process, skilled welders can ensure welding quality through compensating for deviation in the process. This can be achieved just by observing the weld pool surface, because the surface contains sufficient information about the weld quality. More rapid and accurate welding can be got by automatic welding machine. Besides it, it is more convenient for machine to adjust several welding parameters. And welding manipulator and robot can substitute for workers to finish the job in some severe environments. But the machine lacks the visual feedback ability welders possess. So the capable sensing system is the first and most

X. Wang (✉)
School of Information Science and Engineering, East China University of Science and Technology, Shanghai 200237, China
e-mail: wangxuew@ecust.edu.cn

Z. Sun and Z. Deng (eds.), *Proceedings of 2013 Chinese Intelligent Automation Conference*, Lecture Notes in Electrical Engineering 255,
DOI: 10.1007/978-3-642-38460-8_82,

important step to realize intelligent penetration control. Some previous works were done to meet this demand. These methods can be divided into indirect method, two dimensional vision method, and three dimensional vision method.

Indirect methods include line scanner [1], ultrasonic [2], pool oscillation [3], infrared [4, 5], X-ray [6], and non-transferred arc methods [7], etc. Information acquired by above indirect methods can only reflect one characteristic of weld pool, and it is often not sufficient. Besides, high temperature, contamination, extensive arc light, and noise in welding process will make most above sensing methods invalid in some conditions. Vision sensing method is more similar to the welder's visual system, and can reflect the status of the welding process from the weld pool. In addition, it includes sufficient information about weld pool which is useful for process modeling and control. Many researches were conducted to apply vision method in welding process monitoring.

In Refs. [8–16], two dimensional measurements of weld pool were obtained, and model establishments, penetration control were conducted. By sensing two dimensional weld pool images, certain characteristic parameters and penetration control can be achieved. Furthermore, three dimensional weld pool shapes presents more sufficient information for determining the state of weld pool, analyzing pool defect, conducting penetration control, and studying pool physical process.

The measurement of three dimensional weld pool shapes presents more sufficient information for determining the state of weld pool, analyzing pool defect, conducting penetration control, and studying pool physical process. Hence, the weld pools were monitored and reconstructed in three dimensional [17–23]. Besides, multi-variable process model were established, and advanced control were conducted to obtain desired penetration control. Recently, a novel three-dimensional monitoring system was developed and the observed three-dimensional weld pool was characterized by its width, length and convexity [24]. An optimal linear model was also obtained in steady state.

Based on the above sensing methods, some information about weld pool can be obtained, and process modeling and control are conducted to realize process control. In this paper, a 3D sensing system and weld pool reconstruction are proposed in Sect. 82.2. Some researches on weld pool description model are presented in Sect. 82.3. Besides, a 3D weld pool description model based on artificial neural network is established in Sect. 82.3. Then, Sect. 82.4 gives the conclusion.

82.2 3D Weld Pool Reconstruction

82.2.1 System Structure

The proposed three dimensional weld pool sensing system is shown in Fig. 82.1. Gas tungsten arc welding (GTAW) process without filler is exploited in the

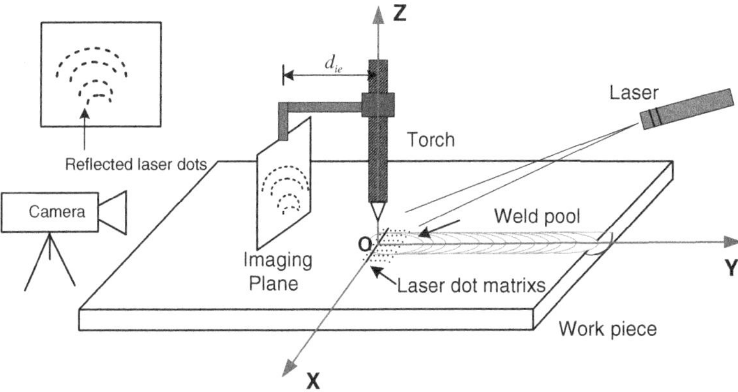

Fig. 82.1 Observation mechanism of weld pool surface

system. A low power structure light projects laser dot-matrix on weld pool surface in the pipe. An imaging plane is fastened coaxially to the laser to intercept the reflected laser rays. The intensity of laser almost remains as the same when the rays intercepted by the imaging plane, on the contrary, the intensity of the arc light decays significantly by distance. So the clear image captured by imaging plane due to the difference between propagation in laser and arc. Moreover, after distorted by a weld pool, either a convex or concave one, the image of reflected points will be reflected to the imaging plane. Thus, the geometry of weld pool surface can be extracted from reflected points in captured images.

82.2.2 Image Processing and Reconstruction

After the reflected points were captured, the captured two dimensional images on the plane were required to be processed to obtain three dimensional weld pool geometries. The process includes calibration, image processing, and reconstruction. The original reflected image, processed image, reconstructed pool shape, and entire weld pool shape are given in the Fig. 82.2.

82.3 ANN Weld Pool Description Model

Weld pool geometry is important to analyze weld pool characteristic and achieve desired weld pool penetration control. Methods to establish the weld pool geometry models can be classified as mechanism modeling and regression modeling. To

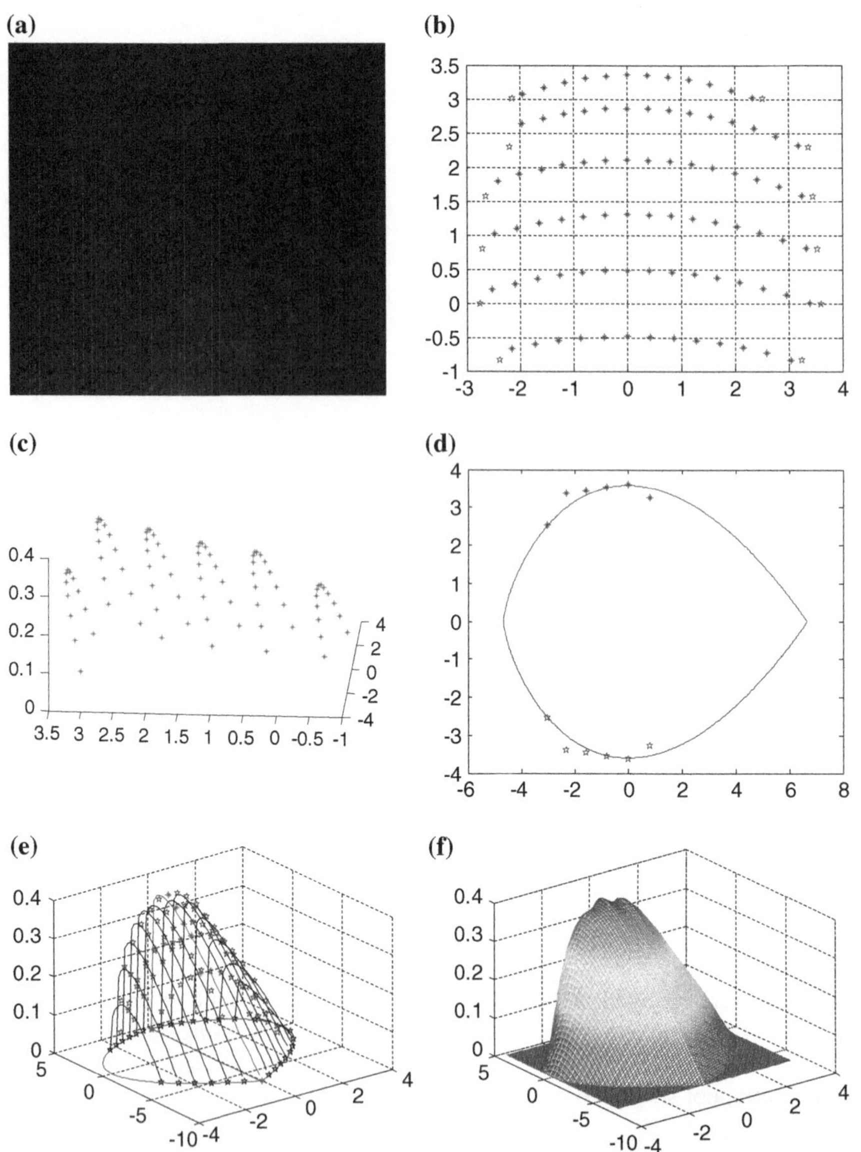

Fig. 82.2 Weld pool reconstruction results. **a** Original reflected image. **b** Points after image processing. **c** Reconstructed points. **d** *Pool shape* in horizontal plane. **e** Calculated pool points. **f** Three dimensional weld *pool shape*

adopt mechanism modeling method, profound and comprehensive process mechanism knowledge must be known in advance. Based on the mechanism knowledge, the mechanism model could be established based on the physical relationship of the primary and auxiliary variables which may be controlled in the process.

Some previous study based on analytical approach was done to get weld pool model [25–27]. Above three dimensional models based on the heat transfer and fluid flow can be used to investigate the dynamic characteristic of weld pool, and explicit physical meaning is contained in the models, but it is complicated, and not convenient to study the geometry of the weld pool directly. In this case, regression model has been widely adopted as simpler yet effective method. Regression model can be established based on regression analysis of process experimental data. Aiming to study weld pool geometry directly, some regression method were adopted to get weld pool model [12, 14, 28].

Above regression models presented weld pool geometries in two dimensional, which can't reflect the weld pool characteristic fully. As a nonlinear regression method, artificial neural network (ANN) is extensively applied due to its nonlinear mapping ability. Hence, BPNN model will be established to reflect three dimensional weld pool geometries.

Based on the proposed sensing system, image processing and reconstructed program, three dimensional shape of the weld pool can be got. The shapes of weld pool are different in some aspects. The height, length, and width of the pool are various, besides it, there are a concave on the top of the convex pool sometimes. The shape can be described by ANN because of the nonlinear mapping ability. Hence, data of various weld pool shape should be selected to train the ANN, and then it can simulate all kinds of weld pool shape. Then ANN model was established to describe the weld pool shape, which is useful for weld pool penetration prediction and control.

In Fig. 82.2, the images were drawn based on the data, which was obtained after the reconstructed image was processed. The data was calculated as follows. Figure 82.2b presents processed points reflected from the weld pool, the boundary points and the bottom boundary curve can be got. Then, points in the bottom curve were calculated through interpolation method. Based on Fig. 82.2e, two points connect the middle upper curve and boundary curve were calculated, and the middle upper curve was regressed. Then, all points in the middle upper curve were calculated based on interpolation method. As to other curves, points on these curves were calculated as the middle upper curve.

Based on the above data, the weld pool was separated into two parts to establish the ANN model. One part is the bottom plane, where Z is zero, and the other part is upper part in the weld pool shape. The original weld pool surface and regression weld pool surface based on the ANN model were given in Fig. 82.3.

From the regression pool surface, it can be seen that the ANN model can describe the welding pool surface well. The MSE of the model is 0.0981, and it means the regression model has enough accuracy. Regression result of the model is given in Fig. 82.4.

The weights of the BPNN model are shown as follows, and the superscript of the weights denotes its layer,

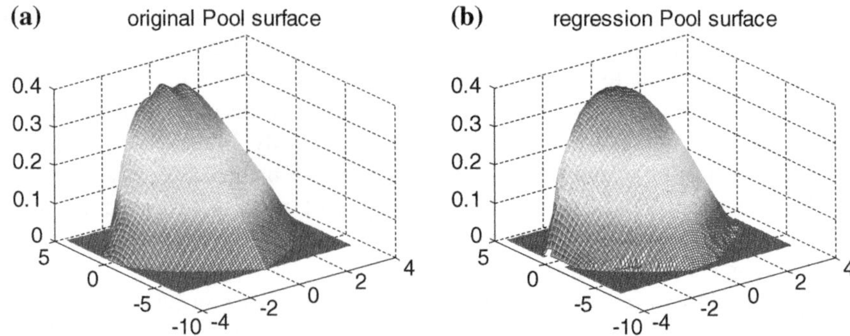

Fig. 82.3 Weld pool surface regression. **a** Original pool surface. **b** Regression pool surface based on model

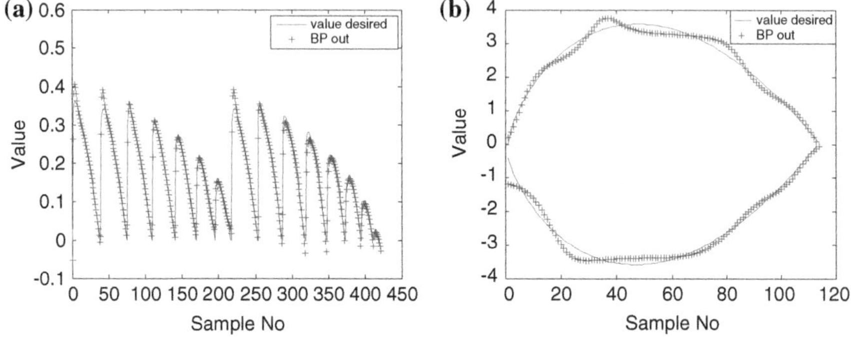

Fig. 82.4 Regression result of model 3. **a** Z > 0. **b** Z = 0

$$[w^1_{1,1}, w^1_{1,2}, \ldots, w^1_{1,n}, b^1_1, w^1_{2,1}, w^1_{2,2}, \ldots, w^1_{2,n}, b^1_{12}, \\ \ldots, w^1_{m,1}, w^1_{m,2}, \ldots, w^1_{m,n}, b^1_m, w^2_1, w^2_2, \ldots, w^2_m, b^2]$$

The number of the weights for the BPNN model is

$$M = m \times (n + 1) + m + 1 \quad (82.1)$$

where n is the number of input-layer neuron nodes, and m is the number of hidden-layer neuron nodes.

$$y = \sum\nolimits_{m=1}^{4} w^2_m \times f\left(\sum\nolimits_{n=1}^{2} w^1_{mn} x_n + b^1_m\right) + b^2 \quad (82.2)$$

$$f(x) = 1/(1 + e^{-x}) \quad (82.3)$$

Structure of the BPNN model is 2-4-1. From the regression results, it can be seen that the model can describe the weld pool geometry well. With the geometry changes, the BPNN model parameters will change too. Hence, the weights of the BPNN model hold information which describes the weld pool geometry when the structure of BPNN model remains same. At the same time, the geometry characteristic parameters relate with the welding parameters and penetration status [29]. After the relationships between the model parameters, weld penetration and welding parameters are found, it can be used to realize penetration estimation and control.

82.4 Conclusion

Weld pool contains important information for welding process analysis, modeling, and control. Based on the three dimensional vision monitoring system for the GTAW process, three dimensional weld pools were reconstructed through image processing. Then, artificial neural network (ANN) models were established to describe the weld pool geometry. Regression results show that the ANN model can reflect weld pool geometry accurately, which will benefit the GTAW process analysis and control.

Acknowledgments The assistance from PhD students WeiJie Zhang in experiments and help from Professor YuMing Zhang at the University of Kentucky Welding Research Laboratory are greatly appreciated.

References

1. Vorman AR, Brandt H (1976) Feedback control of GTA welding using puddle width measurement. Weld J 55(9):742–749
2. Graham GM, Ume IC (1997) Automated system for laser ultrasonic sensing of weld penetration. Mechatronics 7(8):711–721
3. Ju JB, Suga Y, Ogawa K (2002) Penetration control by monitoring molten pool oscillation in TIG arc welding. In: Proceedings of the twelfth international offshore and polar engineering conference, Kitakyushu, Japan, 26–31 May 2002, pp 241–246
4. Hartman DA, DeLapp DR, Cook GE, Barnett RJ (1999) Intelligent fusion control throughout varying thermal regions. In: Proceedings of the IEEE industry applications conference, no 1, Phoenix, AZ, 3–7 Oct, pp 635–644
5. Beardsley HE, Zhang YM, Kovacevic R (1994) Infrared sensing of full penetration state in gas tungsten arc welding. Int J Mach Tools Manuf 34(8):1079–1090
6. Rokhlin SI, Guu AC (1993) A study of arc force, pool depression, and weld penetration during gas tungsten arc welding. Weld J 72(8):381s–390s
7. Lu W, Zhang YM (2006) Robust sensing and control of the weld pool surface. Meas Sci Technol 17:2437–2446
8. Luo H, Devanathan R, Wang J, Chen X, Sun Z (2002) Vision based neurofuzzy logic control of weld pool geometry. Sci Technol Weld Joining 7(5):321–325

9. Wang JJ, Lin T, Chen SB (2005) Obtaining weld pool vision information during aluminum alloy TIG welding. Int J Adv Manuf Technol 26(3):219–227
10. Fan CJ, Lv FL, Chen SB (2009) Visual sensing and penetration control in aluminum alloy pulsed GTA welding. Int J Adv Manuf Technol 42(1–2):126–137
11. Chen B, Wang JF, Chen SB (2010) Prediction of pulsed GTAW penetration status based on BP neural network and D-S evidence theory information fusion. Int J Adv Manuf Technol 48(1–4):83–94
12. Zhang YM, Li L, Kovacevic R (1997) Dynamic estimation of full penetration using geometry of adjacent weld pools. J Manuf Sci Eng Trans ASME 119(4):631–643
13. Kovacevic R, Zhang YM (1997) Neurofuzzy model-based weld fusion state estimation. IEEE Control Syst Mag 17(2):30–42
14. Zhang YM, Kovacevic R (1998) Neurofuzzy model based control of weld fusion zone geometry. IEEE Trans Fuzzy Syst 6(3):389–401
15. Kovacevic R, Zhang YM, Li L (1996) Monitoring of weld penetration based on weld pool geometrical appearance. Weld J 75(10):317s–329s
16. Wu CS, Gao JQ (2006) Vision-based neuro-fuzzy control of weld penetration in gas tungsten arc welding of thin sheets. Int J Model Ident Control 1(2):126–132
17. Zhao DB, Chen SB, Wu L, Dai M, Chen Q (2001) Intelligent Control for the Shape of the Weld Pool in Pulsed GTAW with Filler Metal. Weld Res Suppl 80(11):253s–260s
18. Chen SB, Zhao DB, Lou YJ, Wu L (2004) Computer vision sensing and intelligent control of welding pool dynamics. Rob Weld Intell Autom, LNCIS 299:25–55
19. Zhang YM, Wu L, Walcott BL, Chen DH (1993) Determining joint penetration in GTAW with vision sensing of weld-face geometry. Weld J 72(10):463s–469s
20. Zhang YM, Kovacevic R, Wu L (1996) Dynamic analysis and identification of gas tungsten arc welding process for full penetration control. J Eng Ind Trans ASME 118(1):123–136
21. Zhang YM, Kovacevic R, Li L (1996) Adaptive control of full penetration GTA welding. IEEE Trans Control Syst Technol 4(4):394–403
22. Saeed G, Zhang YM (2007) Weld pool surface depth measurement using calibrated camera and structured-light. Meas Sci Technol 18:2570–2578
23. Song HS, Zhang YM (2008) Measurement and analysis of three-dimensional specular gas tungsten arc weld pool surface. Weld J 87(4):85s–95s
24. Zhang WJ, Liu YK, Wang XW, Zhang YM (2012) Characterization of three dimensional weld pool surface in GTAW. Weld J 91(7):195s–203s
25. Kovacevic R, Cao ZN, Zhang YM (1996) Roles of welding parameters in determining the geometrical appearance of weld pool. J Eng Mater Technol Trans ASME 118(4):589–596
26. Jou M (2003) Experimental study and modeling of GTA welding process. J Manuf Sci Eng 125:801–808
27. Zhao PC, Wu CS, Zhang YM (2004) Numerical simulation of dynamic characteristics of weld pool geometry with step-changes of welding parameters. Modell Simul Mater Sci Eng 12(5):765–780
28. Zhang YM, Kovacevic R, Li L (1996) Characterization and real-time measurement of geometrical appearance of weld pool. Int J Mach Tool Manuf 36(7):799–816
29. Wang XW (2013) Analysis and modeling of GTAW weld pool geometry. CCIS 355:284–293. Springer, Heidelberg

Chapter 83
Design of Intelligent Vertical Axis Turbofan Wind-Driven Generator

Zhenjun He, Fengying Ji, Jian Zhang and Jianrong Lu

Abstract The paper introduces the principle of intelligent vertical axis turbofan wind-driven generator, analyzes the concrete structure of the device and designs the control system. The scheme of controlling the fan blades flexibility is proposed and the methods on solving wire coil in operation is worked out as well. Owe to the unique turbofan structure, the device can adapt to wind in all directions so as to make the wind-driven generator keep constant counterclockwise rotation. The intelligent control of fan blades flexibility realizes the automatic open or close of blades according to the power of nature wind for device's adjustment and protection. The design of the hub solves the problem of wire coil in the process of wind-driven generator operation.

Keywords Wind-driven generator · Intelligent · Vertical axis · Turbofan · Design

83.1 Introduction

As far as scientific estimation, the mineral resources on the earth are available for human beings only in about 60 years. It is urgent to find efficient alternative source of energy. Wind energy is a great choice for it is an inexhaustible treasure in nature. And human beings had made use of it for thousands of years, which were applied in various aspects of life as an important motive power before the invention of steam engine.

After the outbreak of oil crisis in 1973, the developed countries like the United States, Western Europe, invested a lot of funds and encouraged high-tech industry to seek an alternative to oil fuel. They committed to the wind-driven generators

Z. He (✉) · F. Ji · J. Zhang · J. Lu
School of Mechanical Engineering, Nantong Broadcast & Television University,
No.10 Wai Huan West road, Nantong, China
e-mail: nthzj@163.com

Z. Sun and Z. Deng (eds.), *Proceedings of 2013 Chinese Intelligent Automation Conference*, Lecture Notes in Electrical Engineering 255,
DOI: 10.1007/978-3-642-38460-8_83,

research by taking advantage of latest technology in the field of computer, air dynamics, structural mechanics and material sciences, ushering in a new era of wind power utilization. Vertical axis windmills were applied into use a long time ago. In China, the earliest use of wind was in the form of vertical windmills. While the vertical axis wind turbine (VAWT) didn't appear until the 1920s, when is much later than that of horizontal axis wind turbine (HAWT) [1]. As it is widely believed that in no circumstances can the tip speed ratio of vertical axis rotor be more than 1, and that its utilization ratio is lower than HAWT, the VAWT failed to get attention in long term. This paper presents a new intelligent vertical axis turbofan wind-driven generator whose blades are able to automatically adjust to wind speed and apply to wind in all directions.

83.2 Comparison Between Horizontal Axis and Vertical Axis

First, the blades in a HAWT are operated by both inertia force and gravity during one revolution. The direction of inertia force is changeable while the direction of gravity is constant, which leads to an alternating load on blades and then enhances the fatigue strength. In addition, it is inconvenient to repair or maintain HAWT for it is usually placed in dozens of meters high. On the contrary, the blades in a VAWT are under permanent load for the directions of inertial force and gravity are both stable. That is to say, blades in VAWT have longer fatigue life. And VAWT can be set in ground or at the bottom of wind rotor to be easily reached.

Second, as is known, horizontal axis wind rotor is excellent in start performance, but based on the wind tunnel test by Chinese Air Force Research and Development Center, the result is not satisfactory. In the experiment, starting wind velocity is usually between four and five m/s with the maximum incredibly reaching 5.9 m/s. On the other hand, we all know that vertical axis wind rotor is poor in start performance, especially for Darrieus O without self-starting ability, which is a big obstacle in the development of VAWT. However, Darrieus H wind rotor is absolutely different.

Third, although wind power is claimed to be clean and green energy, with the establishment of more and more large wind farms, two major environmental problems have been raised—noise and ecological damage. The tip speed ratio of horizontal axis wind rotor is generally between 5 and 7. The blades cut air in such a high speed that it causes a great pneumatic noise. It is chance that many birds die from blades at high speed. The tip speed ratio of vertical axis wind rotor is relatively much slower, from 1.5 to 2, which reaches the mute effect. So many places like publish facilities, houses, where HAWT couldn't be used before for noise, now are able to install VAWT. In a word, compared with the traditional HAWT, the VAWT is superior in advanced design, higher utilization rate of wind energy, lower starting wind velocity and mute effect, therefore, has broader application prospect in the market [2].

83.3 Intelligent Vertical Axis Wind Turbine

In accordance with advantages and disadvantages between HAWT and VAWT, the paper designs a new intelligent vertical axis turbofan wind-driven generator whose blades are able to automatically adjust to wind speed and apply to wind in all directions. In Fig. 83.1, structure of turbofan deals with the variation of wind direction; Single chip microcomputer (SCM) is applied to intelligently control blades to start, stop, open and close. The weaker the wind is, the bigger the blades

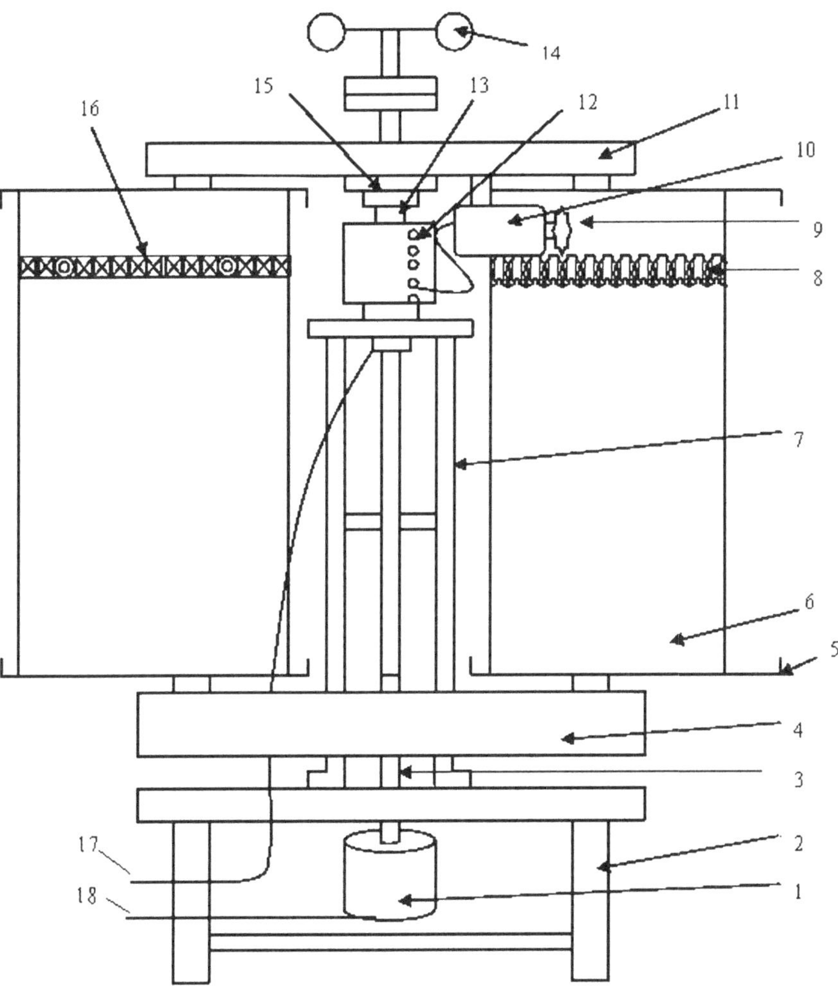

Fig. 83.1 Intelligent vertical axis turbofan wind-driven generator. (*1* Generator, *2* Base, *3* Axle, *4* Planetary gear train, *5* Blades rail, *6* Blades, *7* Tripod, *8* Chain, *9* Sprocket, *10* Stepping motor, *11* Roof, *12* Hub, *13* Main axle, *14* Anemometer, *15* Bearing, *16* Drawstring, *17* Control line, *18* Power line)

opening is, vice versa. SCM is capable of adjusting motor torque to keep the generating power relatively stable.

83.3.1 Structural Design of Speed Up Gears of VAWT

The planetary gear trains driving design is adopted to raise the generator speed so as to achieve speeding up. In the planetary gear mechanism, set the speed of sun wheel, ring gear and planet carrier as n1, n2 and n3, and the number of teeth of which is z1, z2 and z3. The gear driving is chiefly characterized by its high efficiency, compact structure [3], long service life and stable transmission ratio [4]. And the main features of planetary gear driving lie in small size, carrying capacity and stability [5]. The principal advantage of planetary gear speed increaser is to amplify the speed increasing ratio, speed and power from small-sized overall dimensions. Planetary gear driving is widely used in field of high speed transmission, like high-speed steamship, whose transmission power is getting bigger and bigger. In medium–low speed and heavy load transmission, planetary gear driving for big torque and large size has made great progress [6] (Fig. 83.2).

Introduce the relative rest tie rod H, formula is:

$$i_{13}^{H} = \frac{n_1 - n_H}{n_3 - n_H} = \frac{z_3}{z_1}$$

Design results:

[Internal gear parameter] Number of teeth z1 = 90 mm, Module m = 5, Reference circle diameter d1 = 450 mm, Excircle diameter Φ 540 mm;
[Planetary gear parameter] Number of teeth z2 = 36, Module m = 5, Reference circle diameter d2 = 180 mm, Axle hole diameter = Φ 30 mm, Keyway width = 8 mm, Addendum circle diameter = Φ 190 mm;

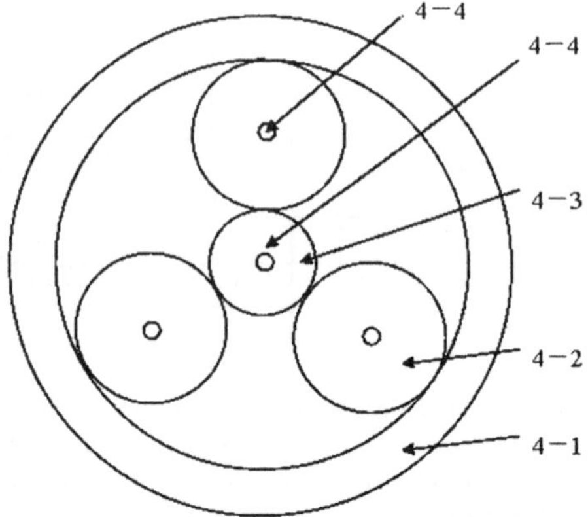

Fig. 83.2 Intelligent vertical axis turbofan wind-driven generator planetary gear trains. (*4-1* Internal gear, *4-2* Planetary gear, *4-3* Central gear, *4-4* Fixed axis)

[Central gear parameter] Number of teeth z3 = 18, Module m = 5, Reference circle diameter d3 = 90 mm, Axle hole diameter = Φ 30 mm, Keyway width = 8 mm, Addendum circle diameter = Φ 100 mm.

83.3.2 Design of Hub of VAWT

It is demanded that multiple wires shouldn't wind in rotating around. The intelligent VAWT invented multiple subdivisions castellated hub device. It can guarantee the signal transmission to control signal through soft contact between slip-ring and spring, namely both of them are in touch with each other while rotating.

The multiple subdivisions castellated hub is composed of axle, slip-ring, loop, frame and spring. Each slip-ring is fixed on excircle of hollow step axle through tight fit. The number of slip-ring is same as wire linked to. Wires come out of hollow step axle with the punched holes distributing evenly around circumference so that leakage won't happen for the contact of wires. The slip-ring engages with loops in which spring pieces keep touch with slip-ring and can be connected with the external together with wires. Two semi-circles castellated loops made of insulating material are set with slip-ring in axle, serving to separate and insulate. During operation, lifting lever drives frame and loop to rotate to prevent wire coil in the process of hub's working (Fig. 83.3).

83.3.3 Principle of Intelligent Control

The wind-driven generator can switch on–off automatically under electromechanical control device. And it also can adjust the diverging size of blades according to the power of nature wind. Once the speed of nature wind reaches the start wind one, generator spreads the blades to work. After receiving a pulse signal,

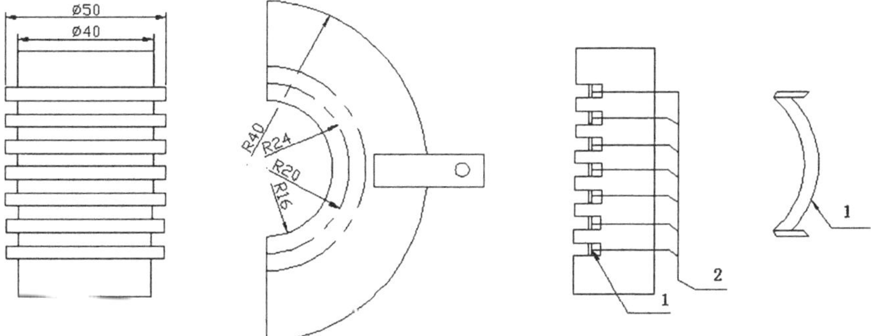

Fig. 83.3 Structure diagrams of multiple subdivisions castellated hub. (*1* spring chip *2* line)

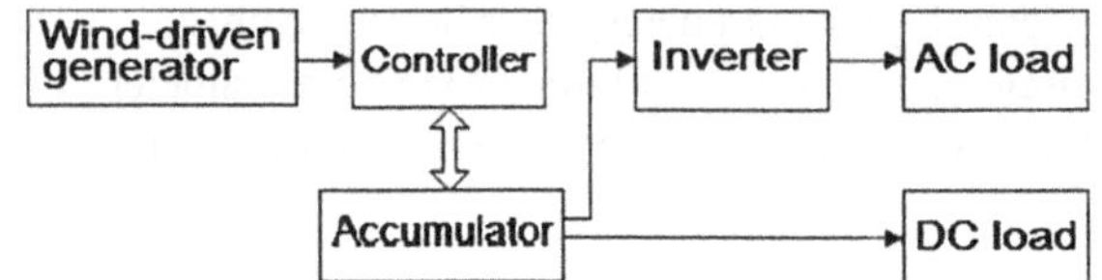

Fig. 83.4 Flow chart of wind power generation system

stepping motor will be launched to spin at a fixed angel and in a set direction, spinning as a fixed angel step by step turning into rotation. Accurate positioning can be achieved by controlling the number of pulses. Speed and acceleration is realized by controlling pulse frequency to ensure the speed in a certain range, avoiding the extremes, so as to improve the working efficiency, reduce excessive wear and avert the spare part overheating. If the speed of nature wind is too fast, generator will shut off by itself, and then blades back in position.

83.3.4 Flow Chart of Intelligent Control

The stepping motor will be driven to spin at a fixed angel (Step angel) an in a set direction after receiving a pulse signal, spinning as a fixed angel step by step gradually turning into rotation. Angular displacement can be achieved by controlling the number of pulses. Speed and acceleration is realized by controlling pulse frequency to adjust speed.

On receiving the wind signals, the anemometer sends pulse signal to single chip microcomputer (SCM) 51 which provides the strength of wind ranging from 3 to 12. Suppose the wind is in this range, a pulse signal will be sent out by anemometer to SCM 51 and then be transformed into program to drive stepping motor. The starting of motor will drive the sprockets, putting the chains fixed in major blade to motion, finally pulling the other three blades, while the extending angel is due to wind speed. Four blades are fastened to external gear in planetary gear trains, blades driving external gear. The power will be expanded by gear trains to increase transmission ratio of internal gear where main axle held on will drive generator [7]. The generator outputs three-phases AC which will be rectified into DC. At last, the output voltage is up to 220 V through inverter (Figs. 83.4 and 83.5).

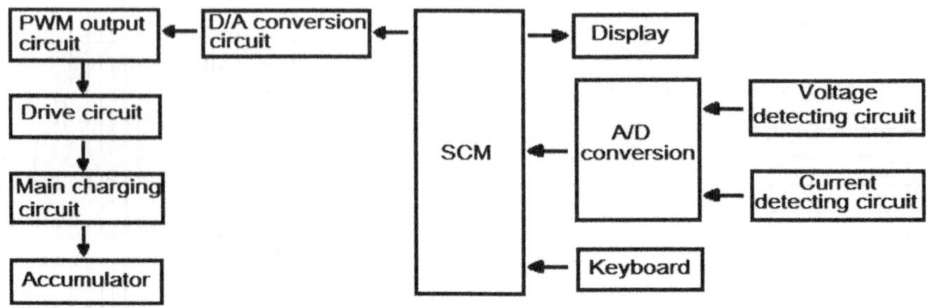

Fig. 83.5 Flow chart of intelligent control

83.4 Conclusion

Intelligent vertical axis wind-driven generator is distinctive in characteristic for the application of mechanical transmission theory, air dynamics and automatic control theory [8]. Owe to the unique turbofan structure, the device can adapt to wind in all directions so as to make the wind-driven generator keep constant counter-clockwise rotation. The intelligent control of fan blades flexibility realizes the automatic open or close of blades according to the power of nature wind for device's adjustment and protection. The design of the hub solves the problem of wire coil in the process of wind-driven generator operation.

References

1. Zhang GM (2001) On the rationality of megawatt vertical axis wind turbine. Wind Power Gener 4:55 (in Chinese)
2. Wu SQ, Zhao DP (2011) Principle of wind power generation. Beijing University Press (in Chinese)
3. Hand MM, Balas MJ (2000) Systematic controller design methodology for variable-speed wind turbines. Wind Eng 24(3):169–187
4. Zhou BY (2005) Design of industrial robot. Machine Industry Press (in Chinese)
5. Ji WF (2005) Mechatronical technology. Publishing House of Electronics Industry (in Chinese)
6. Zhao L (2005) Modern power electronic technology. Qinghua University Press. (in Chinese)
7. Zhang YX (2003) Principle and application of single chip microcomputer, Interface Technology. National Defense Industry Press (in Chinese)
8. Abdel Azimel-sayed AF, Hirsch C, Derdelinckx R (1995) Dynamics of vertical axis wind turbines (Darrieus type). Int J Rotating Mach 2(1):33–41

Chapter 84
Differencial-Clustering: Mining Bicluster Based on Weighted Graph in Microarray Dataset

Jingni Diao, Cuifang Zheng, Jilan Zhang and Jiaju Wu

Abstract In this study, we discuss methods of mining differential biclusters on gene expression data. A differential bicluster is a set of genes corresponding to samples which have substantially different expression values in two sample classes. These differential clusters may have meaningful biology significance that some specified experimental conditions are key factor to gene expression values and certain genes are sensitive to these conditions. In general, most of bicluster algorithms follow a respective clustering framework that they mine clusters in two classes. However, this strategy leads to low efficiency in term of time. We proposed a novel algorithm: Differencial-Clustering with the strategy which mine differential biclusters directly on weighted graph corresponding to two sample classes. The main contribution of our algorithm is that we avoid the respective cluster generation in two classes and comparison strategy. We compare our algorithm with standard subspace clustering method, the experimental result analysis demonstrate our main contribution.

Keywords Differential bicluster · Subspace clustering · Differential graphs

84.1 Introduction

Gene expression data are being generated by DNA chips and other microarray techniques, they are often presented as matrices of expression levels of genes under different conditions (including environments, individuals and tissues). One of the usual goals in expression data analysis is to group genes according to their expression under multiple conditions, or to group conditions based on the expression of a number of genes.

J. Diao (✉) · C. Zheng · J. Zhang · J. Wu
Institution of Computer Application, China Academy of Engineering Physics, Beijing, China
e-mail: diaojn@icaep.cn

Z. Sun and Z. Deng (eds.), *Proceedings of 2013 Chinese Intelligent Automation Conference*, Lecture Notes in Electrical Engineering 255,
DOI: 10.1007/978-3-642-38460-8_84, © Springer-Verlag Berlin Heidelberg 2013

Clustering [1, 2] is a widely used method to reveal the relationship among genes. In recent years, a new field of clustering analysis termed biclustering has gained increasing popularity in the analysis of gene expression data and other biological data. Unlike traditional clustering algorithm, in which each cluster is generated across all columns, biclustering denote groups of items showing co-expression values across certain columns of a data matrix. As a research field in data mining, biclustering has emerged as a possible solution to find co-expression pattern in certain conditions. A typical utility of biclustering is the discovery of functional modules in gene expression data. The outcome reveal that group of genes are only sensitive to a subset of control experimental constrains.

The bicluster can be generally classified into four categories [3]: (1) constant value biclusters, (2) constant row or column biclusters, (3) biclusters with coherent values, where each row and column is obtained by addition or multiplication of the previous row and column by a constant value, and (4) biclusters with coherent evolutions, where the direction of change of values is important rather than the coherence of the values. Each of these types of biclusters holds different types of significance for discovering important knowledge from real-valued data sets. Given the importance of biclusters, many studies have been developed to find them. A variety of algorithms are proposed to mine these types of bicluster. For instance, Cheng and Church's method [4] can naturally find both constant value and constant row or column biclusters. Similarly, OPSM [5] is meant to find biclusters with coherent trends of up or down-regulation in biclusters while xMotifs [6] is designed to find biclusters with constant columns in a gene expression data matrix. Despite the differences in all these biclustering methods in terms of the type of biclusters they seek, they suffer from some common defect. One of the main strategy which these algorithms followed is that start with either all rows or columns, and then iteratively eliminate them to optimize the objective function. Another strategy is that they mine biclusters respectively in two classes and then compare the outcome [7–9]. However, the two strategies both lead to low efficiency in term of time.

The MicroCluster [10] proposed by Lizhuang Zhao and Mohammed J. Zaki is a novel method, it use a weighted graph to mine scaling bicluster, On the contrary, the weighted graph method is computationally efficient compared to respective clustering methods and also produce good clustering of microarray data. The microarray data is represented as a $n \times m$ matrix D which rows correspond to genes and columns correspond to samples. MicroCluster defines a ration range for a pair of columns and summarizes all the ration range, then stat all the valid ratio to get the valid ratio range set R_{ab} and construct a weighted, directed, range multigraph $M = (V, E)$, where $V = S$ (all columns), and for each $R_i^{ab} \in R_{ab}$, there exists a weighted, directed edge $(S_a, S_b) \in E$ with weight w. In addition, each edge in the range multigraph has an associated gene set corresponding to the range on that edge. The experimental evaluation shows that MicroCluster has high performance in mining microarray data. However, there still exists a limitation in MicroCluster that it is designed to mine scaling biclusters. In our study, we proposed a novel algorithm Differencial-Clustering (mining differential bicluster on weighted graphs). In Differencial-Clustering framework, we follow the weighted graph strategy to keep the time efficiency; the weighted between a pair of sample is the genes which are co-expressed in both samples instead of range ratio. The redefined conception of weight makes Differencial-Clustering feasible for biclusters other than scaling type. Several experiments are performed on real microarray dataset to show that Differencial-Clustering is more efficient than the existing algorithms.

The layout of the paper is organized as follows. Some preliminaries and problem definitions are discussed in Sect. 84.2. In Sect. 84.3, we focus on Differencial-Clustering algorithm. Systematic experimental result analysis is reported in Sect. 84.4. In Sect. 84.6, we finished the paper with a conclusion.

84.2 Problem Definition

Before the introduction of our algorithm, some preliminaries and problem definitions are discussed in this section. Generally, the microarray data can be represented as a matrix which rows correspond to genes and columns correspond to experimental samples. We preprocessed the real data by apply a $\log_2$ normalizing function on the original matrix. The normalized matrix can be presented as Table 84.1. The expressing values 0, 1, -1 in the matrix respectively representing the biological significance non-expressed, expressed and depressed. There are three types of relationships between genes in the matrix; these types are defined as following:

1. X and Y are positive co-expressed, denoted as XY, if X and Y are both expressed; or both depressed;
2. X and Y are negative co-expressed, denoted as –XY, if X and Y have the inverse expressing value.
3. X and Y is non co-expressed if X is expressed or depressed, then Y shows both expressed and depressed state, or remains unchanged.

In our method, we denote the microarray dataset by an undirected sample relational graph, the definition of the sample relational graph is as following:

Table 84.1 An example normalized microarray dataset for two classes

	a. Class A							b. Class B				
	S_1	S_2	S_3	S_4	S_5	S_5		S_1	S_2	S_3	S_4	S_5
g_1	1	1	1	1	0	S_5	g_1	1	1	1	0	0
g_2	-1	-1	-1	1	0	S_5	g_2	-1	-1	-1	0	1
g_3	1	1	-1	1	0	S_5	g_3	1	1	-1	0	1
g_4	0	0	0	0	0	S_5	g_4	0	1	-1	0	1

Definition 1: The sample relational graph G = {E, V, W}, every vertex V_i in the graph represent a unique sample, there exist an edge E_i between a pair of samples only if the two samples has co-expressed genes and weighted item set W_i is the genes which are co-expressed in both vertexes

Definition 2: Given A and B be two sample relational graph, the subtraction between A and B can be defined as follow:

$$\begin{aligned}\mathrm{Sub}(A-B) = \{(u_i,v_i,w_i)|[(u_i \in V_a) \cap (v_i \in V_A) \cap (u_i \in V_B) \\ \cap (v_i \in V_B)(v_i \in V_B) \cap (w_i = w_{Ai} - W_{Bi})] \cup (u_{i*} \in V_A) \\ \cap (v_{i*} \in V_A) \cap (u_i \in V_B) \cap (v_i \in V_B) \cap (w_i = W_{Ai}))\}\end{aligned} \quad (84.1)$$

Definition 3: The union between two sample relational graph A and B is defined as follow:

$$\begin{aligned}\mathrm{Union}(A \cup B) = \{(u_i, v_i, w_i)|[(u_i \in V_a) \cap (v_i \in V_A) \cap (u_i \in V_B) \cap (v_i \in V_B)(v_i \in V_B) \\ \cap (w_i = w_{Ai} \cup W_{Bi})] \cup (u_i \in V_A) \cap (v_i \in V_A) \cap (u_i \in V_B) \cap (v_i \in V_B) \\ \cap (w_i = W_{Ai}))] \cup (u_i \in V_b) \cap (u_i \in V_A) \cup (v_i \in V_A) \cap (W_{Ai} = W_{bi})\}\end{aligned} \quad (84.2)$$

Definition 4: The differential sample relational graph between two sample relational graph A and B can be defined as follow:

$$G_{diff}(AB) = \mathrm{Sub}(A - B) \cup \mathrm{Sub}(B - A) \quad (84.3)$$

In order to describe our definitions more clearly, we use Fig. 84.1 to illustrate the process of differential sample relational graph construction. In Fig. 84.1a and b are the sample relational graphs of two datasets A and B in Table 84.1 which respectively correspond to different classes. C represents Sub(A–B) and d represents Sub(B–A). Graph e corresponds to the differential sample relational graph between the two dataset.

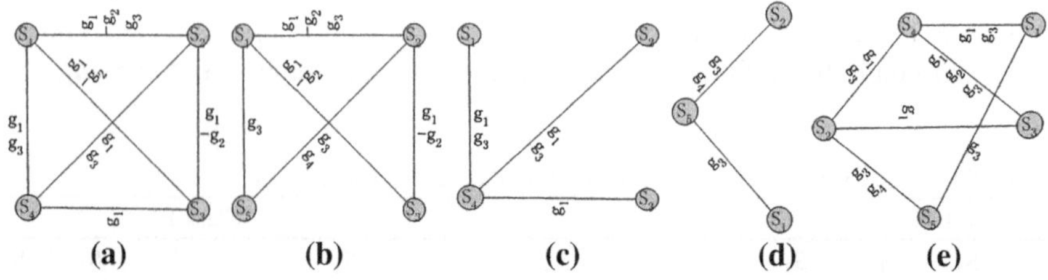

Fig. 84.1 The process of constructing the differential sample relational graph (**a**) the sample relational graphs of class A dataset; (**b**) the sample relational graph of class B dataset; (**c**) sub(A–B); (**d**) Sub(B–A); (**e**) differential sample relational graph

84.3 The Differencial-Clustering Algorithm

This section describes the Differencial-Clustering Model in detail for mining differential cluster in two different classes without respective cluster generation. The Differencial-Clustering framework has three main steps as following:

1. Find co-express genes in each pair of samples in each class, and construct the relational sample graph respectively for each dataset of two classes.
2. Compute the subtraction between the two relational sample graphs and construct the differential sample relational graph.
3. The Differencial-Clustering algorithm

This section describes the Differencial-Clustering Model in detail for mining differential cluster in two different classes without respective cluster generation. The Differencial-Clustering framework has three main steps as following:

4. Find co-express genes in each pair of samples in each class, and construct the relational sample graph respectively for each dataset of two classes.
5. Compute the subtraction between the two relational sample graphs and construct the differential sample relational graph.
6. Mining maximal bicluster on the differential sample relational graph with sample-growth strategy.

84.3.1 Construct the Relational Sample Graph

Let S_a and S_b be any two columns in the dataset, they can be denoted as two vertex V_a and V_b in the relational sample graph. If there exist co-express genes which include both positive co-expressed and negative co-expressed between the two samples, an edge E_{ab} will connect the pair of vertexes. The weighted item W_{ab} is the co-expressed genes between them.

84.3.2 Construct the Differential Sample Relational Graph

According to Definition (1), we get the subtraction between two dataset from two classes. The next step, we construct the differential sample relational graph according to Definition (3). The sample relational graph can filter many unrelated genes and sample which will reduce the time and space redundancy.

84.3.3 Mining Clusters

Construction of the differential sample relational graph can eliminate most unrelated genes and then Differencial-Clustering use a depth-first sample extension framework to mining bicluster.

The Differencial-Clustering Algorithm
Input: the minimum row threshold r_{min}; the minimum column threshold c_{min}; differential sample relational graph G_{diff}; the set of samples and genes S and G; candidate cluster C.
Output: the set of biclusters C_{final}.
Initialization: C=ϕ; Call Differencial-Clustering(C=ϕ)
Differencial-Clustering(C);
1. If C=ϕ, for each $S_i \in G_{diff}$ (i=1,2…n) do
if the weight item in the edge E_{ik}(k=1,2..m) satisfy the r_{min}, then
C$\leftarrow S_i$, Call Differencial-Clustering(C);
else if k<m; k=k+1;
else if i<n, i=i+1, Call Differencial-Clustering(C);
else return;
2. If C$\neq\phi$, for each $S_i \in G_{diff}$ (i=1,2…) ,do
3. for each S_j (j=i+1,i+2,..n) do
4. if weight item between samples in C and S_j satisfy the r_{min}, then
C$\leftarrow S_j$, j=j+1 Call Differencial-Clustering(C)
5. else j=j+1 Call Differencial-Clustering(C)
6. if the number of samples in C satisfy the c_{min}, C $C_{final;}$
7. i=i+1; Call Differencial-Clustering(C)
8. return;

Fig. 84.2 The Differencial-Clustering's mining steps

Figure 84.2 outline the main routine of the Differencial-Clustering algorithm implementation. It takes as input a set of parameters r_{min}, c_{min}, M, G, S and will output the final set of all biclusters. In our implementation, we made the cluster more efficient by using a sample extension strategy since the number of genes far exceeds the number of samples. The initial call of Differencial-Clustering is made by a candidate cluster C = Ø.

The Differencial-Clustering algorithm follows a hierarchical structure; we extend the vertex in the differential sample relational graph to the candidate cluster C, our goal is to mine maximal cluster on the differential sample relational graph. Step 1 constructs the first candidate cluster with one sample in G_{diff}; the follow-up steps mine the clusters with a depth-first frame work. The samples which are already explored can be skipped in the next steps.

84.4 Experimental Result Analysis

In this section, we present extensive experiments to evaluate the effectiveness of our method on gene expression data. We demonstrate the effect of computation efficiency comparison between the Differencial-Clustering algorithm and SDC method (subspace differential algorithm [11]). All experiments were performed on a PC with a 2.53 GHz Intel(R) Core(TM) 2 Duo CPU and 2G RAM running Windows 7 and algorithms were coded in Visual C++ 6.0.

84.5 Dataset

In this study, our dataset come from the AGEMAP [12] gene expression database, which is a resource that catalogs changes in gene expression as a function of age in mice. The AGEMAP database includes expression changes for 8,932 genes and a total of 16,896 cDNA clones in 16 tissues as a function of age. We deal with the mice which were of ages 6 and 16 month, with five male mice and five female mice per age cohort. We analyze the gonad tissue try to figure out the relations between related genes and aging factors.

84.5.1 Experimental Result Analysis

We demonstrate the result of Differencial-Clustering in comparison with SDC method. The Differencial-Clustering algorithm mine clusters on the differential sample relational graph and in [11], the algorithm obtain the differential pattern with the respective result generation and comparison scheme. The differential support used in SDC method is defined as Definition 5. In Differential-Clustering, the differential support corresponds to the parameter c_{min}. Some result analyses are illustrated in Fig. 84.3.

Definition 5: The differential support SDC (α) is computed as the difference between the percent of samples in class A on which pattern α is co-expressed and the maximal the percent of samples in class B on which a size −2 subset of α is co-expressed

$$\mathrm{SDC}(\alpha) = R_A(\alpha) - \max_{i,j \in \alpha}(R_B(\{i,j\})) \tag{84.4}$$

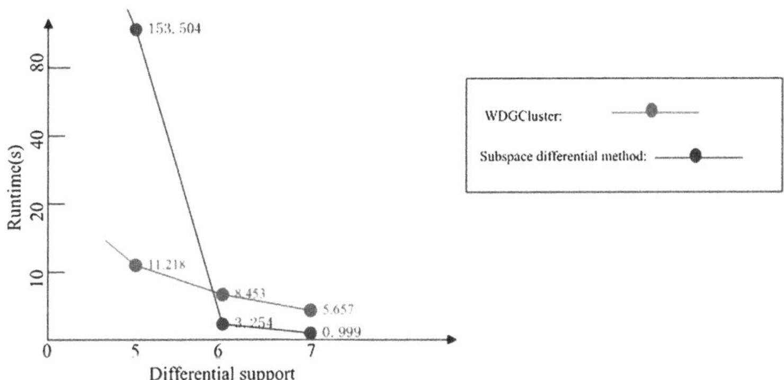

Fig. 84.3 The experimental result of WDGCluster and differential subspace method

In our dataset, the datasets from ages 6 and 16 month are denoted as two classes. Differencial-Clustering constructed a sample relational graph from the two classes. On the counterpart in [11], the outcome is generated respectively from the two classes. From Fig. 84.3, we can see the time efficiency differ a lot. With the differential support of 5 out of 12 samples, the runtime of SDC method is 153.04 s and the Differencial-Clustering algorithm is only 11.128 s. This contrast presents the efficiency of Differencial-Clustering under the low differential support threshold condition. The runtime of SDC method falls sharply when we set the differential support as 6. Under the differential support of 6 and 7, the time efficiency of SDC method is higher than Differencial-Clustering since we considered the combination of genes and samples in Differencial-Clustering which will increase the complexity.

In comparison of two experimental result, we find some interesting genes such as −BG067346, BG077727 and −BG071866 which are differential in one result and normal in another. The reason for this phenomenon is that some patterns in one class cannot satisfy the frequent support threshold which mean non-expressed. However, they showed differential characteristic in two classes more than the user defined threshold. We believe that this kind of genes are easily omitted in the respective result generation framework, the significance in this pattern will be meaningful to biologist. This interesting result motivated our thinking on the frequent bicluster mining in the further study.

84.6 Conclusion

Bicluster is an emerging method in the area of microarray dataset analysis. It provided biologist with a possible solution to find co-expression pattern in certain conditions which is with great significance to the gene analysis. In this paper, we developed a novel method called Differencial-Clustering to mining differential biclusters. Differencial-Clustering defined a differential sample relational graph to get the bicluster with high efficiency. We use sample extension strategy to improve the algorithm's efficiency. The efficiency comparison shows our algorithm is efficient and effective. However, more techniques remain unexplored in our further research such as frequent bicluster mining. We will plan to study the cluster method to find more efficient way to analyze the microarray dataset.

Acknowledgments The work is supported by research and Implement the Technology of Security Retrival based on the ciphertext.

References

1. Tavazoie S, Hughes JD, Campbell MJ, Cho RJ, Church GM (1999) Systematic determination of genetic network architecture. Nat Genet 22:281–285
2. Ramoni M, Sebastiani P, Kohane I (2002) Cluster analysis of gene expression dynamics. PNAS 99:9121–9126
3. Madeira SC, Oliveira AL (2004) Biclustering algorithms for biological data analysis: a survey. IEEE/ACM TCBB 1(1):24–45
4. Cheng Y, Church GM (2000) Biclustering of Expression Data. In: Proceedings of the 8th international conference on intelligent systems for molecular biology (ISMB'00), ACM Press, pp 93–103
5. Ben-Dor A, Chor B, Karp R, Yakhini Z (2003) Discovering local structure in gene expression data: the order-preserving submatrix problem. J Comput Biol 10(3–4):373–384
6. Murali TM, Kasif S (2003) Extracting conserved gene expression motifs from gene expression data. In: Proceedings of the Pacific Symposium Biocomputing, 2003, pp 77–88
7. Leyfer D, Weng Z (2005) Genome-wide decoding of hierarchical modular structure of transcriptional regulation by cis-element and expression clustering. Bioinformatics 21(Suppl 2):ii197–ii203
8. Prelic A, Bleuler S, Zimmermann P, Wille A, Buhlmann P, Gruissem W, Hennig L, Thiele L, Zitzler E (2006) A systematic comparison and evaluation of biclustering methods for gene expression data. Bioinformatics 22(9):1122–1129
9. Gao BJ, Griffith OL, Ester M, Jones SJ (2006) Discovering significant OPSM subspace clusters in massive gene expression data. In: Proceedings of the 12th ACM SIGKDD international conference on knowledge discovery and data mining. ACM Press, Philadelphia, USA, 2006, pp 922–928
10. Zhao L, Zaki MJ (2005) MicroCluster: efficient deterministic biclustering of microarray data. J Data Min Bioinfromatics
11. Fang G, Kuang R, Pandey G, Steinbach M, Myers CL, Kumar V (2010) Subspace differential co-expression analysis problem definition and a general approach. In: Pacific symposium on biocomputing, 2010, pp 145–156
12. Southworth LK, Owen AB, Kim SK (2009) Aging mice show a decreasing correlation of gene. PLoS Genet 5(12):e1000776

Chapter 85
Design and Implementation of an Ultralow Power Data Acquisition System

Chuan Shi, Yang Zhang, Weirong Nie, Liaoliao Yan and Jian Jiang

Abstract A kind of low power data collecting system based on MSP430 singlechip is introduced in this paper. The system which makes use of powerful processing ability and a lot of on-chip external components of the singlechip, is available in batter power supply and portable equipment situations. It can record real-time data of the dynamic information of the object, even if the power is shut off. So it is convenient for data playback and analysis. Practices prove that the design performed effectively.

Keywords MSP430F449 · Data acquisition · Low power · Serial port communication

85.1 Introduction

Microprocessor, the choice of which determines intellectualized level and structure of the signal acquisition system, is the main device for data collecting, processing and transmission. Along with the development of electronic technology, microcomputer and semiconductor, the direction of microprocessor will be smaller volume, lower power consume and higher computation speed. Data acquisition system which based on the microprocessor possesses the characteristics of intelligent automation, high accuracy and low cost [1].

The series of MSP430 singlechip are mixed signal micro-controllers, with the characteristics of low voltage, ultra-low consumption and powerful processing

---

C. Shi (✉) · Y. Zhang · L. Yan · J. Jiang
Unit 63892 of PLA, Mailbox 085-5#, Luoyang 471003, China
e-mail: shi__chuan@163.com

W. Nie
NanJing Science and Technology University, NanJing 210094, China

Z. Sun and Z. Deng (eds.), *Proceedings of 2013 Chinese Intelligent Automation Conference*, Lecture Notes in Electrical Engineering 255,
DOI: 10.1007/978-3-642-38460-8_85,

ability. They also have plenty of on-chip external components and convenient development environment [2–4]. A kind of low power data collecting system based on MSP430F449 singlechip is introduced in this paper, which can collect data information more times. Negative delay function is introduced, which ensures the integrality of the signal acquisition. Thanks to the FLASH storage module of MSP430F449, the data can be recorded even the power is off. Due to the low volume (3 cm^2), high over loading (40,000–60,000 g), lower power consume, high antijamming ability and reliability, the system performs very well in practice.

85.2 Design Scheme of System

The data acquisition system is composed of the signal processing unit, CPU, series communication unit, power supply management unit etc. The basal configuration is given in Fig. 85.1.

The original signal exported form the sensor is amplified and filtered by signal processing unit first. Then it is sent into CPU to process and analyze. In the end the signal is transferred into PC through series communication unit. Considering the factors of improving signal noise ratio, reducing electromagnetic interference, volume and power consume, this data acquisition system also need some indispensable peripheral circuits.

85.2.1 Signal Processing Unit

Because sensibility of the sensor using in the system is microvolt, small signal will easily be covered in noise. So it needs to be pretreated, such as pre amplification and filtering before sent into the CPU.

Amplifier for instrument is suitable to high-accuracy measuring of weak signal, so AD620 with the characteristics of low noise, low bias current and low power dissipation [5], is chosen in pre amplification module. It can adjust gain (1–1,000) conveniently. Gain equation of AD620 is:

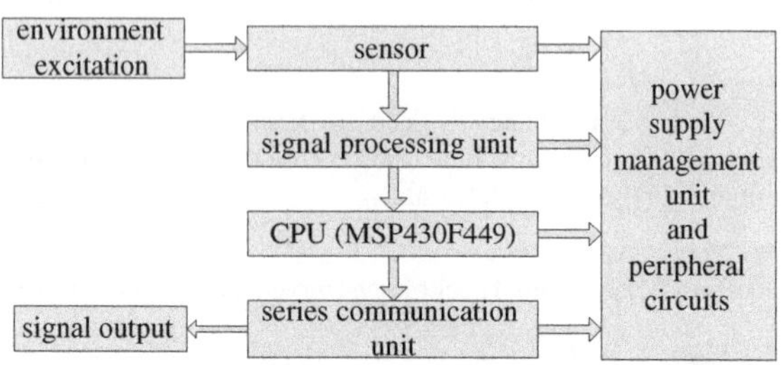

Fig. 85.1 The basic structure of data acquisition system

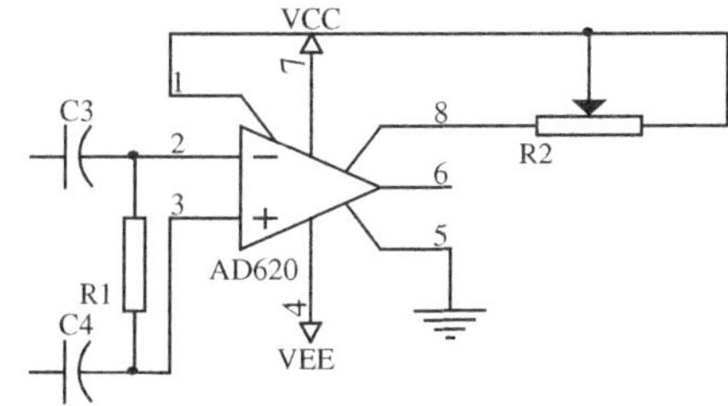

Fig. 85.2 Amplifier unit

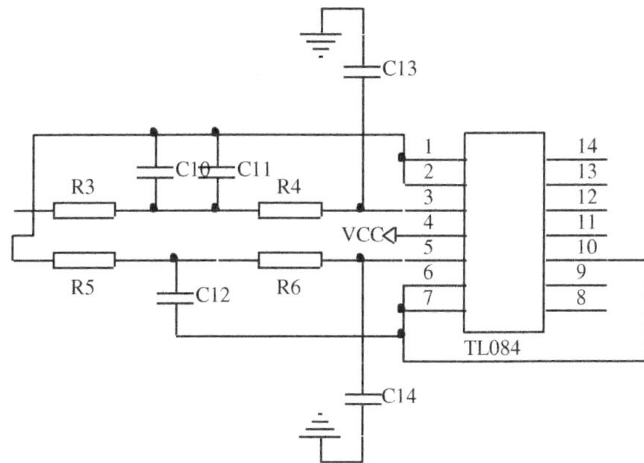

Fig. 85.3 Fourth-order Butterworth low-pass filter

$$G = \frac{49.4K\Omega}{R_2} + 1 \quad (85.1)$$

Gain G can easily be changed by adjusting R_2 between pin1 and pin2. Amplification unit circuitry constructed with AD620 is given in Fig. 85.2.

The system is designed to collect signal whose frequency below 10 K. The filter in hardware of the system uses two operational amplifiers of the integrated amplifier TL084 (the other two are used to drive up the signal) to build a fourth-order Butterworth low-pass filter [6, 7]. Cutoff frequency is 10 K. The circuitry is given in Fig. 85.3.

85.2.2 CPU

MSP430F449 is control kernel of the system, which accomplishes collection, processing and transmission of data. It is the highest grade product in the series of MSP430 singlechip. Besides the common characteristics of MSP430, it has superiority itself: owning inside reference source, sampling holding, 12-bit A/D which can scan automatically; the largest capability of FLASH memory reaches to 60 KB and RAM 2 KB; it is convenient to debug and simulate that using the FLASH combining with the on-chip JTAG interface [2, 4].

Singlechip F449 using its own 12-bit A/D to collect analog quantity, which avoids the complicacy of interfaces and improves the reliability of the system. The conversion mode of A/D is single-channel and multiple conversions. Multiple conversions are carried out in one channel chosen by the system. After conversion accomplishes once, the corresponding interrupt identification bit is set to indicate the end of this conversion. The singlechip, the power supply of which is +3.3V, uses 8 M oscillator joining pin XT2OUT and XT2IN to start up. In the series communication, 32.768 K oscillator joining pin XOUT and XIN is used as the clock root of LFXT1CLK (low frequency clock root generated from low frequency clock crystal). For the collection function of the system, one reset switch is designed to control the work state of the system by controlling the level of digital I/O pin P2.2—when high level, the system is in collection state; when low level, in series communication.

Making use of on-chip JTAG interface and FLASH memory which can be erased and wrote through electricity, storing information in power-off condition is accomplished in this system. The program is downloaded in FLASH by JTAG interface first, then the system control program run by software. In the acquisition process, the data collected are stored in FLASH (keep data when power off). After acquisition is accomplished, the information is transferred into PC through series communication unit. One point is emphasized here, FLASH memory can't be visited when erasing or writing. There are five special JTAG pins, as RST/NMI, TCK, TMS, TDI and TDO/TDI in MSP4430F449. The JTAG interface of the system is given in Fig. 85.4. Hereinto S1 is reset switch to control system state.

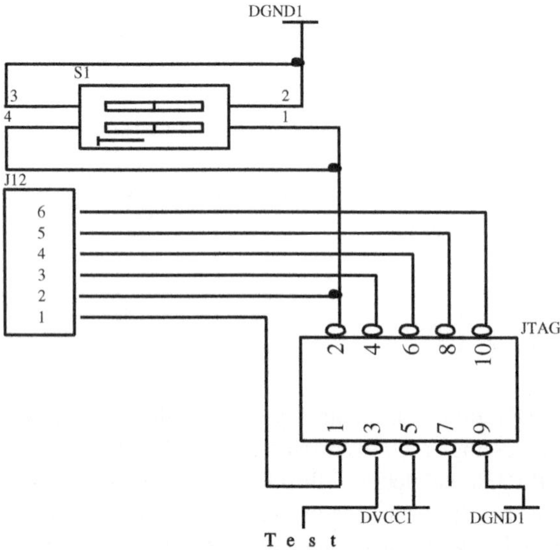

Fig. 85.4 JTAG interfaces of the system

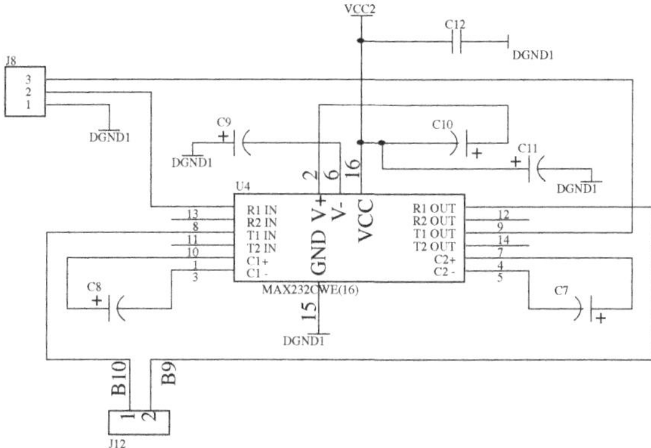

Fig. 85.5 Series communication unit circuit

85.2.3 Series Communication Unit

MSP430F449 has its own predominance in transmission—full-duplex asynchronism series all-purpose module (USART), which is used in series communication unit. The port baud is 9,600 bps. After setting a series of registers, the hardware can shift in or out data automatically to complete serial communication. The transmission or reception of data stream depends on one shift register. Under reception state, data stream are stored in reception buffer when it reaches one byte. Under transmission state, data in transmission buffer is sent to port one bit by one bit.

The system adopts RS232C standard [8] and chooses the MAX232CWE chip to achieve the conversion between TTL level exported form singlechip and CMOS level of computer. The circuitry is given in Fig. 85.5.

85.2.4 Power Supply Management

The working voltage of MSP430F449 is 1.8–3.6 V. In order to ensure low power consumption, all electron device used in the system can work well under ± 3 V.

Three fastener batteries are connected in series, used as the chief power supply source of the system, which ensures stability of the system. The characteristics of Linear voltage regulation chip TPS76930 which made by company TI are ultra-low consumption and low voltage difference, It is used for transforming 9 V input voltage to 3 V output voltage as power supply for MSP430F449. Because both the sensor and AD620 used in the system are ambipolar, CMOS monolithic integrated voltage inverse chip MAX660 is chosen to transform +3 V input voltage to -3 V output voltage as its power supply.

As the amplitude of analog signal imported in AD is small, digital switch noise brought by digital circuit in the system will seriously affect the precision of AD conversion. So power supply should be divided into analogue and digital. Analogue ground and digital ground need to be separated strictly, and they have only one common ground.

85.2.5 Power Consumption Design

In order to reduce the consumption of the system, several measures are adopted, besides choosing low power dissipation components.

1. If the basic function of the system is satisfied, reduce the number of the interface circuit, high performance device and capacitance on the circuit node as many as possible [9].
2. Modularization design is used. The system is divided into three parts, such as power supply module, collection/processing module and series communication module. When data collecting, only power supply module and collection/processing module are used. After collection is completed, series communication module is connected to them to transmit data to PC.
3. Low power consumption modes offered by MSP430F449 are took full advantage of. CPU enters suitable lower consumption mode when its setting of tasks are finished in main program. Most works which need to be completed by CPU will finished in interrupt service routine. Then CPU is awaked by corresponding interrupt to accomplish interrupt service, after which CPU enters power consumption mode again [2].
4. The power consumption of system is directly proportional to clock frequency of CPU in singlechip, so clock frequency of CPU will be reduced as much as possible when the control task and calculating aren't heavy.

85.3 Software of the System

The software program is compiled in integration development environment IAR Embedded Workbench, and debugged using debugger C-SPY. If the program succeeds in debugging, executable codes are downloaded in the FLASH of MSP430F449 to run [10]. Main functions of software are initialization of MSP430F449, data collection, filtering in software etc.

The initialization of MSP430F449 is the important step to set up operating mode of the chip. When electrified, the processor enters presetting state because of reset signal. After choosing clock root and setting high level of P2.2, the A/D module is initialized. Then AD conversions start, waiting for threshold voltage (given beforehand). When the signal reaches threshold, data collection start, and data acquired are stored in FLASH after processing.

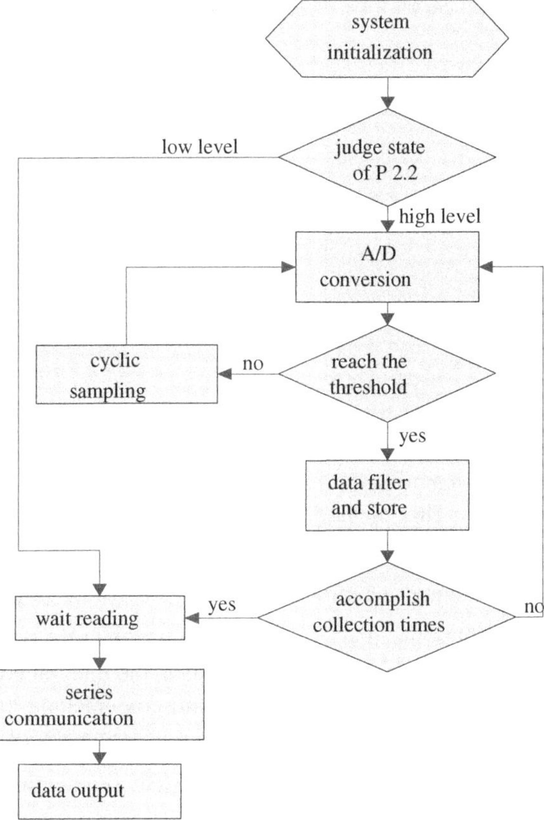

Fig. 85.6 Software flow pattern of the system

After resetting, state of pin P2.2 changes to lower level. At this time, data in FLASH will be sent into PC by serial port transmission program for further analysis. Flow diagram of program is given in Fig. 85.6.

Three points are explained here:

1. In order to ensure integrality of the signal, negative delay method is used in collection.

Figure 85.7 shows the signal acquired not using negative delay design. In this figure the highest wave crest is deficient, so the value of signal around the crest can't be observed and calculated clearly. The design doesn't achieve prospective aim. Two methods can be used to improve it, one is reducing trigger level of collection circuit, and the other one is using negative delay method in hardware of software. The first method is easily to achieve, but considering the high noise of power supply, signal collected can be submerged in the noise when trigger level is reduced. At the same time, considering the complexity of hardware design, this system chooses method that using negative delay in software.

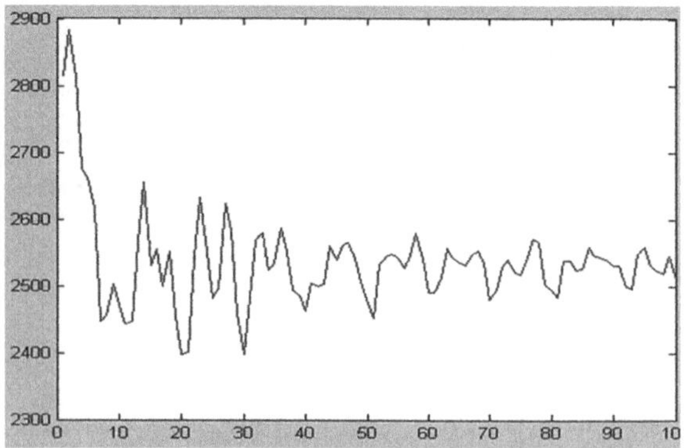

Fig. 85.7 Signal wave

One hundred datum are sampled circularly and stored in RAM till the signal reaches the threshold. Then these one hundred datum are added to the data collected later to acquire integrated signal and both two parts data are stored into FLASH. This method is easy and convenient, which ensures integrality of the signal when powered off.

2. The signal is reprocessed through software filter.

Considering that singlechip MSP430F449 can achieve the highest most efficiency when dealing with 16-bit data, the integral coefficient digital filter with low characteristic is constructed, using the method that the pole point and zero counteract each other [11, 12]. Cutoff frequency of the filter is 1 K.

3. The system can carry out multiple segmenting collections by setting times through software.

85.4 Result and Analysis of Test

Signal collection experiments using vibration sensor were carried out to prove system performance. The sensor was fixed on the table. Vibration signal was generated when knocking the table. Setting times once and twice, data were

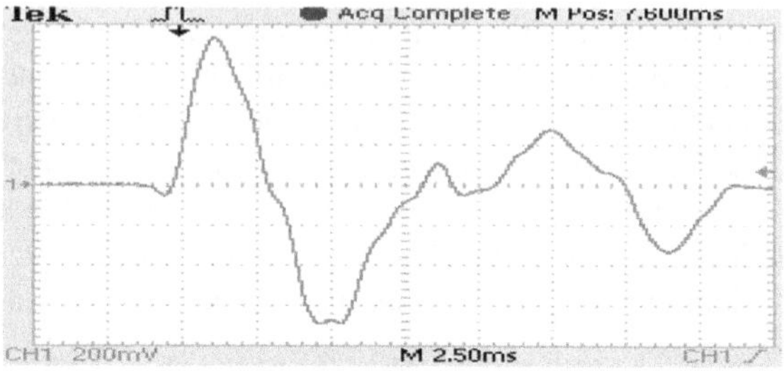

Fig. 85.8 Signal save collected by digital oscilloscope

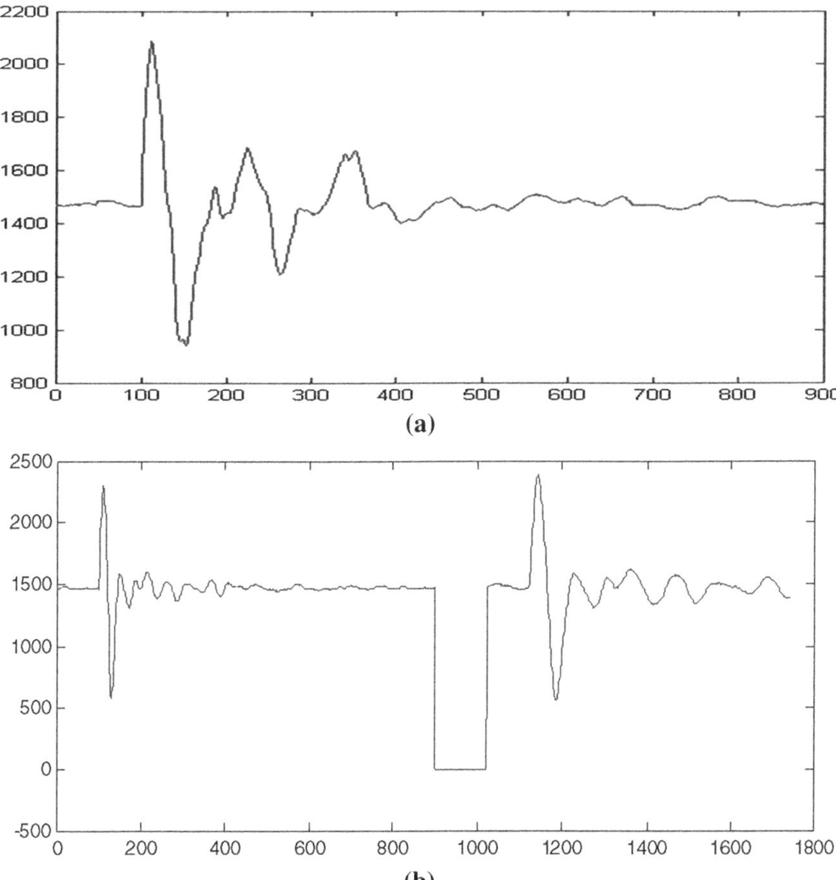

Fig. 85.9 Signal waves collected by system designed

collected by both data acquisition system and digital oscilloscope at the same time. Data acquired are showed in Figs. 85.8, 85.9. Figure 85.8 shows signal wave collected by oscilloscope. Figure 85.9 shows signal waves collected by the system designed (a) and (b) show signal collected once and twice.

Compared the two signal waves collected by the system and oscilloscope, it is obvious that the system can catch the real time signal exported from the sensor well and truly. Data acquired reproduce vibration wave entirely, so this system achieves the requirements. One point need to explain, in order to observe twice signal collection clearly, data between these two collections are set zero.

85.5 End Statement

Data acquisition system designed in this paper possesses the characteristics with simple hardware structure, little volume, lower power consume, high antijamming ability. Negative delay method and the integral coefficient digital filter are used in the software, which ensure the integrality of signal and improve the real time characteristics of data processing. Connected with different type of sensors, the system can collect data from many sources. Results about collection test prove that the design achieves the requirements. It is convenient and steady to collect data with the system, which has already been used in vibration and impact signal acquisition outdoor many times.

References

1. Hu Y, Su Y, Zhang J (2007) Design of low power data collecting system. China Meas Test Technol 33(5):121–123 (in Chinese)
2. Wei X (2002) The series of MSP430 singlechip interface technique and system design. Beijing University of Aeronautics and Astronautics Press, Beijing (in Chinese)
3. Dake H (2002) The series of MSP430 ultra-low power 16-bit singlechip. Beijing University of Aeronautics and Astronautics Press, Beijing (in Chinese)
4. MSP430x44x Family User's Guide(Rev.C) (2003) Texas instruments literature number SLAU049C
5. Analog Denices (2004) Low cost power instrumentation amplifier AD620
6. Williams AB (2008) Design of electronic filter. Science Press, Beijing
7. Chang X, Wang H, Yu Y, Ma Q (2009) Design of a programmable amplifier and filter. J Sichuan Normal Univ 32(3):381–385 (in Chinese)
8. Zhang B, Chen X, Lv X (2007) Design and realization of in system programmable outer-chip flash of DSP. Foreign Electron Meas Technol 26(10):39–42 (in Chinese)
9. Ma X, Sun C (2008) Multicenter data collection and transmission system based on low-power-consumption design. Mach Electron 7:16–18 (in Chinese)
10. Dake H (2003) The series of MSP430 singlechip C++ programing design and development. Beijing University of Aeronautics and Astronautics Press, Beijing (in Chinese)
11. Hu G (2003) Digital signal processing-theoretics, arithmetic and implementation. Tsinghua University Press, Beijing
12. Press WH, Saul C (2004) Numerical algorithm. Publishing House of Electronics Industry, Beijing

Chapter 86
Research on Temperature Optimal Control for the Continuous Casting Billet in Induction Heating Process Based on ARX Model

Zhe Xu, Xulong Che, Bishi He, Yaguang Kong and Anke Xue

Abstract In the hot rolling line continuous casting billet electromagnetic induction heating process, it's generally difficult to make steel billet temperature reach the technological requirement of the target temperature by the conventional voltage control curve. This paper firstly analyzes the mechanism of continuous casting billet electromagnetic induction heating, and establishes ARX model through system identification method. Based on the ARX model, it uses random search algorithm to optimize the output voltage curve. Simulation results show that the optimization strategy can effectively improve the temperature control precision in the continuous casting billet electromagnetic induction heating.

Keywords Electromagnetic induction heating · Model identification · ARX model · Optimal control

86.1 Introduction

Induction heating furnace is the important equipment of the hot rolling production line. It is the intermediate link between continuous casting and continuous rolling heating, and its role is to reheat the continuous casting billet or the roughing billet to the rolling requirement of the target temperature distribution. The billet discharging temperature in heating furnace will directly affect the next step continuous rolling production status, and the heating quality of heating furnace will directly affect the performance of finished steel. Therefore, to seek optimal voltage control curve that makes the billet discharging temperature as far as possible to

Z. Xu (✉) · X. Che · B. He · Y. Kong · A. Xue
Institute of Information and Control, Hangzhou Dianzi University,
Xiasha Higher Education Zone, Hangzhou City, China
e-mail: xuzhe@hdu.edu.cn

Z. Sun and Z. Deng (eds.), *Proceedings of 2013 Chinese Intelligent Automation Conference*, Lecture Notes in Electrical Engineering 255,
DOI: 10.1007/978-3-642-38460-8_86,

achieve the target temperature, is the current hot rolling electromagnetic induction heating problem that urgently need to be solved [1].

In the continuous casting billet hot rolling process, some temperature prediction models and optimal control theories have been discussed at home and abroad [2–5], but temperature optimal control study on casting billet hot rolling production line by electromagnetic induction heating method is also very few [6]. Induction heating has some characteristics, such as serious interference, time variation, non-linearity, coupling, inertia and time delays [7, 8]. Therefore, for such a complex industrial process, it's hard to get ideal control effect through the conventional modeling and control method.

At present, there are still some problems in induction heating optimal control. Firstly, because steel billet temperature fields in the furnace can't be measured online in the actual production process, especially steel billet internal temperature can't be directly measured, we need to build the mathematical model to forecast the heating billet temperature distribution accurately. Secondly, combining forecasting model, we need to choose suitable optimization algorithm and optimal target function to make the optimization effect achieve the best.

Combined with the actual engineering facts, this paper analyzes the heat transfer mechanism of the billet induction heating process and establishes the ARX model through system identification to realize billet temperature accurate prediction. On the basis of the forecast model, it uses random search algorithm to adjust output voltage curve, in order to make the billet discharging temperature achieve target and realize temperature optimal control of continuous casting billet induction heating process.

86.2 Billet Induction Heating ARX Model

86.2.1 Induction Heating System

A hot continuous rolling production line adopts electromagnetic induction heating equipment, which is composed by a 6,000 kw, 1,000 Hz medium frequency source, two induction furnaces and a PLC controller, etc. Two infrared thermometers are set up before the first induction furnace and behind the second one. These two temperatures are recorded as the charging temperature (Tmp2) and the discharging temperature (Tmp3). Detailed parameters of the induction heating equipment: each furnace is 0.8 m long with an interval of 0.67 m, Tmp2 temperature measuring point from the first furnace inlet is 0.56 m, Tmp3 temperature measuring point from the second furnace export is 0.37 m. The steel billet length is 50($\pm$2) m. Tmp2 is 900–1,000 °C, heated to 1,050 °C or so. The roller driving speed is about 1 m/s. The induction heating equipment is shown in Fig. 86.1.

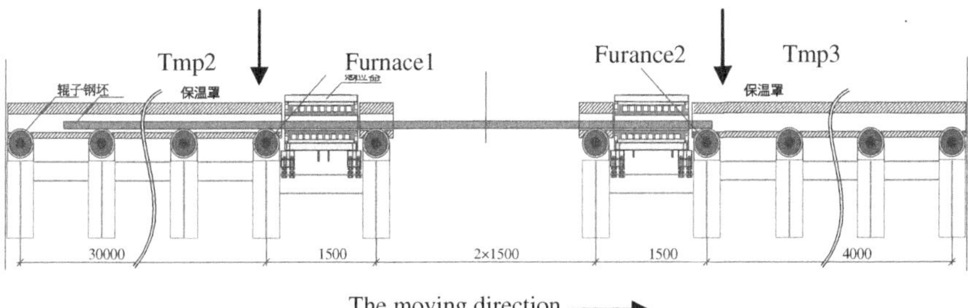

Fig. 86.1 The schematic drawing of the induction heating equipment

86.2.2 Identification of ARX Model

Because it is difficult to realize online measurement for billet temperature especially internal temperature in the heating process, we use the billet surface temperature before and after heating instead. According to the actual length of the billet, this paper divides billet into 250 segments along the length, so the collected temperature data should resample, and each billet has 250 sampling points. We consider that every segment of the temperature is uniform and can be replaced by one temperature which is known as temperature point. Similarly, the output voltage and current of medium frequency source is measured from the billet head passing through the temperature measuring point Tmp2 to the billet tail passing through the temperature measuring point Tmp3, we obtained 266 sampling points in the same sampling time, known as voltage point and current point. So a sample has 250 temperature points, 266 voltage points and current points.

Based on the analysis of the electromagnetic induction heating mechanism, the energy consumption of the billet temperature rise is from electrical work of power supply used in induction heating process, so the temperature rise is mainly affected by the absorbed work in induction heating furnace and the charging temperature Tmp2. Instead of the work, the billet power sum is used to simplify the calculation. We must know the current distribution of each furnace before calculating power, but the above-mentioned current points is the output current of medium frequency source, namely the total current flowing through two heating furnaces, so we must first obtain the current distribution of each heating furnace according to the different positions of the billet. The product of this current and voltage points is the power we need. The heating furnace, whose length is 0.8 m, is divided into four segments. We consider that power allocates to each segment of billet in heating furnace averagely. We should calculate power coefficient first. Taking the calculation method of the first segment billet power coefficient as an example, the first segment enters into the heating furnace and absorbs all the power, so its power coefficient is 1; When the second segment enters into the heating furnace, the first one shares 1/2 power in the furnace, so the power coefficient is 1/2, Similarly we can calculate and gain another two power coefficients. Therefore, a piece of steel billet is divided into 250 segments, four power coefficients of the first segment billet are [1 1/2 1/3 1/4], the power coefficients of the second one are [1/2 1/3 1/4 1/4], the power coefficients of the third one are [1/3 1/4 1/4 1/4]…The final power sum is the sum of products of power coefficients and the four powers in each heating furnace. Thus, we realize the linearization of the non-linear model and get a simple ARX model of double input and single output. The concrete form is as follows:

$$y(t) = au_1(t) + bu_2(t) + e(t) \tag{86.1}$$

where y(t) is temperature rise, $u_1(t)$ is power sum, $u_2(t)$ is Tmp2, e(t) is the system error, a and b are parameters to be identified.

The model parameters are identified by the least-squares method, so the formula (86.1) is translated into least-squares regression model:

$$z(k) = h^T(k)\theta + n(k) \tag{86.2}$$

where z(k) is the output vector of the object, h(k) is the observable data vector, θ is the parameter to be identified, n(k) is the zero mean random noise. The concrete values are:

$$\begin{cases} z(k) = y(t) \\ h(k) = [u_1(t), u_2(t)]^T \\ \theta = [a, b]^T \end{cases} \tag{86.3}$$

The optimizing index of parameters is:

$$J(\theta) = \sum [z(k) - h^T(k)\theta]^2 \tag{86.4}$$

Obtain the estimate $\hat{\theta}$ by making J = min, $\hat{\theta}$ is the least-squares estimate of θ, i.e., we complete the model parameters identification.

86.2.3 ARX Model Validation

Based on the example of the above steel mill's hot continuous rolling induction heating process, it can heat the continuous casting billet whose temperature is 900–1,000 °C (Tmp2) to 1,050 °C or so (Tmp3) by electromagnetic induction heating method. In order to improve the accuracy of the model, we take twenty samples to identify the model. After data pretreatment, each billet sample data contains 250 power sums, Tmp2 and Tmp3. By taking the power sum and the Tmp2 as input and the temperature rise as output, the ARX model parameters in the formula (86.1) can be get by identification method: a = 1.81e-5, b = −3e-3.

The sum of the Tmp2 and the model output temperature rise is the Tmp3 which we need. The identification result is shown in Fig. 86.2.

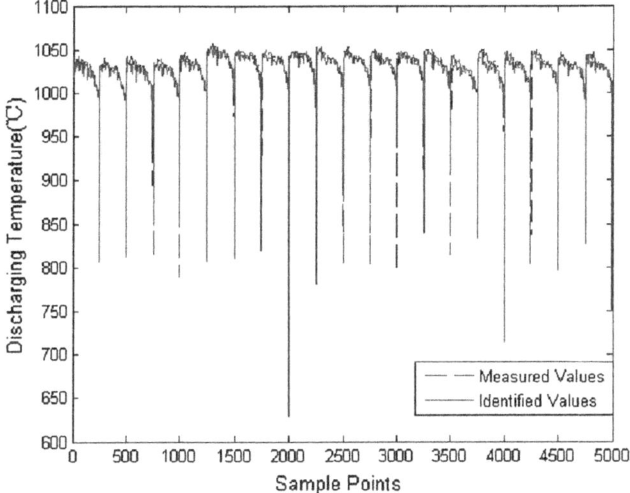

Fig. 86.2 The identification result

Then, we take other ten samples which are not identified to carry on the forecast, and verify the accuracy of the ARX model, the result shown in Fig. 86.3. Table 86.1 lists the prediction errors of samples which are within 2 %. Therefore, the result indicates that the prediction model has high precision and can achieve the requirement of the next step temperature optimal control.

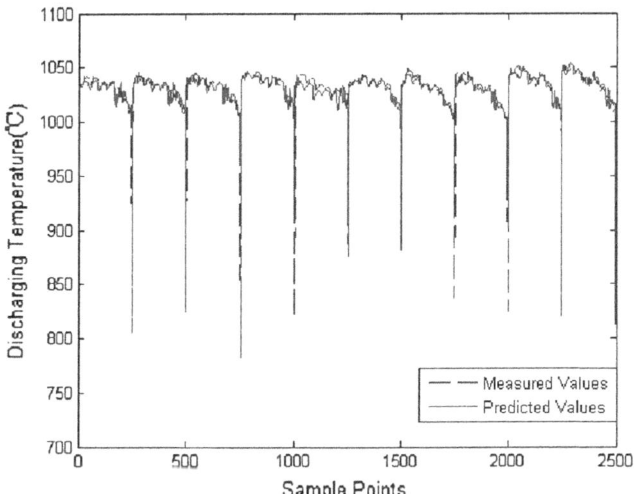

Fig. 86.3 The prediction result

Table 86.1 The prediction error of samples

Sample	1	2	3	4	5
Error (%)	1.15	1.85	1.59	1.10	1.89
Sample	6	7	8	9	10
Error (%)	0.36	0.84	1.43	1.37	0.80

86.3 Billet Temperature Optimal Control

86.3.1 The Necessity of Optimal Control

The work target of heating furnace is to make billets' surface temperature distribution curve approach to the target temperature when they discharge from the furnace, it can guarantee the reliability of the next step rolling production. In this steel mill, the discharge target temperature of continuous casting billets is 1,050 °C, first we select a random sample as the research object, such as choose the eighth sample of the identification process, and then, forecast the Tmp3 with the above ARX model and the original voltage control curve. The prediction result is shown in Fig. 86.4. From the graph, it is discovered that there is some deviation between the billet discharging temperature distribution and the target temperature, especially in tail position, so it is necessary to optimize the control of Tmp3.

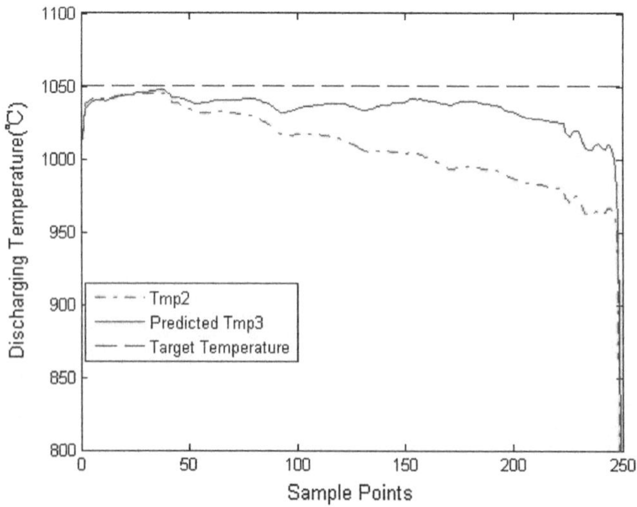

Fig. 86.4 The Tmp3 prediction result of the 8th example

86.3.2 The Realization of the Optimal Algorithm

According to the analysis of induction heating mechanism and actual working condition, the billet temperature rise is mainly affected by the absorbed power in the furnace and the charging temperature Tmp2. We can adjust the output voltage curve to control the billet discharging temperature Tmp3, meanwhile, the current is determined by the voltage and the position of billet, and the Tmp2 can be measured before the billet enters into the furnace. Combined with the prediction model of the billet, we optimize the billet discharging temperature Tmp3 to improve the quality of products.

In order to evaluate the optimization effect after adjusting the output voltage, we give the optimization target function:

$$J_{target} = |T_1 - T_{tar1}| + |\mathrm{T}_2 - T_{tar2}| + \cdots + |T_N - T_{tarN}| \tag{86.5}$$

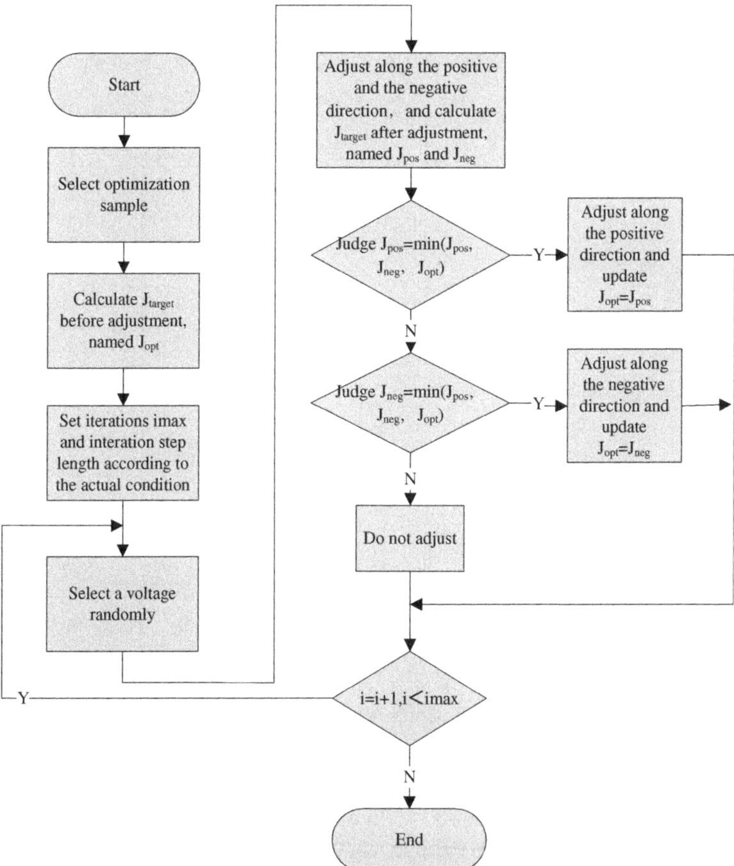

Fig. 86.5 Random search algorithm flow chart

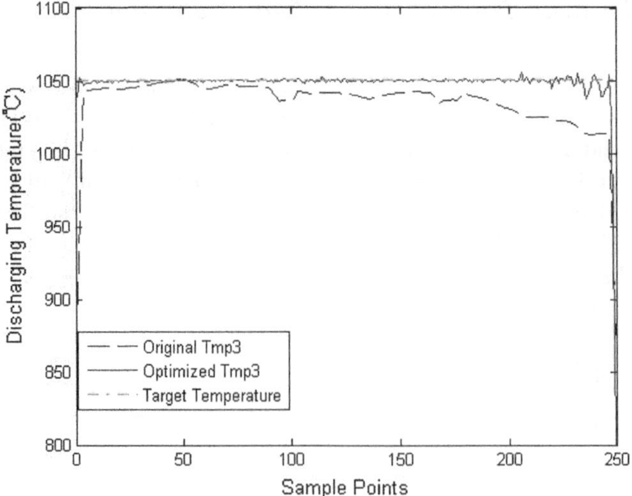

Fig. 86.6 Tmp3 before and after optimization

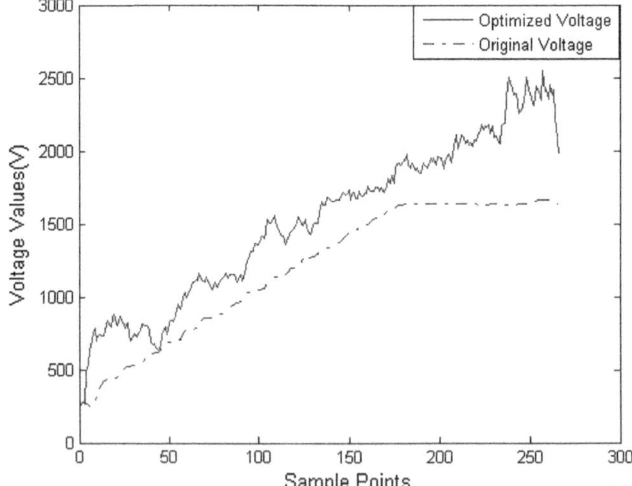

Fig. 86.7 Output voltage before and after optimization

where the subscript N is the length of Tmp3 in one sample, T_1, $T_2 \ldots T_N$ are the Tmp3 series predicted by the above prediction model after optimizing the control voltage, and T_{tar1}, $T_{tar2} \ldots T_{tarN}$ are the Tmp3 target temperature series.

The optimization process adopts the random search algorithm. This method first randomly adjusts the original voltage curve and invokes the billet temperature prediction model to get the prediction Tmp3, and then it calculates the optimization target function and judge the optimization effect, at last the process will be

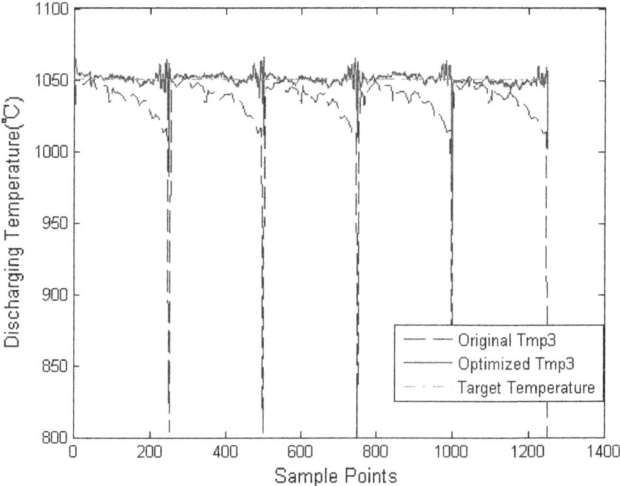

Fig. 86.8 Optimization results of other samples

repeated until the end of iteration. The adjustment strategy of the optimization algorithm is the key process in the optimization, and the merits or defects of the adjustment strategy itself will relate to the algorithm convergence speed and the accuracy of the final optimization results. This paper, from the classical optimization control theory, randomly selects a voltage at a certain time and adjusts it along the positive and the negative with fixed step length respectively. For example, we take the nth voltage point of the original voltage curve, and set the optimization step length to be 100 V, so the adjustment will be as the following:

The Positive Adjustment $Vout_{n,pos} = Vout_{n,orig} + 100$
The Negative Adjustment $Vout_{n,neg} = Vout_{n,orig} - 100$

The voltages at other time remain unchanged. Through this prediction model, we calculate the positive and the negative optimization target function J_{target} respectively, and determine the voltage adjustment scheme based on J_{target}. The specific algorithm flow chart is as shown in Fig. 86.5.

After optimizing, the billet discharging temperature Tmp3 is shown in Fig. 86.6, the optimized voltage curve is shown in Fig. 86.7. Then, the optimized voltage applies to other similar samples, Tmp3 can be predicted by the ARX identification model, as shown in Fig. 86.8.

Figure 86.6 shows that the billet discharging temperature Tmp3 is close to the target temperature by optimizing the output voltage, so this method can achieve the purpose of optimization. Finally, through practical testing, adjusting the output voltage curve by random search algorithm can be effective in optimizing the billet discharging temperature distribution, which realizes the induction heating process temperature control of the hot rolling production line.

86.4 Conclusion

Induction heating furnace is a new kind of heating equipment, so the research on its optimal control model is still in its infancy. This paper, according to a large number of field data, establishes ARX model by system identification, which not only has high prediction accuracy but also can guarantee the speed of operation. Then, based on the ARX model, it optimizes the temperature control of the induction heating furnace and obtains the optimized output voltage curve. But due to the irregular control voltage curve, which cannot satisfy the requirement of practical engineering, we still need to do corresponding treatment of the voltage. The simulation results show that the billet induction heating temperature optimal control method can considerably improve the induction heating furnace heating quality, and realize the induction heating temperature optimal control in the actual production of the hot rolling production line.

Acknowledgments This work is supported by Science and Technology Planning Project of Zhejiang Province #2010C01018 and National 973 Project #2009CB320602.

References

1. Wen H, Xiang S, Zhou Y, Zhou Y (2008) A study of electromagnetic induction heating of steel billet. The 3rd Asian workshop and summer school on electromagnetic processing of materials. The Chinese Society for Metals, Shanghai, 2008. (In Chinese)
2. Wang X, Li N, Li S, Xi Y (2001) Modeling and temperature optimal setpoint strategy for walking beam reheating furnace. J Shanghai Jiaotong Univ 35(9):1306–1309 (In Chinese)
3. Bi CC, Li N, Huang D (2004) Study on billet temperature prediction and furnace temperature optimal setting of regenerative reheating furnace. Acta Autom Sin 30(3):476–480 (In Chinese)
4. Zhang K, Shao C, Zhu H (2006) Improved optimal algorithm and simulation of temperature distribution for reheating furnace. J Syst Simul 18(3):794–796 (In Chinese)
5. Fang X, Yu L, Wang Q, Wang J (2012) Establishment and optimization of heating furnace billet temperature model. The 24th Chinese control and decision conference (CCDC), Haerbin, 2012
6. Kranjc M, Županič A, Jarm T, Miklavčič D (2009) Optimization of induction heating using numerical modeling and genetic algorithm. IEEE, pp 2104–2108. 978-1-4244-4649-0/09 ©2009
7. Shen H, Yao ZQ, Shi YJ, Hu J (2006) Study on temperature field induced in high frequency induction heating. Acta Metall Sin 19(3):190–196 (In Chinese)
8. Zhang K, Shao C, Yin F (2006) On a comprehensive optimal control method based on optimization of hot rolling steel slab temperature. Inf Control 35(6):755–758

Chapter 87
LED Intelligent Dimming System Based on Data Fusion Technology

Yu-jie Fang, Yu Su, Hui-yuan Zhao, Jia-feng Chen and Lian-zhong Qi

Abstract For the purpose of energy saving and energy efficiency, the LED intelligent dimming system making use of solar power supply is designed in this paper. The system is based on STC12C5A60S2 microcontroller with multiple pyroelectric infrared detectors and light intensity sensors, in which LED dimming will be achieved according to the analysis and computation results through Multi-sensor data fusion technology. PWM dimming is adopted in the system depending on presetting programmable counter array (PCA) module of microcontroller in pulse width modulation (PWM) operation mode. Following the discussion about principle, composition and achievement, the Multi-sensor data fusion analysis is stated in accordance with different space distribution of light intensity sensors. Experiments show that the system is in simple structure and stable and reliable with the features of energy saving and intelligent.

Keywords LED · Data fusion · PWM · Sensor · Solar energy

87.1 Introduction

Energy conservation and emission reduction is already a common view of every country. The issue that is the usage of renewable and environmental-friendly energy becomes a focus of concern and study. As the most common nature resources, the inexhaustible solar energy will be fully utilized, which has become a sustainable development energy strategy decision in the world. Recently years, Light Emitting Diode (LED) with its advantages has been used wildly in the field

Y. Fang (✉) · Y. Su · H. Zhao · J. Chen · L. Qi
Department of Automation, School of Information and Technology, Zhuhai, China
e-mail: fyj0202@126.com

Y. Fang · Y. Su · H. Zhao · J. Chen · L. Qi
Beijing Institute of Technology, 6th. Jin Feng Road, Tang Jia Wan,
Beijing, GuangDong, China

Z. Sun and Z. Deng (eds.), *Proceedings of 2013 Chinese Intelligent Automation Conference*, Lecture Notes in Electrical Engineering 255,
DOI: 10.1007/978-3-642-38460-8_87,

of display and lighting. Moreover the new semiconductor materials technology is bound to promote the further development of solar panels and LED [1, 2]. In this paper, we propose a smart lighting scheme that utilizes solar energy as a power source and developed an intelligent dimming system based on STC12C5A60S2 microcontroller with Multi-sensor data fusion technology [3].

87.2 System Components and Working Principle

The whole system are made up of solar cell panels (SCP), battery, Maximum Power Point Tracking controller (MPPT), pyroelectric infrared detectors, light intensity sensors, STC12C5A60S2 microcontroller, LEDs, wireless transmission module (WTM). The structure is showed in Fig. 87.1. Parallel or series solar panels convert solar energy to electric energy for LED illumination. When people enter the room where the pyroelectric infrared detectors are installed in advance, the system will turn on the LED automatically and gradually. At the same time, the system will survey the room's light intensity by three light intensity sensors BH1750FVI. The illumination intensity will be identified based on Multi-sensor data fusion technology through processing data from different light intensity sensors. Then the system will automatically adjust the LED illumination relying on the processed result that is important foundation of generating PWM signals.

87.3 Hardware Design

87.3.1 Solar Panels, Battery and MPPT Controller

Solar panels use light energy from the sun to generate electricity through the photovoltaic effect. A solar panel is a packaged, connected assembly of

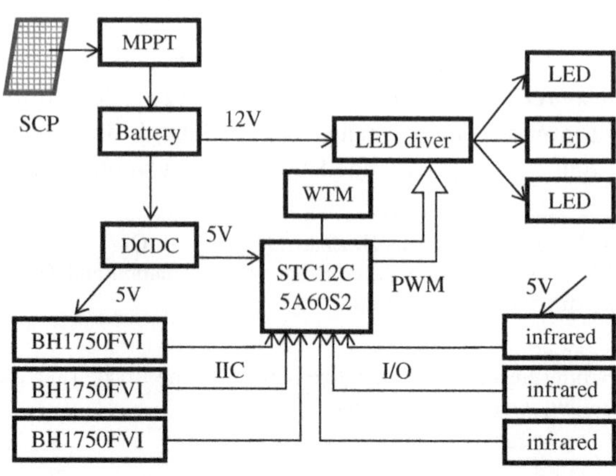

Fig. 87.1 System architecture

photovoltaic cells. It can be used to generate and supply electricity in illumination applications. Because a single solar panel can produce only a limited amount of power, multiple panels would be chosen in most of the cases. Batteries operate in a fairly simple manner. Direct current electricity runs out of the solar panels into batteries for storage. When electricity is needed it runs out of the batteries for use. The most important feature of MPPT controller is intelligent tracking input voltage from solar panel, which could be let solar panel always working at maximum power point of V-A curve. At the same time it includes completely protecting and controlling functions, such as overcharge protection, short circuit protection, and battery reverse current protection.

87.3.2 Light Intensity Sensor Design

In this system, we use three light intensity sensors BH1750FVI. BH1750FVI is a digital ambient light sensor IC for I^{2}C bus interface. This IC is suitable to obtain the ambient light data. It is possible to detect wide range at high resolution by programming. An important thing must be considered is that the I^{2}C bus driving capability is always weak. So we add IC 82B715 to enhance the bus driving capability and prolong the communication distance. All the data operation is completed by imitating I^{2}C communication protocol by microcontroller STC12C5A60S. The circuit diagram is showed in Fig. 87.2.

As slaves, BH1750FVI should be configured their addresses firstly by the ADDR pin for I^{2}C serial communication. When this pin is connected to high level, the slave address is 10111001. To the contrary, when this pin is connected to low

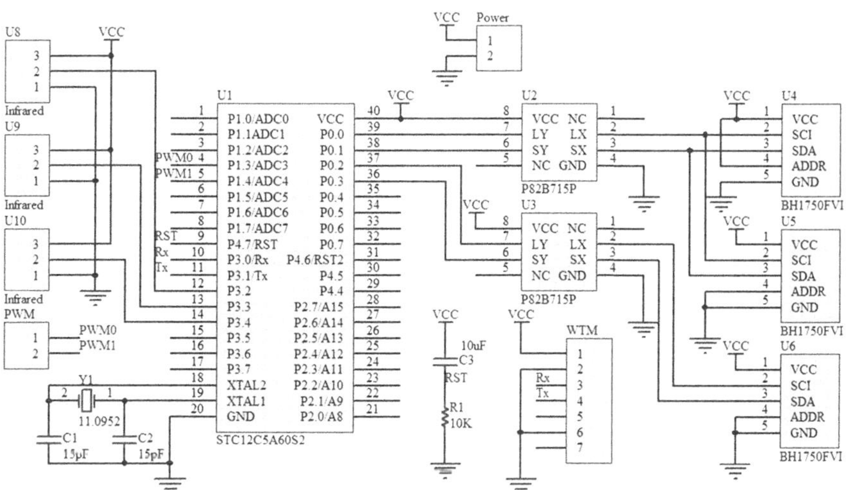

Fig. 87.2 System circuit diagram

level, the slave address is 01000111. In the I^{2}C bus protocol, microcontroller sends address data at first and sends operating code (operating code for high resolution mode is 00100000) after response. Then microcontroller should wait 120 ms in order to convert light intensity to digital data. The result is a 16 bit binary number. Finally achieving the light intensity data need to convert binary number to its decimal number and divide by 1.2.

87.3.3 Pyroelectric Infrared Detectors Design

Pyroelectric infrared detectors could produce weak voltage signal once receiving infrared radiation signal and enlarge it by transistor for output. HC-SR501 with an automatic induction function is used in this system. Its ultra-low voltage operation mode is widely used in various types of automatic sensing in electrical equipment. When people enter the detection range, HC-SR501 will output a high level. A transition from high level to low level will be happened automatically after a short delay once people leave the detection range. It has two trigger mode can be selected by jumpers on the board. In this system, we chose the repeatable trigger mode. Its detection range like a conical area is limited, which is showed in Fig. 87.3. Obviously one detector can not meet requirements of the system. As a result, we add two detectors to enhance accuracy of detection. There output pins connect to P3.2, P3.3 and P3.4 respectively on STC12C5A60S.

87.3.4 PWM Generator

The core of this control system is the STC12C5A60S2 microcontroller. It is a single-chip microcontroller based on a high performance architecture 80C51 CPU and has a 2-channel PCA modules. The reason for selecting is it has integrated PWM generator. The PCA is a special 16-bit timer that connects with two 16-bit capture/compare modules. It can be programmed to work in different modes including PWM output mode. PWM frequency can be calculated by formula (87.1).

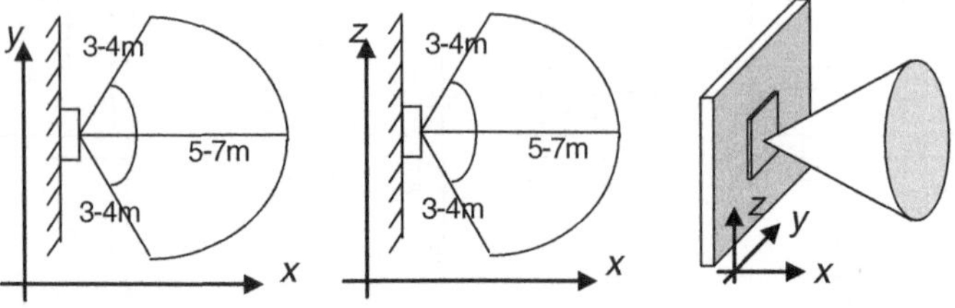

Fig. 87.3 Detection range of HC-SR501 (space coordinates xyz)

$$f(PWM) = \frac{clock(PCA)}{256} \tag{87.1}$$

We should set the register EPCnL and CCAPnL in order to adjust the duty cycle. When the value of register CL is less than [EPCnL, CCAPnL], the output is low level. If the value is equal to or more than [EPCnL, CCAPnL], the output is high level. Changing from FF to 00, the value of CL will overflow. Then [EPCnH, CCAPnH] value will load in [EPCnL, CCAPnL], which will refresh PWM signal without any disturb. The duty cycle can be set by register EPCn and CCAPn for the purpose of LED dimming.

87.3.5 Wireless Transmission Module

The system integrates wireless transmission module that is used for communicating with a special terminal depending on serial ports. Some functions could be completed such as remote control and status monitoring.

87.4 Software Design

In view of automatically detection and dimming, the running system program should be fully considered its reliability and accuracy. So watchdog program is a necessary part of this system. In the first step, the system will initialize and detect all the units. The detail program flowchart is showed in Fig. 87.4. After preparing well, MCU begin to read the status of infrared detectors. LED will not turn on, if HC-SR501 output is low level. It means no person in detection range and will maintain this status till anyone enters in this area. Then the light intensity sensors will begin to work. LED dimming will be determined by the PWM signal, according to the light intensity data which is processed by relative algorithm.

87.5 Algorithm of Data Fusion

Data fusion also called multi-sensor data fusion. Recently, with the development of information technology, there have been a lot of researches about data fusion. Making good use of different kinds of sensors and utilizing computers to analyze or synthesis this data by some algorithm for making decision and estimating would make the system more reliable and accurate. Nowadays there are some algorithms such as weighted algorithm, adaptive algorithm, fuzzy algorithm [4–6] and so on. In this paper, we design an algorithm based on multi-sensor spatial distribution. Sensors data fusion diagram is showed in Fig. 87.5.

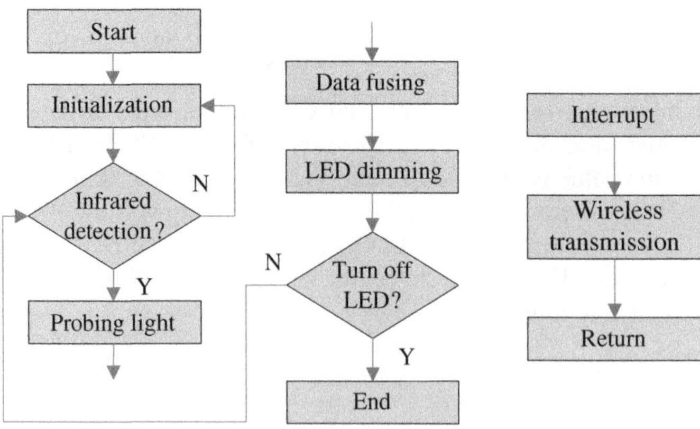

Fig. 87.4 Program flowchart

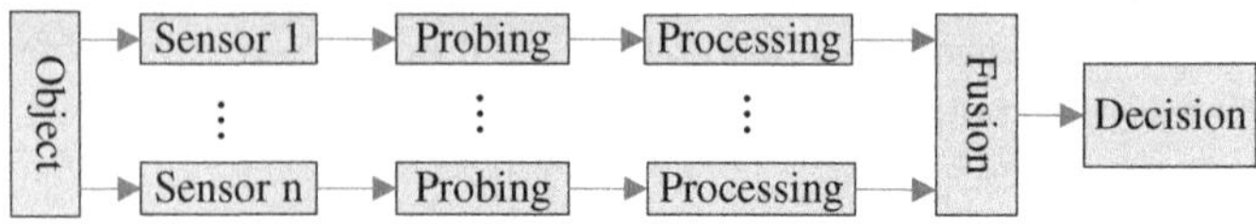

Fig. 87.5 Sensors data fusion diagram

In generally, indoor illumination intensity is asymmetrical. Sensors placed in different position will detect different light intensity. For the sake of optimizing the whole system, we distribute light intensity sensors in different position and applies average algorithm based on credibility [7–9]. While w_i is trust degree of sensor i and d_i is measured data of sensor i, the fusion result can be expressed as

$$\hat{x} = \sum_{i=1}^{n} w_i d_i \tag{87.2}$$

where, w_i must be satisfied the following formula

$$\sum_{i=1}^{n} w_i = 1 \tag{87.3}$$

In our system, the weight value from light intensity sensor can be obtained by experience or experiment.

87.6 Summary

The near window light, indoor activity area light and corner light in the room at dusk are collected respectively. The three position light will be regarded as intensity gradient of test environment. Their values are respectively 356, 287 and

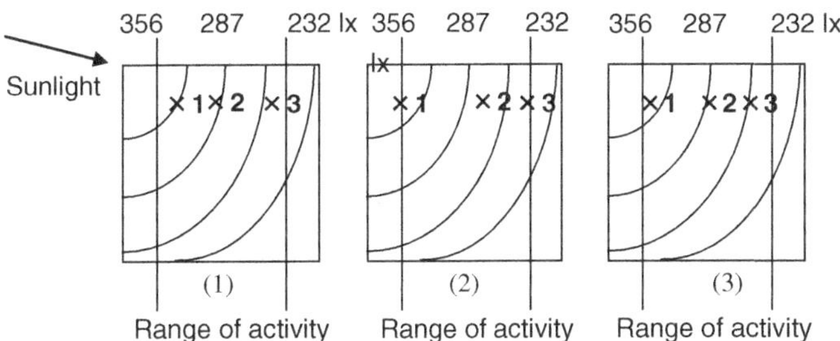

Fig. 87.6 The sensors distribution scheme

Table 87.1 Experimental data

w_1	w_2	w_3	Fusion of distribution (1) (324, 289, 254)	Fusion of distribution (2) (338, 270, 245)	Fusion of distribution (3) (329, 275, 251)
0.2	0.6	0.2	289	278.6	281
0.15	0.6	0.25	285.5	273.95	277.1
0.1	0.6	0.3	282.3	269.3	273.2
0.15	0.65	0.2	287.25	275.2	278.3
0.1	0.7	0.2	285.5	271.8	275.6
0.15	0.7	0.15	289	276.45	279.5
0.2	0.7	0.1	292.5	281.1	283.4

232 lux. The Fig. 87.6 shows the different position and distribution of light intensity sensor, and Table 87.1 shows the fusion results with different weight calculation under this circumstances (The X1, X2, X3 are the position of three light intensity sensor). The system can judge accurately if anyone is in the detection area, use multi-sensors to detect light intensity and complete LED dimming through PWM. Furthermore the real-time information communication with a terminal is realized, including the remote monitoring and manual dimming.

Acknowledgment The author would like to acknowledge the partial financial support of Zhuhai science and technology planned project, contract number 2011B050102005.

References

1. Wang J, Liu TZ, Yang X (2008) The LED lighting system driven by solar energy. J Process Autom Instrum 29(12):5–7, 10 (in Chinese)
2. Jing LXZ (2011) Design of the remote monitoring system for solar LED street lights. J China Illum Eng 22(6):105–107, 110 (in Chinese)
3. Yang YR (2010) Brightness control of led lamps using fuzzy logic controllers. In: The 5th IEEE Conference on industrial electronics and applications (ICIEA), Taipei, Taiwan
4. Guangfu B, Zhinong J, Xuejun L, Dhillon BS (2011) Weighted multi-sensor data level fusion method of vibration signal based on correlation function. J Chin J Mech Eng 24(5):899–904

5. Wu X, Zhou Y, Cai Y, Yang L (2010) Research actualities and problems on multisensor target recognition system model. J Astronaut 5:1413–1420 (in Chinese)
6. Ran C, Deng Z (2010) Self-tuning weighted measurement fusion Kalman filter and its convergence. J Control Theory Appl 8(4):435–440
7. Lei Y, Feng X, Zhu C, Li B (2012) Geometric location algorithm for multi-sensor networking based on data fusion theory. J Infrared Laser Eng 41(5):1339–1344 (in Chinese)
8. Wei L, Pengju H, Shesheng G (2010) Applying random weighted information estimation to implementing a new fusion algorithm for multi-sensors. J Northwest Polytech Univ 5:674–678 (in Chinese)
9. Lin JZXWZ, Baoguo X (2010) Multi-sensor data fusion method based on belief degree and its applications. J SE Univ (Nat Sci Ed) 38(1):253–257 (in Chinese)

Chapter 88
Dual Networks Model for Lower Error and Delay Using RS-CRC Encoding

Yong Li, Rong Zong, Jiang Yu, Ling Zhao and Peng Li

Abstract A key challenge for power distribution network communication is to transmit control and management information with very high reliability and low delay. However, the existing wireless sensor networks (WSN) or power line carrier communication (PLC) networks is not easy to resolve the issue due to the fading and interference properties of power distribution network environment. In this paper, we work out a model of WSN and PLC using RS-CRC (Reed-Solomon code and cyclic redundancy check code) encoding based on the idea of cooperative communication, in which the networks are independent in fading and interference. Simulation results show that the proposed dual networks of WSN and PLC with RS-CRC encoding have lower error bit rate and time delay compared with single network and existing research papers conclusion. In power distribution network communication using the dual networks might solve the communication issue in power industrial environment.

Keywords Wireless sensor networks · Power line carrier communication networks · Cooperative communication · Dual networks · RS-CRC encoding · High reliability and low error

Y. Li · R. Zong (✉) · J. Yu
School of Information, Yunnan University,
Cuihu north road no.2, Kunming 650091, China
e-mail: zongrong@ynu.edu.cn

L. Zhao
Communication Branch of Yunnan Power Grid Corporation,
Kunming 650217, China

P. Li
Queshan Shanhe Fluorite Company, Quesha, China

Z. Sun and Z. Deng (eds.), *Proceedings of 2013 Chinese Intelligent Automation Conference*, Lecture Notes in Electrical Engineering 255,
DOI: 10.1007/978-3-642-38460-8_88,

88.1 Introduction

In the 1980s, along with the rapid development of Ethernet LAN, wireless networks is extended and complementary for wired LAN, and widely applied in the area that needs for mobile data processing or un-allowed to configured by physical transmission medium. Meanwhile, the potential application of wireless communication in power system is attracting more and more attention. For instance, in 2009, National Institute of Standards and Technology (NIST), promulgated the smart grid interoperability framework [1], which proposed the application of wireless communication in the intelligent power grid. In January 2010, Pacific Northwest National Laboratory [2] released a report about wireless communications for the electric power system, which focus on how, where and what type of wireless transmission satisfy requirements of power communication system. A new framework of the next generation power communication transmission network based on Packet Transport Network (PTN) and Optical Transport Network (OTN) is proposed [3], which can solve the problems of traditional MSTP and DWDM network. Urban power distribution network based on WiMax and its automatic meter reading was proposed by our group [4]. Study on ZigBee, WSN [5] are also carried. However, the wireless technologies existed and its standard protocol [6] are impossible to meet the application requirements of field layer in control system, since it has some shortcomings with real-time requirement. What's more, the performance of bit error rate (BER) in wireless networks can hardly meet to the requirement of 10^{-9} in industrial control [7]. So the research of reliability and real-time characteristics during power distribution network communication system is very important.

The multiple path protection mechanism [8] in address issue of two fiber links are simultaneously broken, which can support eight protection paths to more ensure communication reliability and quality. Multiple Relay Selection Scheme [9] and Reed Solomon (RS) code [10] is concatenated as an inner code to improve the communication performance. Inspired by these ideas, combined with a variety of effective network routing algorithm, in this paper, a scheme of dual networks of WSN and PLC based on Reed-Solomon (RS) code and Cyclic Redundancy Check (CRC) code (RS-CRC) encoding is proposed by our team based on the idea of cooperative communication. The dual networks have higher reliability and better real-time characteristics than single network, which are suitable for the requirement of power distribution network environment. The dual networks communication operation model with RS-CRC coding is presented in Sect. 87.2. Section 87.3 describes the simulations and testing results. Finally, a few concluding remarks are given in Sect. 87.4.

88.2 Communication System Model

In this proposed communication system model, both the two networks are introduced with noises of different distribution, so that one of the networks is independent fading network. We consider that two binary symmetric channels as wireless sensor networks (WSN) and power line carrier communication (PLC) network, the parameters of which are different. The communication system logic model structure mainly divides into three parts: signal source terminal, transmitter terminal and receiver terminal. The dual networks logic structure is shown in Fig. 88.1. We assume that two networks transmit same information m(t) at the same time. The communication system model simulation based on the Matlab software platform.

At the transmitter terminal, random binary numbers is used as the signal source m(t), which are generated by Bernoulli Binary Generator block. When some message come from the source, each data frame (or packet) to which a multi-bit Cyclic Redundancy Check (CRC) codes (Bit-16 or Bit-32) is appended will be encoded into one RS codeword. Reed-Solomon (RS) (n, k) code is concatenated as an inner code. Binary Symmetric Channel (BSC) is an important class of channel model, which can reflect the characteristics of two different networks, if set different parameters. At the receiver side, CRC-N Syndrome Detectors check error frame and output check result. When a data frame error, the port of Err outputs 1. Otherwise, it will output 0, i.e., there are not errors in the data frame. The output values are stored in the cache apparatus. Then, we get transmission results of dual channels by doing logic operation between output values of channel1 and channel2. Display1 and Display2 block display the successfully transmitted packets by channel1 and channel2 respectively. Successfully transmitted packets for the dual channels are recorded by Display3. The Bit Error Rate (BER) of dual-channel model could be calculated with logic XOR, Error Rate Calculation and Cumulative Sum logic module with the Matlab platform.

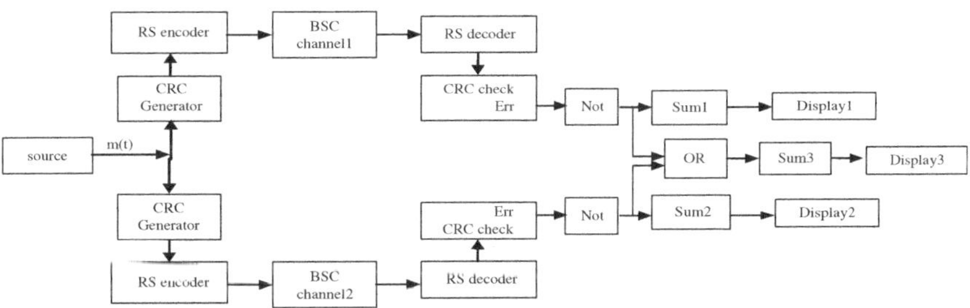

Fig. 88.1 Logic structure of dual-network model

88.3 Communication System

Power distribution network communication requirements high reliability and low error rate, therefore, if the contacting link is suddenly cut or channel is serious interfered by kinds of unknowable channel fading and interference, it would be a serious influence for communication quality. The BER of dual-network proposed less than 10^{-10}, when data packet size is set to 1,484 bits, which solves the problem very well. When sudden interrupt or unknowable error happens in one of the networks, data packets will be transmitted successfully with high probability.

88.3.1 RS(31,25) Probability of Successfully Transmission

There is RS-CRC encoding to simulate the communication system. The CRC generate 16 bits checking information for every data packet. The Reed-Solomon code has message length K (K = 25) and code word length N (N = 31). The data packets' probability of successfully transmission (PST) in single network and dual-network are shown in Fig. 88.2. The packet size is set as 2,984 bits. We can see that the PST for dual networks is higher than the PST for single network, i.e., at the same bit error rate (BER), probability of successfully transmission for dual networks is better than that for single network. PST for dual network is less than

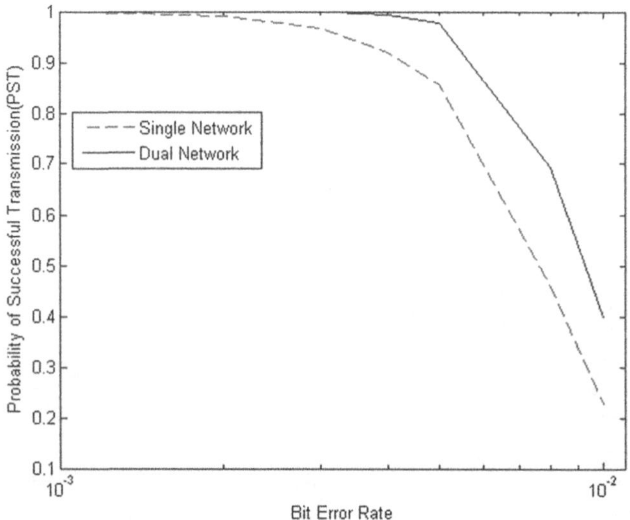

Fig. 88.2 PST for single network and dual-network under different channel BER (packet size = 2,984 bits)

$1-10^{-9}$, while the PST for single network is about 0.95, when BER of the channel is 3×10^{-3}. When the BER $> 5 \times 10^{-3}$, PST of single network starts falling rapidly with the increase of the BER.

Figure 88.3 shows the change of single network and dual-network's PST under different packet size, and the channel BER is set to be 9×10^{-3} in the simulation. It can be seen that the single network and dual-network's PST decreases with the increase of packet size, and the PST of the single network decreases faster than that of the dual-network. The PST of dual-network is very high when the packet size is smaller than 1,109 bits. At the lowest level power distribution network control communication system, communication equipments only need small packet data rapidly transfer between the information terminal and controller center. The packet size may be about several hundred bits or more than one thousand bits. So the dual-network system can improve the quality of control and management information transmission obviously in power distribution network.

If the channel BER is set to be 1×10^{-2} in Fig. 88.4, i.e., the noise interference of channel environment is increased, which compared with Fig. 88.3. The curves of single network and dual-network's PST drop rapidly, the probability of successfully transmission is lower than that in Fig. 88.3 when the packet size is same. But the PST for dual-network is still higher than that in single network. We can see both the networks environment (BER) and data packet size affect the reliability of transmission from Figs. 88.2, 88.3, and 88.4.

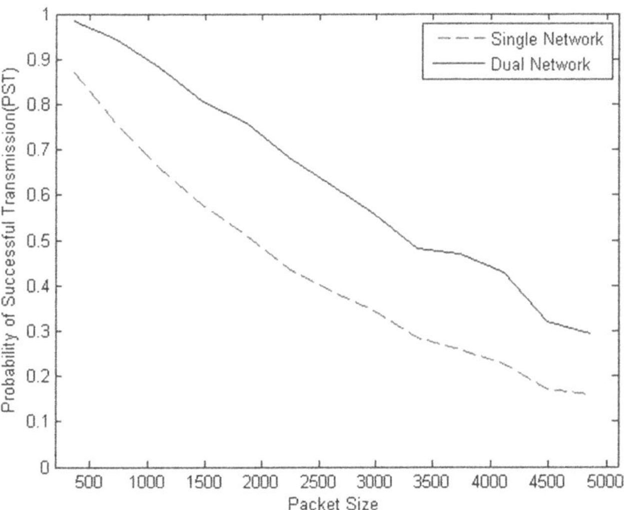

Fig. 88.3 PST for single network and dual-network under different packet size (Network-BER $= 9\times10^{-3}$)

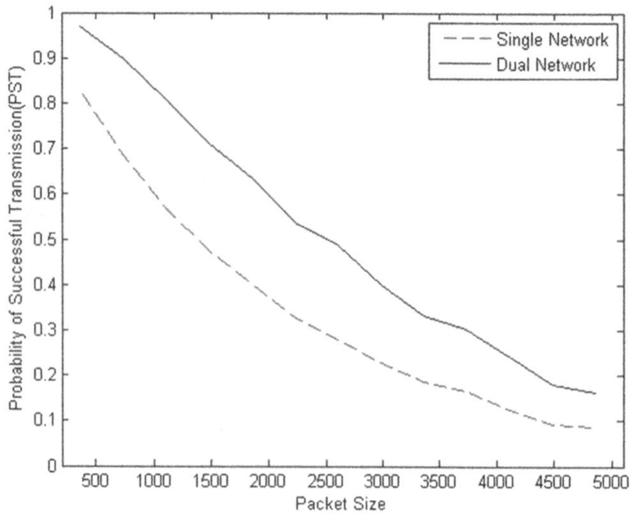

Fig. 88.4 PST for single network and dual-network under different packet size (Network-BER = 1×10^{-2})

88.3.2 RS(31,15) Probability of Successfully Transmission

In this part, RS(31,15) are Selected as inner code. The corresponding code rate is 0.48. Figure 88.5 delivers the relationship between the channel BER and the probability of successfully transmission. In case of constant packet size (2,984 bits), PST decreases with the increase of channel BER. Thus, high PST can be achieved when the BER of channel is small. When BER of channel is 3×10^{-3}, PST in single network is 0.97, while that in dual-network's PST can attain about $1-10^{-10}$, that's very small. So the reliability performance of communication in dual network with RS(31,15) is better than dual-network with higher code rate in RS(31,25) code, in which the code rate is 0.8.

The change of single network and dual-network's PST under different packet size are shown in Fig. 88.6, and the BER is set to be 3×10^{-2} in the simulation. It can be seen dual networks' PST are bigger than single network at different packet size. RS(31,15) can significantly improve the performance. For example, in dual network, when packet size = 1,109 bits, PST = 0.9; the packet size is 4,484 bits, the PST is about 0.45. However, when the network's BER = 9×10^{-3} in Fig. 88.3, packet size = 1,109 bits, PST = 0.88, and packet size = 4,484 bits, the PST is only 0.32.

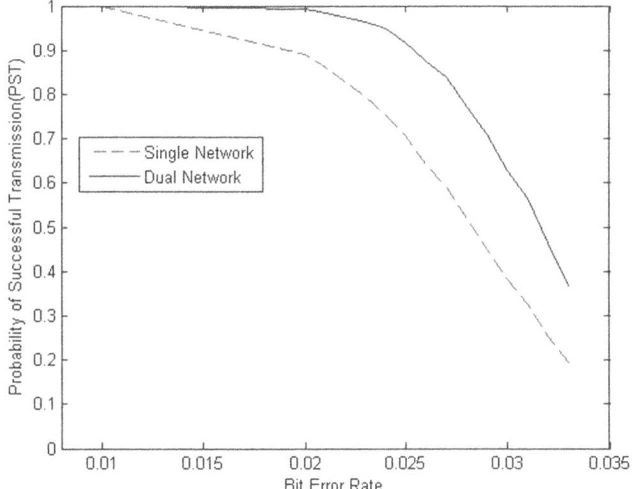

Fig. 88.5 PST for single network and dual-network under different channel BER (packet size = 2,984 bits)

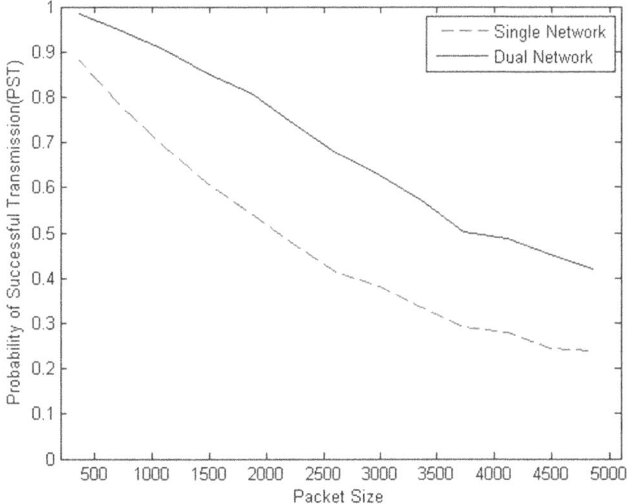

Fig. 88.6 PST for single network and dual-network under different packet size (channel BER = 3×10^{-2})

88.3.3 Using RS(31,25) and RS(31,15)

The power distribution network's environment is complex and changeable. So there are some differences in communication condition of two networks. If we only use a coding scheme might not achieve a good successful transmission

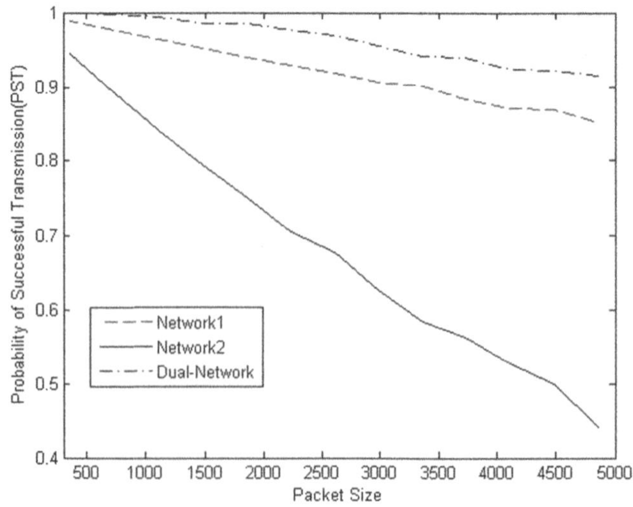

Fig. 88.7 Network relationship between PST and packet size (Network1 BER = 2×10^{-2}, Network2 BER = 7×10^{-3})

requirements. In this section, we consider two different networks, network2 is poor with higher channel BER = 2×10^{-2}, and the other relatively better with BER = 7×10^{-3}. Figure 88.7 illustrates the PST of two networks with different BER. Because of existing serious noise interference, the curves of network2 drop rapidly, but the curves of dual-network drop slowly, that is better than network1 and network2.

Industrial control and automation system for the communication require the fairly stringent response time. The control process is usually completed in milliseconds or even delicate time. Congestion control and automatic retransmission request (ARQ) strategy can be used to reduce the transmission delay in wired communication, but in the WSN and PLC communication, these strategies are ineffective. Need to find a strategy or build a new system. In this paper, dual-network model can effectively solve the problem of transmission delay. This system only need add adaptive coding selection strategy at the transmitting end. Due to the higher probability of successfully transmission, it avoids a lot of time cost for many retransmissions.

88.3.4 RS(31,25) and RS(31,15) Bit Error Rate Simulation Analysis

In order to measure the dual-network's reliability, we introduce some logic modules at the receiver end. Transmitting signal m(t) and received message m′(t) are both sent to Error Rate Calculation module, to calculate the dual-network's

BER. The simulation for the packet size is 1,484 bits and the BER for network1is set to BER 2×10^{-2}, network2 is set to 7×10^{-3}, we can get network1's BER is 1.4×10^{-4}, network2's BER is 6×10^{-4}, but dual-network's BER slightly lower than 10^{-10}; When packet size is 1,859 bits, the channel BER of network1 is set to 2×10^{-2}, network2 is set to 7×10^{-3}, we can get network1's BER is about 1.4×10^{-4}, network2's BER is about 6×10^{-4}, BER of dual-network slightly lower than 10^{-10}. Such a low BER can meet the requirements in industrial communication.

88.4 Conclusion

WSN technology has been gradually used in some industrial area. But how to meet the strict requirements of low BER (less than 10^{-9}), short delay in power distribution network control environment is still a highlighted problem. This paper presented an effective solution through developing a novel dual-network of WSN and PLC communication model with RS-CRC coding, which is an improved method based on physical layer. The simulation results indicate that the reliability in dual-network is better than that in single-network, and transmission with RS-CRC dual-network could effectively reduce the packet transmission error rates and delay. Adaptive code word selection strategy can be set at the transmitting end using the dual-network.

We can fully use above characteristics of the two networks, with adaptive coding selection mechanism: (1) Both the BER of network1 and network2 are more than 2×10^{-2} and less than 2.5×10^{-2}, using dual-network with RS(31,15) to communication, in order to get high PST; (2) Both the BER of the two network's BER are slightly less than 4×10^{-3}, using dual-network with RS(31,25) to communicate, so that access to higher PST and transmission rate. (3) One of the BER of two network's BER are about 7×10^{-3}, but the other network's BER is about 2×10^{-2}, using dual-network with RS(31,25) and RS(31,15) to communication, in order to get high PST. The dual-network obviously improves the service quality of power distribution network communication.

Acknowledgments This work was supported by National Natural Science Foundation (No. 61162004).

References

1. National institute of standards and technology (2010) Online available: http://www.nist.gov/public_affairs/releases/smartgrid_interoperability.pdf
2. Pacific Northwest National Laboratory (2010) Online available: http://www.pnl.gov/main/publications/external/technical_Reports/PNNL-19084.pdf

3. Feng B, Fan Q, Li Y (2012) Research on framework of the next generation power communication transmission network. Int Conf Comput Sci Electron Eng 3:414–417
4. Li T, Huang M, Shi J, Chang J (2010) Communication scheme for urban power distribution network and its application on meter reading based on WiMax. Telecommun Electr Power Syst 31(214) (in Chinese)
5. Li J, Liu S, Wu S (2012) A design of remote computer house monitoring and control system based on ZigBee WSN. IJACT: Int J Adv Comput Technol 4(12):233–240
6. IEEE Std 802.11b-1999, Part 11 (1999) Wireless LAN Medium Access Control (MAV) and Physical Layer (PHY) specifications
7. Goldsmith AJ (2005) Wireless communications. Cambridge University Press, UK
8. Chen WP, Shih FH, Hwang WS (2009) The multiple path protection of DWDM backbone optical networks. J Inf Sci Eng 25:733–745
9. Xu J, Zhang H, Yuan D (2012) E-global optimal multiple relay selection scheme in cognitive relay networks. AISS: Adv Inf Sci Serv Sci 4(4):218–229
10. Wang D, Soleymani MR (2012) Cooperative communication system with systematic raptor codes. In: 25th IEEE Canadian conference on electrical and computer engineering (CCECE), pp 1–6

Chapter 89
Aggregation-Based Privacy-Preservation Approximate Query Protocol in Wireless Sensor Networks

Yongjian Fan, Xiaoying Zhang and Hong Chen

Abstract Privacy preservation in wireless sensor networks has attracted more and more attentions. Answering generic query in wireless sensor networks while preserving data privacy is a challenge. In this paper, we present a Aggregation-based Privacy-preservation Approximate Query Protocol in Wireless Sensor Networks (APAQ). APAQ provides approximate results for multiple types of query, such as Top-k query, range query, SUM, MAX/MIN, etc. APAQ adopts in-network aggregation to reduce energy consumption. Theoretical analysis and simulation results by using real-world data confirm the high efficacy and efficiency of APAQ.

Keywords Wireless sensor network · Privacy preservation · Approximate query · Data aggregation

89.1 Introduction

Privacy problem in WSNs has been paid close attention to in recent years, and data privacy preservation techniques has become a hot spot of research. At present, the studies mainly focus on two fields: privacy-preserving data aggregation and privacy-preserving data query. About privacy-preserving data aggregation, schemes proposed in [1–5] support data aggregation functions such as SUM, MAX/MIN or Median, but cannot support complicated data queries such as Top-k

Y. Fan (✉) · X. Zhang · H. Chen
School of Information, Renmin University of China, Beijing, China
e-mail: fanyj_ruc@ruc.edu.cn

Y. Fan
School of Information and Electrical Engineering, Hebei University of Engineering, Handan, China

Z. Sun and Z. Deng (eds.), *Proceedings of 2013 Chinese Intelligent Automation Conference*, Lecture Notes in Electrical Engineering 255,
DOI: 10.1007/978-3-642-38460-8_89,

query and rang query. About privacy-preserving data query, [6–8] proposed verifiable privacy-preserving range query protocol for in two-tiered sensor networks, and verifiable privacy-preserving Top-k query protocol in two-tiered sensor networks is proposed. Nevertheless, none of these schemes is generic query schemes for multiple query types.

In this paper, we propose a privacy-preserving generic approximate query protocol using in-network aggregation. In the protocol, We product a random vector used to hide information such as sensor IDs and sensory data value; intermediate sensor nodes aggregate these vectors; the base station use these vectors to construct linear equations and draws a histogram with global statistical information, then to compute the results of approximate query according to this histogram.

The rest of this paper is organized as follows: Sect. 89.1 summaries related work. Section 89.2 gives the background and model assumptions of our research. Section 89.3 presents APAQ protocol in details. Subsequently, Sects. 89.4 and 89.5 evaluate the performance of APAQ via detailed theoretical analysis and simulation study, respectively. Finally, Sect. 89.6 concludes this paper.

89.2 Models and Background

89.2.1 Query Model

In this paper, query model falls into two categories: generic approximate query model and in-network query model. Using generic approximate query model, users deliver query commands to the base station. While using in-network query model, the base station deliver query commands to sensors.

Generic approximate query model is described as:

$$\begin{aligned} \mathcal{Q}_G = &(\text{query region} = \text{G}) \wedge (\text{epoch} = \text{t}) \\ &\wedge (\text{query type} = \text{T}) \wedge (\text{query parameter} = \text{P}), \end{aligned}$$

where G is the query region required by users in epoch t, T is the set of required query type, P is the set of all required parameters for T.

APAQ supports multiple query types, including Top-k query, range query, MAX/MIN, SUM, Median, Histogram, etc. T may contain only one type, but also may contain several types that APAQ supports. Parameters in P should be given correctly according to types in T: if the type is Top-k query, parameter in P should be an integer k which means the k-maximal (minimal) values; if the type is range query, parameter in P should be a range $[a, b]$.

In-network query model is given as:

$$\mathcal{Q}_t = \{t, (H_1, \ldots, H_h)\}$$

where $(H_1, \ldots, H_h)$ is a set of regions calculated from ϵ in $\mathcal{Q}_G$, it represents the granularity of histogram. If ϵ equals its default, the base station will not deliver $(H_1, \ldots, H_h)$.

89.2.2 Attack Model and Security Goals

There may have two main kinds of attack models that threat private information in WSNs: external attacks and internal attacks. Sensors communicate with each other through wireless technique. For external attacks, an adversary attempts to gain sensitive information by overhearing the wireless links. For internal attacks, an adversary attempts to gain sensitive information by compromising or copying a sensor node.

In APAQ, we aim to provide a new approach to realize generic approximate query while preventing both external attacks and internal attacks from gaining collected data.

89.3 APAQ Protocol

We propose a privacy-preserving generic approximate query protocol using in-network aggregation. In the protocol, We product a random vector used to hide information such as sensor IDs and sensory data value; intermediate sensor nodes aggregate these vectors; the base station use these vectors to construct linear equations and draws a histogram with global statistical information, then to compute the results of approximate query according to this histogram.

We suppose sensor networks have n sensors denoted as $s_i(i = 1, 2, \ldots, n)$, each sensor node is assigned a unique ID. The domain of sensory data is denoted as $[v_{\min}, v_{\max}]$, where $v_{\min}$ and $v_{\max}$ are respectively the minimum and maximum of sensory data values.

APAQ consists of two phases: system initialization phase and query process phase.

1. System Initialization Phase

In this phase, the base station randomly selects $m \times n$ distinct positive integers $a_{ij}(i = 1, 2, \ldots, m;\ j = 1, 2, \ldots, n)$ to construct a $m \times n$ matrix G as follows:

$$\begin{pmatrix} a_{11} & a_{12} & \cdots & a_{1n} \\ a_{21} & a_{22} & \cdots & a_{2n} \\ \cdots & \cdots & \cdots & \cdots \\ a_{m1} & a_{m2} & \cdots & a_{mn} \end{pmatrix}$$

Sensor node $s_i (i = 1, 2, \ldots, n)$ corresponds to the i-th column vector of matrix G, that is $(a_{1i}, a_{2i}, \ldots, a_{mi})^T$, $(a_{1i}, a_{2i}, \ldots, a_{mi})^T$ is called as the reference vector of sensor node s_i.

Sensor node s_i is preloaded with a unique key k_i only shared between s_i itself and the base station. $[v_{\min}, v_{\max}]$ is partitioned as h consecutive non-overlapping regions $(H_1, \ldots, H_h)$.

The base station encrypts the following message with k_i and sends it to sensor node.

$$Sink \rightarrow s_i (i = 1, 2, \ldots, n):$$
$$i, \mathrm{E}_{k_i} \left\{ \begin{array}{l} (a_{1i}, a_{2i}, \ldots, a_{mi})^T, \\ (H_1, \ldots, H_h) \end{array} \right\}$$

where $\mathrm{E}_{k_i}\{\}$ denote the encryption process.

2. Query Process Phase

In this phase, every sensor node maps its sensory data to corresponding region label, and conceals region labels and its ID into its $m -$ length response vector, add secret vector to its response vector, and then sends the response vector to aggregator. Aggregator sum all received vectors together with its own vector to calculate its $m -$ length response vector and uploads this vector. The base station sums received all vectors and subtracts all secret vector, figures out a histogram that includes statistic information of sensory data and corresponding sensor IDs. It is easily to calculate approximate query results according to the histogram. The detailed process is as follows:

A. **Data Replying at Leaf Node**

Leaf nodes hide its region labels and their IDs into a $m -$ length response vector $(r_{1i}, r_{2i}, \ldots, r_{mi})^T$, using following formula:

$$(r_{1i}, r_{2i}, \ldots, r_{mi})^T = N_i \times (a_{1i}, a_{2i}, \ldots, a_{mi})^T$$

where N_i is the region label that $v_{i,t}$ falls into, i.e. $v_{i,t} \in H_{N_i}$, $(a_{1i}, a_{2i}, \ldots, a_{mi})^T$ is the reference vector of sensor node s_i.

Leaf node s_i computes its response vector as: $(a_{1i} \times N_i, a_{2i} \times N_i, \ldots, a_{mi} \times N_i)^T$.

Then leaf node s_i uploads the response vector $(r_{1i}, r_{2i}, \ldots, r_{mi})^T$ to its parent node that is aggregator.

B. **Data Aggregation at Aggregator**

Using the additive aggregation function, aggregator s_j directly aggregate these received response vectors with its own vector, get response vector $(r_{1j}, r_{2j}, \ldots, r_{mj})^T$ of aggregator s_j. Aggregator node s_j uploads its response vector

$(r_{1j}, r_{2j}, \ldots, r_{mj})^T$ to its parent node. If its parent node isn't the base station, its parent node continues to add these received response vectors with its own vector.

C. **Processing at the Base Station**

After the base station has received all response vectors of its children, it processes these vectors in the following steps:

Compute the vector called as $(b_1, b_2, \ldots, b_m)^T$ by summing all received vectors and gain the following vector:,and get to the following linear equations:

$$\begin{pmatrix} b_1 \\ b_2 \\ \vdots \\ b_m \end{pmatrix} = \begin{pmatrix} \sum_{i=1}^{n} a_{1i} \times N_i \\ \sum_{i=1}^{n} a_{2i} \times N_i \\ \vdots \\ \sum_{i=1}^{n} a_{mi} i \end{pmatrix} \tag{89.1}$$

If $m(m \leq n)$ is large enough, we will work out the result of linear equations, the result is $\{N_1, N_2, \ldots, N_n\}$. If $m = n$, we can solve linear equations (89.1) with an exact and unique result. N_i is the label of region that the data of sensor s_i belong to. Sensory data value of sensor node $s_1, s_2, \ldots, s_n$ correspond to respectively region label $H_{N_1}, H_{N_2}, \ldots, H_{N_n}$. We statistic the number of sensory data in each region and derive a histogram based on $(H_1, \ldots, H_h)$ and correlative sensor IDs, and then deduce the approximate query results according to the histogram.

89.4 Privacy Analysis

Definition 1 Efficient privacy-preservation in privacy-preserving approximate query

For a query $Q_G = \{G, t, T, P, \epsilon\}$, privacy-preservation of query results is efficient, if and only if adversaries are not able to obtain sensitive information as following: (1) sensory data $v_{i,t}$ of any individual node $s_i (i = 1, 2, \ldots, n)$ in area G; (2) the histogram with statistics information; (3) sensor node IDs relevant with query result.

We assume that adversaries may capture aggregator to launch internal attack. We focus on two types of attack: internal attack and external attack, which are mentioned in Sect. 89.2.2.

Theory 1 Supposing aggregator cannot gain keys shared by the base station and sensor nodes, APAQ satisfies efficient privacy-preservation (Definition 1).

Proof During system initialization, the base station encrypts reference vector of each sensor node $s_i (i = 1, 2, \ldots, n)$ with the key k_i shared with sensor node s_i, and then deliver it to sensor node s_i. Hence, if aggregator cannot get the shared keys, they cannot decrypt to get reference vectors. Each node conceals its sensory data and its ID into its reference vector to form response vectors, and upload response vector. Without reference vectors, aggregators are unable to infer private information. Meanwhile, it is response vectors but not sensory data that are transmitted through wireless links, so outside attacker are also unable to obtain sensitive information.

89.5 Experimental Evaluation

We implement experiment by using real-world data in simulation platform OMNeT++. We use publicly available real-world data set LUCE, which consists of the temperature, humidity and other attribute data collected during 2006–2007. We use part temperature data to experiment, which attribute domain is [−20, 30]. We suppose the communication radius of sensor node is 60 m and create a routing by TAG (tiny aggregation service for ad hoc sensor networks) algorithm. Using to sensor location in LUCE, we have accomplished experiments with network sizes: 50 nodes are distributed in 350 × 250 m area, the average of neighbors number is 6.4 and the average of hops number is 2.9.

For better comparison in query process phase, we experimented with APAQ, and KIPDA [5]. Both KIPDA and APAQ adopt a non-cryptographic method to protect sensitive information. KIPDA provides aggregation results for MAX/MIN functions, while APAQ provides generic approximate query result for Top-*k* query, range query, MAX/MIN, etc. we choose parameter *k* of KIPDA to be 5.

In query process phase of three groups of experiments, we respectively investigated the impact of the number of regions *h* and the number of collecting data round on communication cost (C_Q). For ease of description, we assume that each node collects data only one time at each epoch, that is, the number of collecting data round is *t* in epoch *t* We set $m = n$ to guarantee that linear Eq. (89.1) have an exact and unique result.

Figure 89.1 shows the impact of region number *h* on communication cost C_Q when the number of collecting data round is 100. APAQ consumes a little less

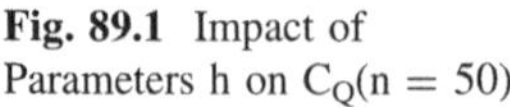

Fig. 89.1 Impact of Parameters h on $C_Q(n = 50)$

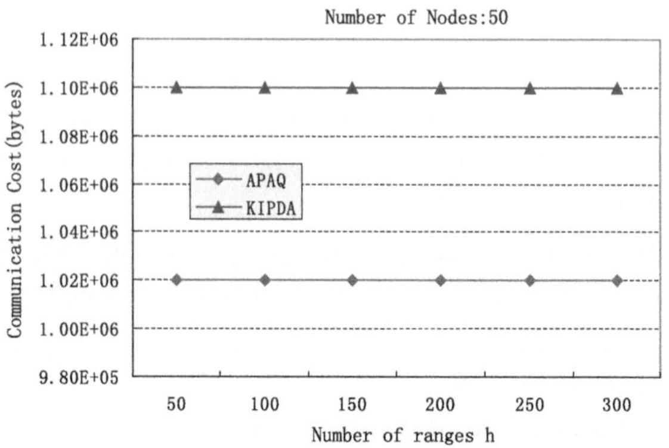

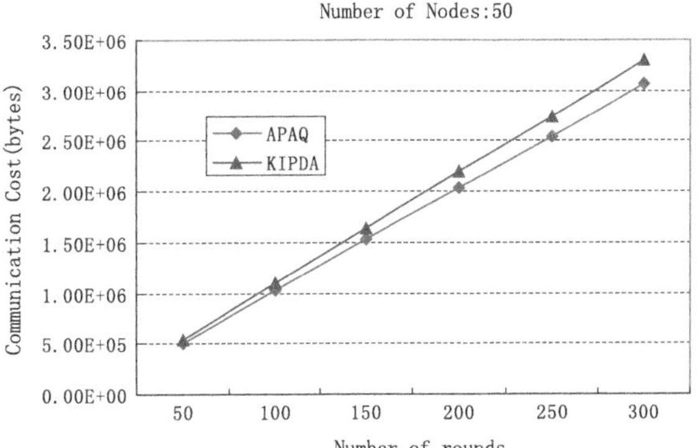

Fig. 89.2 Impact of Parameters t on $C_Q(n = 50)$

communication cost than KIPDA. In addition, the results also display that region number h almost does not affect both APAQ and KIPDA.

Figure 89.2 shows the impact of the number of round on communication cost C_Q when region number h is 100. APAQ consumes a little less communication cost than KIPDA. Furthermore, communication cost of APAQ and KIPDA grows linearly with t.

89.6 Summary

In our paper, we propose a privacy-preserving generic approximate query protocol in wireless sensor networks (APAQ). APAQ adopts in-network aggregation to save energy and strike a balance between accuracy and energy consumption by adjusting the granularity of histogram. Theoretical analysis and simulation results by using real-world data confirm the high efficacy and efficiency of APAQ.

Acknowledgments This research was supported by the National Natural Science Foundation of China (61070056, 61033010), the Natural Science Foundation of Hebei Province, China (F2013402031) and the HGJ Important National Science and Technology Specific Projects of China (2010ZX01042-001-002-002).

References

1. Yang G, Wang A, Chen Z, Xu J, Wang H (2011) An energy-saving privacy-preserving data aggregation algorithm. Chin J Comput 34(5):792–800 (in Chinese)
2. Feng T, Wang C, Zhang W, Ruan L (2008) Confidentiality protection for distributed sensor data aggregation. In: IEEE INFOCOM'08, Phoenix, USA, pp 56–60
3. Stavros P, Aggelos K, Dimitris P (2011) Secure and efficient in-network processing of exact SUM queries. In: Proceedings of the 27th international conference on data engineering (ICDE), Hannover, Germany, pp 517–528

4. Groat MM, He W, Forrest S (2011) KIPDA: k-indistinguishable privacy-preserving data aggregation in wireless sensor networks. In: IEEE INFOCOM'11. Shanghai, China, pp 2024–2032
5. Zhang W, Wang C, Feng T (2008) GP^{2}S: generic privacy-preserving solutions for approximate aggregation of sensor data. In: Proceedings of the 6th annual IEEE intelligent conference on pervasive computing and communications (PerCom). Hong Kong, China, pp 179–184
6. Sheng B, Li Q (2008) Verifiable privacy-preserving range query in two-tiered sensor networks. In:IEEE INFOCOM'08, Phoenix, USA, pp 46–50
7. Chen F, Liu AX (2010) SafeQ: secure and efficient query processing in sensor networks. In: IEEE INFOCOM'10, San Diego, USA, pp 2642–2650
8. Fan Y, Chen H (2012) Verifiable privacy-preserving top-k query protocol in two-tiered sensor networks. Chin J Comput 35(3):423–433 (in Chinese)

Chapter 90
Safety Evaluation of Dike with Cracks

Xizhong Shen, Haobing Li and Ming Zhang

Abstract Dilapidation or cracks existed in dike, only few influencing factors were considered, and systematic quantitative results lacked, safety character of dike was not clarity under the work of cracks. Based on theory of unsaturated soil mechanics and analysis model of cracks, influencing factors on safety of dike such as depth of cracks, status of cracks filled with water, changing of water level, rainfall, seepage coefficient and so on were analyzed, systematic quantitative results of dike under the work of cracks were achieved. Analysis shows that the depth of cracks is deeper, the safety coefficient of dike is less, safety status of dike becomes bad while cracks filled with water. Moreover, the endangering degree of dike with cracks keeps close correlation with characteristic of soil, the chance of accident keeps direct proportion to duration of rainfall, intension of rainfall, depth of cracks, decreasing velocity of water level about dike-slope at near-river. The results can be applicable as a reference in flood prevention, emergency dealing with, maintenance and reinforce.

Keywords Cracks · Dike · Stabilization · Seepage · Unsaturated soil

---

X. Shen (✉) · H. Li (✉)
Yellow River Institute of Hydraulic Research, No 45 Shenhe Road, Zhengzhou, China
e-mail: shenxz@126.com

H. Li
e-mail: shenxz007@163.com

M. Zhang (✉)
Key Research Center on Levee Safety and Disaster Prevention Ministry of Water Resources, Shenhe 45, Zhengzhou 450003, China
e-mail: zhangming2000203@163.com

Z. Sun and Z. Deng (eds.), *Proceedings of 2013 Chinese Intelligent Automation Conference*, Lecture Notes in Electrical Engineering 255,
DOI: 10.1007/978-3-642-38460-8_90,

90.1 Introduction

After rainfall suffered for a long time or dike is soaked by flood with high water level, slide of slope or collapse of bank may often happen in flood dike while water level decreasing [1]. Two types of effect on stability of slope are unsteady seepage aroused for water level increasing of river and decreasing of downstream of dike, unsaturated seepage aroused for rain or evaporation of dike-body, and these are the main factors aroused for breakage of dike. Especially for dike with cracks, the chance of slope slide will increase sharply with changing of water level [2–4].

Safety status of dike without cracks was studied, lots of influencing factors were analyzed, thus the results could be use for reference in safety evaluation of dike with cracks [4–6]. Safety evaluation of dike and dam with cracks were discussed by Fredlund and Rahardjo [5], some methods of analysis were provided, these were dike with cracks was analyzed qualitatively, theory basis on stabilization analysis of dike with cracks, but it was a pity that few application examples were researched and it was difficult to use these methods in safety evaluation of dike with cracks [5]. Danger of safety on and some influencing factors were carried out such as cracks, status of cracks filled with water, changing of water level, rainfall, but few quantitative results were achieved [6–8]. Danger of cracks of dike was analyzed, but farther research was not carried out [9–11]. Dilapidation or cracks existed in dike, there were not systematic quantitative results, flood prevention and dealing with of dike with cracks were restricted. Therefore, it is necessary to evaluate logically the safety character of dike under the work of cracks.

Take some general dike as the example, danger of dike with cracks was discussed under the condition of complex environment based on theory of unsaturated soil mechanics and analysis model of cracks, thus it would provide technical safeguard for flood prevention, emergency and reinforce.

90.2 Model of Analysis

90.2.1 Analysis Methods of the Influence of Dike Safety on Cracks

The influence of dike safety on cracks discussed, safety state of dike-slope with cracks under dangerous hydraulic condition such as slope of back-river for flood level of design, slope of near-river for water level decreasing, rainfall and so on need be analyzed [1].

Distributing of pore water pressure can be achieved from analysis of seepage, and the influence of factors such as changing of water level and rainfall can be considered in safety evaluation of dike [12]. When unsaturated characteristic of soil considered, function of shear strength can be expressed as fellows [5].

$$\tau_f = c' + (\sigma_n - u_a)\tan\phi' + (u_a - u_w)\tan\phi^b \tag{90.1}$$

where, τ_f is shear stress at failure, c' is effective cohesion, σ_n is normal stress, u_a is pore air pressure, u_w is pore water pressure, ϕ' is effective internal friction angle, ϕ^b is internal friction angle of matric suction.

90.2.2 Model and Parameters of Calculating

Convenient in analysis and good in typical, taking some general dike as the example shown as Fig. 90.1. Slope ratio of dike-body is 1:3 in Fig. 90.1, there are 9,860 nodes and 9,500 elements in the model. The curves of coefficient of infiltration and volumetric water content were selected as Figs. 90.2 and 90.3, and parameters of calculating were shown as Table 90.1 based on correlative results [5, 12].

90.2.3 Boundary Condition and Initial Condition

At surface and slope of dike, boundary condition is defined as boundary of flux or fixed water head. The relation between intension of rainfall with permeability of

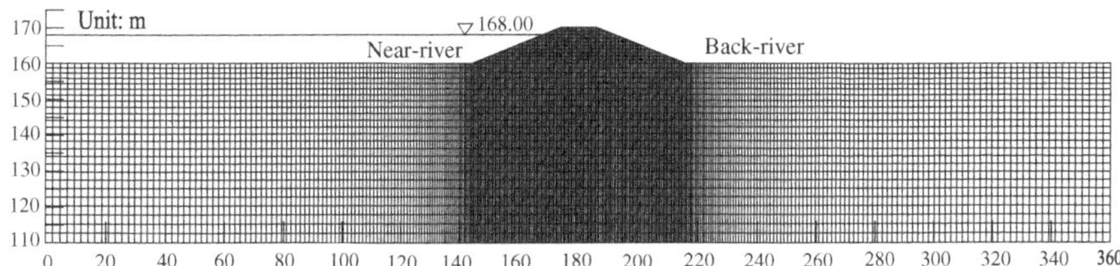

Fig. 90.1 Meshes of analysis about finite element method

Fig. 90.2 *Curves* of seepage coefficient and matric suction

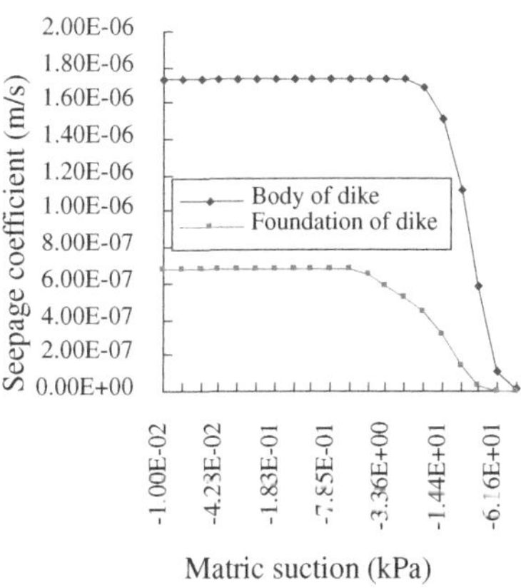

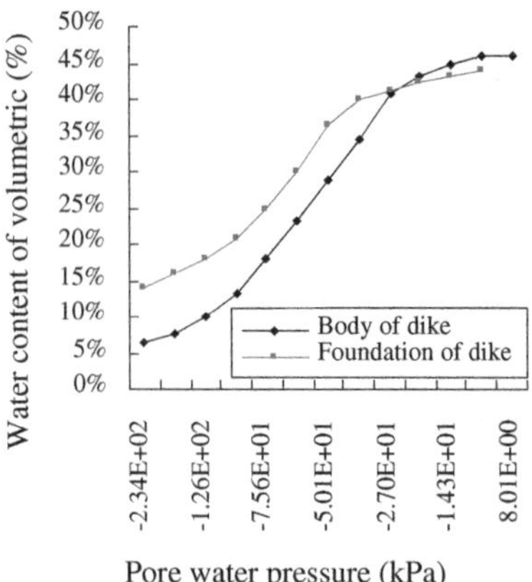

Fig. 90.3 *Curves* of water content of volumetric and pore water pressure

Table 90.1 Parameters of numeration

Type of soil	Thickness (m)	Density (kN/m^3)	Cohesion (kPa)	Internal friction angle (Degree)
Silt (body of dike)	10	17.5	8	18
Clay (foundation of dike)	50	19.0	20	25

soil can be estimated automatically with program [12]. If intension of rainfall is less than coefficient of infiltration of exterior soil, it is defined as flux boundary, and the value is equal to intension of rainfall. If intension of rainfall is more than coefficient of infiltration of exterior soil, boundary condition is defined as fixed water head boundary, and the value is equal to the altitude of exterior soil.

Left side and right side of model and under groundwater level can be defined as fixed water head boundary, it can be defined as flux boundary under groundwater level, and the flux value is equal to 0. Bottom boundary of model can be defined as imperviable one. Initial condition of groundwater level is situated below 0.5 m above exterior foundation of dike. Without evaporation of water considered, suction distributing of soil changes in linear above groundwater level.

90.3 Analysis of the Influence of Dike Safety on Cracks

90.3.1 Analysis of the Influence on Complexion of Cracks

In order to analyze the influence of dike safety on cracks filled with water and depth of cracks, taking the influence of no rainfall as the example, the influence of safety at back-river with different status of cracks filled with water is expressed as

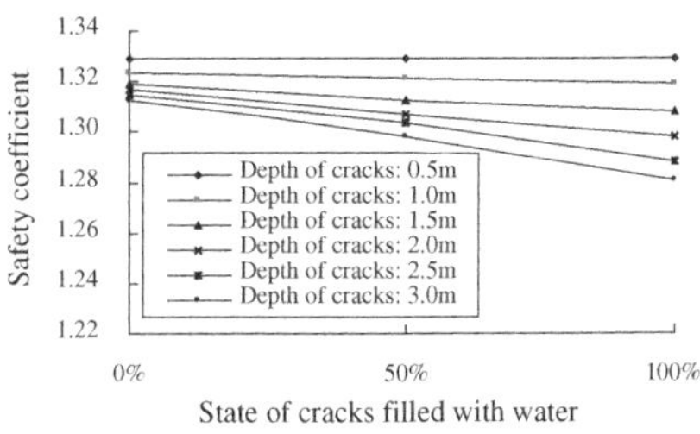

Fig. 90.4 Safety effect on dike of cracks filled with water

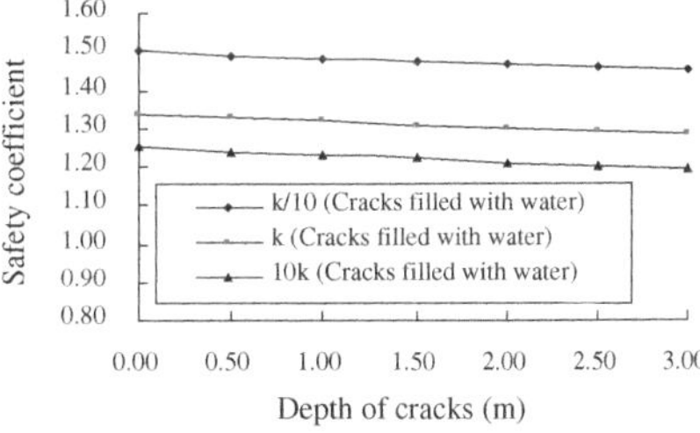

Fig. 90.5 Safety effect on dike about depth of cracks

Figs. 90.4 and 90.5 (where, k is seepage coefficient) when the altitude of water level at near-river arrives at 158 m.

From Fig. 90.4 we can see, it carries out disadvantage effect on dike safety at back-river despite of cracks filled with water or not, and safety coefficient of dike at back-river keeps continually decreasing with the depth of cracks. Safety coefficient of back-dike keeps lower relatively with different depth of cracks while cracks filled with water. Moreover, it is higher in the degree filled with water, safety coefficient of dike-body at back-river is higher accordingly. Finally, safety coefficient of dike-body with cracks can decrease obviously when it rains.

From Fig. 90.5 we can see, safety coefficient of dike-body at back-river keeps inverse proportion to depth of cracks, it is deeper in cracks, and safety coefficient of dike-body at back-river is lower. Cracks can affects safety of dike obviously, thus especial recognition should be aroused.

90.3.2 Analysis of the Influence on Decreasing Velocity of Water Level at Near-River

In order to analyze the influence of dike safety on decreasing velocity of water level at near-river, the relation between decreasing velocity of water level with safety coefficient of dike-slope at near-river is expressed as Fig. 90.6.

From Fig. 90.6 we can see, safety coefficient of dike-slope of near-river keeps inverse proportion to decreasing velocity of water level at near-river. Quickness in decreasing velocity of water level at near-river, water cannot discharged from

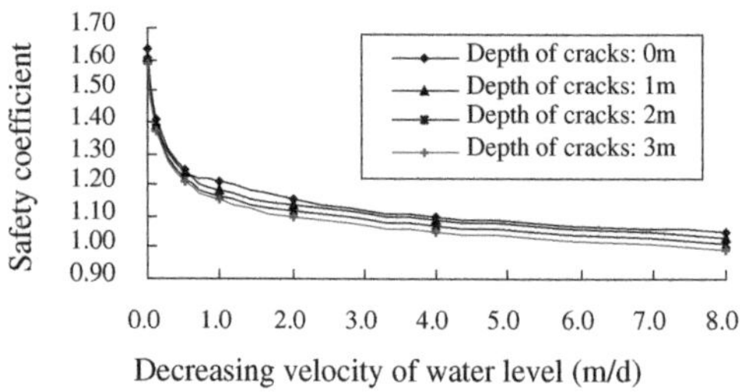

Fig. 90.6 Safety effect on dike about decreasing velocity of water level at near-river

dike-body at near-river quickly, thus pore water pressure is higher in dike-body. Especially when decreasing velocity of water level arrives at 8 m/d, safety coefficient of dike-slope at near-river can become lower than 1.0, thus it can lead to slide of slope and bank collapse of bank.

90.3.3 Analysis of the Influence on Rainfall

90.3.3.1 Intension of Rainfall

When the altitude of water level at near-river is 158 m, the influence of safety of dike without cracks on rainfall is expressed as Fig. 90.7. When it rains for 24 h, the relation between intension of rainfall with safety coefficient of dike-slope at back-river is expressed as Fig. 90.8.

From Fig. 90.7 we can see, safety coefficient of dike-slope at back-river keeps inverse proportion to duration of rainfall. The influence of safety of dike-body on rainfall turns lower after 24 h. Safety coefficient of back dike-slope keeps inverse proportion to intension of rainfall. Especially to super rainstorm (for example: intension of rainfall is 100 mm/d), safety coefficient of back dike-slope can decrease from 1.4 to 0.98, thus it can lead to slide of slope and collapse of bank when super rainstorm lasts for a long time.

From Fig. 90.8 we can see, it is higher in intension of rainfall, and safety coefficient of dike-body at back-river is lower.

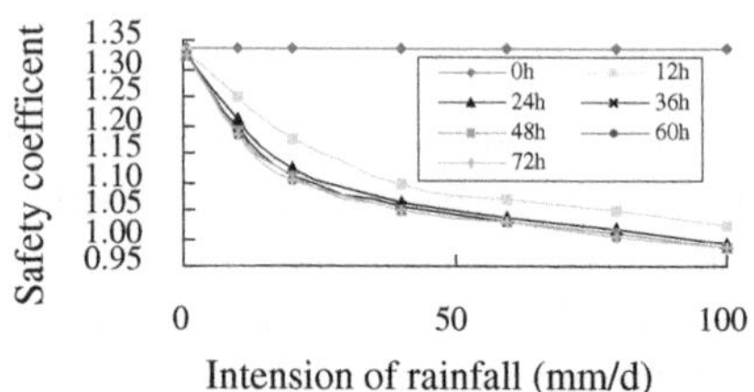

Fig. 90.7 Safety effect on dike about rainfall without cracks

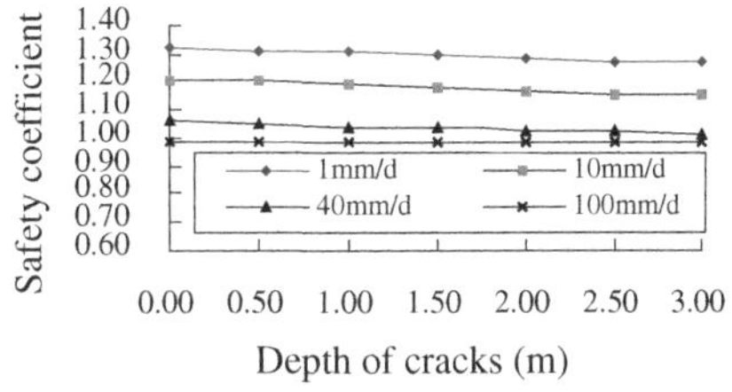

Fig. 90.8 Safety effect on dike about intension of rainfall

90.3.3.2 Duration of Rainfall

When the altitude of water level at near-river is 158 m and depth of cracks is 2 m, the influence of dike safety at back-river on duration of rainfall is expressed as Fig. 90.9. When altitude of water level at near-river is 158 m and intension of rainfall is 10 mm/d, the influence of dike safety at back-river on duration of rainfall with different depth of cracks is expressed as Fig. 90.10.

From Fig. 90.9 we can see, safety coefficient of dike-slope at back-river keeps close correlation with duration of rainfall, it is longer in duration of rainfall, and safety coefficient of dike-body at back-river is lower under the condition of the same intension of rainfall. The influence of safety of dike-body turns lower after 24 h for the type of soil, and the changing trend turns steady.

From Fig. 90.10 we can see, it is longer in duration of rainfall, safety coefficient of dike-body at back-river is lower with depth of cracks, and the influence of safety of dike-body on rainfall turns steady after 24 h. It is higher in depth of cracks,

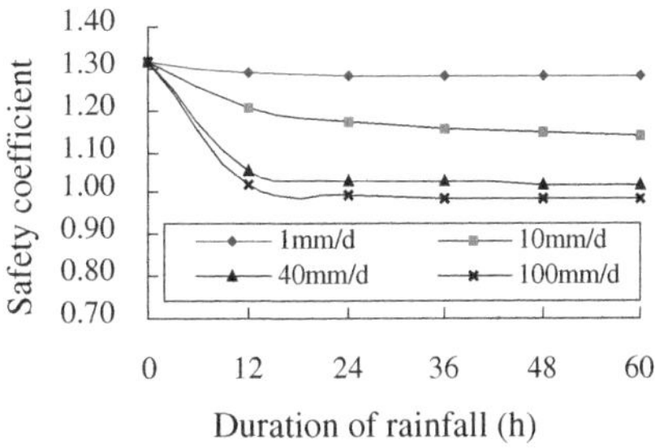

Fig. 90.9 Safety effect on dike about by rainfall

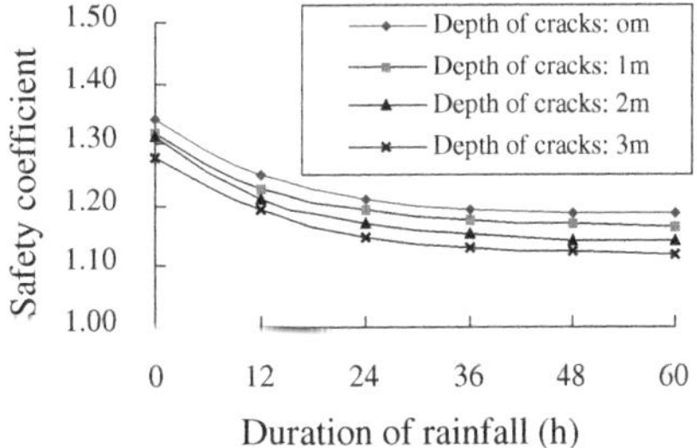

Fig. 90.10 Safety effect on dike about period of rainfall

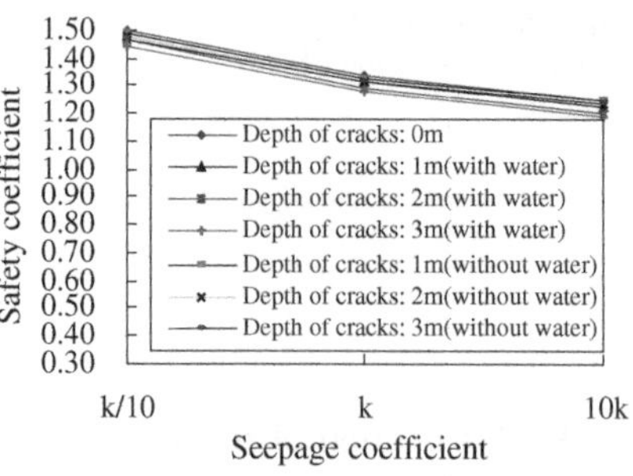

Fig. 90.11 Safety effect on slope of dike at back-river by suction

safety coefficient is lower. It is more in cracks or destroy of dike, safety accident comes forth more easily, and the chance of accident keeps direct proportion to duration of rainfall, intension of rainfall and depth of cracks.

90.3.4 Analysis of the Influence on Seepage Coefficient of Soil

In order to analyze the influence of dike safety on seepage coefficient of soil, the relation between seepage coefficient with safety coefficient is shown as Fig. 90.11.

From Fig. 90.11 we can see, it is higher in safety coefficient of dike-slope at back-river. Because the seepage surface is lower relatively for lower seepage coefficient of dike. Hidden trouble of safety is serious for higher seepage coefficient.

90.4 Conclusions

Because of theory of unsaturated soil mechanics and analysis model of cracks, unsaturated mechanics of soil and cracks of dike can be considered, lots of influencing factors such as depth, status of cracks filled with water, change of water level, rainfall and so on can be analyzed, influencing factors can be achieved.

It can easily lead to slope slide and collapse when decreasing velocity of water level at near-river is quick and duration of rainstorm is long for dike with cracks. Hidden trouble of safety is serious for higher seepage coefficient. The chance of accident keeps direct proportion to duration of rainfall, intension of rainfall, depth of cracks, decreasing velocity of water level about dike-slope at near-river.

Safety of dike can be influenced by status of cracks, changing of water level, rainfall and so on, thus safeguard should be reinforced in flood season, problem should be managed in due course, lash-up measures should be taken necessarily for safe in flood season.

Acknowledgments This work was supported by Research Foundation of Yellow River Institute of Hydraulic Research (No. HKY-JBYW-2013-17), "948" Item of Ministry of Water Resources (No. 201124) and the National Natural Science Foundation of China (No. 60934009).

References

1. Chen ZY (2003) Soil slope stability analysis-theory methods and programs. China Water Power Press, Beijing (in Chinese)
2. Du YH (2002) Research and practice on technique of reinforcing dike with silt in lower Yellow River. Yellow River Water Power Press, Zhengzhou (in Chinese)
3. Fan YW, Zhang Y, Jin XL, Ma YQ (2006) Effect of rainfall on stability of anisotropic soil slope. Rock Soil Mech 26(S2):1097 (in Chinese)
4. Lu TH, Gao GQ, Chen J (2006) Effective stress method of stress and deformation for soil-rock fill dam after water impounding. Rock Soil Mech 26(2):247 (in Chinese)
5. Fredlund DG, Rahardjo H (1993) Soil mechanics for unsaturated soils. John Wiley and Sons Inc., New York
6. Li QY, Zhang JH, Wang HJ, Zhang MG, Li MD (2002) Stability analysis of Tongma Levee with construction cracks. J Yangtze River Sci Res Inst 19(1):25 (in Chinese)
7. Shen JB, Wang ZS (2003) Analysis of cracks in the 1st dam of Sheshang reservoir. Hydro-Sci Eng (2):41 (in Chinese)
8. Xu GM, Gao CS, Zhang L, Yang SH (2005) Preliminary study of stability of levee on soft ground. Chin J Rock Mech Eng 24(13):2315 (in Chinese)
9. Shi ZB, Wei Q, Fu YG, Ge JA (2004) Forming cause and methods of preventing and dealing with of cracks about dike and road in Yellow River. J Yellow River Conservancy Tech Inst 16(1):16 (in Chinese)
10. Wang XY (2006) Simple analysis on the longitudinal cracks of dike about left bank at Jiaozuo section of Yellow River. Yellow River 28(5):75 (in Chinese)
11. Lu KF, Ma LY, Hou XX, Wang SP, Liu GQ (2007) Yellow River main dike cracks origin analysis and preventing and controlling measure. China Water Transp 7(4):82 (in Chinese)
12. GEO-SLOPE International Ltd (2001) GEO-SLOPE user's manual. GEO-SLOPE International Ltd., Canada, Alberta

Chapter 91
Simulation of Water and Floating Body with SPH Method

Zhisheng Li, Ao Sun and Xin Zhao

Abstract A real-time simulating method based on smoothed particle Hydrodynamics (SPH) for water and floating body interaction was proposed in this paper. The force analysis and state update models were established for floating body and water particles, and a collision detecting method between particles and obstacles was proposed also. The water surface was extracted from water particles based on Marching Cubes for final rendering. Experimental results showed that our method was realistic, and could achieve real-time frame rate when particle amount was less than 6,000.

Keywords SPH · Floating body and water interaction · Simulation modeling

91.1 Introduction

As an ordinary phenomenon, the interaction between water and floating body seems simple, but the realistic modeling and rendering of it is a challenging problem in computer graphics. There are two approaches have been used for simulating water in computer graphics nowadays, one is the Euclidian approach, and the other is the Lagrange approach.

Euclidian approach is a grid/mesh based numerical method, which has been applied to create models of water motion for computer graphics [1–5]. In recent

Z. Li (✉)
Post-Doctoral Research Station, The Unit 92493 of PLA, Dalian, China
e-mail: zhisheng_li@126.com

Z. Li · A. Sun
The Unit 91550 of PLA, Dalian, China

X. Zhao
Dalian university, Dalian, China

Z. Sun and Z. Deng (eds.), *Proceedings of 2013 Chinese Intelligent Automation Conference*, Lecture Notes in Electrical Engineering 255,
DOI: 10.1007/978-3-642-38460-8_91, © Springer-Verlag Berlin Heidelberg 2013

years, fluid simulation based on Lagrange approach has got a research focus in computer graphics. It is a mesh free method, which provides accurate and stable numerical solutions for integral equations and partial differential equations (PDEs) with all kinds of possible boundary conditions using a set of arbitrarily distributed nodes or particles. Among Lagrange approaches, Smoothed Particle Hydrodynamics (SPH) is a new one, in which the state of a system is represented by a set of particles, and fluid advection is calculated by particle movement, so that those drawbacks in Euclidian approaches, such as numerical dissipation, mesh deformation, and the limitation of Courant–Friedrichs–Lewy condition (CFL condition), can be avoided. When particle number is not large, real time simulation can be achieved. In 1995, [6] first used SPH in computer graphics to simulate fire and gaseous phenomena. Then [7] simulated bubble rising motion in water with SPH. In 2003, [8] used SPH to simulate pouring water into glass. [9] simulated river flowing in ravine based on the modeling method of [9]. However, the object is only a still obstacle in [8] and [9], so their methods couldn't simulate the interaction between water and floating body.

Here we propose a modeling method based on Smoothed Particle Hydrodynamics to animate the interaction between water and floating body.

91.2 SPH Summary

As a Lagrange approach, SPH is an interpolation method for particle systems. With SPH, field quantities that are only defined at discrete particle locations can be evaluated anywhere in space. For this purpose, SPH distributes quantities in a local neighborhood of each particle using radial symmetrical smoothing kernels. According to SPH, continuous physical field is approximated in Eq. (91.1), and a discrete quantity $f(x_i)$ and its derivative are interpolated at location x by a weighted sum of contributions from all particles:

$$f(x) = \int_{\Omega} f(x')W(x - x', h)\mathrm{d}x' \tag{91.1}$$

$$f(x_i) = \sum_{j=1}^{N} \frac{m_j}{\rho_j} f(x_j)W(x_i - x_j, h) \tag{91.2}$$

$$\nabla f(x_i) = -\sum_{j=1}^{N} \frac{m_j}{\rho_j} f(x_j)\nabla W(x_i - x_j, h) \tag{91.3}$$

$$\nabla^2 f(x_i) = -\sum_{j=1}^{N} \frac{m_j}{\rho_j} f(x_j)\nabla^2 W(x_i - x_j, h) \tag{91.4}$$

where, x is position, Ω is integral volume, x' is a position in Ω, h is the support domain radius, $W(x - x', h)$ is the smoothing kernel function, $f(x_i)$ is the quantity of particle i at position x_i, x_i is the particle position in the supporting domain of particle i.

91.3 Modeling Water and Floating Body with Particles

91.3.1 Grid-Based Neighbour Particles Search

In SPH, effective area of the smoothing function for a particle at some position is usually called as the support domain of that particle, the range of which is defined by smoothing kernel radius, and the particles in support domain is called as neighbor particles. Particle properties are calculated by interpolation of neighbor particles. Therefore, searching neighbor particles of a particle is important step for its properties calculation. In order to search neighbor particles conveniently, we firstly divide simulation space into a 3D grid, of which the side length of grid cell equals to smoothing kernel radius. Figure 91.1 is diagram of the 3D grid, where deep color dots represent floating body and light color dots is water particles. The distribution range of particles are changing with the movement of particles, so the grid should be invidided and reset particles into corresponding grid cell at the beginning of every time step.

We create a list of neighbor particles for every particle by searching particles in its support domain, see Fig. 91.2. Searching progress is limited in cells up, down, left, right, front, and back the current operating particle in, because the grid cell

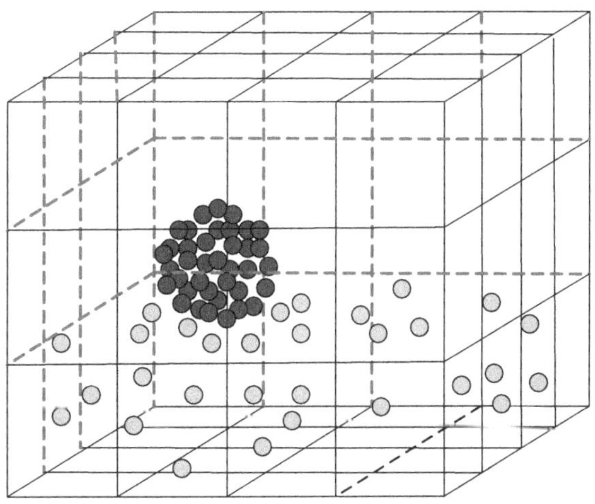

Fig. 91.1 Particles distribution in 3D grid

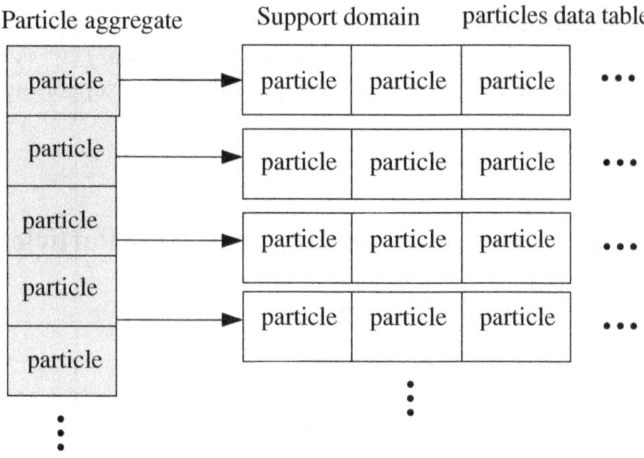

Fig. 91..2 Data structure of neighbor particles list

length is set as the smoothing kernel radius, and the particles in beyond cells don't belong to current support domain. This decreases the sum of searching particles and raise efficiency.

91.3.2 Force Analysis and State Computation of Water Particle

Updating the state of water particles (velocity, acceleration etc.) is important for modeling water flow movement. Every particle is controlled by Lagrange fluid dynamics function as follows:

$$\rho \frac{Du}{Dt} = -\nabla p + \mu \nabla^2 u + \rho g \tag{91.5}$$

where u is the viscosity of water, u is the velocity of water particle, ρ its density, p its pressure, g is the unit force at gravitational direct.

If we see the right expression of the equation as the resultant force acted on a water particle, up equation can be simplified as:

$$a_i = \frac{Du_i}{Dt} = \frac{f_i}{\rho_i} \tag{91.6}$$

where $\boldsymbol{f}_i$ is resultant force, a_i is acceleration, p_i is pressure, ρ_i is density.

Water particle i changes its position and velocity continually under the action of force $\boldsymbol{f}_i$. In order to analyze the force $\boldsymbol{f}_i$ conveniently, we decompose $\boldsymbol{f}_i$ into pressure force, gravity, and viscous force. These forces can be computed as follow equations, which are deduced from Eqs. (91.2)–(91.4).

$$\begin{aligned} f_i &= f_i^{pressure} + f_i^{vis\cos ity} + f_i^{gravit} \\ f_i^{pressure} &= -\nabla p_i = \sum_{j=1}^{N} m_j \frac{(p_i + p_j)}{2\rho_j} \nabla W(x_i - x_j, h) \\ f_i^{vis\cos ity} &= \mu \nabla^2 u_i = -\mu \sum_{j=1}^{N} m_j \frac{(u_j - u_i)}{\rho_j} \nabla^2 W(x_i - x_j, h) \\ f_i^{gravit} &= \rho_i g = g \sum_{j=1}^{N} m_j W(x_i - x_j, h) \end{aligned} \tag{91.7}$$

91.3.3 Force Analysis and State Computation of Floating Body

There are gravity and buoyancy force acted on body when floating body dropping into water.

$$F_j = f_j^{float} + f_j^{gravity} \tag{91.8}$$

According to the principle of mechanical energy conservation, when system is not affected by external force, the resultant force acted on system should be zero. Therefore, there is following equation between adjacent particle i and particle j.

$$f_{ij} + f_{ij} = 0 \tag{91.9}$$

where f_{ij} is force on particle i produced by particle j, and f_{ij} is force on particle j produced by particle i. Equation (91.9) indicates that interaction force between particle i and particle j is equal but opposite direction. The buoyancy on a floating body particle is produced by all water particles in its support domain. If we find all these water particles we can calculate the buoyancy of floating body particle by summating all interaction force between floating body particle and water particles as follows:

$$f_j^{float} = -\sum_{i=1}^{N} f_{ij}^{pressure} + f_{ij}^{vis\cos ity} + f_{ij}^{gravity} \tag{91.10}$$

where f_j^{float} is buoyancy on floating body particle j, $f_{ij}^{pressure}$ $f_{ij}^{vis\cos ity}$, $f_{ij}^{gravity}$ are respectively pressure force, viscous force and gravity, which are produced by water particle i. N is the number of water particles effected on floating body particle j. The gravity $f_j^{gravity}$ is a constant.

91.3.4 Collision Detection Between Water Particle and Obstacle

Moving water will collide with its surrounding environment, such as rigid wall of container, hills, solid rocks etc., which can be called as obstacle of moving water. For the sake of reality, the collision between water and obstacle must be taken into account in water simulation. We can simplify the collision as water particles with obstacle, because water has been considered composing with particles. Obstacle can be modeled with polygons, and every polygon is one of the surfaces of obstacle, so the collision detection between water and obstacles turns into particles and polygons.

We can get acceleration of water particles and floating body particles through methods in Sections 91.3.2 and 91.3.3, and calculate the velocity at next time step of them with the acceleration, and further, through which we can calculate the positions at next time step in advance. If the connection line between current position and next position pass through any obstacle polygons, which indicate that the particle moving with the new calculated velocity will collide with obstacle and come into it, then we should change the velocity to avoid this happening.

As we know, if the particle and polygon are rigid body, the particle will bounce at the polygon surface when they colliding with each other, see Fig. 91.3, where N is the normal of the polygon, point A is the position of particle at current time, point B its next position at the next time. The line between point A and point B pierces through the obstacle surface, which means the particle will penetrate obstacle if it moves at velocity v_{i+1}. Therefore, we should change the velocity v_{i+1}. We suppose the particle bounces at the collision point C, the arrow of line CE is its bounce direction, and the point E is the ideal position at next time after bounce. However, the particle can't arrive E for the energy loss after collision, and can only arrive point D, v_{i+1} is its velocity calculated as following,

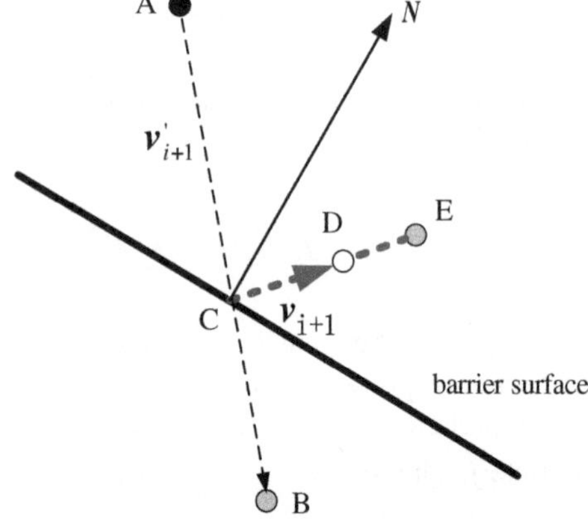

Fig. 91.3 Collision rebound of particle at flat surface

$$v_{i+1} = k_{damp}(v'_{i+1} \times 2N(N \cdot v'_{i+1})) \tag{91.11}$$

where k_{damp} is the collision attenuation coefficient. Acceleration of water particle is linked with the force on particle, collision force will produce collision acceleration, which can be calculated with $a_N = \alpha N$, where α is the transfer coefficient from collision force to collision acceleration. So the next time acceleration of water particle can be calculated with $a_{j+1} = a_j + a_N$.

91.4 Water Surface Modeling Base on Marching Cubes

Water surface can set water apart from surrounding environment, and then be imposed lighting and drawing. Marching Cubes is classical algorithm for surface extraction in three-dimensional data field. It is simple with fast drawing speed, so we use Marching Cubes algorithm to extract water surface from water density field. Several details aspects involved as follow:

1. Density field data sampling base on spatial grid node
   In Sect. 91.3.2, we have built the three dimensional grid according to the spatial distribution of particle. The grid node is considered as the sampling point in flow field. Node sampling value is calculated by the SPH interpolation. First, find all the volume element that use this node as vertex, for example, the nodes located in the boundary of 3D grid match 4 volume elements. Secondly, calculate density at the sampling node by SPH interpolation of each particle in the found volume element as follows:

$$\rho_i = \sum_{j=1}^{N} m_j W(x_i - x_j, h) \tag{91.12}$$

where, j is the particle in the found 3D grid cells.

2. The threshold value of I so surface on water surface in the volume density field
   Extract water surface from water density field with marching cubes algorithm. The density of water surface must be known. According to the water pressure is equal to atmospheric pressure and the ideal gas state equation, the density of water surface can be obtained as follows:

$$\rho_{surface} = \frac{p_{\text{气}}}{k} \tag{91.13}$$

where $\rho_{surface}$ is the density of water surface, 气 represents atmospheric pressure, k is the gas constant.

3. Extracting isosurface of water surface base on marching cubes algorithm We can extract the isosurface base on Marching Cubes algorithm after the up two steps. Because of the limitation of space, this article does not do detailed introduction, you can obtain the detail method in the reference literature [10].

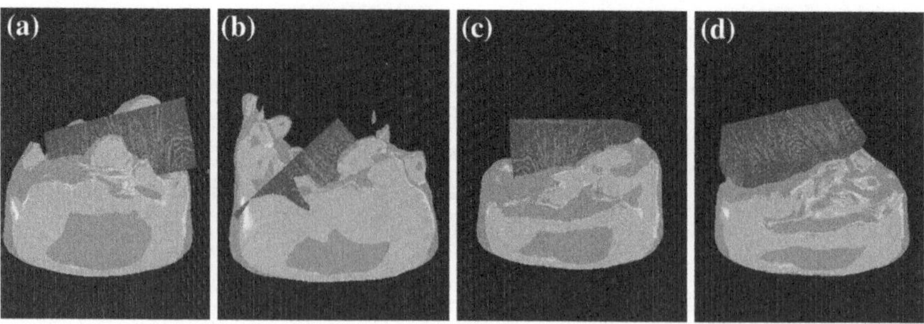

Fig. 91.4 Simulation result

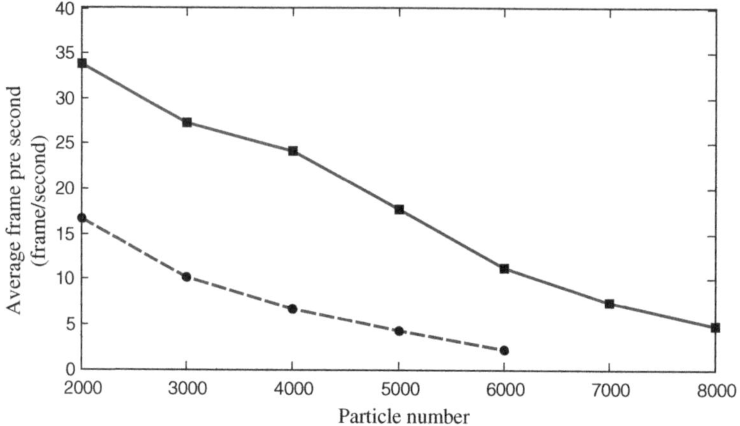

Fig. 91.5 Relationship between average frame pre second and particle numbers

91.5 Implementation

Experiments are carried out with VC++6.0 and OpenGL on the Windows XP platform. And with Intel CPU PIV 3.0 GHz, 512 MB RAM, GeForce 6,600 display card and 128 M display memory.

In our experiments, a block of wood is poured into a glass of water, and then shape the glass. To make it simple, water and the block of wood are drew, ignoring the glass. Figure 91.4a–d show the surface shaking when thrown a block of wood. Solid line in Fig. 91.5 describes the relationship between average rending frame pre second and particle number, which shows that it can reach interaction performance when the number of particles below 6,000.

91.6 Results and Discussion

A real-time simulating method based on SPH for water and floating body interaction proposed in this paper. The force analysis and state update models were established for floating body and water particles, and a collision detecting method between particles and obstacles was proposed also. The water surface was extracted from water particles based on Marching Cubes for final rendering. Experimental results showed that our method was realistic, and could achieve real-time frame rate when particle amount was less than 6,000.

References

1. Chen J, Lobo N (1995) Toward interactive rate simulation of fluids using Navier-Stokes equations. Graph Models Image Process 57(2):107–116
2. O'Brien JF, Hodgins JK (1995) Dynamic simulation of splashing fluids. In: Proceedings of computer animation, IEEE Computer Society, Washington, pp 198–208
3. Foster N, Metaxas D (1996) Realistic animation of liquids. Graph Models Image Process 58(5):471–483
4. Forester N, Fedkiw R (2001) Practical animation of liquids. In: Computer graphics proceedings, annual conference series, ACM SIGGRAPH, Los Angeles, California, pp 23–30
5. Enright D, Marschner S, Fredkiw R (2002) Rendering of complex water surface. ACM Trans Graph 21(3):736–744
6. Stam J, Fiume E (1995) Depicting fire and other gaseous phenomena using diffusion processes. In: Computer graphics proceedings, annual conference series, ACM SIGGRAPH, Los Angeles, California, pp 129–136
7. Muller M, Solenthaler B, Keiser R, Gross M (2005) Particle-based fluid–fluid interaction. In: Computer graphics proceedings, annual conference series, ACM SIGGRAPH, Los Angeles, California, pp 237–244
8. Muller M, Charypar D, Gross M (2003) Particle-based fluid simulation for interactive applications. In: Computer graphics proceedings, annual conference series, ACM SIGGRAPH, Aire-la-Ville, pp 154–159
9. Kipfer P, Westermann R (2006) Realistic and interactive simulation of rivers. Graphics interface. In: Proceedings of graphics interface, Quebec, Canada, pp 41–48
10. Lorensen WE, Cline HE (1987) Marching cubes: a high resolution 3D surface construction algorithm. Comput Graph 21(4):163–169

Chapter 92
A Jacket Robot and its Human–Robot Interacting Technology

Huailin Zhao, Yulong Xia and Yi Liu

Abstract A special jacket actuated by Mckibben pneumatic artificial muscles is developed, which is called as "jacket robot". It can be put on the human upper body and the two arms to help them act. To satisfy the requirement for perceiving the arm actions automatically, an action intention recognition technology based on the acceleration signal is investigated. The paper introduces the constitution of the jacket robot, its main functions and its main technical specification. The designation idea about the recognition method is emphasized. The related experiment shows that the method is efficient.

Keywords Jacket robot · Mckibben muscle · Action intention recognition · Acceleration signal

92.1 Background

In the human-powered robot field, both the wearable robot and the intelligent system of human body exoskeleton are the hot subjects [1–3]. Putting on the wearable robot on the body, the elder, the weaker, the patient and the disable can bear a burden or walk so that their life quality can be improved very much. Putting on it, the nurse will have enough physical strength to carry patients. A special intelligent system of human body exoskeleton was developed by America Defense Department together with some related labs and some companies, which could help the soldiers carry a heavy burden when they did the troops march and completed some difficult tasks [4]. Based on its commerce value and the military application, the wearable robot has become one of the key subjects in the robot

H. Zhao (✉) · Y. Xia · Y. Liu
School of Electrical and Electronic Engineering, Shanghai Institute of Technology, Room A417, the 1st Science Branch Building, No. 100, Haiquan Rd., Shanghai City 201418, China
e-mail: zhao_huailin@yahoo.com

Z. Sun and Z. Deng (eds.), *Proceedings of 2013 Chinese Intelligent Automation Conference*, Lecture Notes in Electrical Engineering 255,
DOI: 10.1007/978-3-642-38460-8_92, © Springer-Verlag Berlin Heidelberg 2013

field. In America and Japan, the related research and developing centers have been applying the wearable robot to the hospitals, the recuperation centers and the armed forces [5, 6]. Besides, Department of Science and Technology of China has put forward the key subject "medical treatment robots for the limb recuperation" in the key program "disable's recuperation technologies and the equipment development", which refers to the wearable robot too.

In this research, the developed artificial assistant jacket, which is called as the jacket robot, is right just a wearable robot which can improve the power of the wearer's two arms. It is actuated by Mckibben pneumatic artificial muscles (simply call it as Mckibben muscle) by which the additional power is provided. Mainly, the functions of both the shoulder joint and the elbow joint of the jacket robot are considered and designed. It's easier for the jacket robot to generate the soft touch and relative safe operation because the Mckibben muscle actuator is similar to the biological muscle. One of the key problems is the human–machine interaction because that the jacket robot is to be put on the human body. How does it can sense automatically the action intention of the human's two arms so that the Mckibben muscles can be controlled to contract in real time? No other than this, the jacket robot can act coordinately with human's action intention and achieve assisting to the human arm action. Besides, the wearer will feel comfortable no other than good human-robot coordination.

92.2 System Introduction

The whole system of the jacket robot includes three parts: the jacket robot itself, pneumatic control subsystem and the measurement subsystem.

The jacket robot mainly consists of the thickened vest, arm-protecting plastic cover and the Mckibben muscles. Each arm of the jacket robot is actuated by 6 pieces of Mckibben muscles. The distribution of them is a little like the human's arm muscles, which stimulates the anatomization structure of the human arm. Based on the above consideration, we studied the anatomization of the human arm. The 6 pieces of Mckibben muscles stimulate separately the main muscles of the human arm. Figure 92.1 shows its appearance. The later experiments show that it didn't feel uncomfortable when the wearer firstly put on it. On whole, the materials used for making the jacket robot are very light. Its weight is less than 2.5 kg.

The jacket robot needs the support of the pneumatic control subsystem, because that the Mckibben muscle is pneumatic. The subsystem mainly includes the compressor, the electro-pneumatic ratio valves, the computer and the I/O card. The compressor provides the pressed air. The electro-pneumatic ratio valve generates the air pressure proportional to the electric voltage output by the computer. And the pressed air is used to actuate the Mckibben muscle.

About the measurement subsystem, we designed one detection method for the arm action intention recognition which is based on the acceleration signal. Later we will explain the idea of the method in detailed.

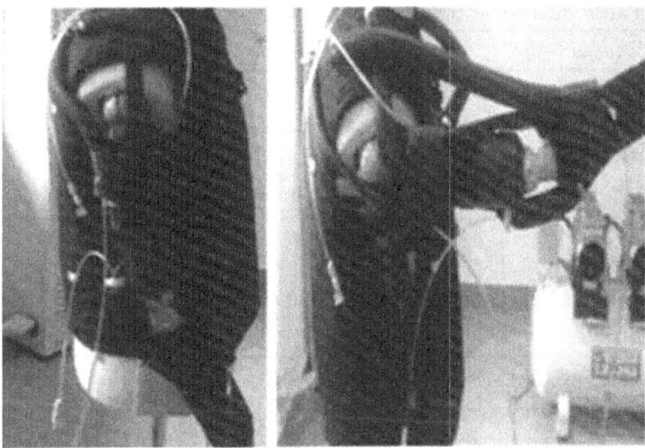

Fig. 92.1 The jacket robot

When the detection method is applied, the whole system is always controlled by the computer.

The jacket robot can realize the basic actions of both the human shoulder joint and the human elbow joint. After analyzing, we sum up four basic actions which are shown in Fig. 92.2. And Table 92.1 shows the detailed testing result.

Zhejiang University has developed a recuperation equipment for the human arm. Table 92.2 shows the actions it can realize and the related angle change field with both the shoulder joint and the elbow joint [1].

Comparing Tables 92.1 and 92.2 with each other, the angle change field of the jacket robot is larger than that of the one developed by Zhejiang University.

On the other hand, we get the related data about the human arm, shown in Table 92.3 [7].

Comparing Tables 92.1 and 92.3 with each other, the difference between them are found. The angle change field is clearly less than the human arm's. To the jacket robot, some angle change fields can get to the 70 % of the corresponding human arms, but some of them can't, which can only get to its 50–60 %. What's the reason? After analyzing, we think there are two main factors: One is that the mechanical structure is actually different from the human's; another one is that the contraction ratio of the Mckibben muscle is much less than the human's (The Mckibben muscle's is less than 30 %, the human's is up to 40 %) [8].

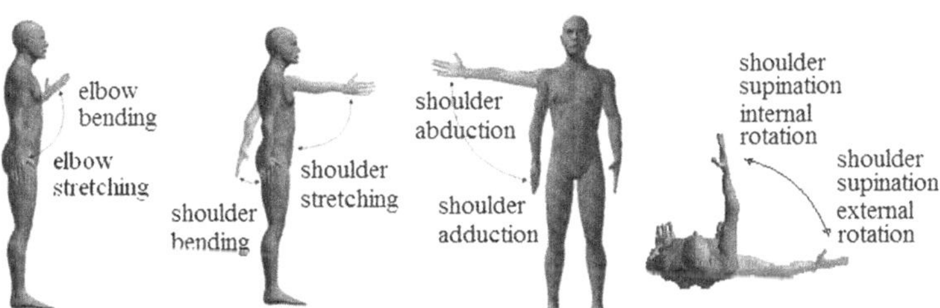

Fig. 92.2 The basic arm actions

Table 92.1 The basic actions and their angle change fields when put on by the wearer

Joint action	Angle field (°)
Bending and stretching of the elbow	5 – (80–85)
Bending and stretching of the shoulder	0 – (90–105)
Abduction and adduction of the shoulder	0 – (80–85)
Supination external rotation and supination internal rotation of the shoulder	0 – (75–80)

Table 92.2 The basic actions and their angle change fields of the recuperation equipment

Joint action	Angle field (°)
Bending and stretching of the elbow	0–90
Bending and stretching of the shoulder	−60–60
Abduction and adduction of the shoulder	−50–60
Supination external rotation and supination internal rotation of the shoulder	–

Table 92.3 The basic actions and their angle change fields of the human arm

Joint action	Angle field (°)
Bending and stretching of the elbow	(0–4) – (135–150)
Bending and stretching of the shoulder	−(30–80) – (130–180)
Abduction and adduction of the shoulder	(30–75) – (150–180)
Supination external rotation and supination internal rotation of the shoulder	−(40–90) – (40–90)

92.3 Recognition Method of the Action Intention of the Human Arms

How does the jacket robot act correctly to help the human's arm? Whenever acting, it must know firstly what action the human arm wants to do. This means that the jacket robot must perceive the action intention of the human arm. Based on the above analysis, we use the 3-axis accelerator to sense the arm actions. Because there are both forearm and rear arm, it needs at least two accelerators for each arm. In another word, it needs the distribution of multi-accelerators. We only used one 3-axis accelerator to study the method because of the cost. Figure 92.3 shows the accelerator which can connect with the computer by USB port. The experiment result shows that it's very efficient.

There are also some problems with this method. The key one of them is how to consider the action sensitivity. Or we say how to set the threshold used for the action intention judgment. The accelerator is very sensitive and therefore the threshold has to be larger than zero. But the appropriate threshold is not easy to be determined, because that the speeds of the actions' generating even though by the same person are not always same. This is why that sometimes the response of the jacket robot delays a little long.

Fig. 92.3 The 3-axis accelerator

We tested and validated the cognition method by two different experimental methods:

The experimental method 1: The jacket robot is put on the wooden model, and the 3-axis accelerator is connected at the forearm or the rear arm of a real person. When the person acts arbitrarily based on the actions listed in Table 92.1, the computer can sense efficiently the human's action intention and then control the related Mckibben muscles to contract, at last the jacket robot helps the human arm act. Figures 92.4, 92.5, 92.6 and 92.7 show the experiment. The sensitivity of the action intention judgment can be regulated by the computer program (changing the threshold is OK). The test results are as the following:

The action delay of the jacket robot: 600–3,500 ms (when the sampling period of the acceleration is 500 ms).

The experimental method 2: Directly putting the jacket robot on a person and attaching the 3-axis accelerator at the forearm or the rear arm of the person. When the wearer does arbitrarily the actions listed in Table 92.1, the computer can sense the action intention of the person's arm and control the related Mckibben muscles to contract in real time, so that the jacket robot achieves the arm action assisting. Of course, the sensitivity of the action intention judgment can be regulated by the

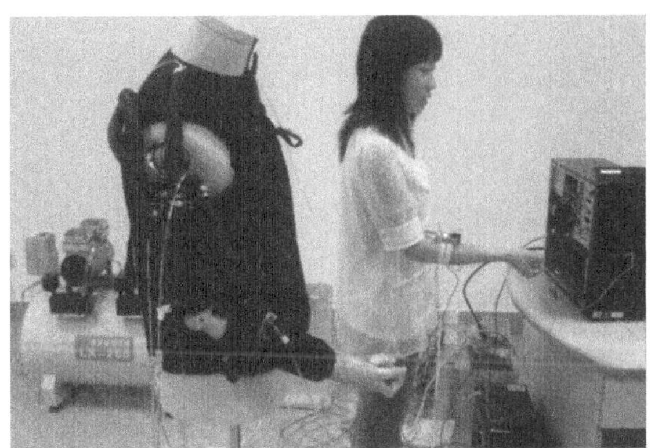

Fig. 92.4 Perceiving the elbow action

Fig. 92.5 Perceiving the shoulder action

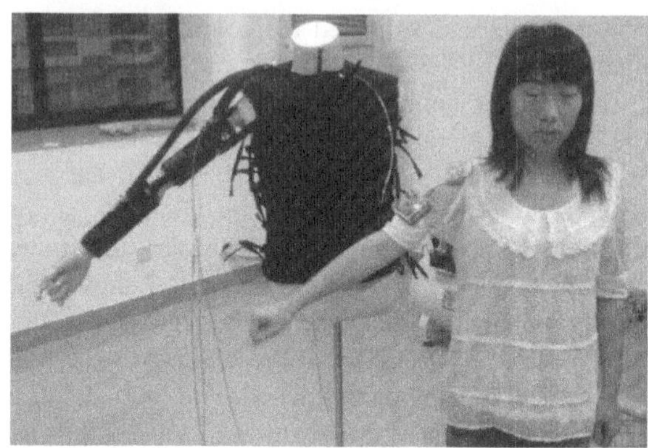

Fig. 92.6 Perceiving the shoulder action

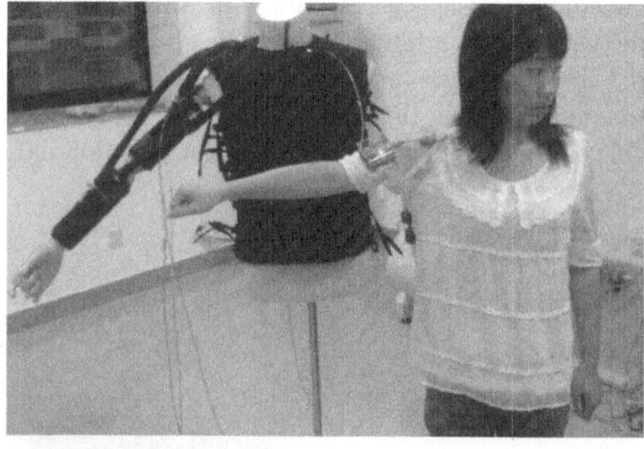

Fig. 92.7 Shoulder supination rotation action

computer program (changing the threshold is OK). The experimental result is as the same as the experimental method 1:

The action delay of the jacket robot: 600–3500 ms (when the sampling period of the acceleration is 500 ms).

92.4 Conclusion

This paper describes an initial developed jacket robot and its supporting subsystems. The jacket robot can realize the basic actions of the human arm. But the activity sphere is not as large as the human's. It emphases on the recognition method based on the acceleration signal for the action intention of the human arms. On whole, the method is simpler because the requirement for the positing of the accelerator is not too high, and the key is the determination of the judgment threshold.

Acknowledgments This research is partly supported by the fund of the outstanding youth of Shanghai City (yyy11075).

References

1. Zhang J, Chen Y, Yang C (2011) The human-machine intelligent system of the soft exoskeleton. Science Publisher, Beijing (In Chinese)
2. http://kobalab.com, 05 2012
3. Zhang J (2009) The basic theory and the application technology research based on the soft exoskeleton human-machine intelligent system. Doctor Dissertation, Zhejiang University
4. http://www.me.berkeley.edu/hel/, 05 2012
5. Chen F (2007) Research on the technologies of the wearable assistant robot. Doctor Dissertation, University of Science and Technology China
6. Kobayashi H, Aida T, Hashimoto T (2009) Muscle suit development and factory application. Int J Autom Technol 3(6):709–715
7. Ryuichi N, Hiroshi S (2000) Fundamental kinesiology, 5th edn. Ishiyaku Publishers, Inc., Tokyo
8. Chou C (1996) Study of human motion control with a physiology based robotic arm and spinal level neural controller. Doctor Dissertation, Washington University

Chapter 93
Time-Delay Estimation Based on Cross-Correlation and Wavelet Denoising

Hua Yan, Yepeng Zhang and Qi Yang

Abstract A new time-delay estimation method based on the combination of wavelet transform and cross-correlation is proposed. This method uses compromise threshold function and Birge-Massart threshold selection strategy to denoise signals, and then determines the time-delay between the signals by computing the cross-correlation of the denoise signals. Comparative study of the new method, cross-correlation method and second correlation method were carried out. Results show that these three methods can give good time-delay estimations when the SNR is not low; however, only the new method can give good time-delay estimations if the SNR is very low and the noises are strongly correlated.

Keywords Time-delay estimation · Cross-correlation · Compromise threshold function · Wavelet denoising

93.1 Introduction

Acoustic temperature measurement method, which uses the dependence of sound speed in materials on temperature along the sound propagation path, has many advantages such as non-contact, wide measurement range and suitability for large size. Acoustic travel-time tomography is an effective tool for the measurement of 2D or 3D temperature distribution. Acoustic pyrometry is its representative application in industry, whereas monitoring the temperature distribution of stored

---

H. Yan (✉) · Y. Zhang
School of Information Science and Engineering, Shenyang University of Technology, Shenyang, China
e-mail: yanhua_01@163.com

Q. Yang
Liaoning Administration College of Police and Justice, Shenyang 110161, China

Z. Sun and Z. Deng (eds.), *Proceedings of 2013 Chinese Intelligent Automation Conference*, Lecture Notes in Electrical Engineering 255,
DOI: 10.1007/978-3-642-38460-8_93,

grain by acoustic travel-time tomography is a new application research being explored [1–4]. The key to acoustic temperature measurement is to measure acoustic travel-time accurately. Grain is a highly absorbing acoustic medium, therefore the acoustic measurement of stored grain temperature has a higher requirement on the accuracy of travel-time measurement [1, 2].

Acoustic travel time measurement is actually a time-delay estimation problem. Time-delay estimation has been an active topic in the signal processing field. Cross-correlation (CC) is the most basic and common time-delay estimation method. Some correlation methods, such as second correlation (SC) method, are based on CC method, and have been widely used [5–8].

In practical measurement, various noises and interferences are inevitable, so getting an accurate time-delay estimation is not easy. In order to satisfy the accuracy requirement of time-delay estimation in acoustic measurement of stored grain temperature, a new time-delay estimation method based on the combination of wavelet compromise threshold denoising and cross-correlation (CC&WCTD) is proposed. Comparative studies on CC, SC and CC&WCTD are given.

93.2 Time-Delay Estimation Methods

93.2.1 Cross-Correlation Method: CC Method

Supposing $x_1(t) = s(t) + n_1(t)$ and $x_2(t) = s(t - D) + n_2(t)$ are the signals received by the two sensors respectively, where s(t) is the source signal, $n_1(t)$ and $n_2(t)$ represent the additive noises, D is the time delay between the two signals. The cross-correlation of $x_1(t)$ and $x_2(t)$ can be expressed as:

$$R_{x_1x_2}(\tau) = E[x_1(t)x_2(t+\tau)] = R_{ss}(\tau - D) + R_{n_1s}(\tau - D) + R_{sn_2}(\tau) + R_{n_1n_2}(\tau) \tag{93.1}$$

where $R_{ss}(.)$ is the auto-correlation of the source signal s(t). $R_{ss}(\tau - D)$ reaches maximum when $\tau = D$. The CC method identifies the abscissa of $R_{x_1x_2}(\tau)$ corresponding to the peak of $R_{x_1x_2}(\tau)$ as the-delay estimation [5–8]. When the SNR is high and there is no correlation between the signal and noise, noise and noise, CC method can get good results of time delay estimation, but in the case of the low SNR or correlated noise, this method cannot be used for time delay estimation [8].

93.2.2 Second Correlation Method: SC Method

The auto-correlation of $x_1(t)$ can be expressed as

$$R_{x_1x_1}(\tau) = E[x_1(t)x_1(t+\tau)] = R_{ss}(\tau) + R_{n_1s}(\tau) + R_{sn_1}(\tau) + R_{n_1n_1}(\tau) \tag{93.2}$$

$R_{x_1x_1}(t)$ and $R_{x_1x_2}(t)$ can be regarded as time signals. Ignoring the correlation of signal and noise, their cross-correlation function can be expressed as:

$$RR(\tau) = E[(R_{x_1x_1}(t)R_{x_1x_2}(t+\tau)] = R_{Rs}(\tau - D) + R_{Rn}(\tau) \tag{93.3}$$

where $R_{Rs}(.)$ is the auto-correlation of the $R_{SS}(t)$, when $\tau = D$ $R_{Rs}(\tau - D)$ reaches maximum. $R_{Rn}(.)$ is the cross-correlation of $R_{n_1n_1}(t)$ and $R_{n_2n_2}(t)$. $R_{n_1n_1}(t)$ and $R_{n_2n_2}(t)$ are respectively the auto-correlation of $n_1(t)$ and $n_2(t)$.

SC method identifies the abscissa of $RR(\tau)$ corresponding to its peak as the time-delay estimation. It reduces the noise influence on estimation by multiple correlation calculations. Compared with CC method, it can be used in a lower SNR environment. But if the noises are correlated, both CC and SC methods can not give accurate time-delay estimation in a very low SNR environment [8].

93.2.3 A Method Based on the Combination of Wavelet Compromise Threshold De-noising and Cross-Correlation: CC&WCTD Method

The idea of CC&WCTD is as follows. First process the received signals by wavelet compromise threshold denoising, then calculate the cross-correlation of the denoised signals, and finally acquire the time-delay estimation by finding the peak position of the cross-correlation function.

The wavelet threshold denoising method [9–12] proposed by Donoho is widely used in engineering. It can be divided into three steps.

1. Wavelet decomposing. Choose an appropriate wavelet mother function, determine a proper decomposition level N, and calculate the wavelet decomposition coefficient of the noisy signal in each layer.
2. Wavelet coefficient threshold processing: Select a proper threshold and conduct threshold processing on the high frequency coefficient of each layer.
3. Wavelet reconstruction. Reconstruct the signal by using the low frequency coefficient of the Nth layer and the high frequency coefficients of all the N layers.

The threshold function can be divided into hard threshold and soft threshold [8–11]. Assuming the threshold is λ, the high frequency coefficient of wavelet decomposition before and after threshold processing are $\omega_{j,k}$ $\hat{\omega}_{j,k}$ respectively, the hard threshold function is defined as:

$$\hat{\omega}_{j,k} = \begin{cases} \omega_{j,k}, & |\omega_{j,k}| > \lambda \\ 0, & |\omega_{j,k}| < \lambda \end{cases} \tag{93.4}$$

and the soft threshold function is defined as:

$$\hat{\omega}_{j,k} = \begin{cases} \text{sign}(\omega_{j,k})(|\omega_{j,k}| - \lambda), & |\omega_{j,k}| > \lambda \\ 0, & |\omega_{j,k}| < \lambda \end{cases} \tag{93.5}$$

Above two threshold functions are often used in practice. But some disadvantages exist in them. The $\hat{\omega}_{j,k}$ determined by the hard threshold is discontinuous at $|\omega_{j,k}| = \lambda$, which will result in oscillation of the reconstruct signal; the $\hat{\omega}_{j,k}$ obtained by the soft threshold has good continuity, but there exists a constant deviation in the case of $|\omega_{j,k}| > \lambda$, which will affect the reconstructed signal directly [11, 12].

In this paper, a compromise threshold method is used, which is defined as:

$$\hat{\omega}_{j,k} = \begin{cases} sign(\omega_{j,k})(|\omega_{j,k}| - \alpha\lambda), & |\omega_{j,k}| > \lambda \\ 0, & |\omega_{j,k}| < \lambda \end{cases} \tag{93.6}$$

where α is adjustment factor ($0 \leq \alpha \leq 1$). When $\alpha = 0$ and 1, Eq. (93.6) are hard and soft threshold function, respectively. When $0 < \alpha < 1$, the value of $\hat{\omega}_{j,k}$ is a value between the values derived from soft and hard threshold function, thus better denoising effect can be obtained.

In this paper, level-dependent thresholds are selected by using Birge-Massart strategy [12]. The use of level-dependent threshold is conducive to maintaining the characteristics of wavelet coefficients and reducing the errors of the reconstructed signal. Daubechies wavelet belongs to compact support wavelet. High order Daubechies wavelet presents an obvious linear FM feature in time domain, therefore, db18 is used as a wavelet basis since the source signal in this paper is a linear swept-frequency cosine signal (chirp signal). The number of wavelet decomposition layers (N) is 5.

93.3 Experiments and Result Analysis

Experiments were carried out in MATLAB. The source signal s(t) was a chirp signal, with a frequency range of 500–2,000 Hz. The signal instantaneous frequency at the initial moment is 500 Hz, and reaches 2,000 Hz in 0.3 s. The received signal $x_1(t)$ and $x_2(t)$ were, respectively, s(t) and $s(\tau - D)$ ($D = 2.8820e - 2$ s), added with Gaussian white noise. The noises in $x_1(t)$ and $x_2(t)$ can be uncorrelated or strongly correlated. Strong correlated noises were generated by adding same noise to the signals. $x_1(t)$ and $x_2(t)$ were sampled 30,000 points respectively at a sampling rate of 100 k.

Wavelet threshold denoising is applied to a noisy signal with SNR (signal to noise ratio) of -5 dB for comparing the denoising effects of hard, soft and compromise threshold functions. In order to estimate the quality of the reconstructed signal quantitatively, signal to noise ratio R_{SN} and mean square error (RMSE) of the reconstruct signal are defined as follows.

$$\mathrm{RMSE} = \sqrt{\frac{1}{N_{sample}} \sum_{n=1}^{N_{sample}} [s(n) - \hat{s}(n)]^2} \tag{93.7}$$

$$R_{SN} = 10\lg\left[\frac{\sum_{n=1}^{N_{sample}} s^2(n)}{\sum_{n=1}^{N_{sample}} [s(n) - \hat{s}(n)]^2}\right] \tag{93.8}$$

where N_{sample} is the length of the discrete sampled signal, s(n) and $\hat{s}(n)$ are, respectively, the source signal and the reconstructed signal. Larger R_{SN} and smaller RMSE give a better reconstructed signal. Table 93.1 shows the denoising effect of hard, soft and compromise ($\alpha = 0.5$) threshold functions. It can be seen from Table 93.1 that the compromise threshold function has better denoising effect compare with the hard and the soft threshold functions.

Figure 93.1 shows the curves of R_{SN} versus adjustment factor α when the SNR of the noisy signal are 6 and −8 dB, respectively. Following can be seen from Fig. 93.1:

1. When the SNR of the noisy signal is 6 dB, the denoising effect is best at $\alpha = 0.1$;
2. When the SNR of the noisy signal is −8 dB, the denoising effect is best at $\alpha = 0.8$.

That means a small adjustment factor should be used in high SNR condition, and a larger adjustment factor should be used in low SNR condition.

Using CC, SC and CC&WCTD methods, the-time delay estimations are calculated in different SNR. The coefficient of variance cv(%) and the average relative error e (%) are used to assess the stability and accuracy of the estimation values.

$$\mathrm{cv} = \frac{\sqrt{\frac{\sum_{j=1}^{100} (t_j - t)^2}{100-1}}}{\bar{t}} \times 100\,\% \tag{93.9}$$

$$e = \frac{\sum_{j=1}^{100} \frac{|t_j - t|}{t}}{100} \times 100\,\% \tag{93.10}$$

Table 93.1 Denoising effects of hard, soft and compromise threshold functions

Evaluation parameters	Hard threshold	Soft threshold	Compromise threshold
R_{SN}	12.2795	13.0376	14.5479
RMSE	0.3827	0.3684	0.3417

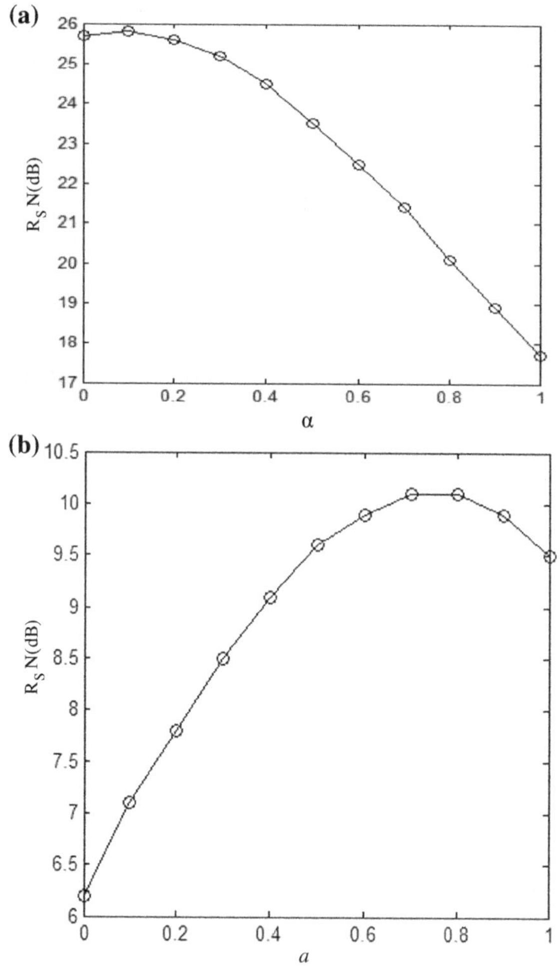

Fig. 93.1 R_{SN} versus α when the SNR of the noisy signal is 6 and −8 dB **a** 6 dB, and **b** −8 dB

where t_j is the *j*th time-delay estimation value, t is the real value of time delay (2.8820e-2 s).

The assessment results are given in Tables 93.2 and 93.3. When the SNR of $x_1(t)$ and $x_2(t)$ are 15, 10, 5 and 0 dB, the adjustment factor of CC&WCTD is set to 0.1; when the SNR of $x_1(t)$ and $x_2(t)$ is −5, −10 and −15 dB, the adjustment factor of CC&WCTD is set to 0.9. Following can be found in Tables 93.2 and 93.3.

1. When the noises are uncorrelated each other, these three methods can obtain good time-delay estimation as long as the SNR of the signals is higher than −10 dB.
2. When the noises are strongly correlated each other, these three methods can obtain good time-delay estimations too as long as the SNR of the signals is higher than 5 dB.

Table 93.2 Estimation results when uncorrelated Gaussian white noise added

	Methods	SNR of $x_1(t)$ and $x_1(t)$					
		10 dB	5 dB	0 dB	−5 dB	−10 dB	−15 dB
cv	CC	1.33e − 13	1.16e − 2	2.56e − 2	4.08e − 2	6.56e − 2	1.32e − 1
	SC	1.33e − 13	1.33e − 13	1.33e − 13	2.23e − 2	5.11e − 2	1.19e − 1
	CC&WCTD	1.33e − 13	1.33e − 13	1.33e − 13	1.47e − 2	1.70e − 2	2.88e − 2
e	CC	5.40e − 3	8.57e − 3	1.65e − 2	3.33e − 2	5.50e − 2	1.10e − 1
	SC	5.40e − 3	5.40e − 3	5.40e − 3	1.76e − 2	4.17e − 2	9.67e − 2
	CC&WCTD	5.40e − 3	5.40e − 3	5.40e − 3	1.09e − 2	1.36e − 2	2.32e − 2

Table 93.3 Estimation results when correlated Gaussian white noise added

	Methods	SNR of $x_1(t)$ and $x_1(t)$					
		15 dB	10 dB	5 dB	0 dB	−5 dB	−10 dB
cv	CC	1.33e − 13	1.33e − 13	1.29e − 2	5.96e + 1	NaN	NaN
	SC	1.33e − 13	1.33e − 13	1.33e − 13	1.33e − 13	3.20e − 2	NaN
	CC&WCTD	1.33e − 13	1.33e − 13	1.33e − 13	1.33e − 13	1.37e − 2	1.75e − 2
e	CC	5.40e − 3	5.40e − 3	9.18e − 2	2.60e + 1	100	100
	SC	5.40e − 3	5.40e − 3	5.40e − 3	5.40e − 3	2.45e − 2	100
	CC&WCTD	5.40e − 3	5.40e − 3	5.40e − 3	5.40e − 3	9.94e − 3	1.46e − 2

3. In view of anti-noise ability, CC method is the worst and CC&WCTD method is the best. When the noises are strongly correlated each other and the SNR of the signals is low as −10 dB, only CC&WCTD method can obtain good time-delay estimations.

93.4 Conclusion

Wavelet compromise threshold denoising is an effective denoising method. The comparative study of CC, SC and CC&WCTD methods show that the time-delay estimation obtained by CC&WCTD is obviously better than those obtained by CC and SC methods. When the noises are strongly correlated each other and the SNR of the signals is low as −10 dB, only CC&WCTD method can give good time-delay estimation. Therefore, CC&WCTD has strong ability to adapt to a noisy environment, and can be expected to be used in an actual acoustic temperature monitoring system for stored grain.

Acknowledgments The work is supported by the National Natural Science Foundation of China (60772054) and the Specialized Research Fund for the Doctoral Program of Higher Education of China (20102102110003).

References

1. Yan H, Chen GN, Zhou YG et al (2012) Primary study of temperature distribution measurement in stored grain based on acoustic tomography. Exp Thermal Fluid Sci 42:55–63
2. Yan H, Chen GN, Zhou YG et al (2012) Experimental study of sound travel-time estimation method in stored grain. J Comput 7:947–953
3. Holstein P, Raabe R, Müller et al (2004) Acoustic tomography on the basis of travel-time measurement. Meas Sci Technol 15:1240–1248
4. Bramanti M, Salerno EA, Tonazzini A et al (1996) An acoustic pyrometer system for tomographic thermal imaging in power plant boilers. IEEE Trans Instrum Meas 45:159–167
5. Gedalyahu H, Eldar CY (2010) Time-delay estimation from low-rate samples: a union of subspaces approach. IEEE Trans Signal Process 58:3017–3031
6. Dhull S, Arya S, Sahu OP (2010) Comparison of time-delay estimation techniques in acoustic environment. Int J Comput Appl 8:29–31
7. Kan Z, Shao FQ, Ding L (2010) Correlative flow velocity measurement based on electrostatic sensor. J Shenyang Univ Technol 32:90–94 (in Chinese)
8. Tang J, Xing HY (2007) Time delay estimation based on second correlation. Comput Eng 27:265–267 (in Chinese)
9. Cai M (2011) Improvement on threshold de-noising based on wavelet analysis. Inf Electron Eng 9:211–214 (in Chinese)
10. Liang L, Suo L (2010) Improving accuracy of delay estimation in wavelet de-noising. J Appl Opt 32:65–69 (in Chinese)
11. Donoho DL (1995) Denoising by soft-thresholding. IEEE Trans Inf Theory 41:613–627
12. Ger ZX, Sha W (2007) Wavelet analysis theory with MATLAB R2007. Publishing House of Electronics Industry, BeiJing (in Chinese)

Chapter 94
3D Temperature Field Reconstruction: A Comparison Study of Direct and Indirect Method

Hua Yan, Hongzheng Lin and Shanhui Wang

Abstract The reconstructions of 3D temperature fields by acoustic tomography are studied. Three cross-sections of the cylinder space to be measured are selected as typical planes; each plane has 8 acoustic sensors amounted on its periphery. Using the reconstruction algorithm based on Markov radial basis function and Tikhonov regularization we proposed, the 3D temperature distributions in the space are obtained in two ways. The first, called indirect 3D, is to reconstruct the 2D temperature fields of the three planes by 2D reconstruction algorithm and to interpolate them into a 3D distribution. The second, called direct 3D, is to reconstruct the 3D temperature fields directly by 3D reconstruction algorithm. Reconstructions of three temperature field models demonstrate that the direct method has better ability in complex 3D temperature field reconstruction at the expense of more travel-times to be measured.

Keywords Acoustic tomography · 3D temperature field · Reconstruction algorithm · Direct 3D reconstruction · Indirect 3D reconstruction

94.1 Introduction

Temperature field reconstruction based on acoustic tomography [1–4] uses the dependence of sound speed in materials on temperature along the sound propagation path. It has many advantages such as non-destructive, noncontact sensing and quick in response. Acoustic pyrometers are its representative application in industry [4], whereas monitoring the temperature distribution of stored grain by acoustic tomography [5, 6] is a new application research being explored.

H. Yan (✉) · H. Lin · S. Wang
School of Information Science and Engineering,
Shenyang University of Technology, Shenyang, China
e-mail: yanhua_01@163.com

Z. Sun and Z. Deng (eds.), *Proceedings of 2013 Chinese Intelligent Automation Conference*, Lecture Notes in Electrical Engineering 255,
DOI: 10.1007/978-3-642-38460-8_94,

Reconstruction of 2D temperature fields by acoustic tomography on a typical plane is often reported. But in many cases, for example, monitoring the temperature distribution of stored grain, reconstruction of 3D temperature fields is needed.

In this paper, the reconstructions of 3D temperature fields by acoustic tomography are investigated. Using the reconstruction algorithm based on Markov radial basis function and Tikhonov regularization we proposed, the 3D temperature distributions are obtained by indirect 3D method and direct 3D method, respectively.

94.2 Theory of Temperature Field Reconstruction Based on Acoustic Tomography

Temperature measurement by acoustic method is based on the principle that the sound velocity in a medium is a function of the medium temperature. The sound velocity c in a gaseous medium at an absolute temperature T is given by [3–6].

$$c = z\sqrt{T} \tag{94.1}$$

where z is a constant decided by gas composition. The value of z for air is 20.05.

To survey the temperature distribution in a space, several acoustic sensors (sources/receivers) should be installed on its periphery. A sound signal emitted by one transceiver can be received by all the transceivers. Using a proper time delay estimation method, for example, a method based on cross-correlation [6, 7], the sound travel-times along the sound path specially selected can be measured. Then, the temperature distribution can be reconstructed by an appropriate reconstruction algorithm.

94.3 The Reconstruction Algorithm Based on Markov Radial Basis Function and Tikhonov Regularization

If the reciprocal distribution of sound velocity in medium is f(x, y) in 2D or f(x, y, z) in 3D, the sound travel-time t_j of the jth sound path p_j can be express as

$$\begin{aligned} t_j &= \int_{p_j} f(x, y)ds, \quad j = 1, 2, \ldots, m \quad (2D) \\ t_j &= \int_{p_j} f(x, y, z)ds, \quad j = 1, 2, \ldots, m \quad (3D) \end{aligned} \tag{94.2}$$

where m is the number of sound path, s is the path length.

Markov function is used to approximate the sound velocity distribution. Markov function can be expressed as

$$\begin{aligned} \varphi_i(x, y) &= e^{-a\sqrt{(x-x_i)^2+(y-y_i)^2}} \quad (2D) \\ \varphi_i(x, y, z) &= e^{-a\sqrt{(x-x_i)^2+(y-y_i)^2+(z-z_i)^2}} \quad (3D) \end{aligned} \tag{94.3}$$

where $a > 0$ is the shape parameter of radial basis function; x_i and y_i are the coordinate of the ith data point in function space. In this paper $a = 0.0002$.

If(x, y) in 2D or f(x, y, z) in 3D can be expressed as the linear combination of number of Markov radial basis function.

$$\begin{aligned} f(x, y) &= \sum_{i=1}^{n} \beta_i e^{-a\sqrt{(x-x_i)^2+(y-y_i)^2}} \quad (2D) \\ f(x, y, z) &= \sum_{i=1}^{n} \beta_i e^{-a\sqrt{(x-x_i)^2+(y-y_i)^2+(z-z_i)^2}} \quad (3D) \end{aligned} \tag{94.4}$$

where β_i is the coefficient to be determined, x_i, y_i and z_i are the coordinate of the ith original pixel. Usually, the denser the original pixels are, the larger the condition number of the interpolation matrix (hereinafter matrix A), which will result in serious ill-posedness of the inverse problem. Thus, the number of the original pixels cannot be too large.

Substituting Eq. (94.4) into Eq. (94.2), we have

$$\begin{aligned} t_j &= \sum_{i=1}^{n} \beta_i \int_{p_j} \varphi_i(x, y)ds = \sum_{i=1}^{n} \beta_i A_{ji} \quad (2D) \\ t_j &= \sum_{i=1}^{n} \beta_i \int_{p_j} \varphi_i(x, y, z)ds = \sum_{i=1}^{n} \beta_i A_{ji} \quad (3D) \end{aligned} \tag{94.5}$$

where $A_{ji} = \int \varphi_i(x, y)ds$ or $A_{ji} = \int \varphi_i(x, y, z)ds$. Defined $A = (A_{ji})_{j=1,\ldots}$, m, $i = 1, \ldots, n$, $b \overset{p_j}{=} (\beta_1, \ldots \beta_n)^T$, $t = (t_1^{p_j} \ldots t_m)^T$, the model of the forward problem of acoustic CT can be expressed as

$$\mathbf{A\beta} = \mathbf{t} \tag{94.6}$$

Using the Singular Value Decomposition of matrix A and Tikhonov regularization [8], the regularization solution of Eq. (94.6) can be expressed as

$$\boldsymbol{\beta} = \sum_{i=1}^{p} \left(\frac{\sigma_i^2}{\sigma_i^2 + \mu} \right) \frac{\mathbf{u}_i^T \mathbf{t}}{\sigma_i} \mathbf{v}_i = \sum_{i=1}^{p} \frac{\mathbf{u}_i^T \mathbf{t}}{\sigma_i + \mu/\sigma_i} \mathbf{v}_i \tag{94.7}$$

where μ is regularization parameter, $\sigma_1 \geq \sigma_2 \geq \ldots \geq \sigma_p > 0$ are the singular values of matrix A, p is the number of non-zero singular values, $\mathbf{u}_i$ and $\mathbf{v}_i$ are the left and right singular value vectors of $\mathbf{A}$. Because of the ill-posedness of the inverse problem, the condition number of **A** is very large, that means very small singular values exist. It can be known from Eq. (94.7) that, if σ_i is very small and $\mu = 0$, the errors in sound travel-times could be magnified many times and make the solution far away from the truth. A proper small positive regularization parameter can restrain errors of the solution effectively. In fact, μ controls the weight of measured data and experience in solution. Sound velocity is bounded and fluctuant is used as experience information in Eq. (94.7). If μ is too small, the error of measured data can't be well restrained, while if μ is too large, the solution will lose much detailed information. Usually, the regularization parameter is chosen by experience.

As the location of sound transceivers and original pixels are decided, matrix **A** and its singular value decomposition can be obtained. Then the parameter vector $\boldsymbol{\beta}$ can be calculated by using measured or simulated sound travel-time vector t and Eq. (94.7). Substituting β into Eq. (94.4), the reciprocal distribution of sound velocity can be calculated, and the temperature distribution in free space can be obtained from Eq. (94.1).

94.4 3D Temperature Field Reconstruction by Indirect or Direct Way

In this paper, the space to be measured is a cylinder with a height of 10 m and a radius of 5 m. Three cross-sections of the space are selected as typical planes. Eight acoustic sensors are amounted on the periphery of each plane, see Fig. 94.1a.

If we want to reconstruct the 3D temperature fields indirectly, we subdivide the plane to be reconstructed into 131 pixels shown in Fig. 94.1b, and measure the travel-times along the $20 \times 3 = 60$ effective sound wave paths shown in Fig. 94.1c. After we reconstruct the 2D temperature fields in the three typical planes, we can obtain the 3D temperature field by interpolation method. In this paper, 'linear' interpolation method is used.

If we want to reconstruct the 3D temperature fields directly, we subdivide the space to be reconstructed into 131×10 volume pixels and measure the travel-times along the 196 effective sound wave paths shown in Fig. 94.1d. Using these travel-times, we reconstruct the 3D temperature fields directly.

94.5 3D Temperature Field Reconstructions from Simulated Data

In order to compare the reconstruction performance of direct and indirect ways, four typical temperature field models are reconstructed. They are

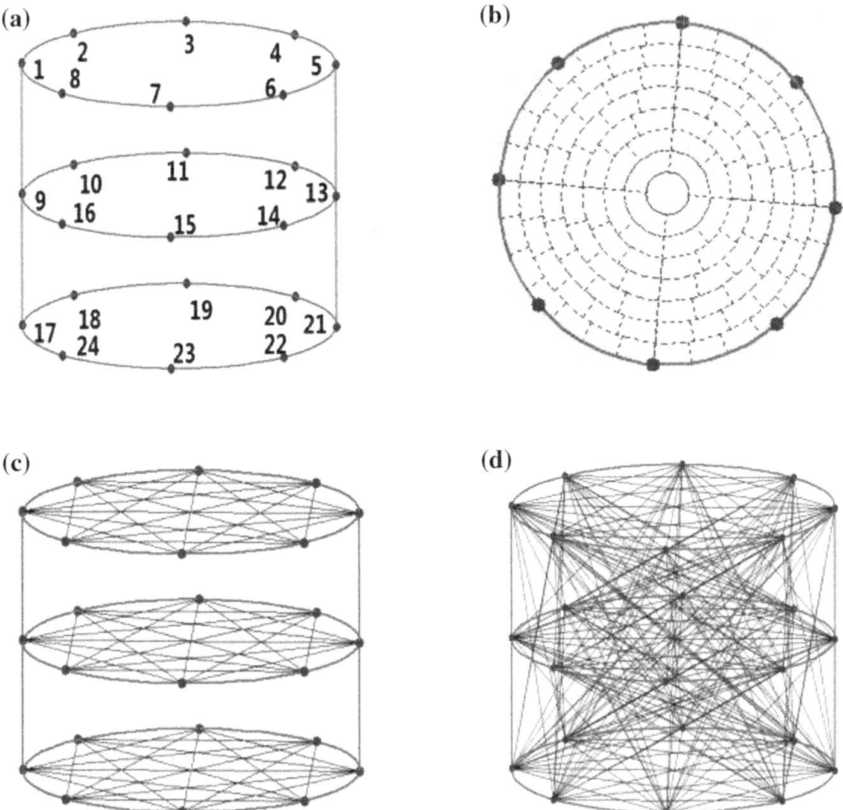

Fig. 94.1 Sensors, pixels and effective sound wave paths. **a** A space surrounded by 24 sensors. **b** Pixels in a typical plane. **c** The 60 effective sound wave paths for indirect 3D reconstruction. **d** The 196 effective sound wave paths for direct 3D reconstruction

Model 1. Symmetry one-peak model defined as:

$$T(x, y, z) = \frac{350}{0.05(x^2 + y^2 + z^2) + 1} \tag{94.8}$$

Model 2. Symmetry two-peak model defined as:

$$T(x, y, z) = \frac{350}{0.2[(x+3)^2 + y^2 + (z-1.6)^2] + 1} + \frac{350}{0.2[(x-3)^2 + y^2 + (z-1.6)^2] + 1} \tag{94.9}$$

Model 3. Symmetry four-peak model defined as:

$$T(x, y, z) = \frac{350}{0.5\left[(x+2.5)^2 + y^2 + (z-2.5)^2\right] + 1} + \frac{350}{0.5\left[(x-2.5)^2 + y^2 + (z-2.5)^2\right] + 1} + \frac{350}{0.5\left[(x+2.5)^2 + y^2 + (z+2.5)^2\right] + 1} + \frac{350}{0.5\left[(x-2.5)^2 + y^2 + (z+2.5)^2\right] + 1} \quad (94.10)$$

To evaluate the quality of the reconstructions, following reconstruction errors are used. The root-mean-squared error E_{rms} of the reconstructed field is defined as

$$E_{rms} = \frac{\sqrt{\frac{1}{M}\sum_{j=1}^{M}\left[T(j) - T(\hat{j})\right]^2}}{T_{ave}} \times 100\,\% \quad (94.11)$$

The relative error E_{max} in the maximum reconstructed temperature is

$$E_{max} = \left|\frac{T_{max} - \widehat{T}_{max}}{T_{max}}\right| \times 100\,\% \quad (94.12)$$

The relative error E_{mean} in the mean reconstructed temperature is

$$E_{mean} = \left|\frac{T_{mean} - \widehat{T}_{mean}}{T_{mean}}\right| \times 100\,\% \quad (94.13)$$

In Eqs. (94.11–94.13), $\hat{T}(j)$ and $T(j)$ are the reconstructed and the true temperature of jth pixel, respectively; T_{max} and T_{mean} are the true maximum temperature and the true mean temperature; $\hat{T}_{max}$ and $\hat{T}_{mean}$ are the reconstructed maximum temperature and the reconstructed mean temperature; M is the number of pixels.

Using the MTR algorithm, temperature field model 1–3 are reconstructed in direct way and indirect way, respectively. The travel-times are computer simulation values calculated using temperature field models, the locations of the sensors

Table 94.1 2D reconstruction errors of the three typical planes

Error criteria		Model 1 (%)	Model 2 (%)	Model 3 (%)
E_{mean}	Upper plane	0.095	0.0987	0.096
	Middle plane	0.073	0.109	0.055
	Lower plane	0.095	0.0908	0.096
E_{max}	Upper plane	0.733	0.686	1.300
	Middle plane	1.226	1.200	1.562
	Lower plane	0.733	0.594	1.300
E_{rms}	Upper plane	1.357	2.010	2.348
	Middle plane	1.983	3.067	2.822
	Lower plane	1.357	1.124	2.348

Table 94.2 3D reconstruction errors

Error Criteria	Model 1		Model 2		Model 3	
	Direct 3D (%)	Indirect 3D (%)	Direct 3D (%)	Indirect 3D (%)	Indirect 3D (%)	Direct 3D (%)
E_{mean}	0.446	5.726	0.056	9.4202	0.004	22.012
E_{max}	0.451	5.624	1.058	34.183	4.795	48.817
E_{rms}	1.054	6.316	1.856	24.921	3.006	30.9691

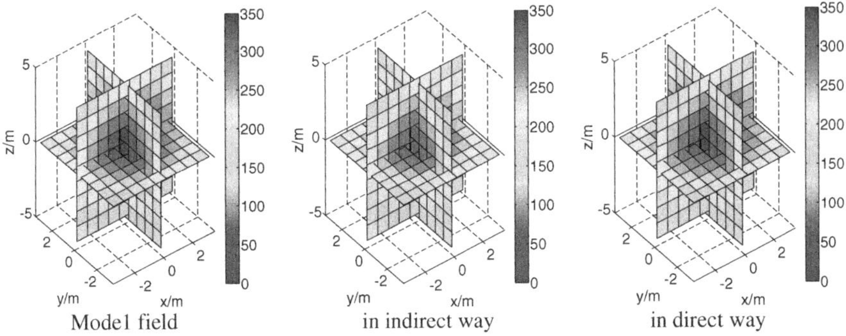

Fig. 94.2 Model 1 field and fields reconstructed in indirect way and direct way

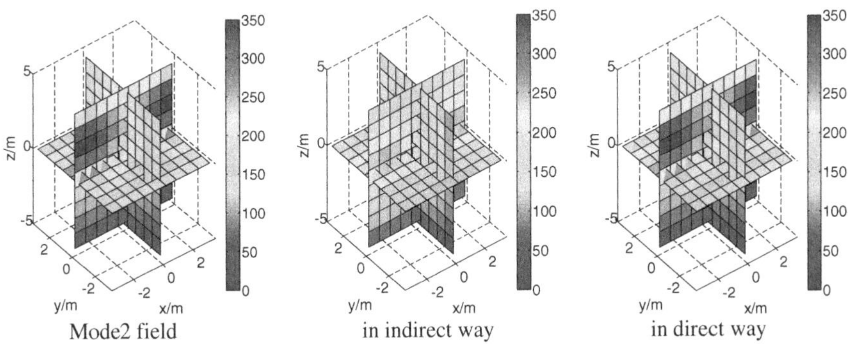

Fig. 94.3 Model 2 field and fields reconstructed in indirect way and direct way

and the relationship between the sound velocity and the temperature in gas. The 2D reconstruction errors of the three typical planes and the 3D reconstruction errors of the space are given in Tables 94.1 and 94.2, respectively. The model temperature fields and the temperature fields reconstructed are shown in Figs. 94.2, 94.3 and 94.4. Since the simulation data used in this paper are exact, without noise added, the regularization parameter μ is set to zero.

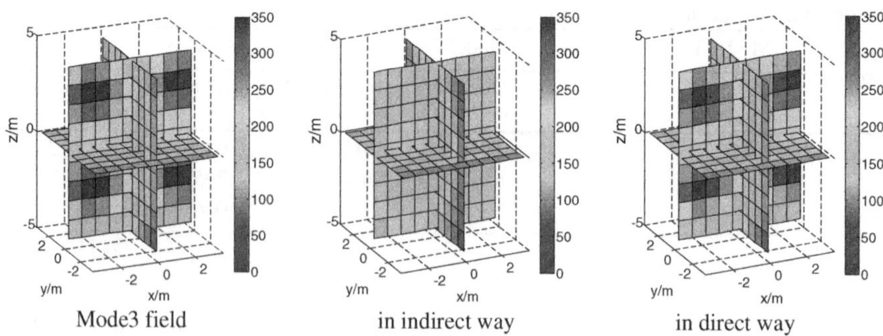

Fig. 94.4 Model 3 field and fields reconstructed in indirect way and direct way

94.6 Conclusion

Following can be concluded from the comparison study of the direct 3D and indirect 3D reconstruction methods.

1. Compared with the indirect 3D method, the direct 3D method needs more travel-times to be measured.
2. Although the reconstruction errors of the three typical planes are small, the reconstruction errors generated by the indirect 3D method are much bigger than those generated by the direct 3D method. The more complex the temperature field is, the large the difference becomes. Thus the direct 3D method is a better way for constructing complex 3D temperature fields.

Acknowledgments The work is supported by the National Natural Science Foundation of China (60772054) and the Specialized Research Fund for the Doctoral Program of Higher Education of China (20102102110003)

References

1. Holstein P, Raabe R, Müller R et al (2004) Acoustic tomography on the basis of travel-time measurement. Meas Sci Technol 15:1240–1248
2. Barth M, Armin R (2011) Acoustic tomographic imaging of temperature and flow fields in air. Meas Sci Technol 22. doi:10.1088/0957-0233/22/3/035102
3. Wei F, Chen Y, Pan HC et al (2010) Experimental study on underwater acoustic imaging of 2-D temperature distribution around hot springs on floor of Lake Qiezishan China. Exp Thermal Fluid Sci 34:1334–1345
4. Bramanti M, Salerno EA, Tonazzini A et al (1996) An acoustic pyrometer system for tomographic thermal imaging in power plant boilers. IEEE Trans Instrum Meas 45:159–167
5. Yan H, Chen GN, Zhou YG et al (2012) Primary study of temperature distribution measurement in stored grain based on acoustic tomography. Exp Therm Fluid Sci 42:55–63

6. Yan H, Chen GN, Zhou YG et al (2012) Experimental study of sound travel-time estimation method in stored grain. J Comput 7:947–953
7. Kan Z, Shao FQ, Ding L (2010) Correlative flow velocity measurement based on electrostatic sensor. J Shenyang Univ Technol 32:90–94 (in Chinese)
8. Li ZC, Huang HT, Wei Y (2011) Ill-conditioning of the truncated singular value decomposition, Tikhonov regularization and their applications to numerical partial differential equations. Numer Linear Algebra Appl 18:205–221

Chapter 95
Real Time Simulation of Ship Wake Based on Particle System

Xin Zhao and Zhisheng Li

Abstract A new ship wake simulation method based on particle systems was proposed in this paper. We firstly modeled 2-D ship waves by using Kelvin waves theory, and then extended it to 3-D model by using particle systems. After that, we simulated bow wave and stern wave with experiential models through actual observation on their shape characteristics, range and stochastic. Experimental results show that the modeling methods are efficient and realistic for ship wakes simulation.

Keywords Ship wakes · Bow wave · Stern wave · Ship wave · Kelvin waves · Particle systems

95.1 Introduction

Ship wake is a complex water flow phenomenon, which comprises bow wave, stern wave and ship wave. Due to its shape irregularity and complex causes, real-time simulation of ship wakes in virtual ocean scene has always been a difficult problem in computer graphics. The particle system theory is a commonly used method to describe dynamic irregular objects, which suitable for natural phenomena simulation [1–3]. Therefore, some researchers introduced it to ship wake simulation [2–7]. We deeply studied on the causes of ship wake, and a novel ship wake simulation method based on particle system is proposed in paper, which absorbs the advantages of previous method.

X. Zhao (✉)
Information Engineering College, Dalian University, Dalian, China
e-mail: zx38610@yeah.net

Z. Li
The unit 91550 of PLA

Z. Sun and Z. Deng (eds.), *Proceedings of 2013 Chinese Intelligent Automation Conference*, Lecture Notes in Electrical Engineering 255,
DOI: 10.1007/978-3-642-38460-8_95,

95.2 Ship Wave Modeling

95.2.1 Two Dimensional Ship Wave Modeling Based on Kelvin Wave Theory

According to the Kelvin wave theory in ship fluid dynamics, Disturbance waves formed by a motive point on deep water surface is called the Kelvin wave [6]. The crest line of Kelvin wave can be calculated as following [8]:

$$x = X_1 \cos\theta\left(1 - \frac{1}{2}\cos^2\theta\right), y = \frac{1}{2}X_1\cos^2\theta\sin\theta \tag{95.1}$$

where X_1 is the relative distance between motive disturbance point and origin, indicating the wave interval size of disturbance waves. θ is the angle between moving direction and line connecting motive disturbance point and the point on crest line $\theta \in [-\pi/2, \pi/2]$. For different constant X_1, gives a group of θ, we can get a group of (x, y) by Eq. (95.1).

Kelvin wave theory is in accord with the characteristics of ship wave, so we simulate ship wave based on Kelvin wave theory. For meeting the simulation requirements, we adjust Eq. (95.1).

First, transverse wave is not obvious in the actual ship sailing because of the ship overlying, but the longitudinal diffusion wave is very consistent with the V-shaped wave on both sides of actual ship on sailing, so we need to eliminate the transverse wave in crest line and retain the longitudinal wave through adjustment of θ range. The greater θ value is, the more obvious Kelvin wave tends to longitudinal wave, so we set $[-\pi/2, -\pi/8]$ and $[\pi/8, \pi/2]$ as the range of θ. Its wave shape is shown in Fig. 95.1.

Second, ship speed affects the force and the velocity of water drainage, when ship speed increases, the force on water and the velocity of water drainage increase, ship wake becomes longer, and wave interval becomes bigger; when ship speed reduces, the force on water reduces, drainage velocity slower, ship wake shorter, and wave interval smaller. Therefore, the ship wake is deeply influenced by ship speed. We can set $X_1 = V \cdot t$, because ship speed represents distance of

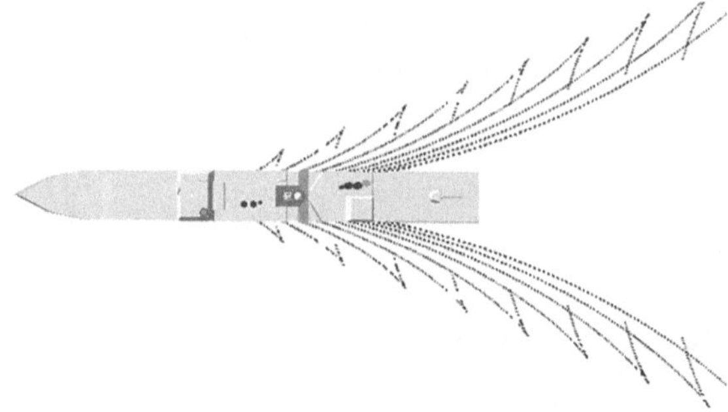

Fig. 95.1 Crest line of Kelvin wave after adjustment

disturbance point movement in unit time, where V is ship speed; X_1 is the movement distance of disturbance point in t seconds. We can fix the value of t, then the wake size is only related with the speed of V.

We use the adjustment Kelvin wave crest line model the ship wave crest line, and modeling equation as follow:

$$\begin{aligned} x &= V \cdot t \cdot \cos\theta\left(1 - \frac{1}{2}\cos^2\theta\right) \\ y &= \frac{1}{2}V \cdot t \cdot \cos^2\theta \sin\theta \end{aligned} \tag{95.2}$$

95.2.2 Particle System Modeling of Ship Waves

The ship wave crest line equation mentioned in 2.1 is a two-dimensional equation. In order to obtain 3D effect, we let water particles spray out from the two-dimensional crest line, which can form wave spray clouds.

1. Initial position

As mentioned up words, particles spray out from the two-dimensional crest line, so the crest line is the basic framework of ship waves. In order to ensure the spray has the stereoscopic effect, the initial position of particle ejection should be randomly selected in a certain range around crest line, and each position ejects certain number of particles outward. The initial position can be set as follows:

$$\begin{aligned} x_0 &= V \cdot t \cdot \cos\theta\left(1 - \frac{1}{2}\cos^2\theta\right) + Rand() \\ y_0 &= Rand() \\ z_0 &= \frac{1}{2}V \cdot t \cdot \cos^2\theta \sin\theta + Rand() \end{aligned} \tag{95.3}$$

where (x_0, y_0, z_0) is the initial position of particle, $Rand()$ is a random offset.

2. Initial velocity

We assume that particles do free falling after ejecting from initial positions. In order to get scattered ejection effect, movement direction should be set in a certain angle range randomly. The initial velocity can be set as follows:

$$
\begin{aligned}
V &= \langle V_x, V_y, V_z \rangle \\
\theta &= C_1 \cdot Rand() \\
\beta &= C_2 \cdot Rand() \\
V_x &= S \cdot \cos(\theta) \cdot \cos(\beta) \\
V_y &= S \cdot \sin(\theta) \\
V_z &= S \cdot \cos(\theta) \cdot \sin(\beta)
\end{aligned}
\tag{95.4}
$$

where θ is the angle between direction vector and the horizontal plane, C_1 is the maximum value of θ, β is the angle between horizontal projection of direction vector and Z axis, C_2 is the maximum value of β, $Rand()$ is the random number in $0 \sim 1$, S is speed rate.

95.3 Bow Wave and Stern Wave Modeling

95.3.1 Bow Wave Modeling

Bow wave comprises clouds of sprays and droplets generated from collision between sea water and bow plate. We set a sector around the bow plate on the sea level, and the sector center coincides with the bow front, as shown in Fig. 95.2. Particles eject out from the sector region randomly and then doing uniform deceleration parabolic motion without considering particles collision with each other.

1. Initial position

The initial position of bow wave particle should be set in sector region randomly as following:

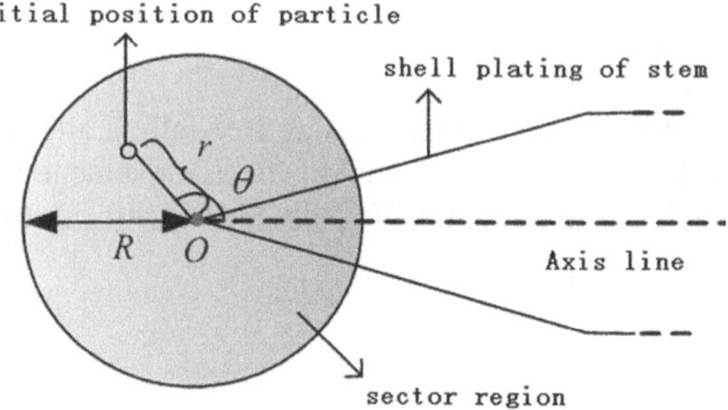

Fig. 95.2 Particle generation area of bow wave

$$\begin{aligned} x_0 &= O_x - r \cdot \cos\theta \\ y_0 &= 0 \\ z_0 &= O_z - r \cdot \sin\theta \\ r &= Rand(0, R) \end{aligned} \quad (95.5)$$

where (x_0, y_0, z_0) is the initial position of particle, r is the distance from sector center to initial position of particle, its value is a random in 0 ~ R, and R is the sector radius. $O(O_x, O_z)$, θ is the center. θ is the angle between axis and initial position, its value within the sector angle range.

2. Initial velocity and acceleration

The closer to bow, the bigger the ship's motion impact to particle is, so the initial rate of particle is relevant to its initial position. We assume that the closer to center, the bigger the initial rate of particle is, and vice versa, meanwhile, the closer to center, the smaller the ejection angle and initial acceleration of particle are. We calculate the initial velocity and acceleration as following:

$$\begin{aligned} V_x &= v_{\max} \cdot \cos\beta \cdot \cos\theta \cdot \frac{(R-r)}{R} \\ V_x &= v_{\max} \cdot \sin\beta \cdot \frac{(R-r)}{R} \\ V_x &= v_{\max} \cdot \cos\beta \cdot \sin\theta \cdot \frac{(R-r)}{R} \\ \beta &= \frac{3\pi}{4} \cdot \frac{(R-r)}{R} \\ a_x &= 0 \\ a_y &= a_{\max} \cdot \frac{(R-r)}{R} \\ a_z &= 0 \end{aligned} \quad (95.6)$$

where (V_x, V_y, V_z) is the initial velocity of particle. β is the initial ejection angle. (a_x, a_y, a_z) is the initial acceleration, $a_{\max}$ is the maximum initial acceleration in vertical direction. $v_{\max}$ is the maximum initial rate of particles.

3. Location update

Assuming particles do uniform motion in the horizontal direction after ejection, and do uniform deceleration motion in the vertical direction until ending its life cycle and sinking into sea. After ejects t seconds, particle location (x, y, z) can be calculated as follows:

$$
\begin{aligned}
x &= x_0 + V_x \cdot t + \frac{1}{2} a_x \cdot t^2 \\
y &= y_0 + V_y \cdot t + \frac{1}{2} a_y \cdot t^2 \\
z &= z_0 + V_z \cdot t + \frac{1}{2} a_z \cdot t^2
\end{aligned} \tag{95.7}
$$

95.3.2 Stern Wave Modeling

Stern wave is produced by propeller fluttering and sea water filling, which cause vortex and circulation in the ship tail, and vortex diffusion causes spray, that is stern wave. According to the observation, the area of stern wave area is approximate trapezoid, therefore, we can set a trapezoidal shape in the stern, and let stern wave particle eject from the middle point at the trapezoid front and move in the trapezoidal region, as shown in Fig. 95.3.

1. Initial velocity and acceleration

The ejection velocity and acceleration of particles should be randomly set as following:

$$
\begin{aligned}
V_x &= v_{\max} \cdot \cos\beta \cdot \cos\alpha \cdot Rand() \\
V_x &= v_{\max} \cdot \sin\beta \cdot Rand() \\
V_x &= v_{\max} \cdot \cos\beta \cdot \sin\alpha \cdot Rand() \\
\beta &= \beta_{\max} \cdot Rand() \\
\alpha &= \pm\,\alpha_{\max} \cdot Rand() \\
a_x &= 0 \\
a_y &= a_{\max} \cdot Rand() \\
a_z &= 0
\end{aligned} \tag{95.8}
$$

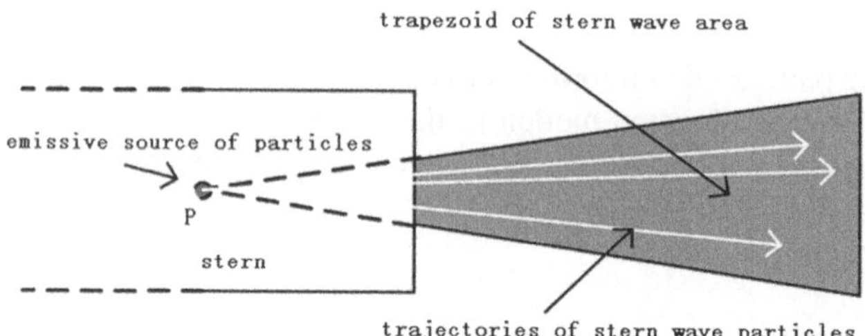

Fig. 95.3 Particle generation area of stern wave

where (V_x, V_y, V_z) is the initial velocity of particle. V_{max} is the maximum initial particle Rate. β is particle ejection angle. β_{max} is maximum ejection angle; α particle ejection angle between horizontal direction and trapezoidal axis. α_{max} is maximum value of α. $Rand()$ is a random number in the $0 \sim 1$. (a_x, a_y, a_z) is the initial acceleration, a_{max} is the maximum initial acceleration in the vertical direction.

2. Initial position

Stern wave particle ejects from the point at which the two trapezoidal sides intersect each other, so the initial position of a particle (x_0, y_0, z_0) is equal to the intersection point coordinate (p_x, p_y, p_z).

95.4 Implementation

In order to verify the effectiveness of algorithm mentioned above, we do simulation experiments with VC++6.0 and OpenGL on Windows XP platform. The hardware environment are Intel CPU P4 3 GHz, 256 MB RAM, GeForce 6600 graphics card, and 128 MB video card. We use the Point Sprites function in OpenGL Expansion for particle system rendering. The rendering number of point sprites is about 54000 each frame, and the average rendering frame rate is about 57 frames per second. Figure 95.4 shows the ship wake simulation, from which we can clearly see the bow wave, stern Wave and ship wave effect generated by our algorithm. We also simulated ship wake with Goss's algorithm, with 32000 line-shape particles and about 64 frames per second. The simulation with our algorithm is more comprehensive and true through comparison simulation results.

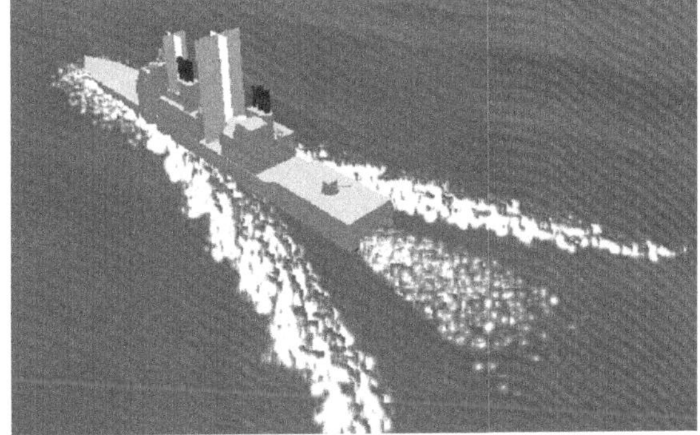

Fig. 95.4 Ship wake simulation using our algorithm

95.5 Conclusion

We simulated ship wakes including bow wave, stern wave and ship wave in this paper using particle system technology. Kelvin wave theory is used to model ship wave. Experiential models are used to model bow wave and stern wave through actual observation. Simulation results show that our modeling method is more efficient and realistic than Gross method, which is more suitable for real-time simulation of virtual ocean environment.

References

1. Reeves WT (1983) Particle systems-a technique for modeling a class of fuzzy objects. Comput Graphics 17(3):359–376
2. Muller M, Solenthaler B, Keiser R, Gross M (2005) Particle-based fluid–fluid interaction. In: Proceedings of computer graphics proceedings, annual conference series, ACM SIGGRAPH, Los Angeles, California, pp 237–244
3. Muller M, Charypar D, Gross M (2003) Particle-based fluid simulation for interactive applications. In: Proceedings of computer graphics proceedings, annual conference series, ACM SIGGRAPH, Aire-la-Ville, pp 154–159
4. Michael E, Goss A (1990) real time particle system for display of ship wakes. IEEE Comput Graphics Appl 272(7):30–35
5. YIN Y, Ren H, Zhang X et al (2002) Wave simulation technology applied to virtual environment of navigation simulation. J Syst Simul 14(3):313–315 (in Chinese)
6. Hu Y, Jiang Y (2008) Simulation o f 3D ship waves. Comput Appl 28 (S1):247–249 (in Chinese)
7. Wang Y, Wang Z (2006) Simulation of ship wake in complex environment. J Syst Simul 18(11):3247–3249 (in Chinese)
8. Xia G (2003) Ship hydrodynamics. HUA Zhong University of Science and Technology press, WuHan, China (in Chinese)

CPSIA information can be obtained at www.ICGtesting.com
Printed in the USA
LVOW01*2039270713

344969LV00005BA/158/P